MW00845480

HAZARDOUS MATERIALS TECHNICIAN

Chris Weber

PEARSON

Boston Columbus Indianapolis New York San Francisco Upper Saddle River
Amsterdam Cape Town Dubai London Madrid Milan Munich Paris Montreal Toronto
Delhi Mexico City São Paulo Sydney Hong Kong Seoul Singapore Taipei Tokyo

Publisher: Julie Levin Alexander
Publisher's Assistant: Regina Bruno
Senior Acquisitions Editor: Stephen Smith
Associate Editor: Monica Moosang
Development Editor: Eileen M. Clawson
Editorial Assistant: Samantha Sheehan
Director of Marketing: David Gesell
Marketing Manager: Brian Hoehl
Marketing Specialist: Michael Sirinides
Marketing Assistant: Crystal Gonzalez
Managing Production Editor: Patrick Walsh
Production Liaison: Julie Boddorf
Production Editor: Lisa S. Garboski, bookworks editorial services
Senior Media Editor: Amy Peltier
Media Project Manager: Lorena Cerisano
Manufacturing Manager: Ilene Sanford
Creative Director: Jayne Conte
Cover Designer: Sue Behnke
Cover Photo: Chris Weber
Composition: S4Carlisle Publishing Services
Printing and Binding: R.R. Donnelley/Willard
Cover Printer: Lehigh-Phoenix Color/Hagerstown

Credits and acknowledgments borrowed from other sources and reproduced, with permission, in this textbook appear on appropriate pages within the text. Unless otherwise stated, all photos have been provided by the author.

The information contained in this book is believed to be accurate at the time of publication. Changing standards and regulations may periodically warrant significant changes and updates to the contents. No warranty, express or implied, is given to the accuracy or applicability of the contents of this book to a given hazardous materials or weapons of mass destruction incident. This textbook is designed to be used during an instructor-led hazardous materials technician class in a controlled environment. THIS TEXTBOOK IS NOT DESIGNED TO BE USED DURING EMERGENCY INCIDENT RESPONSE. Every acceptable procedure has not been included, and listed procedures may warrant incident-specific modification. Multiple reference sources should be used at all hazardous materials incidents. To safely and effectively operate at a hazmat incident the reader must be a member of an organized hazardous materials response team that trains regularly and adheres to all applicable portions of 29 CFR 1910.120 and NFPA parts 471, 472, and 473. Hazardous materials response entails risk that must be effectively managed through proper training and pre-incident planning. IMPROPER USE OF THIS TEXTBOOK MAY RESULT IN DEATH, SERIOUS INJURY, AND/OR PROPERTY DAMAGE.

Library of Congress Cataloging-in-Publication Data
Weber, Chris, Ph.D.
Hazardous materials technician / Chris Weber.
p. ; cm.
Includes index.
ISBN-13: 978-0-13-172019-0
ISBN-10: 0-13-172019-8
I. Title.
[DNLM: 1. Disasters—prevention & control. 2. Hazardous Substances—toxicity.
3. Chemical Warfare—prevention & control. 4. Emergency Responders. WA 295]
363.17'91—dc23

2012014181

10 9 8 7 6 5 4 3 2 1

ISBN 10: 0-13-172019-8
ISBN 13: 978-0-13-172019-0

CONTENTS

Chapter 2 Regulations and Standards 37

Chapter 3 Toxicology 59

Chapter 6 Information Resources 146

Chapter 13 Incident Command 423

PREFACE

This *Hazardous Materials Technician* textbook is designed primarily for emergency responders who will respond in an offensive manner to assess and control hazardous materials releases and other chemical emergencies. These emergency responders may hail from the fire service, law enforcement, emergency medical services (EMS), military, or industry.

Firefighters have traditionally responded to hazardous materials releases and have been at the forefront of shaping modern hazardous materials response—in large part owing to injuries and deaths suffered during responses to chemical emergencies. Law enforcement personnel have more recently been faced with clandestine drug labs and WMD incidents that they must characterize and render safe and from which they must collect and process evidence. Unfortunately, a lack of hazardous materials training has led to significant injuries to law enforcement personnel in the recent past. Likewise, EMS workers are called on to decontaminate and treat victims as well as respond to industrial accidents involving chemical releases. Many cleanup contractors operate at the scene of emergencies, often before the incident is downgraded to a nonemergency situation, because the first responders lack training or experience at the scene of the hazardous materials incident. These industry employees should be trained to paragraph Q (emergency response operations) of 29 CFR 1910.120 (HAZWOPER), not just to paragraph E (uncontrolled hazardous waste sites), which is intended for waste site cleanup workers.

Emergency operations near the site of a chemical release are inherently more dangerous than defensive operations performed at a distance. This text is designed to comply with the 2008 edition of NFPA 472, Chapter 7: *Competencies for Hazardous Materials Technicians,* and HAZWOPER (29 CFR 1910.120) promulgated by the Occupational Safety and Health Administration (OSHA).

The text is organized in a sequence that follows the logical progression of a hazardous materials response. Chapter 1 introduces the concept of hazardous materials incidents and WMD incidents as well as reviews awareness and operations level concepts that are critical to successfully completing the hazardous materials technician competencies that follow. Chapter 2 lays the legal foundation of offensive hazardous materials response and answers many of the questions regarding why we do the things the way we do. Chapter 3 describes how hazardous materials affect the human body as well as how these effects are measured and quantified—both in epidemiological studies as well as on the scene of a hazardous materials release. Chapter 4 explains the chemistry of hazardous materials. Chapter 5 describes how hazardous materials behave in the context of their chemical and physical properties and the environmental conditions at the site of release. Chapter 6 explains how to determine key characteristics of hazardous materials, such as their chemical and physical properties, using reference sources. Chapter 7 describes the proper use of personal protective equipment, including chemical protective clothing and respiratory protection. Chapter 8 explains the use of monitoring equipment to assess the atmospheric hazards at a hazardous materials release. Chapter 9 describes how to conduct solid, liquid, and gas sampling and sample identification procedures for unknown chemical releases. Chapter 10 describes containers—including their features, design parameters, construction methods and materials, and their failure modes—and explains the damage assessment process. Chapter 11 discusses product control options for hazardous materials releases, including plugging and patching, drum patching and overpacking, grounding and bonding, product transfers, and venting and flaring. Chapter 12 describes decontamination procedures, and Chapter 13 explains the incident command system. Chapter 14 ties all the previous components of hazardous materials response together and describes the hazardous materials response process and organization.

This textbook may be used for a 40-hour class up to a 240-hour class by choosing the appropriate chapters and depth of coverage. For example, a 40-hour class would not cover many advanced applications such as cold tapping, transfer operations, and flaring found in Chapter 11 (Product Control), but would cover drum patching, plugging, and overpacking as well as chlorine kit application. Likewise, a 40-hour hazmat technician class would likely completely omit Chapter 4 (Chemistry) and Chapter 9 (Sample Collection and Identification). Therefore the following syllabi are suggested for the following hazardous materials technician class lengths:

40 hours:

- **Chapter 1** Introduction (complete coverage)
- **Chapter 2** Regulations and Standards (complete coverage)
- **Chapter 3** Toxicology (complete coverage)
- **Chapter 5** Chemical and Physical Properties (complete coverage)
- **Chapter 6** Information Resources (partial coverage)
- **Chapter 7** Personal Protective Equipment (complete coverage)
- **Chapter 8** Detection and Air Monitoring (complete coverage)
- **Chapter 10** Containers and Damage Assessment (complete coverage)
- **Chapter 11** Product Control (partial coverage based on agency needs)
- **Chapter 12** Decontamination (complete coverage)
- **Chapter 13** Incident Command System (coverage based on agency need and previous instruction)
- **Chapter 14** Response Organization (complete coverage)

80 hours: All the preceding chapters with the addition of significantly more hands-on activities, skill stations, and exercises and the following chapter:

- **Chapter 9** Sample Collection and Identification (partial coverage)

160 hours: All chapters should be included (either partially or entirely depending on agency needs), as well as hands-on exercises and skill stations that reinforce the competencies presented and explained in each chapter with the use of agency equipment and standard operating procedures (SOPs).

240 hours: All chapters in their entirety as well as in-depth functional exercises to reinforce hazardous materials response concepts.

Irrespective of class length, the information in this book must be complemented with hands-on activities designed to reinforce key response concepts. The online component, offered through Resource Central on www.bradybooks.com, is available to both instructors and students and contains suggested hands-on activities and exercises for each chapter. Resource Central also contains web resources as well as step-by-step visual guides for hands-on activities such as decontamination, donning and doffing of personal protective equipment, and product control procedures. No technician-level class is complete without hands-on skills stations and exercises with the actual equipment that responders will be using in the field.

Additionally, the authority having jurisdiction (AHJ) must ensure that the appropriate equipment and refresher training are available to their personnel. This textbook provides technical information that can be used for initial training and periodic recurrent training. HAZWOPER law requires that emergency response personnel operating under paragraph Q receive annual refresher training. This training must be "of sufficient content and duration to maintain their competency." OSHA did not assign a length of time to refresher training. Remember that the more complicated the equipment and procedures that personnel are required to carry out, the more periodic training will be required to maintain proficiency. Stay safe!

Chris Weber

ACKNOWLEDGMENTS

I would first and foremost like to thank my family for their patience and support during all my writing projects, including this book. Long writing projects are difficult not only for authors but also for their families. Sometimes the project appears all-consuming to everyone involved. I deeply appreciate their understanding and help throughout the duration!

This book has been the work of the better part of 5 years as I have obsessed about what to include and what not to include, and spent days trying to ensure that what I was trying to explain was actually getting explained. We all stand on the shoulders of giants, and I have had a lot of help and support from many friends and colleagues over the years, not to mention the accumulated work of the legends in the field of hazardous materials response.

I would like to thank the Washtenaw County Hazardous Materials Response Team for making me a hazardous materials responder. Over the years, we as a team honed our skills through education, training, and much practice. We learned many lessons the easy way, during classes, training, and exercises; and we learned some the hard way, but we always strived to learn them well and not repeat our mistakes. I received much support, especially from past and present team members, including the following: Chief Jim Roberts, Chief Bill Steele, Chief Bill Wagner, Chief Mike Skrypec, Chief Tom Edman, Chief Darwin Loyer, Chief Mark Nicholai, Captain Dan Cain, Captain Max Anthouard, Lieutenant Scotty Maddison, Supervisors Chuck Rork and Dean Lloyd, Captain Craig Liggett, Brad Tanner, Don Dettling, Tim Andrews, Doug Armstrong, Ernie Close, Jim Rachwal, Jeff French, Lieutenant Shaun Bach, Fred Anstead, Mike Loria, Andy Box, Chief Chris Bishop, and all the other hardworking and dedicated team members.

The Michigan State Police Emergency Management and Homeland Security Training Center asked me to become an adjunct instructor in 2003, and what an eye-opening experience that was. There I had the opportunity to polish my instructing skills with sage advice from program managers and mentors Bob Cook, Tony Garcia, and Ed Halcomb. The experience and advice of the following center instructors have proved invaluable in framing my understanding of the hazardous materials world: Brian Stults, Dean Blauser, Gary Brannock, Dennis Herbert, Chuck Doherty, Don Tillery, Pat Nael, Jeremy Connell, Tim James, Don Owsley, Larry Kosmalski, Chad Tackett, Phil Salinas, Gregg Ginebaugh, Dennis Reilly (now a program manager himself), Sergeant Don Tillery, Sergeant Emmitt McGowan, and Matt Ratliff. The center would not run like a well-oiled machine without the tireless efforts of Sergeant Dennis Harris, Lieutenant Mike Johnson, Stacy Theis, Renee Osborne, Rudy Walsdorf, Wendy Galbreath, Lieutenant Dave Wood, and Jim Porcello.

Over the years, I have had the honor and privilege of training thousands of students nationally and internationally. The most valuable feedback for an instructor and an author comes from the students themselves. The questions they ask steer the class and make it better. Students, please ask questions! This gives your instructor valuable information regarding your background, your interests, your temporary gaps in knowledge, and consequently, what areas need to be addressed. Questions also guide the discussions and spur the instructor to add information that may be missing from the syllabus, or information and anecdotes of which he or she is reminded through the question. I would like to thank every student who asked a question during one of my classes, who asked questions during break that I could address immediately after break, and who honestly filled out the evaluation at the end of class. You have played a large role in shaping not only this text but also this instructor.

The Longmont Fire Department, Longmont Police Department Special Enforcement Unit, Denver Fire Department, Michigan State Police Emergency Management and Homeland Security Training Center, Colorado WMD-CST, Colorado State Patrol, University of Colorado at Boulder Environmental Health and Safety, University of California San Diego Environmental Health and Safety, Las Vegas Metro Police ARMOR, Oakland County Regional Response Team, Pittsfield Township Fire Department, and Ypsilanti City Fire Department were kind enough to provide me with opportunities to take photographs at exercises and training scenarios.

Some individuals helped by discussing aspects of the book, others starred in photos or provided photos and case histories, and still others read and edited chapters to make sure I didn't leave anything out and to make sure the copy was understandable. I would like to thank the following colleagues for their generous help during this project: Mike Becker, Jeff Peterson, Martin McFarland, Michele Goldman, Todd Feaster, Craig Paull, Jeff Moll, Jared Mannering, Ethan Unwin, Emily Allen, Nate Rea, Micah Holmes, and Brian Jackson (Longmont FD); SSG Jay Lemons (8th WMD-CST); Brian Wagner (National Institutes of Health Fire Department); Doug Huffmaster (Las Vegas Metro PD/ARMOR); Steve Curry (Columbia, SC FD); Dean DeMark (Fort Bragg); Robert Kaminky (Colorado State Patrol); Tony Garcia, Dennis Harris, and Chad Tackett (Michigan State Police Emergency Management and Homeland Security Training Center); Dean Blauser and Sue Blauser (HazMat Solutions, Inc); Jay Daum (U.S. Coast Guard); MSgt William Bennett (51st WMD-CST); Doug Rohn (Madison Fire Department); Richard Dufek (Carmel Fire Department, retired); Gary Brandt (Carmel Fire Department); Bill Hand (Houston Fire Department, retired); Ralph Bogle (University of Colorado, Boulder); Tod Ferguson (University of California, San Diego); Nick Vent (County of San Diego); Tom Clawson (Technical Resources Group, Inc); Eugene Ngai (Chemically Speaking, LLC); Dr. Derek Gill (Airgas); Bill Cullen and Keith Silverman, PhD, MPH (GoldShield TEAM, LLC); Brian Heinz (CST); Jerry Grey (San Francisco Fire Department, retired); Chantel Javois, Mark Norman, Pauline Leary, and John Corbett (Smiths Detection); Jim Westgate and Kerstin Barr (Thermo Fisher Scientific); Lisa Mork Davis and Lonnie Toby (RAE Systems); Chris Wrenn (Environics USA); and Mike Everett (Pall Corporation).

I would like to thank the following individuals for reviewing this book. Their comments and suggestions were invaluable throughout the process.

Lieutenant Douglas Rohn
Madison Fire Department
Madison, WI

Anthony R. DeAngelo, BS, ME, CHP
Adjunct Instructor, Schenectady County Community College
Schenectady, NY

Division Chief Matthew W. Knott, MS, CFO, CEM, MIFireE
Rockford Fire Department
Rockford, IL

Deputy Fire Chief Gerard E. Mahoney
MA Public Administration, Framington State College; Cambridge Fire Department
Cambridge, MA

Assistant Fire Chief Jeff Travers
Director of Public Safety, Great Oaks Institute; Mariemont Fire Department
Cincinnati, OH

Battalion Chief David C. Harrington, AA, TN-EMT-P/IC, NR-EMT-P
City of Oak Ridge Fire Department
Oak Ridge, TN

Professor Shankara Babu, PhD, CHMM, CEM
College of Southern Nevada
Las Vegas, NV

Hazmat Chief Richard J. Lenius
BS Fire Science, DLS HAZMAT; Former Fire Chief, City of Katy;
Hazmat Instructor, Angelina College Fire Academy
Longview, Texas

Christopher Gilbert
MS Safety Engineering, Senior Environmental Specialist–Hazardous Materials
ACEPD, Gainesville, FL

Lieutenant Mike Becker
Longmont Fire Department–Station 5
Longmont, CO

Tom Clawson
Technical Resources Group
Idaho Falls, ID

Robert Cook
Hazard Management
Mount Pleasant, MI

Bill Cullen
GoldShield TEAM
Somerset, NJ

Jay Daum, Response Management School
United States Coast Guard
Yorktown, VA

Dean DeMark
Emergency Response Training
Fayetteville, NC

Richard Dufek
Firehouse Training & Consulting
Westfield, IN

Tod Ferguson
Environmental Health & Safety, University of California
San Diego, CA

Dr. Derek Gill
Airgas-SAFECOR
Albuquerque, NM

Bill Hand
Hazardous Materials Response Team
Harris County, Texas

Lieutenant Scott Maddison
Ypsilanti Fire Department
Ypsilanti, MI

John Meyers
JAM Consulting
Chicago, IL

Eugene Ngai
Chemically Speaking
Whitehouse Station, NJ

Dr. Keith Silverman
GoldShield TEAM
Somerset, NJ

Nick Vent
Department of Environmental Health
County of San Diego, CA

I would like to thank David Heskett, an aspiring artist and graphic designer, for most of the artwork in this book. He has done a tremendous job. His other work can be viewed at the website www.DHdesigns.com. I would also like to thank all of the hardworking people behind the scenes at Pearson on this book project: Eileen Clawson, Lisa Garboski, Barbara Liguori, Monica Moosang, Samantha Sheehan, and Stephen Smith.

Chris Weber
Longmont, CO

ABOUT THE AUTHOR

Chris Weber teaches hazardous materials topics internationally, actively develops and instructs numerous hazardous materials and counterterrorism courses, and runs full-scale training exercises through his training and consulting firm, Dr. Hazmat, Inc. (www.drhazmat.com). Chris regularly instructs classes in illicit laboratory response for the U.S. military; tactical chemistry classes for hazmat technicians; 24-hour Hazmat Medic classes for EMS personnel; regulatory classes such as HAZWOPER, RCRA, and DOT courses and refreshers for industry; and numerous other courses for hazardous materials responders including the following:

Hazardous Materials Operations Train-the-Trainer Course (www.hazmatoperations.us)
Hazardous Materials Technician Train-the-Trainer Course (www.hazmattechnician.us)
24-, 40-, 80-, 160-, and 240-hour Hazardous Materials Technician Courses
Hazardous Materials Technician Refresher Course
Air Monitoring and Sample Identification Course
Tactical Chemistry Course
Illicit WMD and Drug Laboratory Recognition and Response Course (www.illicitlabs.com)

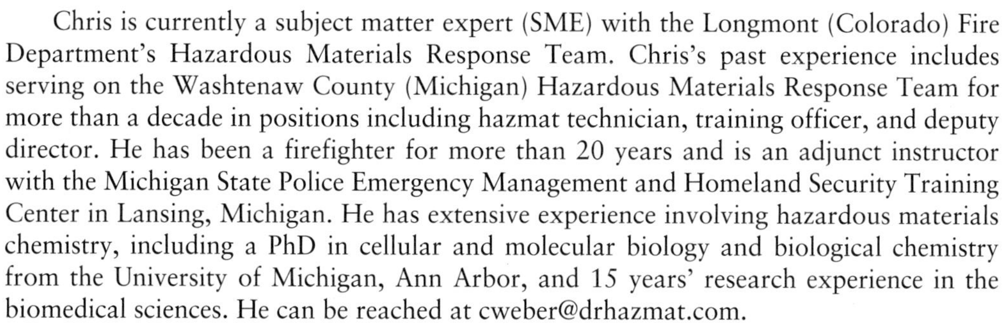

Chris is currently a subject matter expert (SME) with the Longmont (Colorado) Fire Department's Hazardous Materials Response Team. Chris's past experience includes serving on the Washtenaw County (Michigan) Hazardous Materials Response Team for more than a decade in positions including hazmat technician, training officer, and deputy director. He has been a firefighter for more than 20 years and is an adjunct instructor with the Michigan State Police Emergency Management and Homeland Security Training Center in Lansing, Michigan. He has extensive experience involving hazardous materials chemistry, including a PhD in cellular and molecular biology and biological chemistry from the University of Michigan, Ann Arbor, and 15 years' research experience in the biomedical sciences. He can be reached at cweber@drhazmat.com.

Chris's other publications include the following:

Pocket Reference for Hazardous Materials Response
Hazardous Materials Operations

CHAPTER 1

Introduction

Courtesy of Chris Weber, Dr. Hazmat, Inc.

KEY TERMS

biological agent, *p. 2*
cold zone, *p. 18*
corrosive material, *p. 3*
cryogenic liquid, *p. 20*
dangerous goods, *p. 3*
explosion, *p. 20*
explosive, *p. 20*
flammable liquid, *p. 3*
flashpoint, *p. 22*
hazardous material, *p. 2*
hot zone, *p. 18*
illicit laboratory, *p. 7*
olfactory fatigue, *p. 19*
oxidizer, *p. 22*
pathogens, *p. 24*
prion, *p. 24*
pyrophoric material, *p. 22*
radiological dispersal device (RDD), *p. 16*
terrorism, *p. 11*
toxin, *p. 6*
warm zone, *p. 18*
weapon of mass destruction (WMD), *p. 6*

OBJECTIVES

After reading this chapter, the student should be able to:

- Describe six common types of hazardous materials incidents.
- Describe five signs of a hazardous materials release.
- List four types of weapons of mass destruction (WMDs).
- Name the nine DOT hazard classes.
- Use the most recent version of the *Emergency Response Guidebook* (*ERG*).
- Describe eight characteristics common to progressive hazmat teams.

ResourceCentral For additional review and practice tests, visit **www.bradybooks.com** and click on Resource Central to access book-specific resources for this text! To access Resource Central, follow directions on the Student Access Card provided with this text. If there is no card, go to **www.bradybooks.com** and follow the Resource Central link to Buy Access from there.

You are dispatched to an incident at the district courthouse. According to the caller, a letter-sized envelope is leaking a white powder in the mail room of the building. As you respond, several questions go through your mind:

- Will this be a false alarm, or is this the real deal?
- What types of clues should I be looking for?
- How should I approach the scene?
- Who will be my contact at the courthouse?
- How many people have been exposed to this white powder?
- What other resources might be helpful?

Let's see if we can get some answers to these questions and others.

hazardous material ■ A substance that, because of its quantity, concentration, corrosiveness, flammability, reactivity, toxicity, infectiousness, or radioactivity, constitutes a threat to human health, public safety, or the environment.

What are **hazardous materials**? We should formulate a definition before we try to respond to releases of hazardous materials so that we can understand the scope of the problem and minimize the danger to ourselves and to the public. We can surmise that a hazardous material might be dangerous, but we must know precisely what we are dealing with when we respond to incidents involving hazardous materials.

There are many different definitions of hazardous materials, from the technical bureaucratic definitions to the colloquial. Let's start with the official versions:

> Hazardous material means a substance or material that the Secretary of Transportation has determined is capable of posing an unreasonable risk to health, safety, and property when transported in commerce. . . . The term includes hazardous substances, hazardous wastes, marine pollutants, elevated temperature materials, materials designated as hazardous in the Hazardous Materials Table . . ., and materials that meet the defining criteria for hazard classes and divisions. . . .
>
> —U.S. Department of Transportation (49 CFR 171.8)

Possibly the most far-reaching definition has been put forth by the Occupational Safety and Health Administration (OSHA):

biological agent ■ A disease-causing organism or toxic substance of living origin, such as a bacterium, virus, or toxin. Examples include anthrax, smallpox, and ricin.

> *Hazardous substance* means any substance designated or listed under paragraphs (A) through (D) of this definition, exposure to which results or may result in adverse affects on the health or safety of employees:
>
> (A) Any substance defined under section 101(14) of the Comprehensive Environmental Response, Compensation, and Liability Act (CERCLA);
>
> (B) Any **biological agent** and other disease-causing agent which after release into the environment and upon exposure, ingestion, inhalation, or assimilation into any person, either directly from the environment or indirectly by ingestion through food chains, will or may reasonably be anticipated to cause death, disease, behavioral abnormalities, cancer, genetic mutation,

physiological malfunctions (including malfunctions in reproduction) or physical deformations in such persons or their offspring;

(C) Any substance listed by the U.S. Department of Transportation as hazardous materials under 49 CFR 172.101 and appendices; and

(D) Hazardous waste as herein defined.

—U.S. Occupational Safety and Health Administration (29 CFR)

Hazardous waste is defined by the U.S. EPA and 40 CFR 261.3. Internationally, hazardous materials are known as **dangerous goods**, and with increasing international standardization of transportation regulations through global harmonization, you'll likely be hearing this term more often in the future.

dangerous goods ■ Internationally accepted term for hazardous materials.

Well over 50 million different chemicals are used, transported, and manufactured, or have been registered in the Chemical Abstracts Service (CAS) database. According to the U.S. Department of Transportation, the following were the top 10 chemicals involved in hazardous materials transportation incidents in 2009:

1. Paint or Paint Related (Flammable-Combustible Liquid) 9.14% of incidents
2. Paint Related Material (Flammable-Combustible Liquid) 7.22%
3. Flammable Liquids N.O.S. (Flammable-Combustible Liquid) 5.40%
4. Paint (Combustible Liquid) 4.13%
5. Corrosive Liquids N.O.S. (Corrosive Material) 4.05%
6. Isopropyl Alcohol (Flammable-Combustible Liquid) 3.25%
7. Sodium Hydroxide Solution (Corrosive Material) 2.78%
8. Fire Extinguishers (Nonflammable Compressed Gas) 2.16%
9. Hydrochloric Acid Solution (Corrosive Material) 2.15%
10. Corrosive Liquid Basic Inorganic (Corrosive Material) 2.15%

Statistically, this means that in your career you are much more likely to encounter **flammable liquids** and **corrosive materials** than other classes of hazardous materials. However, you still need to be prepared to deal with any of those 50 million chemicals!

flammable liquid ■ A liquid having a flashpoint below 100°F (37.8°C) (per OSHA), or a liquid having a flashpoint of 141°F (60.5°C) or below (per DOT).

corrosive material ■ A substance that causes destruction of human skin at the site of contact or reacts with steel or aluminum at a specified rate.

Now that we have a better understanding of what hazardous materials are, let's look at an excellent colloquial definition of a hazardous material:

> A hazardous material is any substance that jumps out at you when something goes wrong, and hurts or harms the things it touches when released.
>
> —Ludwig Benner, *Hazardous Materials Emergencies*, 1976

This is a very practical definition. We need to be prepared for anything that can come out and bite us. We must have the appropriate equipment and the ability to use it safely in dealing with a hazardous material, and we must have sufficient knowledge to predict accurately how the hazardous material will behave. That is, we need to have both good equipment and, even more important, good training.

A hazardous material (hazmat) incident is any unintended release of a hazardous material from its container. A well-engineered chemical process, or hazardous materials that are properly stored in compatible containers, pose no immediate hazard. However, an unintended release, such as by human error, container failure, natural disaster, or criminal act, poses a risk. A hazmat response is necessary when control over a chemical is lost through container breach, pipe rupture, runaway chemical reaction, or multiple other means. Some hazardous materials incidents are relatively minor and can be cleaned up safely and easily. In industry these are called *incidental spills*, and the emergency response community usually never even becomes aware of them. Other incidents are larger and pose a greater risk to people and to the environment. Such hazardous materials incidents require an emergency response and often pose a significant risk to responders, to the public, and to the environment. These are the types of incidents we will consider.

Location of Hazardous Materials

Hazardous materials are located throughout our communities. They can be found at large industrial facilities, water treatment plants, research universities, warehouses, and even local schools. They are also found in grocery stores, home improvement stores, beauty salons, auto parts stores, and most retail establishments. Unexpectedly large quantities of hazardous materials may even be found at private residences producing their own biodiesel—for example, in the garage or in an outside shed. However, the quantities of a hazardous material found in large fixed-site storage tanks, railroad tank cars, or highway tankers typically pose a much larger risk than the contents of smaller containers such as pint containers, 5-gallon pails, or even 55-gallon drums.

Common Hazardous Materials Incidents

Just as in fire department, emergency medical system (EMS), or law enforcement daily operations, there are "bread-and-butter" hazardous materials runs—those incidents that occur with some frequency and that a competent hazmat team will be prepared for and train for regularly (Figure 1-1).

SICK BUILDING SYNDROME

Acute building-occupant chemical exposures, often called *sick building* responses, are routine calls for the typical hazardous materials response team. A sick building call can range from reports of strange smells in the building to sickened occupants inside the building. It is the hazmat team's responsibility to identify the problem. Sometimes a chemical or a mixture of chemicals is the root cause, such as roofing solvents or an incompatible mixture of cleaning chemicals. At other times the cause may remain a mystery, such as when a transient source disappears before the investigation can be started (as when a delivery

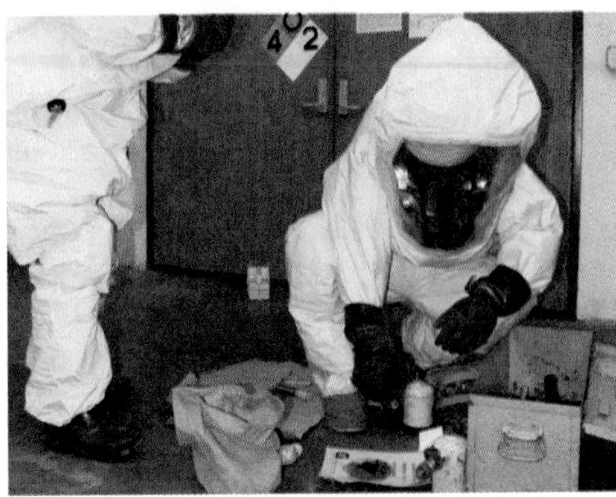

FIGURE 1-1 Four hazardous materials incidents. Clockwise from top left: a clandestine drug lab dump site, chemical sampling during an illicit lab investigation, abandoned gold mine effluent caused by illegal mining, and application of a chlorine B-kit to stop a chlorine leak in a 1-ton container. These are hazardous materials incidents that hazardous materials technicians may encounter and must be prepared to handle. *Photos courtesy of Chris Weber, Dr. Hazmat, Inc.*

truck idled too long near a building air intake, then drove away). At yet other times there may not even be a legitimate cause, for example, on a Friday afternoon at a school when the students merely want the afternoon off. As a hazmat technician who is part of a hazmat response team, it will be your responsibility to determine what is happening. These types of calls require excellent training, good air monitoring skills, and good discipline. These calls can quickly escalate out of control for an unprepared hazmat team.

MERCURY SPILLS

Mercury is a toxic liquid metal found in many consumer and industrial items, including some thermometers, certain switches in cars and household thermostats, and fluorescent bulbs. Mercury is not particularly toxic if an individual is exposed for a single short-term exposure; however, long-term exposure can lead to severe health effects. It is therefore very important to detect and thoroughly clean up spilled mercury in enclosed occupied areas such as residential homes and businesses. The number of mercury responses should decrease over the next several years as mercury thermometers are phased out in favor of other types, including alcohol-filled and electronic thermometers.

FIRES AT HAZARDOUS MATERIALS OCCUPANCIES

Fires in occupancies that store quantities of hazardous materials must be dealt with very carefully. Emergency responders—including not only firefighters but also other agency personnel such as law enforcement officers and emergency medical technicians (EMTs)—at the scenes of these fires should be monitored for immediate and delayed symptoms of exposure to chemicals and toxic products of combustion.

Environmental concerns are a significant issue at such fires as well. In the Midwest several years ago, a fire at an agricultural supply store was aggressively fought using great amounts of water. The runoff from this fire contaminated the groundwater in a large area. Often, it is better to let these fires burn and to protect exposures only as necessary, taking care to contain any runoff. However, downwind populations must be protected from potentially toxic smoke and/or any runoff from firefighting operations. It is very important that the hazmat team advise the incident commander at fire scenes involving such occupancies to prevent dire long-term consequences. It is also prudent to have the incident commander notify local and state public health agencies and environmental protection agencies.

OVERTURNED HIGHWAY CARGO TANKERS

Gasoline tankers, propane tankers, and other chemical cargo tanks are ubiquitous in our communities. Periodically, motor vehicle accidents involve these vehicles. Often, the cargo tank overturns and starts leaking. As a hazardous materials technician you will be tasked with confining spilled material to the area of the release and containing the leak. With a gasoline tanker this may involve diking and damming sewer openings and applying dome clamps. These types of cargo tanks must be unloaded before they can be righted. Perhaps your hazardous materials response team will perform this task, or the team may stand by as a mitigation company does so. In Chapter 11 we will see how a gasoline tanker is drilled, and gasoline is transferred to a waiting cargo tanker. It is only a matter of time before you are called to one of these incidents.

DERAILMENTS

Fortunately, derailments are significantly less common than highway cargo tanker accidents, but when they do occur, they are significantly more complicated and dangerous. Cargo trains can have dozens to hundreds of railcars, many of which will be carrying significant quantities of hazardous materials—typically, 10,000 to 40,000 gallons of product. These larger quantities of chemicals dictate larger downwind protective action distances. A leaking or breached railcar is a very difficult problem for any hazardous materials response team. The best way to deal with a train derailment is to plan with the rail industry for such an eventuality during your training.

VICTIM RESCUE AND RECOVERY

Sometimes, routine medical emergency calls turn into hazardous materials incidents. For example, dispatch information indicates a person with difficulty breathing, and upon arrival at the scene you find a male patient lying unconscious next to a 55-gallon drum of muriatic acid. How should you handle this situation? Your awareness and operations level training tell you to stay away and call the hazardous materials response team. But now you are in the process of becoming a hazardous materials technician and must act as a member of the hazmat team.

Quick action is important at EMS calls involving hazardous materials. There are two possibilities: the first is that this is a routine medical emergency, and the muriatic acid is safely contained in the drum. Perhaps the patient collapsed owing to a diabetic emergency, for example, and this is not a hazardous materials incident. However, the second possibility is that muriatic acid is leaking, and the patient was overcome by hydrogen chloride vapors. He may be contaminated, and if you enter the area without respiratory protection you will become exposed to the same dangerous vapors. At these types of incidents, it is essential that victim rescue be completed rapidly, but hazmat technicians must survey the scene using air monitoring equipment and wear the appropriate level of personal protective equipment (PPE) while performing the rescue.

CRIME SCENES

Hazardous materials incidents can become crime scenes for many different reasons. For example, if hazardous materials were transported or stored improperly, laws may have been violated, and a criminal investigation will ensue. If hazardous materials were dumped illegally, environmental crimes may have been committed. Hazardous materials may also be used in the process of committing a crime, such as synthesizing illegal drugs in an illicit laboratory.

Criminals or terrorists may target various locations in your community. The location will vary based on the intent, ideology, and ultimate goal of the perpetrator. Criminal incidents may include chemical releases used as diversionary tactics during crimes such as bank robberies, toxic chemical releases during a murder or attempted murder, or terrorist attacks using hazardous materials or **weapons of mass destruction (WMDs)**. Incidents involving WMDs are by definition a crime scene and fall under the jurisdiction of the Federal Bureau of Investigation (FBI).

weapon of mass destruction (WMD) ▪ Any weapon composed of explosives, chemical agents, biological agents, or radioactive material designed to inflict large numbers of casualties and/or large amounts of property damage.

Suspicious Letters and Packages

Suspicious letters and packages may take many forms. These types of responses pose a real danger to responders and must be carefully managed. The most significant risk of suspicious letters is that they may contain biological agents that pose a grave inhalation hazard. Suspicious letters may include other dangerous hazardous materials or WMDs, such as the readily available **toxin** ricin.

toxin ▪ A poisonous substance produced by a living organism or bacterium. Examples include ricin, botulinum toxin, and mycotoxins.

Other suspicious letters or packages may be bombs. Letter bombs may contain explosives such as Primasheet, a commercial plastic explosive, and be designed to kill or injure individuals. Packages may contain larger quantities of explosives and create significant structural damage when detonated. If a letter bomb has traveled in the mail, the explosive it contains is typically fairly stable, since it must withstand the rigors of transport. Letter bombs may be triggered by many different means: they may be rigged to explode on opening, remotely, on a time delay, or at a precise time. These types of suspicious letters are extremely dangerous owing to the energy released by the explosives. You should always call an explosive ordnance disposal (EOD) team, or bomb squad, when dealing with a potential letter bomb.

Suspicious letter calls are still fairly common for most communities. The vast majority of these suspicious letters are harmless. The hazmat team must determine which ones may be credible and which ones are almost certainly hoaxes or misperceptions. Remember: always assume the threat is real until it can be ruled out.

Illicit Laboratories

Illicit laboratories are a very common crime scene in many communities. The most common **illicit laboratories**, or clandestine laboratories, you will respond to are methamphetamine laboratories (meth labs). These labs are becoming increasingly common throughout North America and are extremely dangerous because they contain a witch's brew of hazardous materials, including corrosives, flammables, water-reactive and/or air-reactive substances, as well as many toxic materials. In addition, the individuals making the drugs (the "cooks") often deploy booby traps to dissuade would-be thieves, do not possess proper laboratory or chemistry training, have poor housekeeping skills, and dispose of hazardous waste improperly. In the case of illegal drug labs, the "cooks" are often users of the drugs they are producing and can be quite dangerous and unpredictable.

illicit laboratory ■ A location where a final product is made using precursor chemicals, reagents, and solvents. Illicit labs can be configured to make illegal drugs, chemical warfare agents (CWAs), biological warfare agents (BWAs), or explosives. Also known as clandestine laboratory, clan lab, or illicit lab.

Clandestine explosive laboratories have been common throughout parts of the Middle East, North Africa, and Asia for several decades and are becoming more common in the United States. Over the years these laboratories have become very sophisticated and can produce large quantities of powerful and sensitive explosives. In the United States terrorists have traditionally used commercially acquired and homemade ammonium nitrate/fuel oil (ANFO) mixtures in their attacks. ANFO is a very simple explosive to produce. Recently, foreign terrorists have begun to use organic peroxide–based explosives such as triacetone triperoxide (TATP) and hexamethylene triperoxide diamine (HMTD). These explosives are a little bit more complicated to produce but are a lot more dangerous owing to their sensitivity to heat, shock, and friction. You should always call an EOD team when dealing with a suspected illicit explosives laboratory.

Environmental Crimes

Given the large amount of chemicals used annually in North America, it is not surprising that environmental crimes are quite common. Environmental crimes include illegally dumping chemical waste and discharging untreated chemical waste into waterways. An example of the latter is introducing chemicals into sanitary sewers or storm drains. For example, a plant manager at a metal plating facility directs employees to bypass the wastewater treatment unit to save money. This shortcut sends untreated wastewater directly into the sewer system in violation of the permit issued by the municipal sewer authority. The plant manager is guilty of a criminal violation of the Clean Water Act (CWA).

Examples of environmental crimes involving illegal dumping include abandoning 55-gallon drums full of waste chemicals by the side of the road or disposing of hazardous chemicals in the municipal waste stream. For example, a cleaning products manufacturer decides to dump several buckets of corrosive waste into its municipal waste dumpster. The responsible party within the company is guilty of a violation of the Resource Conservation and Recovery Act (RCRA).

Environmental crime investigations often involve tracing pollutants to the originator (responsible party). This may involve the analysis of soil, water, or air samples that contain low pollutant concentrations and so requires sensitive laboratory equipment. Proper sample collection and rapid transport to the testing laboratory are essential for successful prosecution of many environmental crimes. The discharge site, such as a storm sewer outflow into a lake, may be distant from the source of the pollutant. Good detective work is necessary to trace the path of the pollution to its source and to identify the responsible party.

WMD Release or Attack

Weapons of mass destruction may be released or used in an attack in an almost endless variety of ways. WMD releases and attacks are therefore more common than you might

SOLVED EXERCISE 1-1

You have been dispatched to a reported clandestine methamphetamine laboratory. Prior to entering the area, what are some of your concerns?

Solution: Is the scene secure? Has law enforcement swept the area for suspects? Are there any booby traps? Has the air been monitored, and is the atmosphere safe? Have any chemicals been spilled, or are any containers leaking? You need answers to all these questions before you enter the area. As you can imagine, representatives from multiple agencies—or cross-trained personnel—will typically be required to provide the answers.

initially think. In 1983 cyanide was placed in packages of Tylenol and caused the death of several people. This incident spurred the creation of the tamperproof closures for pharmaceutical and food containers with which we are all familiar. Ricin, a toxin that can be isolated from the castor bean, has been used in numerous attacks throughout the United States. The FBI must be notified as soon as possible in cases of crime involving WMDs.

Hazardous Materials Accidents That Shaped Modern Hazmat Response

One of the best approaches to preparing for future events is to look at past events:

> Those who cannot remember the past are condemned to repeat it.
>
> —George Santayana, *The Life of Reason*, 1905

Over the years many chemical releases have shaped the development of hazardous materials response at the public safety, industry, and military levels. Some of these accidents caused serious environmental damage, and yet others caused significant injury and death. Many hazardous materials incidents involved all the preceding elements. Accidents such as those at Texas City, Texas; Waverley, Tennessee; Kingman, Arizona; and Kansas City, Kansas, were instrumental in shaping modern hazmat response. Not surprisingly, there have been many significant international hazardous materials incidents as well, the most significant of which occurred in Bhopal, India—a tragedy we will examine in more detail in the next chapter. We will study all the aforementioned incidents and many other significant hazardous materials incidents through the course of this textbook, to help us learn to use risk-based response at hazmat incidents.

SEVESO, ITALY (1976)

The Seveso chemical reactor accident, named after the municipality most affected, tough the reactor was located in Meda, was a watershed event for European industrial safety laws that led to the Seveso II Directive, as European Union (EU) industrial safety regulations are now commonly known. On July 10, 1976, more than 6 tons of chemicals containing highly toxic and environmentally damaging dioxins were released over approximately 7 square miles (18 km^2) about 10 miles (16.1 km) north of Milan, Italy.

This accident bore a causative failure chain of multiple errors common to most large-scale disasters: (1) an atypical process was performed, (2) inadequate monitoring devices were in place, and (3) the process was interrupted by unforeseen events. The manufacturing plant was attempting to create a chemical intermediate necessary for both herbicide 2,4,5-T and hexachlorophene production. The reaction required a higher temperature than was available from the normal utilities. To achieve this elevated temperature, personnel opted to route high-pressure steam generated from an electrical turbine into the external heating coils of the chemical reactor. The temperature of the steam could not be monitored, even though it was known that violent decomposition occurs at elevated temperatures. Because Italian law mandated that plant operations be shut down over the weekend, the steam generated by the turbine became hotter as the load on the turbine decreased. This increased temperature started a slow decomposition reaction that rapidly turned into a runaway exothermic (heat-generating) reaction several hours later. A pressure relief valve finally opened, causing the widespread dioxin contamination.

Almost 100,000 people lived near the plant. There were 736 people in the most highly contaminated area, almost 5000 in a secondary zone affected, and more than 30,000 in a region of lower contamination. The dioxins killed 3300 animals within days. To prevent food chain contamination, more than 80,000 domesticated animals were slaughtered and disposed of. Almost 500 people developed skin lesions, a condition known as chloracne. In the ensuing years, several pivotal epidemiological studies showed that 2,3,7,8-tetrachlorodibenzodioxin (TCDD), one of the chemicals that was produced in the reaction, is carcinogenic and causes cardiovascular and endocrine-related health effects. TCDD was also a significant contaminant of Agent Orange used during the Vietnam War defoliation program.

In a soap opera–like twist, a hazardous waste that was originally improperly disposed of by a cleanup contractor in France was ultimately incinerated by the parent company of the chemical plant. Several plant managers were initially sentenced to prison terms, although only two sentences were upheld, and much less time was actually served.

MEXICO CITY, MEXICO (1984)

On November 19, 1984, a large liquefied petroleum gas (LPG) terminal in San Juanico near Mexico City was completely destroyed by fire and multiple explosions. More than 500 people were killed, including scores of emergency responders, and more than 5000 others suffered severe burns. The terminal contained more than 50 LPG storage tanks, six large spherical tanks, and almost 50 horizontal storage tanks. The terminal was being supplied from a refinery 249 miles (400 km) away through an approximately 12-inch (30.5-cm) pipeline. A large LPG leak developed, whose location the terminal operators could not immediately identify. Therefore, highly flammable gas continued to leak for 5 to 10 minutes until an ignition source in the form of a flare stack ignited the large gas cloud. More than 11,000 cubic meters of gas was consumed in the inferno. The large fire subsequently caused multiple boiling-liquid, expanding-vapor explosions (BLEVEs) over the course of an hour and a half. A large number of casualties occurred in residential areas immediately adjacent to the terminal. Contributing factors to this accident included the close proximity of tanks to one another, the lack of flammable gas detection, the ineffectiveness of emergency shutoff valves, and the destruction of the fire protection systems during the first explosion. In addition, emergency response was delayed by heavy traffic in the area immediately after the accident.

AKRON, OHIO (1989)

This CSX train derailment in Akron, Ohio, graphically illustrates the need for complete and accurate communications between emergency responders and railroad officials. The derailment of 21 railcars—including nine loaded with butane—occurred on February 26, 1989, very close to a B. F. Goodrich Chemical Company polymer manufacturing facility that was actively using the monomers butadiene and acrylonitrile. Two of the tank cars released their butane, which ignited almost immediately. This explosion prompted the evacuation of 1750 people within approximately one square mile. The fire damaged the chemical facility, and chemical facility representatives later noted that had cooling systems failed inside the building, a large toxic cloud would have been released, significantly compounding the emergency.

Initially, communication between emergency responders and the train crew was spotty at best, and there was confusion as to which railcars were involved in the derailment. Because a printed consist was not up to date, the incident commander was unsure whether several chlorine tank cars and a butadiene tank car were also involved in the derailment. This confusion took over an hour to resolve. Eventually it was determined that only butane tank cars were involved. The tank cars were kept cool with master streams, and the fires were controlled.

Communications between emergency responders and railroad representatives continued to be suboptimal during clearance of the wreckage. Railroad representatives determined that the best plan of action would be to rerail the one damaged butane car—despite extensive damage to load-bearing structural components—and transport it approximately 4 miles to another facility. Despite internal disagreement among CSX representatives about the safety of rerailing the damaged butane railcar, these concerns were not communicated to the incident commander. Neither alternative response plans nor information about the hazards of rerailing and transporting the damaged butane tank car was relayed to emergency responders or to city representatives. Approximately 3.5 miles (5.6 km) into the journey the damaged butane tank car rolled off its trucks (wheel assemblies) owing to a damaged bolster assembly. This second hazmat incident prompted a second, although smaller, evacuation of 25 families.

The National Transportation Safety Board (NTSB) determined that a number of factors were problematic in this incident:

1. The absence of an updated and accurate train consist
2. Poor communication between railroad representatives and emergency responders, including lack of consist information from the train crew, and a lack of hazard communication during wreckage clearance operations
3. The lack of qualified technical experts to independently advise local emergency responders during wreckage clearance operations
4. The proximity of the chemical facility to the mainline railroad track
5. Maintenance issues with the railcars
6. Maintenance issues with the railroad track

Hazardous materials technicians responding to train derailments should always immediately attempt to locate the train crew and obtain the consist. During wreckage clearance operations it is imperative that emergency responders ask the right questions, including the following: What are alternatives to the plan you are proposing? What are the hazards associated with your plan? How do you propose to minimize these hazards?

BALTIMORE, MARYLAND (2001)

On July 18, 2001, a 60-car CSX freight train derailed inside the Howard Street tunnel, sparking a fire that burned for five days and incapacitated downtown Baltimore. The fire reached peak temperatures of 1800°F (980°C), ruptured a 40-inch (102-cm) water main over the tunnel, disrupted the Internet along the East Coast by damaging fiber optic cables, and forced countless trains to be rerouted. The chemical-fueled fire was caused by the rupture and ignition of a tank car carrying tripropylene. The heat of the fire, which averaged about 900°F (482°C) throughout the length of the tunnel, also caused a hydrochloric acid tank car to fail. The fire was extremely difficult to fight from either end of the tunnel owing to the intense heat. Eventually, a manhole above the tunnel was used to access the seat of the fire and extinguish it. It has been reported that delayed notification may have led to the inability to fight the fire from the tunnel entrance, because the fire had time to build up intense heat.

TOULOUSE, FRANCE (2001)

A catastrophic explosion occurred on September 21, 2001, in a fertilizer factory containing 300 tons of ammonium nitrate in Toulouse, France. The blast destroyed the factory and killed 29 people, including one school-aged boy. The force of the explosion seriously wounded 2500 people and caused an additional 8000 relatively minor injuries. Structural damage displaced more than 40,000 people—roughly 10% of the population—for several days. There are conflicting theories about the cause of this catastrophic event. Official experts believe that an exothermic chemical reaction caused by the mixing of incompatible chemicals in a waste stream resulted in a heat and pressure buildup that detonated the stored ammonium nitrate. An alternative theory advanced by the French Minister of the Environment is that a plant subcontractor purposefully caused the explosion owing to "possible Islamic fundamentalist sympathies." This conjecture highlights the fact that disgruntled employees can do significant damage at chemical facilities.

Terrorism

Terrorism is an emotionally and politically charged word. International law enforcement agencies and governments have not been able to agree on a conclusive definition, perhaps because several governments carry out terrorism or sponsor it indirectly. In military circles **terrorism** has been defined as a form of asymmetric warfare. This means that a weaker enemy uses subterfuge, sneak attacks, and improvised weapons to attempt to defeat a more powerful foe. The U.S. Department of Justice has promulgated a useful definition of terrorism in the Code of Federal Regulations:

terrorism ■ The unlawful use of force or violence to intimidate or coerce the government, the civilian population, or any segment thereof in furtherance of political or social objectives.

> Terrorism includes the unlawful use of force and violence against persons or property to intimidate or coerce a government, the civilian population, or any segment thereof, in furtherance of political or social objectives.
>
> —28 CFR 0.85(l)

The U.S. Department of State defines terrorism in the United States Code as

> premeditated, politically motivated violence perpetrated against non-combatant targets by subnational groups or clandestine agents
>
> —22 USC 2656f(d)

The U.S. Department of Defense defines terrorism as

> the unlawful use of violence or threat of violence to instill fear and coerce governments or societies. Terrorism is often motivated by religious, political, or other ideological beliefs and committed in the pursuit of goals that are usually political.
>
> —Joint Publication 3-07.2, "Antiterrorism", 24 November 2010

It is important to examine the motive behind terrorism to be able to predict terrorist targets, the timing of terrorism, and terrorist methods.

RECOGNIZING TERRORIST INCIDENTS AND CRIMINAL EVENTS

A terrorist may target locations based on their high profile, function, high occupancy load, or significance to the community. Thus, transportation hubs such as airports and train stations, national monuments and symbols, federal buildings, critical infrastructure such as power plants and water treatment plants, abortion clinics, schools, stadiums, malls, courthouses, religious establishments, and many other locations may be targeted. WMD incidents may involve military or improvised chemical or biological agents deployed to hurt or kill civilians and responders alike (Figure 1-2).

The acronym CBRNE stands for Chemical, Biological, Radiological, Nuclear, and Explosives. These five categories are collectively termed weapons of mass destruction, but they are not the only tools criminals and terrorists may use. There are a host of very deadly chemicals in homes, stores, and workplaces that also may be misused.

Explosives and Incendiary Devices

By far the most commonly employed criminal and terrorist weapons are explosives and incendiary devices, considered in both historical and modern perspectives, chiefly because

4TH GRADE
GREENDALE SCHOOL
FRANKLIN PARK NJ 08852

SENATOR DASCHLE
509 HART SENATE OFFICE
BUILDING
WASHINGTON D.C. 20510

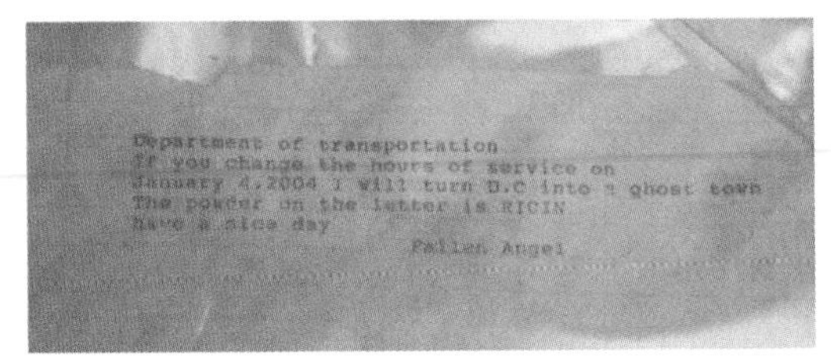

Department of transportation
If you change the hours of service on
January 4,2004 I will turn D.C into a ghost town
The powder on the letter is RICIN
Have a nice day

Fallen Angel

FIGURE 1-2 Four terrorist incidents. Clockwise from top left: the Alfred P. Murrah Federal Building after detonation of the truck bomb, smoke rising from Ground Zero at the World Trade Center site, a ricin threat letter, and an anthrax threat letter addressed to Senator Tom Daschle. Fortunately, terrorist incidents are relatively uncommon, but hazardous materials technicians must be prepared to assist with these incidents.

they are very destructive and are widely available owing to their industrial and military use. Explosives are commonly used in the construction industry, by miners, by farmers, and in a wide range of other legitimate businesses. Unfortunately there is a long history of the deliberate misuse of explosives in the United States and around the world.

Bath, Michigan (1927) On May 18, 1927, multiple bombings in a school in Bath Township, Michigan, killed 38 school children and seven adults, and injured at least 58 additional people. The attack on the school was perpetrated by Andrew Kehoe, a disgruntled school board member who had access to the school. He was allegedly upset at having to pay higher taxes owing to the consolidation of several smaller schools in the area. Kehoe carried out the attack in three phases. Initially he killed his wife and burned his farm. While firefighters were busy fighting the blazes at his home he detonated explosives he had previously placed in the basement of the school building over a period of months under the ruse of performing building maintenance. Then, after first responders arrived at the school, Kehoe drove up and detonated his shrapnel- and explosive-filled vehicle, killing himself and the school superintendent, and grievously injuring several others.

This was the first and deadliest attack on a school in U.S. history and also bore many hallmarks of modern terrorist attacks. This bombing marked the first terrorist attack to specifically target first responders trying to help victims at the scene. Kehoe also acquired his explosives in small quantities to avoid detection.

He used the explosives pyrotol and dynamite. Pyrotol was an incendiary explosive reprocessed from surplus World War I cordite and smokeless powder. It was commonly used by farmers in conjunction with dynamite to remove tree stumps. Investigators later found an additional 500 pounds of explosives planted in the basement of a different wing of the school that had not detonated because the firing train was interrupted by the first explosion. The government response to this incident was to remove pyrotol from the market.

Chesterton, Indiana (1933) The first bombing of a commercial airliner, United Airlines flight 23, occurred on October 10, 1933, over Chesterton, Indiana, and killed seven people. Witnesses described hearing an explosion and seeing the plane in flames shortly after 9 PM at an altitude of about 1000 feet. The United Airlines flight was allegedly brought down by a nitroglycerin bomb that was placed in the baggage compartment at the rear of the airplane.

The Unabomber (1978–1995) The Unabomber, the alias given by authorities to Ted Kaczynski, went on a 17-year bombing spree from May 1978 until he was apprehended in 1995. The Unabomber was a lone-wolf terrorist, a neo-Luddite who disapproved of modern technology. Most of the bombs were delivered through the postal service or placed at specific locations. In 1979 he placed a bomb in the cargo hold of American Airlines flight 444. Although the bomb started smoking, and some passengers were treated for smoke inhalation, a faulty timing mechanism luckily prevented the bomb from detonating. Experts estimated that it would have destroyed the plane had it detonated. Over the 17 years three people were killed and at least 24 were injured.

Oklahoma City, Oklahoma (1995) On April 19, 1995, Timothy McVeigh and Terry Nichols detonated almost 7000 pounds of explosives in front of the Alfred P. Murrah Federal Building in Oklahoma City, Oklahoma. This explosion killed 168 people, including 19 children, and injured almost 700. The explosion damaged or destroyed hundreds of buildings and dozens of cars and shattered glass in hundreds of nearby buildings. Property damage exceeded $650 million. The attack occurred on the second anniversary of the deaths in Waco, Texas, at the end of a fiery standoff with federal agents. McVeigh and Nichols were motivated by antigovernment sentiment. This attack, the second deadliest after the September 11, 2001, World Trade Center disaster, spurred legislation designed to increase isolation distances and barriers around federal buildings to limit the damage from any future terrorist attacks.

The bomb consisted of 5000 pounds of ammonium nitrate fertilizer mixed with 1200 pounds of nitromethane and 350 pounds of Tovex in a rental truck. On September 30, 1994, Terry Nichols bought 2000 pounds of ammonium nitrate from a farm supply company in Kansas in the form of forty 50-pound bags. This was an excessive amount of ammonium nitrate even for farming operations and should have raised red flags. It is very important for first responders, especially hazardous materials technicians and law enforcement, to have good relationships with local chemical suppliers and to be involved in their education concerning signs of terrorism.

September 11th Attacks (2001) The September 11 (9/11) attacks on the twin towers of the World Trade Center in New York City, the Pentagon in Washington DC, and United Airlines flight 93 were the deadliest terrorist attacks on U.S. soil, claiming almost 3000 victims, including 343 firefighters, 60 police officers, and another eight paramedics from private EMS. Four separate flights were hijacked by 19 Islamist terrorist hijackers in a coordinated effort. Two flights were crashed into the twin towers, one flight was crashed into the Pentagon, and one flight that was presumably bound for the U.S. Capitol or White House was seized by passengers and crashed near Shanksville, Pennsylvania. Essentially, these planes were incendiary bombs, each estimated to have been carrying more than 20,000 gallons of jet fuel, or the energy equivalent of 600 tons of TNT. In New York City, the heat generated by the fuel was enough to catastrophically weaken the steel support members of the twin towers and cause the floors above the fire at the impact zone to collapse in a pancake fashion, leading to a complete collapse of the buildings.

Tragically, the death toll continues to rise, since many first responders that worked in the toxic dust cloud are dying from respiratory ailments and other related diseases. Advocacy groups believe that more than 800 responders involved in rescue and/or cleanup work have died; however, it cannot be conclusively proven how many deaths were due to the work carried out at Ground Zero. It is critical that first responders wear proper PPE in the wake of terrorist attacks and at the release site of any hazardous material or WMD.

You may be asking yourself how these attacks could have happened. The terrorists took advantage of the airline industry policy of obeying hijackers. This policy originated in an attempt to minimize casualties, since hijackers had never before been suicidal and were almost always apprehended at their final destination. The Islamist terrorist organization al-Qaeda took advantage of this knowledge in planning the 9/11 attacks. The hijackers did not need any weapons or bombs on the plane, and compliance was almost guaranteed by prevailing industry procedures; the fuel load on the planes themselves constituted their explosive, or incendiary, devices. This tragedy illustrates the importance of careful consideration in setting agency policy and response procedures. When these procedures are publicized or become self-evident from response patterns, terrorists can easily take advantage of this information. Operational security is very important for first responders.

Madrid, Spain (2004) On March 11, 2004, during morning rush hour, an al-Qaeda–inspired terrorist cell carried out a coordinated attack on the commuter train system in Madrid, Spain, that killed 191 people and wounded at least 1800. Thirteen bombs were placed aboard four separate commuter trains, and 10 of those bombs were remotely detonated. Examination of an unexploded device revealed that its core comprised approximately 22 pounds of Goma-2 ECO, a Spanish explosive, with about 2.2 pounds of nails and screws added as shrapnel. One of the devices reportedly contained dynamite as the primary explosive. Spanish miners sold the explosives to the terrorists. If the Spanish miners knew the intent of their customers, this would be the first and only case of collaboration between Islamist terrorists and non-Muslims.

The Madrid train bombings were symbolically planned for a total of 911 days after September 11, 2001. Terrorists often choose symbolic dates or anniversaries of events of importance to their cause.

London, England (2005) On July 7, 2005, during morning rush hour, another al-Qaeda–inspired terrorist cell carried out a coordinated suicide attack on the commuter bus and underground rail (the Tube) systems in London, England, that killed 52 people in addition to the four terrorists, and injured more than 700. The explosives used by the attackers were homemade organic peroxide–based explosives. On July 21, 2005, a second attack took place on the London bus and Tube systems. In this attack none of the main charges detonated, and there were no injuries. All the terrorists were later arrested owing to the excellent video surveillance system deployed in the United Kingdom, especially in London.

Moscow, Russia (2010) On March 29, 2010, during morning rush hour, two women who were part of an Islamist Chechen separatist terrorist group carried out two suicide bombings on the Moscow, Russia, metro that killed at least 40 people and injured more than 100. The explosive they used is believed to be a common commercial and military explosive called RDX. Both bombs were packed with shrapnel consisting of nuts, bolts, and screws.

Biological Agents

Biological agents were used in warfare even before pathogens— the microscopic causative agents of disease—were discovered. As early as 400 BC, Scythian archers used arrows dipped in the blood of decomposing bodies and in manure to increase the death toll by infection. In the mid-fourteenth century, Tartars catapulted bodies of bubonic plague victims over the walls of the besieged city of Kaffa, Ukraine, and the British gave smallpox-infested blankets to the enemy during the French and Indian War in the mid-eighteenth century.

In the modern era, there have been two confirmed terrorist attacks with biological agents in the United States (not involving toxins). One involved the dispersal of salmonella bacteria to influence a local election, and the second case involved the release of anthrax spores through the U.S. Postal Service. Biological agents are readily available and can be prepared relatively easily with rudimentary knowledge. Terrorist attacks involving biological agents are also very difficult to detect, since victims may not come forward for several days to weeks after the attack owing to the extended incubation period of biological organisms. Epidemiological surveillance activities performed by public health authorities are therefore very important in detecting a biological attack as early as possible to give victims the best possible chance for survival and to give law enforcement the best possible chance of preserving evidence and prosecuting the suspect.

Salmonella: The Dalles, Oregon (1984) In September 1984, followers of the Bhagwan Shree Rajneesh poisoned salad bars in 10 different restaurants in The Dalles, Oregon, using *Salmonella enterica* Typhimurium obtained from the American Type Culture Collection. More than 750 people became ill, and 45 were hospitalized. Fortunately, no one died. The followers of Rajneesh were attempting to influence a Wasco County election by sickening the local population to ensure that their candidates would win.

Residents of Antelope, Oregon, and The Dalles, Oregon, almost immediately suspected cult members of the attack. Indeed, public health officials investigated this outbreak, since it was the largest food-borne outbreak in the United States that year. They could not rule out intentional dissemination but blamed poor personal hygiene of the food workers, since they were stricken ill first. Oregon Democratic Congressman James H. Weaver suspected the cult members were to blame and on February 28, 1985, voiced a persuasive circumstantial case on the floor of the U.S. House of Representatives. In September 1985 Rajneesh himself came forward and implicated several of his followers of involvement in the salmonella poisoning. A federal investigation ensued that ended in the conviction of two Rajneeshpuram officials, who served 29 months in federal prison.

Anthrax: New York City, Washington DC, and Boca Raton, Florida (2001) In September and October of 2001, members of the media and two U.S. senators received letters containing powdered anthrax. According to the FBI, the anthrax was of the Ames strain and appeared to have originated at the U.S. Army Medical Research Institute for Infectious Diseases (USAMRIID) in the laboratories of a well-respected microbiologist. The FBI investigation became known as the Amerithrax case. Ultimately, five people died, and 17 others became ill. Eleven of those cases involved inhalational anthrax, which has a very high mortality rate. All five of the victims died of inhalational anthrax. Antibiotics are very effective at both preventing and treating anthrax infections when they are discovered in the early stages. However, once an anthrax infection has progressed, the production of anthrax toxin and its resultant edema and immune system toxicity generally leads to a poor outcome for the patient. Anthrax toxin is a combination of three proteins produced by *Bacillus anthracis* to evade the host immune system. One component, the cell binding protein (PA), is capable of attaching to and gaining entry into targeted immune cells, which allows the other two toxic components to act: the edema factor (EF), which causes edema, as the name implies; and the lethal factor (LF), which kills the immune cells.

The first wave of anthrax attacks were letters sent from Trenton, New Jersey, postmarked September 18, 2001. Five letters were mailed to ABC news, CBS news, NBC news, the *New York Post*, which are all in New York City, and the fifth one was mailed to the *National Enquirer* at the American Media, Inc. (AMI), offices in Boca Raton, Florida. This anthrax was a coarse brown material and caused mainly cutaneous anthrax, with a few exceptions. The first victim, from AMI, died on October 6, 2001, from inhalational anthrax. In this case the diagnosis of anthrax was overlooked early on, and inhalational anthrax is almost impossible to successfully treat in its later stages.

The second wave of anthrax letters were sent to Democratic Senators Tom Daschle of South Dakota and Patrick Leahy of Vermont. This anthrax was a more highly refined powder and was nearly pure. None of the Amerithrax anthrax spores appeared to have been weaponized. *Weaponization* refers to the application of an antistatic coating to the outside of the spores to increase their dispersal and capability for inhalation. Radiocarbon dating showed that the spores were produced within two years of dissemination and could have been produced in a relatively cheap and simple homemade illicit laboratory.

Chemical Agents

Chemicals have been employed in warfare since as early as 1000 BC in China, where arsenic smoke was used. Modern chemical warfare agents were first employed in World War I. Initially, toxic industrial chemicals such as chlorine were used militarily, and eventually, agents such as sulfur mustard were specifically developed for warfare. Nerve agents were developed during World War II. Most military-style chemical warfare agents, such as sarin, VX, and mustard, are almost impossible to obtain commercially and are very difficult to obtain on the black market. However, chemical agents can be prepared in illicit laboratories by persons with even a limited knowledge of chemistry. In recent years several illicit laboratories have been discovered set up to produce chemical agents. Other chemical agents, such as hydrogen cyanide, unfortunately are extremely easy to obtain and use.

Sarin: Tokyo, Japan (1995) On March 20, 1995, during morning rush hour, Aum Shinrikyo cult members punctured plastic bags containing sarin on five trains of the Tokyo subway system, killing 12 people and injuring thousands more. When law enforcement officers investigated the headquarters of the cult, they found explosives, other chemical weapons, biological warfare agents, and weapons caches. The biological agents discovered included anthrax and Ebola cultures obtained from Zaire. In addition, officers found laboratories set up to manufacture illegal drugs such as methamphetamine and LSD. This is an example of a very sophisticated illicit laboratory operation in which the terrorists were making multiple categories and types of illegal substances. It is important to call the appropriate resources quickly when illicit laboratories are found.

In 1993 the Aum Shinrikyo cult manufactured sarin and VX nerve agents. Members tested their final product on sheep in a remote location in Western Australia. Over the next several years they used the nerve agents in several assassination attempts.

On May 5, 2005, the cult initiated another attack on the Tokyo subway system, this time in a bathroom of the Shinjuku station. In this case cult members used a hydrogen cyanide–generating device to release the gas into the ventilation system. Several other hydrogen cyanide–generating devices were found in other locations in the subway system. Reportedly, they had enough hydrogen cyanide to kill 20,000 people.

Radioactive Materials

Radioactive materials are relatively easy to obtain commercially and on the black market. Such materials can be incorporated in or strapped to explosives and dispersed in what is known as a *dirty bomb* or a **radiological dispersal device (RDD)**. Typically, the radioactivity levels of the materials found in an RDD will be very low, and the primary hazard will be a psychological effect. However, this psychological effect may be very powerful. The American public is extremely wary of radiation. Therefore, any release of radioactive material, or even a perceived release of radioactive material, will likely cause extensive economic damage to the region. To date an RDD has not been used. These devices would likely not produce many casualties and would primarily be weapons of mass disruption, causing economic damage. The largest number of casualties would result from the explosion itself, as it would be very difficult to incorporate enough radioactive material to cause significant morbidity or mortality from radiation exposure.

radiological dispersal device (RDD) ▪ An explosive device that contains a radioactive material that is disseminated into the environment as the bomb detonates. Also known as a dirty bomb.

Most radioactive materials are not fissile. A fissile material is capable of sustaining a nuclear chain reaction such as in a nuclear reactor or atomic weapon. Fissile radioactive materials can be used to make a nuclear weapon; however, nuclear weapons are technically very difficult to construct, for two primary reasons: (1) a relatively large quantity of highly purified uranium or plutonium is required to achieve critical mass, and (2) the critical mass of fissile material must be very rapidly and symmetrically compressed for a nuclear detonation to occur. This requires very precisely engineered explosive shape charges. Do not confuse an RDD with a nuclear weapon.

A nuclear incident also may occur if a nuclear power plant has an accident or is attacked by terrorists. Two examples of serious nuclear incidents involving nuclear power plants are the core meltdown at Chernobyl in the Soviet Union during the 1980s, and the core meltdown in March 2011 at Fukushima, Japan, caused by the magnitude 9.0 earthquake and subsequent tsunami. Both incidents destroyed the nuclear reactor itself, made a wide area around the nuclear power plant uninhabitable, and spread radioactivity around the globe. These incidents could be considered RDDs on a massive scale. Thousands of tons of radioactive material were dispersed in each incident, but there were no reported short-term deaths from radiation exposure due to the radioactive materials in the cloud.

Moscow, Russia (1995) In November 1995, Chechen terrorists contacted Russian media and stated that they had buried a dirty bomb in a public park in Moscow, Russia. When authorities investigated the report, they did indeed find a container of radioactive cesium-137 partially buried in the park. The source of the cesium and the identity of the perpetrators have never been determined.

José Padilla (2002) In 2002, José Padilla, a former Chicago gang member, was arrested at the airport after returning from central Asia, where he allegedly trained at al-Qaeda–sponsored training camps. Padilla allegedly received training on how to assemble a dirty bomb in Lahore, Pakistan. He was allegedly instructed to carry out an attack on U.S. soil and was planning to detonate a radioactive dirty bomb in a large city in the United States. Padilla was held as an enemy combatant for 3-1/2 years and in August 2007 was convicted in a civilian court of conspiring to kill people in an overseas jihad and funding and supporting terrorist organizations.

INDICATORS OF A TERRORIST INCIDENT

Signs of a terrorist incident involving hazardous materials or WMDs may include any of the following:

- reports of an unexplained explosion
- many sick people with unusual symptoms, complaining of similar symptoms, and/or at an unusual time of year
- many injured people with an unexplained cause
- crop dusting or spraying activities in unusual locations or at unusual times (for example, in the evening or early morning)
- unusual devices or equipment near building air intakes
- an unusual number of dead animals, insects, or vegetation in an area

If an incident is suspected to be of a criminal or terrorist nature, several extra steps and precautions should be taken. Often, secondary devices targeted at responding personnel are deployed by the perpetrator. Extreme caution should be used in and around the release site. In addition, the release site is now a crime scene and should be disturbed as little as possible. Evidence preservation should be one of the top priorities right after life safety and incident stabilization. Almost everything at the scene is potential evidence—including tire tracks, footprints, trash, and other more obvious items. Disturbing, moving, or walking over and/or through these pieces of evidence must be avoided. The local law enforcement agency and the FBI should be notified as soon as possible.

Awareness Level and Operations Level Review

Awareness level personnel are not typically emergency responders. They are any employees who may be first on the scene of a hazardous materials or WMD release. They may encounter hazardous materials or WMDs during the course of their normal job duties. The primary role of awareness level personnel is to recognize a dangerous situation in the form of a hazardous materials release, avoid it, isolate the area so others are not hurt or killed, and notify the appropriate personnel or agencies that can handle a hazardous materials release of that magnitude. This process is captured in the acronym RAIN: Recognize, Avoid, Isolate, and Notify.

Awareness level personnel will notify 911, thereby summoning emergency response employees of law enforcement organizations, fire departments, or EMS organizations who should be trained to the hazmat operations or technician level. Operations level responders are members of public safety agencies that respond to hazardous materials or WMD incidents to protect nearby persons, the environment, and property from the effects of the hazardous materials release.

You should approach hazardous materials incidents from upwind, uphill, and upstream: from upwind to avoid accidentally entering a vapor or gas cloud; from uphill to avoid the flow of liquid releases and heavier-than-air gases, especially during a catastrophic container rupture that may occur after arrival; and from upstream, when applicable, to avoid hazardous materials brought along with the current of the falling water or the contour of the water channel. Additionally, gas and vapor clouds—such as chlorine—have a tendency to travel along bodies of water.

Once operations level personnel have ensured their own safety, they need to ensure the safety of the public by isolating the area and setting up control zones. There are three control zones: the **hot zone**, also known as the *exclusion zone*; the **cold zone**, also known as the *support zone*; and the **warm zone**, also known as the *contamination reduction zone* or *decontamination area*. When you arrive at a hazardous materials incident, it will typically be divided only into a hot zone and a cold zone initially. The hot zone is any area that is already contaminated or can be reasonably expected to become contaminated—in other words the initial isolation area. When setting up the hot zone always remember that gases and vapors travel. Gas and vapor clouds are obviously considered contamination. The *Emergency Response Guidebook (ERG)* is an excellent initial source for determining the size of the hot zone. It tells you how far to isolate the area in all directions in either the orange section or the green section of the guide, depending on whether the product is highlighted. If there are contaminated victims, or if more highly trained entry teams start entering the hot zone, a warm zone or decontamination area is carved out of the cold zone on the upwind and/or uphill side. The warm zone is the location of most decontamination activities. We will discuss decontamination in detail in Chapter 12.

hot zone ▪ The exclusion zone in which hazardous materials are located or could reasonably be expected to migrate toward, or the area in which effects may be felt upon catastrophic container failure (such as from a boiling-liquid, expanding-vapor explosion, or BLEVE).

cold zone ▪ The clean, uncontaminated area of the hazardous materials incident. Also known as the support zone. It is where the incident command post and hazardous materials team support functions are located.

warm zone ▪ The control zone where decontamination is performed.

One of the most important functions that operations level responders perform is to assess the severity of the incident. This will be very helpful to you—the hazardous materials technician—on your arrival. To assess the severity of the incident, hazardous materials operations personnel must determine whether anyone has been injured or killed, the number of victims, the type and quantity of released material, and the properties of the released material. As a hazardous materials technician it will be important for you to interface with other first responders who arrived at the scene before you, especially those that have been trained to the hazardous materials operations level, to hear their scene assessment.

RECOGNITION OF HAZARDOUS MATERIALS INCIDENTS

It is imperative that hazardous materials spills, terrorist incidents, or other events involving the release of substances that may pose a risk to people or property be recognized early, especially by the first responders who discover the release. Typically, the people who discover a hazardous materials incident become the first victims if they do not recognize the situation, as many in the fire, law enforcement, EMS, and military communities are tragically aware.

There are many cues to a release of a dangerous substance. These range from those detected by the senses, such as sight, sound, and smell; to labels and placards; to situational awareness. Naturally, the farther away you are from the spill when you recognize it, the safer you will be. The following are several clues, listed in the approximate order from lower risk to higher risk:

- occupancy and location information (pre-incident planning information)
- container shape and size (using binoculars)
- container and building markings (using binoculars)
- shipping papers
- air monitoring readings
- use of the senses

If you can smell or taste the chemical, you have already been exposed! Sight and sound may be very effective for early detection and warning—as long as they are used from a distance or from witness accounts. Hazardous materials spills may be visible from a distance owing to smoke, a vapor cloud, or visible liquid or solid spilled material. If something, especially a hazardous material looks out of place, stop and retreat until you can determine that the situation is indeed safe. Hazardous materials spills may be heard from a distance owing to the high-pressure squeal given off by a leaking gas cylinder, the sound of materials burning, or the sound of chemicals reacting. The sense of smell is not the ideal tool for early detection, since many materials already have a degree of toxicity by the time you smell them. Many chemicals do not have any odor at all, while others cause olfactory fatigue. **Olfactory fatigue** refers to the temporary deadening of the sense of smell by some chemicals, such as hydrogen sulfide. However, some chemicals have an odor threshold that is below the toxicity level. Witness accounts of unusual odors may still be a good indication of the presence of a hazardous materials release and its nature (Table 1-1).

olfactory fatigue ■ The loss of ability to identify an odor after prolonged exposure. An example of a chemical that causes olfactory fatigue is hydrogen sulfide.

TABLE 1-1 Representative Odor Thresholds

CHEMICAL	ODOR THRESHOLD	PERMISSIBLE EXPOSURE LIMIT (PEL)
Acetic acid	0.1–24 ppm	10 ppm
Acetone	0.1–699 ppm	1000 ppm
Ammonia	0.04–55 ppm	50 ppm
Benzene	1.4–120 ppm	1 ppm
Carbon monoxide	Odorless	50 ppm
Chlorine	0.01–5 ppm	1 ppm
Ethanol	1–5100 ppm	1000 ppm
Ethyl ether	0.1–9 ppm	400 ppm
Ethyl mercaptan	0.00051–0.075 ppm	10 ppm
Gasoline	0.005–10 ppm	None (however, listed as carcinogen by NIOSH)
Hydrogen chloride	1–10 ppm	5 ppm
Hydrogen cyanide	0.00027–5 ppm (not everyone is capable of detecting odor)	10 ppm
Hydrogen sulfide	0.00001–1.4 ppm (caution! olfactory fatigue)	20 ppm
Methane	Odorless	None
Methyl ethyl ketone (MEK)	0.25–85 ppm	200 ppm
Propane	Odorless	1000 ppm
Styrene	0.001–200 ppm	100 ppm
Toluene	0.02–70 ppm	200 ppm

MARKING SYSTEMS

Marking systems are designed to make employees, the public, and emergency responders aware of hazardous materials in hazardous situations. Many different marking systems are in use throughout North America and the world. These have been devised by the military, the federal government, private organizations such as gas companies, and safety organizations such as the National Fire Protection Association (NFPA).

The U.S. Department of Transportation (U.S. DOT) has devised a system of placards and labels for the quick and early identification of hazardous materials containers in the U.S. transportation system. Likewise, the NFPA has devised the NFPA 704 marking system for fixed-site facilities. We will cover each of these marking systems in detail, since they can provide useful information about a hazardous materials release.

DOT Hazard Classes and Divisions

Nine hazard classes have been defined by the U.S. Department of Transportation:

Class 1	Explosives
Class 2	Gases
Class 3	Flammable liquids (and combustible liquids in the United States)
Class 4	Flammable solids and water reactive materials
Class 5	Oxidizers
Class 6	Toxic and infectious materials
Class 7	Radioactive materials
Class 8	Corrosive materials
Class 9	Miscellaneous hazardous materials

Vehicles that transport hazardous materials are placarded with the hazard class and, if appropriate, a four-digit UN/NA number. The placards are diamond shaped and are required to be 10-3/4" × 10-3/4". Packages containing hazardous materials that are transported are labeled with the hazard class marking. The labels are 4" × 4". Figure 1-3 shows the U.S. DOT approved placard and label designs. These hazard classes are often further subdivided into divisions based on more precisely defined properties within a given class.

explosion ■ An extremely rapid release of energy in the form of gas and heat.

explosive ■ A substance or article capable of detonation.

Class 1: Explosives An **explosion** is an extremely rapid release of energy in the form of gas and heat. **Explosives** are substances and devices capable of yielding an explosion. Explosives can lead to devastating injuries from burns, pressure waves, and shrapnel. Class 1 explosives are subdivided into six divisions in descending order based on how rapidly energy is released: Division 1.1 is the most dangerous, and Division 1.6 is the least dangerous (Table 1-2).

cryogenic liquid ■ Liquefied compressed gas having a boiling point below –130°F (–90°C) at atmospheric pressure (per DOT).

Class 2: Gases Class 2 consists of materials that are gases at standard temperature and pressure. In this case, standard temperature and pressure refers to approximately room temperature (68°F/20°C) and normal atmospheric pressure (14.7 psi). However, Class 2 materials may be found in the solid, liquid, or gas state during transport and storage. For example, carbon dioxide may be a solid in the form of dry ice, a liquid in the form of a cryogenic liquid, or a gas in a pressurized compressed gas cylinder. A **cryogenic liquid** is a gas that has a boiling point below −130°F (−90°C). Cryogenic liquids pose a severe frostbite hazard when they are released from their container. Gases are subdivided into three divisions, depending on their properties (Table 1-3).

Materials placarded as Class 2 may have a wide range of chemical and physical properties. They may react with each other and form polymers, such as butadiene; they may be

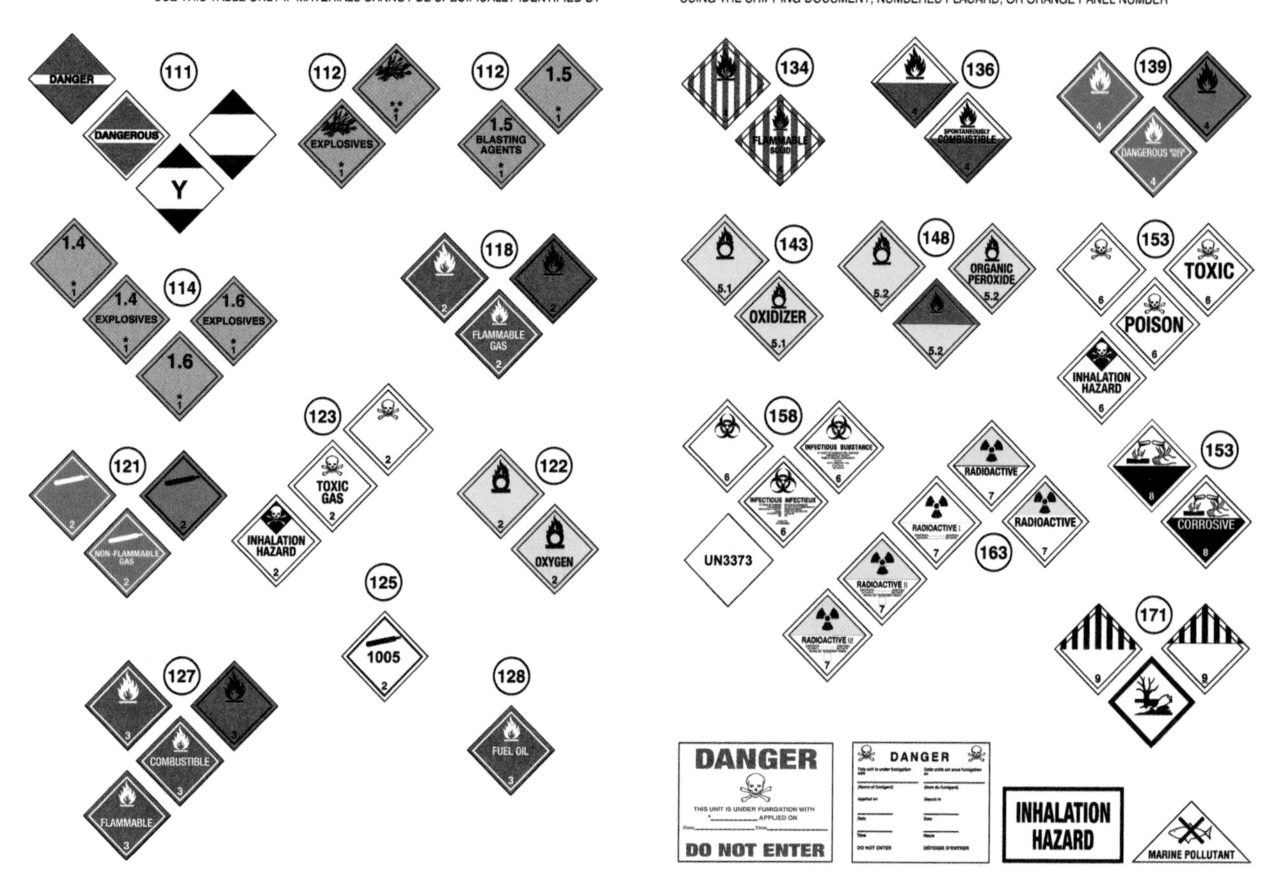

FIGURE 1-3 The U.S. DOT hazardous materials placards as found in the 2012 *Emergency Response Guidebook*. Placards must be displayed on vehicles carrying sufficient quantities of hazardous materials. *Courtesy of the U.S. Department of Transportation Pipeline and Hazardous Materials Safety Administration.*

TABLE 1-2 Explosives

DIVISION	DEFINITION	EXAMPLES
Division 1.1	Explosives with a mass explosion hazard	Black powder, TNT, dynamite, and other high explosives
Division 1.2	Explosives with a projection hazard	Detonating cord, flares
Division 1.3	Explosives with predominately a fire hazard	Propellants, liquid-fueled rocket mortars, professional fireworks
Division 1.4	Explosives with no significant blast hazard	Practice ammunition, signal cartridges, line throwing rockets, consumer fireworks
Division 1.5	Very insensitive explosives with a mass explosion hazard	Blasting agents such as ammonium nitrate and fuel oil mixtures (ANFO)
Division 1.6	Extremely insensitive articles	Certain military ammunition, bombs and warheads

TABLE 1-3 Gases

DIVISION	DEFINITION	EXAMPLES
Division 2.1	Flammable gases	Methane, propane, methyl chloride, inhibited butadienes
Division 2.2	Nonflammable and nontoxic gases	Anhydrous ammonia (U.S. domestic transport), nitrogen, carbon dioxide, helium, argon
Division 2.3	Toxic gases	Anhydrous ammonia (international transport), chlorine, methyl bromide, hydrogen fluoride, arsine, hydrogen sulfide

corrosive, such as anhydrous ammonia and hydrogen fluoride; they may be toxic, such as carbon monoxide and phosphine; or they may be simple asphyxiating gases like nitrogen and argon. Their common property is that they are gases at room temperature and atmospheric pressure. Because gases have a wide range of chemical and physical properties, they pose a wide range of hazards. In addition to their chemical hazards, Class 2 materials are stored in containers under pressure that may rupture violently under fire conditions or when damaged.

Class 3: Flammable Liquids Flammable liquids are defined by the U.S. DOT as liquids with a flashpoint below 141°F (60.5°C). **Flashpoint** is the temperature at which a liquid gives off enough vapors to ignite but not necessarily to sustain combustion (continue to burn). Some examples of flammable liquids are gasoline, acetone, methanol, and toluene. In the United States, Class 3 also includes combustible liquids. *Combustible liquids* are defined by the U.S. DOT as liquids with a flashpoint between 141°F (60.5°C) and 200°F (93°C), and in some cases, liquids with a flashpoint between 100°F and 141°F (exceptions such as diesel fuel). Some examples of liquids classified as combustible are diesel fuel, kerosene, and mineral oil.

flashpoint ■ The temperature of a liquid at which it gives off enough vapors to ignite but not necessarily to sustain combustion.

Class 4: Flammable Solids and Dangerous When Wet Materials Hazard Class 4 comprises a wide range of materials within its three divisions. Division 4.1 encompasses three distinct types of materials:

1. Desensitized explosives, such as wetted explosives
2. Self-reactive materials, such as thermally unstable substances
3. Readily combustible solids that may ignite through friction

Spontaneously combustible materials of Division 4.2 consist of pyrophoric materials and self-heating materials. A **pyrophoric material** is a liquid or solid that can quickly ignite when it comes into contact with air, even in the absence of an external ignition source. A self-heating material increases in temperature on contact with air, without input of any external energy. Division 4.3 consists of dangerous when wet materials. Dangerous when wet materials tend to ignite or give off flammable vapors when in contact with water (Table 1-4).

pyrophoric material ■ A substance capable of igniting on exposure to air at temperatures below 130°F.

Class 5: Oxidizing Substances and Organic Peroxides Oxidizing materials increase the rate of combustion of other materials. These materials can cause ordinary combustibles to burn at a much more rapid rate and spontaneously ignite without an ignition source. Some **oxidizers** liberate oxygen, thereby accelerating combustion. Yet other materials, such as fluorine and chlorine, are more powerful oxidizers than oxygen and increase the rate of combustion by themselves. The organic peroxides are such strong oxidizers that they can even have explosive properties under certain conditions.

oxidizer ■ A substance that enhances or supports the combustion of other substances.

TABLE 1-4 Flammable Solids

DIVISION	DEFINITION	EXAMPLES
Division 4.1	Flammable solids	Magnesium metal, nitrocellulose, wetted titanium powder
Division 4.2	Spontaneously combustible materials	Charcoal briquettes, phosphorus, aluminum alkyls and magnesium alkyls, dry titanium powder
Division 4.3	Water reactive substances and dangerous when wet materials	Calcium carbide, magnesium powder, sodium hydride, potassium metal alloys

According to the U.S. DOT, organic peroxides are classified as types A through G:

Type A can detonate or deflagrate rapidly as packaged for transport (forbidden in transport)

Type B as packaged for transport, neither detonates nor deflagrates rapidly, but can undergo a thermal explosion

Type C as packaged for transport, neither detonates nor deflagrates rapidly and cannot undergo a thermal explosion

Type D is an organic peroxide which—
- (i) Detonates only partially, but does not deflagrate rapidly and is not affected by heat when confined;
- (ii) Does not detonate, deflagrates slowly, and shows no violent effect if heated when confined; or
- (iii) Does not detonate or deflagrate, and shows a medium effect when heated under confinement.

Type E neither detonates nor deflagrates and shows low, or no, effect when heated under confinement

Type F will not detonate in a cavitated state, does not deflagrate, shows only a low, or no, effect if heated when confined, and has low, or no, explosive power

Type G will not detonate in a cavitated state, will not deflagrate at all, shows no effect when heated under confinement, and shows no explosive power. A type G organic peroxide is not subject to the requirements of this subchapter for organic peroxides of Division 5.2 provided that it is thermally stable (self-accelerating decomposition temperature is 50°C (122°F) or higher for a 50 kg (110 pounds) package). An organic peroxide meeting all characteristics of type G except thermal stability and requiring temperature control is classed as a type F, temperature control organic peroxide.

—49 CFR 173.128

Type A peroxides are so dangerous that they are forbidden in transport. Less dangerous organic peroxides are usually shipped under refrigeration in liquid form and contain inhibitors (Table 1-5).

Class 6: Toxic Substances and Infectious Substances Poisonous materials of Division 6.1 consist of materials other than gases that are toxic to humans or presumed toxic to humans, and may pose a danger during transportation. These materials pose a wide variety of hazards including causing immediate and/or delayed illness, cancer, birth defects, and even death. Division 6.2 consists of infectious substances that contain or are

TABLE 1-5 Oxidizers

DIVISION	DEFINITION	EXAMPLES
Division 5.1	Oxidizing substances	Hydrogen peroxide, ammonium nitrate, calcium hypochlorite, bromine trifluoride
Division 5.2	Organic peroxides	Dibenzoyl peroxide, methyl ethyl ketone peroxide, peroxyacetic acid

TABLE 1-6 Poisonous Materials

DIVISION	DEFINITION	EXAMPLES
Division 6.1	Toxic substances	Aniline, arsenic compounds, carbon tetrachloride, hydrocyanic acid, tear gas
Division 6.2	Infectious substances	Medical waste and laboratory waste containing pathogens such as anthrax, botulism, rabies, tetanus

pathogens ▪ Microorganisms (such as viruses and bacteria) or infectious substances (such as prion proteins) that have the potential to cause disease in humans or animals. Examples of pathogens include the smallpox virus, *Bacillus anthracis*, and the causative agent of mad cow disease (a prion).

prion ▪ An infectious protein. Examples of prion diseases include kuru, mad cow disease, and scrapie.

suspected to contain human or animal pathogens. **Pathogens** are bacteria, viruses, **prions**, or microorganisms that have the potential to cause human or animal disease (Table 1-6).

Class 7: Radioactive Materials Radioactive materials are dangerous because they emit ionizing radiation, which is harmful to the human body. High doses of radiation can cause a condition known as acute radiation syndrome (ARS), which is characterized by loss of appetite, fatigue, fever, nausea, vomiting, diarrhea, and skin damage. Extremely high radiation doses can lead to seizures and coma, and even death. Since radiation is a known human carcinogen, even exposures below the level of ARS will increase an individual's risk for cancer. Ionizing radiation can be extremely dangerous because it may go unnoticed since it cannot be seen, tasted, or smelled. Unless specialized detection equipment is used, it will not be noticed until significant damage has occurred. Some examples of radioactive materials include cobalt-60, uranium hexafluoride, radioactive waste, and medical and laboratory isotopes such as phosphorus-32, sulfur-35, and iodine-131. We will discuss radioactivity in greater detail in Chapter 3.

Class 8: Corrosive Materials Corrosive materials are liquids and solids that can cause tissue destruction, and corrode steel or aluminum at a specified rate. Some examples of corrosives include sulfuric acid, hydrochloric acid (muriatic acid), hydrogen fluoride, sodium hydroxide (lye), phosphoric acid, and aluminum trichloride. There are two types of corrosives: acids and bases. Acidic burns are usually detected quickly owing to intense pain at the contact site, whereas caustic burns may initially be undetected because of a lack of immediate pain at the contact site.

Class 9: Miscellaneous Hazardous Materials, Products, Substances, or Organisms Class 9 materials present a hazard during transportation but do not meet the criteria of any other hazard class. Some examples of miscellaneous hazardous materials are molten sulfur, hazardous substances (such as polychlorinated biphenyls, or PCBs), and some hazardous waste shipments. Class 9 materials have a wide range of hazards.

The NFPA 704 Marking System

The NFPA 704 marking system is designed for fixed-site facilities to warn occupants, employees, and responding emergency services personnel of chemical hazards located at that facility. Buildings or rooms containing hazardous materials will be designated with a diamond with four quadrants (Figure 1-4). Each quadrant is a different color:

SOLVED EXERCISE 1-2

Determine from the figure the hazard class of the material in the drum.

Solution: The material in the drum belongs to hazard Class 4: flammable solids and/or dangerous when wet materials. Based on the blue color of the label we can even determine the division: The only labels and placards that are a characteristic blue color are those for water reactive substances/dangerous when wet materials (Division 4.3).

Courtesy of Chris Weber, Dr. Hazmat, Inc.

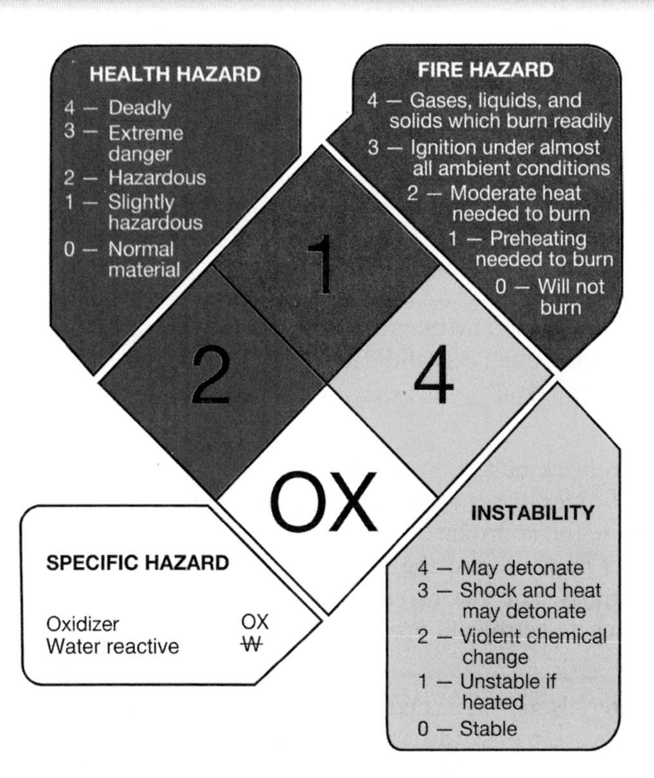

FIGURE 1-4 The NFPA 704 marking system. This voluntary marking system is displayed at participating facilities that store and/or use hazardous materials. *Adapted with permission from NFPA 704-2007,* System for the Identification of the Hazards of Materials for Emergency Response, *Copyright © 2007, National Fire Protection Association. This reprinted material is not the complete and official position of the NFPA on the referenced subject, which is represented solely by the standard in its entirety. The classification of any particular material within this system is the sole responsibility of the user and not the NFPA. NFPA bears no responsibility for any determinations of any values for any particular material classified or represented using this system.*

- Blue signifies health hazards.
- Red signifies fire hazards.
- Yellow signifies degree of instability (reactivity hazards).
- White signifies specific hazards.

SOLVED EXERCISE 1-3

Courtesy of Chris Weber, Dr. Hazmat, Inc.

Determine from the figure the type of material stored in this building.

Solution: The NFPA 704 marking system indicates that the material or materials stored in this building are highly flammable, very toxic, quite reactive, and water reactive. The red quadrant indicates that at least one material in this building has the highest degree of flammability hazard. The blue quadrant indicates that at least one material in this building has a health hazard of 3, where 4 is the highest health hazard. The yellow quadrant indicates that at least one material in the building is moderately unstable at 2 on a scale of 0 to 4. The white quadrant showing the W with a horizontal slash through it indicates that at least one material in the building is water reactive.

The red, blue, and yellow quadrants will each contain a number from 0 to 4. The number 4 indicates a high degree of hazard, and 0 indicates a low degree of hazard. The white quadrant will contain markings indicating specific types of hazards such as oxidizer (OX) or water reactivity (₩). When you see the water reactive symbol, remember not to use water during fire suppression operations. Some facilities will also include markings for corrosives (COR), acids (ACID), bases (BASE), radioactive materials (RAD or the trefoil propeller symbol), or other symbols in this quadrant. Even though these symbols are not officially NFPA sanctioned, the information can still be very useful.

Hazardous Materials Identification System (HMIS)

The Hazardous Materials Identification System (HMIS) is a marking system designed for chemical workplace safety. It combines a chemical hazard ranking similar to the NFPA 704 marking system with personal protective equipment recommendations based on a system of pictograms (Figure 1-5). The HMIS marking system is very common in industries utilizing potentially hazardous chemicals and can be a good starting point for assessing appropriate personal protective equipment.

Pipeline Markings

Pipeline markers are usually used to mark underground pipelines, but not always. There should be a pipeline marker on both sides of roadways, railways, and waterways crossed by the pipeline. The sign will indicate the approximate location of the pipeline, the pipeline contents, an emergency contact number, and the pipeline operator. Pipelines often carry many products at different times, and the pipeline marker may indicate several materials or a class of materials, such as petroleum products (Figure 1-6).

Container Markings

Containers carrying hazardous materials are often marked with labels containing virtually the same information as transportation placards. The shipping label may also contain additional manufacturer information. The hazardous materials containers, including cardboard boxes, totes, drums, and others, may be regulated by the U.S. DOT and may be built to specific standards. These containers will have special markings indicating their design parameters (Figure 1-7). We will discuss containers and their features in much greater detail in Chapter 10.

HAZARDOUS MATERIALS IDENTIFICATION SYSTEM

HMIS®

HAZARD INDEX

4 = SEVERE HAZARD
3 = SERIOUS HAZARD
2 = MODERATE HAZARD
1 = SLIGHT HAZARD
0 = MINIMAL HAZARD

An asterisk(*) or other designation corresponds to additional information on a data sheet or separate chronic effects notification

Additional Information

PERSONAL PROTECTION EQUIPMENT

A Safety Glasses	n Splash Goggles	o Face Shield & Eye Protection	p Gloves
q Boots	r Synthetic Apron	s Full Suit	t Dust Respirator
u Vapor Respirator	w Dust & Vapor Respirator	y Full Face Respirator	z Airline Hood or Mask

PERSONAL PROTECTION INDEX

A
B
C
D
E
F
G
H
I
J
K
X Consult your supervisor or S.O.P. for "*SPECIAL*" handling directions

FIGURE 1-5 The Hazardous Materials Identification System is used to label containers that hold hazardous chemicals. *Courtesy of the American Coatings Association (ACA).*

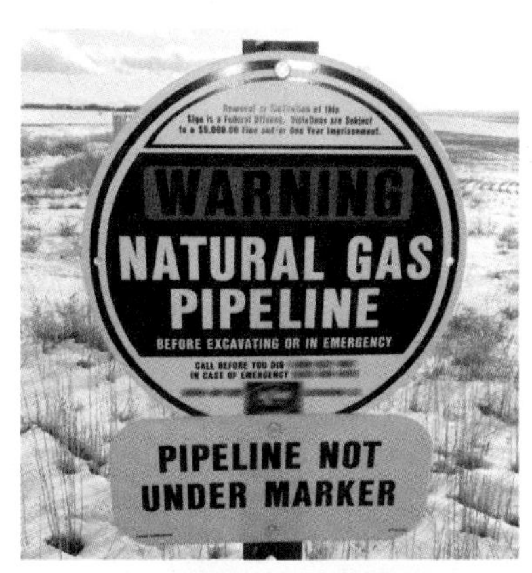

FIGURE 1-6 Pipeline markers are used to mark the location of underground pipelines containing hazardous materials. They contain useful information regarding the owner and contents of the pipeline. Note that the pipeline may not be directly underneath the marker. Although intentionally blurred in these images, the pipeline marker should always clearly state the pipeline owner and a telephone number to reach them. *Photos courtesy of Chris Weber, Dr. Hazmat, Inc.*

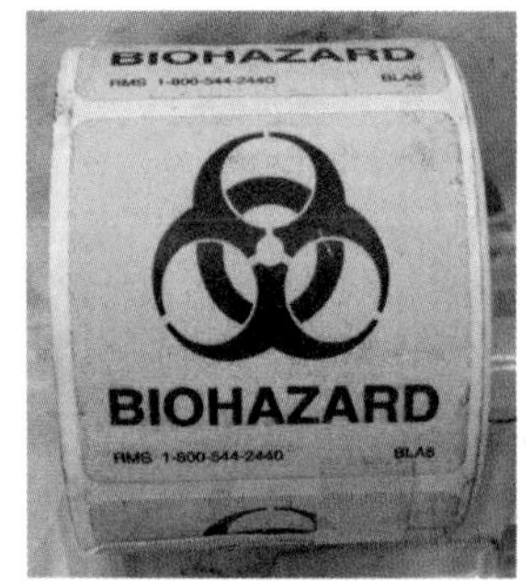

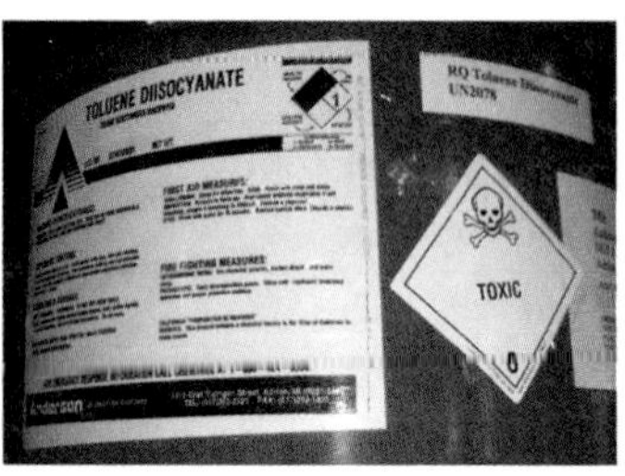

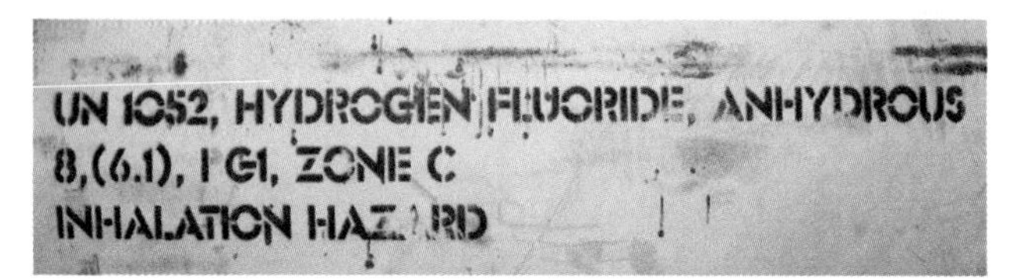

FIGURE 1-7 Many different types of labels may be found on packages containing hazardous materials. Clockwise from the top left are a DOT radioactive label, a biohazard label, an HMIS marking, a shipping name stenciled on a hydrogen fluoride 1-ton container, a manufacturer's label including the NFPA 704 diamond, and a DOT hazard class label. *Photos courtesy of Chris Weber, Dr. Hazmat, Inc.*

Container shape may indicate the presence of hazardous materials. Cylinders often contain compressed gases, dewars contain cryogenic liquids, metal drums may contain liquids, plastic drums may contain corrosive liquids, and cardboard drums may contain a solid material. Heavy packaging often indicates that a pressurized material or dangerous material is being transported. But don't jump to conclusions based solely on container shape. Always check the container label and shipping papers!

Military Markings

Military organizations, including the U.S. military, have their own hazard identification marking systems, especially as they relate to explosives recognition. Figure 1-8 shows some of the U.S. military identification marks for explosives, chemicals, and chemical warfare agents.

The U.S. DOT *Emergency Response Guidebook*

The *Emergency Response Guidebook (ERG)* gives general information about hazardous materials and is primarily designed for use in the first 15 minutes of an incident, although it may be useful during later stages of an incident, for example, for determining downwind evacuation distances. The information in the *ERG* is very general; however, it does initially point you in the right direction. The guidebook contains much useful information about potential hazards, initial isolation distance, evacuation information, actions to take in case of a fire, spill, or leak, as well as first aid procedures in case of injuries. The *ERG* is quite easy to use by following the directions on page 1.

If you know the name of the chemical involved, you can look it up alphabetically in the blue section. Or if you know the four-digit UN/NA number, you can look it up numerically in the yellow section (Figure 1-9). If all you can see is a placard, you can

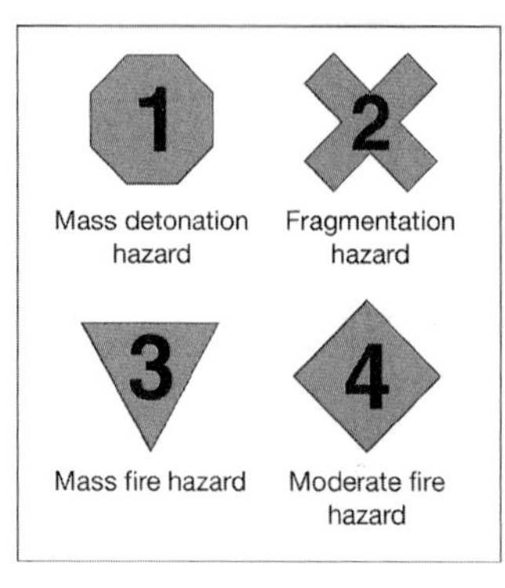

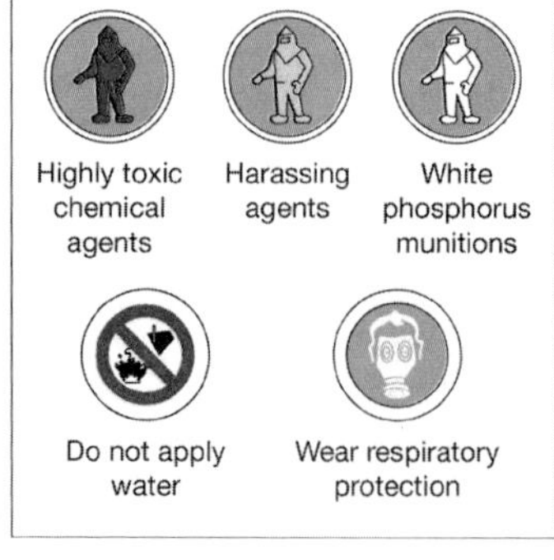

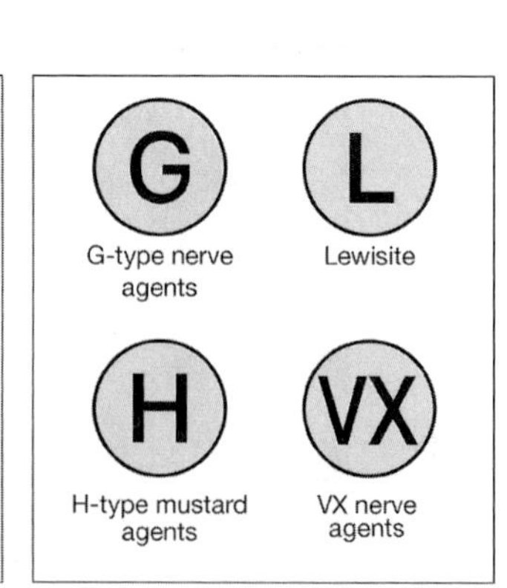

FIGURE 1-8 The U.S. military uses its own marking system to identify explosive hazards (left), chemical hazards (middle), and chemical warfare agents (right). *Art by David Heskett.*

ID No.	Guide No.	Name of Material
1392	**138**	Alkaline earth metal amalgam, liquid
1393	**138**	Alkaline earth metal alloy, n.o.s.
1394	**138**	Aluminum carbide
1395	**139**	Aluminum ferrosilicon powder
1396	**138**	Aluminum powder, uncoated
1397	**139**	Aluminum phosphide
1398	**138**	Aluminum silicon powder, uncoated
1400	**138**	Barium
1401	**138**	Calcium
1402	**138**	Calcium carbide
1403	**138**	Calcium cyanamide, with more than 0.1% Calcium carbide
1404	**138**	Calcium hydride
1405	**138**	Calcium silicide
1407	**138**	Caesium
1407	**138**	Cesium
1408	**139**	Ferrosilicon
1409	**138**	Hydrides, metal, n.o.s.
1409	**138**	Metal hydrides, water-reactive, n.o.s.
1410	**138**	Lithium aluminum hydride
1411	**138**	Lithium aluminum hydride, ethereal
1413	**138**	Lithium borohydride
1414	**138**	Lithium hydride
1415	**138**	Lithium
1417	**138**	Lithium silicon
1418	**138**	Magnesium alloys powder
1418	**138**	Magnesium powder
1419	**139**	Magnesium aluminum phosphide
1420	**138**	Potassium, metal alloys
1420	**138**	Potassium, metal alloys, liquid
1421	**138**	Alkali metal alloy, liquid, n.o.s.
1422	**138**	Potassium sodium alloys
1422	**138**	Potassium sodium alloys, liquid
1422	**138**	Sodium potassium alloys
1422	**138**	Sodium potassium alloys, liquid
1423	**138**	Rubidium
1423	**138**	Rubidium metal
1426	**138**	Sodium borohydride
1427	**138**	Sodium hydride
1428	**138**	Sodium
1431	**138**	Sodium methylate
1431	**138**	Sodium methylate, dry
1432	**139**	Sodium phosphide
1433	**139**	Stannic phosphides
1435	**138**	Zinc ashes
1435	**138**	Zinc dross
1435	**138**	Zinc residue
1435	**138**	Zinc skimmings
1436	**138**	Zinc dust
1436	**138**	Zinc powder
1437	**138**	Zirconium hydride
1438	**140**	Aluminum nitrate
1439	**141**	Ammonium dichromate
1442	**143**	Ammonium perchlorate
1444	**140**	Ammonium persulfate
1444	**140**	Ammonium persulphate
1445	**141**	Barium chlorate
1445	**141**	Barium chlorate, solid
1446	**141**	Barium nitrate
1447	**141**	Barium perchlorate
1447	**141**	Barium perchlorate, solid

Page 30

Name of Material	Guide No.	ID No.
Allyl bromide	**131**	1099
Allyl chloride	**131**	1100
Allyl chlorocarbonate	**155**	1722
Allyl chloroformate	**155**	1722
Allyl ethyl ether	**131**	2335
Allyl formate	**131**	2336
Allyl glycidyl ether	**129**	2219
Allyl iodide	**132**	1723
Allyl isothiocyanate, stabilized	**155**	1545
Allyltrichlorosilane, stabilized	**155**	1724
Aluminum, molten	**169**	9260
Aluminum alkyl halides	**135**	3052
Aluminum alkyl halides, liquid	**135**	3052
Aluminum alkyl halides, solid	**135**	3052
Aluminum alkyl halides, solid	**135**	3461
Aluminum alkyl hydrides	**138**	3076
Aluminum alkyls	**135**	3051
Aluminum borohydride	**135**	2870
Aluminum borohydride in devices	**135**	2870
Aluminum bromide, anhydrous	**137**	1725
Aluminum bromide, solution	**154**	2580
Aluminum carbide	**138**	1394
Aluminum chloride, anhydrous	**137**	1726
Aluminum chloride, solution	**154**	2581
Aluminum dross	**138**	3170
Aluminum ferrosilicon powder	**139**	1395
Aluminum hydride	**138**	2463
Aluminum nitrate	**140**	1438
Aluminum phosphide	**139**	1397
Aluminum phosphide pesticide	**157**	3048
Aluminum powder, coated	**170**	1309
Aluminum powder, pyrophoric	**135**	1383
Aluminum powder, uncoated	**138**	1396
Aluminum processing by-products	**138**	3170
Aluminum remelting by-products	**138**	3170
Aluminum resinate	**133**	2715
Aluminum silicon powder, uncoated	**138**	1398
Aluminum smelting by-products	**138**	3170
Amines, flammable, corrosive, n.o.s.	**132**	2733
Amines, liquid, corrosive, flammable, n.o.s.	**132**	2734
Amines, liquid, corrosive, n.o.s.	**153**	2735
Amines, solid, corrosive, n.o.s.	**154**	3259
2-Amino-4-chlorophenol	**151**	2673
2-Amino-5-diethylaminopentane	**153**	2946
2-Amino-4,6-dinitrophenol, wetted with not less than 20% water	**113**	3317
2-(2-Aminoethoxy)ethanol	**154**	3055
N-Aminoethylpiperazine	**153**	2815
Aminophenols	**152**	2512
Aminopyridines	**153**	2671
Ammonia, anhydrous	**125**	1005
Ammonia, solution, with more than 10% but not more than 35% Ammonia	**154**	2672
Ammonia, solution, with more than 35% but not more than 50% Ammonia	**125**	2073
Ammonia solution, with more than 50% Ammonia	**125**	3318
Ammonium arsenate	**151**	1546
Ammonium bifluoride, solid	**154**	1727
Ammonium bifluoride, solution	**154**	2817
Ammonium dichromate	**141**	1439
Ammonium dinitro-o-cresolate	**141**	1843

Page 93

FIGURE 1-9 The yellow, blue, and green sections of the *Emergency Response Guidebook* (*ERG*). The yellow section lists hazardous materials in numerical order according to their UN number. The blue section lists hazardous materials in alphabetical order according to their shipping name. The green section lists downwind evacuation distances for highlighted materials. *Courtesy of the U.S. Department of Transportation Pipeline and Hazardous Materials Safety Administration.*

(continued)

1251	**131P**	Methyl vinyl ketone, stabilized	100 m	(300 ft)	0.3 km	(0.2 mi)	0.8 km	(0.5 mi)	800 m	(2500 ft)	1.5 km	(1.0 mi)	3.0 km	(1.9 mi)
1259	**131**	Nickel carbonyl	100 m	(300 ft)	1.4 km	(0 9 mi)	5.4 km	(3.4 mi)	1000 m	(3000 ft)	11.0+ km	(7.0+ mi)	11.0+ km	(7.0+ mi)
1295	**139**	Trichlorosilane **(when spilled in water)**	30 m	(100 ft)	0.1 km	(0.1 mi)	0.3 km	(0.2 mi)	60 m	(200 ft)	0.7 km	(0.4 mi)	2.2 km	(1.4 mi)
1298	**155**	Trimethylchlorosilane **(when spilled in water)**	30 m	(100 ft)	0.1 km	(0.1 mi)	0.2 km	(0.1 mi)	60 m	(200 ft)	0.6 km	(0.4 mi)	1.6 km	(1.0 mi)
1305	**155P**	Vinyltrichlorosilane **(when spilled in water)**	30 m	(100 ft)	0.1 km	(0.1 mi)	0.2 km	(0.2 mi)	60 m	(200 ft)	0.6 km	(0.4 mi)	2.0 km	(1.3 mi)
1305	**155P**	Vinyltrichlorosilane, stabilized **(when spilled in water)**												
1340	**139**	Phosphorus pentasulfide, free from yellow and white Phosphorus **(when spilled in water)**	30 m	(100 ft)	0.1 km	(0.1 mi)	0.2 km	(0.1 mi)	60 m	(200 ft)	0.4 km	(0.2 mi)	1.4 km	(0.9 mi)
1340	**139**	Phosphorus pentasulphide, free from yellow and white Phosphorus **(when spilled in water)**												
1360	**139**	Calcium phosphide **(when spilled in water)**	30 m	(100 ft)	0.2 km	(0.1 mi)	0.7 km	(0.4 mi)	300 m	(1000 ft)	1.1 km	(0.7 mi)	3.8 km	(2.4 mi)
1380	**135**	Pentaborane	60 m	(200 ft)	0.6 km	(0.4 mi)	2.0 km	(1.2 mi)	200 m	(600 ft)	2.7 km	(1.7 mi)	8.2 km	(5.1 mi)
1384	**135**	Sodium dithionite **(when spilled in water)**	30 m	(100 ft)	0.2 km	(0.1 mi)	0.6 km	(0.4 mi)	60 m	(200 ft)	0.8 km	(0.5 mi)	2.7 km	(1.7 mi)
1384	**135**	Sodium hydrosulfite **(when spilled in water)**												
1384	**135**	Sodium hydrosulphite **(when spilled in water)**												
1397	**139**	Aluminum phosphide **(when spilled in water)**	60 m	(200 ft)	0.2 km	(0.2 mi)	0.9 km	(0.6 mi)	500 m	(1500 ft)	2.1 km	(1.3 mi)	7.5 km	(4.7 mi)

Page 295

"+" means distance can be larger in certain atmospheric conditions

FIGURE 1-9 *(Continued)*

find the three-digit guide number for various placards on pages 16 and 17. If a railcar is involved, and you can recognize only the railcar shape, you can compare it with silhouettes on page 18. Likewise, if an over-the-road trailer or tanker is involved and all you can see is the shape of the vehicle, you can refer to silhouettes on page 19. In all cases the goal is to find a three-digit guide number in the orange section that will give guidance in responding to an emergency (see Solved Exercise 1-4).

The title of the guide immediately gives some information. For example, Guide 140 is titled "Oxidizers." This immediately alerts you to some of the potential hazards, which are listed under that heading and include two categories: "FIRE OR EXPLOSION" and "HEALTH." The one listed first is the primary hazard. In the case of Guide 140, "FIRE OR EXPLOSION" is the primary hazard. Under the "PUBLIC SAFETY" heading you can find such useful information as an immediate isolation distance, appropriate protective clothing, and evacuation guidelines for large spills or fire. On the facing page, you will find guidance for handling fires, spills or leaks, and basic first aid advice for exposure victims. Thus, you can quickly gather a lot of critical information from the *ERG*.

With regard to evacuation, if an entry is highlighted in the yellow or blue sections, and it is not on fire, you should first consult the green section (Figure 1-9). If an entry is highlighted, it means that the chemical is a toxic inhalation hazard (TIH) and poses a downwind health hazard. The green section is organized in numerical order, by the four-digit UN/NA number. This section contains more specific information, since it deals with a single chemical, and is organized into a "SMALL SPILLS" and a "LARGE SPILLS" column. A small spill is less than 55 gallons. Under each heading is listed an isolation distance and a downwind protective action distance for daytime and nighttime. Typically, the isolation distance will be larger than that found in the respective guide number. At the end of the green section is a table of water-reactive materials that also specifies which toxic gas is produced when the material reacts with water.

Advancing from the Operations Level to the Technician Level

As a hazardous materials technician you will be your community's primary defense once a release has occurred. Response to a hazmat incident typically starts at the local level with a call to 911. The local fire department, law enforcement agency, and, possibly, emergency medical services normally respond. Depending on their initial assessment, they will call a local or regional hazmat response team, probably the one of which you are, or will become, a member. If the situation is beyond the capability and training of local or regional resources, state and federal hazardous materials response assets will be called. One of the defining characteristics of hazardous materials incidents is that they typically are multijurisdictional and multiagency and/or involve several levels of government, often up to and including the federal level, represented most commonly by the Environmental Protection Agency (EPA). The number of resources needed at a hazardous materials incident depends on a competent analysis of hazard versus risk.

HAZARD VERSUS RISK

The differences between a hazard and a risk form the core of any emergency response planning—in other words, of a risk-based response. *Hazard* is the characteristic of having the capability to cause harm. Just about everything is hazardous to a degree. Driving your car to work is hazardous. According to the Bureau of Transportation Statistics, a total of 29,086 motor vehicle occupants were killed in car accidents in the United States in 2009. Walking or bicycling to work is hazardous is well: 4092 pedestrians and 630 bicyclists were killed in that same year in the United States. But only 18 airline passengers were killed while flying worldwide according to the IATA 2009 Air Travel Safety Report.

SOLVED EXERCISE 1-4

Consult the current DOT *Emergency Response Guidebook* to find the recommended isolation distance for a cargo tanker carrying allyl chloroformate that is on fire (which directs you to Guide 155).

GUIDE 155 SUBSTANCES - TOXIC AND/OR CORROSIVE (FLAMMABLE/WATER-SENSITIVE) ERG2012

POTENTIAL HAZARDS

FIRE OR EXPLOSION

- **HIGHLY FLAMMABLE: Will be easily ignited by heat, sparks or flames.**
- Vapors form explosive mixtures with air: indoors, outdoors and sewers explosion hazards.
- Most vapors are heavier than air. They will spread along ground and collect in low or confined areas (sewers, basements, tanks).
- Vapors may travel to source of ignition and flash back.
- Those substances designated with a **(P)** may polymerize explosively when heated or involved in a fire.
- Substance will react with water (some violently) releasing flammable, toxic or corrosive gases and runoff.
- Contact with metals may evolve flammable hydrogen gas.
- Containers may explode when heated or if contaminated with water.

HEALTH

- **TOXIC**; inhalation, ingestion or contact (skin, eyes) with vapors, dusts or substance may cause severe injury, burns or death.
- **Bromoacetates and chloroacetates are extremely irritating/lachrymators.**
- Reaction with water or moist air will release toxic, corrosive or flammable gases.
- Reaction with water may generate much heat that will increase the concentration of fumes in the air.
- Fire will produce irritating, corrosive and/or toxic gases.
- Runoff from fire control or dilution water may be corrosive and/or toxic and cause pollution.

PUBLIC SAFETY

- **CALL EMERGENCY RESPONSE Telephone Number on Shipping Paper first. If Shipping Paper not available or no answer, refer to appropriate telephone number listed on the inside back cover.**
- As an immediate precautionary measure, isolate spill or leak area in all directions for at least 50 meters (150 feet) for liquids and at least 25 meters (75 feet) for solids.
- Keep unauthorized personnel away.
- Stay upwind.
- Keep out of low areas.
- Ventilate enclosed areas.

PROTECTIVE CLOTHING

- Wear positive pressure self-contained breathing apparatus (SCBA).
- Wear chemical protective clothing that is specifically recommended by the manufacturer. It may provide little or no thermal protection.
- Structural firefighters' protective clothing provides limited protection in fire situations ONLY; it is not effective in spill situations where direct contact with the substance is possible.

EVACUATION

Spill

- See Table 1 - Initial Isolation and Protective Action Distances for highlighted materials. For non-highlighted materials, increase, in the downwind direction, as necessary, the isolation distance shown under "PUBLIC SAFETY".

Fire

- If tank, rail car or tank truck is involved in a fire, ISOLATE for 800 meters (1/2 mile) in all directions; also, consider initial evacuation for 800 meters (1/2 mile) in all directions.

Page 248

ERG2012 SUBSTANCES - TOXIC AND/OR CORROSIVE (FLAMMABLE/WATER-SENSITIVE) **GUIDE 155**

EMERGENCY RESPONSE

FIRE

- Note: Most foams will react with the material and release corrosive/toxic gases.

CAUTION: For Acetyl chloride (UN1717), use CO_2 or dry chemical only.

Small Fire

- CO_2, dry chemical, dry sand, alcohol-resistant foam.

Large Fire

- Water spray, fog or alcohol-resistant foam.
- **FOR CHLOROSILANES, DO NOT USE WATER**; use AFFF alcohol-resistant medium expansion foam.
- Move containers from fire area if you can do it without risk.
- Use water spray or fog; do not use straight streams.

Fire involving Tanks or Car/Trailer Loads

- Fight fire from maximum distance or use unmanned hose holders or monitor nozzles.
- Do not get water inside containers.
- Cool containers with flooding quantities of water until well after fire is out.
- Withdraw immediately in case of rising sound from venting safety devices or discoloration of tank.
- ALWAYS stay away from tanks engulfed in fire.

SPILL OR LEAK

- ELIMINATE all ignition sources (no smoking, flares, sparks or flames in immediate area).
- All equipment used when handling the product must be grounded.
- Do not touch damaged containers or spilled material unless wearing appropriate protective clothing.
- Stop leak if you can do it without risk.
- A vapor suppressing foam may be used to reduce vapors.
- **FOR CHLOROSILANES**, use AFFF alcohol-resistant medium expansion foam to reduce vapors.
- **DO NOT GET WATER on spilled substance or inside containers.**
- Use water spray to reduce vapors or divert vapor cloud drift. Avoid allowing water runoff to contact spilled material.
- Prevent entry into waterways, sewers, basements or confined areas.

Small Spill

- Cover with DRY earth, DRY sand or other non-combustible material followed with plastic sheet to minimize spreading or contact with rain.
- Use clean non-sparking tools to collect material and place it into loosely covered plastic containers for later disposal.

FIRST AID

- Move victim to fresh air. • Call 911 or emergency medical service.
- Give artificial respiration if victim is not breathing.
- **Do not use mouth-to-mouth method if victim ingested or inhaled the substance; give artificial respiration with the aid of a pocket mask equipped with a one-way valve or other proper respiratory medical device.**
- Administer oxygen if breathing is difficult.
- Remove and isolate contaminated clothing and shoes.
- In case of contact with substance, immediately flush skin or eyes with running water for at least 20 minutes.
- For minor skin contact, avoid spreading material on unaffected skin.
- Keep victim warm and quiet.
- Effects of exposure (inhalation, ingestion or skin contact) to substance may be delayed.
- Ensure that medical personnel are aware of the material(s) involved and take precautions to protect themselves.

Page 249

Courtesy of the U.S. Department of Transportation Pipeline and Hazardous Materials Safety Administration.

Solution: If a tank truck is involved in a fire, isolate for 800 meters (1/2 mile) in all directions.

Just because these activities are hazardous does not mean we avoid them completely. We try to make them as safe as possible. We maintain our cars. We drive carefully and keep an eye out for careless drivers. We try to minimize the hazards of driving, thereby minimizing the risk to ourselves and our family. *Risk* can be defined as the likelihood that a hazardous activity will actually cause harm—that is, it is the probability of harm. Stated in a different way, these statistics show that in 2008 there were 310 deaths per 100 billion passenger vehicle miles traveled, whereas there were only 5 deaths per 100 billion miles traveled by air. Travel by air is less risky, but nevertheless passengers have died in plane crashes. In our daily lives as well as in our careers we are constantly managing the risks posed by hazards.

Hazardous materials response is no different. The hazard of concentrated phenol is much higher than that of a 2% phenol solution (cough suppressant throat spray). However, the risk each poses will depend on the situation, quantity, container integrity, and many other factors.

Three key attributes of the hazardous materials release drive the risk assessment process—the 3 C's:

- Chemical
- Container
- Conditions

The specific hazards of the chemical are a good starting point for assessing the risks the chemical may pose (Chapter 5). The container will dictate the quantity of the material present and the state of matter of the chemical, and a damage assessment of the container will reveal an estimate of the likelihood of container failure (Chapter 10). The conditions under which a hazardous materials release occurs, such as temperature, precipitation, and environment—such as other nearby containers containing hazardous materials, exposed populations, and natural features such as rivers, lakes, and streams—will affect the severity of the incident. Evaluation of the 3 C's is a good start to a risk-based response.

DEFENSIVE VERSUS OFFENSIVE OPERATIONS

One of the biggest differences between the hazardous materials operations level and the hazardous materials technician level is an offensive versus a defensive mindset. What does this mean? *Defensive* tactics consist primarily of preventing a hazardous material release from becoming worse while operating mainly outside the primary hazard zone. *Offensive* tactics are used to stop the release and require entry into the hot zone, thus placing a hazmat technician in a more hazardous situation. This requires much greater training and experience on the part of hazmat technicians to minimize the risk to themselves, their crew, and the public.

THE HAZARDOUS MATERIALS RESPONSE TEAM OR COMPANY

One way to reduce the risk of a hazard is to operate as part of a well-trained and well-equipped hazardous materials response team. A well-trained hazmat team knows that safety is the number one priority. Your safety is the first priority, then that of your team members and other first responders, and finally that of the public. This may seem counterintuitive, but if you get hurt, you can't help anyone else, and the majority of incident resources will be diverted to helping you, thereby making the situation much worse and much more dangerous for everyone involved. The following are several key characteristics common to progressive hazardous materials response teams:

- Regular training
- Standard operating procedures
- Pre-incident planning
- Appropriate response equipment

- Exercises
- Working with mutual-aid hazmat teams
- Working with partner agencies
- Regular evaluations to document competency
- Risk-based emergency response

Training is a critical part of forming a cohesive hazmat team that has the ability to operate safely and efficiently at hazardous materials releases. Training should be conducted on a regular basis with all members of the hazmat response team. It is very important that team members understand one another, be familiar with the team's standard operating procedures, and know one another's strengths and weaknesses to form a cohesive group. Training should follow the logical progression outlined in Figure 1-10.

The foundation of the training pyramid is a solid understanding of the core concepts underlying hazardous materials response. This includes scientific concepts such as chemistry and toxicology, as well as how hazardous materials are containerized, transported, and used in industry. Next comes recognition of hazards, which includes placarding and labeling, signs of hazardous materials releases, and container recognition. In other words, when you arrive on the scene, you should quickly have a reasonable idea of the scope of the problem. The next stage of the pyramid is gas and vapor detection through air monitoring. Air monitoring and radiation detection are extremely important in protecting first responders and the public from unseen chemical hazards. The next stage is substance identification, which is crucial to performing the final step of the pyramid—mitigation of the incident. Hazmat mitigation involves containing a chemical release and stabilizing the situation until cleanup contractors can arrive to remove the hazardous materials. Hazmat team training should start at the base and progress to the top. A hazmat team that is great at putting on a chlorine kit or transferring liquids but has no idea how to perform air monitoring or how to deal with pressure effects on a container is a catastrophic accident waiting to happen.

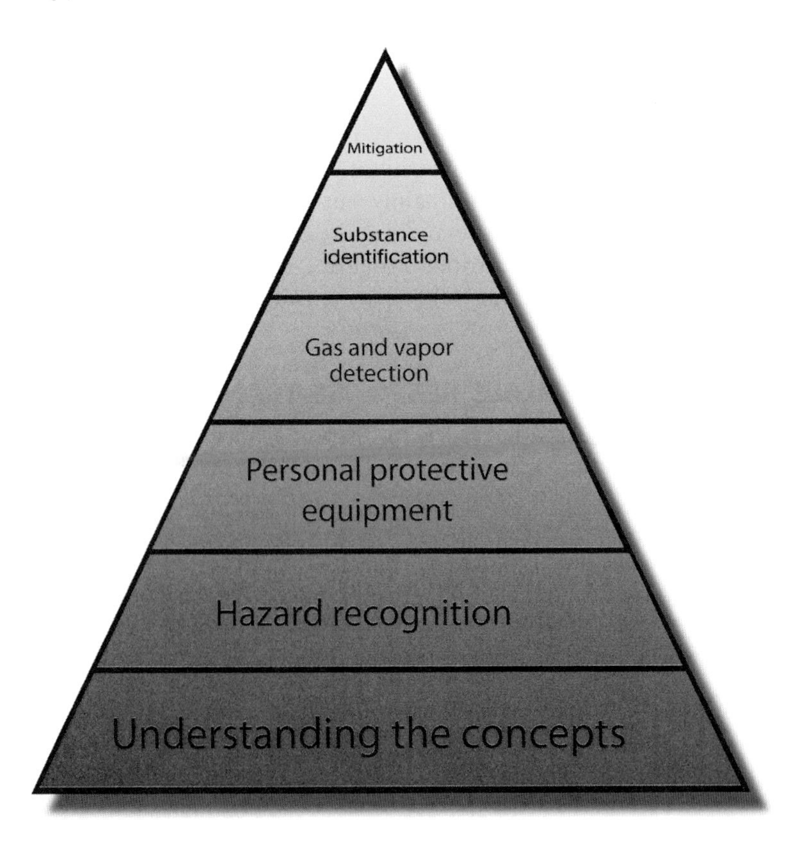

FIGURE 1-10 The hazmat training pyramid. It is important to train from the ground up, starting with a firm grasp of the concepts. *Art by David Heskett.*

Two other important components of hazardous materials response team operations are pre-incident planning and conducting exercises. It is important to identify the hazardous materials facilities in your jurisdiction as well as to know what types of hazardous materials move through your area on roads, railways, pipelines, and waterways, and through the airports so that you can select and purchase the appropriate equipment and focus on the applicable training.

Because hazardous materials incidents can be complex and resource intensive it is important to assess local, regional, state, federal, and tribal resources in your area. First and foremost, get to know any mutual-aid hazmat response teams. What are their capabilities? How good is their training? Is their equipment compatible with yours? Is their equipment complementary to yours? Joint exercises at industrial facilities are a great way to answer these questions and more.

Hazardous materials response is an interdisciplinary field. Illicit laboratory incidents will require close cooperation with local, state, and federal law enforcement agencies; WMD incidents will require close cooperation not only with local law enforcement agencies but also with many federal agencies such as the FBI; bioterrorism incidents will require close cooperation with public health agencies; and hazardous materials releases into the environment will require close cooperation with county, state, and federal environmental protection agencies. It is especially important that hazardous materials response teams train with their local hazmat operations level responders—such as fire departments, EMS agencies, police departments, and emergency management agencies—that will be supporting their operations, as well as with special response teams such as urban search and rescue teams, bomb squads, and SWAT teams. Your hazmat team should ensure that standard operating procedures among agencies are compatible and, ideally, should train together on high-risk operations that may involve them, such as bomb scares and barricaded gunmen at industrial facilities.

Where Do We Go from Here?

This book covers the key concepts outlined in the training pyramid described previously. The text follows the National Fire Protection Association's response philosophy outlined in NFPA 472 (2008):

1. Analyze the incident.
2. Plan a response.
3. Implement the planned response.
4. Evaluate the progress.
5. Terminate the incident.

The first half of the book covers analyzing the incident and lays the foundation for planning a response. The second half of the book focuses on planning and implementing the planned response. Evaluating the progress is a key component of any safe emergency response operation and is stressed throughout the book. Terminating the incident is covered in the last chapter (Chapter 14: Response Operations).

CHAPTER **REVIEW**

Summary

You are now training to be one of the primary first responders that will be notified by awareness level personnel and called on to enter the hot zone and mitigate a hazardous materials release. This entails switching from a defensive mindset to an offensive mindset and learning how to use the necessary tools and the appropriate techniques to remain safe. This book focuses on providing the knowledge and expertise to safely accomplish mitigation techniques in a potentially contaminated hot zone to protect life, property, and the environment.

Review Questions

1. What is the most common WMD used by terrorists?
2. Which section of the *ERG* would you consult to look up a chemical by name?
3. Which section of the *ERG* would you consult to look up a chemical by its four-digit UN/NA number?
4. What type of information can you find in the orange section of the *ERG*?
5. What should you do when dealing with a hazard whose entry in the *ERG* is highlighted?
6. What are the nine hazard classes as defined by the U.S. DOT?
7. How does the NFPA 704 marking system for fixed-site facilities work?
8. What are the indications that a pipeline is located nearby?

References

Callan, Michael. (2001) *Street Smart Hazmat Response.* Chester, MD: Red Hat Publishing Company.

Hawley, Chris. (2003) *Hazardous Materials Incidents*, 2nd Ed. United States: Delmar Learning.

International Fire Service Training Association (2004) *Hazardous Materials for First Responders.* Stillwater, Oklahoma: Fire Protection Publications.

Lesak, David. (1999) *Hazardous Materials Strategies and Tactics.* Upper Saddle River, NJ: Pearson/Brady.

Oldfield, Kenneth W. (2005) *Emergency Responder Training Manual for the Hazardous Materials Technician.* Hoboken, New Jersey: Wiley – Interscience.

Schnepp, Rob (2010) *Hazardous Materials Awareness and Operations.* Sudbury, Massachusetts: Jones and Bartlett Publishers.

U.S. Department of Transportation. (2012). *2012 Emergency Response Guidebook.* Washington, DC: Pipeline & Hazardous Materials Safety Administration.

Weber, Chris. (2011) *Hazardous Materials Operations.* Upper Saddle River, NJ: Pearson/Brady.

Weber, Chris. (2007). *Pocket Reference for Hazardous Materials Response.* Upper Saddle River, NJ: Pearson/Brady.

CHAPTER 2

Regulations and Standards

Courtesy of Chris Weber, Dr. Hazmat, Inc.

KEY TERMS

buddy system, *p. 43*
consensus standard, *p. 38*
decontamination, *p. 42*
emergency response plan (ERP), *p. 42*
engineering control, *p. 41*
global harmonization, *p. 53*
gross negligence, *p. 57*
hazardous waste, *p. 49*
immediately dangerous to life and health (IDLH), *p. 41*
incident command system (ICS), *p. 43*
negligence, *p. 56*
permissible exposure limit (PEL), *p. 41*
personal protective equipment (PPE), *p. 41*
respiratory protection, *p. 45*

OBJECTIVES

After reading this chapter, the student should be able to:

- Describe the requirements of HAZWOPER with regard to hazardous materials emergency response at the hazardous materials technician level.
- List the five levels of hazardous materials training defined in HAZWOPER.
- Describe the training requirements at the hazardous materials technician level according to HAZWOPER.
- Summarize the requirements of the Resource Conservation and Recovery Act (RCRA).
- Name the key EPA regulations that affect hazardous materials response and training, and summarize their requirements.
- Summarize the requirements of the Hazardous Materials Regulations (HMR).
- Describe how the global harmonization effort affects, and will affect, response to hazardous materials releases at the hazardous materials technician level.

- Describe the differences between the HAZWOPER regulation and the NFPA 472 (2008) standard regarding hazardous materials technician proficiencies and training.
- Describe the role of voluntary consensus standard organizations and their relationship to hazardous materials response.

After bouncing from station to station, you've landed at station 6—on the edge of the industrial section of town. Over the last year you've been dispatched to your share of motor vehicle accidents in the area, several of which involved tanker trucks from a local refinery. Now you've been assigned to the hazmat team and have a few questions:

- Who regulates the refinery industry?
- What type of hazardous materials training do the drivers of the tanker have?
- Who needs be notified if the tanker leaks at one of these motor vehicle accidents?
- What type of hazardous materials training am I required to have to respond to these accidents?

Let's see if we can get some answers to these questions and more.

consensus standard ▪ Guidelines created by industry experts through a collaborative process as members of nongovernmental organizations.

Federal, tribal, state, and local regulations along with **consensus standards** make up the rules that govern hazardous materials technicians and hazardous materials response teams. You might be thinking: What is the difference between the two and how does this affect me?

Laws are passed by Congress and direct federal agencies to promulgate regulations. The regulations are written by experts in the subject matter and generally have industry and public input through a review and commenting process. Voluntary consensus standards are created by organizations made up of industry experts. These consensus standards may be adopted by reference at the federal, state, or local level, whereby they attain the force of law. It is also important to know your local ordinances and state laws, as these can have a profound effect on how you will be required to respond to hazardous materials incidents. Essentially, these standards and regulations mandate a risk-based response to hazardous materials incidents.

Legal Foundation of Hazardous Materials Response

One reason your employer is putting you through this training program, besides being concerned for your safety, is that several laws and federal regulations require it. State and federal governments regulate hazmat response owing to the danger hazardous materials and weapons of mass destruction (WMD) incidents pose to responders and the impact they have on people, property, and the environment. State regulations may be stricter than federal regulations, so you should also be familiar with your state's laws and regulations. Good resources are your state Department of the Environment (DOE), Department of Environmental Quality (DEQ), or Department of Environmental Protection (DEP).

Three agencies are primarily responsible for hazardous materials regulation at the federal level: the Environmental Protection Agency (EPA), the Occupational Safety and Health Administration (OSHA), and the Department of Transportation (DOT). Congress authorized these agencies to promulgate regulations by passing legislation that mandated the agencies to regulate how hazardous materials are disposed of and affect the environment, how they are handled in the workplace, and how they are transported.

The Superfund Amendments and Reauthorization Act (SARA) is a key law in the legal framework of hazardous materials response. SARA Title I directed OSHA and the EPA to develop safety regulations to protect workers at hazardous waste sites. The resulting regulations, found in 29 CFR 1910.120 and 40 CFR 311, are commonly known as HAZWOPER (HAZardous Waste OPerations and Emergency Response). HAZWOPER is the federal statute that covers public and private sector response to hazardous materials incidents, hazardous waste sites, and the operation of transport, storage, and disposal facilities. SARA Title III is commonly known as EPCRA, which stands for Emergency Planning & Community Right to Know Act. It requires facilities handling hazardous materials to provide information to the local community to support effective emergency response planning. Let's explore each of these in turn, since together they lay the legal foundation for hazardous materials response in the United States.

U.S. Occupational Safety and Health Administration

We start our discussion of regulations with OSHA, since we must always remember to protect ourselves first. If we get hurt, we cannot help anyone else. OSHA is tasked with protecting workers in the workplace. As employees of an emergency response organization, we fall under many OSHA regulations. We're going to discuss three of them here: the hazardous waste operations and emergency response regulation (HAZWOPER), the respiratory protection regulation, and the hazard communication regulation (HAZCOM). By far the most important one to us as hazmat technicians, which also references the respiratory protection regulation, is HAZWOPER, since it directly regulates our actions prior to, during, and after emergency responses to hazardous materials releases.

HAZARDOUS WASTE OPERATIONS AND EMERGENCY RESPONSE (HAZWOPER) (29 CFR 1910.120)

HAZWOPER is a comprehensive hazardous materials worker regulation developed in the late 1980s to improve worker safety during hazmat cleanup response. The EPA adopted an almost identical version of HAZWOPER, found in 40 CFR 311, since local and state government employees are not subject to federal OSHA regulation and are therefore not subject to HAZWOPER. Personnel such as local and state government workers who may not be covered by OSHA's regulation are covered by the EPA's (Figure 2-1). Therefore, EPA specifically included public sector employees under 40 CFR 311. Planned states—those states that have state-run occupational safety and health plans—have adopted a version of HAZWOPER that is identical with or more stringent than the federal version (Table 2-1). Thus irrespective of your state, your actions as a hazardous materials technician will be governed by the HAZWOPER regulation, so let's explore this regulation in detail.

See Hazardous Waste Operations and Emergency Response (HAZWOPER) from the Code of Federal Regulations for more information.

Paragraph A: Scope

HAZWOPER applies to

1. Cleanup operations required by a governmental body at uncontrolled hazardous waste sites;
2. Corrective actions involving cleanup operations at Resource Conservation and Recovery Act (RCRA) sites;

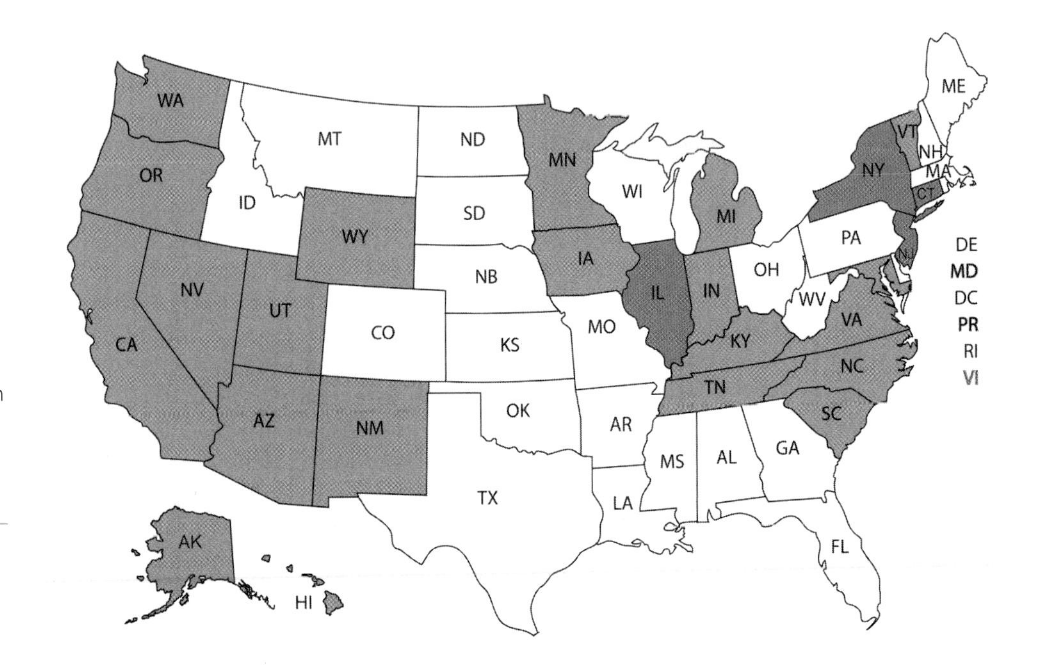

FIGURE 2-1 States and territories that have adopted their own version of HAZWOPER for all workers are shown in red, states that have adopted their own version of HAZWOPER for government workers only are shown in blue, and states that have not adopted their own regulations are shown in white. Your state jurisdiction can have a significant impact on how you respond to hazardous materials emergencies.

3. Voluntary cleanup operations at governmentally recognized uncontrolled hazardous waste sites;
4. Hazardous waste operations at treatment, storage and disposal (TSD) facilities;
5. *Emergency response operations for releases of, or substantial threats of releases of, hazardous substances without regard to the location of the hazard.*

except in cases where the employer can demonstrate that the operation does not involve or have the reasonable possibility to involve employee exposure to safety or health hazards.

Paragraphs B–O: Waste Site Worker Regulations

Do not confuse waste site worker requirements with emergency response worker requirements. The regulations for the former differ substantially in certain areas from those for emergency response workers, described in paragraph Q, which we will examine in detail later.

TABLE 2-1 The OSHA Plan States

OSHA PLAN STATES THAT COVER ALL EMPLOYEES		OSHA PLAN STATES THAT COVER ONLY PUBLIC SECTOR EMPLOYEES
Alaska	New Mexico	Connecticut
Arizona	North Carolina	Illinois
California	Oregon	New Jersey
Hawaii	Puerto Rico	New York
Indiana	South Carolina	Virgin Islands
Iowa	Tennessee	
Kentucky	Utah	
Maryland	Vermont	
Michigan	Virginia	
Minnesota	Washington	
Nevada	Wyoming	

Paragraph F: Medical Surveillance

HAZWOPER requires medical exams prior to your assignment to a hazmat team, at specific times during your hazmat career, and after you leave your hazmat team assignment. Emergency response employers and employees must comply with this regulation, including public safety agencies operated by local and state governments. Covered employees are

1. Employees who are, or may become, exposed to hazardous substances at or above permissible exposure levels (PEL) for 30 or more days per year;
2. Employees who wear respirators for 30 or more days per year;

3. Employees who are injured, become ill, or develop signs and symptoms due to a job-related exposure;
4. *Members of hazardous materials response teams.*

Medical exams must include a medical and work history that emphasizes hazardous substance exposure and fitness to wear **personal protective equipment (PPE)**. All exams must be performed by or under the supervision of a licensed physician and must be provided without cost to the employee. Information that must be provided to the physician by the employer includes the following:

personal protective equipment (PPE) ▪ A combination of chemical protective clothing (CPC) and respiratory protection designed to prevent hazardous materials from contacting and harming the wearer.

1. A copy of the HAZWOPER standard
2. A description of the employee's duties
3. The employee's actual or anticipated exposure levels
4. A description of the PPE to be used by the employee
5. Information pertaining to previous medical exams that are unavailable to the physician
6. Any other information required by 29 CFR 1910.134 (Respiratory Protection)

HAZWOPER also stipulates when a medical exam must be performed, and the frequency of follow-up medical exams and consultations:

1. Prior to assignment.
2. At least once every 12 months, unless the attending physician believes a longer interval (not to exceed 24 months) is appropriate.
3. At termination of employment or assignment (if there has not been an exam within the last six months).
4. As soon as possible if the employee has been exposed above the **permissible exposure limit (PEL)** or has developed signs and symptoms consistent with a job-related exposure to hazardous substances. We will discuss permissible exposure limits in Chapter 3.
5. More frequently if the examining physician believes it medically necessary.

permissible exposure limit (PEL) ▪ The OSHA enforced time-weighted average threshold limit value to which workers can be exposed continuously during an 8-hour work shift and a 40-hour work week without suffering ill effects.

The employer must give you, as the employee, a copy of the physician's written opinion, and the physician must inform you of the findings of the medical exam. *The physician may not inform your employer of specific findings or diagnoses unrelated to occupational exposure.* The employer must keep most medical records for at least the duration of employment plus 30 years, as specified in 29 CFR 1910.1020. The records shall contain the name and social security number of the employee, the physician's written opinion(s), employee medical complaints, and a copy of the information supplied to the physician by the employer.

engineering control ▪ Mechanical or structural devices or equipment that reduces the risk or extent of injury or exposure to a hazardous material or atmosphere.

Paragraph G: Engineering Controls, Work Practices, and PPE

Engineering controls, administrative controls, personal protective equipment (PPE), or a combination of these must be implemented to protect employees from exposure to hazardous substances and safety and health hazards. Whenever feasible, engineering controls and administrative controls need to be instituted to reduce and maintain employee exposure below the PEL. Examples of engineering controls include pressurized cabs or control booths on equipment and the use of remotely operated material handling equipment. Examples of administrative controls are staying upwind of the release, keeping nonessential personnel out of the hot zone, and wetting down dusty operations.

When engineering controls and changes in work practices are not sufficient to eliminate the health and safety hazards, PPE must be selected and used that will protect the employee from the hazards and potential hazards they are likely to encounter (as identified during the site characterization and analysis). PPE selection must be based on hazard- and site-specific performance criteria—a core component of a risk-based response to hazardous materials releases. Furthermore, supplied-air breathing apparatus must be used in **immediately dangerous to life and health (IDLH)** conditions. The level of PPE

immediately dangerous to life and health (IDLH) ▪ The atmospheric concentration of any substance that poses an immediate threat to life, causes an irreversible or delayed adverse health effect, or interferes with an individual's ability to escape during a 30-minute period. IDLH atmospheres may not be entered unless supplied-air respiratory protection is used.

must be increased if and when additional information on site conditions indicates an exposure above the PEL is possible. Conversely, the level of PPE may be decreased if site conditions improve and an exposure above the PEL will not result from downgrading to a lower level of PPE. A written PPE program must be part of the employer's safety and health program and contain the following elements:

1. PPE selection criteria based on site hazards
2. Instructions on use of the PPE and limitations of the equipment
3. Work mission duration
4. PPE maintenance and storage
5. PPE **decontamination** and disposal
6. PPE training and proper fitting
7. PPE donning and doffing procedures
8. PPE inspection prior to, during, and after use
9. Evaluation of the effectiveness of the program
10. Limitations during temperature extremes, heat stress, and other appropriate medical considerations

decontamination ▪ The process of removing unwanted contaminants from personnel and equipment.

Paragraph P: Treatment, Storage and Disposal Facility Regulations (RCRA)

This paragraph does not pertain to emergency response at the hazardous materials technician level. These regulations differ substantially in certain areas with respect to those for emergency response workers, which are described in paragraph Q, discussed in detail next.

Paragraph Q: Emergency Response to Hazardous Substance Releases

Paragraph Q specifies how public and private emergency response teams must handle hazardous materials incidents. First, it stipulates that an emergency response plan must be developed to address anticipated emergencies (transportation, pipeline, fixed-site facilities, etc.). As a hazardous materials technician, you should become familiar with your agency's **emergency response plan (ERP)**, the city or county ERP, and the ERPs developed by fixed-site facilities in your jurisdiction.

emergency response plan (ERP) ▪ A guidance document that explains how an entity or agency will handle conceivable emergencies such as natural disasters and hazardous materials releases. Required by several laws and regulations such as SARA Title III, RCRA, and HAZWOPER.

The ERP should address the following key points if they are not addressed elsewhere, such as in standard operating procedures or guidelines, in the local or state ERP, or in any applicable SARA Title III facility plans:

1. Preemergency planning and coordination with outside agencies
2. Personnel roles, lines of authority, training, and communication
3. Emergency recognition and prevention
4. Safe distances and places of refuge
5. Site security and control
6. Evacuation routes and procedures
7. Decontamination
8. Emergency medical treatment and first aid to exposed victims
9. Emergency alerting and response procedures
10. Critique of any previous hazardous materials incident responses and any follow-up actions
11. Personal protective equipment (PPE) and other emergency equipment

HAZMAT HANDLE

OSHA considers personal protective equipment (PPE) the last line of defense against hazardous materials, not the first! Always attempt to minimize the effects of hazardous materials using engineering controls and safe work practices first, then use PPE as a final precautionary measure.

Procedures should be developed for handling an emergency response. These response procedures should contain the following key points:

1. The senior response official will become the incident commander under an **incident command system (ICS)**. The incident command system should be National Incident Management System (NIMS) compliant. We will cover NIMS-compliant ICS in Chapter 13.
2. All hazardous substances and conditions present must be identified to the extent possible.
3. Appropriate emergency operations must be implemented using the correct PPE.
4. If the possibility of an inhalation hazard exists, self-contained breathing apparatus (SCBA) must be worn until appropriate air monitoring has been performed and a decreased level of respiratory protection is warranted.
5. The **buddy system** must be used in the hot zone. Responders should never enter the hot zone by themselves. Only personnel actively performing emergency operations should be permitted into the hot zone.
6. Backup personnel must be available for rescue of hot zone personnel. A minimum of two trained hazmat technicians should serve as the backup team (two-in/two-out rule).
7. Advanced first aid support personnel must be on standby with medical equipment and transport capability.
8. An experienced and knowledgeable safety officer must be appointed.
9. If IDLH or imminent danger conditions exist, the safety officer has the authority to terminate or alter emergency activities immediately and unilaterally.
10. After emergency operations have terminated, appropriate decontamination procedures must be implemented.
11. When it is deemed necessary for meeting the tasks at hand, approved SCBA bottles from different manufacturers may be interchanged provided they are of the same capacity and pressure rating.

incident command system (ICS) ■ A system designed to optimally use and direct resources at emergency incidents.

buddy system ■ The procedure that requires the hot zone to be entered in groups of at least two. For safety, the buddy system is mandated by OSHA's respiratory protection regulation.

Skilled support personnel, such as heavy equipment and crane operators, who do not possess HAZWOPER training, may be used at emergency incidents involving hazardous materials releases. During an emergency, these personnel may perform only the task(s) that cannot be safely performed by hazmat team members, and such operators must be briefed prior to entry. The briefing must encompass the following:

1. Instruction on the use of appropriate PPE
2. Chemical hazards involved
3. Duties to be performed

Specialist employees are those employees who work with and are trained in the hazards of specific hazardous substances as part of their routine duties and who will be called on to provide technical advice or assistance at a hazardous substance release. Specialist employees must receive training and demonstrate competency annually in their field of specialization.

HAZWOPER defines five levels of training:

- Awareness
- Operations
- Technician
- Specialist
- Incident commander

Awareness level personnel should understand and recognize hazardous materials incidents and their associated risks, with the ultimate goal to self-evacuate and notify the appropriate authorities who can take further action. They should be able to function within the agency's ERP, understand the use of the DOT *Emergency Response Guidebook*, and be able to implement site security and control.

The role of operations level personnel is to protect nearby persons, property, and the environment in a defensive fashion without coming into direct contact with the hazardous substance. In addition to possessing the awareness competencies, they must know how to execute basic hazard and risk assessment techniques, how to select and use PPE, how to perform basic product control procedures, how to carry out containment and/or confinement operations, and how to implement decontamination procedures. The operations level training must be a minimum of 8 hours in length.

Technician level personnel respond to hazardous materials emergencies for the purpose of stopping the release. They must be able to do the following:

1. Implement the employer's emergency response plan.
2. Use field survey instruments to verify and/or determine the nature of the release.
3. Function within the ICS.
4. Select and use PPE. The chemical protective clothing must meet the requirements of paragraph G, as explained earlier.
5. Understand hazard and risk assessment techniques.
6. Perform advanced product control, containment, and/or confinement techniques.
7. Understand and implement decontamination procedures.
8. Understand termination procedures.
9. Understand basic chemical and toxicological terminology and behavior.

The technician level training must be a minimum of 24 hours in length.

Specialist level personnel respond with and support hazmat technician level personnel. They must know how to implement the local emergency response plan; be able to use advanced survey instruments and equipment; know how to select and use PPE; understand in-depth hazard and risk assessment techniques; be able to perform specialized control, containment, and/or confinement techniques; know how to determine and implement decontamination procedures; be able to develop a site safety and control plan; and understand chemical, radiological, and toxicological terminology and behavior. The specialist level training must be a minimum of 24 hours in length.

Incident commander level personnel assume control of hazmat incidents beyond the awareness level response. They must be able to implement the employer's ICS and both the local and employer's emergency response plans, know of the state emergency response plan, know federal response capabilities, understand the risks associated with working in PPE, and understand the importance of decontamination procedures. They must receive a minimum of 24 hours of training equivalent to the hazardous materials operations level.

Trainers for hazardous materials emergency response classes must have taken an appropriate train-the-trainer class or possess training and/or academic credentials and instructional experience to be competent. Refresher training shall consist of receiving annual training of sufficient content and duration to maintain competency; alternatively, competency must be demonstrated at least annually. The employer must make a statement of competency and keep a record of the methodology used to demonstrate competency. This is a very important requirement to keep all hazmat team members safe. As a hazardous materials technician you should ensure that all members of the hazmat team are competent to perform their duties. The hazmat team is only as strong as its weakest link.

HAZMAT HANDLE

Although HAZWOPER specifies a minimum of 24 hours of training, this is almost certainly not enough time to become proficient in all the tasks required to be performed by hazardous materials technicians today.

HAZMAT HANDLE

Personnel trained to the awareness, operations, technician, specialist, and incident commander levels must receive at least as much annual training to remain competent. A mere 8 hours of annual training is *not* sufficient for technician level personnel to remain competent!

Medical surveillance and consultation is required for hazardous materials response team members. Members of a hazmat team and hazmat specialists must receive a baseline physical exam and be provided with medical surveillance according to paragraph F. Anyone who exhibits signs or symptoms that may have been caused by a job-related exposure to hazardous materials must be provided with medical consultation consistent with paragraph F, as explained earlier.

Post–emergency response operations, otherwise known as *cleanup*, are addressed in HAZWOPER as well. If hazardous materials or contaminated materials need to be removed from the site, the employer conducting the cleanup must comply with one of the following: (1) meet all the requirements of paragraphs B through O (the HAZWOPER hazardous waste site worker regulations) or (2) if cleanup is done on plant property by plant employees, they must complete the following training:

1. 29 CFR 1910.38 (a) (Emergency Action Plans)
2. 29 CFR 1910.134 (Respiratory Protection)
3. 29 CFR 1910.1200 (Hazard Communication)
4. Other appropriate task-specific safety and health training

Thus, HAZWOPER provides a comprehensive framework for hazardous materials response. The NFPA 472 (2008 edition) standard is fully consistent with paragraph Q of HAZWOPER. The HAZWOPER standard is currently in the process of being revised and will reportedly be updated and aligned with current editions of NFPA standards 472 and 473 (which we discuss in greater detail later in the chapter).

RESPIRATORY PROTECTION (29 CFR 1910.134)

The OSHA respiratory protection regulation, which is referenced by HAZWOPER, is designed to protect employees from hazardous atmospheres. It requires the employer to provide workers with the appropriate **respiratory protection**, training in how to use the respirators, and a medical evaluation to ensure the employees are fit to wear the respirator. Depending on the type of respirator and whether the use is voluntary or mandated, a written respiratory protection plan must be in place that covers respirator selection, use, maintenance, care, fit testing, training, respiratory protection program evaluation, and record keeping. Persons using a respirator should understand their employer's respiratory protection plan and be trained in the selection and use of the respiratory protection they will be wearing.

respiratory protection ■ PPE that prevents inhalation of airborne contaminants.

HAZARD COMMUNICATION (HAZCOM) (29 CFR 1910.1200)

The hazard communication regulation, or HAZCOM, was enacted in 1983. This regulation requires employers to inform employees of any hazardous chemicals with which

SOLVED EXERCISE 2-1

According to HAZWOPER, what is the minimum level to which personnel performing decontamination at a hazardous materials release (i.e., operating in the warm zone) need to be trained?

Solution: Operations level.

Your respiratory protection is your last line of defense, keeping hazardous materials from entering your lungs. Make sure you understand how to use your respirator well!

they will be working and to identify their specific hazards. Thus this law is also called the employee *right-to-know* law. HAZCOM requires chemical manufacturers to inform their customers of the chemical and health hazards associated with their products. This information must be detailed enough to allow the customer to develop worker protection and safety plans for its employees. Out of this law was born the Material Safety Data Sheet (MSDS). You should have received HAZCOM training from your employer informing you where to find MSDS describing the hazards of any chemicals stored or used at your workplace, and the storage locations for those chemicals. We will discuss what information may be found in MSDS and how to read and interpret them in detail in Chapter 6, Information Resources.

HAZCOM also includes employee training in how to handle incidental spills and health emergencies with these chemicals. HAZCOM is useful to you as a hazmat technician at the scene of a hazardous materials release because each employer must keep copies of MSDS that are readily available to employees. Therefore, they are often a convenient source of specific chemical information at the scene of the hazardous materials release at a fixed-site facility.

U.S. Environmental Protection Agency

The EPA was established in 1970. Its mission is to protect human health and to safeguard the natural environment—air, water, and land—on which we depend.

FEDERAL INSECTICIDE, FUNGICIDE, AND RODENTICIDE ACT (FIFRA)

Before the EPA had even been established, pesticides were already being regulated through FIFRA to protect consumers from ineffective products. FIFRA was established in 1947 to regulate pesticides—primarily, their distribution, sale, and use. At that time the U.S. Department of Agriculture was primarily responsible for its enforcement. As part of the newly formed EPA, FIFRA was amended in 1972 by the Federal Environmental Pesticide Control Act (FEPCA) and subsequently was focused more on the protection of human health and the environment. These amendments specifically included provisions to

1. Strengthen the pesticide registration process by shifting the burden of proof to the chemical manufacturer;
2. Enforce compliance against banned and unregistered products;
3. Promulgate the regulatory framework missing from the original law.

The EPA formulated a Pesticide Product Labeling System (PPLS) to increase consumer safety by ensuring that pesticide labels are consistent and have thorough safety and use directions. The pesticide label must be securely attached and include the following information:

- name and address of the producer, registrant, or person for whom produced
- restricted use statement (when required)
- product name, brand, or trademark
- ingredient statement
- signal word—including skull and crossbones (when required)

- "Keep Out of Reach of Children" (KOOROC)
- cautionary statements—including hazards to humans and domestic animals
- EPA registration number
- EPA establishment number
- storage and disposal statements
- referral statement to direction for use in booklet (when supplemental labeling is used)
- net weight or measure of contents

The EPA registration number (EPA Reg. No.) for pesticides is composed of two parts separated by a hyphen, such as 123-456. The first half (123) is the company registration number, and the second half (456) is the product registration number, which uniquely identifies that pesticide. Both parts together form the pesticide registration number. An example of a pesticide label is shown in Figure 2-2. A product labeled with a three-part number, such as 123-456-789, is a *distributor product*, that is, a product manufactured by one company but marketed, usually under a different name, by another company. In this case, the last third of the number (789) refers to the company doing the distributing.

The signal word describes the degree of hazard the pesticide poses. The signal word must be highlighted in a different color, be in bold print, be outlined, or draw attention to itself in some other way on the label. Three signal words are used:

- DANGER is used for the most highly toxic pesticides.
- WARNING is used for moderately toxic pesticides.
- CAUTION is used for less toxic pesticides.

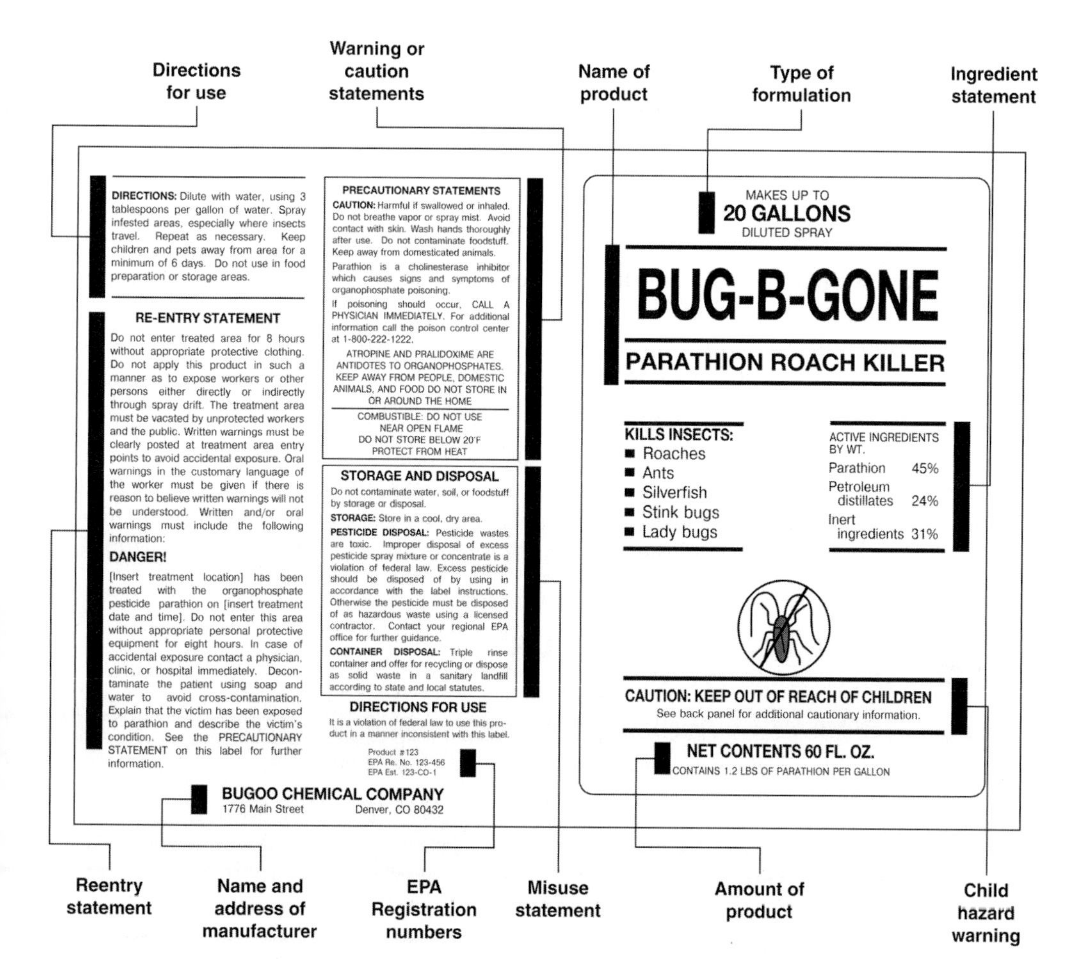

FIGURE 2-2 An example of the key features of pesticide labels. The EPA requires the inclusion of specific information that can be very helpful during hazardous materials emergencies.

CLEAN AIR ACT OF 1970 (CAA)

The Clean Air Act required the EPA to set national ambient air quality standards for certain pollutants deemed especially dangerous to human health, called *criteria air pollutants*. Currently there are six such pollutants: sulfur dioxide, nitrogen dioxide, ozone, carbon monoxide, lead, and particulate matter. Under EPA's Clean Air Act contaminant concentrations are regulated for the general welfare of the public, not for employees in particular, as with OSHA regulations. Almost 200 additional air pollutants that may cause significant adverse health effects in the public are regulated under the National Emission Standards for Hazardous Air Pollutants. These chemicals include asbestos and mercury.

The CAA also established the U.S. Chemical Safety and Hazard Investigation Board, also known as the Chemical Safety Board (CSB), which investigates major chemical accidents to provide recommendations to improve safety. These reports can be very useful to hazmat response teams for training and learning more about the hazards certain industries located in a local response area may pose.

CLEAN WATER ACT OF 1972 (CWA)

The Clean Water Act established a federal response system for hazardous materials spills in waterways. Although the legislation was created primarily for oil spills, it was expanded to include the response to and cleanup of emergency releases of chemicals in general. This act established the National Contingency Plan (NCP) and established the National Response Team (NRT).

National Contingency Plan (NCP)

The National Oil and Hazardous Substances Pollution Contingency Plan, commonly called the National Contingency Plan (NCP), describes the federal response structure for significant hazardous materials releases in the United States. The NCP provides guidance for federal agencies involved in hazardous materials response under CERCLA. The industry Tier I and Tier II plans received by the local emergency planning committees (LEPCs) and reviewed by the state emergency response commission (SERC) must follow the NCP.

Three fundamental kinds of activities are performed pursuant to the NCP:

1. Preparedness planning and coordination for response to a discharge of oil or release of a hazardous substance, pollutant, or contaminant
2. Notification and communications
3. Response operations at the scene of a discharge or release

National Response Team (NRT)

National planning and coordination is accomplished through the National Response Team (NRT), which consists of representatives from many agencies involved in hazardous materials response, from boots on the ground to financing and logistics. The NRT is chaired by a representative from the EPA, and the vice chair is a representative of the U.S. Coast Guard.

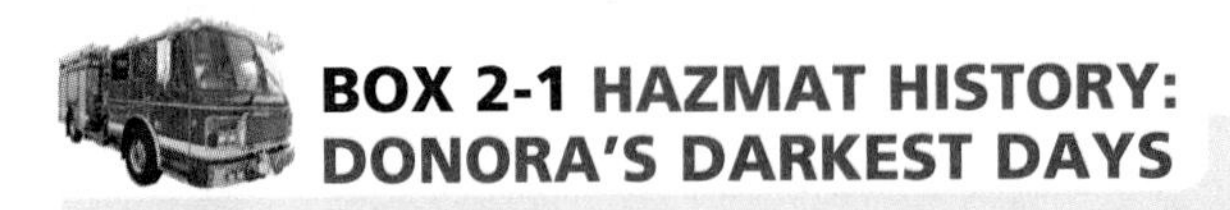

BOX 2-1 HAZMAT HISTORY: DONORA'S DARKEST DAYS

In the 1940s Donora, Pennsylvania, was a steel mill town with a population approaching 14,000. At that time, there were no regulations governing pollution, and the town's numerous mills produced large amounts of toxic yellow smog during normal operations. In the fall of 1948 an air inversion caused a toxic cloud to be trapped low to the ground for an extended period. It was so dark during the middle of the day that residents needed flashlights just to navigate the streets. By the time the incident ended almost a week later, 20 people had died and almost seven thousand had become ill. Several similar incidents eventually prompted the Clean Air Act of 1970.

TOXIC SUBSTANCES CONTROL ACT OF 1976 (TSCA)

The Toxic Substances Control Act (TSCA) was passed in 1976 to regulate the manufacture, distribution, use, and disposal of chemicals. This law arose in response to many accidents and injuries from chemicals routinely used in the workplace and the home, as well as improper disposal of these chemicals. Many chemicals are tightly regulated, and some are even banned outright, such as polychlorinated biphenyls (PCBs), chlorofluorocarbons (CFCs), asbestos, and lead paint. One result of this act is the warning labels we've come to rely on found on many common consumer commodities. More than 84,000 chemicals are regulated under the TSCA, and they are listed in the Chemical Substance Inventory. When products contain one or more of these chemicals, a warning label listing the following information must be included on the product:

- name and address of the manufacturer
- common name of the chemical
- health hazards
- environmental hazards
- exposure hazards
- actions to minimize any of the aforementioned hazards

RESOURCE CONSERVATION AND RECOVERY ACT OF 1976 (RCRA)

The Resource Conservation and Recovery Act of 1976, commonly referred to as RCRA, was passed by Congress in response to several significant environmental disasters, such as the Valley of the Drums in Kentucky. This regulation imposed "cradle-to-grave" regulation of **hazardous waste** and made the generator of hazardous waste responsible for its safe and legal disposal. Any entity that generates hazardous waste must properly store, transport, and dispose of it. Proper disposal may include recycling, such as redistillation of waste solvent; incineration, such as burning of waste hydrocarbons in specially licensed kiln ovens; or disposal in a licensed and specially engineered hazardous waste landfill. RCRA has significantly reduced the amount of illegal dumping owing to stiff civil monetary penalties and criminal penalties including jail time for willful acts.

hazardous waste ■ A waste material designated by EPA and listed in the CFR, or a waste material exhibiting one or more hazardous waste characteristics (ignitability, corrosiveness, reactivity, or toxicity).

The RCRA

- requires companies that generate hazardous waste to properly characterize the waste;
- limits the time hazardous waste may be stored on their premises before disposal (90–270 days depending on the size of the generator);
- dictates safe storage conditions during temporary hazardous waste storage;
- requires the use of hazardous waste manifests during transport;
- requires the licensing of hazardous waste haulers;
- regulates disposal facilities;
- requires that the hazardous waste manifest be returned to the original generator indicating how the hazardous waste was legally disposed of.

State and federal environmental agencies regularly inspect RCRA-licensed facilities, and employees of such facilities must receive initial RCRA training and annual refresher training.

It is a good idea to periodically tour RCRA facilities in your response area. Dealing with hazardous waste releases can often be more difficult than dealing with virgin chemicals, since detailed toxicological and chemical and physical property information about the waste often does not exist. Sometimes, hazardous waste is merely a used solvent that is just a dirtier version of the virgin material; however, hazardous waste also can be a complicated mixture of several different chemicals that have been used in a complicated industrial process. There are no practical ways accurately to identify all hazards of these mixtures in the field. The EPA does require hazardous waste generators to classify the waste into three categories:

- universal
- characteristic (flammability, reactivity, corrosiveness, and toxicity)
- listed

BOX 2-2 HAZMAT HISTORY: VALLEY OF THE DRUMS

The A.L. Taylor Site, also known as "Valley of the Drums," covers 13 acres in Brooks, Kentucky, 12 miles south of Louisville. It was used as a refuse dump, drum recycling center, and chemical waste dump for a decade, from 1967 to 1977. The chemical waste was largely from the paint and coatings industries and contaminated the air, surface water, groundwater, and soil with organic and inorganic chemicals.

The EPA conducted emergency response activities in March 1979 under Section 311 of the Clean Water Act, and in September 1981 under CERCLA, at a total cost of $650,000. Through these response activities and voluntary removal of wastes by known generators, a majority of the surface wastes (about 17,000 drums) were removed. A system was installed to control and treat contaminated run-off from the site.

Source: U.S. Environmental Protection Agency, National Priorities List

SOLVED EXERCISE 2-2

You have been dispatched to a hazardous materials incident at a chemical manufacturing plant. On your arrival the plant manager tells you that a 55-gallon drum of hazardous waste has been punctured by a forklift on the loading dock. How would you determine the contents of the drum?

Solution: Consult the copy of the universal hazardous waste manifest the facility is required to have on site.

These are very broad classifications and do not provide detailed toxicological information, chemical reactivity information, or chemical and physical property information.

COMPREHENSIVE ENVIRONMENTAL RESPONSE, COMPENSATION, AND LIABILITY ACT OF 1980 (CERCLA)

The Comprehensive Environmental Response, Compensation, and Liability Act (CERCLA), commonly known as Superfund, provides federal authority to respond to releases or threatened releases of hazardous materials that may endanger public health or the environment. Additionally, CERCLA provides for liability of persons responsible for releases of hazardous waste at closed and abandoned hazardous waste sites and establishes a trust fund to provide for cleanup when a responsible party cannot be identified. The trust fund is funded by a tax on the chemical and petroleum industries. The actions taken by local and state governments at hazardous materials incidents are exempted from liability under CERCLA. The law authorizes short- and long-term removal actions based on the EPA's National Priorities List (NPL).

SUPERFUND AMENDMENTS AND REAUTHORIZATION ACT OF 1986 (SARA) AND THE EMERGENCY PLANNING & COMMUNITY RIGHT-TO-KNOW ACT (EPCRA)

In 1986, CERCLA was amended by the Superfund Amendments and Reauthorization Act (SARA). SARA comprises four titles:

Title I	Response and liability
Title II	Miscellaneous provisions
Title III	Emergency planning and community right to know (EPCRA)
Title IV	Interior air quality research and radon

Title III of SARA, also called the Emergency Planning & Community Right-to-Know Act (EPCRA), mandates that communities plan for emergencies involving hazardous

BOX 2-3 HAZMAT HISTORY: BHOPAL, INDIA

On the evening of December 2, 1984, the world's worst industrial accident occurred at a Union Carbide subsidiary in Bhopal, India. Water entered an industrial tank containing 42 tons of water-reactive methyl isocyanate (MIC), which caused an exothermic reaction that resulted in the release of a large toxic gas cloud. Almost four thousand people were killed in the immediate aftermath of the accident, and estimates indicate that up to fifteen thousand died within weeks of the accident. In fact, people continue to suffer ill effects to this day in and around Bhopal. The cause of the accident has been disputed, and scenarios range from poor maintenance that allowed water to seep into the tank, to a cleaning operation using high-pressure water, to sabotage. Several factors contributed to the severity of the accident:

- Extremely large quantities of the reactive methyl isocyanate were stored in a single, overfilled storage tank.
- Reports indicated that maintenance on the storage tank, pipes, and valves was substandard.
- Reports also indicated that facility personnel were not properly trained, and safety systems were disabled owing to economic considerations. Supposedly, additional safety systems that required manual intervention were not activated.
- Large population densities, in the form of slums, were allowed to build up around the industrial facility.
- There were no community emergency response plans to handle a toxic release or a mass evacuation.

This incident directly led to the implementation of Sara Title III legislation in 1986 to prevent a similar incident in the United States.

materials and requires local industrial facilities to report the hazardous materials they store on-site to local first responders. Facilities that use and store large amounts of hazardous materials are therefore often called SARA Title III sites. The following are the four primary provisions of EPCRA:

- emergency planning (sections 301–303)
- emergency release notification (section 304)
- hazardous chemical inventory (sections 311 and 312)
- toxic chemical release inventory (section 313)

Exemptions from reporting requirements are permitted for certain trade secrets per section 322. This law is designed to help local communities protect public health, safety, and the environment from chemical hazards. The chemical inventory report, along with the site maps and the emergency response procedures that are contained in the plans, is a great source of information for hazardous materials technicians and should be used for training and planning by the hazmat team.

This legislation also requires the formation of State Emergency Response Commissions (SERCs) and Local Emergency Planning Committees (LEPCs) to plan for and react to hazardous materials incidents. LEPCs are usually formed at the municipal or county level and are responsible for the development of hazardous materials emergency response plans. LEPCs comprise local stakeholders, such as industry, public safety, and nonprofit organizations, and they review the ERP on an annual basis. The local emergency management agency and fire department usually work very closely with the LEPC to ensure that the hazardous materials response in their community will be adequate. You should become familiar with your role as a hazardous materials technician in the local ERP. By law, communities—primarily represented by their hazardous materials response team—must have a response plan, must exercise that plan, and must follow that plan during emergency responses.

OIL POLLUTION ACT OF 1990 (OPA)

The Oil Pollution Act (OPA) regulates the petroleum industry, including producers, carriers, and refineries. Under OPA, companies must develop emergency response plans, training programs, and exercises, and have adequate spill resources available. A cleanup trust was established that is funded by a tax on oil. Alledgedly violations of this act by British Petroleum (BP) and others in their management of the Deepwater Horizon oil rig contributed to the disastrous explosion in April 2010.

U.S. Department of Transportation

The U.S. Department of Transportation (U.S. DOT) has been tasked by Congress to maintain safe interstate transportation, including the transportation of hazardous materials by road, rail, air, and water. Air transportation is regulated by the Federal Aviation Administration (FAA), which is a part of the department. As hazardous materials responders we must be familiar with many of these laws, since they regulate packaging and transportation as well as accident reporting. Although accident reporting is typically the responsibility of the carrier (the responsible party), we should ensure that the appropriate authorities have been notified. Let's start by examining the hazardous materials regulations and get a broad overview of what they contain.

PLACARDING TABLES

[§172.504(e)]

TABLE 1

Category of material (Hazard Class or division number and additional description, as appropriate)	Placard name
1.1	EXPLOSIVES 1.1
1.2	EXPLOSIVES 1.2
1.3	EXPLOSIVES 1.3
2.3	POISON GAS
4.3	DANGEROUS WHEN WET
5.2 (Organic peroxide, Type B, liquid or solid, temperature controlled)	ORGANIC PEROXIDE
6.1 (materials poisonous by inhalation (see §171.8 of this subchapter))	POISON INHALATION HAZARD
7 (Radioactive Yellow III label only)	RADIOACTIVE[1]

[1] RADIOACTIVE placard also required for exclusive use shipments of low specific activity material and surface contaminated objects transported in accordance with §173.427(a)(6).

TABLE 2

Category of material (Hazard Class or division number and additional description, as appropriate)	Placard name
1.4	EXPLOSIVES 1.4
1.5	EXPLOSIVES 1.5
1.6	EXPLOSIVES 1.6
2.1	FLAMMABLE GAS
2.2	NON-FLAMMABLE GAS
3	FLAMMABLE
Combustible Liquid	COMBUSTIBLE
4.1	FLAMMABLE SOLID
4.2	SPONTANEOUSLY COMBUSTIBLE
5.1	OXIDIZER
5.2 (Other than organic peroxide, Type B, liquid or solid, temperature controlled)	ORGANIC PEROXIDE
6.1 (Other than materials poisonous by inhalation)	POISON
6.2	(None)
8	CORROSIVE
9	Class 9 (see §172.504(f)(9))
ORM-D	(None)

FIGURE 2-3 DOT placarding requirements: Table 1 materials must be placarded in any quantity, and Table 2 materials must be placarded when more than 1001 pounds of the material is being transported. *Courtesy of the U.S. Department of Transportation Pipeline and Hazardous Materials Safety Administration.*

THE HAZARDOUS MATERIALS REGULATIONS (HMR)

The U.S. DOT regulates hazardous materials during transport. These regulations are known as the Hazardous Materials Regulations (HMR) and can be found in 49 CFR, Parts 100–185. The HMR require that hazardous materials shipments are properly characterized, safely packaged, appropriately marked and labeled, and accompanied by shipping papers during transportation.

Labels, placards, and shipping papers are an important source of information for first responders at incidents involving hazardous materials releases during transportation. Placarding and labeling requirements are summarized in Figure 2-3. According to 49 CFR 172.504, any quantity of material listed in DOT Table 1 must be placarded (Figure 2-3, top). DOT Table 2 materials must be placarded when transported by highway or rail when the aggregate gross weight of all hazardous materials in the nonbulk packages is over 454 kg (1001 lb) (Figure 2-3, bottom). It is also worth noting that it is illegal to place a label on a package or a placard on a vehicle when it does not contain that hazardous material during transportation. However, it is permissible to placard a vehicle that contains a Table 2 material even when the placard is not required by the quantity present, as long as any quantity of that hazardous material is in the vehicle.

You may also encounter a white square placard with a black border that is required for certain highly dangerous shipments, such as highway route controlled quantities of radioactive material, rail shipments of specified explosives and poisons, and flammable gases in DOT 113 tank cars. We will study the different types of packages and containers in which hazardous materials are shipped and used in Chapter 10, Containers and Damage Assessment. Most of the specifications and features we will discuss come directly from the Hazardous Materials Regulations.

Any employees that offer hazardous materials for transport must receive initial training that at a minimum includes

- general awareness/familiarization training (such as recognition of hazardous materials);
- function-specific training (such as proper packaging);
- safety training (such as use of personal protective equipment);
- security awareness training (such as understanding how hazardous materials shipments may be stolen);
- in-depth security training (when applicable, on the written transportation security plan of the employer).

Hazmat employees must receive refresher training at least every 3 years. These training requirements are meant to reduce the number of transportation accidents such as the ValuJet flight 592 crash in Florida, which killed 110 people on May 11, 1996. This passenger aircraft was brought down by the improper packaging of oxygen generators, which resulted in a fire in the cargo compartment. Had the employees of the company that shipped the oxygen generators received better training, this accident and loss of life could have been prevented.

GLOBAL HARMONIZATION

The hazardous materials regulations are modified on an annual basis, but they have recently been changing at a more rapid rate and will continue to do so over the next several years owing to global harmonization. **Global harmonization** refers to the worldwide development of consistent hazardous materials classification and transportation regulations. The official name for this scheme as designated by the United Nations is the Globally Harmonized System of Classification and Labelling of Chemical Substances (GHS). One of the aims of GHS is to reduce inconsistencies such as the labeling of anhydrous ammonia between the United States and the rest of the world. For example, in the United States, anhydrous ammonia is considered a Class 2 nonflammable gas. Internationally,

global harmonization ■ The worldwide development of consistent hazardous materials classification and transportation regulations.

BOX 2-4 HAZMAT HISTORY: THE EARLY YEARS OF HAZMAT TRANSPORTATION

Nitroglycerin is an extremely sensitive explosive that was commonly used in the construction and mining industries in the nineteenth century. Because nitroglycerin is sensitive to heat, friction, and shock, many accidents occurred during its transportation, storage, and use. In fact, the degree of hazard associated with nitroglycerin was the driving factor behind the invention of dynamite by Alfred Nobel. Let's take a look at a nitroglycerin accident from the nineteenth century to show how differently hazardous materials were transported, stored, and used then.

SAN FRANCISCO, CALIFORNIA (1866)

In April 1866 an unmarked shipment of nitroglycerin that originated in Germany arrived in the Port of San Francisco. One of the crates was leaking and was taken to the Wells Fargo & Co. office, where curious employees pried open the box, causing an explosion that leveled everything within a 50-foot radius, killed many people, and shattered windows within a half-mile radius. Such a catastrophe illustrates why safety regulations governing hazardous materials awareness training, marking and labeling requirements, and employee right-to-know laws are important.

FIGURE 2-4 An example of the sometimes drastic differences in placarding regulations in the United States versus in Europe. In the United States anhydrous ammonia is placarded as a nonflammable gas (left); in Europe it is placarded as a poisonous and corrosive gas (right). *Photos courtesy of Chris Weber, Dr. Hazmat, Inc.*

anhydrous ammonia is considered a Class 2 poison gas with subsidiary hazard of Class 8, corrosive (Figure 2-4).

In addition to consistent labeling and placarding categories, a consistent set of pictograms, labels, and placards is being devised (Figure 2-5). Global harmonization should make it easier for manufacturers, shippers, and carriers to understand the dangerous goods regulations, in turn making it safer for hazardous materials emergency responders. As an example, in January 2011 the Pipeline and Hazardous Materials Safety Administration (PHMSA) authorized changes to the following areas of the HMR relating to global harmonization:

- proper shipping names
- hazard classes
- packing groups
- special provisions
- packaging authorizations
- air transport of limited quantities
- vessel stowage requirements

A final ruling regarding some of these changes is expected in 2012. In expectation of these changes several manufacturers have already started using the GHS sanctioned safety data sheet (SDS) format for their hazard communication sheets. SDS contains virtually the identical information as MSDS, but the order and format has been standardized.

FIGURE 2-5 Nine labels used to identify hazardous materials in the Global Harmonization System (GHS). Top row: explosive (left), flammable (middle), and oxidizer (right). Middle row: compressed gas (left), corrosive (middle), and poison (right). Bottom row: the cautionary symbol used for multiple hazards (left), the health hazard symbol used for such materials as carcinogens (middle), and environmental hazards (right). As a hazardous materials technician you will soon encounter these labels much more often as global harmonization becomes more widespread. *Art by David Heskett.*

FEDERAL AVIATION ADMINISTRATION (FAA)

At large hazardous materials incidents it can be helpful to temporarily close the airspace above and near the location of the hazardous materials release for the safety of the aircraft and that of responders on the ground. Obviously, a sudden explosion at the site could cause aircraft to lose control and crash. This actually

SOLVED EXERCISE 2-3

What type of training would you expect an employee at a trucking company that ships corrosive materials to have?

Solution: The training must include general awareness/familiarization training, function-specific training, safety training, security awareness training, and in-depth security training.

happened in 1947 at the Texas City, Texas, ammonium nitrate explosion (see Box 4-2); two aircraft were blown out of the sky. In addition, low-flying aircraft such as helicopters can cause problems for emergency responders and the public, as the downwash from the rotors can affect gas and vapor movement and can cause other unintended consequences on the ground. Therefore, under the FAA Temporary Flight Restrictions in the Vicinity of Disaster/Hazard Areas (14 CFR 91.137), the incident commander can request a temporary flight restriction near hazardous materials incidents, which the FAA would implement. Accredited news representatives have been excepted from this rule; however, they must file a flight plan.

National Fire Protection Association (NFPA)

The National Fire Protection Association (NFPA), a nongovernmental organization, has promulgated a set of hazardous materials response standards to which most emergency response agencies try to adhere. Standards typically do not have the force of law, but they may be enforced through civil lawsuits. Thus, NFPA standards provide a "standard of care" that hazmat responders may be held to in a court of law. Furthermore, standards may be adopted by reference in a regulation, effectively making the standard a regulation that carries the force of law.

NFPA 472 (2008 EDITION)

NFPA 472, *Standard for Competence of Responders to Hazardous Materials/Weapons of Mass Destruction Incidents*, gives explicit guidance for fire service and law enforcement personnel responding to hazardous materials and weapons of mass destruction incidents. NFPA 472 follows the five levels of training stipulated by OSHA in HAZWOPER: awareness, operations, technician, specialist, and incident commander.

The standard provides specific competencies, both knowledge-based and performance-based, that personnel certified to any given level must be able to master. The standard is divided into 15 chapters, and annexes A through K. Key chapters are the following:

- Chapter 4: Competencies for Awareness Level Personnel
- Chapter 5: Core Competencies for Operations Level Responders
- Chapter 6: Competencies for Operations Level Responders Assigned Mission Specific Responsibilities
- Chapter 7: Competencies for Hazardous Materials Technicians
- Chapter 8: Competencies for Incident Commanders
- Chapter 9: Competencies for Specialist Employees
- Chapter 10: Competencies for Hazardous Materials Officers
- Chapter 11: Competencies for Hazardous Materials Safety Officers
- Chapter 12: Competencies for Hazardous Materials Technicians with a Tank Car Specialty
- Chapter 13: Competencies for Hazardous Materials Technicians with a Cargo Tank Specialty
- Chapter 14: Competencies for Hazardous Materials Technicians with an Intermodal Tank Specialty
- Chapter 15: Competencies for Hazardous Materials Technicians with a Marine Tank Vessel Specialty

Each chapter outlines the knowledge- and performance-based competencies required for the appropriate level of hazardous materials response. In addition, Annex A contains detailed explanatory material for selected topics. NFPA 472 is therefore an excellent document outlining the training requirements for hazardous materials responders and is used as a training guide for developing hazardous materials training curricula.

This text covers the competencies for hazardous materials technicians as outlined in Chapter 7 of the 2008 edition of NFPA 472. All training to the NFPA 472 standard must demonstrate a student's ability to perform required tasks. The class you are taking part in will therefore have not only classroom portions but also hands-on exercises based on the information provided in this text.

NFPA 473 (2008 EDITION)

NFPA 473, *Standard for Competencies for EMS Personnel Responding to Hazardous Materials Incidents*, gives specific guidance to emergency medical service personnel responding to hazardous materials and weapons of mass destruction incidents. The standard defines where EMS personnel may perform patient care:

- Level 1 EMS personnel may perform patient care in the cold zone.
- Level 2 EMS personnel may perform patient decontamination and perform patient care in the warm zone (at triage points and at the decontamination line).

ASTM, ANSI, and Other Standards Organizations

Several other standards organizations have been formed to increase safety, to standardize certain procedures, and generally to improve the efficiency of business and industry. The American National Standards Institute (ANSI) provides a venue for interested parties to formulate voluntary consensus standards. Throughout this book you will periodically encounter specific standards from various organizations that affect how you respond to hazardous materials incidents as hazmat technicians. For example, ASTM International (which was formerly known as the American Society for Testing and Materials) has issued standards ranging from personal protective equipment testing to sample collection of suspected biothreat agents. The American Petroleum Institute (API) is an advocate for safety standards and voluntary consensus standards in the petroleum industry and has been very influential since it promulgated its first standards in 1924.

Liability

Liability may arise out of your actions at a hazardous materials emergency when your actions are negligent or criminal in nature. When you use poor judgment and your actions cause injury or damages, you have been negligent. **Negligence** can occur before, during, and after hazardous materials responses. Failure to adequately plan for facility emergencies, train and equip personnel, and keep equipment in working order before an emergency occurs could all be construed as negligence if they subsequently cause injury or damage

negligence ■ The use of poor judgment in which the consequent actions cause injury or damages to others.

SOLVED EXERCISE 2-4

You are having difficulties finding formal guidance documents relating to sampling biological agents in the federal regulations. What other resources are available to you?

Solution: Voluntary consensus standards published by such organizations as the NFPA, ANSI, or ASTM may be helpful. Indeed, ASTM has just such a document: E2458-10: *Standard Practices for Bulk Sample Collection and Swab Sample Collection of Visible Powders Suspected of Being Biothreat Agents from Nonporous Surfaces.*

during an emergency response. Failing to take action, or taking the wrong action, during an emergency could also be construed as negligence. Failure to medically monitor exposed personnel, failure to correct ineffective standard operating procedures, or failure to repair and replace damaged equipment after an emergency response could also be construed as negligence. **Gross negligence** occurs when an individual acts in an outrageously negligent manner. To prove negligence in a civil suit, the following three components must be present:

gross negligence ■ Occurs when an individual acts in an outrageously negligent manner.

1. There was a duty to act, and a failure to do so—hazardous materials technicians as part of the hazardous materials response team automatically have a duty to act.
2. There was a failure to conform to the generally accepted standard of care (such as HAZWOPER or NFPA 472).
3. A loss or damage was suffered owing to the allegedly negligent actions.

The answer to the question of who was responsible for the negligence will determine who has the liability for the negligent actions that caused the injury and/or damage. In the case of hazardous materials emergency response, this can be a complicated question. There are four general types of liability:

- **Vicarious liability:** One person or agency is responsible for the actions of another person.
- **Respondeat superior:** The employer is responsible for the actions of the employee; however, if the employee acts outside the scope of employment, the employee may be personally liable for his or her negligent actions.
- **Joint and several liability:** The victim is allowed to recover damages from all responsible parties.
- **Landowner liability:** Property owners are responsible for damages arising out of the use of that property, such as poorly maintained or defective equipment.

Criminal liability arises when laws are violated, and a criminal act is thereby committed. This would be a very unusual circumstance for emergency response personnel. Civil liability arises when accepted codes of conduct are not followed. There are three types of civil liability:

- **Common law nuisance:** An activity interferes with the use of private property (private nuisance) or interferes with public health and safety (public nuisance).
- **Common law negligence:** A wrongful act causes damage or injury; the act can be willful or negligent.
- **Statutory liability:** Accountability is imposed through federal, state, or local laws and ordinances.

Public sector emergency responders have a certain amount of immunity depending on their actions and their physical location (state and municipality). Most states provide qualified immunity from liability as long as there was no gross negligence involved. However, as hazardous materials technicians you still have the duty to act in a competent and professional manner, through pre-incident planning, rigorous training, exercising, and maintaining your equipment in working order.

CHAPTER **REVIEW**

Summary

Responding to hazardous materials incidents can be dangerous if not properly performed. Therefore, there are numerous regulations and standards involving hazardous materials response that apply to both private and public sector employees. The Occupational Safety and Health Administration (OSHA) regulates hazardous materials response through the HAZWOPER regulation as well as the respiratory protection and Hazard Communication (HAZCOM) regulations. Other federal agencies that are heavily involved in hazardous materials response regulation are the Environmental Protection Agency (EPA) and the Department of Transportation (DOT). The National Fire Protection Association (NFPA) has promulgated many consensus standards that apply directly to hazardous materials response. The most significant NFPA standard that affects hazardous materials technicians is NFPA 472 (2008). It is important to follow all applicable regulations and consensus standards for your own safety and that of your team members and the public.

Review Questions

1. What is the difference between a law, a regulation, and a standard?
2. What regulation covers emergency response to hazardous materials incidents?
3. Which standard covers emergency response to hazardous materials incidents?
4. What are the five levels of hazardous materials response training?
5. Who receives the SARA Title III plans and paperwork provided by industry?
6. What is the function of an LEPC?
7. What is global harmonization, and what is its significance to the emergency response community?

References

Department of Transportation. (1998) 49 CFR. Hazardous Materials Regulations. Washington, DC: U.S. Department of Transportation.

Environmental Protection Agency (1970) 40 CFR. Protection of Environment. Washington, DC: U.S. Environmental Protection Agency.

National Fire Protection Association. (2008). NFPA 472, *Standard for Competence of Responders to Hazardous Materials/Weapons of Mass Destruction Incidents.* Quincy, MA: Author.

National Fire Protection Association. (2008). NFPA 473, *Standard for Competencies for EMS Personnel Responding to Hazardous Materials/Weapons of Mass Destruction Incidents.* Quincy, MA: Author.

National Fire Protection Association. (2008). NFPA 1999, *Standard on Protective Clothing for Emergency Medical Operations.* Quincy, MA: Author.

National Fire Protection Association. (2007). NFPA 1994, *Standard on Protective Ensembles for First Responders to CBRN Terrorism Incidents.* Quincy, MA: Author.

National Fire Protection Association. (2005). NFPA 1991, *Standard on Vapor-Protective Ensembles for Hazardous Materials Emergencies.* Quincy, MA: Author.

National Fire Protection Association. (2005). NFPA 1992, *Standard on Liquid Splash-Protective Ensembles and Clothing for Hazardous Materials Emergencies.* Quincy, MA: Author.

Occupational Safety and Health Administration. (1990). 29 CFR 1910.120, *Hazardous Waste Operations and Emergency Response (HAZWOPER).* Washington, DC: U.S. Department of Labor.

Occupational Safety and Health Administration. (1984). 29 CFR 1910.134, *Respiratory Protection Standard.* Washington, DC: U.S. Department of Labor.

CHAPTER 3

Toxicology

Art by David Heskett.

KEY TERMS

absorption, *p. 76*
acute effects, *p. 77*
asphyxiant, *p. 61*
bacteria, *p. 66*
carcinogen, *p. 61*
ceiling, *p. 80*
chemical agent, *p. 63*
chronic effects, *p. 77*
convulsant, *p. 61*
dose, *p. 77*
dose-response curve, *p. 77*
dose-response relationship, *p. 77*
effective concentration low (EC_{LO}), *p. 78*
effective dose low (ED_{LO}), *p. 78*
excretion, *p. 75*
external contamination, *p. 83*
half-life, *p. 72*
incapacitating concentration 50% (IC_{50}), *p. 78*
incapacitating dose 50% (ID_{50}), *p. 78*
incubation period, *p. 66*
infectious dose, *p. 66*
ingestion, *p. 76*
inhalation, *p. 75*
injection, *p. 76*
internal contamination, *p. 83*
ionizing radiation, *p. 70*
irritant, *p. 61*
latent effect, *p. 61*
lethal concentration 50% (LC_{50}), *p. 78*
lethal dose 50% (LD_{50}), *p. 78*
metabolism, *p. 75*
potentiated effect, *p. 77*
quality factor (Q), *p. 74*
radionuclide, *p. 69*
recommended exposure limit (REL), *p. 80*
secondary contamination, *p. 83*
sensitizer, *p. 61*
short-term exposure limit (STEL), *p. 80*
target organ effect, *p. 62*
teratogen, *p. 61*
threshold effect, *p. 77*
threshold limit value (TLV), *p. 80*
time-weighted average (TWA), *p. 79*
toxic concentration low (TC_{LO}), *p. 78*
toxic dose low (TD_{LO}), *p. 78*
toxic industrial chemical (TIC), *p. 62*
toxicity, *p. 60*
toxicology, *p. 60*

OBJECTIVES

After reading this chapter, the student should be able to:

- Name 10 organs of the body that may be affected by hazardous substances and how they are affected.
- Name four categories of chemical agents.

- Name five chemical agents and their signs and symptoms.
- Name four categories of biological agents.
- Name five biological agents and their signs and symptoms.
- Name four types of ionizing radiation.
- Name three ways to protect yourself from radiation.
- List the four routes of entry.
- Explain the difference between acute and chronic effects.
- Explain the significance of the dose-response relationship.
- Explain the interrelationship between the permissible exposure limit (PEL), the short-term exposure limit (STEL), and the ceiling (C).
- Explain the difference between workplace exposure limits and public safety exposure limits.
- Explain the difference between internal contamination and external contamination.

Your agency's hazardous materials response team is dispatched to the chemistry building of the local community college for a puncture wound to the hand with uncontrolled bleeding. When you arrive you are escorted to the organic chemistry lab, where a student has a jagged piece of glass protruding from his hand. You notice that the glass is hollow, and there are various jugs, beakers, and flasks full of chemicals nearby.

- How do you handle this patient?
- What are your concerns?
- Could the patient be contaminated?
- Could any chemicals have been injected into his body through the puncture wound?

Let's see if we can get some answers to these questions and more.

What makes a hazardous material a health hazard? In other words, how will chemical releases affect people? In this chapter we will explore how chemicals interact with the human body and exert deleterious effects.

toxicology ▪ The study of poisons and their effect on living organisms.

Toxicology is broadly defined as the study of poisons. Modern toxicology is the science that studies how chemicals interact with biological systems. For hazardous materials technicians, toxicology can be defined as the study of the effect of chemicals—specifically, industrial and household chemicals to which victims have been or may become exposed—on the human body.

toxicity ▪ The degree of damage a chemical does to the body.

Toxicity is the degree of harm a chemical does to the body. All substances are toxic at some dose. Even ingesting water in large enough amounts can be lethal. A famous quote by Paracelsus (1493–1541), considered the father of toxicology, captures this idea quite nicely:

> All substances are poisons; there is none which is not a poison. The right dose differentiates a poison and a remedy.

This is an important point to keep in mind when conducting a risk-based response and considering hazard versus risk at hazmat incidents. Many aspects of hazardous materials response are affected by an understanding of toxicology. For example, a responder is subjected to an additional hazard (heat stress) when forced to wear a high level of personal protective equipment (PPE) for a low-risk chemical release. There are also risks (such as travel hazards and economic loss) associated with unnecessary evacuations and road closures that may arise from overreacting to hazardous materials incidents.

Health Effects of Hazardous Materials

The health effects of hazardous materials vary greatly. The health effects of any given hazardous material depend on a number of factors including the chemical and physical properties of the substance, the route of exposure, the health and physiology of the exposed individual, storage conditions, ambient temperature, and weather conditions. For example, sulfuric acid evaporates extremely slowly. Therefore, a liquid spill in a room full of people will have minimal health effects as long as no one touches the spill. However, if the liquid is aerosolized, or skin contact is made with the liquid, the health effects can be quite severe.

MECHANISMS OF HARM

Hazardous materials may harm individuals by several different mechanisms. **Asphyxiants** interfere with breathing, oxygen transport in the body, or oxygen uptake in the cells. Simple asphyxiants displace oxygen, while chemical asphyxiants interfere with oxygen transport or uptake in the body. Examples of simple asphyxiants include nitrogen, argon, carbon dioxide, and helium; examples of chemical asphyxiants include carbon monoxide, hydrogen sulfide, and hydrogen cyanide. Carbon monoxide is a blood asphyxiant; it prevents oxygen from binding to hemoglobin in the red blood cells. Hydrogen cyanide is a tissue asphyxiant; it prevents the cells in the body from using oxygen to create energy. Hydrogen sulfide is a respiratory paralyzer; it paralyzes the phrenic nerve as well as attacks the olfactory nerve (thereby causing olfactory fatigue).

Some materials are **convulsants**; they cause the body to have convulsions and seizures. Some materials are **irritants**; they can cause reversible skin, eye, or respiratory inflammation. Skin or eye irritation can be redness, swelling, or a rash at the site of contact. Respiratory irritation can lead to coughing, sneezing, swelling of the respiratory passages, and difficulty breathing. Irritants are often corrosive materials. Examples of irritants include ammonia, chlorine, and hydrogen chloride.

Some materials are allergens that produce allergic reactions. An allergic reaction is a heightened immune system response to a chemical that leads to tissue swelling and skin rash. Severe allergic reactions can lead to anaphylactic shock, which can rapidly lead to respiratory arrest and death. Some materials are **sensitizers**. This means that after repeated exposure, or even a single exposure, they can cause severe allergic reactions up to and including anaphylactic shock. Isocyanates, which are used in the plastics industry, can sensitize workers, making them subject to severe asthma attacks if they are exposed again. Another example is formaldehyde.

Yet other materials are **carcinogens**, which are capable of causing cancer. Examples of carcinogens include benzene, asbestos, coal tar, and ethidium bromide. Carcinogens often exhibit a long latency period (latent effect). **Latent effects** occur a significant time after exposure—months, years, or even decades after exposure. **Teratogens** are agents that when administered to the mother cause birth defects in the unborn child. Examples of teratogens include thalidomide and ionizing radiation.

asphyxiant ■ A gas or vapor that dilutes or displaces air or interferes with breathing, oxygen transport in the body, or oxygen uptake in the cells.

convulsant ■ A material that causes the body to have convulsions and seizures.

irritant ■ A material that causes reversible skin, eye, or respiratory inflammation.

sensitizer ■ A substance that after the initial exposure causes a more severe allergic reaction on subsequent exposures. Examples of sensitizers are isocyanates.

carcinogen ■ A substance that causes cancer (tumor formation).

latent effect ■ Disease or impairment that is manifested after an extended time period following exposure to the hazardous substance.

teratogen ■ A substance that causes birth defects.

TABLE 3-1 Target Organ Effects

TARGET ORGAN EFFECT	ORGAN AFFECTED	SIGNS AND SYMPTOMS	SELECTED EXAMPLES OF CAUSATIVE AGENTS
Cutaneous hazards	Skin	Defatting of the skin, rashes, irritation	Ketones, chlorinated compounds
Eye hazards	Eyes	Conjunctivitis, corneal damage	Organic solvents, acids
Central nervous system (CNS)	Brain and spinal cord	Drooping of upper eyelids, slurred speech, respiratory difficulty, seizures, unconsciousness	Heavy metals (such as lead, mercury, and thallium)
Peripheral nervous system (PNS)	Nerves	Numbness, tingling, decreased sensation, change in reflexes, decreased motor strength	Arsenic, lead, toluene, styrene
Neurotoxins	Nervous system	Varied (see CNS and PNS)	Acrylamide, nerve agents (such as sarin and VX), many venoms
Pulmonary toxins	Lungs	Cough, chest tightness, shortness of breath	Silica, asbestos, hydrochloric acid
Hematopoietic toxins	Red blood cells	Cyanosis, loss of consciousness	Carbon monoxide, benzene
Nephrotoxins	Kidneys	Edema, proteinuria	Halogenated hydrocarbons, uranium
Hepatotoxins	Liver	Jaundice, liver enlargement	Carbon tetrachloride, nitrosamines
Reproductive toxins	Reproductive organs	Birth defects, sterility	Lead, DBCP

TARGET ORGAN EFFECTS

target organ effect ■ Tissue damage caused by hazardous materials that affect specific body systems.

Some materials have target organ effects. A **target organ effect** is tissue damage caused by hazardous materials that affect specific organs. Table 3-1 lists some of these target organ effects, the signs and symptoms of exposure, and examples of chemicals that cause them.

TOXIC INDUSTRIAL CHEMICALS

toxic industrial chemical (TIC) ■ Poisonous substances used widely in industry that pose a great risk if accidentally or intentionally released.

Toxic industrial chemicals (TIC), also called *toxic industrial materials* (TIM), are extremely dangerous chemicals used in industry that may also be used as weapons by terrorists or criminals. Examples of toxic industrial chemicals include chlorine, anhydrous ammonia, arsine, phosgene, phosphine, hydrogen cyanide, hydrogen chloride, and hydrogen fluoride. Currently, toxic industrial chemicals are a major concern in counterterrorism strategy. Industrial facilities that use these chemicals in quantity are required to have security plans in place. Some of these materials pose such a severe risk that the U.S. DOT has classified them as *toxic inhalation hazards*, often abbreviated TIH. A TIH is defined as a gas or volatile liquid that is toxic when inhaled, typically with an LC_{50} value below 5000 ppm. We will discuss what LC_{50} means and the units of measurement shortly. The DOT Inhalation Hazard Zones are defined as follows:

HAZARD ZONE A: *Gases:* LC_{50} of less than or equal to 200 ppm
Liquids: Saturated vapor concentration in air (at standard temperature and pressure [STP]) equal to or greater than 500 LC_{50}; and LC_{50} less than or equal to 200 ppm

HAZARD ZONE B: *Gases:* LC_{50} greater than 200 ppm and less than or equal to 1000 ppm

Liquids: Saturated vapor concentration in air (at STP) equal to or greater than 10 LC_{50}; and LC_{50} less than or equal to 1000 ppm and criteria for Hazard Zone A are not met

HAZARD ZONE C: LC_{50} greater than 1000 ppm and less than or equal to 3000 ppm

HAZARD ZONE D: LC_{50} greater than 3000 ppm and less than or equal to 5000 ppm

CHEMICAL AGENTS

Chemical warfare agents, often abbreviated CWA, are chemicals that were used, and in some cases are still being used, in military warfare. Shortly after World War II the Geneva Conventions banned the use of chemical warfare agents by signatory nations. However, some rogue nations still synthesize, stockpile, and even use CWAs in military applications. In addition, CWAs can be synthesized in industrial facilities and smaller illicit laboratories by terrorists. **Chemical agents** are classified into four broad categories based on their biological effect: choking agents, blister agents or vesicants, blood agents, and nerve agents. The physiological effect of chemical agents is generally immediate, appearing within seconds to minutes, but in some cases the signs and symptoms may be delayed for hours. Table 3-2 summarizes the toxic properties of the most common chemical agents, including the irritants. Irritants such as Mace and pepper spray are still used by the military and by law enforcement agencies for crowd control and are even commercially available in many cases.

chemical agent ■ An extremely toxic chemical substance that has the capability to cause mass casualties when disseminated. Examples include sarin, hydrogen cyanide, and vesicants such as sulfur mustard.

Choking Agents

Choking agents are respiratory irritants that produce hydrochloric acid on contact with the moisture in the respiratory tract. Choking agents are gases and are thus easily inhaled. Hydrochloric acid burns the respiratory tract tissue and causes fluid secretion into the lungs and leads to what is commonly called "dry land drowning." The medical term for this is pulmonary edema. The two most common choking agents are phosgene and chlorine, both of which were used during World War I. Both agents are commonly used in industry and are also considered toxic industrial chemicals.

Blister Agents

Blister agents, also called vesicants, are chemicals that produce blisters and skin lesions at the site of contact. There are three subcategories of blister agents: the mustard family, lewisite, and phosgene oxime. The mustard family contains sulfur mustards, such as distilled mustard (HD), and nitrogen mustards, such as HN. The mustard family causes blister formation because its members are alkylating agents. These are reactive chemicals that can interfere with biologically important chemicals in the body, including the genetic material, or DNA. Its alkylation can result in mutations that cause birth defects and cancer. Blister agents were also first used in World War I. Most blister agents are liquids, with the exception of phosgene oxime, which is a solid. Although HD is often called "mustard gas," this is a misnomer. It is a liquid at room temperature. The primary route of exposure is skin absorption, although in enclosed areas inhalation may also be a significant route of entry. Symptoms of blister agent exposure are usually delayed by several hours. Lewisite—an arsenical—is the exception to this rule, and symptoms occur within minutes of exposure. The antidote to lewisite is the chelating agent British anti-lewisite (BAL).

Blood Agents

Blood agents interfere with the body's ability to take in or use oxygen. They also were first used in World War I. The two most common blood agents are hydrogen cyanide and cyanogen chloride. These chemicals are very volatile liquids and are therefore both inhalation and contact hazards. Hydrogen cyanide exerts its most potent effect in cells by binding to and irreversibly inactivating the key protein cytochrome oxidase in the biochemical pathway known as oxidative phosphorylation. This prevents the cells in the body from using oxygen and making energy. Two antidotes to hydrogen cyanide are manufactured. The older of the two antidotes consists of a two-part cocktail, a nitrite and a thiosulfate,

TABLE 3-2 Examples of Chemical Agents

COMMON NAME	MILITARY ABBREVIATION	DOT HAZARD CLASS	SIGNS AND SYMPTOMS
Nerve Agents			SLUDGEM (salivation, lacrimation [tearing], urination, defecation, gastrointestinal upset, emesis [vomiting], miosis [pinpoint pupils])
Tabun	GA	6.1	
Sarin	GB	6.1	
Soman	GD	6.1	
VX	VX	6.1	
Blister Agents			Skin lesions, skin redness, skin irritation, blistering
Mustard	H	6.1	
Distilled mustard	HD	6.1	
Nitrogen mustard	HN	6.1	
Lewisite	L	6.1	
Blood Agents			Difficulty breathing, reduced level of consciousness
Hydrogen cyanide	AC	6.1	
Cyanogen chloride	CK	2.3	
Cyanogen bromide		6.1	
Choking Agents			Difficulty breathing, throat irritation, respiratory distress
Chlorine	CL	2.3	
Phosgene	CG	2.3	
Irritants			Difficulty breathing, throat irritation, coughing, lacrimation
Tear gas	CS	6.1	
Dibenzoxazepine	CR	6.1	
Chloroacetophenone	CN	6.1	
Pepper spray, Oleoresin capsicum, capsaicin	OC	2.2 (subsidiary hazard 6.1)	
Phenylchloromethylketone, chloropicrin	PS	6.1	

Source: Adapted from NFPA 472 (2008) Annex A.

that binds the cyanide, and then converts it to the relatively nontoxic and excretable by-product thiocyanate ion. The second antidote is hydroxocobalamin (vitamin B_{12a}), which both binds and eliminates the cyanide ion. This antidote is considered to be much safer than the older one and has fewer side effects.

Nerve Agents

Nerve agents, members of the organophosphate family, were developed in Germany in the 1930's and weaponized during World War II but were never used at that time. Nerve agents interfere with the body's ability to transfer an electrochemical signal through the neurological system to the muscles, the pathway called *nerve cell signaling*. Acetylcholinesterase is an enzyme that normally breaks down the neurotransmitter acetylcholine and terminates the nerve signaling process.

$$\text{Acetylcholine} \rightarrow \text{Acetate} + \text{Choline}$$

Nerve agents bind to and inactivate acetylcholinesterase. When acetylcholinesterase is inactivated at the neuromuscular junction, the nerve signal continues to be transmitted, and the muscle continues to be stimulated. This is analogous to having a radio keyed up, which renders the whole radio frequency useless. An antidote to nerve agents is a two-part sequence consisting of atropine and pralidoxime chloride (such as commercially available DuoDote). Atropine reduces secretions in the airway and respiratory passages, while pralidoxime reactivates the acetylcholinesterase. However, some nerve agents undergo a process called *aging*, whereby they become irreversibly bound to the acetylcholinesterase, and pralidoxime no longer works. Soman is the most common example of a nerve agent that ages rapidly (within minutes).

There are many nerve agents, but the most common ones are sarin (GB), soman (GD), tabun (GA), and VX. All the nerve agents are liquids at room temperature and atmospheric pressure, but some are more volatile than others. Therefore "nerve gas" is another misnomer. However, nerve agents can be both an inhalation hazard and a contact hazard owing to their varying degrees of volatility. The G series of nerve agents are more volatile than the V series. V series nerve agents such as VX are primarily a contact hazard, in other words, a skin absorption hazard. Nerve agents were used by Saddam Hussein against the Kurds in Iraq in the 1980s with devastating effect.

BIOLOGICAL AGENTS

Biological warfare agents (BWAs) are living organisms, or material derived from living organisms, that can be used in warfare or terrorism. BWAs have been used in warfare for millennia. In the Middle Ages enemies besieging cities would hurl the bodies of plague victims over the city walls to cause disease outbreaks. During the French and Indian War the British gave blankets contaminated with smallpox virus to the Indians. In the modern age of biological warfare genetic engineering makes it possible to create greatly enhanced biological agents such as bacteria that are resistant to multiple antibiotics as well as agents that have had toxin genes transferred from one organism to another. Indeed, in 2011, researchers investigating influenza transmission created a highly contagious H5N1 bird flu through genetic engineering.

There are three types of biological agents: bacteria, viruses, and toxins. Table 3-3 lists the toxicological properties of selected biological agents. Biological agents differ from chemical agents in that the signs and symptoms of biological agents *typically* appear days to weeks after exposure. The exceptions to this rule are toxins, for which signs and symptoms may appear as early as minutes after exposure but are commonly delayed for several hours. For this reason a terrorist attack using biological agents will likely have occurred days prior to detection of the incident. In 2001, anthrax spores were mailed to Florida, New York City, and Washington, DC, and infected more than a dozen people and killed two before the anthrax attack was even discovered (see Chapter 1 for a detailed description of this incident). Anthrax can be effectively treated if an antibiotic regimen is started early enough. Owing to the relatively generic signs and symptoms of the initial stages of anthrax infection, this treatment unfortunately was not administered quickly enough in this case.

Biological agents typically initiate an immune response in the body. The immune response causes most of the generic signs and symptoms: fever, headache, body aches, and the like, that we typically experience when we get sick.

SOLVED EXERCISE 3-1

You are dispatched to the local convention center because multiple people are complaining of vomiting and stomach upset. On arrival you notice at least a dozen patients, many of whom have pinpoint pupils. With what chemical agent exposure are these signs and symptoms consistent?

Solution: These signs and symptoms are consistent with nerve agent exposure. Pinpoint pupils, or *miosis*, and copious secretions are a very characteristic and unique early-stage sign of nerve agent exposure and can be used for differential diagnosis.

TABLE 3-3 Examples and Properties of Biological Agents

NAME	DOT HAZARD CLASS	SIGNIFICANT SIGNS AND SYMPTOMS
Anthrax	6.2	Possible skin lesions (cutaneous), possible gastrointestinal upset (ingestion), severe flulike symptoms (inhalation)
Mycotoxin	6.1 or 6.2	Vary in detail depending on type of mycotoxin: Skin pain and lesions, necrosis (cutaneous), abdominal pain, diarrhea, prostration (ingestion)
Plague	6.2	Skin lesions (necrotic)
Viral hemorrhagic fevers	6.2	Bleeding
Smallpox	6.2	Skin lesions (pustules)
Botulinum toxin	6.1 or 6.2	Flaccid paralysis
Ricin toxin	6.1 or 6.2	Loss of consciousness, coma

Source: Adapted from NFPA 472 (2008) Annex A.

Biological agents are generally transmitted through inhalation or ingestion. Skin absorption is typically not a very effective route of entry for biological agents. The exceptions to this rule are some of the smaller toxins, such as the T-2 mycotoxins, which are skin absorptive. For biological agents to cause disease a minimum quantity of infectious particles, called the **infectious dose**, must be inhaled or ingested. Following infection there is an extended period during which a victim exhibits no signs and symptoms while the biological organism is reproducing inside the body. This is the **incubation period**, and it varies depending on the life cycle of the biological agent, the physical condition of the victim, and the dose of biological agent received.

infectious dose ▪ The minimum quantity of infectious particles that must be inhaled or ingested to cause disease.

incubation period ▪ The period after infection during which a victim exhibits no signs and symptoms while a biological organism is reproducing inside the body.

Biological agents are comparatively large and are therefore nonvolatile in their natural state. Therefore, a terrorist attack using biological weapons will likely involve a dissemination device. Many biological agents can be processed into either liquid or solid aerosols that can readily be inhaled. Because biological agents are living organisms, they tend to be more fragile than chemical agents and will likely need to be disseminated at night or indoors to avoid UV radiation (a component of sunlight), which destroys biological agents.

Bacteria

bacteria ▪ Single-celled living organisms.

Bacteria are single-celled living organisms. Most naturally occurring bacteria are harmless, and some are even beneficial. Anyone who has taken too many antibiotics and has contracted a gastrointestinal yeast infection knows how important *Escherichia coli* are to

BOX 3-1 HAZMAT HISTORY: ANTHRAX ATTACK IN JAPAN

In June 1993 the Aum Shinrikyo cult released aerosolized liquid *Bacillus anthracis* cultures over Kameido, Japan, from the roof of its headquarters. The attack went largely unnoticed because the cult used the Sterne strain of anthrax, which is used for veterinary purposes as a vaccine and does not contain the business end of anthrax—the anthrax lethal factors. Residents complained of foul odors emanating from the building at that time, and public health officials investigated and collected a liquid sample from the outside of the building. Several years later scientists confirmed that anthrax had indeed been released in that area after cult members admitted to the crime after the 1995 Tokyo sarin subway attack (see Chapter 1 for more information about this attack). This example illustrates how difficult it can be to detect biological agent attacks.

the digestive system. However, many bacteria are human pathogens, which means they cause human disease. Examples of disease-causing bacteria include *Bacillus anthracis*, which causes anthrax, and *Yersinia pestis,* which causes the plague.

B. anthracis is a spore-forming bacterium widely distributed in the environment anywhere ungulates (hoofed animals) are found. When bacteria produce spores, they change from a growing, vegetative state to a dormant, environmentally hardy state. Spores allow bacteria to lie dormant in the soil for years, if not decades. There are three forms of anthrax: inhalational anthrax, gastrointestinal anthrax, and cutaneous anthrax. Inhalational anthrax is by far the most dangerous form of the disease and occurs when *Bacillus* spores are inhaled. Signs and symptoms of inhalational anthrax include flulike symptoms with onset 1 to 7 days after exposure. Anthrax treatment consists of antibiotics and supportive care. Anthrax is not highly contagious; that is, it is not easily transmittable from person to person.

The plague, also known as the black death, is caused by the bacterium *Yersinia pestis.* This bacterium has a complicated life cycle and spends time in rodents such as rats; in fleas, which transmit the bacterium between different hosts; and, of course, in humans. The plague bacterium is endemic to much of the world, including the United States. Many prairie dogs in the Southwest are infected with *Y. pestis.* There are two forms of the plague: bubonic plague and pneumonic plague. Bubonic plague and pneumonic plague differ in their route of entry. Pneumonic plague is the more severe form of the disease and occurs when an infectious dose is inhaled. Bubonic plague is a cutaneous form of the disease. The incubation period for pneumonic plague is 2 to 3 days; signs and symptoms include flulike symptoms, shortness of breath, and coughing up blood. Both forms of the disease can be treated successfully with antibiotics. Pneumonic plague can be highly contagious.

Some bacteria require a host to survive, as do viruses. Q fever is caused by such an obligate organism, *Coxiella burnetti,* which was developed for use as a biological warfare agent in the past. It is a naturally occurring pathogen of sheep, goats, and cows, and is sometimes transmitted to humans when dust is inhaled that contains bacterial particles or when contaminated milk is ingested. Q fever produces flulike symptoms with an incubation period of 2 to 10 days. It has a very low mortality rate and was used in military programs as a biological incapacitating agent designed to tie up resources in caring for sick soldiers. Antibiotics can be used to shorten the duration of illness.

Viruses

Viruses are organisms that require a living host to reproduce. They do not possess all the biochemical machinery needed to survive and so must hijack the infrastructure of the host to reproduce. Examples of viruses include smallpox; Ebola; the rhinovirus family, which causes the common cold; and the influenza virus family.

Smallpox, which lives only in humans, was eradicated through a ring vaccination strategy in the late 1970s. Theoretically, it is now stored in only two places in the world, Russia and the United States. However, it is likely that there are other smallpox stores in existence. The incubation period is 10 to 12 days, by which time the victim is already contagious. Signs and symptoms of infection include flulike symptoms with a synchronous, centrifugal rash. A *synchronous* rash is one in which all the lesions are in the same stage of development. A *centrifugal* rash first appears at the extremities and head and then moves to the center of the body. The rash consists of pustules similar to those of chickenpox, but chickenpox causes an asynchronous, *centripetal* rash (which first appears on the thorax, or trunk, of the body). Smallpox is of grave concern because it is very lethal, with a mortality rate of up to 30%, and highly contagious, spreading rapidly from person to person before signs and symptoms appear. There is no known cure for smallpox, but vaccines do exist. Fortunately, it is fairly unlikely that terrorists will be able to obtain the smallpox virus.

Ebola and Marburg viruses, two of the causative agents of hemorrhagic fevers, are relatively new viruses that were discovered in the last several decades. These viruses cause flulike symptoms and very characteristic subcutaneous bleeding, including in the eyes. The incubation period is 3 to 21 days, and the mortality rate can be as high as 90% with Ebola, depending on the viral strain. There is no cure or vaccine for viral hemorrhagic fevers; treatment includes supportive care and antiviral treatments such as Ribavirin (a type of antiviral medication).

Toxins

Toxins, in the context of biological agents, are proteins or chemicals produced by living organisms—including plants, animals, and bacteria. Keep in mind the two definitions of toxin: toxin as we define here in the context of biological agents, and toxin as a more generic synonym for poison. Examples of toxins include snake, scorpion, and spider venoms; ricin from the castor bean plant; and staphylococcal enterotoxin B (SEB) from the bacterium *Staphylococcus aureus*. Because toxins do not reproduce in the body, they produce signs and symptoms much more quickly than bacteria or viruses.

There are two major categories of toxins: neurotoxins and cytotoxins. Neurotoxins attack the nervous system and typically, in contrast with chemical nerve agents, *prevent* nerve cell signaling. Symptoms include respiratory difficulty, mental confusion, vision problems, and loss of muscle control, as in flaccid paralysis. Cytotoxins destroy tissue and generally cause tissue necrosis. Tissue necrosis leads to a festering wound at the site of toxin introduction, as through an envenomated snake bite. Cytotoxins introduced orally or through inhalation may lead to systemic necrosis.

Botulinum toxin is a neurotoxin produced by the bacterium *Clostridium botulinum*. It is a naturally occurring bacterium in the soil and often causes illness when improperly canned items are consumed. On average, one or two botulism outbreaks occur each year owing to improper food handling and packaging. Botulinum toxin can also be intentionally produced, purified, and then used as a biological weapon. The toxin produces symptoms 1 to 3 days after ingestion or inhalation. Signs and symptoms of botulinum poisoning include weakness, dizziness, dry mouth, blurred vision, and flaccid paralysis. Flaccid paralysis is one of the hallmarks of botulinum poisoning. In fact, the common antiaging treatment Botox uses botulinum toxin to paralyze the muscles that cause wrinkles. A botulinum antitoxin is available.

SEB, a cytotoxin produced by *Staphylococcus* bacteria, is also a naturally occurring toxin that causes periodic outbreaks of food poisoning when food is not properly handled, refrigerated, and stored. SEB produces symptoms in about 4 to 6 hours. Signs and symptoms of SEB ingestion include vomiting, intense abdominal cramps, and diarrhea. Exposure to the causative bacteria via inhalation produces very different signs and symptoms, including fever, chills, cough, and prostration, which can last 1 to 2 weeks.

Other toxins act through different mechanisms, such as the plant toxin ricin, which inhibits protein synthesis. The normal route of entry for ricin is ingestion, but ricin can also be made into an aerosol powder. Signs and symptoms of ricin poisoning may appear as quickly as within 4 to 8 hours and include difficulty breathing, nausea, vomiting, bloody diarrhea, and abdominal cramps. The castor bean plant is found throughout the United States, and it is relatively easy to extract the toxin from the plant. Therefore, it is not surprising that there have been many ricin incidents in the United States over the last

SOLVED EXERCISE 3-2

You are dispatched to the county courthouse cafeteria because several people are complaining of vomiting and stomach upset. On arrival you notice they are exhibiting signs of flaccid paralysis. With what biological agent exposure are these signs and symptoms consistent?

Solution: These signs and symptoms are consistent with botulinum toxin poisoning. Flaccid paralysis is a very characteristic sign of botulinum toxin exposure.

several decades. In one case, a man almost died in Las Vegas, Nevada, after he became exposed to a sample of ricin in his possession. There is no ricin antidote available, and treatment is primarily supportive.

See Hazmat History on Gambling with Ricin.

EXPLOSIVES

Explosives are commonly used in industry and are the most common WMD agents used in the United States and throughout the world, largely owing to their ease of use and wide availability. An ammonium nitrate/fuel oil mixture (ANFO) was used to bring down the Alfred P. Murrah Federal Building in Oklahoma City in 1995. Explosives lead to a variety of injuries including lung damage from the initial overpressure wave; thermal burn injuries from the heat of the rapid exothermic chemical reaction of the explosion; shrapnel and trauma injuries from the explosive itself and from added shrapnel (such as nuts, nails, screws, and ball bearings); as well as trauma from the surrounding room contents and destroyed building components.

RADIOLOGICAL MATERIALS

Radioactive materials, also known as **radionuclides** or radioisotopes, are all around us—in equipment we use as well as naturally occurring in our bodies and in the environment. Some commonly used radionuclides are americium-241 (in smoke detectors); cesium-137 (in radiography instruments and nuclear density gauges); tritium (hydrogen-3) (in gun scopes, emergency exits, and airport runway lights); and cobalt-60 (to sterilize equipment and food). X-rays are used in medicine, and various other radioisotopes are used in cancer treatments (Table 3-4). Radioactive materials are also the by-products of industry. You're

radionuclide ■ An unstable atom capable of emitting radiation from the nucleus. Also known as a radioisotope or radioactive isotope.

TABLE 3-4 Table of Common Radioactive Isotopes

ISOTOPE	EMISSION	HALF-LIFE	COMMERCIAL USES
Americium-241	Alpha, gamma	458 years	Smoke detectors, gauges
Californium-252	Alpha, neutron, gamma	2.6 years	Cancer therapy, nuclear reactors, density gauges
Carbon-11	Beta, gamma	20.4 minutes	PET imaging
Cesium-137	Beta, gamma	30.1 years	Gauges, tracer
Cobalt-60	Beta, gamma	5.3 years	Cancer therapy, sterilization, radiography cameras, density gauges
Iodine-125	Gamma	60 days	Cancer therapy, medical research
Iodine-131	Beta, gamma	8 days	Cancer therapy
Iridium-192	Beta, gamma	74 days	Cancer therapy, density gauges, radiography cameras
Krypton-85	Beta, gamma	10.8 years	Density gauges
Nickel-63	Beta	100 years	Power source, APD2000 detector
Phosphorus-32	Beta	14 days	Medical research, medical treatment
Strontium-90	Beta	28.8 years	Radioisotope Thermal Generators (RTG)
Sulfur-35	Beta	87.3 days	Medical research
Technetium-99m	Gamma	6 hours	Medical imaging, tracer
Thallium-201	Gamma	73 hours	Medical imaging (stress test)
Tritium (hydrogen-3)	Beta	12.5 years	Medical research, tracer
Uranium-238	Alpha	4.5 billion years	Naturally occurring uranium

probably familiar with the nuclear waste generated by nuclear power plants, but the oil industry also often generates radioactive sludge from drilling operations when naturally occurring radioactive isotopes are brought to the surface.

Radiological materials may be combined with explosives to form radiological dispersal devices (RDDs). Radiological dispersal devices have the same primary hazard as explosive devices, with the added psychological and health hazard of disseminating radioactive materials into the environment.

Radioactivity

Radiation is all around us, and most of it is harmless and even useful. For example, visible light is a form of nonionizing radiation and is obviously very useful, and nonionizing microwave radiation is very useful for cooking food. Figure 3-1 shows the electromagnetic spectrum. When we speak of radiation in the context of hazardous materials, we mean **ionizing radiation.**

ionizing radiation ■ High-energy emissions from radionuclides capable of causing harm to the human body. Examples of ionizing radiation include alpha radiation, beta radiation, gamma radiation, x-rays, and neutron radiation.

IONIZING RADIATION

Ionization is the loss or gain of one or more electrons by an atom or molecule. There are several different types of ionizing radiation: alpha, beta, gamma, X-ray, and neutron radiation. Gamma radiation and X-rays are purely electromagnetic ionizing radiation, whereas alpha, beta, and neutron radiation are particulate. Radiation cannot be seen, smelled, or heard. Therefore, detection equipment is extremely important. Table 3-5 lists the attributes of the different types of radiation.

Radiation is emitted from the nucleus of radioactive materials, or radionuclides, in the form of waves or particles at a very high rate of speed. This high rate of speed means

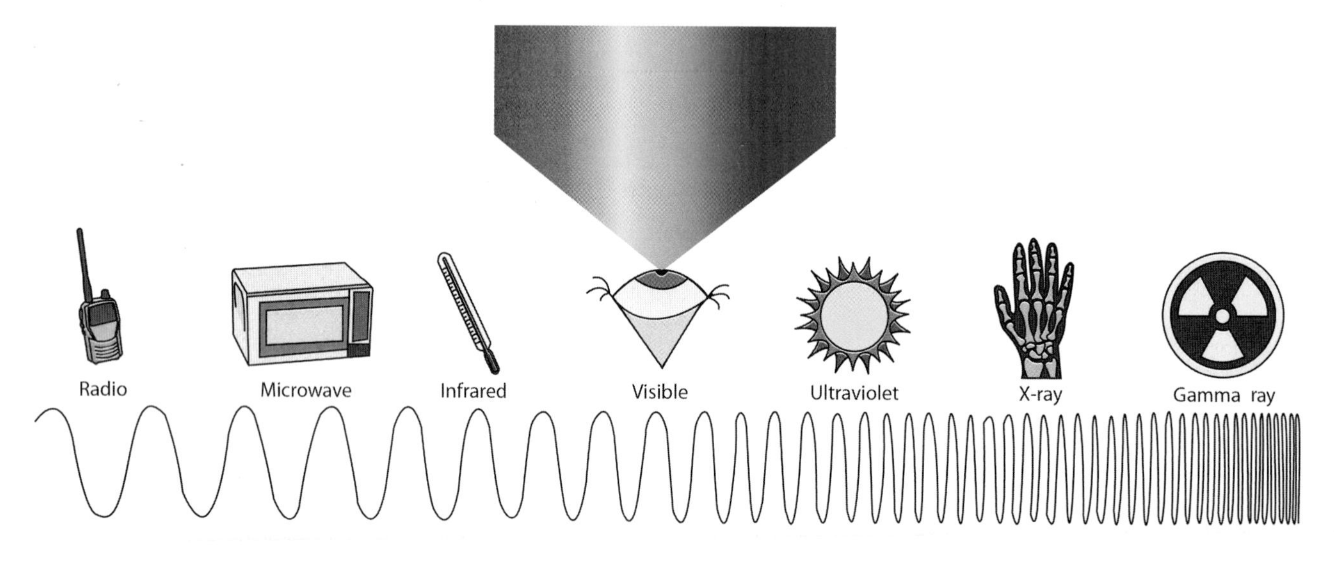

FIGURE 3-1 The electromagnetic spectrum. Ionizing radiation is only a small portion of the electromagnetic spectrum. *Art by David Heskett.*

TABLE 3-5 Types of Radiation

RADIATION	MASS	CHARGE	SHIELDING
Alpha	4 amu*	+2	Paper
Beta	1/1836 amu	−1	Plexiglas or aluminum
Gamma/X-ray	None	0	Lead
Neutron	1 amu	0	Concrete

*atomic mass unit

that radiation has a lot of energy. For example, purely electromagnetic radiation travels at the speed of light, while alpha particles travel at about 1/20th the speed of light. This is still extremely fast and energetic!

Ionizing radiation contains significant amounts of energy and is very dangerous to humans, because the energy is transferred to tissues and causes extensive damage to DNA and proteins. This is the mechanism by which radioactive materials that emit ionizing radiation at high levels damage our bodies. Radiation exposure produces a characteristic set of signs and symptoms that has become known as *radiation sickness*. Initial signs and symptoms of radiation sickness include nausea, vomiting, and possibly a sunburnlike rash. Signs and symptoms of higher-level radiation exposure include gastrointestinal upset and diarrhea. Extremely high radiation exposures will lead to coma and death. Table 3-6 lists some common signs and symptoms of radiation exposure based on the dose received. Radiation sickness occurs hours to days after exposure.

Alpha Particles

Alpha particles are the largest form of ionizing radiation. Each alpha particle is composed of two protons and two neutrons (4 atomic mass units, or amu) and has a net charge of +2, so it is identical with the nucleus of the helium atom. Because of their heavy size and charge, alpha particles do not travel very far in air—only a few inches—since they tend to interact very rapidly with the molecules in air. Therefore, you can protect yourself from alpha radiation sources quite easily. For example, you just need to stay back a few inches or use a piece of paper for shielding (Figure 3-2). In fact, alpha particles will not even penetrate the layer of dead skin, but they may damage the living cells on the corneal surface of your eyes. However, if radioactive materials that emit alpha radiation are ingested, inhaled, or otherwise enter the body, they are extremely dangerous. Once the alpha emitter (the radioactive material

TABLE 3-6 Signs and Symptoms of Radiation Exposure

RADIATION DOSE	DELAY IN TIME OF ONSET	SIGNS AND SYMPTOMS
<100 rem*	1–3 hours	Mild weakness
100–200 rem	3–6 hours	Reduced white blood cell count, vomiting
200–600 rem	2–3 hours	Vomiting, white blood cells completely destroyed
600–1000 rem	1 hour	Vomiting, white blood cells completely destroyed
>1000 rem	30 minutes Hours to days	Diarrhea and fever Death

*roentgen equivalent man

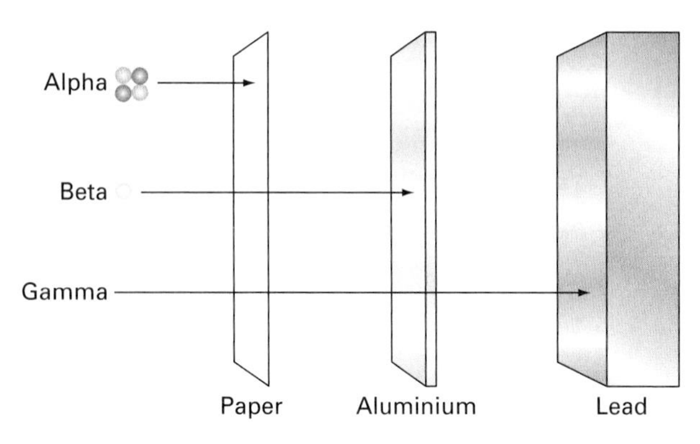

FIGURE 3-2 The relative penetrating power of alpha, beta, and gamma radiation. Gamma and neutron radiation are very penetrating and extremely difficult to block.

or radionuclide) is inside the body, all the alpha particles transfer their energy to the body's tissues and cause extensive damage.

Beta Particles

Beta particles are the smallest form of particulate ionizing radiation. They have the same mass and charge as an electron but are emitted from the nucleus at nearly the speed of light. Beta particles are produced when a proton spontaneously decays into a neutron and an electron. We will discuss the atom in greater detail in the next chapter. Beta particles have a mass of approximately 1/1800th amu and a net charge of -1. Although beta particles are relatively light, they are strongly charged. Owing in large part to their charge, beta particles travel only 10 to 20 feet in air, but they can be an external hazard to the skin and eyes, and may cause surface burns similar to a sunburn, called a *beta burn*. Aluminum or thicker plastic sheeting is required to effectively shield against radioactive materials that emit beta radiation (beta emitters) (Figure 3-2). Such materials are also an internal hazard if they are ingested or inhaled.

Gamma Radiation

Gamma rays and X-rays are purely electromagnetic radiation, so they have no mass or charge. They travel at the speed of light and can travel several hundred feet in the air. Gamma radiation is so energetic that approximately three quarters of the gamma rays pass through the body without doing any damage. However, the other 25% make up for the rest! X-rays and gamma rays have almost identical properties, but gamma radiation is produced by radioactive decay, whereas X-rays are generated when a metal anode, usually copper, is bombarded with electrons (an X-ray generator as found in a hospital or dental office). For our purposes, we lump them together. Because gamma radiation travels hundreds of feet in the air, shielding consists of dense material that is several inches thick (such as lead or steel) (Figure 3-2). This makes shielding rather impractical for dealing with hazardous materials incidents involving gamma-emitting radionuclides (gamma emitters). Such materials are also an internal hazard if they are ingested or inhaled.

Resource Central

See Hazmat History for the story of a Boy Scout who tried to build a small breeder reactor near Detroit, Michigan, and in the process created a federal Superfund site.

Neutron Radiation

Neutrons are relatively massive at 1 amu; however, they have no charge (as their name implies). Because neutrons are uncharged, they can also travel hundreds of feet in the air like gamma radiation. Neutron radiation requires massive amounts of water or concrete for shielding, making shielding impractical as a primary mode of protection. Neutron radiation is typically encountered in nuclear processes, such as operating nuclear reactors, high-level nuclear waste, and detonated nuclear weapons. Neutron emitters are also an internal hazard if they are ingested or inhaled. An example of a neutron emitter is californium-252.

HALF-LIFE

half-life ▪ The time period during which the specific activity of a radioisotope is reduced by half. After one half-life there will be half as much radiation present as before.

Radioactive materials spontaneously emit radiation through a process of disintegration or decay of their atomic nuclei. After they emit radiation, the materials are theoretically no longer radioactive, although the radionuclides may decay into other radioactive isotopes in the decay process. The time it takes for half the atoms in a quantity of radioactive material to emit radiation and decay—in other words, become half as radioactive—is called the **half-life** of the radioisotope. Some radioactive materials have very short half-lives, such as technetium-99m with a half-life of 6 hours. Other radioactive materials have extremely long half-lives, such as uranium-238 with a half-life of 4.5 billion years; and yet others have moderately long half-lives, such as cesium-137 with a half-life of 30 years. An important point to remember is that the shorter the half-life of the radioactive material, the more radiation it emits per unit time. Therefore, standing next to a pound of pure technetium-99m (which emits gamma rays and has a half-life of 6 hours) is much more dangerous than standing next to a pound of pure cesium-137 (which also emits gamma rays and has a half-life of 30 years), because radioisotopes decay to background levels in 7 to 10 half-lives. In fact, radioisotopes with short half-lives are often stored for 10 half-lives and then disposed of as nonradioactive waste (which is much cheaper). Thus, the technetium-99m will have decayed to background levels in 60 hours. This means that if you had stood next to the technetium-99m for 60 hours, you would have received virtually 100% of the possible gamma radiation emission. In contrast, if you had stood next to the cesium-137 for 60 hours you would have received less than 0.02% of the possible gamma radiation emission.

MEASUREMENT OF RADIATION LEVELS

There are several different units of measurement associated with radiation, which can be confusing (Table 3-7) because there are two sets of units—conventional units and the International System of Units (SI units). Depending on the document, the regulation, or the radiation detection equipment you are using, you may see any number of these units at a hazardous materials incident involving radiation.

The quantity, or amount, of the radioactive material is measured in curies (Ci) or becquerels (Bq). Shipping papers are required to list the quantity of radioactive materials in becquerels, while the packaging may list the quantity of radioactive material in curies.

Dose refers to an amount of radiation received and can be measured using roentgens (R), rads, or rems in conventional units; or grays (Gy) and sieverts (Sv) in SI units. The unit roentgen (R) applies only to gamma radiation and X-rays. Rad—which stands for radiation absorbed dose—is the amount of radiation absorbed by a *material*. Rem—which stands for roentgen equivalent man—is the amount of radiation absorbed by *biological tissue*. The amount of radiation absorbed by biological tissue depends on the energy transfer of the ionizing radiation. The radiation that is most efficient at transferring its energy to biological tissue is alpha radiation. For alpha radiation, 1 rad equals approximately 20 rem; for gamma radiation 1 rad equals 1 rem. This conversion number is called

TABLE 3-7 Units of Radiation Measurement

TYPE OF MEASUREMENT	DEFINITION	CONVENTIONAL UNITS	SI UNITS	NOTES
Quantity (activity level)	The quantity of radioactive material	curie (Ci)	becquerel (Bq)	1 Bq = 1 disintegration per second
Exposure	Amount of ionization in air from gamma and X-ray radiation only	roentgen (R)		
Absorbed dose	Amount of radiation absorbed by a material	radiation absorbed dose (rad)	gray (Gy)	1 Gy = 100 rad = 1 J/kg 1 rad = 100 erg/g of material
Dose equivalent	Amount of radiation damage in human tissue	roentgen equivalent man (rem)	sievert (Sv)	1 Sv = 100 rem rem = rad × Q
Exposure rate (Dose rate for gamma radiation)	Amount of X-ray and gamma radiation received per hour	roentgen per hour (R/hr)		Also referred to as field strength measurement
Dose rate	Amount of radiation absorbed per hour	rad/hr rem/hr		

quality factor (Q) ▪ The relative effectiveness of a given type of radiation in transferring energy to biological tissue.

the quality factor. **Quality factor (Q)** refers to how well the particular type of radiation transfers energy to biological tissue. The quality factor for gamma radiation, X-rays, and beta particles is 1; for neutrons it is 2 to 10; and for alpha particles, it is 20. This means that alpha particles do about 20-fold more tissue damage than does gamma radiation.

Dose rate is usually measured in roentgen per hour (R/hr) or sieverts per hour (Sv/hr). Dose rate in R/hr is applicable only to gamma radiation and X-rays. We will discuss dose rate measurements in greater detail in Chapter 8 when we discuss monitoring for radiation.

MINIMIZING RADIATION EXPOSURE

As mentioned earlier, it is impossible to reduce your radiation exposure to zero. Therefore, in the workplace, radiation exposure is kept to "as low as reasonably achievable" (ALARA). ALARA is the mantra used in workplaces where radiation is routinely used, such as hospitals, research laboratories, nuclear reactors, and doctor and dentist offices that use X-ray machines.

The most effective way to protect yourself from radiation is to use time, distance, and shielding (TDS). The less time you spend near a radioactive source, the less radiation you'll receive. Thus, minimizing time spent near a source is a very effective way of reducing radiation exposure. The more shielding you place between yourself and the radioactive source, the less radiation can reach your body. However, shielding is often not a practical option with highly penetrating radiation like gamma and neutron radiation owing to the weight of the shielding needed. The most effective way to protect yourself from radiation is to keep your distance. According to the inverse-square law:

$$R_1 \times (D_1)^2 = R_2 \times (D_2)^2$$

where R is radiation intensity and D is distance.

This equation means that every time you double the distance away from the source, you decrease by fourfold the amount of radiation you receive (Figure 3-3). For example, if you move twice the distance away from the source, let's say from 1 foot to 2 feet, you reduce the radiation received from 100 mrem at 1 foot away to 25 mrem at 2 feet away. This is an important point to remember when establishing the isolation distance for a release involving radioactive materials. If you have indications that a strong radioactive source is present, isolate the area a sufficient distance away.

At hazardous materials incidents involving radiation it is important to track how much radiation each individual receives to compare those levels with the action levels recommended by federal agencies (Table 3-8).

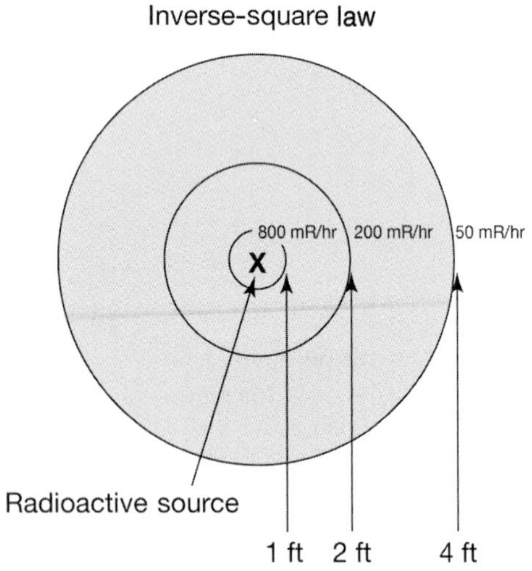

FIGURE 3-3 The inverse-square law and its effects. Doubling your distance away from a radioactive material reduces your exposure fourfold.

TABLE 3-8 Guidelines for Radiation Exposure

DOSE LIMIT	ACTIVITY	CONDITIONS
5 rem	All	Using ALARA principle
10 rem	Protection of major property	Where lower doses are not practical
25 rem	Lifesaving or protection of large populations	Where lower doses are not practical
>25 rem	Lifesaving or protection of large populations	Only on a voluntary basis by personnel fully aware of the risks

Source: Data courtesy of the U.S. Department of Energy, Transportation Emergency Preparedness Program (TEPP) and the U.S. EPA *Manual of Protective Action Guides and Protective Actions for Nuclear Incidents* (EPA 400-R-92-001).

SOLVED EXERCISE 3-3

You are dispatched to a construction site along an interstate highway for a report of a radioactive material release. On your arrival you are told that a radiography camera (pictured on top) was run over by a bulldozer and completely destroyed, releasing the radioactive pigtail source (pictured at bottom) from the protective housing. It reportedly contains iridium-192. At 30 feet away your Geiger counter is reading 100 mR per hour. What would the reading be if you approached to within 3 feet of the radionuclide?

Solution: According to the inverse-square law, the reading would be 10,000 mR, or 10 R, per hour. This is a very large dose rate and could be life threatening. It is calculated as follows:

$$(100 \text{ mR/hr}) \times (30)^2 = (\text{field strength at 3 ft}) \times (3)^2$$

$$(100 \text{ mR/hr}) \times 900 = (\text{field strength at 3 ft}) \times 9$$

$$\text{field strength at 3 ft} = (100 \text{ mR/hr}) \times 900 / 9 = 10{,}000 \text{ mR/hr} = \mathbf{10\ R/hr}$$

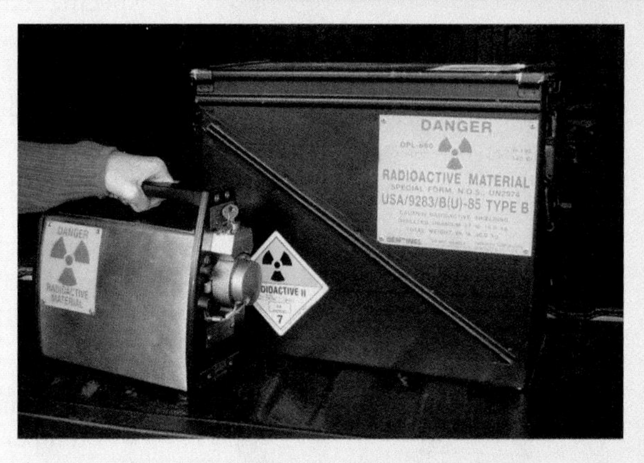

Courtesy of Tom Clawson TRG, Inc.

HAZMAT HANDLE

Reduce your exposure to radiation to as low as reasonably achievable (ALARA) by using time, distance, and shielding (TDS). Remember, increasing your distance from the radioactive material is the most effective way to limit exposure!

Elimination of Poisonous Substances (Detoxification)

The body processes poisons in three different ways: via excretion, storage, or metabolism. **Excretion**, or elimination, refers to the removal of a poison by the body as a liquid via urine or sweating, as a solid via fecal matter, or as a vapor by exhalation. Other toxic materials may be stored in the body for longer periods of time. Examples of poisons that are stored in the body include fat-soluble chemicals such as polychlorinated biphenyls (PCBs) and heavy metals such as lead and mercury that bind to proteins in the body. Yet others are metabolized. **Metabolism**, or biotransformation, involves the breakdown of the chemical or conversion of the chemical into other substances—possibly poisons—by the body. For example, ethanol is metabolized into nontoxic breakdown products that contribute to energy production in the body (and can make you overweight). However, another alcohol, methanol, is metabolized into toxic breakdown products including formaldehyde, which may cause blindness. How the body eliminates a poison directly contributes to its toxicity.

excretion ■ A process by which the body removes a poison. Also called elimination.

metabolism ■ Transformation of one chemical into another by the body.

Routes of Entry

Hazardous materials can enter the body through four routes of entry: inhalation, absorption, ingestion, and injection (Figure 3-4). **Inhalation** is the most effective route of entry and the most common source of exposure for emergency responders, partly because it

inhalation ■ The introduction of a substance in the form of a gas, vapor, aerosol, fume, mist, or dust into the body through the respiratory tract.

FIGURE 3-4 The four routes of entry into the human body for hazardous materials. *Art by David Heskett.*

is an involuntary act. The lungs are designed to exchange gases (absorb oxygen and excrete carbon dioxide), but unfortunately, the lungs are not specific for these two gases and efficiently absorb many potentially toxic chemical vapors and gases. Materials that can be inhaled include gases, vapors produced by volatile liquids, vapors sublimating from solids, aerosols, dusts, and fumes.

The second most common route of entry is **absorption**, since we tend to touch materials with our hands if we do not perceive a hazard. This is another reason early recognition of hazardous materials releases is so important. Chemicals can be absorbed either through the skin (percutaneous absorption) or through the eyes (ocular absorption). The following factors influence the effectiveness of absorption:

- contact time with the body
- concentration of the material
- temperature of the material
- state of matter
- surface area affected
- integrity of the skin (thickness and wounds)
- location on the body

absorption ■ (1) One of the four routes of entry of toxic substances into the body by passage through the skin, mucous membranes, or eyes. (2) The process of using a material or media to soak up a spilled hazardous material through incorporation.

ingestion ■ The introduction of a substance into the digestive tract of the body.

injection ■ The introduction of a substance into the body through a skin puncture or transdermal wound.

Ingestion can be an effective route of entry, since the gastrointestinal (GI) tract is designed to absorb nutrients; however, eating and drinking are voluntary activities. The GI tract includes the mouth, esophagus, stomach, and small and large intestines. Accidental ingestion due to smoking or by placing dirty fingers in the mouth (such as to turn a page) can lead to exposure. You should be careful not to eat, drink, or smoke at hazardous materials incidents (or carry contaminants home with you through poorly performed decontamination).

Injection is probably the least common route of entry, but if a hazardous material is injected directly into the body it can enter the bloodstream and be distributed throughout

BOX 3-2 HAZMAT HISTORY: FLUOROSULFONIC ACID RELEASE IN BRIDGMAN, MICHIGAN

At 5:07 AM on August 7, 1981, the residents of Bridgman, Michigan, awoke to the unmistakable screeching of grinding and bending steel. Fourteen cars of a northbound C&O train with a total of 79 cars had derailed at the 60th car. This car belonged to Dow Chemical of Midland, Michigan, and contained 9655 gallons of fluorosulfonic acid. A large white plume began escaping from the overturned car almost immediately. A total of 3200 gallons of liquid escaped from the tank that day before the leak was eventually plugged.

Various police, fire, and EMS agencies responded to the derailment. Evacuations were started, a command post was set up, the scene was secured, and eventually, mitigation efforts were started. A law enforcement officer positioned at the inner perimeter of the release scene, approximately 100 yards from the overturned railcar was responsible for keeping motorists, the media, and curious onlookers away from the leaking tank car. He spent a total of 13 hours in proximity to the white plume. Several other law enforcement officers complained of breathing discomfort and skin irritation. A few of them returned to their station with weapons tarnished by the corrosive plume. On the morning of August 29, 1981—almost 2 weeks later—the 37-year-old law enforcement officer who had been closest to the plume died of a massive heart attack after a coughing fit at his home. This death was subsequently ruled a line-of-duty death caused by one component of the noxious fumes: hydrogen fluoride. Hydrogen fluoride very effectively binds calcium in the body, including calcium in the bones and calcium in the heart—the same calcium that keeps it beating.

This incident illustrates the importance of proper site assessment and perimeter air monitoring.

the body very quickly. Intravenous (IV) injection refers to introduction of a chemical directly into the bloodstream. Intramuscular (IM) injection refers to introduction of the chemical into a muscle. Victims who have been involved in an industrial accident may have puncture wounds from contaminated shrapnel or equipment. Compressed air and compressed liquids in high-pressure hydraulic lines can also be very effective injection routes of entry. In fact, industrial facilities that use compressed air have long educated employees not to remove dust and dirt by blowing compressed air over themselves to avoid inadvertently injecting contaminants into their bodies.

Signs and Symptoms

Hazardous materials may have **acute effects** or **chronic effects**. Acute effects occur quickly on exposure and may have either short- or long-term effects on the victim. Chronic effects occur over a longer period of time, and signs and symptoms of exposure develop slowly. Exposures to hazardous materials also can be acute or chronic. An acute exposure is an exposure to a hazardous material over a short time—on the order of minutes to days, while a chronic exposure is a long-term exposure—on the order of months to many years. For example, you respond to a hazardous materials incident and unwittingly get too close. You smell a pungent, sweet odor. You immediately realize that you are too close and back away to a safe distance. You've just received an acute exposure to that hazardous material. However, if you've worked in a factory for 5 years that uses benzene as a solvent without adequate ventilation, you have had a chronic exposure to benzene.

Dose-Response Relationship

The amount, or dose, of a toxin or medicine that is received determines the result on the human body. The **dose** is the amount of a poison or drug delivered to a person via a specific route of entry. For example, a person may receive 500 milligrams of aspirin orally. The dose depends on several factors, including route of entry, duration of exposure, and frequency of exposure. Generally, the smaller the dose is, the smaller the effect will be. The **dose-response relationship** describes how a given material affects the body in relation to the quantity administered (Figure 3-5). The dose-response relationship can describe the dose-dependent healing effects of a medicine, or the dose-dependent damage caused by a toxic substance. A graph of these values is a **dose-response curve** that can be consulted to determine the expected physiological response after exposure to a certain amount of material.

Some chemicals have a **threshold effect** below which no ill effects are seen or felt. Other poisons can act synergistically, which means that the effect of the two chemicals together is worse than the effect of either chemical separately. Yet other chemicals may be harmless by themselves yet cause damage when combined with other materials. This is called a **potentiated effect**, an example of which is the accentuated liver damage produced by the combination of acetaminophen and alcohol. The combined effects of different chemical mixtures can be summarized as follows:

Additive effect: Two different toxic materials combine and have a toxic effect that is the sum of their individual toxic effects: (1 + 1 = 2).

Synergistic effect: Two different toxic materials combine, and they enhance each other's toxic effect: (1 + 1 = 3).

acute effects ▪ Signs and symptoms that appear rapidly on exposure to a hazardous substance.

chronic effects ▪ Illness resulting from exposure to a hazardous substance that appears slowly over an extended period (weeks to decades).

dose ▪ The amount of a poison or drug delivered to a person via a specific route of entry.

dose-response relationship ▪ The effect of a given material on the body in relation to the quantity of the material administered.

dose-response curve ▪ The graphic representation of the effect of a given material on the body in relation to the quantity of material administered.

threshold effect ▪ The dose of a material below which no ill effects are seen or felt.

potentiated effect ▪ The combination of a nontoxic material and a toxic material that produces a greater toxic effect on the body than does the toxic material alone.

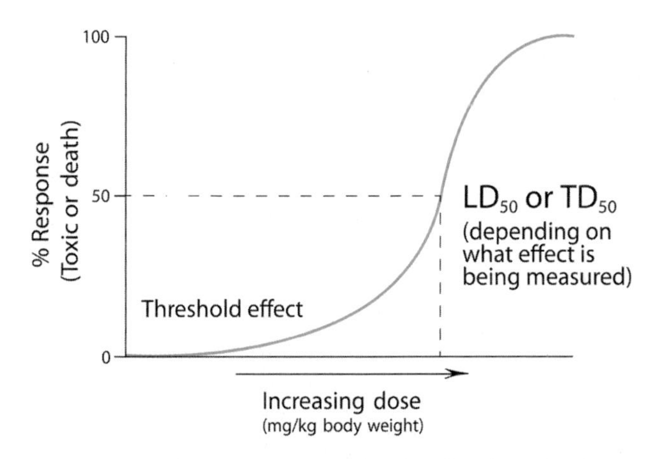

FIGURE 3-5 A dose-response curve describes the relationship between the quantity of the chemical entering the body and its effect on the body, including death at a high enough dose.

lethal concentration 50% (LC_{50}) ▪ The amount of a substance that kills half the test population by the route of inhalation.

lethal dose 50% (LD_{50}) ▪ The amount of a substance that kills half the test population by ingestion, absorption, or injection.

incapacitating concentration 50% (IC_{50}) ▪ The concentration of a chemical that will incapacitate 50% of exposed people by inhalation.

incapacitating dose 50% (ID_{50}) ▪ The concentration of a substance that will incapacitate 50% of the exposed population by a route of entry other than inhalation.

effective concentration low (EC_{LO}) ▪ The lowest amount of the substance required to reach the desired effect of a medicine by inhalation.

effective dose low (ED_{LO}) ▪ The lowest amount of the substance required to reach the desired effect of a medicine by skin absorption or ingestion.

toxic concentration low (TC_{LO}) ▪ The lowest amount of the material that produces signs and symptoms of poisoning by inhalation.

toxic dose low (TD_{LO}) ▪ The lowest amount of the material that produces signs and symptoms of poisoning by skin absorption or ingestion.

Potentiated effect: A nontoxic material and a toxic material combine and produce a more toxic effect than does the toxic material alone: $(0 + 1 = 2)$.

Antagonistic effect: Two different toxic materials combine and have a lower combined toxicity—they effectively cancel or neutralize each other: $(1 + 1 = 0)$.

Antidotes act as antagonists to poisons. Antidotes may increase the rate of elimination of a toxin, or they may block the toxic effects of a poison. Many antidotes by themselves are toxic, such as amyl nitrite, which is used as an antidote to cyanide poisoning. Amyl nitrite causes methemoglobinemia, which reduces the oxygen-carrying ability of blood.

The cumulative effects of some chemicals may be latent, so it is extremely important to avoid exposure to chemicals whenever possible, since signs and symptoms can be delayed from hours to decades.

The dose-response relationship is often difficult to determine. Toxicity information is determined using animal studies, epidemiological data, case reports, and molecular structure comparisons. Because the dose-response relationship for toxic substances can't be tested on humans directly, other species, such as pigs, rabbits, and rats, are often used as test subjects. However, the physiology of different species does not always mirror human response. Thus, a toxic effect seen in pigs may not occur in humans, and more disconcertingly, a toxic effect that occurs in humans may not necessarily be seen when tested on another species. Arsenic is a perfect example. Animal studies consistently do not show a carcinogenic effect in animals, yet there is convincing evidence using epidemiological studies of humans exposed to arsenic that it does cause cancer in humans.

The different effects of substances based on dose can also be described by threshold effects. Two of the most common measures of the toxicity of a substance are the LD_{50} and the LC_{50}. LC_{50} stands for the **lethal concentration 50%** of a substance, that is, the amount of a substance that kills half the test population by the inhalation route. Similarly, LD_{50} stands for the **lethal dose 50%** of a substance, in this case the amount of a substance that kills half the test population by ingestion, absorption, or injection. The important distinction between LC and LD is the route of entry.

Other effects that can be quantified are the **incapacitating concentration 50% (IC_{50})**, which is the concentration of a chemical that will incapacitate 50% of exposed people by inhalation. Similarly, the **incapacitating dose 50% (ID_{50})** is the concentration of a substance that will incapacitate 50% of the exposed population by a route of entry other than inhalation. The **effective concentration low (EC_{LO})** or **effective dose low (ED_{LO})** is the lowest amount of the substance required to reach the desired effect of a medicine by the respective route of entry. The **toxic concentration low (TC_{LO})** or **toxic dose low (TD_{LO})** of a material is the lowest amount of the material that produces signs and symptoms of poisoning.

Measurement of Health Effects

It is important to be able to measure the health effects of hazardous materials on persons exposed to them. Several different metrics have been created to measure the toxicity of different chemicals. You might ask yourself why there are so many different metrics. One reason is the type of exposure that may occur. For example, a factory worker could be subjected to a chronic exposure lasting over a career, for 40 hours a week, for 8 hours a day. In contrast, a member of the public might have an acute exposure of an unpredictable length once in a lifetime if a hazardous materials release occurs nearby, or a stay-at-home parent may be exposed to a hazard 24 hours a day. Clearly, worker exposure guidelines are not applicable to the public. This is an important concept to remember, especially when considering evacuation versus in-place sheltering for downwind protection at hazardous materials releases.

Health effects of chemical exposure are notoriously difficult to measure owing to variations in response among species and races, and between sexes; genetic disposition; age; degree of health; and previous chemical exposures. This is why it is so difficult to predict allergies to foods or bee stings. For example, some people may be stung multiple different times and show no allergic effects, then suddenly may experience a severe allergic reaction.

UNITS OF MEASUREMENT

Concentration values of materials can be reported in many different ways. Airborne concentrations of gases and vapors are often reported in percentage (%) or parts per million (ppm). One percent (1%), or one part per hundred, refers to one molecule of gas or vapor for every 100 molecules of air. This value can be converted to parts per million: 1% equals 10,000 ppm. You may also see concentrations of parts per billion (ppb) and parts per trillion (ppt) for highly toxic materials. Chemical warfare agents often have exposure limits in the part per billion or part per trillion range.

Airborne concentrations of dusts and mists are usually reported in units of milligrams per cubic meter (mg/m^3). Because dust particles and mist droplets are not individual molecules, concentration values are reported as a weight of material in a given volume of air. Parts per million and milligrams per cubic meter can be interconverted using the following formulas, which take into account the molecular weight of the substance:

$$\text{ppm} = \text{mg/m}^3 \times 24.45 \text{ / molecular weight}$$
$$\text{mg/m}^3 = \text{ppm} \times \text{molecular weight / } 24.45$$

This formula is temperature- and pressure-dependent and is based on standard conditions of 25°C and 1 atmosphere of pressure. The conversion factor 24.45 is derived from the volume of one mole (mol) of air expressed in liters (L) and the molecular weight of air, which is measured in grams per mole (g/mol). In this way we can convert the weight per volume into parts per million assuming that the gas and vapor molecules do not interact with one another. This assumption does not apply to some chemicals, such as hydrogen fluoride, that form strong interactions with one another in the gas phase (see Box 5-3 for more detailed information). This formula also does not apply exactly under conditions of high humidity (generally, greater than 60%) to some chemicals that interact strongly with water.

As an example of these units of measurement, toxicity has been defined by the Occupational Safety and Health Administration (OSHA) using LD_{50} and LC_{50} values in albino rat test populations:

Toxic: An LD_{50} value between 50 mg/kg and 500 mg/kg orally, or an LD_{50} value between 200 mg/kg and 2000 mg/kg by contact, or an LC_{50} value between 200 ppm and 3000 ppm by inhalation

Highly toxic: An LD_{50} below 50 mg/kg orally, or an LD_{50} less than 200 mg/kg by contact, or an LC_{50} below 200 ppm by inhalation

EXPOSURE LIMITS

Different airborne exposure limits are determined based on the target population, such as workers or the public, and the amount of time that the target population will be exposed to the chemical, such as an 8-hour workday or continuously for a prolonged period at a hazardous materials release.

time-weighted average (TWA) ▪ The average concentration of a chemical to which a person has been exposed over a given period of time (usually 8 or 10 hours).

Time-Weighted Averages (TWAs)

A **time-weighted average (TWA)** is the average concentration of a chemical to which a person may be exposed over a given period of time. For example, the permissible exposure limit (PEL) is an 8-hour time-weighted average, while the recommended exposure limit (REL) is a 10-hour time-weighted average. A TWA value may be exceeded for a period of time as long as the average value over the entire period remains below the TWA, with some caveats that we will discuss shortly. For example, if a TWA is 100 ppm for an 8-hour time period, then a person can potentially be exposed to 200 ppm for the first 2 hours, 100 ppm for the next 2 hours, and 25 ppm for the last 4 hours because the TWA for the entire 8-hour period is 87.5 ppm, which is below the 100 ppm TWA. The TWA is calculated as follows:

$$(200 \text{ ppm} \times 2 \text{ hr}) + (100 \text{ ppm} \times 2 \text{ hr}) + (25 \text{ ppm} \times 4 \text{ hr}) = 700 \text{ ppm} / 8 \text{ hr} = 87.5 \text{ ppm}$$

Workplace Exposure Limits

It is important to know the level at which exposure to a hazardous material becomes dangerous to be able to choose the appropriate engineering controls, to decide on appropriate work practices, and to choose the appropriate level of PPE at hazardous materials releases. Many different exposure limits have been set by government and industry. Two of the most important are the permissible exposure limit (PEL) and the immediately dangerous to life and health (IDLH) level.

Permissible Exposure Limit (PEL) The permissible exposure limit (PEL) is set by OSHA and has the force of law. The PEL is the maximum airborne concentration of a chemical to which a healthy worker can be exposed 8 hours a day, 40 hours a week, throughout his or her career without experiencing any permanent health problems. The PEL is an 8-hour TWA, which means that the PEL may be temporarily exceeded under certain conditions. Generally, if the airborne concentration of the chemical, as determined by air monitoring, is below the PEL, no respiratory protection is required. Above the PEL, respiratory protection is required. Permissible exposure limits are typically measured in parts per million.

recommended exposure limit (REL) ▪ The NIOSH-determined time-weighted average threshold limit value to which workers can be exposed continuously during a 10-hour time period and a 40-hour workweek without suffering ill effects.

Recommended Exposure Limit (REL) The **recommended exposure limit (REL)** is set by the National Institute for Occupational Safety and Health (NIOSH). NIOSH can be considered the research arm of OSHA. The REL is the maximum airborne concentration of a chemical to which a healthy worker can be exposed 10 hours a day, 40 hours a week, throughout his or her career without experiencing any permanent health problems. The REL is a 10-hour TWA.

threshold limit value (TLV) ▪ An exposure guideline established by the American Conference of Governmental Industrial Hygienists (ACGIH) for airborne concentrations to which an average worker may be exposed for 8 hours per day during a 40 hour workweek without experiencing adverse health effects.

Threshold Limit Value (TLV) The **threshold limit value (TLV)** is the airborne concentration of a chemical to which nearly all workers may be repeatedly exposed without exhibiting adverse health effects. TLVs are set by the American Conference of Governmental Industrial Hygienists (ACGIH) based on industrial experience, human studies, animal studies, or a combination of these. The TLV-TWA is the time-weighted average concentration to which a healthy worker may be exposed 8 hours a day, 40 hours a week, throughout his or her career without experiencing any adverse health effects. These values are updated annually based on the latest research. Threshold limit values are also typically measured in parts per million.

short-term exposure limit (STEL) ▪ The airborne concentration to which workers can be exposed to for up to 15 minutes, four times per day without suffering irritation, chronic, or irreversible tissue damage.

Short-Term Exposure Limits (STEL) To prevent inordinately high short-term exposures during a longer time-weighted average, the **short-term exposure limit (STEL)** was devised. The STEL is the chemical concentration to which a worker may be exposed for up to 15 minutes without suffering adverse affects such as irritation or irreversible tissue damage. There may be no more than four 15-minute STEL exposures during the day, and they must be at least 1 hour apart. In addition, the sum of the STEL exposures may not exceed the 8-hour TLV-TWA.

ceiling (C) ▪ The concentration of a chemical that should not be exceeded even instantaneously.

Ceiling (C) Likewise, the ceiling was devised to avoid inordinately high exposures during a short-term exposure limit. The **ceiling (C)** is the concentration of a chemical that should not be exceeded even instantaneously. To comply with ceiling values, real-time continuous air monitoring instrumentation must be available with extremely rapid readouts. When this is not feasible, the law makes specific exceptions to ceiling values for specified chemicals listed in 29 CFR. The interrelationship of PEL, STEL, and ceiling values is illustrated in Figure 3-6.

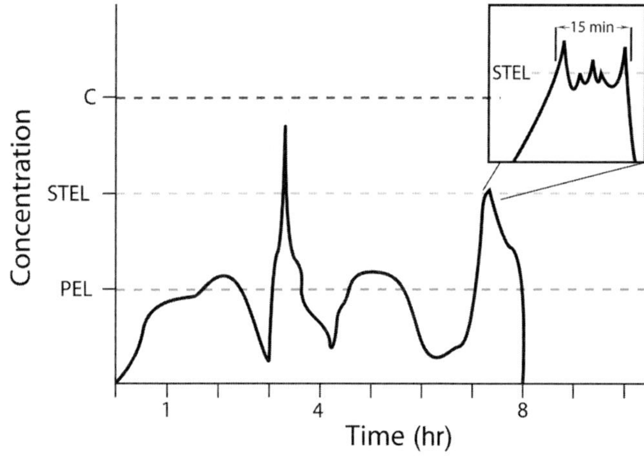

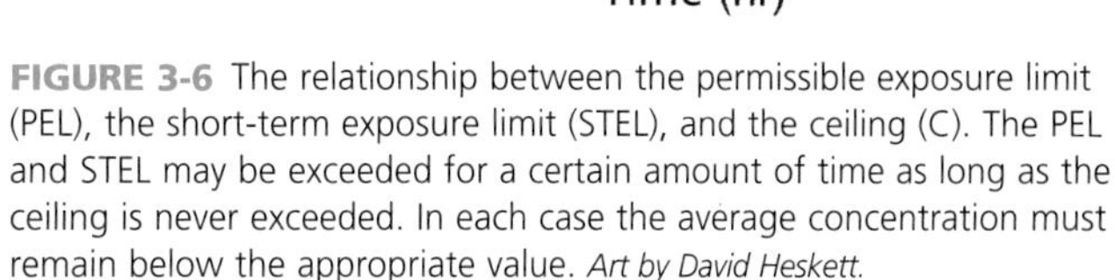

FIGURE 3-6 The relationship between the permissible exposure limit (PEL), the short-term exposure limit (STEL), and the ceiling (C). The PEL and STEL may be exceeded for a certain amount of time as long as the ceiling is never exceeded. In each case the average concentration must remain below the appropriate value. *Art by David Heskett.*

Immediately Dangerous to Life and Health (IDLH) The "immediately dangerous to life and health (IDLH)" level is set by NIOSH. The IDLH is specifically mentioned in the confined-space regulations and thus also carries the force of law. Any exposure above the IDLH requires supplied-air respiratory protection. Exposures below the IDLH, but above the PEL, require some type of respiratory protection, but not necessarily supplied-air respiratory protection. In this case the respiratory protection may be either supplied-air respiratory protection or an air purifying respirator. We will discuss respiratory protection at great length in Chapter 7. Thus, understanding the significance of exposure limits, especially the PEL and IDLH, is extremely important to the hazmat technician for selecting the appropriate personal protective equipment.

These values cannot be used alone to compare the toxicity of different chemicals, since the values can be misleading. For example, let's try to compare the toxicity of acetic acid, which is the primary ingredient in household vinegar, and hydrogen cyanide, which is a highly toxic pesticide and fumigant. The IDLH of both of these materials is 50 ppm, and the PEL for both of these materials is 10 ppm. How can that be? The answer is that the IDLH and PEL values are based on short-term exposure to a corrosive vapor. Both acetic acid and hydrogen cyanide are acidic and corrosive, and damage the upper respiratory tract. The toxicity of hydrogen cyanide as a chemical asphyxiant is manifested over time and at a higher concentration. The body can effectively metabolize low levels of hydrogen cyanide. Thus, it is advisable to consult an industrial hygienist or toxicologist at complex hazardous materials incidents.

Public Safety Exposure Limits

Public safety exposure limits—also known as *protective action criteria* (PAC)—are different from worker exposure limits primarily because each group comprises different types of people, and the potential period of exposure is different for each group as well. Workers tend to be healthy men and women at the prime of their lives who are exposed to a chemical for a relatively limited time period (the 8-hour workday).

Public exposure limits must take into account the young and the old as well as the sick and injured. In addition, during an emergency the public may be exposed to the chemical for an unknown length of time—from hours to days at a time. Therefore, public safety exposure limits should be more protective than worker exposure limits. Often, 1/10 of the IDLH value is initially chosen as a prudent public safety exposure limit for a short time (30 minutes to an hour). This value has been suggested by the EPA, the Federal Emergency Management Agency (FEMA), and DOT as an initial public safety exposure limit in the document "Technical Guidance for Hazard Analysis."

Two commonly accepted public exposure limits are the Acute Exposure Guideline Levels (AEGLs) and the Emergency Response Planning Guidelines (ERPGs), which are discussed in the next two sections. There is also a third public exposure limit called the Temporary Emergency Exposure Limit (TEEL). TEELs were developed by the U.S. Department of Energy and are designed to be used when ERPGs and AEGLs are not available. There are currently TEELs for more than 3000 chemicals.

Acute Exposure Guideline Levels (AEGLs) Acute Exposure Guideline Levels (AEGLs) were developed by the National Research Council's Committee on Toxicology to develop uniform emergency response standards for general public exposures. The AEGLs (pronounced "eagles") take into account the most sensitive members of a given population. This is a more complicated system but is still based on a three-tiered system of health effects, with five additional time lengths of exposure. The three AEGL health effect tiers are as follows:

> *AEGL-1:* The airborne concentration of a substance above which it is predicted that the general population, *including susceptible individuals*, could experience notable discomfort, irritation, or certain asymptomatic non-sensory effects. However, the effects are not disabling and are transient and reversible upon cessation of exposure.
>
> *AEGL-2:* The airborne concentration of a substance above which it is predicted that the general population, including susceptible individuals, could experience irreversible or other serious, long-lasting adverse health effects or an impaired ability to escape.
>
> *AEGL-3:* The airborne concentration of a substance above which is predicted that the general population, including susceptible individuals, could experience life-threatening health effects or death.
>
> —U.S. Environmental Protection Agency

For each of these health effect levels there are five predicted exposure periods: 10 minutes, 30 minutes, 1 hour, 4 hours, and a maximum of 8 hours. This system has two advantages over ERPGs: (1) it takes into account susceptible individuals, and (2) values are published for various chemical exposure time frames. This information can be very helpful in the event of a continuous hazardous materials release. There are currently AEGLs for 66 chemicals.

Emergency Response Planning Guidelines (ERPGs) Emergency Response Planning Guidelines (ERPGs) were developed by the American Industrial Hygiene Association (AIHA) for planning purposes in the event of hazardous materials releases. This is a three-tiered system based on 1-hour contact times:

> *ERPG-1:* The maximum airborne concentration below which it is believed that nearly all individuals could be exposed for up to one hour without experiencing other than *mild transient adverse health effects* or perceiving a clearly defined, objectionable odor.
>
> *ERPG-2:* The maximum airborne concentration below which it is believed that nearly all individuals could be exposed for up to one hour without experiencing or developing irreversible or other serious health effects or symptoms which could *impair an individual's ability to take protective action.*
>
> *ERPG-3:* The maximum airborne concentration below which it is believed that nearly all individuals could be exposed for up to one hour without *experiencing or developing life-threatening health effects.*
>
> —American Industrial Hygiene Association (AIHA)

ERPGs are not designed to protect everyone, so individuals with preexisting conditions such as asthma or congestive heart failure may still be at risk under these guidelines. These guidelines also do not incorporate safety factors. There are currently ERPGs for approximately 150 chemicals.

Enforcement by OSHA

OSHA enforces airborne exposure limits in four separate tiers. The OSHA published PELs are enforced first, followed by the NIOSH published RELs, followed by the TLVs published by the ACGIH, and, finally, if none of these published values exist for a particular

SOLVED EXERCISE 3-4

Your hazmat team is overseeing the removal of more than two hundred drums from a warehouse, many of which are leaking. These drums all contain the same material, which is toxic and has a permissible exposure limit of 100 ppm. Continuous air monitoring is being performed. The contractor who is performing the work says that no respiratory protection is necessary because the PEL has not been exceeded. He presents you with the following list of readings:

Time	Reading
0800–1000	120 ppm
1000–1100	178 ppm
1100–1200	54 ppm
1300–1600	113 ppm
1600–1700	67 ppm

Is he right or wrong?

Solution: He is wrong. The average concentration over the 8-hour time period was 110 ppm, while the PEL is 100 ppm. Respiratory protection is certainly necessary. The TWA is calculated as follows:

2 hr × 120 ppm = 240 ppm
+ 1 hr × 178 ppm = 178 ppm
+ 1 hr × 54 ppm = 54 ppm
+ 3 hr × 113 ppm = 339 ppm
+ 1 hr × 67 ppm = 67 ppm
878 ppm / 8 hr = **110 ppm**

chemical, any toxicological values found in the scientific literature or published by the manufacturer may be enforced. It is therefore very important that hazardous materials response teams consult the appropriate airborne exposure limits when deciding which level of personal protective equipment is appropriate to use. The PPE ensemble must have the necessary chemical protective clothing and respiratory protection components to adequately protect the wearer. The air monitoring equipment must be capable of detecting the chemical at the airborne exposure limit in real time. If these criteria cannot be met, incident objectives may have to be modified. When in doubt it is essential to consult a qualified industrial hygienist. We will discuss PPE selection at great length in Chapter 7, and air monitoring in Chapter 8.

Chemical Contamination

You should be careful to avoid contact with hazardous materials. An individual is contaminated with a hazardous material if it is on or inside the body. If the material is on the outside of the body, it is **external contamination.** If the material has entered the body, either through ingestion, inhalation, or injection, it is **internal contamination.** Often, patients become both internally and externally contaminated. It is very difficult to remove internal contamination. **Secondary contamination** may occur if a contaminated person tracks the hazardous material beyond the immediate area of release, and others become contaminated. The health effects of hazardous materials may linger if proper decontamination is not performed. Decontamination is the removal of hazardous materials from people and equipment. We will discuss decontamination at great length in Chapter 12.

external contamination ▪ The unintentional transfer of a material to the outside surface of the body.

internal contamination ▪ The unintentional transfer of a material to the inside of the body, either through absorption, inhalation, ingestion, or injection.

secondary contamination ▪ The transfer of unwanted material (contamination) from one location to another.

CHAPTER **REVIEW**

Summary

It is important for the hazardous materials technician to understand how chemicals affect the human body. Entry team members going into the hot zone should be familiar with the signs and symptoms of exposure to the released chemical in case of a suit breach, which may expose them to the toxic effects of the chemical. Airborne exposure limits have been determined for various populations, including workers and the general public. The appropriate airborne exposure limit should be consulted for the task at hand, such as downwind protection of the public or selection of the appropriate level of respiratory and skin protection. Understanding the fundamentals of toxicology is critical to correctly selecting and using both air monitoring equipment and personal protective equipment, and to initiating appropriate emergency medical treatment.

Review Questions

1. What are the signs and symptoms of blister agent exposure?
2. What are the signs and symptoms of ricin exposure?
3. What three methods can be used to protect oneself from radiation exposure?
4. What is the most effective route of entry and why?
5. What is the difference between acute health effects and chronic health effects?
6. Which federal agency enforces the permissible exposure limit (PEL)?
7. What is the definition of IDLH?
8. Why is it important to know the PEL and IDLH of a released chemical?
9. What is the difference between a threshold limit value (TLV), a short-term exposure limit (STEL), and the ceiling (C)?
10. Why are there both workplace exposure limits and public safety exposure limits?

References

American Council of Governmental Industrial Hygienists (2011) Guide to Occupational Exposure Values-2011. Cincinnati, Ohio: ACGIH, Inc.

Centers for Disease Control and Prevention. (2007). *NIOSH Pocket Guide to Chemical Hazards.* Washington, DC: U.S. Government Printing Office.

National Nuclear Security Administration. (2005) Handbook for Response to Suspect Radioactive Materials. Washington DC: Department of Energy.

Occupational Safety and Health Administration. (1990). 29 CFR 1910.120, *Hazardous Waste Operations and Emergency Response (HAZWOPER).* Washington, DC: U.S. Department of Labor.

Ottoboni, M. Alice (2011) The Dose Makes the Poison: A Plain-Language Guide to Toxicology, 3rd edition. Wiley

US Army Medical Research Institute of Chemical Defense. (1996) *Medical Management Of Chemical Casualties.* Aberdeen Proving Ground, Maryland: U.S. Army.

US Army Medical Research Institute of Infectious Diseases. (1996) Medical Management of Biological Casualties. Fort Detrick, Maryland: U.S. Army.

US Department of Energy National Nuclear Security Administration. (2005) Handbook For Response to Suspect Radioactive Materials. Washington DC: US Department of Energy.

Silverstein, Ken. (2004). *The Radioactive Boy Scout.* New York: Random House.

en	oxygen 8 O 16.00	fluorine 9 F 19.00
orus	sulfur 16 S 32.07	chlorine 17 Cl 35.45
ic	selenium 34 Se 78.96	bromine 35 Br 79.90

CHAPTER 4

Chemistry

Art by David Heskett.

KEY TERMS

anion, *p. 94*
atomic number, *p. 87*
atomic weight, *p. 87*
cation, *p. 93*
concentration, *p. 113*
covalent bond, *p. 94*
covalent compound, *p. 96*
double bond, *p. 94*
electron, *p. 86*
halogenated hydrocarbon, *p. 106*
hydrocarbon, *p. 103*
hydrocarbon radical, *p. 105*
ionic bond, *p. 93*
ionic compound, *p. 96*
maximum safe storage temperature (MSST), *p. 107*
metal, *p. 90*
mixture, *p. 112*
molecule, *p. 96*
neutron, *p. 86*
nonmetal, *p. 91*
nucleus, *p. 86*
polyatomic ion, *p. 94*
proton, *p. 86*
self-accelerating decomposition temperature (SADT), *p. 107*
single bond, *p. 94*
slurry, *p. 112*
solute, *p. 112*
solution, *p. 112*
solvent, *p. 112*
strength, *p. 113*
triple bond, *p. 95*

OBJECTIVES

- Name the three components of an atom.
- Describe the significance of the periodic table of the elements.
- Describe the difference between metals and nonmetals.
- Name the primary hazard of alkali metals and alkaline earth metals.
- Name the primary hazards of the halogens.
- Describe the significance of electrons to bonding.
- Describe the differences between ionic bonds and covalent bonds and their significance.
- Name six different types of salts and their primary hazards.
- Name the primary hazard of hydrocarbons and hydrocarbon derivatives.

- Describe the difference between saturated and unsaturated hydrocarbons.
- Name 10 different hydrocarbon derivatives and their primary hazards.
- Describe the significance of the name of a chemical with respect to its identification, classification, and reactivity.

You are dispatched to the local rail yard with reports that an overfilled tank car is leaking product. On arrival you are met by rail yard representatives who tell you a railcar full of styrene monomer is leaking from a top fitting. They also advise you that styrene is highly reactive and may polymerize if the inhibitor breaks down. Right about now you're thinking, why did I take this overtime shift? but there are also other questions running through your mind:

- What does polymerize mean?
- What is an inhibitor?
- What is going to happen now?
- What isolation distance should be enforced?
- What other resources will I need?

Let's see if we can get some answers to these questions and more.

nucleus ▪ The positively charged core of an atom composed of protons and usually neutrons and that constitutes the vast majority of the atomic mass.

proton ▪ A positively charged particle in the nucleus of an atom whose charge is equal to and opposite that of an electron.

neutron ▪ A particle located in the nucleus of an atom that has no charge. Neutrons that are violently ejected from the nucleus of radioisotopes are known as neutron radiation.

electron ▪ A negatively charged particle whose charge is equal to and opposite that of a proton. Electrons exist in a cloud surrounding the atomic nucleus. The outermost electrons are responsible for bonding between atoms.

What makes a hazardous material—a chemical—behave the way it does? Why are some materials radioactive? Why do some burn? Why are yet others oxidizers? The properties of chemicals are determined by their makeup—their atoms and how they are bonded together. This determines their chemistry.

We need to focus our attention on chemistry—the study of matter—and explore the origins of chemical hazards and how they affect our response procedures. To spot trends and form a deeper understanding of hazardous materials we have to delve into some of the nuts and bolts of chemistry. The more chemistry you learn the more "common sense" you will develop about hazardous materials. Chemistry also forms the basis of detection and identification of chemicals. Understanding chemistry allows you to optimally use your detection and chemical identification equipment and troubleshoot it in the field. To mount a risk-based response to a hazardous materials incident you must understand the risk any given chemical poses.

Atoms

Atoms are the most basic form of matter. Matter is anything that takes up space and has mass, such as metals and chemicals. The building blocks of all chemicals and hazardous materials are atoms. Atoms consist of a **nucleus** containing **protons** and **neutrons**, and an electron cloud around the nucleus (Figure 4-1). Table 4-1 summarizes the key properties of protons, neutrons, and **electrons**. The nucleus of an atom is very dense and contains most of the mass of the atom. Protons and neutrons weigh one atomic mass unit (amu) each. The **atomic weight** of an atom is the sum of the masses of the protons, neutrons, and electrons—although the electrons weigh very little.

The number of protons in the nucleus determines the identity of the element and is called the **atomic number**. For example, if the nucleus of an atom contains one proton, it is the element hydrogen; if it contains two protons, it is helium, and so on. Protons are positively charged and have a charge of +1. Most nuclei also contain neutrons, which are uncharged. The more protons in the nucleus, typically the more neutrons it will have to distribute and dilute the positive charges.

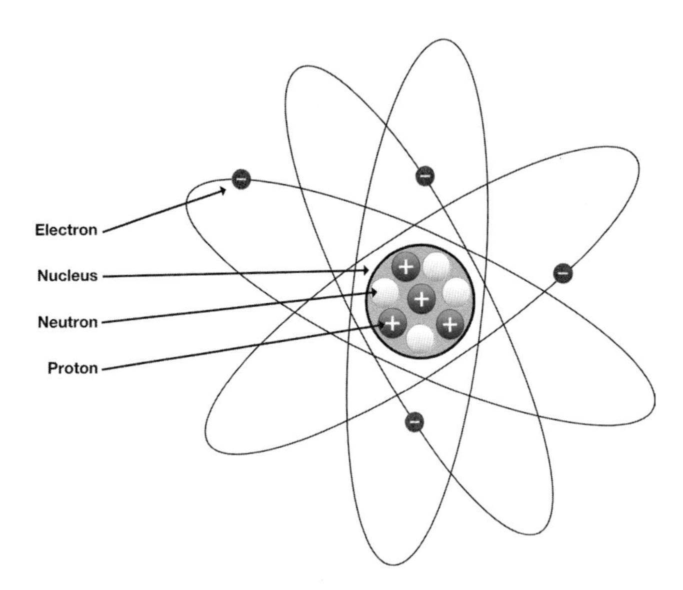

FIGURE 4-1 The Bohr model of the atom. The nucleus contains protons and neutrons, and electrons orbit around the nucleus. Radioactivity is a property of the nucleus, while reactivity is a property of the electrons of an atom. Not drawn to scale. *Art by David Heskett.*

The number of neutrons in the nucleus determines which isotope of an element it is. Isotopes of an element have the same number of protons in their nucleus but differ in the number of neutrons. Every element has several different isotopes, even the smallest one. Hydrogen has three isotopes—hydrogen, deuterium, and tritium. The most common isotope is what we call "hydrogen" and has one proton in the nucleus and no neutrons. The second isotope of hydrogen is *deuterium,* which contains one proton and one neutron in the nucleus. The third isotope is radioactive and is called *tritium.* Tritium contains one proton and two neutrons in the nucleus.

Electrons are located in the electron cloud at a comparatively great distance from the nucleus, which means that most of the volume that an atom occupies is actually empty space. This phenomenon explains a number of interesting facts, especially as they relate to radioactivity (which was discussed in the previous chapter). An electron has approximately 1/1800th the mass of a proton or neutron and carries a negative charge of −1. Elements have equal numbers of electrons and protons, so the charges cancel, and the atoms are neutral.

atomic weight ■ The sum of the mass of the protons, neutrons, and electrons.

atomic number ■ The unique number of protons in the nucleus of an atom; the atomic number determines the identity of the element.

Elements

Elements are pure substances and contain only a single type of atom. The majority of elements do not consist of solitary atoms; most exist as molecules. The only elements that consist of single atoms are the noble gases, such as helium (He) and argon (Ar). All other elements are either metals, compounds, such as table salt (NaCl), or molecules such as oxygen (O_2).

TABLE 4-1 Characteristics of Subatomic Particles

	MASS (AMU)	CHARGE	LOCATION	CHARACTERISTICS
Proton	1.0000000	+1	Nucleus	The number of protons determines the identity of the element (atomic number).
Neutron	1.0013811	0	Nucleus	The number of neutrons determines the isotope of the element.
Electron	1/1836	−1	Electron cloud	Electrons determine the reactivity of the element.

Elements are symbolized by either a single capital letter—such as carbon (C), oxygen (O), and nitrogen (N)—or a two-letter symbol consisting of a capital letter followed by a lowercase letter—such as chlorine (Cl), sodium (Na), and aluminum (Al). This distinction will become important when we discuss chemical formulas that contain multiple atoms later in this chapter. For example, it will be extremely important to recognize the difference between the element cobalt (Co) and the molecule carbon monoxide (CO). Your incident response would be very different for each of these materials! Some of the symbols were derived from Latin or German names for the elements, e.g., *natrium* for sodium.

Metals, such as copper, nickel, gold, iron, and silver, contain conductance band electrons in their outermost shell. Conductance band electrons can be considered a communal pool of electrons that give metals some of their unique characteristics such as high electrical conductivity.

The Periodic Table

In the late 1800s Dmitri Mendeleev, a Russian chemist, arranged all the elements that were known at that time based on their size (essentially atomic number) and their observable properties (reactivity). While doing so he noticed that the elements formed a repeating, or periodic, pattern. He also noticed there were some holes in his newly formed periodic table. Mendeleev correctly theorized that these holes must be missing elements, and he predicted the existence of gallium, scandium, and germanium years before they were actually discovered. Because of the periodicity of the elements major advances were made in understanding how reactivity, bonding, and the number of electrons are related. This relationship is called *orbital theory*, but we will touch on it only briefly later.

The periodic table can be divided into two distinct parts: the metals on the left, and the nonmetals at the top right (Figure 4-2). The staircaselike line on the table denotes the boundary between metals and nonmetals. Nonmetals are "on the top of the stairs," while metals are located "under the stairs." It might help you to remember this relationship by thinking about "metal" tools stored in a closet or a nook under the basement stairs—metals are under the stairs. In contrast, people—composed of nonmetals like carbon—walk on top of the stairs.

When a metal bonds with a nonmetal, an ionic bond is formed in most cases. When nonmetals bond with other nonmetals, covalent bonds are generally formed. We will discuss chemical bonding in much more detail later, since it determines how hazardous materials behave and how we detect them. For example, both Raman and FTIR spectrometers rely on the properties of covalent bonds to identify hazardous materials. For now let's take a closer look at the periodic table.

FAMILIES OR GROUPS

The periodic table is useful because elements with similar properties are arranged in columns, called *families* or *groups*. Elements in the same family share similar chemical and physical properties. Let's explore some of these families and some of their members you may encounter on the job.

The Noble Gases

The noble gases are the last column of the periodic table on the right, starting with helium (He). The noble gases were once called the *inert* gases because it is virtually impossible to get them to react with other materials. Later we will explore why this is the case. For now, let's just consider a few of them. Helium is probably the most well known of the inert gases. Helium is what makes balloons float. But helium also has many other uses in industry. In industry you will see two noble gases in particular, helium and argon, stored as compressed gases and as cryogenic liquids. Helium has a boiling point of −452°F, and argon has a boiling point of −302°F. The only real dangers the noble gases pose are

1 IA	2 IIA	3 IIIB	4 IVB	5 VB	6 VIB	7 VIIB	8 VIII	9 VIII	10 VIII	11 IB	12 IIB	13 IIIA	14 IVA	15 VA	16 VIA	17 VIIA	18 VIIIA
hydrogen 1 H 1.01																	helium 2 He 4.00
lithium 3 Li 6.94	beryllium 4 Be 9.01											boron 5 B 10.81	carbon 6 C 12.01	nitrogen 7 N 14.01	oxygen 8 O 16.00	fluorine 9 F 19.00	neon 10 Ne 20.18
sodium 11 Na 22.99	magnesium 12 Mg 24.30											aluminum 13 Al 26.99	silicon 14 Si 28.09	phosphorus 15 P 30.97	sulfur 16 S 32.07	chlorine 17 Cl 35.45	argon 18 Ar 39.95
potassium 19 K 39.10	calcium 20 Ca 40.01	scandium 21 Sc 44.96	titanium 22 Ti 47.87	vanadium 23 V 50.94	chromium 24 Cr 52.00	manganese 25 Mn 54.94	iron 26 Fe 55.85	cobalt 27 Co 58.93	nickel 28 Ni 58.70	copper 29 Cu 63.55	zinc 30 Zn 65.39	gallium 31 Ga 69.72	germanium 32 Ge 72.61	arsenic 33 As 74.92	selenium 34 Se 78.96	bromine 35 Br 79.90	krypton 36 Kr 83.80
rubidium 37 Rb 85.47	strontium 38 Sr 87.62	yttrium 39 Y 88.91	zirconium 40 Zr 91.22	niobium 41 Nb 92.91	molybdenum 42 Mo 95.94	technetium 43 Tc [98]	ruthenium 44 Ru 101.07	rhodium 45 Rh 102.91	palladium 46 Pd 106.42	silver 47 Ag 107.87	cadmium 48 Cd 112.41	indium 49 In 114.82	tin 50 Sn 118.71	antimony 51 Sb 121.76	tellurium 52 Te 127.60	iodine 53 I 126.91	xenon 54 Xe 131.29
cesium 55 Cs 132.91	barium 56 Ba 137.33	lutetium 71 Lu 174.97	hafnium 72 Hf 178.49	tantalum 73 Ta 180.95	tungsten 74 W 183.84	rhenium 75 Re 186.21	osmium 76 Os 190.23	iridium 77 Ir 192.22	platinum 78 Pt 195.08	gold 79 Au 196.97	mercury 80 Hg 200.59	thallium 81 Tl 204.38	lead 82 Pb 207.2	bismuth 83 Bi 208.98	polonium 84 Po [209]	astatine 85 At [210]	radon 86 Rn [222]
francium 87 Fr [223]	radium 88 Ra [226]	lawrencium 103 Lr [262]	rutherfordium 104 Rf [261]	dubnium 105 Db [262]	seaborgium 106 Sg [263]	bohrium 107 Bh [264]	hassium 108 Hs [265]	meitnerium 109 Mt [266]	darmstadtium 110 Ds [269]	roentgenium 111 Rg [272]	copernicium 112 Cn [277]						

*Lanthanides														
lanthanum 57 La 138.91	cerium 58 Ce 140.12	praseodymium 59 Pr 140.91	neodymium 60 Nd 144.24	promethium 61 Pm [145]	samarium 62 Sm 150.36	europium 63 Eu 151.96	gadolinium 64 Gd 157.25	terbium 65 Tb 158.93	dysprosium 66 Dy 162.50	homium 67 Ho 164.93	erbium 68 Er 167.26	thulium 69 Tm 168.93	ytterbium 70 Yb 173.04	lutetium 71 Lu 174.97

**Actinides														
actinium 89 Ac [227]	thorium 90 Th 232.04	protactinium 91 Pa 231.04	uranium 92 U 238.03	neptunium 93 Np 237.05	plutonium 94 Pu [244]	americium 95 Am [243]	curium 96 Cm [247]	berkelium 97 Bk [247]	californium 98 Cf [251]	einsteinium 99 Es [252]	fermium 100 Fm [257	mendelevium 101 Md [258]	nobelium 102 No [259]	lawrencium 103 Lr [262]

Note: For radioactive elements, the atomic mass number of the common isotope is shown in parentheses; for thorium and uranium, the mass of the naturally occuring radioisotope is listed.

FIGURE 4-2 The periodic table of the elements. *Art by David Heskett.*

simple asphyxiation and cryogenic burns (when they are encountered as liquefied compressed gases)—although their containers may violently rupture if mishandled. The noble gases include helium, neon, argon, krypton, xenon, and radon (which is radioactive).

METALS

metal ■ An element found in the left two thirds of the periodic table below the stairs. Metals are good conductors of heat and electricity, are malleable and ductile, and are typically solids at room temperature.

The **metals** are found on the left side of the periodic table—below the stairs. Generally they are good conductors of heat and electricity, are malleable and ductile, and are solids at room temperature. Some metals are extremely reactive, and others can be extremely toxic at low concentrations. In simple terms, metals like to lose one or more electrons and form ionic bonds with nonmetals. This property affects how they behave and how we can detect them.

Alkali Metals

The alkali metals are the first column of the periodic table, the one that starts with hydrogen (H)—although lithium (Li) is the first true alkali metal. The alkali metals include lithium, sodium, potassium, rubidium, cesium, and francium. The alkali metals are extremely water reactive, as evidenced by lithium strips from batteries that you might encounter in a clandestine methamphetamine lab, or the chunk of sodium metal you may have heard about that was thrown into a toilet and blew it up. Larger quantities of alkali metals can be very dangerous in the presence of even the relatively small amount of water vapor contained in the air. When alkali metals react with water they create an extremely flammable gas and a corrosive solution by forming a metal hydroxide salt:

$$2Na^0(s) + 2H_2O(l) \rightarrow 2NaOH(aq) + H_2(g) + \text{heat}$$

This chemical equation means "Solid elemental sodium metal and liquid water REACT TO FORM an aqueous solution of sodium hydroxide with the evolution of hydrogen gas and heat." The numbers in front of the chemical symbols mean that two atoms of elemental sodium (the superscript zero indicates the elemental form) react with two molecules of water to form two molecules of sodium hydroxide and one molecule of hydrogen gas. The letters in parentheses indicate the solid (s), liquid (l), or gaseous (g) state.

BOX 4-1 HAZMAT HISTORY: SODIUM EXPLOSION IN NEWTON, MASSACHUSETTS (1993)

On October 25, 1993, eleven firefighters were injured—eight of whom were very seriously burned—fighting a fire at a metal processing facility in Newton, Massachusetts. An explosion occurred as they were putting water on a sodium metal fire at the facility. The fire occurred when employees improperly disposed of a dangerously large amount of residual sodium metal (approximately 100 pounds), which contacted water when it overflowed, causing a fire and an initial explosion. The facility used a single common enclosure to both wet-wash drums and equipment and burn off the excess sodium metal left inside the 55-gallon steel drums used to ship solid sodium metal—the half-inch heel left at the bottom of the drum, which normally weighs 2 to 5 pounds.

A second explosion occurred when firefighters attempted to extinguish the sodium metal fire with water. The firefighters were splattered with molten, burning sodium. Initial arriving companies were not informed that the fire involved water-reactive sodium metal. Therefore, they had no idea that firefighter protective equipment would be inadequate to fight this fire. Structural firefighter protective equipment does not protect against molten metals such as sodium, potassium, magnesium, and lithium metals. The manufacturer had large containers of sodium chloride (table salt) available on-site as the designated sodium fire extinguishing agent. Other extinguishing agents that may be used on sodium metal fires are dry soda ash, graphite, diatomaceous earth, and other inert class D extinguishing agents. This incident underscores the importance of recognizing the hazards an emergency poses—such as the chemical and its hazards—early on.

Because of their extreme reactivity, the most common way you will encounter alkali metals is as part of a salt. For example, sodium constitutes half of common table salt, known to chemists as sodium chloride. It is also half of sodium hydroxide, otherwise known as lye, the very corrosive alkaline solid in the equation.

Alkaline Earth Metals

The alkaline earth metals are the second column of the periodic table, starting with beryllium (Be). The alkaline earth metals include beryllium, magnesium, calcium, strontium, barium, and radium. You may be familiar with magnesium metal, a flammable solid often used in fireworks and as the component of road flares that makes them burn so brightly. You may even have fought a magnesium fire before—such as the stubborn engine fire of an old VW car. The alkaline earth metals are less reactive than the alkali metals; however, they are still reactive enough that you will usually encounter them as part of a salt as well.

Transition Metals

The transition metals are located in the middle of the periodic table, primarily columns 3 through 12, starting with scandium (Sc). There are too many transition metals to mention all of them, but they include elements such as iron and chromium, which are part of steel. They also include precious metals such as silver, gold, and platinum. Many of the transition metals are very toxic and are often encountered as environmental pollutants, examples of which are lead, thallium, and mercury. Mercury is the only metallic element that is a liquid at room temperature, although the relatively uncommon elements cesium, francium, and gallium have melting points only slightly above room temperature.

NONMETALS

Nonmetals are found at the top right-hand corner of the periodic table—above the stairs. Nonmetals are poor conductors of heat and electricity. Nonmetals may be solids, liquids, or gases at room temperature. It is important to note that the noble gases are not considered nonmetals, especially for our considerations—namely, reactivity. Carbon, one of the primary constituents of hydrocarbons, is a nonmetal. Nitrogen gas, which constitutes roughly 78% of the atmosphere, is also a nonmetal. Some nonmetals are flammable solids and pyrophoric, such as red phosphorus and white and yellow phosphorus, respectively. The reactivity of the nonmetals varies widely. The halogens and the chalcogens are two especially reactive families within the nonmetals.

nonmetal ▪ An element that is found in the top right-hand corner of the periodic table over the stairs. Nonmetals are poor conductors of heat and electricity and may be solids, liquids, or gases at room temperature.

Halogens

The halogens are the second-to-last column of the periodic table, starting with fluorine (F). The halogens include fluorine, chlorine, bromine, iodine, and astatine. Halogens are very strong oxidizers, and are extremely corrosive and toxic. The halogens are very common industrial chemicals (with the exception of astatine). They are not typically found in residential applications owing to their strong reactivity. Fluorine and chlorine are gases. Fluorine is a common reagent used in many industries including the plastics manufacturing and petroleum refining industries. Chlorine is widely used in wastewater treatment and water purification. Bromine, a volatile liquid, is used as a disinfectant. Iodine, which is a volatile solid, may be encountered in clandestine methamphetamine labs. You've probably noticed that as the halogens become bigger they go from gases (fluorine and chlorine) to liquid (bromine) to solid (iodine).

Chalcogens

The chalcogens are the third-to-last column of the periodic table, starting with oxygen (O). The chalcogens include oxygen, sulfur, selenium, and tellurium. Members of the chalcogen family are also oxidizers but significantly weaker oxidizers than the halogen family. Because oxygen is a principal component of water, the atmosphere, and many salts in the earth's crust, it is the most abundant element on earth. Oxygen is a gas, while sulfur, selenium, and tellurium are solids.

SOLVED EXERCISE 4-1

Complete the following table:

Name	Metal or Nonmetal	Family	Hazardous Characteristics
Chlorine			
Sulfur			
Magnesium			
Potassium			
Argon			

Solution:

Name	Metal or Nonmetal	Family	Hazardous Characteristics
Chlorine	Nonmetal	Halogen	Strong oxidizer; toxic and corrosive gas
Sulfur	Nonmetal	Chalcogen	Weakly oxidizing solid; burns in the presence of stronger oxidizers (such as oxygen and fluorine)
Magnesium	Metal	Alkaline earth	Flammable solid
Potassium	Metal	Alkali	Water-reactive and flammable solid
Argon	Nonmetal	Noble gas	Simple asphyxiating gas

Bonding

Atoms combine with other atoms through a process called bonding. Electrons are the key to the bonding process between atoms. Two primary types of bonds can be formed: ionic bonds and covalent bonds. In the real world any given bond is really a combination of both ionic and covalent bonding to various degrees. Any particular bond is classified according to its primary character. Although this discussion may seem like too much detail, it is important to air monitoring and sample identification.

THE OCTET RULE

Elements have different chemical reactivities and tend to bond with particular atoms, because atoms prefer to have full electron shells. In a simple model, electron shells are the spaces around the atoms where electrons spend their time. Each electron shell has a certain number of slots that can be filled with electrons. The innermost slot, or electron shell, has space for two electrons, and the second shell has space for eight electrons. This number is the origin of the term "octet rule." (There is also a "duet rule" for the two electrons of the innermost shell for hydrogen and helium.)

The most stable configuration for all atoms is the one in which the outermost electron shell is completely filled. This is naturally the case for the noble gases and is the reason for their extraordinary stability. When other atoms reach this state—a completely full outermost electron shell—they become stable. The best examples of this concept are the extremely reactive elements chlorine and sodium. Chlorine needs to gain one electron to complete its outermost shell (and become like argon), while sodium needs to lose an electron (to become like neon) to fulfill the octet rule. When atoms of these two elements exchange electrons and form Na^+ and Cl^-, or table salt, they become extremely nonreactive and inert. If you reorganize the periodic table into a "periodic cylinder," this concept becomes more intuitive (Figure 4-3). Atoms try to find the shortest path to gaining a full outer shell of electrons. There are two ways to accomplish this: through ionic bonding,

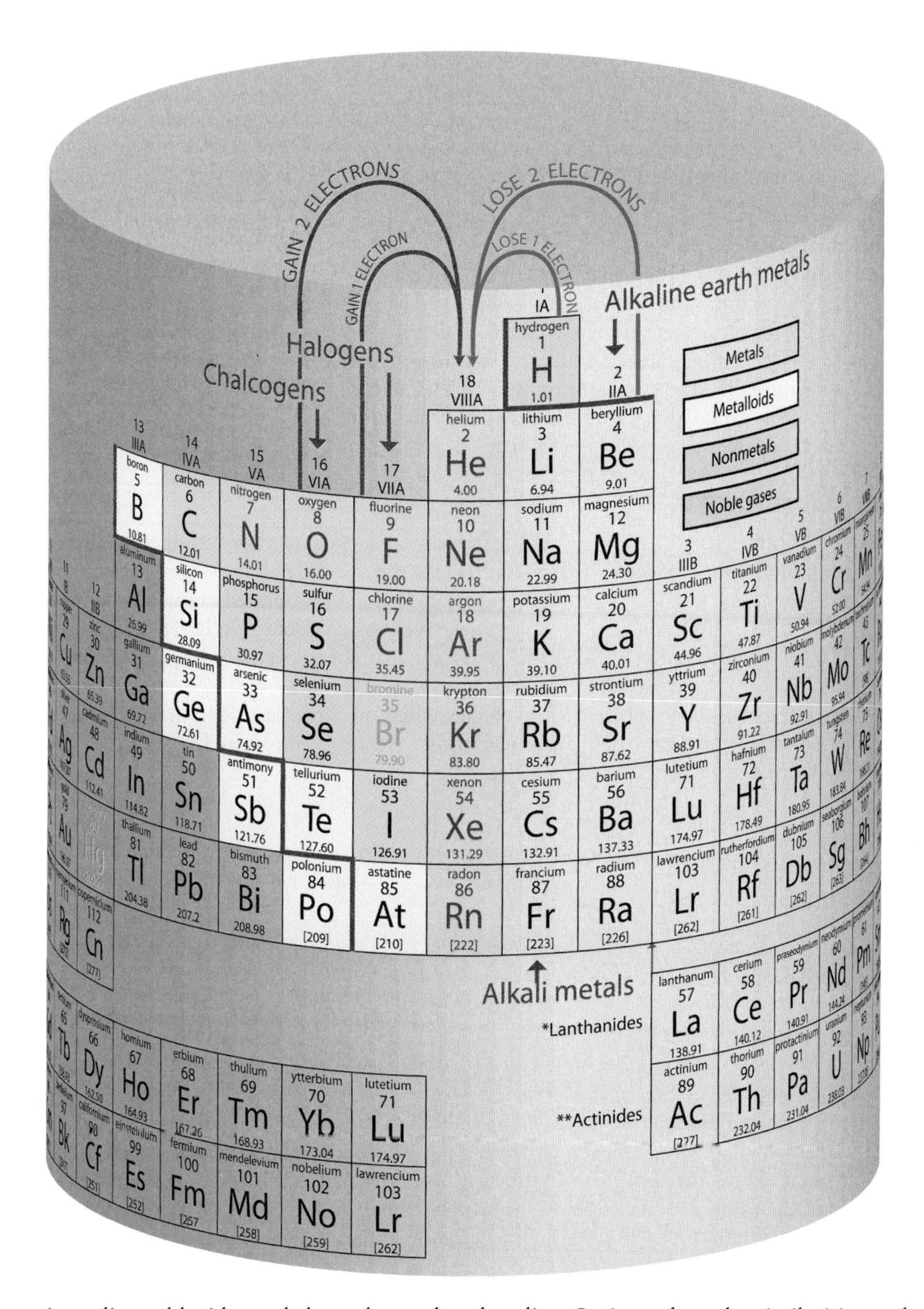

FIGURE 4-3 The periodic cylinder of the elements. When the periodic table is folded in such a way that the noble gases and the alkaline earth metals are brought together, an additional insight into chemical reactivity can be gained. The elements to the left of the noble gases want to gain electrons, while the elements to the right of the noble gases want to lose electrons, to achieve the stable electron configuration of a noble gas. The number of electrons they want to gain or lose is indicated by their proximity to the noble gases. Elements directly next to the noble gases will gain or lose one electron, while elements two rows away from the noble gases will gain or lose two electrons, and so on. *Art by David Heskett.*

as in sodium chloride, and through covalent bonding. Let's explore the similarities and differences between ionic and covalent bonds and learn how to use this knowledge to advantage at hazardous materials incidents.

ionic bond ■ The electrostatic force of attraction between oppositely charged ions that binds them. An example of a compound with ionic bonding is sodium chloride.

IONIC BONDS

One way atoms can achieve a full outer electron shell, thereby satisfying the octet rule, is to gain or lose one or more electrons. **Ionic bonds** are formed when two or more atoms, typically a metal and a nonmetal, combine by transferring one or more electrons between them. This creates one positively charged ion, called a **cation**, and one negatively charged

cation ■ A positively charged atom or group of atoms.

anion ▪ A negatively charged atom or group of atoms.

ion, called an **anion**. The positively charged atom and the negatively charged atom are held together by the attraction of their opposite charges, similar to the attraction between the opposite poles of a magnet. Materials held together by at least one ionic bond are called *salts* and are very often water soluble. Purely ionic compounds, such as sodium chloride (Na^+ and Cl^-) cannot be identified using FTIR or Raman spectroscopy.

Polyatomic Ions

polyatomic ion ▪ A positively or negatively charged group of two or more atoms covalently bonded together.

Polyatomic ions are composed of two or more nonmetal atoms covalently bonded together. Polyatomic ions are sometimes incorrectly referred to as *complex* ions. Technically, complex ions contain a transition metal covalently bonded to one or more nonmetals. Generally, salts containing polyatomic ions are amenable to sample identification using FTIR and Raman instrumentation such as the HazMatID and AhuraFD, since they contain covalent bonds. Table 4-2 lists commonly encountered ions (monatomic and polyatomic ions).

COVALENT BONDS

covalent bond ▪ A shared pair of electrons between two atoms that binds them. Covalent bonds can be seen by Raman and FTIR spectrometers.

Covalent bonds are formed when two or more atoms, typically two nonmetals, combine by sharing one or more electrons. This means that the shared electrons spend time around each of the two atoms that are bonded together. Covalently bonded compounds are also called *molecules*. Materials held together exclusively by covalent bonds are often referred to as *nonsalts*. Covalently bonded materials can for the most part be identified using Raman and FTIR spectroscopy.

Covalent bonds are represented by dashes when the structural formulas of molecules are written. Each dash represents two electrons that are being shared by the two atoms covalently bonded together, as shown here.

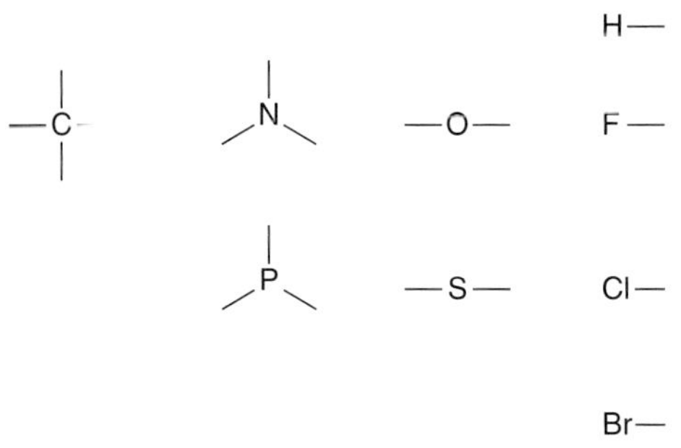

H—

—C—　　N　　—O—　　F—

P　　—S—　　Cl—

Br—

single bond ▪ A chemical bond in which a single pair of electrons is shared by two atoms.

Each of the halogens needs to gain one electron to have a complete outer electron shell, which means it needs to form a **single bond**—represented by one dash in the figure. This means that each halogen atom contributes one electron to the shared bond, and the atom to which it is bonding contributes one electron to the shared bond. Remember, two electrons form a single covalent bond. This is referred to as a shared electron pair.

double bond ▪ A chemical bond in which two pairs of electrons are shared by two atoms in a molecule.

Oxygen and sulfur, members of the chalcogen family, are one column to the left of the halogens. This means that each atom of these elements needs to gain two electrons to have a complete outer electron shell. Therefore each atom needs to form two single bonds, or one **double bond**. Double bonds are made up of four electrons. Double bonds contain two shared electron pairs. A double bond contains more energy than a single bond and is often more reactive. Elemental oxygen, or oxygen gas, contains a double bond between the two oxygen atoms, which gives each a complete outer shell of electrons:

O=O

TABLE 4-2 Monatomic Versus Polyatomic Ions

MONATOMIC IONS (GENERALLY NOT SEEN BY FTIR AND RAMAN SPECTROSCOPY)		POLYATOMIC IONS (GENERALLY SEEN BY FTIR AND RAMAN SPECTROSCOPY)	
CATIONS (POSITIVELY CHARGED)	**ANIONS (NEGATIVELY CHARGED)**	**CATIONS (POSITIVELY CHARGED)**	**ANIONS (NEGATIVELY CHARGED)**
Aluminum (Al^{3+})	Bromide (Br^-)	Ammonium (NH_4^+)	Acetate (CH_3COO^-)
Barium (Ba^{2+})	Chloride (Cl^-)	Mercurous (Hg_2^{2+})	Arsenate (AsO_4^{3-})
Calcium (Ca^{2+})	Fluoride (F^-)	Mercury (I) (Hg_2^{2+})	Azide (N_3^-)
Cesium (Cs^+)	Hydride (H^-)	Pyridinium ($C_5H_5NH^+$)	Benzoate ($C_6H_5COO^-$)
Chromium (II) (Cr^{2+})	Iodide (I^-)		Bicarbonate (HCO_3^-)
Chromium (III) (Cr^{3+})	Nitride (N^{3-})		Bisulfide (SH^-)
Copper (I) (Cu^+)	Oxide (O^{2-})		Bisulfate (HSO_4^-)
Copper (II) (Cu^{2+})	Phosphide (P^{3-})		Bisulfite (HSO_3^-)
Ferrous (Fe^{2+})	Selenide (Se^{2-})		Borate (BO_3^{3-})
Ferric (Fe^{3+})	Sulfide (S^{2-})		Carbide (C_2^{2-})
Hydrogen (H^+)	Telluride (Te^{2-})		Carbonate (CO_3^{2-})
Iron (II) (Fe^{2+})			Citrate ($C_3H_5O(COO)_3^{3-}$)
Iron (III) (Fe^{3+})			Chlorate (ClO_3^-)
Lead (Pb^{2+})			Chromate (CrO_4^{2-})
Lithium (Li^+)			Cyanamide (CN_2^{2-})
Magnesium (Mg^{2+})			Cyanide (CN^-)
Mercuric (Hg^{2+})			Diborate ($B_2O_5^{4-}$)
Mercury (II) (Hg^{2+})			Dichromate ($Cr_2O_7^{2-}$)
Potassium (K^+)			Formate ($HCOO^-$)
Silver (Ag^+)			Hexafluoroaluminate (AlF_6^{2-})
Sodium (Na^+)			Hexafluoroarsenate (AsF_6^{2-})
Stannic (Sn^{4+})			Hexafluorosilicate (SiF_6^{2-})
Stannous (Sn^{2+})			Hydrosulfide (SH^-)
Tin (II) (Sn^{2+})			Hydroxide (OH^-)
Tin (IV) (Sn^{4+})			Metabisulfite ($S_2O_5^{2-}$)
			Molybdate (MoO_4^{2-})
			Nitrate (NO_3^-)
			Nitrite (NO_2^-)
			Oxalate ($C_2O_4^{2-}$)
			Perchlorate (ClO_4^-)
			Peroxide (O_2^{2-})
			Phosphate (PO_4^{3-})
			Pyrophosphate ($P_2O_7^{4-}$)
			Silicate (SiO_3^{2-})
			Sulfate (SO_4^{2-})
			Sulfite (SO_3^{2-})
			Tetraborate ($B_4O_9^{6-}$)
			Triborate ($B_3O_7^{5-}$)

Nitrogen and phosphorus are yet another column to the left, and each atom of these elements needs to gain three electrons to have a complete outer shell. This means that each atom must form three bonds—three single bonds, a single bond and a double bond, or one **triple bond**. A triple bond is made up of six shared electrons. Triple bonds contain even more energy than double bonds and can be even more reactive (especially in the case of carbon-carbon triple bonds). Acetylene contains a reactive triple bond between the two carbon atoms. In contrast, elemental nitrogen, or nitrogen gas, contains a very stable triple bond between the two nitrogen atoms, since it gives each a complete outer shell of electrons.

triple bond ■ A chemical bond in which three pairs of electrons are shared by two atoms in a molecule.

$$N{\equiv}N$$

Carbon is found four columns to the left of the noble gases. Therefore, each atom of carbon requires four extra electrons to complete its outermost electron shell. This means carbon prefers to form four bonds with other nonmetals. It can form four single bonds, two double bonds, a triple bond and a single bond, or many other combinations, so long as the total number of shared electron pairs equals four.

Hydrogen is a very special element. It can lose an electron and act like a metal (forming the H^+ cation), or it can share an electron and act like a nonmetal. Hydrogen can therefore form a covalent single bond by sharing an electron with another atom. In this capacity it acts like a member of the halogen family. However, it is not an oxidizer—far from it. Elemental hydrogen is instead an excellent fuel. Like a diatomic halogen molecule, hydrogen consists of two hydrogen atoms bound together covalently through a single bond.

H—H F—F Cl—Cl Br—Br I—I

We will discuss covalently bonded molecules, especially the hydrocarbons and the hydrocarbon derivatives, in much greater detail later in the chapter.

Compounds

ionic compound ▪ A substance that has at least one ionic bond. Also called a salt.

covalent compound ▪ A substance made up of two or more different types of atoms that contains only covalent bonds. Also called a molecule. Covalent compounds usually can be identified using FTIR and Raman spectroscopy.

molecule ▪ Compounds which exclusively contain covalent bonds.

Compounds are substances made up of two or more different types of atoms. If the compound has at least one ionic bond in it, it is referred to as an **ionic compound** or salt. If the compound is made up entirely of covalent bonds, it is referred to as a **covalent compound** or **molecule.** The vast majority of substances are compounds. Therefore we will spend a great deal of time discussing the composition and properties of compounds and how to best handle them at hazardous materials releases.

RECOGNIZING CHEMICALS BY THEIR NAMES

Determining whether a chemical is a salt or a nonsalt is relatively easy based on the chemical name. If the chemical name starts with the name of a metal—an element found under the stairs—the compound is ionic and likely has the following characteristics:

- It is primarily solid but often is transported and used as a solution in liquid form.
- It has various degrees of water solubility.
- It is nonflammable, although it may produce flammable gases on contact with water.
- It may produce corrosive solutions on contact with water.
- It may produce toxic gases on contact with water.

The first part of the name of an ionic compound (salt) will be a cation. Cation names include any of the metals below the stairs (such as sodium, calcium, or iron), ammonium, and older metal names such as cuprous, cupric, ferrous, ferric, mercurous, mercuric, stannous, and stannic. The last part of the name of a salt will be an anion – a negatively charged ion. See Table 4-2 for a more complete listing of cations and anions.

Of course, there are some exceptions to this general rule. Several transition metals primarily tend to form covalent bonds, especially with carbonyls. Examples of these include chromium carbonyl, iron pentacarbonyl, iron oxide, osmium tetroxide, titanium dioxide, and uranium hexafluoride. In addition, some elements located next to the stairs—either immediately above or below the stairs—also primarily tend to form covalent bonds. Examples of these include aluminum trichloride, aluminum phosphide, tungsten hexafluoride, and germanium tetrahydride. Despite these exceptions, the vast majority of salts you will commonly encounter will adhere to the naming rule explained in the previous paragraph.

SOLVED EXERCISE 4-2

Complete the following table:

Name	Ionic or Covalent (Salt or Nonsalt)
Sodium hydroxide	
Propane	
Titanium dioxide	
Magnesium sulfate	
Methyl alcohol	

Solution:

Name	Ionic or Covalent (Salt or Nonsalt)
Sodium hydroxide	Ionic/salt
Propane	Covalent/nonsalt
Titanium dioxide	Exception (has predominantly covalent characteristics)
Magnesium sulfate	Ionic/salt
Methyl alcohol	Covalent/nonsalt

Conversely, if the name of a chemical does not start with that of a metal, it is a covalent compound and has the following characteristics:

- It is a solid, liquid, or gas.
- It is usually flammable or combustible.
- It has various degrees of water solubility.
- It has various degrees of chemical reactivity.
- It has various degrees of toxicity.

IONIC COMPOUNDS (SALTS)

Salts are ionically bonded compounds. Recall that ionic bond formation involves the transfer of an electron from one atom to another; the result is two oppositely charged ions. Ionic bonds are formed between metals and nonmetals. Let's take table salt as an example again. In this case the sodium atom (metal) gives up one of its electrons to the chlorine atom (nonmetal). The chemist might write this process the following way:

$$Na^0(s) + Cl^0(g) \rightarrow Na^+ \text{ and } Cl^- \rightarrow NaCl$$

Table salt, or sodium chloride, is an ionic compound. However, all ionic compounds are not as innocuous as table salt. Let's take a look at several salts that could cause trouble at a hazardous materials incident. There are six primary categories of ionic salts we will consider:

- binary
- metal oxide
- metal hydroxide
- metal cyanide
- metal peroxide
- metal oxysalt

Binary Salts

Binary salts are made up of a metal and a nonmetal. Sodium chloride, from the preceding example, is an example of a binary salt. There are five types of binary salts that can be extremely dangerous because they are water reactive and produce toxic or flammable gases and/or a corrosive solution (Table 4-3). For example, calcium carbide is used to produce acetylene gas for oxyacetylene welding. In this chemical reaction water is added to chunks of calcium carbide and acetylene gas is produced along with the corrosive calcium hydroxide and a significant amount of heat. In fact, if this reaction proceeds too fast, the heat generated is enough to ignite the acetylene gas.

$$CaC_2(s) + 2H_2O \rightarrow Ca(OH)_2 + C_2H_2(g) + \text{heat}$$

Metal sulfides react with acids to produce the extremely toxic gas hydrogen sulfide. An example is the reaction of sodium sulfide with muriatic acid:

$$Na_2S(s) + 2HCl(l) \rightarrow 2NaCl + H_2S(g) + \text{heat}$$

Some binary salts react with water to produce an acid. An example is the reaction of aluminum trichloride with water to produce hydrochloric acid and aluminum hydroxide:

$$AlCl_3(s) + 3H_2O \rightarrow 3HCl(g) + Al(OH)_3(s) + \text{heat}$$

Metal Oxide Salts

Metal oxide salts are made up of a metal and an oxygen atom. Examples of metal oxides include sodium oxide (Na_2O) and calcium oxide (CaO), which is also known as lime. When metal oxides react with water they produce a corrosive solution and heat. For example, sodium oxide reacts with water to produce sodium hydroxide and heat. Sodium hydroxide, also known as lye, is a strong base. Although their names may sound otherwise, metal oxide salts are not oxidizers.

$$K_2O(s) + H_2O \rightarrow 2KOH + \text{heat}$$

Metal Hydroxide Salts

Metal hydroxide salts are made up of a metal bound to a hydroxide anion (OH^-). The hydroxide anion consists of an oxygen atom covalently bonded to a hydrogen atom. The hydroxide anion has a net charge of -1 and forms an ionic bond with the metal, which is positively charged. The hydroxide anion is a polyatomic ion, since it is made up of two or more nonmetal atoms covalently bonded together. Metal hydroxide salts are an example of ionic compounds that contain covalent bonds. This information will become important when we discuss sample identification and spectroscopic techniques later in the book.

Metal hydroxide salts are very corrosive, since they are mostly strong bases owing to the presence of the hydroxide anion. Examples of metal hydroxides include sodium hydroxide (NaOH), potassium hydroxide (KOH), magnesium hydroxide ($Mg(OH)_2$), and calcium hydroxide ($Ca(OH)_2$), which is also known as slaked lime. Lithium hydroxide (LiOH) and magnesium hydroxide are weaker bases than potassium hydroxide and calcium hydroxide. In fact, magnesium hydroxide—which is used in the antacid milk of magnesia—is such a weak base that it is not even required to be placarded during transport.

Metal Cyanide Salts

Metal cyanide salts are made up of a metal bound to a cyanide anion (CN^-). The cyanide anion consists of a carbon atom covalently bonded to a nitrogen atom. The cyanide anion has a net charge of -1 and forms an ionic bond with the metal that is positively charged. The cyanide anion is also a polyatomic ion, another example of an ionic compound that contains a covalent bond.

When metal cyanide salts react with acids they liberate toxic hydrogen cyanide gas, which can easily be inhaled and absorbed by the body.

$$NaCN(s) + HCl(l) \rightarrow NaCl + HCN(g) + \text{heat}$$

The metal cyanide salts are extremely toxic, because the cyanide anion very effectively binds iron in the body and prevents oxygen from being delivered to the tissues by hemoglobin, and prevents the oxygen from being used in the tissues by the enzyme cytochrome oxidase. Two types of antidotes are available to treat cyanide poisoning (see Chapter 3 for more information).

Metal Peroxide Salts

Metal peroxide salts are made up of a metal ionically bonded to the peroxide radical ($O—O^{2-}$). The peroxide radical is made up of two oxygen atoms covalently bonded together. The peroxide radical has a net charge of −2 and forms two ionic bonds with alkali or alkaline earth metals. In the case of potassium peroxide (K_2O_2), the potassium–oxygen bonds are ionic, and the oxygen–oxygen bond is covalent. The metal peroxide salts are another example of a water-reactive salt. On contact with water, metal peroxides form a caustic solution, and oxygen gas is evolved, which will contribute to combustion. Metal peroxides are strong oxidizers. Examples of metal peroxide salts include sodium peroxide and potassium peroxide.

$$2K_2O_2 + 2H_2O \rightarrow 4KOH + O_2(g) + \text{heat}$$

Metal Oxysalts

Metal oxysalts are made up of a metal, a nonmetal, and one or more oxygen atoms. The nonmetal or transition metal and oxygen are covalently bonded and form a polyatomic anion. The names of metal oxysalts end in "-ate" (e.g., chlorate) or "-ite" (e.g., nitrite) and may have "hypo-" (e.g., hypochlorite) or "per-" (e.g., persulfate) as a prefix. The base state is referred to with the ending "-ate" (chlorate, ClO_3^-), and the oxysalt with the "-ite" ending contains one fewer oxygen than the base state (chlorite, ClO_2^-). The oxysalt containing one more oxygen than the base state has the "per-" prefix (perchlorate, ClO_4^-), while the oxysalt containing two fewer oxygens has the "hypo-" prefix and the "-ite" suffix (hypochlorite, ClO^-).

The metal oxysalts are generally strong oxidizers. They dissolve in water but are not water reactive. Examples of metal oxysalts include sodium hypochlorite, potassium nitrate, sodium chlorate, and potassium perchlorate. You may recognize some of these compounds as key components of explosives.

Metal Thiosulfates

Metal thiosulfates are made up of a metal bonded to a thiosulfate anion [$(S_2O_3)^{2-}$]. Once again, the thiosulfate anion is a polyatomic ion that contains covalent bonds. Metal thiosulfates react with acids to produce toxic sulfur dioxide gas.

$$Na_2S_2O_3 + 2HCl \rightarrow 2NaCl + S^0(s) + SO_2(g) + H_2O$$

Interestingly, thiosulfates are used to treat cyanide poisoning. In the body, thiosulfate recombines with the cyanide to form thiocyanate, which is excreted by the kidneys.

TABLE 4-3 Water-Reactive Salts

DANGEROUS WATER-REACTIVE SALTS	TOXIC AND/OR FLAMMABLE GAS PRODUCED ON CONTACT WITH WATER	CORROSIVE SOLUTION FORMED	EXAMPLE NAME	EXAMPLE FORMULA
Metal carbides	Acetylene	Yes	Calcium carbide	CaC_2
Metal chloride	Hydrogen chloride	Yes	Aluminum trichloride	$AlCl_3$
Metal hydrides	Hydrogen	Yes	Potassium hydride	KH
Metal nitrides	Ammonia	Yes	Aluminum nitride	AlN
Metal phosphides	Phosphine	Yes	Calcium phosphide	Ca_3P_2
Metal peroxide	Oxygen	Yes	Sodium peroxide	Na_2O_2

Metal Thiocyanates

Metal thiocyanates are made up of a metal bonded to a thiocyanate anion $(SCN)^-$. The thiocyanate anion is also a polyatomic ion that contains covalent bonds. Metal thiocyanates react with acids to produce toxic and flammable hydrogen cyanide gas.

COVALENT COMPOUNDS (MOLECULES)

When nonmetals combine with each other they form purely covalent compounds. Covalent compounds commonly exist as gases, liquids, and solids. As increasing numbers of atoms are covalently bonded these compounds tend to progress from gases (such as methane, which contains only five atoms), to liquids (such as pentane, which contains 17 atoms), to solids (such as icosane, which contains 62 atoms).

Inorganic Molecules

Inorganic molecules typically do not contain carbon atoms. However, there are some exceptions to this rule, including carbon monoxide, carbon dioxide, urea, carbon tetrachloride, and phosgene. Typical examples of inorganic covalent compounds are water, ammonia, sulfur dioxide, diborane, arsine, phosphine, and silane.

Nonmetal Hydrides Nonmetal hydrides are made up of a nonmetal and three or four hydrogen atoms covalently bonded together. They behave very differently from their

TABLE 4-4 Acid-Reactive Salts

DANGEROUS ACID-REACTIVE SALTS	TOXIC AND/OR FLAMMABLE GAS PRODUCED ON CONTACT WITH ACID	CORROSIVE SOLUTION FORMED	EXAMPLE NAME	EXAMPLE FORMULA
Metal cyanides	Hydrogen cyanide	No	Sodium cyanide	$NaCN$
Metal sulfides	Hydrogen sulfide	No	Sodium sulfide	Na_2S
Metal thiosulfates	Sulfur dioxide	No	Sodium thiosulfate	$Na_2S_2O_3$
Metal thiocyanates	Hydrogen cyanide	No	Potassium thiocyanate	$KSCN$

BOX 4-2 HAZMAT HISTORY: TEXAS CITY, TEXAS (1947)

On April 16, 1947, the SS *Grandcamp* was being loaded in the Texas City, Texas, harbor with 2300 tons of 32.5% ammonium nitrate fertilizer bound for European farmers. The ammonium nitrate was mixed with clay, rosin, and paraffin to improve its moisture resistance, and it was stored and transported at elevated temperatures, which increased its chemical reactivity. Longshoremen loading the vessel reported that the fertilizer bags were warm to the touch. At about 8 AM smoke and fire were reported in the hold of the ship, and at about 9 AM the captain ordered the hold to be steamed. Steaming the hold was a common firefighting method at the time that involved pumping high volumes of steam into the hold to smother the fire while theoretically preserving the cargo.

Ammonium nitrate is not considered an explosive according to transportation regulations. However, each ammonium nitrate molecule contains both an oxidizer and a fuel. In fact, ammonium nitrate can detonate under heat and pressure, the exact conditions that were created in the hold of the SS *Grandcamp* by a combination of the fire and fire suppression efforts. At 9:12 AM the ship detonated, killing almost six hundred people, leveling almost a thousand buildings, causing many additional residential and industrial fires, and even causing a 15-foot wave detectable 100 miles down the Texas shoreline. All but one member (who didn't respond initially) of the Texas City, Texas, volunteer fire department were killed in the initial blast. The *Grandcamp* explosion caused a fire and subsequent explosion on the SS *High Flyer*, which was carrying almost 1000 tons of ammonium nitrate and 1800 tons of sulfur. Total property loss was estimated at $100 million, which is almost $1 billion in today's currency. This disaster is still considered the worst industrial accident in the United States and led to significant improvements in disaster response planning in U.S. communities.

SOLVED EXERCISE 4-3

Complete the following table:

Name	Chemical Formula	Type of Salt	Primary Hazards
Sodium fluoride			
	$Ca(OH)_2$		
	CaO		
Potassium permanganate			
Magnesium peroxide			
	NaCN		

Solution:

Name	Chemical Formula	Type of Salt	Primary Hazards
Sodium fluoride	NaF	Binary salt	Various; toxicity
Calcium hydroxide	$Ca(OH)_2$	Hydroxide	Corrosiveness
Calcium oxide	CaO	Binary oxide	Corrosiveness
Potassium permanganate	$KMnO_4$	Oxysalt	Oxidizer
Magnesium peroxide	MgO_2	Peroxide	Oxidizer, corrosiveness
Sodium cyanide	NaCN	Cyanide	Toxicity

metal hydride counterparts. Nonmetal hydrides are almost universally pyrophoric, flammable, and toxic—except for ammonia, which is corrosive, flammable, and toxic. They do not evolve hydrogen on contact with water and are all colorless gases, not solids. These properties are the result of the covalent bonds between the nonmetal and the hydrogen atoms, in contrast to the ionic bond between the metal and hydrogen atom in metal hydrides. Examples of nonmetal hydrides include borane, ammonia, phosphine, arsine, and silane (Table 4-5). Nonmetal hydrides such as phosphine, arsine, and silane are often used in the semiconductor industry.

Binary Acids (Acid Gases) Binary acids are composed of a halogen atom covalently bonded to a hydrogen atom. The binary acids are all gases, although hydrogen fluoride is almost a liquid at room temperature owing to extensive hydrogen bonding interactions between HF molecules (see Box 5-3). Binary acids are typically quite soluble in water and form binary acid solutions. For example, a solution of hydrogen chloride in water is called hydrochloric acid, which is also known as muriatic acid. At this point the binary acid acts more like an ionically bonded material rather than a covalently bonded material,

TABLE 4-5 Nonmetal Hydrides and Their Properties

NONMETAL HYDRIDE	FORMULA	HAZARDS	APPEARANCE
Borane	BH_3	Pyrophoric, toxic, flammable	Colorless gas
Ammonia	NH_3	Corrosive, flammable, toxic	Colorless gas
Phosphine	PH_3	Pyrophoric, toxic, flammable	Colorless gas
Arsine	AsH_3	Pyrophoric, flammable, toxic	Colorless gas
Stibine	SbH_3	Unstable, flammable, toxic	Colorless gas
Silane	SiH_4	Pyrophoric, flammable, toxic	Colorless gas

TABLE 4-6 Binary Acids and Their Properties

BINARY ACID	FORMULA	HAZARDS	APPEARANCE
Hydrogen fluoride	HF	Corrosive, toxic, oxidizer	Colorless gas or fuming liquid
Hydrogen chloride	HCl	Corrosive	Colorless gas
Hydrogen bromide	HBr	Corrosive	Colorless gas
Hydrogen iodide	HI	Corrosive	Colorless gas

since the acid dissociates almost completely into hydrogen ions (H^+) and chloride ions (Cl^-). Therefore, binary acid solutions are typically strong acids. The anhydrous binary acids in their pure form are not corrosive. However, they are treated as corrosive, since water is ubiquitous in the environment and can easily enter the container and cause internal corrosion. In fact, this is one possible cause of binary acid releases.

Inorganic Acids (Oxyacids) Inorganic acids, also known as oxyacids, are composed of hydrogen and a polyatomic group of nonmetals including oxygen. Examples of oxyacids include nitric acid, sulfuric acid, and phosphoric acid. Inorganic acids are typically clear liquids with varying volatility. The exception is chromic acid, which is a red crystalline solid in its pure form—although it is often used in aqueous solution. Inorganic acids also dissociate almost completely and liberate hydrogen ions (H^+) in aqueous solution like the binary acids. Some inorganic acids are classified as weak—such as chlorous acid—and others are classified as strong—such as nitric acid and sulfuric acid. Inorganic acids are commonly used in industry. In fact, sulfuric acid is such an important commodity that the gross domestic product (GDP) of countries has been calculated based on its annual production.

Nonmetal Oxides Nonmetal oxides are composed of one or more nonmetal atoms bonded covalently with one or more oxygen atoms. Examples include carbon monoxide, carbon dioxide, sulfur dioxide, and nitrous oxide. Many of these nonmetal oxides form acids on contact with water. For example, carbon dioxide forms carbonic acid, and nitric oxide forms nitric acid.

Organic Molecules

Organic molecules contain the element carbon. Naming organic compounds may seem complicated at first, but it is a systematic method, consisting of two parts. The first part of the name generally gives the number of carbon atoms or halogen atoms, and the second

TABLE 4-7 Inorganic Acids (Oxyacids) and Their Properties

OXYACID	FORMULA	HAZARDS	APPEARANCE
Nitric acid	HNO_3	Corrosive, oxidizer, toxic	Clear liquid
Sulfuric acid	H_2SO_4	Corrosive, oxidizer	Clear liquid
Hypochlorous acid	$HClO$	Corrosive, oxidizer	Clear liquid
Chlorous acid	$HClO_2$	Corrosive, oxidizer	Clear liquid
Chloric acid	$HClO_3$	Corrosive, oxidizer	Clear liquid
Perchloric acid	$HClO_4$	Corrosive, oxidizer	Clear liquid
Chromic acid	H_2CrO_4	Corrosive, oxidizer, toxic	Red solid
Phosphoric acid	H_3PO_4	Corrosive	White solid or viscous liquid (above 42°C)

TABLE 4-8 Nonmetal Oxides and Their Properties

NONMETAL OXIDE	FORMULA	HAZARDS	APPEARANCE
Carbon monoxide	CO	Chemical asphyxiant	Colorless, odorless gas
Carbon dioxide	CO_2	Simple asphyxiant	Colorless gas
Sulfur dioxide	SO_2	Corrosive, toxic	Colorless gas
Nitric oxide	NO	Corrosive, oxidizer, toxic	Colorless gas
Nitrous oxide	N_2O	Oxidizer	Colorless gas
Nitrogen dioxide	NO_2	Corrosive, toxic, oxidizer	Volatile yellowish brown liquid with reddish brown vapors
Dihydrogen monoxide (water)	H_2O	Drowning	Clear liquid

part or the ending of the name indicates the type of organic compound, that is, it identifies the type of hydrocarbon derivative. Table 4-9 illustrates this naming convention. The size of a hydrocarbon is very important to predicting its behavior.

Another naming convention identifies the substituents on a molecule using prefixes that come from Greek and is illustrated in Table 4-10.

Hydrocarbons **Hydrocarbons** are organic compounds that contain both carbon and hydrogen. When a chemical is composed of hydrogen and carbon, it is a good fuel, so it most likely has a flashpoint and a flammable range and can be detected by flame ionization detectors (FIDs). This is one reason why it is important to understand the composition of chemicals. The name of the chemical can tell you a lot about how it will behave during a hazardous materials release and how it can be detected.

hydrocarbon ■ An organic compound that contains both carbon and hydrogen.

The simplest hydrocarbons contain only carbon and hydrogen. Hydrocarbon derivatives contain additional atoms—or groups of atoms—besides carbon and hydrogen. For example, alcohols and ethers are hydrocarbon derivatives that contain oxygen in addition to

TABLE 4-9 Hydrocarbon Naming Convention

NUMBER OF CARBONS	PREFIX	EXAMPLE NAME	EXAMPLE FORMULA	RADICAL (R)
1	Meth-	Methane	CH_4	Methyl ($\bullet CH_3$)
2	Eth-	Ethane	C_2H_6	Ethyl ($\bullet C_2H_5$)
3	Prop-	Propane	C_3H_8	Propyl ($\bullet C_3H_7$)
4	But-	Butane	C_4H_{10}	Butyl ($\bullet C_4H_9$)
5	Pent-	Pentane	C_5H_{12}	Pentyl ($\bullet C_5H_{11}$)
6	Hex-	Hexane	C_6H_{14}	Hexyl ($\bullet C_6H_{13}$)
7	Hept-	Heptane	C_7H_{16}	Heptyl ($\bullet C_7H_{15}$)
8	Oct-	Octane	C_8H_{18}	Octyl ($\bullet C_8H_{17}$)
9	Non-	Nonane	C_9H_{20}	Nonyl ($\bullet C_9H_{19}$)
10	Dec-	Decane	$C_{10}H_{22}$	Decyl ($\bullet C_{10}H_{21}$)
11	Undec-	Undecane	$C_{11}H_{24}$	Undecyl ($\bullet C_{11}H_{23}$)
12	Dodec-	Dodecane	$C_{12}H_{26}$	Dodecyl ($\bullet C_{12}H_{25}$)

TABLE 4-10	Hydrocarbon Numbering Convention		
NUMBER	PREFIX	EXAMPLE NAME	EXAMPLE FORMULA
1	Mono-	Carbon monoxide	CO
2	Di-	Carbon dioxide	CO_2
3	Tri-	Arsenic trichloride	$AsCl_3$
4	Tetra-	Carbon tetrachloride	CCl_4
5	Penta-	Pentachloroethane	Cl_5C_2H
6	Hexa-	Uranium hexafluoride	UF_6
7	Hepta-	Heptahydrate	$(H_2O)_7$
8	Octa-	Octabromodiphenyl ether	$C_{12}H_2Br_8O$
9	Nona- or Ennea-	Potassium nonahydridorhenate	K_2ReH_9
10	Deca-	Decafluorobutane	C_4F_{10}

carbon and hydrogen. In this section we will explore both simple hydrocarbons and more complex hydrocarbon derivatives.

The toxicity of hydrocarbons varies greatly depending on the chemical structure of the hydrocarbon and how it interacts with the human body. For example, the members of the aromatic family can be relatively benign, like xylene and toluene, or they can be extremely toxic, like benzene, which is a potent carcinogen. In the alcohol family, ethanol is imbibed, isopropyl alcohol is relatively nontoxic, while methanol is highly toxic.

Saturated Hydrocarbons (Alkanes) Each carbon atom can form a maximum of four covalent bonds. The members of the alkane class of hydrocarbons contain the maximum number of hydrogen atoms per carbon atom within the molecule. Thus, the carbon in methane (CH_4) is bonded to four hydrogens, and each carbon in ethane (C_2H_6) is bonded to three hydrogens in addition to being bonded to the other carbon atom.

```
   H            H H
   |            | |
 H-C-H        H-C-C-H
   |            | |
   H            H H
Methane        Ethane
```

Alkanes are thus said to be *saturated* hydrocarbons analogous to the saturated fats that you may be familiar with in the context of the healthy food debate. Saturated fats are typically fatty acids that contain the maximum number of hydrogens; in other words, no additional hydrogens can be added. In fact, hydrogenated vegetable oils have been chemically treated to add the maximum number of hydrogens to unsaturated fats.

The generic chemical formula for alkanes is $C_nH_{(2n+2)}$. Alkanes contain only single bonds. Examples of alkanes include the fuels methane, propane, and octane—for which the energy content of gasoline is named.

Unsaturated Hydrocarbons Unsaturated bonds—which for carbon consist of double and triple bonds—are less stable than saturated single bonds. Recall that in a double bond four electrons are shared by two atoms; in a single bond two electrons are shared; and in a triple bond six electrons are shared. Unsaturated bonds are less stable because the extra electrons repel each other and force the bonds away from one another, thereby distorting and straining the molecule. This allows oxygen to more readily react with—or oxidize—these bonds.

Alkenes Alkenes contain at least one double bond between two carbon atoms, so they are unsaturated hydrocarbons. This means that additional hydrogens can be added through chemical reactions (hydrogenation). The double bonds in these materials make these chemicals more reactive. Many monomers used in polymer production are alkenes, such as styrene used in the production of polystyrene.

The generic chemical formula for alkenes is C_nH_{2n}. Examples of alkenes include ethylene, which is used in the food industry to ripen fruit, and propylene, which is used to make polypropylene. Alkenes are prone to polymerization, which is a highly exothermic reaction. *Polymerization* is the formation of a larger molecule from multiple smaller ones, such as the manufacture of polystyrene from a great many styrene monomers.

Alkynes Alkynes contain at least one triple bond between two carbon atoms, so they also are unsaturated hydrocarbons, and they are more reactive than both alkanes and alkenes. The generic chemical formula for alkynes is $C_nH_{(2n-2)}$. An example of an alkyne is acetylene, which is used in oxyacetylene torches for welding. Acetylene, also known as ethyne, is extremely dangerous because of both its flammability and its reactivity. When acetylene is stored as a pure substance under even low pressure it will spontaneously polymerize and explode. This is why acetylene is stored in cylinders filled with a packing material and acetone. The acetylene dissolves in the acetone and thereby stabilizes the acetylene. This is why you should never use an oxyacetylene torch with the tank on its side, since the acetone will bleed out with the acetylene.

Aromatic Hydrocarbons Members of the aromatic family are uniquely stable owing to an alternating pattern of single and double bonds. The position of the double bonds can be described as alternating—or resonating—between adjacent pairs of carbon atoms:

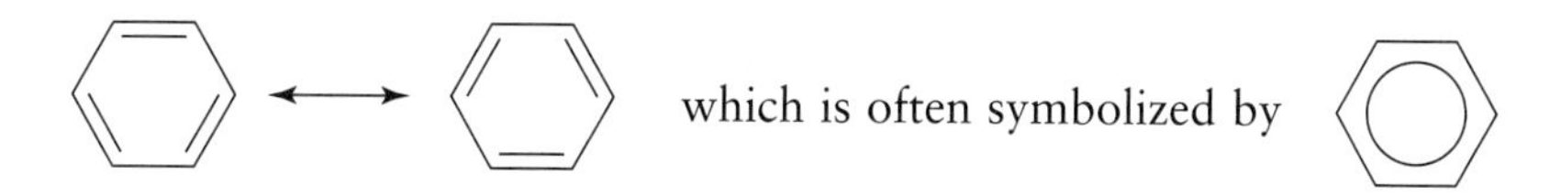

Molecules containing *resonance* bonds, another name for aromatic bonds, can therefore be described as having on average one and a half bonds between atoms. The alternating single and double bonds form a flat ring structure that renders the ring very stable. The most common aromatics are six-membered rings and are known by the acronym BTEX, which stands for benzene, toluene, ethyl benzene, and xylene. The simple members of the aromatic family tend to be relatively nonreactive and burn with a yellow flame and dark, sooty smoke with characteristic cobweblike strands. The distinctive smoke is due to incomplete combustion. The generic chemical formula for simple aromatics is $C_nH_{(2n-6)}$.

Hydrocarbon Derivatives Hydrocarbon derivatives are chemicals that contain a root composed of carbon and hydrogen—referred to as a **hydrocarbon radical** and abbreviated "R"—combined with other nonmetal atoms that collectively form the *functional group* (Table 4-9). The hydrocarbon root of methane (CH_4) is the methyl radical (CH_3•). Similarly, the hydrocarbon root of ethane (C_2H_6) is the ethyl radical (C_2H_5•). Notice that the formula for the radical contains one less hydrogen than the parent molecule. The dot in the formula for the radicals represents the unpaired electron that forms the covalent bond with a functional group which makes it a stable molecule. Radicals cannot exist on their own. You cannot buy a 55-gallon drum of "methyl radical." Radicals are highly reactive species that have a fleeting existence during the chemical reaction that creates any particular hydrocarbon derivative.

hydrocarbon radical ▪ A highly reactive hydrocarbon species that has a fleeting existence during the chemical reaction that creates any particular hydrocarbon derivative.

A functional group may consist of carbon and oxygen, oxygen and hydrogen, nitrogen and oxygen, or a wide range of other combinations. Chemists named them functional groups because it is these novel groups that give many organic compounds their unique chemical reactivity and make them very useful, or functional, for chemical synthesis in

SOLVED EXERCISE 4-4

Complete the following table:

Name	Chemical Formula	Molecular Structure	Family	Primary Hazards
Toluene				
	CH_3CHCH_2			
	C_2H_2			
Cyclohexane				

Solution:

Name	Chemical Formula	Molecular Structure	Family	Primary Hazards
Toluene	$C_6H_5CH_3$		Aromatics	Flammability
Propene	CH_3CHCH_2		Alkenes	Flammability, reactivity
Acetylene	C_2H_2	$H-C\equiv C-H$	Alkynes	Flammability, reactivity
Benzene	C_6H_6		Aromatics	Flammability, toxicity
Cyclohexane	C_6H_{12}		Alkanes	Flammability

industry. Let's examine some of the unique characteristics of the more common hydrocarbon derivatives.

Epoxides Epoxides are triangular "ring" structures made up of two carbon atoms bound to an oxygen atom. Epoxides are extremely toxic and extremely flammable. Ethylene oxide is an example of an epoxide that is used to sterilize hospital equipment and other heat-sensitive equipment that cannot be sterilized in other ways.

halogenated hydrocarbon ■ A hydrocarbon derivative which contains one or more halogen atoms. Liquid halogenated hydrocarbons tend to be heavier than water and sink; they have a specific gravity greater than 1.

Alkyl Halides (R—X) Alkyl halides, also known as **halogenated hydrocarbons**, consist of one or more halogen atoms (fluorine, chlorine, bromine, or iodine) bound to a hydrocarbon radical. This is a very large and diverse family of compounds with many common household and industrial uses. Some members of this family are flammable, such as methylchloride (chloromethane), but many are not, especially as more halogen atoms are added, such as in carbon tetrachloride (CCl_4). "Carbon tet" for short is a single carbon atom with four chlorine atoms attached to it. Alkyl halides are often toxic and volatile, tend not to be water soluble (immiscible or sparingly soluble), and sink in water. The halogen root in the name often ends in "-ide" (for example, chloride or bromide). Examples of alkyl halides include methyl chloride (a common solvent), carbon tetrachloride (a very toxic former fire extinguishing agent), tetrachloroethylene (brake cleaner), vinyl chloride (the monomer used to make PVC plastic), and methylene chloride (furniture stripper). Broad classes of halogenated hydrocarbons include the Freon family, which are used as refrigerants, and the halon family, which are used as fire extinguishing agents.

Ethers (R—O—R) Ethers consist of two hydrocarbon radicals with an oxygen atom in the middle. Ethers are very volatile and flammable—with a wide flammable range—and are not water soluble. Their names end in "ether." Examples of ethers include ethyl ether and methyl tertiary butyl ether (MTBE), a former gasoline additive that is environmentally damaging. Ethers, like many organic solvents, have an anesthetic effect on the body. In fact, diethyl ether was one of the first anesthetics used by doctors. When ethers are exposed to oxygen they tend to form potentially explosive organic peroxides when they dry out and crystallize. Therefore, ether containers—especially old ones—that have crystals within or around the cap are extremely dangerous and should not be handled owing to the risk of explosion. These crystals must be dissolved in water, alcohol, or acetone before they are handled or moved. Disposal should be performed by trained EOD personnel.

Organic Peroxides (R—O—O—R) Organic peroxides consist of two carbon radicals with two oxygen atoms in the middle. Their names end in "peroxide." Organic peroxides are extremely reactive and temperature sensitive, and are explosive when dry. They are particularly dangerous hydrocarbon derivatives because they are unstable. They have a **self-accelerating decomposition temperature (SADT)** at which they will rapidly, exothermically, and, often explosively, disintegrate. Therefore, organic peroxides typically have a **maximum safe storage temperature (MSST)** below which they must be transported and stored at all times to prevent catastrophic decomposition. Under uncontrolled conditions organic peroxides are explosive, and they should be treated as such during an emergency until proven otherwise.

self-accelerating decomposition temperature (SADT) ▪ The temperature at which organic peroxides will rapidly, exothermically, and, often explosively, disintegrate.

maximum safe storage temperature (MSST) ▪ The temperature below which organic peroxides must be transported and stored at all times to prevent catastrophic decomposition.

Organic peroxides are primarily used in industry in the production of plastics as polymerization catalysts. Organic peroxides are typically shipped with inhibitors and under refrigeration or other temperature-controlled conditions. When a hazardous materials release involves organic peroxides, it is extremely important to contact the manufacturer to determine what actions to take. It is especially important to determine how long the inhibitor will remain effective and what the SADT and MSST are. Examples of organic peroxides include methyl ethyl ketone peroxide (MEKP), a common industrial solvent; benzoyl peroxide, the active ingredient in acne medication; and diacetyl peroxide. Organic peroxides can also be used as explosives by terrorists. Triacetone triperoxide (TATP) and hexamethylene triperoxide diamine (HMTD) are organic peroxide–based explosives that may be encountered in WMD incidents or in illicit explosive labs.

Alcohols (R—OH) Alcohols consist of a single hydrocarbon radical attached to a hydroxyl group (which is an oxygen atom bonded to a hydrogen atom). Alcohols are flammable and water soluble, especially when the carbon hydrogen radical is short, as with methanol, ethanol, and isopropanol. Small alcohols tend to have a wide flammable range and may pose a significant flammability hazard during hazardous materials releases. Some alcohols are toxic, such as methanol. Methanol, which is converted to formaldehyde by the liver, can cause blindness. Methanol burns with a barely visible blue flame, while larger alcohols tend to burn with a primarily yellow flame and a blue base. Their names end in "-ol" or "alcohol."

Glycols, such as propylene glycol, contain two hydroxyl groups. These alcohols tend to be less flammable than simple alcohols, and some are toxic, such as ethylene glycol.

Glycols also tend to be more water soluble than simple alcohols. Ethylene glycol, which is toxic, is a common antifreeze that is currently being replaced with the much less toxic antifreeze propylene glycol. Ethylene glycol is metabolized by the liver into uric acid, which causes kidney failure and intense joint pain when uric acid crystals accumulate in the joints. Glycerol, also known as *glycerin*, is also an alcohol and contains three hydroxyl groups. It is a by-product of biodiesel manufacturing and is the precursor of nitroglycerin. Sometimes, small household biodiesel labs will dump their excess glycerol in an inappropriate location, which may cause neighbors to call 911 and trigger a hazmat response.

Carboxylic Acids (Organic Acids) R—COOH Carboxylic acids, also known as *organic acids,* consist of a hydrocarbon radical with a carboxylic acid group. The carboxylic acid group is a combination of a carbonyl group and a hydroxyl group. A carbonyl group consists of a carbon atom double bonded to an oxygen atom. As their name implies, carboxylic acids are corrosive, although they are weaker acids than the inorganic acids. Their names typically end in "-ic acid" or "-oic acid." Examples of carboxylic acids include formic acid, acetic acid, and citric acid. Acetic acid is found in vinegar, and formic acid is excreted by ants as a defense mechanism. Small organic acids are soluble in water, while larger organic acids are not.

Esters (R—(C=O)—O—R) Esters consist of two hydrocarbon radicals and a combination of a carbonyl group bound to an oxygen atom. Esters are flammable and burn with a yellow flame, which has a blue base when the ester contains a shorter-chain hydrocarbon radical. Esters tend to be sweet smelling and are the primary odors associated with bananas, oranges, and other fruits. Their names typically end in "-ate," although this ending can also be associated with organic acids and metal oxysalts. When esters contain double bonds, such as vinyl acetate and methyl acrylate, they can polymerize and are typically quite toxic.

Aldehydes (R—(C=O)—H) Aldehydes consist of a hydrocarbon radical bonded to a carbonyl group and a hydrogen attached to the carbon atom of the carbonyl group. Aldehydes are typically toxic and flammable, with a comparatively wide flammable range (especially the smaller members of the family). Aldehydes burn with a yellow flame, which has a blue base when the aldehyde contains a shorter-chain radical. Common examples of aldehydes include formaldehyde, glutaraldehyde, and acrolein. Aldehydes are commonly used as preservatives, and the shorter aldehydes are water soluble. Formalin is a solution of 37% formaldehyde in water. The names end in "-al" or "aldehyde."

Ketones (R—(C=O)—R) Ketones consist of two hydrocarbon radicals with a carbonyl group in the middle. They typically end in "-one" or "ketone." For example, methyl ethyl ketone (MEK) is the common name of 2-butanone. Ketones tend to be flammable or combustible liquids. They burn with a yellow flame, which has a blue base for the shorter ketones, such as acetone. Smaller ketones tend to be water soluble, owing to the presence of the carbonyl group. Ketones are common industrial solvents, such as acetone and MEK.

Amines (R—N)

Amines consist of one to three hydrocarbon radicals attached to a single nitrogen atom. Nitrogen almost always forms three bonds (or a triple bond):

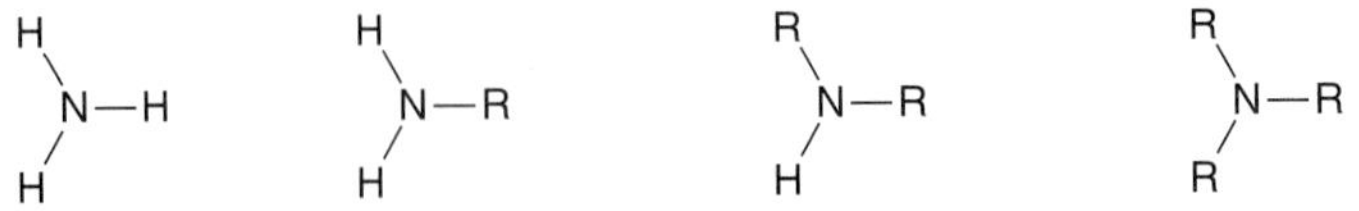

```
H              H              R              R
 \              \              \              \
  N—H            N—R            N—R            N—R
 /              /              /              /
H              H              H              R
Ammonia     Primary amine  Secondary amine  Tertiary amine
```

When only one carbon radical is attached to the nitrogen atom, two hydrogen atoms complete the bonding pattern, and the compound is known as a *primary* amine. When two carbon radicals are attached to the nitrogen, one additional hydrogen is attached to the nitrogen, and the compound is called a *secondary* amine. When three carbon radicals are attached to the nitrogen atom, there are no additional hydrogen atoms, and the compound is called a *tertiary* amine. Their names all end in "amine." Amines are flammable, with a wide flammable range for small-chain amines; burn with a yellow flame; are corrosive (alkaline); and are water soluble (again, the small amines). Amines also tend to be relatively toxic and have very disagreeable, foul odors. Indeed, one amine is aptly named putrescine and is in large part responsible for the characteristic odor of decaying flesh. Owing to the presence of the nitrogen atom, amines liberate toxic gases such as hydrogen cyanide and nitrogen oxide when they burn.

Examples of amines include aniline, methylamine, dimethylamine, and trimethylamine.

Amides Amides consist of a carbonyl group and an amine (nitrogen atom) bound to two hydrocarbon radicals. Small amides tend to be flammable, toxic, and soluble in water. Examples of amides include acrylamide, which is used in the plastics industry to make such common polymers as Plexiglas.

Amides also form the core structure of proteins, the workhorse polymer of cells. In proteins amino acids are connected together by amide bonds to form long chains.

Nitro Groups ($R—NO_2$) The nitro group consists of a single nitrogen atom bound to two oxygen atoms. Hydrocarbon derivatives that contain nitro groups are typically explosive and very dangerous. Nitro groups are strong oxidizers and, when combined with the hydrocarbon radical, which is a fuel, form compounds that burn extremely readily. In fact, the amount of energy released in the form of heat and gases leads to a supersonic shock wave. Supersonic combustion—or extremely rapid burning—is a definition of an explosive.

Examples of nitro compounds include nitromethane, which is a racing fuel; trinitrotoluene (TNT), which is a commercial explosive; and picric acid (trinitrophenol) which is an extremely unstable contact explosive when dried out. Nevertheless, nitro compounds are often used in commercial and military explosives owing to their stability. RDX, HMX, PETN, and many others are very stable to routine heat and shock encountered during handling. They are "cap-sensitive" explosives, which means they require a blasting cap to detonate.

Nitriles ($R—C{\equiv}N$) Nitriles are hydrocarbon derivatives that contain a carbon atom attached to a nitrogen atom via a very reactive triple bond. This is also called a *cyanide* group or *cyano* group. Nitriles tend to be reactive, flammable, toxic, and water soluble. The smallest nitrile is hydrogen cyanide, which is a potent pesticide and a common industrial reagent. Acetonitrile and acrylonitrile are common industrial solvents and reagents. All these compounds are toxic and flammable. Nitriles can polymerize owing to the triple bond and are therefore often used in the plastics industry, such as to produce the nitrile gloves you probably wear on medical calls and as part of your hazmat team PPE.

BOX 4-3 HAZMAT HISTORY: WORLD TRADE CENTER I, NEW YORK CITY (1993)

On February 26, 1993, a truck bomb was detonated in the parking structure below the north tower of the World Trade Center in New York City. The explosive device was a 1500-pound urea nitrate bomb enhanced with hydrogen gas that killed six people and injured more than a thousand. The objective of the attack was to topple the North Tower into the South Tower. Most of the injuries were caused by smoke inhalation and occurred during the evacuation of the WTC through the smoke-filled stairwells. The attack was perpetrated by Islamist terrorists.

The main charge of the device consisted of urea nitrate mixed with powdered aluminum, magnesium, and ferric oxide. The booster charge consisted of nitroglycerin, ammonium nitrate, dynamite, and smokeless powder. Three tanks of hydrogen gas were placed around the main charge to enhance the incendiary effects. This type of configuration is called a gas-enhanced explosive - a type of thermobaric bomb - and was similar to devices used in the 1983 Beirut barracks bombing.

Thiols (R—SH) Thiols are hydrocarbon derivatives that contain a sulfur atom bonded to a hydrogen atom. Thiols are typically very flammable and moderately toxic. Thiols have very distinctive, skunklike odors and are the additive that gives natural gas and propane their distinctive odor. Remember, methane, ethane, propane, and other short-chain hydrocarbons are odorless in their natural state. Thiols are also known as *mercaptans,* which means "mercury seizing." Thiols tend to be effective at binding certain toxic metals such as mercury and can be used as chelating agents. Examples of thiols include beta mercaptoethanol (BME), methanethiol, and ethanethiol.

Carbamates (R—N—CO_2—R) Carbamates can be considered a combination of an ester and an amine. Carbamates are toxic compounds and are common pesticides owing to their toxicity and to their favorable environmental attributes such as biodegradability. Carbamates attack the nervous system, and victims will exhibit signs and symptoms similar to those for organophosphate nerve agent exposure. Examples of carbamates include the pesticides Sevin and Propoxur.

Organophosphates Organophosphates, as their name implies, consist of a hydrocarbon radical bound to a phosphate group and an oxygen atom. However,

SOLVED EXERCISE 4-5

Complete the following table:

<table>
<tr><th>Name</th><th>Chemical Formula</th><th>Molecular Structure</th><th>Family</th><th>Primary Hazards</th></tr>
<tr><td>Nitromethane</td><td></td><td></td><td></td><td></td></tr>
<tr><td></td><td>CH_3OH</td><td></td><td></td><td></td></tr>
<tr><td>Ethylamine</td><td></td><td></td><td></td><td></td></tr>
<tr><td></td><td></td><td>H H
| |
H–C–O–C–H
| |
H H</td><td></td><td></td></tr>
<tr><td></td><td>$CH_3CH_2CH_2CH_2SH$</td><td></td><td></td><td></td></tr>
</table>

Solution:

<table>
<tr><th>Name</th><th>Chemical Formula</th><th>Molecular Structure</th><th>Family</th><th>Primary Hazards</th></tr>
<tr><td>Nitromethane</td><td>CH_3NO_2</td><td>H
| O⁻
H–C–N⁺
| ⩵O
H</td><td>Nitro</td><td>Explosiveness</td></tr>
<tr><td>Methanol</td><td>CH_3OH</td><td>H
|
H–C–H
|
O
|
H</td><td>Alcohol</td><td>Flammability, toxicity</td></tr>
<tr><td>Ethylamine</td><td>$CH_3CH_2NH_2$</td><td>H H H
| | /
H–C–C–N
| | \
H H H</td><td>Amine</td><td>Flammability, toxicity, corrosiveness</td></tr>
<tr><td>Dimethyl ether</td><td>CH_3OCH_3</td><td>H H
| |
H–C–O–C–H
| |
H H</td><td>Ether</td><td>Flammability, peroxide formation</td></tr>
<tr><td>Butanethiol</td><td>$CH_3CH_2CH_2CH_2SH$</td><td>／＼／＼SH</td><td>Thiol</td><td>Flammability, toxicity</td></tr>
</table>

organophosphates are often complex molecules that contain many other atoms such as nitrogen, fluorine, and sulfur in various arrangements. Organophosphates are very toxic pesticides that are typically not soluble in water. Examples of organophosphates include Dursban (chlorpyrifos), malathion, and parathion. Military nerve agents such as sarin and VX belong to the organophosphate family as well.

Isomers

Isomers can be considered identical twins of a chemical family (Table 4-11). Isomers contain the exact same atoms—just as identical twins have the same genetic makeup—but the atoms are arranged differently—just as identical twins have different personalities and

TABLE 4-11 Types of Isomers

TYPE OF ISOMER	DESCRIPTION	STRUCTURAL EXAMPLE
n- (normal)	Straight chain with no branching	*n*-Pentane
iso- (iso)	Single branched chain	isopentane
neo- (neo)	Double branched chain	H_3C, CH_3, C, H_3C, CH_3 neopentane
sec- (secondary)	Two carbon atoms connected to the carbon atom bearing the functional group	OH *sec*-Amyl alcohol
tert- (tertiary)	Three carbon atoms connected to the carbon atom bearing the functional group	OH *tert*-Butyl alcohol
o- (ortho)	A second functional group located next to the first on an aromatic ring (on the number 2 carbon)	CH_3 H_3C *o*-Xylene
m- (meta)	A second functional group located two carbons from the first on an aromatic ring (on the number 3 carbon)	CH_3 H_3C *m*-Xylene
p- (para)	A second functional group located across from the first (on the number 4 carbon)	CH_3 CH_3 *p*-Xylene

Note: The "cyclo-" prefix denotes a completely different chemical (and not an isomer), since the number of hydrogen atoms is different in the molecule.

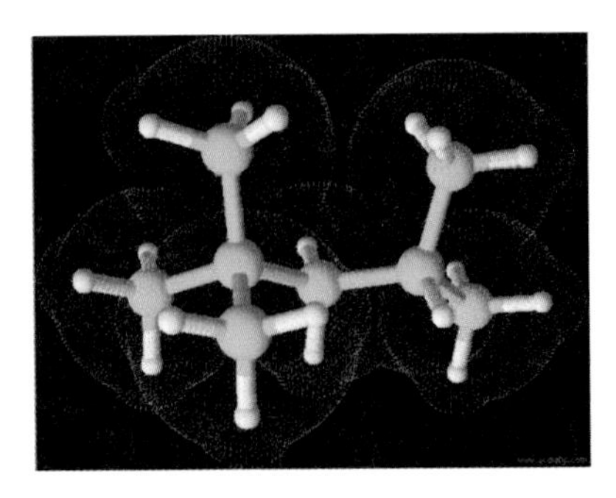

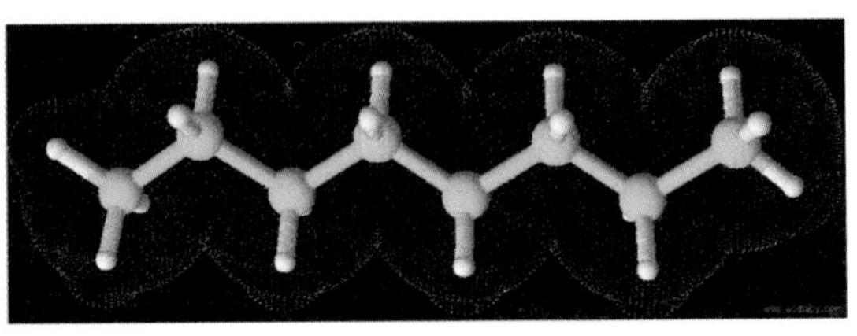

FIGURE 4-4 The molecular structures of isooctane (left) and octane (right) dictate the differing chemical and physical properties of these two isomers. Owing to its branching structure, isooctane has a flashpoint of 10°F, while octane has a flashpoint of 56°F, even though the molecules contain the exact same atoms—they are just arranged differently. *ACD/3D Viewer (Freeware), Version 12.01, Advanced Chemistry Development, Inc., Toronto, ON, Canada, www.acdlabs.com, 2012.*

behaviors. This structural difference can result in some drastically different chemical and physical properties. For example, let's consider octane, which is a significant component of gasoline. There are two primary forms of octane: isooctane, which contains a branch structure, and *n*-octane, which is the straight-chain isomer (Figure 4-4).

The flashpoint of *n*-octane is so high that the gasoline cannot be ignited in a normal internal combustion engine. Only isooctane is useful as an automotive fuel. Properties that are affected by isomer type include vapor pressure and toxicity as well. When we respond to hazardous materials releases it is important to consider if isomers of a given chemical exist, and if so, which isomers are involved in the release.

Mixtures

Pure substances are materials composed of a single chemical. However, you are just as likely to encounter mixtures at hazardous materials incidents. **Mixtures** are a combination of two or more chemicals that may be either homogenous or heterogeneous. Homogenous mixtures are evenly mixed and look like a single, pure material to the naked eye. You will have to use scene information such as placards, labels, and shipping papers, air monitoring, and sample identification clues to determine if you are dealing with a pure substance or a homogenous mixture. A heterogeneous mixture typically looks as if it is composed of two or more different materials. Examples of heterogeneous mixtures include slurries and biphasic liquids—a mixture with two distinct phases, such as oil-and-vinegar salad dressing.

mixture ■ A non-reacting combination of two or more chemicals. It may be either homogeneous or heterogeneous.

SOLUTIONS

A **solution** consists of a **solute** (typically a solid material) dissolved in a **solvent** (typically a liquid). A solution is a uniform, or homogenous, mixture of the two materials. There will not be two layers. A *saturated* solution may have excess solid at the bottom, but this indicates that solute has been added beyond its solubility limit in the solvent.

solution ■ A homogeneous mixture consisting of a solute (typically a solid material) dissolved in a solvent (typically a liquid).

solute ■ A substance that is dissolved in a solution.

solvent ■ A liquid capable of dissolving another material (solid, liquid, or gas) to form a mixture (a solution).

SLURRIES

A **slurry** is a nonuniform mixture of a solid and a liquid. The solid does not completely dissolve in the liquid and is essentially wetted by or suspended in the liquid. If a slurry is left to sit, the solid will eventually separate from the liquid by gravity (and either sink or float depending on its density). Examples of slurries include flour mixed with water, and wood pulp mixed with water. The flour and wood pulp are wetted by the water but do not dissolve and do not form a homogenous solution.

slurry ■ A nonuniform mixture of a solid and a liquid.

CONCENTRATION AND STRENGTH

The term **concentration** refers to the quantity of a solute dissolved in a solution. A concentrated solution has much solute dissolved in the solvent. Concentration is sometimes casually referred to as "strength"; a concentrated solution being referred to as strong, while a dilute solution is referred to as weak. However, **strength** more commonly refers to how readily an acid or a base gives up its proton or hydroxide thereby making the solution more corrosive. A strong acid almost completely ionizes in water (the protons dissociate from the acid), while a weak acid gives up significantly fewer protons (they remain bonded to the acid). An example of a strong acid is sulfuric acid and an example of a weak acid is acetic acid. An example of a strong base is sodium hydroxide and an example of a weak base is ammonia.

concentration ▪ The amount of a material in a mixture.

strength ▪ An expression of how readily an acid or a base dissociates in water.

CHAPTER **REVIEW**

Summary

Chemistry is the science that studies how elements and compounds behave and interact. Hazardous materials technicians who have a sound understanding of chemistry can predict how common hazardous materials may react. Understanding the difference between metals and nonmetals makes it possible for hazmat technicians to quickly interpret chemical and physical properties from the chemical name. Understanding the difference between covalent and ionic bonds makes it possible for the hazmat technician to make sound judgments when choosing detection and chemical identification equipment and interpreting their results. Knowing the characteristics of hydrocarbon derivatives makes it possible for the hazmat technician to quickly and accurately assess hazards versus risks at a hazardous materials release. Chemistry knowledge can therefore greatly facilitate risk-based responses.

Review Questions

1. What are the differences between a metal and a nonmetal?
2. Which hydrocarbon derivatives are most flammable?
3. Which hydrocarbon derivatives may be explosive?
4. Name two types of chemical reactions.
5. What are the differences between a salt and a nonsalt?

References

Centers for Disease Control and Prevention. (2007). *NIOSH Pocket Guide to Chemical Hazards.* Washington, DC: U.S. Government Printing Office.

Dufek, Richard. (2010) *Basic Chemistry for Emergency Responders, Student Manual.* Westfield, Indiana: Author.

Federal Emergency Management Agency (2009) *Chemistry for Emergency Response, Student Manual*, fourth edition. Washington DC: Author.

Fire, Frank (1986) *The Common Sense Approach to Hazardous Materials*, 2nd ed. Tulsa, Oklahoma: Fire Engineering Books and Videos.

Lide, David R. (2008) CRC Handbook of Chemistry and Physics, 89th edition, Boca Raton, Florida: CRC Press.

Meyer, Eugene. (2009). *Chemistry of Hazardous Materials* (5th ed.). Upper Saddle River, NJ: Pearson/Brady.

CHAPTER 5

Chemical and Physical Properties

Courtesy of Chris Weber, Dr. Hazmat, Inc.

KEY TERMS

acid, *p. 120*
autorefrigeration, *p. 126*
base, *p. 120*
boiling-liquid, expanding-vapor explosion (BLEVE), *p. 143*
boiling point, *p. 125*
catalyst, *p. 117*
caustic, *p. 123*
chemical property, *p. 123*
chemical reactivity, *p. 117*
combustion, *p. 118*
compatibility, *p. 117*
condensation point, *p. 125*
corrosive, *p. 120*
critical pressure, *p. 126*
critical temperature, *p. 126*
dust explosion, *p. 124*
endothermic reaction, *p. 117*
exothermic reaction, *p. 117*
expansion ratio, *p. 126*
fire point, *p. 138*
flammable range, *p. 138*
flashpoint, *p. 138*
freezing point, *p. 125*
gas, *p. 124*
heat capacity, *p. 125*
ignition temperature, *p. 141*
inhibitor, *p. 117*
liquid, *p. 124*
lower explosive limit (LEL), *p. 138*
melting point, *p. 125*
miscible, *p. 134*
nonpersistent, *p. 128*
persistent, *p. 128*
physical property, *p. 123*
polymerization, *p. 119*
pyrophoric, *p. 119*
reducing agent, *p. 118*
saponification, *p. 123*
solid, *p. 124*
specific gravity, *p. 134*
standard temperature and pressure (STP), *p. 125*
sublimation, *p. 130*
triple point, *p. 128*
upper explosive limit (UEL), *p. 138*
vapor density, *p. 130*
vapor pressure, *p. 128*
viscosity, *p. 124*
water solubility, *p. 133*

OBJECTIVES

After reading this chapter, the student should be able to:

- Apply the chemical and physical properties of a given hazardous material to a risk-based response at a hazardous materials release.
- Determine the chemical reactivity of a given hazardous material.

- Describe the significance of the pH scale.
- Name five strong acids and five strong bases.
- Describe the significance of the expansion ratio to liquefied compressed gas responses.
- Determine how a given hazardous material will behave on contact with air.
- Name the 13 lighter-than-air gases.
- Determine how a given hazardous material will behave on contact with water.
- Determine the flammability properties of a given hazardous material.
- Explain the mechanics of boiling-liquid, expanding-vapor explosions (BLEVEs).

ResourceCentral For additional review and practice tests, visit **www.bradybooks.com** and click on Resource Central to access book-specific resources for this text! To access Resource Central, follow directions on the Student Access Card provided with this text. If there is no card, go to **www.bradybooks.com** and follow the Resource Central link to Buy Access from there.

You are dispatched to a remote area of your county for a reported train derailment near a river. That particular line often carries cargo trains containing hazardous materials railcars. As you are responding the dispatcher notifies you that several railcars are leaking. The contents of one railcar are reportedly slowly migrating toward the river. A hopper car carrying solid material has overturned and spilled its load on the bridge. As you continue to respond toward the staging area you start to wonder how these chemicals are going to react with the air, the water, and with each other.

- What determines how a chemical behaves on contact with the air?
- What determines how a chemical behaves on contact with water?
- What determines which chemicals react with each other?
- What information tells how flammable a chemical is?
- What information tells how corrosive a chemical is?
- What information tells how reactive a chemical is?
- What information tells which state of matter a material is in and whether it can and/or will change states of matter during a response?
- What information tells if a chemical has the potential to create a plume, and if so how big will the plume be?

Let's see if we can get some answers to these questions and more.

Know your enemy. These words apply to hazardous materials response as much as they apply to war. The best way to understand your enemy—the hazardous materials or dangerous goods that have spilled or been mismanaged—is to understand its behavior. The first step in this process is to gain a fundamental understanding of what the different chemical and physical properties mean and then learn how to use that information to predict how the chemical will behave. In this chapter we explore how to describe the behavioral characteristics of chemicals using chemical and physical properties. We will use and reuse the concepts we learn in this chapter throughout the rest of this textbook in all facets of hazardous materials response.

The many different types of chemical hazards include the following:

- explosivity
- reactivity
- flammability
- radioactivity (see Chapter 3 for more information)
- toxicity (see Chapter 3 for more information)
- corrosivity

Energy-releasing hazards, such as explosivity, reactivity, flammability, and radioactivity, are especially dangerous, because problems involving these hazards can escalate to a lethal level in a very short period of time. Let's start our discussion of chemical and physical properties with energy-releasing hazards.

Reactivity

Chemical reactivity refers to the ability of one substance to interact and combine with another substance to form one or more different substances. This process is called a *chemical reaction*. Some examples of chemical reactions are combustion and polymerization. Chemical reactions can be accompanied by a release of energy, typically in the form of heat, and are sometimes accompanied by light, as in fire. When large amounts of chemicals are involved, or when the chemicals rapidly release a large amount of energy, chemical reactivity can be extremely dangerous—even explosive. Whenever there is a potential for large amounts of incompatible materials to come in contact with one another, a large isolation distance must be enforced.

chemical reactivity ■ The ability of one substance to interact and combine with another substance to form one or more different substances.

Compatibility refers to a lack of reactivity between materials. If two materials do not react with each other, they are considered to be compatible. Incompatible materials react with each other. Some materials react rapidly (a strong oxidizer and a good fuel during combustion), while other materials react slowly (iron and oxygen in air). Some reactions liberate heat (such as combustion) and are called **exothermic reactions**, while other reactions take heat from the environment (as when ammonium nitrate dissolves in water in a cold compress) and are called **endothermic reactions**.

compatibility ■ A lack of reactivity between materials.

exothermic reaction ■ A reaction that liberates heat to the environment.

endothermic reaction ■ A reaction that takes heat from the environment.

CATALYSTS

The rate of a chemical reaction depends on the chemical and physical properties of the materials involved, such as the temperature, amount, and concentration of the materials, the pressure of the materials, and the presence of any catalysts. **Catalysts** increase the rate of a chemical reaction without being consumed in the reaction. An example of a catalyst is the platinum metal that helps remove pollutants in your car's catalytic converter and forms the basis of the sensing element in the combustible gas indicator. The platinum catalyzes the combustion of residual smog components by lowering the amount of oxygen needed to burn them. In the polymer industry, methyl ethyl ketone peroxide (MEKP) catalyzes the curing of fiberglass-reinforced plastics.

catalyst ■ A substance that increases the rate of a chemical reaction without itself being consumed in the reaction.

INHIBITORS

In contrast with catalysts, **inhibitors** slow the rate of reaction or even prevent chemical reactions from occurring. Inhibitors are used to increase the stability of highly reactive chemicals during transport and storage. Different types of inhibitors include rust inhibitors, corrosion inhibitors, oxidation inhibitors, stabilizers, and polymerization inhibitors. For example, liquid organic peroxides are inhibited for transport and storage by diluting them to roughly 25% with a volatile alcohol or ketone. Inhibitors are included in shipments of plastic monomers to prevent runaway polymerization.

inhibitor ■ A substance or mixture of substances used to slow or stop one or more chemical reactions.

Acetylene gas is stored, transported, and used in cylinders containing a porous filling and either the inhibitor acetone or dimethylformamide (DMF), because at pressures over approximately 29 psi acetylene violently decomposes, and the cylinder would explode. Therefore, the acetylene cannot be shipped under greater pressure. This is also the reason why acetylene cylinders should not be used on their side, since the acetone would come out with the acetylene during use.

When responding to hazmat incidents involving polymerization hazards, be aware that inhibitors become less effective over time and with increased heat. It is very important to find out from the manufacturer when the inhibitor was added and how much time you have left to deal with the situation safely.

Organic peroxides are typically shipped with inhibitors and under refrigeration or other temperature-controlled conditions. Organic peroxides are heavily used in industry in the production of plastics as polymerization catalysts. Organic peroxides are extremely reactive and temperature sensitive, and when dry they are explosive. They are particularly dangerous hydrocarbon derivatives because they are unstable. They have a self-accelerating decomposition temperature (SADT) at which point they will rapidly, exothermically, and—often explosively—disintegrate. Therefore, organic peroxides typically have a maximum safe storage temperature (MSST) below which they must be transported and stored at all times to prevent catastrophic decomposition. Under uncontrolled conditions organic peroxides are explosive, and they should be treated as such during an emergency until proven otherwise (Box 5-2).

COMBUSTION

combustion ▪ A very rapid oxidation reduction reaction.

Combustion is a very rapid oxidation–reduction reaction. The oxidation-reduction reactions with which you're probably most familiar are fire and rusting. Fire is an extremely rapid rate of oxidation that releases a lot of energy quickly, in the form of heat and light. Rusting is a very slow rate of oxidation that produces neither a palpable amount of heat nor light. Toxic products of combustion are hazardous materials from which firefighters and other first responders must protect themselves using respiratory protection such as a self-contained breathing apparatus (SCBA).

Reducing Agents

reducing agent ▪ A fuel. Reducing agents donate electrons during chemical reactions, including combustion.

In simple terms, **reducing agents** are fuels. Chemically, reducing agents donate electrons during chemical reactions, including combustion. Combustion is the rapid exothermic reaction between a reducing agent (fuel) and an oxidizing agent (such as oxygen). In rusting, a metal—the reducing agent—slowly donates electrons to an oxidizer.

Oxidizing Agents

Oxidizing ability refers to a chemical's ability to take electrons from another substance. During combustion, oxidizers take electrons from the reducing agent—the fuel. The more vigorously something is burning, the more forcibly the oxidizer is taking the electrons

SOLVED EXERCISE 5-1

A tanker carrying methyl ethyl ketone peroxide (MEKP) has overturned on the interstate and is leaking. What chemical and physical properties are you most concerned about and why?

Solution: MEKP is an organic peroxide. Whenever the name of a chemical ends in "peroxide," think oxidizer. And worse, if it is an organic peroxide—as identified by the "methyl ethyl ketone" portion of the name—it also has a fuel component to go along with the oxidizing component, so think explosive. The most important pieces of information to know when faced with an organic peroxide emergency are the self-accelerating decomposition temperature (SADT) and the maximum safe storage temperature (MSST). The SADT for MEKP is 130°F, and it should be listed in the MSDS or will be available from the manufacturer if you use the emergency response telephone number. You must ensure the temperature of the MEKP never reaches 130°F!

TABLE 5-1	Classification of Oxidizers According to NFPA 400	
CLASSIFICATION	**DEFINITION**	**EXAMPLES**
Class 1 Oxidizer	A material that only slightly increases the rate of combustion and does not cause spontaneous ignition of fuels.	Hydrogen peroxide (8% to 27.5% by weight), nitric acid (40% concentration or less), potassium nitrate, sodium nitrite, potassium dichromate
Class 2 Oxidizer	A material that moderately increases the rate of combustion and may cause spontaneous ignition of fuels.	Calcium chlorate, calcium hypochlorite (50% or less by weight), hydrogen peroxide (27.5% to 52% by weight), nitric acid (concentration greater than 40% but less than 86%), potassium permanganate, sodium peroxide
Class 3 Oxidizer	A material that significantly increases the rate of combustion and will decompose vigorously on contact with a fuel or heat.	Hydrogen peroxide (52% to 91% by weight), nitric acid, fuming (concentration greater than 86%), potassium chlorate
Class 4 Oxidizer	A material that severely increases the rate of combustion and causes spontaneous ignition of fuels. These materials may explode if contaminated or exposed to heat, shock, or friction.	Ammonium perchlorate (particle size greater than 15 microns), hydrogen peroxide (greater than 91% by weight), perchloric acid solutions (greater than 72.5% by weight), tetranitromethane

from the fuel. Examples of oxidizers include oxygen, the halogens such as chlorine and fluorine, peroxides, and oxidizing acids such as sulfuric acid and nitric acid.

Keep in mind that oxidizing power and reducing power are relative. Be acutely aware of this fact when handling or sorting chemicals. If a strong oxidizer is mixed with a weak oxidizer, the weak oxidizer becomes the reducing agent to the strong oxidizer, and a highly exothermic reaction can occur. This reaction may manifest itself as heat, a fire, or even an explosion. This is one reason the NFPA categorizes oxidizers into four classes in the NFPA 400 standard (Table 5-1).

Note that the oxidizer classification is strongly dependent on concentration. For example, 3% household hydrogen peroxide is not even considered an oxidizer by these standards, while hydrogen peroxide concentrations greater than 91% may explode spontaneously under certain conditions. What a world of difference! Always consult a technical expert when sorting and disposing of chemicals.

POLYMERIZATION

Polymerization is the formation of a larger molecule from two or more smaller ones. Typically, polymerization occurs when a relatively homogenous material is able to react with itself to form a continuous chain of a longer polymer. The plastics industry is based on the polymerization reaction. Polymerization is an exothermic reaction, which means it liberates heat during the reaction. Runaway polymerization is therefore extremely dangerous, since large amounts of heat can be generated in an enclosed environment, which may cause the container to explode. Containers of styrene, the monomer used in the polymerization reaction that produces polystyrene, have been involved in unanticipated chemical reactions and explosions during transportation. Materials that polymerize may be more likely to fail catastrophically because as the material polymerizes it may clog the pressure relief device (PRD) of the container, rendering the PRD inoperable.

polymerization ■ The generally exothermic chemical reaction in which monomer molecules react to form larger molecules (polymers).

AIR-REACTIVE MATERIALS

Air-reactive materials undergo a chemical reaction with one or more components of the air when exposed to the atmosphere. **Pyrophoric** materials are one type of air-reactive material. These materials will ignite if they come into contact with the air at temperatures below 130°F. Examples of pyrophoric materials include white and yellow phosphorus,

pyrophoric ■ Capable of igniting on exposure to air at temperatures below 130°F.

BOX 5-1 TACTICS: FIRES AT HAZMAT OCCUPANCIES

Fires at occupancies that use or store hazardous materials such as industrial facilities, agricultural supply stores, pool supply stores, and hardware stores are very dangerous operations for firefighters, the surrounding population, and the environment. It is important that incident commanders consider the pros and cons of letting the fire burn itself out at high temperatures, thereby consuming the hazardous materials, or fighting the fire aggressively.

There have been many incidents that were not handled properly and caused injury to firefighters and the public—such as aggressive firefighting at a chemical facility in Florida that caused a significant number of cancers in firefighters; other incidents have caused significant property damage—such as the ladder truck that was consumed by flammable runoff at a manufacturing facility in Waxahachie, Texas, in October 2011; and yet other incidents have led to massive groundwater contamination that has affected hundreds of thousands of residents for many years afterwards—such as fertilizer and pesticide groundwater contamination caused by aggressive firefighting of agricultural supply stores.

Before aggressively fighting fires at these occupancies several questions must be asked:

- What hazardous materials are currently located at the facility?
- Are the hazardous materials involved in the fire?
- What are the dangers the hazardous materials pose?
- How will the hazardous materials react when water is used during extinguishment?
- If the hazardous materials are flammable, will they create burning runoff that may interfere with firefighting efforts and endanger firefighters?
- How toxic will the firefighting runoff be to people and the environment?
- Where will uncontrolled runoff go?
- Can we control the runoff? If so, how?
- Does the runoff have the potential to contaminate groundwater?
- Does the runoff have the potential to enter sanitary and/or storm sewers?

This is a long list of questions, and the best time to get answers to most of them is during pre-incident planning, inspections, and facility tours. It is imperative that fire departments pre-incident plan firefighting operations at any facility that may contain hazardous materials and include provisions for various fire scenarios. It is a good idea to involve the local hazardous materials response team during the pre-incident planning phase as well as to have it on-site during firefighting operations to perform runoff control operations and air monitoring activities.

titanium chloride, trimethyl aluminum, and tributyl aluminum. Improperly stored air-reactive materials have caused many hazmat emergencies.

WATER-REACTIVE MATERIALS

Water-reactive materials interact with water and release energy or form dangerous products. Typically, such reactions are highly exothermic and can even be explosive. Alkali metals, for example, are water reactive. Even a small chunk of sodium metal when dropped into water produces a sizable explosion. Often, reactions of water-reactive materials will also generate flammable or toxic gases, such as hydrogen in the reaction between alkali metals and water, or hydrogen cyanide in the reaction between cyanide salts and water (as discussed in Chapter 4).

Corrosives and the pH Scale

Corrosives are materials that cause visible destruction of skin tissue or cause significant corrosion of metals, such as steel (Figure 5-1). There are two types of corrosive materials: **acids** and **bases.** The difference between them is measured by their pH. The pH scale is a way to determine the corrosive properties of a material that contains water. A liquid that contains water is called an *aqueous* solution. A solution with a pH of 7 is considered neutral, one with a pH below 7 is considered acidic, and one with a pH above 7 is considered basic.

corrosive ■ A liquid or solid that causes destruction of human skin at the site of contact or reacts with steel or aluminum at a specified rate.

acid ■ A corrosive material that has a pH below 7. A substance that corrodes steel or destroys tissue at the site of contact. A compound that releases hydrogen ions when dissolved in water. Examples of acids include battery acid, muriatic acid (hydrogen chloride), nitric acid, and vinegar.

base ■ A substance that releases hydroxide ions (OH^-) in water. Bases form solutions that have a pH above 7. Also known as alkali, alkaline material, or caustic.

BOX 5-2 HAZMAT HISTORY: BENZOYL PEROXIDE EXPLOSION (2003)

On January 2, 2003, a benzoyl peroxide explosion occurred at the Catalyst Systems Inc. production facility in Gnadenhutten, Ohio, lightly injuring one employee. Employees were drying 200 pounds of 75% benzoyl peroxide in granular form, attempting to produce 98% benzoyl peroxide using a vacuum dryer. The benzoyl peroxide explosively decomposed, almost leveling the production building.

The cause of the accident was most likely that the benzoyl peroxide reached the self-accelerating decomposition temperature (SADT), which is 68°C for a 1-pound bag of 98% benzoyl peroxide. The process was designed to heat the benzoyl peroxide to approximately 42°C using 82°C water circulated through the vacuum dryer jacket. The SADT of organic peroxide is dependent on the size and type of package, and was likely lower than 68°C for the 200-pound batch being tumbled in the dryer. Hotspots in the dryer, contamination, or static electricity may have initiated the explosive self-accelerating decomposition of the benzoyl peroxide.

Courtesy of the U.S. Chemical Safety and Hazard Investigation Board.

Benzoyl peroxide is used in a number of commercial applications including acne medication, dental resins, automotive body putty, food bleach, and some polymer manufacturing (such as silicone rubber and PVC).

The pH scale is based on the dissociation of a water molecule (H_2O) into a proton (H^+)—also called a hydrogen ion—and a hydroxide ion (OH^-):

$$H_2O \leftrightarrows H^+ + OH^-$$

The pH scale measures the concentration of protons in the solution according to the following equation:

$$pH = \log(1 / [H^+])$$

The more protons there are, the more acidic the solution is. The logarithmic nature of the pH scale means that a difference of one pH unit signifies a 10-fold change in the acidity of the solution. Thus, a solution with a pH of 6 is 10-fold more acidic than one with a pH of 7; and a solution with a pH of 1 is a million-fold more acidic than one with a pH of 7, since there is a difference of six pH units ($10 \times 10 \times 10 \times 10 \times 10 \times 10 = 1$ million) between them. Figure 5-2 illustrates the logarithmic nature of the pH scale. Thus, the corrosive properties of a liquid increase the farther its pH value is from 7.

FIGURE 5-1 A valve corroded by nitric acid. Hazardous materials releases are often caused by using incompatible fittings, transfer hoses, or containers. This damage was caused by a less than 30-second exposure to concentrated nitric acid. *Courtesy of Chris Weber, Dr. Hazmat, Inc.*

Although liquids with both high and with low pH values are corrosive, acids and bases have very different properties. Table 5-2 lists examples of both acids and bases. Acids are materials that contribute protons to a solution by dissociation. Dissociation means that a substance breaks apart into its constituent ions in solution. For example, muriatic acid, or hydrogen chloride (HCl), almost completely dissociates into protons and chloride ions in aqueous solution:

$$HCl \rightarrow H^+ + Cl^-$$

This is why muriatic acid is a strong acid; it increases the concentration of protons. Bases contribute hydroxide ions to a solution, which causes the hydroxide ions to combine with

FIGURE 5-2 The pH scale is logarithmic. A low pH indicates acidic conditions, whereas a high pH indicates alkaline conditions.

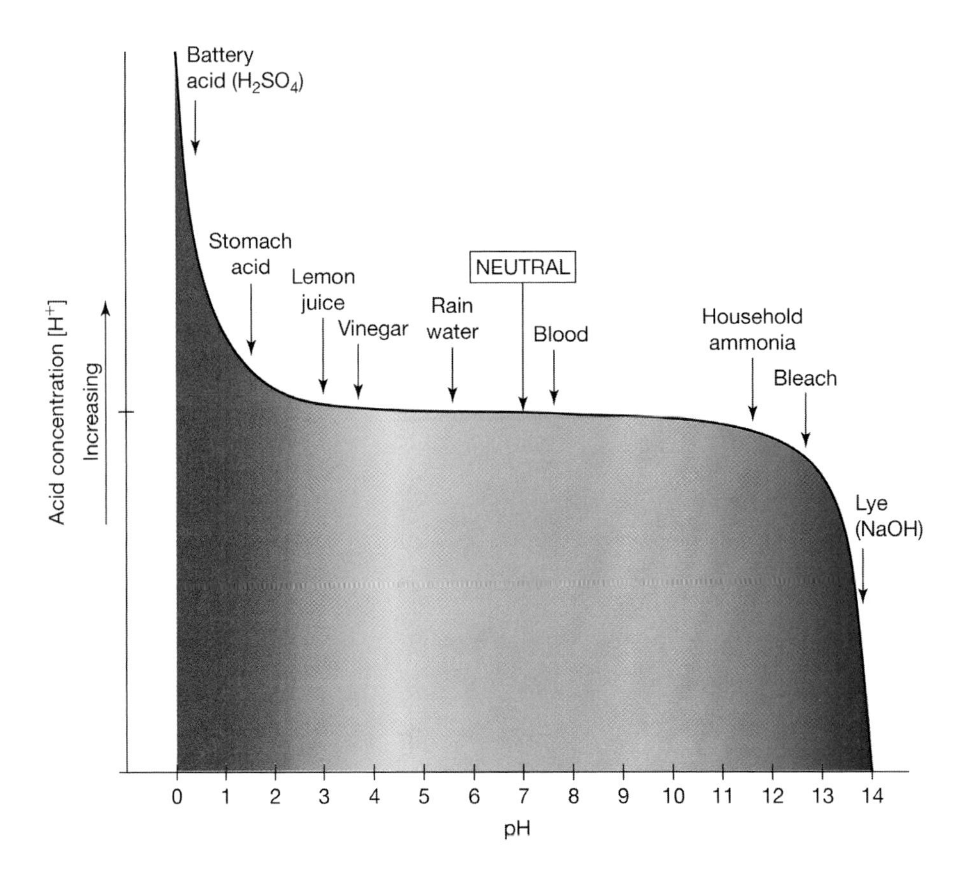

TABLE 5-2	Selected Acids and Bases
ACIDS	**BASES**
Acetic acid (also flammable)	Ammonia (also flammable)
Chromic acid (also an oxidizer)	Ammonium hydroxide
Citric acid	Bleach (also an oxidizer)
Formic acid	Calcium hydroxide
Hydrogen bromide	Lye
Hydrogen chloride	Potassium hydroxide
Hydrogen fluoride (also highly toxic)	Potassium hypochlorite (also an oxidizer)
Hydrogen iodide	Sodium hydroxide
Muriatic acid	Sodium hypochlorite (also an oxidizer)
Nitric acid (also an oxidizer)	
Perchloric acid (also an oxidizer)	
Phosphoric acid	
Sulfuric acid (also an oxidizer)	
Hydrochloric acid	
Hydrofluoric acid (also highly toxic)	

protons in the solution and form water, thereby decreasing the hydrogen ion concentration and increasing the pH:

$$H^+ + OH^- \rightarrow H_2O$$

For example, sodium hydroxide (NaOH), or lye, almost completely dissociates into a hydroxide ion and a sodium ion in aqueous solution:

$$NaOH \rightarrow Na^+ + OH^-$$

This is why lye is a strong base. Bases are also called *alkalis*, *alkaline* materials, or **caustics**.

Acid burns and caustic burns are not the same. Acid burns are typically painful at the site of contact, and acids can usually be washed off relatively easily with water. The most dangerous acid is hydrofluoric (HF) acid. This acid is not only corrosive, but it is also toxic and skin absorptive. The fluoride ion has a strong attraction to calcium. Since bones contain a lot of calcium, hydrofluoric acid is also called the "bone seeker." Hydrofluoric acid can also lead to heart dysrhythmias and was the cause of death of a law enforcement officer who responded to a hazmat incident involving a railcar carrying a product that liberated hydrofluoric acid on contact with water (see Box 3-2 Hazmat History: Fluorosulfonic Acid Release in Bridgman, Michigan). Some acids also have oxidizing potential. This means they can cause combustible and flammable materials to ignite on contact. Some examples are nitric acid, sulfuric acid, hydrofluoric acid, and perchloric acid.

Conversely, caustic burns may not initially be painful at the site of contact. Caustic burns feel slippery on the skin. This is because the hydroxide ion actually dissolves the skin and creates a soaplike substance. **Saponification** is the process of making soap. Soap has traditionally been made from animal fat and lye. Recall that lye is a strong base, and your skin is a fatty tissue. This is why bases are more difficult to decontaminate from the skin; they actually react with the skin and penetrate more deeply. As they penetrate they often inactivate nerve endings as well, deadening the sense of pain. It is very important to thoroughly decontaminate all chemical exposures, especially caustic exposures, using copious amounts of water.

caustic ■ A substance that releases hydroxide ions (OH^-) in water. Bases form solutions that have a pH above 7. Also known as alkali, alkaline material, or base.

saponification ■ The reaction of triglycerides with alkalis to form soap.

physical property ■ A change that does not result in an alteration of the chemical identity of a substance, such as vaporization, sublimation, melting, and boiling.

chemical property ■ A characteristic of a substance involving a molecular change in that substance. Examples of chemical properties include flashpoint, LEL, and ionization potential.

Chemical and Physical Properties

The chemical and physical properties of a material are the roadmap to its behavior. If you understand the behavior of the hazardous material, you can plan for a more positive outcome of the hazmat release. **Physical properties** are characteristics of a substance that do not involve a change in chemical identity. For example, boiling point, melting point, vapor pressure, and solubility are all physical properties. In contrast, **chemical properties** are characteristics of a substance that involve a chemical reaction that changes its chemical identity. Examples of chemical properties include flashpoint, lower explosive limit, upper explosive limit, and ignition temperature.

SOLVED EXERCISE 5-2

An intermediate bulk container (IBC) full of muriatic acid solution fell off of a semi trailer and is leaking at the loading dock of a small business. Approximately 100 gallons of acid leak out of the container before you can plug and patch it. Your crew takes a pH reading of the spilled liquid, and it has a pH of 2. You would like to dilute the acid and flush it into the sanitary sewer system. Before proceeding with your plan you call the supervisor at the wastewater treatment plant, who informs you that the plant can handle the liquid if its pH is raised to 6. Approximately how much water will it take to bring the 100 gallons of spilled acid to a pH of at least 6?

Solution: The pH of the spilled material needs to be raised from 2 to at least 6, which is four pH units. Four pH units means the acid needs to be diluted 10,000-fold ($10 \times 10 \times 10 \times 10 = 10{,}000$). There are 100 gallons of acid, so you will need 1,000,000 gallons of water (100 gallons of acid $\times$ 10,000 = 1,000,000 gallons of water). This is certainly a tall order, and probably not very practical for most jurisdictions. The better option would be to recover as much acid as possible and neutralize or dilute the residual.

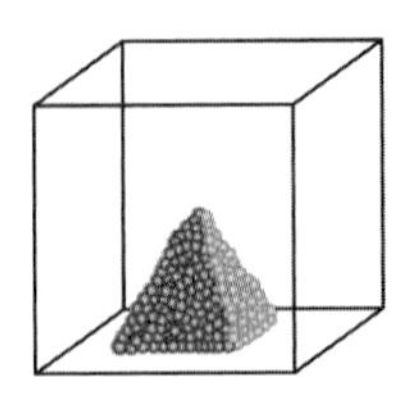
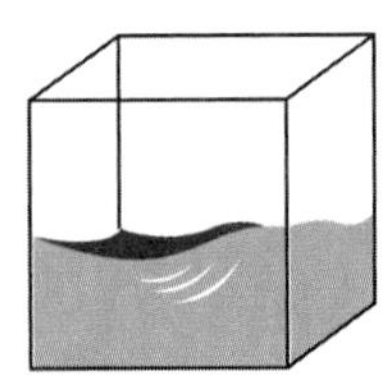
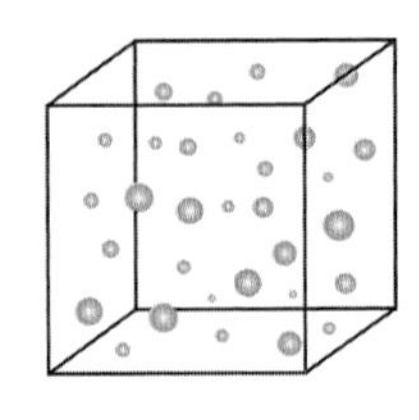

FIGURE 5-3 The three states of matter: solid, liquid, and gas. *Art by David Heskett.*

The different chemical and physical properties should be studied in conjunction with one another; they don't operate in a vacuum. We are going to examine what these properties can tell us about the behavior of a hazardous material in the real world. What is going to happen if the product contacts air? What will happen if the product contacts water? What is going to happen if the product escapes from its container? If you understand the chemical and physical properties of materials and know where to find this information for a particular hazardous material, you can make an informed decision and favorably affect the outcome of the incident—that is, you can plan a risk-based response.

STATES OF MATTER

The state of matter of a hazardous material plays a significant role in how it will be stored, handled, and remediated in case of a release. Hazardous materials are commonly found in one of three states of matter: **solid, liquid**, or **gas** (Figure 5-3). The containers and industrial process systems that handle the various states of matter vary dramatically, and the inherent dangers of the storage and handling systems are very different. Consequently, the ways of mitigating emergencies will also vary considerably. We will explore containers in detail in Chapter 10, and mitigation methods in detail in Chapter 11.

solid ▪ A material with a melting point above 68°F (20°C) at one atmosphere of pressure (101.3 kPa); a material that has a defined shape and volume.

liquid ▪ A material with a melting point of 68°F (20°C) or lower and a boiling point above 68°F (20°C) at one atmosphere of pressure (101.3 kPa). Liquids take on the shape of their container and possess a definite volume.

gas ▪ A material with a boiling point below 68°F (20°C) at one atmosphere. Matter that does not have a definite volume or shape and disperses in three dimensions when released from its container.

dust explosion ▪ The sudden and energetic combustion of dust particles suspended within a confined area.

viscosity ▪ The thickness of a liquid.

Solids

Solids are the least mobile form of matter, tend to stay in one location, and retain their shape and size. They can be considered primarily a one-dimensional hazard. However, solids can melt and are potentially volatile (through sublimation). In addition, finely divided solids can accumulate on surfaces and in the air and cause dust explosions.

Dust explosions cause a significant number of deaths and injuries (and property damage) every year. Equipment that produces a dust or fine powder has the potential to cause a dust explosion—often months or years after the equipment and process are initially started. This happens when combustible dust generated during production accumulates on surfaces, proper housekeeping is not performed to remove this dust, a sudden air current or building vibration dislodges the dust and it becomes suspended in the air, and a nearby ignition source ignites the finely suspended combustible particles, causing an explosion. Examples of materials that can cause dust explosions include sugar, sawdust, flour and other grains, and coal dust. A common rule of thumb is that an explosive atmosphere is likely present when the visibility is less than 5 feet in the area (29 CFR 1910.146). Needless to say, you should immediately evacuate when conditions indicate a dust explosion is possible.

Liquids

Liquids take the shape of their container, or once they are released, tend to flow in two dimensions and follow gravity downhill. Liquids can freeze and are often volatile (through evaporation). The vapors of liquids behave like gases.

Viscosity refers to the thickness of a liquid, which affects its ability to flow. For example, water has a lower viscosity than oil, since water is less thick. As the viscosity of the liquid increases, it tends to move less under gravity. For you as a hazmat technician it means that a very viscous liquid may behave more like a solid and stay put, while a less viscous liquid will flow more readily. This may be an advantage or a disadvantage depending on the situation. Viscous liquids are more difficult to pump but will also not flow downhill as quickly toward a sewer opening.

Gases

Gases are the most mobile of the three states of matter and disperse readily in three dimensions. Gases may be compressed or condensed into liquids with temperature and/or pressure changes. Depending on the scenario with which you are faced, the state of matter can be either an advantage or a disadvantage. For example, if you need to clean up a hazardous material and get it back in its container, it may be an advantage if it is a solid, since you can readily shovel it. However, if it is an extremely dangerous material, and there is a strong wind blowing, it may be an advantage if the material is a gas. The wind will disperse the gas and dilute it to nontoxic levels. Knowing the state of matter of a product is crucial to making sound decisions at the scene of a release.

Any chemical can be found in any of the three states of matter at the appropriate temperature and pressure. For example, when we say gasoline is a liquid, we mean it is a liquid at standard temperature and pressure. **Standard temperature and pressure (STP)** are considered to be 68°F and one atmosphere (760 mmHg) by NIST. At temperatures below its freezing point, gasoline is a solid, and above its boiling point, it is a gas. Each state of matter has its benefits and drawbacks as they relate to hazmat response.

standard temperature and pressure (STP) ■ Considered to be 68°F and one atmosphere (760 mmHg) by NIST.

Phase Transitions

A *phase transition* is the transformation of a chemical from one state of matter to another that occurs at a characteristic temperature and pressure. This is a physical change, since the chemical composition and identity remain the same. There are two primary phase changes: between solid and liquid, and between liquid and gas. Each of these phase changes can be approached from above or below, temperature wise. A substance can pass through a phase transition from higher to lower temperature, such as in freezing; or it can pass through it from lower to higher temperature, such as in melting.

Transitions between states of matter occur through temperature and pressure changes. The transition between a liquid and a solid state is called *freezing*. This process involves cooling. The transition between a solid and a liquid state is called *melting*. This process involves heating. The temperatures of the **freezing point** and the **melting point** are identical; the only difference between the two processes is whether the temperature is increasing or decreasing as the chemical passes through the transition point. The transition between a liquid and a gas state is called *boiling*. The **boiling point** is the temperature at which a liquid transitions to a gas at its most rapid rate. Keep in mind that at almost any temperature liquids transition to vapor through evaporation. The evaporation rate is based on the vapor pressure of the liquid, which we will discuss in more detail. The transition between a gas and a liquid state is called *condensation*. Once again, the temperatures of the boiling point and **condensation point** are identical, and the only difference is the temperature direction in which the substance passes through the transition point.

When you respond to hazmat incidents you must be aware of the possibility of phase transitions that might occur owing to changing environmental conditions or changes in temperature due to chemical reactivity. A situation may become significantly more challenging if a phase transition occurs unexpectedly.

freezing point ■ The temperature at which a liquid turns into a solid at atmospheric pressure.

melting point ■ The temperature at which a solid becomes a liquid.

boiling point ■ The temperature at which the vapor pressure of a substance equals the atmospheric pressure. The temperature at which a liquid evaporates at its most rapid rate.

condensation point ■ The temperature at which a gas or vapor turns into a liquid at its most rapid rate.

Heat Capacity and Heat Transfer

Heat capacity refers to the amount of energy a material can absorb in a given amount of time. Liquids tend to have a higher heat capacity and heat transfer rate than solids, although there are many exceptions, such as aluminum metal. Gases have the lowest heat capacity and heat transfer capability. Often, it is an advantage when materials have a high heat capacity and a fast heat transfer rate. For example, aluminum transfers heat quickly, and gasoline has a relatively high heat capacity. This is a distinct advantage when there is a tire fire under a full DOT 406 highway cargo tanker, as opposed to under an empty cargo tanker that contains only gasoline vapors. The empty cargo tanker will eventually explode in a dramatic fashion when the impinging flames melt the aluminum body of the tank, since the large amount of heat from the fire cannot effectively be dissipated by the small amount of liquid gasoline.

heat capacity ■ The amount of energy a material can absorb in a given amount of time.

Critical Temperature and Pressure

The state of matter of a chemical is determined by temperature and pressure. The temperature above which a gas cannot be liquefied is known as the **critical temperature**. The **critical pressure** is the pressure at which the liquid state, the gaseous state, and the solid state of a material can coexist simultaneously. The **triple point** is the temperature (critical temperature) and pressure (critical pressure) at which all three states of matter coexist simultaneously. This explains why some chemicals behave oddly, such as dry ice (solid carbon dioxide). Under atmospheric pressure carbon dioxide is never found in the liquid state, no matter what the temperature is. Carbon dioxide must be pressurized above 5 atmospheres to be transported as a liquid. The critical pressure and critical temperature are important in accurately calculating a plume model in programs such as ALOHA (which we will discuss in the next chapter).

critical temperature ■ The temperature above which a gas cannot be liquefied by pressure alone.

critical pressure ■ The pressure at which the liquid state, the gaseous state, and the solid state of a material can coexist simultaneously.

triple point ■ The temperature (critical temperature) and pressure (critical pressure) at which all three states of matter coexist simultaneously.

Expansion Ratio of Gases

When a gas is liquefied, using either pressurization or temperature reduction, or a combination thereof, new dangers ensue. Apart from being inherently toxic, the gas may cause freezing injuries (cryogenic burns) when it escapes from the pressurized container. **Autorefrigeration** occurs when a compressed gas or liquefied compressed gas escapes from its pressurized container and loses energy. This energy loss is realized as a lower temperature of the remaining liquid (due to evaporation), and as a lower temperature of the released gas (due to depressurization). In humid environments autorefrigeration leads to a buildup of ice at the release point, whether this is at an open valve or a container breach. An ice plug at the point of release on the container may temporarily slow or stop the leak.

In addition, the expansion ratio of the material must be taken into account. The **expansion ratio** is the comparison between the volume of a given amount of liquefied gas and the volume of that gas at a given temperature and pressure. In practical terms, the expansion ratio of a gas is the n-fold increase in volume that a liquefied compressed gas undergoes as it boils off. For example, 1 cubic foot (ft^3) of liquid oxygen (LOX) turns into 860 ft^3 of oxygen gas. The expansion ratio ranges from a low of 239 for n-butane to a high of 1445 for neon (see Table 5-3). This increase in volume can be especially dangerous in buildings and other relatively confined areas. Even a nontoxic, simple asphyxiating gas such as nitrogen can become deadly as a small liquid nitrogen spill of a few gallons becomes thousands of gallons of nitrogen gas and displaces the oxygen in the room. Plume migrations of toxic, flammable, corrosive, or oxidizing gases are even more dangerous at larger liquefied compressed gas releases.

autorefrigeration ■ The cooling of a liquid, usually cryogenic, through the evaporation of that liquid; also known as evaporative cooling. Containers that carry cryogenic liquids (such as liquefied oxygen and liquefied nitrogen) usually maintain relatively low operating pressures through this process.

expansion ratio ■ The ratio of the increase in volume of a given amount of liquefied gas as it turns into a gas at a given temperature and pressure (usually room temperature and atmospheric pressure).

GAS AND VAPOR BEHAVIOR

How will a hazardous material behave if it is released from its container and it contacts the air? You need to know this information because the material may react violently with the air and explode. Or you may want to know how quickly vapors are generated by a flammable liquid to determine the fire hazard. Or you may want to know how much of a toxic material victims may have inhaled. These are all reasons for understanding gas and vapor behavior.

The first piece of information you need to know is whether the material reacts with the air, that is, whether it is pyrophoric. A pyrophoric material is capable of igniting

SOLVED EXERCISE 5-3

The liquid line on a cryogenic liquid container has ruptured at the local university's chemistry building. Laboratory personnel advise you that the material involved is argon. Argon is a nontoxic inert gas that may pose an asphyxiation hazard in large quantities. Laboratory personnel also advise that not more than 20 liters (L) of liquid argon escaped before a graduate student was able to close the valve. How much gaseous argon, would you estimate, is in the room?

Solution: Argon has an expansion ratio of 842. Therefore you could expect approximately 16,840 L (842 $\times$ 20 L) of gaseous argon in the building.

TABLE 5-3 Expansion Ratios for Liquefied Compressed Gases

LIQUEFIED GAS	EXPANSION RATIO (1 TO)	TEMPERATURE OF GAS AT 1 ATM
Ammonia	947	59°F (15°C)
Argon	842	70°F (21°C)
Arsine	502	59°F (15°C)
n-Butane	239	59°F (15°C)
Carbon dioxide	845	59°F (15°C)
Carbon monoxide	680	70°F (21°C)
Chlorine	521	59°F (15°C)
Diborane	362	59°F (15°C)
Dichlorosilane	290	59°F (15°C)
Ethylene	450	70°F (21°C)
Helium	745	70°F (21°C)
Hydrogen	850	70°F (21°C)
Hydrogen bromide	648	59°F (15°C)
Hydrogen chloride	772	59°F (15°C)
Hydrogen iodide	518	59°F (15°C)
Hydrogen sulfide	638	59°F (15°C)
Krypton	693	70°F (21°C)
LPG (commonly a propane/ butane mix)	270	Not listed
Methane	625	70°F (21°C)
Methanethiol	433	59°F (15°C)
Methylamine	522	59°F (15°C)
Neon	1445	70°F (21°C)
Nitrogen	696	70°F (21°C)
Nitrous oxide	662	68°F (20°C)
Oxygen	860	70°F (21°C)
Phosgene	337	59°F (15°C)
Phosphine	510	59°F (15°C)
Propane	311	59°F (15°C)
Silane	412	59°F (15°C)
Sulfur dioxide	535	59°F (15°C)
Tungsten hexafluoride	270	59°F (15°C)
Vinyl chloride	365	59°F (15°C)
Water	1600	Not listed
Xenon	559	70°F (21°C)

Sources: MSDSs from Air Liquide, Air Product, Linde Gas, and Praxair.

TABLE 5-4 Vapor Pressures of Selected Materials Listed in Order of Increasing Volatility

CHEMICAL	VAPOR PRESSURE
Sodium cyanide	0 mmHg (approximate)
Methylene bisphenyl diisocyanate (MDI)	0.000005 mmHg (at 77°F)
VX	0.0001 mmHg
Sulfuric acid	0.001 mmHg
Sarin	2 mmHg
Kerosene	5 mmHg (at 100°F)
Water	20 mmHg
Ethanol	44 mmHg
Methyl ethyl ketone (MEK)	78 mmHg
Acetone	180 mmHg
Gasoline	38 to 300 mmHg (depending on blend)
Ethyl ether	440 mmHg
Hydrogen fluoride	783 mmHg
Chlorine	6.8 atm
Ammonia	8.5 atm
Hydrogen chloride	40.5 atm

spontaneously when exposed to air at temperatures below 130°F. These types of materials are extremely dangerous and can release large amounts of energy very quickly. Examples of pyrophoric materials include silane, arsine, phosphine, diborane, white phosphorus, iron sulfide, and lithium aluminum hydride.

Vapor Pressure

vapor pressure ■ The pressure exerted on a container by a solid or liquid in equilibrium with its vapors. A measure of a material's tendency to evaporate.

Next, you would like to know how quickly the material evaporates. The evaporation rate of a material is defined by its **vapor pressure**. Vapor pressure is typically measured in millimeters of mercury (mmHg), atmospheres (atm), or pounds per square inch (psi). The higher the vapor pressure, the more rapidly the material evaporates when it escapes from its container (Table 5-4). For example, sulfuric acid has a vapor pressure of 0.001 mmHg, while hydrofluoric acid has a vapor pressure of 783 mmHg at room temperature. But what do these numbers mean?

We need to put these numbers into context. The easiest way to remember the significance of vapor pressure is to remember the vapor pressure of water and acetone, since most of you have a good idea of how quickly these materials evaporate at room temperature. Water has a vapor pressure of roughly 20 mmHg, and acetone has a vapor pressure of 180 mmHg at room temperature. Acetone is a fairly volatile solvent found in nail polish remover and paint thinner.

When we examine these two materials and their relative vapor pressures, we quickly realize that lower numbers mean less evaporation, and higher numbers mean greater evaporation. Thus, in comparing the vapor pressures of sulfuric acid and hydrofluoric acid, we see that sulfuric acid evaporates very slowly. Such slowly evaporating materials are known as **persistent** materials. In contrast, hydrofluoric acid evaporates readily and is a **nonpersistent** material.

persistent ■ Not volatile; not prone to evaporate.

nonpersistent ■ Volatile; prone to evaporate.

Vapor pressure is strongly dependent on both the size as well as the chemical composition of the molecule. Table 5-5 lists the vapor pressures of several members of common

TABLE 5-5 Hydrocarbon and Hydrocarbon Derivative Size Dependence of Vapor Pressure

FUNCTIONAL GROUP	SYMBOL	R=1 C (METHYL-)	2 C (ETHYL-)	3 C (PROPYL-)	4 C (BUTYL-)	5 C (PENTYL-)	6 C (HEXYL-)
Hydrocarbons							
Alkenes	R=R	N/A	81.66 atm (ACT) Ethene	10.3 atm Propene	2.6 atm 1-Butene	637 mmHg (77°F) 1-Pentene	157 mmHg 1-Hexene
Alkanes	R–R	Above critical temperature (ACT) Methane	37.3 atm Ethane	8.4 atm Propane	2.05 atm Butane	420 mmHg Pentane	124 mmHg Hexane
Alkynes	R≡R	N/A	44.2 atm Acetylene	5.2 atm 1-Propyne	1.6 atm 1-Butyne	352 mmHg 1-Pentyne	253 mmHg (38°C) 1-Hexyne
Aromatics	R–C_6H_5	75 mmHg Benzene	21 mmHg Toluene	7 mmHg Ethyl benzene	2 mmHg *n*-Propyl benzene	1.03 mmHg *n*-Butyl benzene	0.4 mmHg *n*-Pentyl benzene
Hydrocarbon Derivatives							
Ethers	R–O–R	5.0 atm Dimethyl ether	440 mmHg Diethyl ether	55 mmHg Dipropyl ether	4.8 mmHg Dibutyl ether		
Primary amines	R–N–H_2	3.0 atm Methylamine	874 mmHg Ethylamine	471 mmHg *n*-Propylamine	82 mmHg *n*-Butylamine		
Aldehydes	R–(C=O)H	>1 atm Formaldehyde	740 mmHg Acetaldehyde	271 mmHg Propionaldehyde	90 mmHg Butyraldehyde	27 mmHg Valeraldehyde	
Ketones	R–(C=O)–R	N/A	N/A	180 mmHg Acetone	78 mmHg 2-Butanone	27 mmHg 2-Pentanone	11 mmHg 2-Hexanone
Thiols	R–S–H	1.7 atm Methyl mercaptan	442 mmHg Ethyl mercaptan	155 mmHg (77°F) 1-Propanethiol	35 mmHg *n*-Butyl mercaptan		
Nitriles	R–C≡N	630 mmHg Hydrogen cyanide	73 mmHg Acetonitrile	35 mmHg Propionitrile	14 mmHg *n*-Butyronitrile		
Esters	R–(C=O)O–R	476 mmHg Methyl formate	173 mmHg Methyl acetate	73 mmHg Ethyl acetate	25 mmHg *n*-Propyl acetate	10 mmHg *n*-Butyl acetate	
Alcohols	R–OH	96 mmHg Methanol	44 mmHg Ethanol	15 mmHg *n*-Propanol	6 mmHg *n*-Butanol	2.2 mmHg Amyl alcohol	0.5 mmHg *n*-Hexanol
Carboxylic Acids	R–(C=O)OH	35 mmHg Formic acid	11 mmHg Acetic acid	3 mmHg Propionic acid	0.4 mmHg Butyric acid		
Nitroalkanes	R–NO_2	28 mmHg Nitromethane	21 mmHg (77°F) Nitroethane	8 mmHg 1-Nitropropane	4.4 mmHg (77°F) 1-Nitrobutane		
Chloroalkanes	R–Cl	5.0 atm Methyl chloride	1000 mmHg Ethyl chloride	344 mmHg 1-Chloropropane	81 mmHg 1-Chlorobutane	20 mmHg 1-Chloropentane	9 mmHg 1-Chlorohexane

Source: Data courtesy of *NIOSH Pocket Guide*, CAMEO, WISER, *CRC Handbook of Chemistry and Physics*, ChemSpider, Sigma-Aldrich, MSDS.

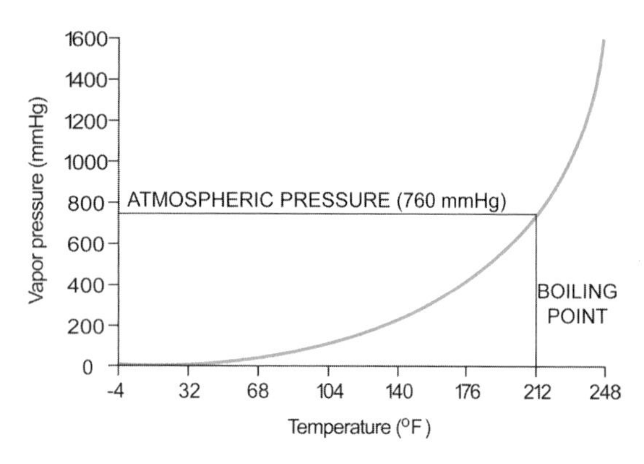

FIGURE 5-4 The relationship between vapor pressure and temperature—in this case for water. As the temperature increases, so does the vapor pressure. When the vapor pressure equals atmospheric pressure (760 mmHg), the material will boil. At sea level, when water boils (212°F), its vapor pressure equals 760 mmHg. *Art by David Heskett.*

hydrocarbon and hydrocarbon derivative families, listed in order of increasing size, from one carbon atom up to six carbon atoms. The table shows that the vapor pressure of family members decreases as their size increases. This can be a very helpful observation if you know the name of the chemical that has been released. If it contains more than five or six carbons, it is most likely not particularly volatile, with the notable exception of the simple hydrocarbons.

Vapor pressure is highly temperature dependent. Most reference sources list the vapor pressure at approximately room temperature, somewhere between 68°F and 75°F. But be aware that some materials with very low vapor pressure have a vapor pressure listed at a much higher temperature—often, the temperature at which that material is used in manufacturing processes. You should get in the habit of looking at not only the vapor pressure value but also the temperature to which that value applies. When you are on the scene of hazardous materials releases keep in mind that lower temperatures will mean lower vapor pressures (in other words, less volatility); and vice versa, higher temperatures will mean a higher vapor pressure (more volatility). In fact, the temperature dependence of vapor pressure is logarithmic. This means that the vapor pressure increases much faster as the temperature rises. In fact, the vapor pressure doubles for approximately every 15°C (27°F) rise in temperature. Figure 5-4 shows the temperature dependence of vapor pressure for water.

sublimation ■ The phase transition of a solid directly to a vapor without first passing through the liquid state.

vapor density ■ The mass of a gas or vapor compared with the mass of an equal volume of air.

The boiling point of a material is closely related to its vapor pressure. In fact, they are inversely related. This means that as the boiling point increases, the vapor pressure decreases. Thus, if you know something has a high boiling point, you also know that it evaporates slowly; and vice versa, if it has a low boiling point, it evaporates quickly.

When a solid becomes a gas directly without first becoming a liquid, that is, evaporates, the process is known as **sublimation**. Examples of materials that sublimate readily include water at subfreezing temperatures (in the form of ice and snow), carbon dioxide (dry ice), naphthalene (mothballs), and iodine. The important thing to remember is that not all solids are nonvolatile. For example, because iodine is extremely toxic, the sublimation of iodine in a closed space, such as an illicit methamphetamine lab, can be very dangerous.

FIGURE 5-5 Vapor density. The top drawing illustrates the movement of a gas with a vapor density less than 1; the bottom drawing illustrates a gas or vapor with a vapor density greater than 1. *Art by David Heskett.*

Vapor Density

Once you know how quickly a material evaporates, the next thing you need to know is where the vapor is going to go. Whether a gas or vapor rises or sinks in air is described by its **vapor density** (Figure 5-5). Vapor density is a comparison between the weight of a volume of gas or vapor and an equal volume of air. The vapor density of air is therefore set at 1.0. The *NIOSH Pocket Guide to Chemical Hazards* refers to vapor density as "RGasD," which stands for relative gas density. Note that the vapor density is a unitless number, since the density of a gas or vapor is being compared with the density of air.

If the vapor density of the material is greater than 1, such as 2.47 for chlorine, the material will eventually sink in air. Conversely, if the vapor density of the material is less than 1, such as 0.6 for ammonia, it will eventually rise. Why does it only "eventually" rise? Ammonia escaping from a liquefied compressed gas cylinder will initially fall owing to its extremely cold temperature. The

BOX 5-3 UNUSUAL PHYSICAL PROPERTIES OF HYDROGEN FLUORIDE

Hydrogen fluoride (HF) is a toxic and corrosive gas that is extensively used in the fluorocarbon manufacturing process and in many other industries including the petroleum industry. Hydrofluoric acid is an aqueous solution of hydrogen fluoride. Hydrogen fluoride is used and shipped primarily in its anhydrous form, as a 70% HF solution, or as a 49% HF solution. Hydrogen fluoride causes deep skin burns and complexes with calcium and magnesium in the body, causing intense pain at the site of contact, deep skin burns, and heart dysrhythmias in systemic poisoning cases. Burns of only a few percent of body surface area are often fatal.

Hydrogen fluoride is a very polar molecule—a similarity it shares with water— which gives it many unique properties. Polarity means one half of the molecule is more positively charged (H) while the other half of the molecule is more negatively charged (F). Owing to its polarity hydrogen fluoride is able to form hydrogen bonds with itself and with water in solution. Hydrogen bonds are similar to ionic bonds in that they are formed by the attraction of opposite charges, but hydrogen bonds are much weaker than ionic bonds.

Hydrogen bonding is the primary reason for the unusually low vapor pressure of hydrogen fluoride. For example, hydrogen fluoride has a vapor pressure of 783 mmHg, while hydrogen chloride has a vapor pressure of 40.5 atmospheres, hydrogen bromide has a vapor pressure of 20 atmospheres, and hydrogen iodide has a vapor pressure of 7.5 atmospheres. Based on this trend of size versus vapor pressure, HF should have a vapor pressure much greater than 40 atmospheres. But the vapor pressure of HF is barely above 1 atmosphere. What makes HF behave so strangely? The answer is that it can form hydrogen bonds, which allows it to interact with itself, thereby dramatically reducing its volatility. The reduction in volatility occurs because HF forms multimers—dimers, trimers, and tetramers.

	Molecular weight	Vapor density
H—F	20	0.69
$^{+}$H—F^{-}···$^{+}$H—F^{-}	40	1.38
(HF trimer, cyclic)	60	2.07
(HF tetramer, cyclic)	80	2.76

Hydrogen bond ···

Art by David Heskett.

The tendency to form multimers is also the root cause of its abnormally high vapor density. HF has a molecular weight of 20, which gives it a theoretical vapor density of approximately 0.7. However, experiments have shown that the vapor density of HF is very temperature dependent and varies between 1.2 and 2.4 at typical environmental temperatures (0°C to 35°C; 32°F to 95°F). The vapor density of HF does not approach the theoretical value of 0.7 until it gets to 60°C (140°F) or warmer. This is explained by the fact that HF dimers have a molecular weight of 40, HF trimers have a molecular weight of 60, and HF tetramers have a molecular weight of 80. Under normal conditions HF consists of varying sizes of multimers, from the monomer to the tetramer. Therefore, expect HF to behave as a heavier-than-air gas under typical release conditions.

cold ammonia gas is actually denser (heavier) than the much warmer ambient air surrounding it. This is the same principle that allows a hot air balloon to float. Heated air is significantly less dense than cold air, which allows the hot air balloon and its loaded basket to rise.

Most gases and vapors are heavier than air. Two mnemonics can be used to help remember the relatively small number of lighter-than-air gases. The first is HAHA MICE, which is incomplete but lists all the significantly lighter-than-air gases:

H—Hydrogen (vapor density = 0.07, the lightest gas)
A—Ammonia (0.59)
H—Helium (0.14)
A—Acetylene (0.91)

TABLE 5-6 Vapor Densities of Selected Chemicals

CHEMICAL	VAPOR DENSITY (RGASD)
Hydrogen	0.07
Helium	0.14
Methane	0.55
Ammonia	0.59
Water	0.62
Nitrogen	0.97
Air	1
Oxygen	1.11
Ethanol	1.59
Hydrogen fluoride	1.9 (approximate)
Chlorine	2.47
Gasoline	3.8 (approximate)
Sulfur hexafluoride	5.11
Tetrachloroethylene (Perk)	5.7
Mercury	6.9

M—Methane (0.55)
I—Illuminating Gases [includes natural gas (0.6) and neon (0.7)]
C—Carbon monoxide (0.97)
E—Ethylene (0.97)

The more complete mnemonic is 4H MEDIC ANNA:

H—Hydrogen (0.07)
H—Helium (0.14)
H—Hydrogen fluoride (0.69) [theoretical; actually approximately 1.9; see Box 5-3]
H—Hydrogen cyanide (0.93)

M—Methane (0.55)
E—Ethylene (0.97)
D—Diborane (0.96)
I—Illuminating gases [in this case natural gas (0.6), which is roughly 90% methane and 10% ethane]
C—Carbon monoxide (0.96)

A—Ammonia (0.59)
N—Neon (0.70)
N—Nitrogen (0.97)
A—Acetylene (0.91)

There are relatively few lighter-than-air gases because the density of gases and vapors is directly proportionate to their molecular weight (size). The molecular weight of air is 29, which is a relatively small number. In comparison, acetone, with a molecular weight of 58.1, is twice as heavy as air (58 divided by 29 equals a vapor density of 2). The molecular weight of a chemical can be found in most reference materials that list chemical and physical properties, such as the *NIOSH Pocket Guide*. Table 5-7 lists the lighter-than-air gases and their principle characteristics.

Be aware that air currents and the wind can have a much greater role than vapor density in determining where gases and vapors will go. This is especially true for gases that are barely lighter than air such as nitrogen, carbon monoxide, and ethylene. Recall that the atmosphere is a mixture of roughly 80% nitrogen and 20% oxygen. All the nitrogen does not rise and leave a 100% oxygen atmosphere on the ground just because its vapor density is slightly less than that of oxygen! However, in an undisturbed, sealed room heavy gases like chlorine will be found near the ground, and lighter gases like ammonia will be found near the ceiling.

SOLVED EXERCISE 5-4

Your local hazmat team is called to a residential neighborhood for a reported hydrogen fluoride release. On arrival you find out that the resident has a 5-gallon jug of hydrofluoric acid that he keeps in his garage and uses for his glass etching hobby. Over the years the container has become brittle, and when the resident bumped it, the bottom cracked. The homeowner was not injured, and first responders decontaminated him as a precaution. What are your concerns regarding the hydrogen fluoride vapors?

Solution: You should have three primary concerns with any gas or vapor: (1) atmospheric reactivity, (2) volatility, and (3) vapor density. Hydrogen fluoride is not pyrophoric, is very volatile with a vapor pressure of 783 mmHg, and is heavier than air with a vapor density of approximately 1.9. Therefore you should expect significant amounts of hydrogen fluoride vapors that will eventually settle near the ground. Hydrogen fluoride is extremely toxic, so this would certainly be a response that requires the highest level of personal protective equipment for responders. The homeowner should be seen at the emergency room to determine if fluoride poisoning has occurred from his exposure.

TABLE 5-7 Lighter-Than-Air-Gases

CHEMICAL NAME	HAZARDS	VAPOR DENSITY/RGASD	CHEMICAL FORMULA (MW)
Acetylene	Flammable	0.91	C_2H_2 (26.0)
Ammonia	Flammable, corrosive, toxic	0.59	NH_3 (17.0)
Carbon monoxide	Flammable, toxic	0.97	CO (28.0)
Diborane	Flammable, toxic	0.96	B_2H_6 (27.7)
Ethene (ethylene)	Flammable	0.97	C_2H_4 (28.0)
Helium	Asphyxiant	0.14	He (4.0)
Hydrogen	Flammable	0.07	H_2 (2.0)
Hydrogen cyanide	Flammable, corrosive, toxic	0.93	HCN (27.0)
Methane	Flammable	0.55	CH_4 (16.0)
Natural gas (10% ethane, 90% methane)	Flammable	0.60	Mixture (17.4)
Neon	Asphyxiant	0.70	Ne (20.2)
Nitrogen	Asphyxiant	0.97	N_2 (28.0)

Vapor density tells you where the safest areas are during a release, how the gas or vapor will spread, and where to test and sample during air monitoring activities.

WATER BEHAVIOR

How will a hazardous material behave if it is released from its container and it contacts water? To understand how a hazardous material will behave when it contacts water, you need to know three things, in the following order:

1. Will it react violently with water?
2. If not, will it dissolve in water?
3. If it doesn't mix with water, will it float or sink?

Let's see how we can determine this information.

Water Reactivity

Some materials are extremely water reactive, such as sodium metal. Sodium metal is stored under oil to prevent contact with even the small amount of moisture in the air. It ignites and will explode on contact with water. A truck transporting sodium metal that has lost its load on a bridge over water presents a serious water reactivity hazard. Thus, it is important to consult reference materials and check for water reactivity! Chemical reactivity, including water reactivity, is listed in the *NIOSH Pocket Guide,* on the MSDS sheet, in WISER and CAMEO, and in many other reference materials. We will cover the use of reference materials in the next chapter.

Water Solubility

After you are confident that the material is not water reactive, the next question to ask is whether the material is water soluble. **Water solubility** describes the extent to which a material will dissolve in water. For example, *n* butanol is 9% soluble in water; however, the percentage is not necessarily what you're used to. Percent water solubility

water solubility ▪ The extent to which a material will dissolve in water.

TABLE 5-8	Water Solubility of Selected Chemicals
CHEMICAL	**WATER SOLUBILITY**
Gasoline	Insoluble
Tetrachloroethylene (Perc)	0.02%
Chlorine	0.7%
n-Butanol	9%
Methyl ethyl ketone (MEK)	28%
Ammonia	34%
Hydrogen chloride	67% (at 86°F)
Sodium hydroxide	111%
Acetone	Miscible
Ethanol	Miscible
Sulfuric acid	Miscible

means that 9 milliliters (mL) of *n*-butanol will dissolve in water to a final volume of 100 mL. A counterintuitive example is sodium hydroxide, which is 111% water soluble. This simply means that 111 g of sodium hydroxide can be dissolved in water to a final volume of 100 mL. So how do we convey that something is completely soluble in water in any and all proportions? The term for complete solubility is **miscible**. For example, ethanol is soluble in all proportions in water. That means we can dissolve a drop of ethanol in a bathtub full of water, or we can dissolve a drop of water in a bathtub full of ethanol, or any ratio of ethanol and water in between. Ethyl alcohol is miscible with water. Table 5-8 lists the water solubility of selected chemicals. Solubility is temperature dependent. Generally, water solubility of liquids and solids increases with increasing temperature. However, most gases become less soluble in water as the temperature increases.

miscible ■ Completely soluble in all proportions.

As with vapor pressure, the water solubility of substances decreases with the size of the hydrocarbon and hydrocarbon derivative. Table 5-9 shows these trends. As the number of carbon atoms in a hydrocarbon derivative increases, its solubility decreases. We typically think alcohols, aldehydes, amines, organic acids, and ketones are fairly water soluble, but this is true only for the smaller members of the family. Hydrocarbons are almost completely insoluble in water. As the size of hydrocarbon derivatives increases, they look less like a derivative and more like a hydrocarbon owing to the increased number of carbon atoms in their makeup.

Specific Gravity

specific gravity ■ The mass of a given volume of material compared with the mass of an equal volume of water.

Specific gravity, akin to vapor density for gases, tells you whether a liquid will float on top of the water or sink to the bottom. Specific gravity is a comparison between the weight of a volume of liquid and an equal volume of water. The specific gravity of water is defined as 1.0. Thus, an insoluble liquid with a specific gravity of 0.8 will float on water, while an insoluble liquid with a specific gravity of 1.8 will sink in water. Note that specific gravity is also a unitless number, just like vapor density, since the density of a liquid is being compared with the density of water. Table 5-10 lists the specific gravity of selected materials.

How does knowing all this information affect your hazardous materials incident? Let's imagine you have a railcar full of butanol that is derailed and is leaking, and product is flowing toward a river. One option you might have is to go downstream, ahead of the release, and place a boom across the river (a technique we will discuss in Chapter 11). Butanol is only 9% soluble, so most of the material should not dissolve. Since the specific gravity of butanol is 0.81, the material should float, and you will be able to capture it behind a boom. Right? Not so fast. What happens after the bulk of the butanol is trapped behind the boom, and the river continues to move underneath the material? Since the butanol is 9% soluble, the water will continue to pick up 9 mL of butanol for every 100 mL of water that flows by. Therefore, if it takes the environmental company 2 hours to arrive and start pumping off the butanol, there will likely not be much left to pump out. However, if you have trapped a completely insoluble material behind the boom, all the material will still be there when the environmental company arrives. This example illustrates the importance of taking into account the chemical and physical properties of the material when determining response tactics.

TABLE 5-9 Hydrocarbon and Hydrocarbon Derivative Size Dependence of Solubility in Water

FUNCTIONAL GROUP	SYMBOL	1 C (METHYL-)	2 C (ETHYL-)	3 C (PROPYL-)	4 C (BUTYL-)	5 C (PENTYL-)	6 C (HEXYL-)
Hydrocarbons							
Alkanes	R–R	0.00227% g Methane	0.00568% g Ethane	0.00669% g Propane	0.00724% g Butane	0.0041% Pentane	0.00098% Hexane
Alkenes	R=R	N/A	0.0131% g Ethene	0.0200% g Propene	0.0222% g 1-Butene	0.0148% 1-Pentene	0.0053% 1-Hexene
Alkynes	R≡R	N/A	2% g Acetylene	0.364% g Propyne	0.287% g 1-Butyne	0.157% 1-Pentyne	0.036% 1-Hexyne
Aromatics	R–C_6H_5	0.178% Benzene	0.0519% Toluene	0.0161% Ethyl benzene	0.0052% Propyl benzene	0.00138% Butyl benzene	0.00105% Pentyl benzene
Hydrocarbon Derivatives							
Alcohols	R–OH	miscible Methanol	miscible Ethanol	miscible *n*-Propanol	7.4% *n*-Butanol	2.20% *n*-Pentanol	0.60% *n*-Hexanol
Aldehydes	R–(C=O)H	miscible Formaldehyde	miscible Acetaldehyde	30.6% 1-Propanal	7.1% 1-Butanal	1.2% 1-Pentanal	
Primary amines	R–N–H_2	miscible Methylamine	miscible Ethylamine	miscible *n*-Propylamine	miscible *n*-Butylamine	soluble *n*-Pentylamine	soluble *n*-Hexylamine
Carboxylic Acids	R–(C=O)OH	miscible Formic Acid	miscible Acetic Acid	miscible Propionic Acid	miscible or 5% Butanoic acid	3.6% Pentanoic acid	0.96% (20°C) Hexanoic acid
Esters	R–(C=O)O–R	30% Methyl formate	24.5% (20°C) Methyl acetate	8.08% Ethyl acetate	2.3% *n*-Propyl acetate	0.68% (20°C) *n*-Butyl acetate	0.17% (20°C) *n*-Pentyl acetate
Ethers	R–O–R	7.1% Dimethyl ether	6.42% Diethyl ether	0.49% Dipropyl ether	0.03% Dibutyl ether		0.019 % (20°C) Dihexyl ether
Ketones	R–(C=O)–R	N/A	N/A	miscible Acetone	28% 2-Butanone	5.5% 2-Pentanone	2% 2-Hexanone
Thiols	R–S–H	2% Methyl mercaptan	0.7% Ethyl mercaptan	slight 1-Propanethiol	0.0597% (20°C) 1-Butanethiol		
Nitriles	R–C≡N	miscible Hydrogen Cyanide	miscible Acetonitrile	11.9% Propionitrile	3% (77°F) *n*-Butyronitrile		
Nitroalkanes	R–NO_2	11.0% Nitromethane	4.4% Nitroethane	1.54% 1-Nitropropane			
Chloroalkanes	R–Cl	0.535% g Chloromethane	0.67% g Chloroethane	0.250% 1-Chloropropane	0.087% 1-Chlorobutane	0.0201% 1-Chloropentane	0.0064% 1-Chlorohexane

Sources: NIOSH Pocket Guide, CRC Handbook of Chemistry and Physics (89th edition), MSDS.

SOLVED EXERCISE 5-5

You find out that the railcar leaking its contents toward the river in the opening scenario contains ethanol. How do you expect the ethanol to behave once it reaches the river?

Solution: Ethanol is not water reactive, is miscible with water, and has a specific gravity of 0.79. Therefore the ethanol will mix with the water and be diluted. In this case the specific gravity is all but irrelevant, since the ethanol mixes with the water and will not float on top, as implied by the specific gravity.

TABLE 5-10 Specific Gravity of Selected Materials

CHEMICAL	SPECIFIC GRAVITY
Ammonia	0.62 (exists as a gas at ambient temperature and pressure)
Gasoline	0.72–0.76 (at 60°F)
Acetone	0.79
Ethanol	0.79
Methyl ethyl ketone (MEK)	0.81
n-Butanol	0.81
Water	1.0
12 M Hydrochloric acid	1.2
Chlorine	1.41 (at 6.86 atm) (exists as a gas at ambient temperature and pressure)
Tetrachloroethylene (Perc)	1.62
Sulfuric acid	1.85

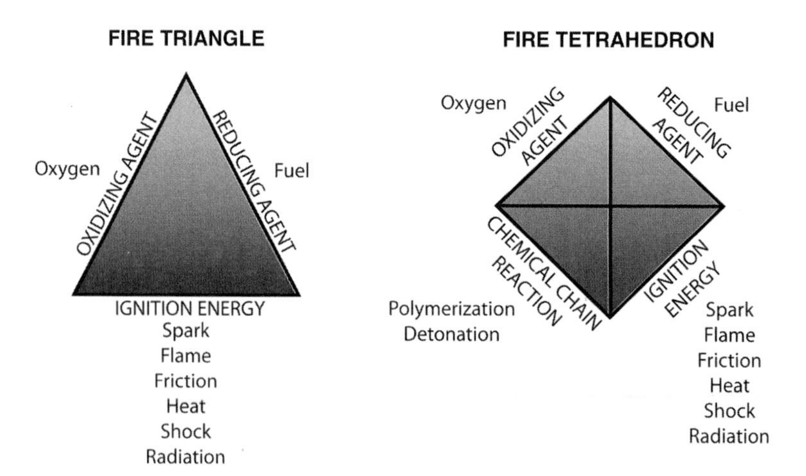

FIGURE 5-6 The fire triangle and the fire tetrahedron. For combustion to occur, three elements must be present—sufficient oxygen, a fuel to burn, and an ignition source. This is the fire triangle. The fire tetrahedron generalizes the fire triangle to include oxidizers in general, fuels, and ignition energy, and expands the triangle to include the chemical chain reaction—such as polymerization. *Art by David Heskett.*

FIRE BEHAVIOR

Flammability is one of the most serious dangers of a hazardous material. All line-of-duty hazardous materials deaths to date have been caused by flammable liquids or flammable gases and the resulting fire and/or explosion (see Boxes 5-4 and 5-5). Three components are needed for combustion: fuel, oxidizer, and an ignition source—or heat. This is referred to as the fire triangle. When we consider runaway chemical reactions, the fire triangle has been expanded to the fire tetrahedron (Figure 5-6). The three most important

TABLE 5-11	Flashpoint, Flammable Range, and Ignition Temperature of Common Materials		
CHEMICAL	**FLASHPOINT**	**FLAMMABLE RANGE (LEL–UEL)**	**IGNITION TEMPERATURE**
Propane	N/A* (gas)	2.1% to 9.5%	842°F (450°C)
Acetylene	N/A (gas)	2.5% to 100%	581°F (305°C)
Hydrogen	N/A (gas)	4% to 74%	1040°F (560°C)
Methylamine	N/A (gas)	4.9% to 20.7%	806°F (430°C)
Methane	N/A (gas)	5% to 15%	999°F (537°C)
Ammonia	N/A (gas)	15% to 28%	1166°F (630°C)
Ethyl ether	−49°F	1.9% to 36%	356°F (180°C)
Gasoline	−45°F	1.4% to 7.6%	495°F (257°C)
Acetone	0°F	2.5% to 12.8%	869°F (465°C)
Hydrogen cyanide	0°F	5.6% to 40%	1000°F (538°C)
Methanol	52°F	6% to 36%	867°F (464°C)
Ethanol	55°F	3.3% to 19%	685°F (363°C)
Styrene	88°F	0.9% to 6.8%	914°F (490°C)
Acetic acid	103°F	4.0% to 19.9% (at 200°F)	865°F (463°C)
Kerosene	100-162°F	0.7% to 5%	410°F (210°C)
Chlorine	N/A (nonflammable, oxidizing gas)	N/A	N/A

*N/A – not applicable

BOX 5-4 HAZMAT HISTORY: COLD STORAGE WAREHOUSE EXPLOSION (1984)

On Monday, September 17, 1984, an anhydrous ammonia explosion occurred at the Dixie Cold Storage Company in Shreveport, Louisiana. Anhydrous ammonia is a toxic, flammable, and corrosive gas that is often used as a refrigerant in cold storage warehouses. Two hazardous materials team members from the Shreveport Fire Department Hazardous Materials Unit were caught in the explosion, one of whom died.

Company employees discovered a small ammonia leak several days before the explosion, and they attempted to isolate and repair the leak. Later, they determined there was a faulty valve at the evaporator in the room. The employees removed the perishable contents from the freezer room, attempted to close an isolation valve, and started repairs. The workers wore only minimal respiratory protection to perform the work and noticed that the ammonia levels were rising steadily. They attempted to neutralize the ammonia with a carbon dioxide fire extinguisher, which reduced the visibility in the room to near zero. Eventually they exited the room and called the fire department for assistance.

On arrival, the fire department first responders assessed the situation and called the hazmat team because of the presence of the anhydrous ammonia. After an initial assessment the hazmat team members entered the warehouse and attempted to repair the faulty valve 17 feet above the floor. They used a forklift truck to access the valve, which caused the spark that ignited the anhydrous ammonia. The ammonia was able to accumulate to flammable levels owing to poor ventilation. The explosion ignited the chemical protective clothing the hazmat team members were wearing, fatally burning the forklift operator and critically injuring the second team member. The fatally injured hazmat team member died approximately 36 hours later in the hospital after receiving burns to over 97% of his body.

Anhydrous ammonia is placarded differently in the United States than it is in almost every other country. In the United States it is considered to be a Division 2.2, a nonflammable and nontoxic gas. In Canada and Europe it is considered a poisonous and corrosive gas. Despite the U.S. DOT classification, anhydrous ammonia is a corrosive, poisonous, and flammable gas—as this incident tragically illustrated.

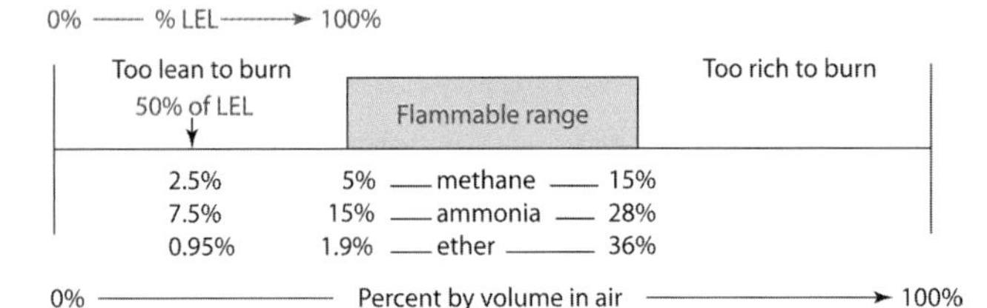

FIGURE 5-7 The flammable range consists of the concentration of fuel that will burn in atmospheric oxygen—the range between the lower explosive limit (LEL) and the upper explosive limit (UEL). Below the flammable range there is not enough fuel to burn in the presence of atmospheric oxygen. Above the flammable range there is too much fuel to burn in the presence of atmospheric oxygen.

chemical properties you need to consider when dealing with flammable materials are the flammable range, flashpoint, and ignition temperature.

Flammable Range

The **flammable range** is the concentration of vapors, usually measured in percent, at which the product will burn. The flammable range is bracketed by the **lower explosive limit (LEL)** and the **upper explosive limit (UEL)** (Figure 5-7). Below the LEL there are not enough vapors to burn, and the mixture is too lean. Above the UEL there is too much fuel and not enough oxygen to burn, and the mixture is too rich. Below the LEL and above the UEL, even if a flammable gas or vapor is present, and there is an ignition source, the mixture will not ignite. You have most likely encountered this problem with your lawnmower. You pull the drawstring and the motor will not turn over. The air/fuel mixture going into the carburetor is too lean, so you push the choke several times, and you start to smell gasoline. Then you pull the drawstring, but the mower still won't start. Now the air/fuel mixture entering the carburetor is too rich—you flooded the carburetor. You passed right through the flammable range. Initially you were below the LEL, and after you operated the choke too much, you ended up above the UEL.

flammable range ▪ The concentration of gas or vapor between the lower and upper explosive limits in ambient air. Also known as the explosive range.

lower explosive limit (LEL) ▪ The concentration of a gas or vapor in air below which it will not ignite.

upper explosive limit (UEL) ▪ The maximum concentration of a gas or vapor in the ambient air that will burn.

A high LEL implies that higher concentrations of a gas or vapor are needed before it will ignite, such as in an enclosed area in a building. A low LEL or a wide flammable range implies that a gas or vapor can ignite readily almost anywhere—even in open areas—and at a significant distance from the release.

Hydrocarbons and hydrocarbon derivatives are typically flammable or combustible materials owing to their carbon, hydrogen, and oxygen content. The most flammable hydrocarbons are the unsaturated hydrocarbons, followed by the saturated and aromatic hydrocarbons, as shown in Table 5-12. The most flammable hydrocarbon derivatives are the aldehydes, ethers, alcohols, carboxylic acids, esters, and ketones. As the size of the molecule increases, the flammable range narrows, the LEL decreases, and the UEL decreases. Thus, it takes less of a larger molecule to reach the flammable range, but once the flammable range is reached, the range is narrower. We tend to say aldehydes have a wide flammable range, but this is true only for formaldehyde (which has one carbon atom), acetaldehyde (which has two carbon atoms), and 1-propanal (which has three carbon atoms). Aldehydes with four carbons or more have flammable ranges of 10% or less. As we discussed in Chapter 4, it is important to look at the complete name of the chemical. Besides knowing to which family the chemical belongs, such as saturated hydrocarbons or alcohols, it is just as important to look at the size of the molecule, in other words, how many carbons it has.

Flashpoint

flashpoint ▪ The temperature of a liquid at which it gives off enough vapors to ignite, but not necessarily continues to burn.

fire point ▪ The temperature of a liquid at which it gives off enough vapors to ignite and continues to burn without an additional external ignition source.

The flashpoint gives a lot of information about a flammable material. It is actually a combination of a chemical property and a physical property: the lower explosive limit and the vapor pressure. The **flashpoint** is defined as the temperature at which a liquid gives off enough vapors to ignite, but not necessarily to sustain combustion. Therefore, the flashpoint gives you information about the evaporation rate (a liquid gives off vapors) and the LEL (enough vapors to ignite). A related property is the fire point. The **fire point** is the temperature at which a liquid gives off enough vapors to ignite and continue to burn without an additional external ignition source. The fire point is typically a few degrees higher than the flashpoint.

Knowledge of the flashpoint can make or break a hazmat incident involving flammable materials. For example, let's take as an example an uninsulated tanker rolling over on the interstate and spilling approximately 250 gallons of methanol. This spill occurs next to a large apartment complex. The temperature is currently 25°F. Do you need to

TABLE 5-12 Hydrocarbon and Hydrocarbon Derivative Size Dependence of Flammable Range

FUNCTIONAL GROUP	SYMBOL	1 C (METHYL-)	2 C (ETHYL-)	3 C (PROPYL-)	4 C (BUTYL-)	5 C (PENTYL-)	6 C (HEXYL-)
Hydrocarbons							
Alkynes	R≡R	N/A	2.4%–100% └97.5%┘ Acetylene	1.7%–16.8% └15.1%┘ 1-Propyne	1.1%–11.7% └10.6%┘ 1-Butyne		
Alkenes	R=R	N/A	2.75%–28.6% └25.85%┘ Ethene	2.0%–11.1% └9.1%┘ Propene	1.6%–10% └8.4%┘ 1-Butene	1.4%–8.7% └7.3%┘ 1-Pentene	1.2%–6.9% └5.7%┘ 1-Hexene
Alkanes	R–R	5%–15 % └10%┘ Methane	3%–12.5% └9.5%┘ Ethane	2.1%–9.5% └7.4%┘ Propane	1.6%–8.4% └6.8%┘ Butane	1.4%–7.8% └6.4%┘ Pentane	1.1%–7.5% └6.4%┘ Hexane
Aromatics	R–C_6H_5	1.2%–8.0% └6.8%┘ Benzene	1.1%–7.1% └6%┘ Toluene	0.8%–6.7% └5.9%┘ Ethyl benzene	0.8%–6.0% └5.2%┘ *n*-Propyl benzene	0.8%–5.8% └5.0%┘ *n*-Butyl benzene	
Hydrocarbon Derivatives							
Alcohols	R–OH	5.5%–38% └32.5%┘ Methanol	3%–19 % └16%┘ Ethanol	2.2%–13.7% └11.5%┘ *n*-Propanol	1.4%–11.2% └9.8%┘ *n*-Butanol	1.2%–10% └8.8%┘ Amyl alcohol	1.2%–7.7% └6.5%┘ *n*-Hexanol
Aldehydes	R–(C=O)H	7.0%–73% └66%┘ Formaldehyde	4.0%–60% └56%┘ Acetaldehyde	2.9%–17.0% └14.1%┘ 1-Propanal	2.5%–12.5% └10%┘ 1-Butanal	2.1%–7.8% └5.7%┘ 1-Pentanal	1%–7.5% └6.5%┘ 1-Hexanal
1° Amines	R–N–H_2	4.2%–20.8% └16.6%┘ Methylamine	3.5%–14.0% └10.5%┘ Ethylamine	2.0%–10.4% └8.4%┘ *n*-Propylamine	1.7%–9.8% └8.1%┘ *n*-Butylamine	2.2%–22% └19.8%┘ *n*-Pentylamine	2.1%–9.3% └7.2%┘ *n*-Hexylamine
Carboxylic Acids	R–(C=O)OH	10%–57% └47%┘ Formic acid	4.0%–19.9% └15.9%┘ Acetic acid	2.1%–12.1% └10%┘ Propionic acid	2%–10% └8%┘ Butyric acid	1.6%–7.6% └5.1%┘ Valeric acid	2%–10% └8%┘ Hexanoic acid
Esters	R–(C=O)O–R	4.5%–23% └18.5%┘ Methyl formate	3.1%–16% └12.9%┘ Methyl acetate	2.0%–11.5% └9.5%┘ Ethyl acetate	1.7%–8% └6.3%┘ *n*-Propyl acetate	1.3%–7.5% └6.2%┘ *n*-Butyl acetate	1.1%–7.5% └6.4%┘ Amyl acetate
Ethers	R–O–R	2%–50% └48%┘ Dimethyl ether	1.7%–36% └34.3%┘ Diethyl ether	1.2%–9.5% └8.3%┘ Dipropyl ether	0.9%–8.5% └7.6%┘ Dibutyl ether	2%–8% └6%┘ Dipentyl ether	

(Continued)

TABLE 5-12 Hydrocarbon and Hydrocarbon Derivative Size Dependence of Flammable Range (*Continued*)

FUNCTIONAL GROUP	SYMBOL	1 C (METHYL-)	2 C (ETHYL-)	3 C (PROPYL-)	4 C (BUTYL-)	5 C (PENTYL-)	6 C (HEXYL-)
Hydrocarbon Derivatives							
Ketones	R–(C=O)–R	N/A	N/A	2.5%–12.8% —10.3%— Acetone	1.8%–10.1% —8.3%— 2-Butanone	1.5%–8.2% —6.7%— 2-Pentanone	1.3%–8.1% —6.8%— 2-Hexanone
Thiols	R–S–H	3.9%–21.8% —17.9%— Methyl mercaptan	2.8%–18.0% —15.2%— Ethyl mercaptan				
Nitriles	R–C≡N	5.4%–46% —40.6%— Hydrogen cyanide	3.0%–16.0% —13%— Acetonitrile	3.1%–? Propionitrile	1.65%–? *n*-Butyronitrile		
Nitroalkanes	R–NO_2	7.3%–63% —55.7%— Nitromethane	3.4%–40% —36.6%— Nitroethane	2.2%–? 1-Nitropropane			
Chloroalkanes	R–Cl	7.1%–19% —11.9%— 1-Chloromethane	3.8%–15.4% —11.6%— 1-Chloroethane	2.4%–11.1% —8.7%— 1-Chloropropane	1.8%–10% —8.2%— 1-Chlorobutane	1.4%–8.6% —7.2%— 1-Chloropentane	1%–9.6% —8.6%— 1-Chlorohexane

TABLE 5-13 NFPA Definitions of Flammable and Combustible Liquids

CLASSIFICATION	DEFINITION	EXAMPLES
Class IA Flammable Liquid	Flashpoint less than 73°F with a boiling point less than 100°F	Diethyl ether, ethylene oxide, acetaldehyde
Class IB Flammable Liquid	Flashpoint less than 73°F with a boiling point equal to or greater than 100°F	Gasoline, ethanol, toluene, methyl ethyl ketone (MEK), diethylamine, acrolein
Class IC Flammable Liquid	Flashpoint equal to or greater than 73°F, but less than 100°F	Xylene, *n*-butanol
Class II Combustible Liquid	Flashpoint equal to or greater than 100°F, but less than 140°F	Diesel fuel, acetic acid
Class IIIA Combustible Liquid	Flashpoint equal to or greater than 140°F, but less than 200°F	Home heating oil, benzyl chloride
Class IIIB Combustible Liquid	Flashpoint equal to or greater than 200°F	Lubricating and cooking oils

evacuate the apartment complex because of the threat of fire? The first chemical property you should investigate is the flashpoint, which turns out to be 52°F. What does this mean? It means the temperature of the methanol needs to be at least 52°F before it will generate enough vapors to ignite. Therefore, as long as the temperature of the methanol is below 52°F an ignition source can be present, and the vapors will not ignite. In fact, you could theoretically even put a cigarette out in the 25°F methanol (but don't try this at home or on your next hazmat incident!).

Flashpoint is used to determine the difference between a flammable liquid and a combustible liquid. The U.S. DOT defines a flammable liquid as having a flashpoint at or below 141°F, while a combustible liquid has a flashpoint above 141°F, but below 200°F. The U.S. DOT allows flammable liquids with flashpoints between 100°F and 141°F to be reclassified as combustible liquids. This is why diesel fuel is treated as a combustible liquid in the United States even though it may have a flashpoint as low as 126°F.

The NFPA has a more nuanced and complex definition of flammable and combustible liquids that also incorporates volatility in the form of boiling point. These definitions are shown in Table 5-13.

Both saturated and unsaturated hydrocarbons have very low flashpoints. In fact, they are all gases until they contain five carbon atoms. Therefore the saturated and unsaturated hydrocarbons pose a significant flammability risk at the scene of hazmat releases. This is one reason oil refineries are such high-risk occupancies. Aromatic hydrocarbons are quite flammable as well, but because the smallest aromatic compound—benzene—contains six carbon atoms, it has a flashpoint of 12°F. In contrast, methane has a flashpoint of −306°F. Many hydrocarbon derivatives are also extremely flammable, but not as flammable as the saturated and unsaturated hydrocarbons. The most flammable hydrocarbon derivatives in descending order are thiols, primary amines, ethers, aldehydes, chloroalkanes, ketones, esters, and nitriles. Counterintuitively, the alcohols (such as methanol and ethanol) and nitroalkanes (such as the racing fuel nitromethane) have comparatively high flashpoints compared with the other classes of hydrocarbon derivatives. The flashpoints of both hydrocarbons and hydrocarbon derivatives up to six carbon atoms in size are listed in Table 5-14.

Ignition Temperature

The **ignition temperature**, also known as the *autoignition temperature*, is the temperature at which a material will initiate self-sustained combustion without an external ignition source. This means that when a material reaches this temperature it will start to burn and release flammable decomposition products. An example is oil-soaked rags that burst into flames after being stuffed into a trash can. The oxidation of oils such as linseed oil

ignition temperature ▪ The minimum temperature at which a material will initiate self-sustained combustion without an external ignition source.

TABLE 5-14 Hydrocarbon and Hydrocarbon Derivative Size Dependence of Flashpoint

FUNCTIONAL GROUP	SYMBOL	1 C (METHYL-)	2 C (ETHYL-)	3 C (PROPYL-)	4 C (BUTYL-)	5 C (PENTYL-)	6 C (HEXYL-)
Hydrocarbons							
Alkenes	R═CH_2		–213°F Ethene	–162°F Propene	–110°F 1-Butene	–56°F 1-Pentene	–15°F 1-Hexene
Alkanes	R–CH_3	–306°F Methane	–211°F Ethane	–156°F Propane	–76°F Butane	–40°F Pentane	–7°F Hexane
Alkynes	R≡CH		N/A Acetylene	–125°F 1-Propyne	–81°F 1-Butyne	–29°F 1-Pentyne	–6°F 1-Hexyne
Aromatics	R–C_6H_5	12°F Benzene	40°F Toluene	55°F Ethyl benzene	86°F *n*-Propyl benzene	138°F *n*-Butyl benzene	149°F *n*-Pentyl benzene
Hydrocarbon Derivatives							
Thiols	R–S–H	N/A Methyl mercaptan	–55°F Ethyl mercaptan	–5°F 1-Propanethiol	35°F *n*-Butyl mercaptan	65°F 1-Pentanethiol	68°F 1-Hexanethiol
Primary amines	R–NH_2	–72°F Methylamine	–51°F Ethylamine	–22°F *n*-Propylamine	10°F *n*-Butylamine	45°F *n*-Pentylamine	81°F *n*-Hexylamine
Ethers	R–O–R	–41.8°F Dimethyl ether	–40°F Diethyl ether	–18.4°F Dipropyl ether	77°F Dibutyl ether	134°F Dipentyl ether	172°F Dihexyl ether
Aldehydes	R–(C═O)H	–64°F Formaldehyde	–36°F Acetaldehyde	–22°F 1-Propanal	10°F 1-Butanal	39°F 1-Pentanal	77°F 1-Hexanal
Chloroalkanes	R–Cl	–87°F Methyl chloride	–58°F Ethyl chloride	–24°F 1-Chloropropane	15°F 1-Chlorobutane	52°F 1-Chloropentane	81°F 1-Chlorohexane
Ketones	R–(C═O)–CH_3			0°F Acetone	16°F 2-Butanone	45°F 2-Pentanone	73°F 2-Hexanone
Esters	R–(C═O)O–CH_3	–2°F Methyl formate	14°F Methyl acetate	24°F Ethyl acetate	55°F *n*-Propyl acetate	72°F *n*-Butyl acetate	98°F *n*-Pentylacetate
Nitriles	R–C≡N	0°F Hydrogen Cyanide	36°F Acetonitrile	42°F Propanenitrile	62°F *n*-Butyronitrile	115°F Valeronitrile	142°F Hexanenitrile
Alcohols	R–OH	52°F Methanol	55°F Ethanol	71°F *n*-Propanol	95°F *n*-Butanol	118°F Amyl alcohol	145°F *n*-Hexanol
Nitroalkanes	R–NO_2	95°F Nitromethane	82°F Nitroethane	96°F 1-Nitropropane	111°F 1-Nitrobutane	153°F 1-Nitropentane	162°F 1-Nitrohexane
Carboxylic acids	R–(C═O)OH	85°F Formic acid	103°F Acetic acid	126°F Propionic acid	161°F Butyric acid	190°F Pentanoic acid	219°F Hexanoic acid

Sources: CHRIS, CAMEO, *NIOSH Pocket Guide*, WISER, CRC Handbook of Chemistry and Physics, MSDS, ChemSpider.

SOLVED EXERCISE 5-6

You've been dispatched to a biodiesel fuel production facility. One of its outdoor storage tanks containing methanol has catastrophically ruptured and spilled methanol out of the dike area. It is winter, and current weather conditions are clear and sunny skies with a temperature of 10°F. What are your primary concerns with this methanol release?

Solution: Methanol is a flammable liquid with a flashpoint of 52°F; a wide flammable range, from 6% to 36%; and an ignition temperature of 867°F. Because the ambient methanol temperature of 10°F is below the flashpoint of 52°F, not enough vapors are being generated to reach the lower explosive limit. Therefore, flammability is not a primary concern. Toxicity is a greater concern at this temperature. You should keep an eye on the weather conditions, and monitor the level of methanol vapors periodically. However, if this were a heated tank, and the methanol spilled outdoors, there would be a flammability risk until the methanol was cooled to below its flashpoint.

generates heat. As the heat builds up and reaches the ignition temperature of the linseed oil (650°F/343°C) it ignites and causes the contents of the trash to catch fire. This situation is often referred to as "spontaneous combustion." Another example is the common kitchen fire caused by burning vegetable oil. When the temperature of the cooking oil in the pan reaches the ignition temperature of the vegetable oil, it spontaneously combusts. Materials with low ignition temperatures therefore pose fire hazards. When these materials are reactive, such as linseed oil, which oxidizes in air, the hazard is significantly increased.

Boiling-Liquid, Expanding-Vapor Explosions (BLEVEs)

Boiling-liquid, expanding-vapor explosions (BLEVEs) occur when a liquid inside a tank or vessel is heated above its boiling point, and the tank catastrophically fails owing to a structural defect or because it has been pressurized beyond its limitations (Figure 5-8). Many people, including first responders, have been killed by BLEVEs caused by brittle metal used in the construction of tanks and the complete lack of pressure relief devices, or inadequate pressure relief devices. Even now, BLEVEs occur at regular intervals when the pressure inside tanks increases faster than the pressure relief devices can dissipate that pressure—and people are injured and killed. This pressure increase is typically caused by fires underneath partially empty or mostly empty tanks carrying flammable liquids or liquefied compressed gases such as propane. When a tank catastrophically fails in this manner the liquid contents inside the tank are instantaneously converted to a gas or vapor and are often ignited almost immediately with an ensuing fireball. The catastrophic failure of the tank sends deadly shrapnel up to a mile and a half away (especially in the

boiling-liquid, expanding-vapor explosion (BLEVE) ■ The process by which the buildup of internal pressure within a container is catastrophically relieved by its violent rupture, accompanied by the release of a vapor or gas and fragmentation of the container.

FIGURE 5-8 A boiling-liquid, expanding-vapor explosion (BLEVE). The burning propane cylinder on the left is not being sufficiently cooled by the hose line and eventually explodes (right) when the vapor pressure of the propane exceeds the bursting pressure of the cylinder. *Courtesy of Mike Becker, Longmont (CO) Fire Department.*

BOX 5-5 HAZMAT HISTORY: BLEVE: THE DEADLY DECADE

Liquid petroleum gas (LPG) and propane are widely used flammable liquid fuels that are transported in bulk amounts on a daily basis. Not surprisingly this hazardous material has been involved in many releases and fatalities over the years. In the 1970s there were several notable cases in which boiling liquid expanding vapor explosions (BLEVEs) were involved in serious incidents that caused fatalities or large property losses. The series of accidents impacted how compressed liquefied gases are stored and transported as well as how emergency responders handle compressed liquefied gas releases.

CRESCENT CITY, ILLINOIS (1970)

In June of 1970 a 108 car Toledo, Peoria & Western Railroad freight train derailed at approximately 6 AM on a Sunday morning in downtown Crescent City, Illinois. A little over an hour later the first of 12 tank cars carrying propane exploded sending a fireball over 1000 feet in the air. This conflagration destroyed the downtown of Crescent city including 15 homes, 70 people were injured, and nine were hospitalized with damage estimated at $2 million.

KINGMAN, ARIZONA (1973)

In July of 1973 a railroad tank car filled with propane was being unloaded in Kingman, Arizona, when it caught fire and eventually exploded. Three firefighters were killed instantly, and there were ultimately six fatalities and at least 70 other people were injured. According to witnesses the tank car exploded minutes after firefighters arrived to battle the flames. The explosion moved the tank car a quarter of a mile and dug a crater 10 feet deep. The BLEVE was blamed upon a relief valve failure during the transfer operation.

WAVERLY, TENNESSEE (1978)

In February of 1978 24 cars of a Louisville and Nashville Railroad freight train derailed in downtown Waverly, Tennessee. Emergency responders initially evacuated the area for a quarter-mile and cooled damaged tanks with streams of water. The initial damage assessment indicated cleanup operations could commence. Crews eventually found another tank car that was buried by debris the day after the derailment which was moved to clear the tracks to partially reopen the rail line. On the third day after the derailment temperatures that had hovered in the mid-20s were now climbing into the mid-50s. On the afternoon of the third day the damaged railcar violently ruptured shortly before offloading operations were to begin. The BLEVE killed six people instantly, including the Waverly police chief and fire chief as well as a state investigator. Ultimately 16 people died, 43 others were injured, and several dozen buildings were damaged or destroyed.

MISSISSAUGA, ONTARIO, CANADA (1979)

In November of 1979 a 106 car Canadian Pacific freight train carrying explosives, flammable liquids, and poisons derailed in Mississauga, Ontario. The derailment was caused when an improperly maintained journal bearing overheated (also known as a hot box), caught fire, and burned through the axle causing a pair of wheels (trucks) to completely fall off. The subsequent derailment ruptured several other tank cars carrying styrene, toluene, propane, caustic soda, and chlorine. The propane tank cars caught fire and exploded sending flames 5000 feet into the air. Concerns about a leaking chlorine tank car caused the evacuation of 200,000 people which remains the second largest peacetime evacuation behind the New Orleans evacuation during Hurricane Katrina in 2005.

case of railcars) and can cause the entire container to rocket. The fireball can engulf everything that is nearby. Box 5-5 describes several deadly BLEVEs that occurred in the 1970s.

It is important to note that BLEVEs also can occur with nonflammable liquefied gases. The container will still catastrophically rupture; there just won't be an ensuing fireball. People can still be killed and injured by the mechanical force of the explosion and the release of toxic or corrosive gases.

Summary

Hazardous materials technicians must be able to interpret chemical information and make sound decisions to protect themselves, the public, and prevent the hazmat incident from becoming worse. Understanding the chemical and physical properties of a hazardous material is crucial. You must understand how the hazardous material will behave on contact with air and with water, and you must know its chemical reactivity, including flammability. Anticipating how a hazardous material will act under the circumstances with which you are faced can make the difference between a safe and successful hazmat response and a disaster. It is impossible to mount a risk-based response to a hazardous materials incident without understanding the chemical and physical properties of the material.

Review Questions

1. What is the temperature at which a liquid changes to a gas at its most rapid rate?
2. At what temperature does a liquid turn to a solid?
3. What physical property describes how quickly a substance evaporates?
4. What is the vapor pressure of water at room temperature?
5. What physical property describes whether a gas or vapor will rise or fall when released into the atmosphere?
6. What are the three properties you must know if a hazardous material threatens to come into contact with water?
7. Why is water solubility important?
8. What physical property describes whether a liquid will sink or float when released into water?
9. Will a vapor burn if its concentration is 80% of the LEL under normal atmospheric conditions?
10. What is the definition of flashpoint?

References

Centers for Disease Control and Prevention. (2007). *NIOSH Pocket Guide to Chemical Hazards*. Washington, DC: U.S. Government Printing Office.

Houghton, Rick. Emergency Characterization of Unknown Materials, CRC Press (2008).

Klem, Thomas J. "Cold Storage Warehouse", NFPA Fire Investigations (1984).

Lide, David R. CRC Handbook of Chemistry and Physics, 89th edition, CRC Press (2008).

Meyer, Eugene. Chemistry of Hazardous Materials, 3rd edition, Pearson-Prentice Hall (2010).

US Chemical Safety and Hazard Investigation Board. (2003) "FIRE AND EXPLOSION: HAZARDS OF BENZOYL PEROXIDE" Washington DC: author.

Weber, Chris. (2007). Pocket Reference for Hazardous Materials Response. Upper Saddle River, NJ: Pearson/Brady.

CHAPTER 6
Information Resources

Courtesy of Chris Weber, Dr. Hazmat, Inc.

KEY TERMS

air bill, *p. 153*
bill of lading, *p. 151*
consist, *p. 151*
dangerous cargo manifest, *p. 153*
manifest, *p. 153*
Material Safety Data Sheet (MSDS), *p. 157*
packing group, *p. 151*
shipping paper, *p. 150*
waybill, *p. 151*

OBJECTIVES

After reading this chapter, the student should be able to:

- List the components of the local emergency response plan.
- List the components of a facility emergency response plan.
- Name the DOT-required components of shipping papers.
- Use an MSDS sheet to determine the chemical, physical, and toxicological properties of a product.
- Use the *NIOSH Pocket Guide* to determine the chemical, physical, and toxicological properties of a chemical.
- Use the CHRIS manual to determine the chemical and physical properties of a chemical.
- Use WISER to determine the chemical, physical, and toxicological properties of a chemical.
- Use CAMEO to determine the chemical, physical, and toxicological properties and recommended PPE of a chemical.
- Use ALOHA and MARPLOT to model the expected plume from a volatile chemical release.

- List the reference materials that can provide chemical reactivity information.
- Determine the current and future weather conditions at the location of a hazardous materials release.

ResourceCentral For additional review and practice tests, visit **www.bradybooks.com** and click on Resource Central to access book-specific resources for this text! To access Resource Central, follow directions on the Student Access Card provided with this text. If there is no card, go to **www.bradybooks.com** and follow the Resource Central link to Buy Access from there.

Your jurisdiction includes the university, the community college, and several industries that have their own small-scale research and analytical laboratories. Therefore, you have known for some time that FedEx and UPS carry many hazardous materials shipments through your jurisdiction almost every day. What you didn't expect was to encounter the mother lode on your shift. You've been dispatched to a motor vehicle accident involving a UPS truck bound for the university and the industrial park. The UPS truck overturned and spilled many packages across the highway, many of which are leaking. Now you're wondering how you're going to make heads or tails out of this mess. Some of the questions running through your mind are the following:

- Where can I find a list of all the chemicals on that truck?
- Where can I determine the chemical and physical properties of those materials?
- Which materials will react with the air?
- Which materials will react with water?
- Which materials will react with one another?
- How can I determine the weather conditions at the scene?
- Where can I get reliable weather forecasts, since this will undoubtedly be a time-consuming incident?

Let's see if we can get some answers to these questions and more.

In the last few chapters we extensively discussed chemical and physical properties and exposure limits. To make sound decisions at a hazmat incident you need to find the values of the chemical and physical properties for the hazardous material involved. You do this by consulting reference sources and technical experts. Let's explore some of these reference materials and learn how to use them. You should also examine general sources of information regarding the hazmat-related hazards present in your jurisdiction as well as your planned role in mitigating those emergencies.

On the other hand, it is also important to avoid information overload at an emergency. There are literally hundreds of pieces of information you could look up, write down, and hand to the incident commander. But which ones are really important? As we discussed in the last chapter, their importance depends on the nature of the release, environmental conditions, weather conditions, and the product.

Some information resources should be consulted well before an actual hazardous materials emergency. These information resources should be used to ensure that resources are spent on the appropriate training and equipment to address all hazards and risks faced by your community. The foundation of a risk-based response is proper planning and preparation.

Government Emergency Response Plans

Government agencies at the federal, tribal, state, and local level have all developed emergency response plans (ERPs) designed to minimize the scope of emergencies as well as the consequences of those emergencies. These plans strive to be all-encompassing, including natural disasters, man-made disasters, terrorist incidents, as well as hazardous materials emergencies. The federal emergency response plans, such as the National Response Framework, are designed to address incidents of national significance, in other words, those emergencies that are so severe they require more resources than the states, tribes, and local governments can reasonably provide. The federal plans also address which federal agencies will provide what resources in case of emergency. In many cases the federal government provides resources during less severe emergencies because they have essential equipment, expertise, or legal jurisdiction, or the emergency has affected or will affect federal property. For this last reason it is quite likely that you will be working with one or more federal agencies at hazardous materials emergencies to which you respond. The most commonly encountered federal agencies during hazmat and WMD response are the EPA and the FBI. Your hazardous materials response team should understand the equipment and resources various federal agencies are able to supply, as well as how to incorporate representatives from these agencies into the incident command structure. This information can be obtained by studying the most current version of the National Response Framework and networking with regional representatives of those agencies.

STATE EMERGENCY RESPONSE PLANS

The state ERP addresses severe natural and man-made disasters that rise to the level of incidents of state significance, that is, those emergencies that are so severe they require more resources than local governments can reasonably provide. As is the case with the federal emergency response plan, state resources will be available during less severe emergencies when the state has essential equipment, expertise, or legal jurisdiction, or the emergency has affected or will affect state property. Local hazardous materials response teams often work with state agencies such as the state police or highway patrol, the state's Department of Environmental Quality or Protection, and the state public health department. Once again, your hazardous materials response team should understand the equipment and resources these various state agencies have available. It is also very important to network with the local and regional representatives of these state agencies to facilitate cooperation during an emergency.

LOCAL EMERGENCY RESPONSE PLANS

The local ERP is the most important one to you as a hazardous materials technician assigned to a hazmat response team. The local ERP will detail the types of incidents to which you are expected to respond, and what your role at those incidents is expected to be. It also details what type of support the hazardous materials response team can expect from other local government agencies such as the fire department, the police department, the emergency medical services, the public works department, the health department, and the emergency management agency. The leadership of your hazardous materials response team should be intimately familiar with the local ERP and ensure that it acquires the necessary equipment, resources, and training to fulfill the response functions expected by the community. You should also be familiar with the local ERP and ensure that you are personally capable of carrying out your expected role. It is always a good idea to familiarize yourself with the responsibilities and ask yourself whether you have the appropriate training and are acquiring the requisite experience.

The local ERP is typically organized by type of emergency, including winter weather, tornadoes, flooding, hurricanes, hazardous materials releases, utility failures, and terrorist incidents. The plan typically also includes a detailed risk assessment based on previous incidents, quantities of hazardous materials used in and moving through the community, and high-hazard occupancies.

The risk assessment comprises a hazard assessment, a vulnerability assessment, and a threat assessment. The hazard assessment identifies the dangers that occupancy or piece of infrastructure may pose during a natural or man-made disaster, including the consequences of an unintended release. The vulnerability assessment identifies how easily that particular hazard may be released. A high-hazard occupancy that has very little vulnerability may actually pose a much lower risk to the community than a low to moderate hazard that has much vulnerability. The threat assessment consists of the types of dangers posed to that occupancy, such as the threat of terrorism or severe weather. The risk assessment should drive the prioritization of your hazmat response team's training and equipment.

This analysis of your jurisdiction can also be very helpful in formulating an idea of where you personally should focus your resources with regard to training. The hazard identification and risk analysis found in the local ERP will also guide the training program established by the leadership of your hazmat team. It doesn't make much sense to spend an inordinate amount of time on radiation when you live in a community that has three refineries and no nuclear power plants.

Facility Emergency Response Plans

Facility ERPs are extremely helpful both before hazardous materials incidents occur—such as during training, planning, and exercise events—as well as during emergencies involving hazardous materials. There are many different types of ERPs mandated by many different agencies. For example, OSHA mandates ERPs for SARA Title III facilities that use threshold quantities of hazardous chemicals (Figure 6-1). The EPA, through RCRA, requires facilities that generate, store, or dispose of hazardous waste to have contingency plans. The DOT requires facilities that ship or receive hazardous materials to have security plans. All these plans contain helpful information for the first responder. Some information may be readily available—even mandated to be made publicly available—before a hazardous materials incident occurs, while others may not (such as the DOT mandated security plan). Typical ERPs contain the following information:

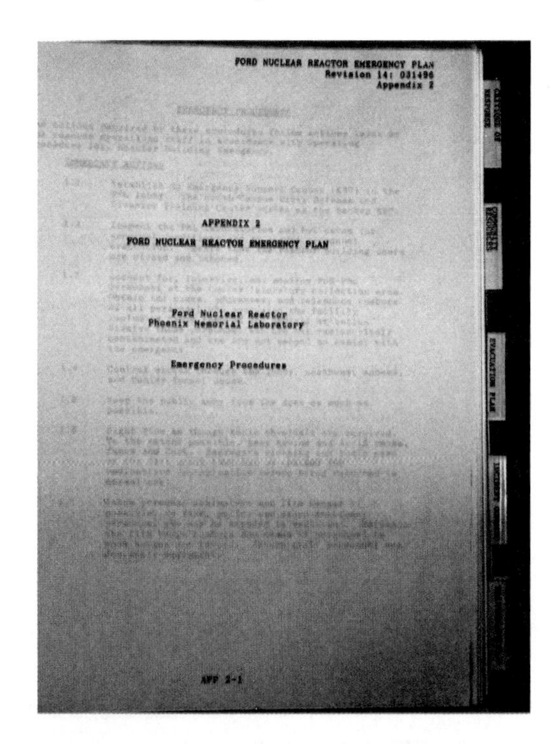

FIGURE 6-1 An example of a facility emergency response plan. Emergency response plans contain detailed information about the facility, including the locations of hazardous materials, safety data sheets, and recommended emergency response procedures. *Courtesy of Chris Weber, Dr. Hazmat, Inc.*

- physical hazard evaluation
- chemical hazard evaluation
- hazard communication
- facility emergency response procedures
- emergency response training and equipment
- emergency contact numbers
- facility maps
- MSDS sheets
- chemical storage information

At hazardous materials releases at fixed-site facilities, the facility ERP should be one of the first reference materials you consult. The facility ERPs for all the SARA Title III facilities in your jurisdiction should already be in the possession of the hazmat response team and located in the emergency response vehicle—either electronically or in several binders.

SOLVED EXERCISE 6-1

Using the facility map shown, determine what chemical hazards are present at this SARA Title III facility.

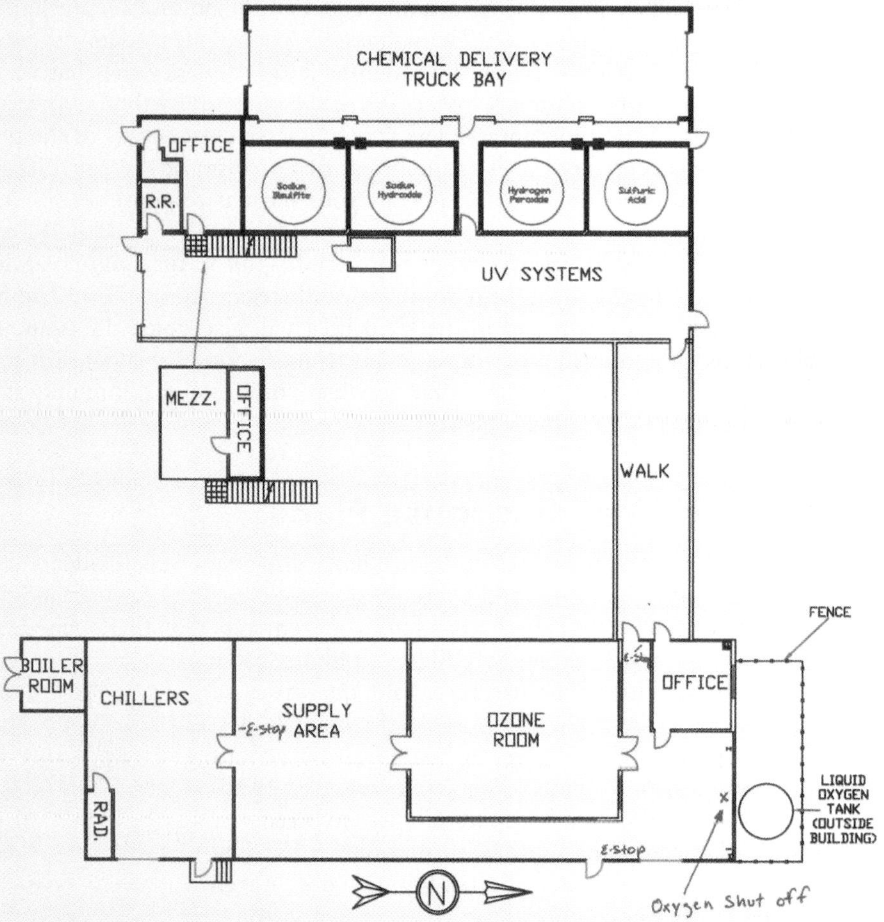

Courtesy of Chris Weber, Dr. Hazmat, Inc.

Solution: There is a liquid oxygen tank located outside the building at the northeast corner. There are also four fixed-site storage tanks containing sodium bisulfite, sodium hydroxide, hydrogen peroxide, and sulfuric acid.

Further reading of the ERP (not shown) would reveal that the liquid oxygen tank has a capacity of 9000 gallons and the four liquid tanks have a capacity of 6000 gallons each. Also, the concentrations of the chemicals are 50% hydrogen peroxide, 93% sulfuric acid, 50% sodium hydroxide, and the sodium bisulfite solution has a pH of 3.

Transportation Shipping Papers

Bulk containers carrying hazardous materials encountered during transportation will usually be marked on the outside with a placard on all four sides, stenciling, and/or a UN/NA number. This is still very limited information, even if you have the *Emergency Response Guidebook* at your disposal. How can you get more detailed information at the scene of a hazmat incident involving transportation vehicles?

shipping paper ■ The document required by DOT that describes the contents, quantity, and hazards of the materials contained in a shipment.

All modes of commercial transportation are required to carry **shipping papers.** Shipping papers will give you much more detailed information regarding contents of the shipment, including the name of the product, the quantity of the material, origination and destination locations, an emergency contact number, the hazard class, and often, packaging requirements. Obtaining the shipping papers early in the incident is very beneficial. The following is the type of information you can expect to find in the shipping papers:

1. UN/NA number: the four-digit United Nations or North American identification number.
2. Proper shipping name: the shipping name as stated in the Hazardous Materials Regulations (HMR).
3. Hazard class and division: the DOT hazard class and division.
4. **Packing group**: classification of the material based on its danger in transportation and the consequent packaging requirements:
   I—great danger
   II—medium danger
   III—minor danger
5. Type of hazard and other notations: such as "CORROSIVE" or "Marine Pollutant."
6. Total quantity (weight).
7. Number of packages and package type(s).
8. Reportable quantity (RQ) if applicable: The reportable quantity is the amount of material that if spilled must be reported to the National Spill Response Center (NRC).
9. Inhalation hazard or poison inhalation hazard (PIH):
   A—LC_{50} = 0–200 ppm
   B—LC_{50} = 200–1000 ppm
   C—LC_{50} = 1000–3000 ppm
   D—LC_{50} = 3000–5000 ppm
   As explained in Chapter 3, the LC_{50}, which stands for lethal concentration 50%, is the concentration at which 50% of the test population (such as laboratory animals) would be expected to die.
10. EPA waste stream number: The waste stream number must be included for RCRA hazardous waste shipments. This number will be required for some Class 9 materials per 40 CFR 262.20.

packing group ■ Per DOT, a classification based on the degree of danger a hazardous material presents. Packing Group I indicates great danger; Packing Group II, medium danger; and Packing Group III, minor danger.

But remember, the information contained in the shipping papers is dependent on the person completing the form! Shipping papers have been known to contain significant errors. Each year, fines are issued to the transportation industry for improper shipping papers. Additionally, there may be many shipments on any given transportation vehicle, and some may already have been delivered. More disturbingly, the driver may be doing a favor to a friend and may have accepted a shipment that is not listed on shipping papers. This is exceedingly rare, but has been known to happen.

HIGHWAY: BILL OF LADING

The **bill of lading**, or freight bill, is the shipping paper used in the highway transportation industry and must remain in the cab of the vehicle within reach of the driver at all times. Two common locations you will find the bill of lading are the front seat of the cab and in a driver's-side door compartment. On occasion, an extra copy of the shipping papers may be found located in a tube on the chassis of the trailer or within the valve compartment. Figure 6-2 shows an example of typical shipping papers found in highway transportation.

RAILROAD: WAYBILLS AND TRAIN CONSIST

Every railcar within a freight train will have an individual **waybill** indicating the shipper, consignee, and the contents of the railcar. The train **consist** is the compilation of

bill of lading ■ The shipping papers used in the road transportation industry.

waybill ■ The shipping papers most commonly used for individual railcars in the rail industry. The consolidated waybills are part of the train consist.

consist ■ The consolidated shipping papers used in the rail industry. It comprises all the individual railcar waybills and usually includes other administrative paperwork such as the order of railcars in the train.

SHIPPER PLEASE NOTE FREIGHT CHARGES ARE PREPAID ON THIS BILL UNLESS MARKED COLLECT Page 1

STRAIGHT BILL OF LADING–NOT NEGOTIABLE

Carrier: ETOH HAULERS

PLACE PRO LABEL HERE

Shipper's Bill or Lading No.

Consignee's Reference/PO No.

To: CANADIAN FUEL MIXERS
CONSIGNEE
STREET 1 E85 BLVD
DESTINATION TORONTO ON M3R 1A1 CA

TRAILER | B/L DATE
Route
SPECIAL INSTRUCTIONS

FROM: IOWA BIOFUELS
SHIPPER
STREET 1 GRAIN ALLEY
ORIGIN CITY/ST/ZIP DES MOINES IA 50307 US

FOR PAYMENT SEND BILL TO:
NAME
STREET
CITY/ST/ZIP

Collection Delivery __________________ and mail to ________________
Street__________________________ City ____________ State__________________
Carrier must collect cash, money order, bank cashier's check, or bank certified check unless shipper signs here to accept company check:
Signed:____________________

C.O.D. Charge to be paid by — Shipper ☐ Consignee ☐

Units No. Type	Packages No. Type	Kind of Package, Description of Articles, Special Marks and Exceptions (Subject to Correction)	Weight Subj. to Correction) (LBS)	Class or Rate Per. (For info. only)	Cube (Optional) Cuft
1	1 TL	UN 1170, ETHANOL, 3, BULK PACKAGE EMERGENCY CONTACT: (800) 424-9300 Ethanol Additional Services: Arrival Notification Customs Administrative Fee Single Shipment	56000		

Total Handling Pieces: 1	Individual Pieces: 1	Weight: 56000	Cube:

NOTE(1) Where the rate is dependent on value, shippers are required to state specifically in writing the agreed or declared value of the property as follows:
"The agreed or declared value of the property is specifically stated by the shipper to be not exceeding ______________________ per ______________________."
Note (2) Liability Limitation for loss and damage on this shipment may be applicable. See 49 U.S.C 14708 (c)(1)(A) and (B).
Note (3) Commodities requiring special or additional care or attention in handling or stowing must be so marked and packaged as to ensure safe transportation with ordinary care. See Sec. 2(e) of NMFC item 380.
Freight Charges are PREPAID unless marked collect

SHIPPER IOWA BIOFUELS

PER (SIGNATURE REQUIRED)

CARRIER

PER DATE

FIGURE 6-2 An example of the shipping paper used in over-the-road transportation.

the individual railcar waybills. It will be with a member of the train crew—most commonly the conductor or the engineer. The conductor is ultimately responsible for the train consist.

The consist commonly comprises four parts: the tonnage graph, the position-in-train document, the train listing, and the emergency handling instructions. The tonnage graph shows the lading of each railcar in a train. Railcars containing hazardous materials will be specially indicated—often by marking the lading in tons with "HHHH." The position-in-train document specifically lists railcars containing hazardous materials and their position in the train. The train listing contains information about each railcar in the train.

Railcars that contain hazardous materials contain extra information and are highlighted. This extra information includes the chemical name, the UN/NA identification number, the standard transportation commodity code (STCC) number, the origin and destination locations, and emergency contact information. STCC codes are unique seven-digit identification numbers. STCCs for hazardous materials start with "49," while those for hazardous waste start with "48." The emergency handling instructions will be included for all hazardous materials that are carried on the train. Different railroads may have slightly different formats but typically include all the aforementioned information. Figure 6-3 shows a typical consist used by CSX Transportation.

AIR: AIR BILL

The **air bill** is the shipping paper used in the air transportation industry. The air bill, or shipper's declaration for dangerous goods, contains much of the same information as in other shipping papers, including the following:

air bill ▪ The shipping papers used in the airline industry.

- shipper
- consignee
- airport of departure
- permitted aircraft (cargo and/or passenger aircraft)
- proper shipping name
- hazard class
- identification number
- packing group
- quantity and type of packaging
- emergency contact number

The air bill will be found in the cockpit, and the pilot is the responsible party. Copies of individual waybills will usually also be attached to the outside of the respective packages in the cargo hold. Solved Exercise 6-2 shows an example of an air dangerous goods declaration.

MARINE: DANGEROUS CARGO MANIFEST

The **dangerous cargo manifest**, also called the *IMO dangerous goods declaration,* will be found in the wheelhouse, or in a pipe-like container on a barge. The captain or master of the vessel is responsible for the dangerous cargo manifest. On large vessels it may be almost impossible to match individual waybills to cargo. Be aware that there is a difference between individual bills of lading and the ship's **manifest**. There is one bill of lading per client. The ship's manifest consolidates all these individual bills of lading and lists the entire ship's cargo. Thus, when you are involved in a shipping incident make sure to obtain the ship's manifest with a complete listing of the hazardous materials on board. Figure 6-4 shows an example of a marine bill of lading as found in a dangerous cargo manifest.

dangerous cargo manifest ▪ The shipping papers used in marine transportation. The manifest lists the hazardous materials and their locations aboard watercraft.

manifest ▪ In marine transportation, the consolidated shipping papers of ship's cargo.

SOLVED EXERCISE 6-2

You are dispatched to the local airport for a passenger suffering burns to his hands. On your arrival, the passenger states that his hands started burning after he retrieved his luggage. Determine from the air bill shown what chemicals may have leaked onto the passenger's luggage.

SHIPPER'S DECLARATION FOR DANGEROUS GOODS

Shipper
Acme Chemical Co.
23451 Main St.
Industry, MI 48564

Air Waybill No. 456-67-8973452

Page 1 of 1 Pages

Shipper's Reference Number
(optional)

Consignee
John Doe
765 Washington St.
Anytown, GA 30237

Two completed and signed copies of this Declaration must be handed to the operator.

TRANSPORT DETAILS

This shipment is within the limitations prescribed for:
(delete non-applicable)

PASSENGER AND CARGO AIRCRAFT

Airport of Departure: DTW

Airport of Destination: ATL

WARNING

Failure to comply in all respects with the applicable Dangerous Goods Regulations may be in breach of the applicable law, subject to legal penalties.

Shipment type: *(delete non-applicable)*
NON-RADIOACTIVE

NATURE AND QUANTITY OF DANGEROUS GOODS

Dangerous Goods Identification						
UN or ID No.	Proper Shipping Name	Class or Division (Subsidiary Risk)	Pack-ing Group	Quantity and type of packing	Packing Inst.	Authorization
UN1830	Sulfuric acid with more than 51 percent acid	8	II	1 Liter		

Additional Handling Information

24 Hour Emergency Contact Telephone Number: 1-800-424-9300

I hereby declare that the contents of this consignment are fully and accurately described above by the proper shipping name, and are classified, packaged, marked and labelled/placarded, and are in all respects in proper condition for transport according to applicable international and national governmental regulations. I declare that all of the applicable air transport requirements have been met.

Name/Title of Signatory
David Smith

Place and Date
8/24/2012

Signature
(see warning above)

Courtesy of Chris Weber, Dr. Hazmat, Inc.

Solution: The air bill states that commercial cargo containing sulfuric acid was being shipped in the cargo hold of this passenger plane. The passenger's signs and symptoms are consistent with corrosive material contact.

Part A: Tonnage Graph

TONNAGE GRAPH
================

TRAIN Q40629 CR TRN# ORIG: CFP 1 TIME: 0408290245 CONSIST# 991199

C = SHIPMENT GOVERNED BY CLEARANCE BUREAU INSTRUCTION
R = POTENTIAL RESTRICTED EQUIPMENT - SEE "RESTRICTED AND SPECIAL HANDLING LIST".
REFERENCE TIMETABLES, OPERATING RULES AND MECHANICAL INSTRUCTIONS FOR HANDLING IN YOUR TRAIN.
L = LONG CAR
S = SHORT CAR
* = EMPTY TOFC CAR

PLATE SIZE----------------|
TYPE----------------------|
LOAD/EMPTY----------------|

CSRR 7749 L D
CSRR 8866 L D

Hazardous Material

RES KEY	TRN POS	==CAR== NUMBER	LCP ETS	UN/NA NUMB	=======TONS===== 20	40	60	80	100	120	140	EST TONS	TRN LGTH	A X
R 9040	045-	CNW 490978	LCC		*****	*****	*****	*****	*****	*****	***	3910	2920	4
R 9040	046-	ATSF315464	LCC		*****	*****	*****	*****	*****	*****	***	4040	2975	4
	047-	RBOX553070	LAE		*****	*****	*****	*****	*****	**		4150	3025	4
R 9050	048-	CLC 10002	LFC		*****	*****	*****	*****	*****	***		4290	3070	4
R 9050	049-	TTZX 84081	LFC		*****	*****	*****	*****	*****	*****	**	4420	3130	4
R 9250	050-	CORX 4325	LTC	1824	HHHHH	HHHHH	HHHHH	HHHHH	HHHHH	HHHHH	HHH	4550	3175	4
	051-	RBOX400602	LAE		*****	*****	*****	*****	*****	*****	***	4680	3225	4
	052-	RBOX400501	LAE		*****	*****	*****	*****	*****	*****	***	4810	3275	4
	053-	DBCX 11098	LTC		*****	*****	*****	*****	*****	*****	***	4940	3325	4
R 9250	054-	DBEX123456	LTC	2448	HHHHH	HHHHH	HHHHH	HHHHH	HHHHH	HHHHH		5060	3370	4
R 9040	055-	GRAX 3333	LCC		*****	*****	*****	*****	*****	*****	*	5185	3425	4
	056-	CSXT777898	ECC		*****	***						5220	3470	4
R 9250	057-	ABCX 002	LTC	1075	HHHHH	HHHHH	HHHHH	HHHHH	HHHHH	HHHHH		5340	3525	4
R 9250	058-	CDUX543210	LTC	1075	HHHHH	HHHHH	HHHHH	HHHHH	HHHHH	HHHHH	HHH	5470	3585	4
R 9250	059-	ABCX 001	LTC	1830	HHHHH	HHHHH	HHHHH	HHHHH	HHHHH	HHHHH	HHH	5605	3645	4
	060-	CSXT887700	ECC		*****	***						5636	3725	4
	061-	DBCX109876	LTC		*****	*****	*****	*****	*****	*****	*	5761	3780	4
R 9040	062-	CNW 490835	LCC		*****	*****	*****	*****	*****	*****		5891	3840	4

GRAND TOTAL L	E	MERCHAN L	E	PIGFLAT L	E	COAL L	E	PERISH L	E	AUTORAK L	E	TOTAL TONS	TOTAL LENGTH	TOTAL AXLE
51	24	51	24	0	0	0	0	0	0	0	0	7985	4250	300

RESTRICTION KEYS***
9040 = RE RULE 7 - HIGH CUBE CAR
9250 = HAZARDOUS -SEE HAZ MATERIAL DESCRIPTION
9050 = LONG CAR
9160 = RESTRICTED TO 40 MPH - COAL WEIGHT RESTRICTION.

ENGINEER IS TO BE PROVIDED WITH TONNAGE GRAPH OR A LEGIBLE COPY OF THE TONNAGE GRAPH. THE UN/NA NUMBER COLUMN CAN BE USED TO REFERENCE THE PRODUCT NAME OF THE HAZARDOUS MATERIAL IN THE EMERGENCY RESPONSE GUIDEBOOK.

Part B: Position-in-Train Document

CT-168 REPORT - NOTICE OF RAIL CARS/UNITS CONTAINING HAZARDOUS MATERIALS

TRAIN Q40629 CR TRN# ORIG: CFP 1 TIME: 0408290245 CONSIST# 991199

THE FOLLOWING RAIL CARS/UNITS CONTAINING HAZARDOUS MATERIALS ARE LOCATED IN YOUR TRAIN. THEY MUST BE POSITIONED IN YOUR TRAIN IN ACCORDANCE WITH FEDERAL REGULATIONS. WHENEVER THERE IS A CHANGE IN THE POSITION OF ANY HAZARDOUS MATERIAL CAR IN THE TRAIN, THE CONDUCTOR (OR DESIGNEE) MUST IMMEDIATELY UPDATE THIS DOCUMENT TO SHOW THE NEW POSITION OF ALL HAZARDOUS MATERIAL CARS.

KEY FIELD CODES:
P=POISON INHALATION HAZARD ZONE A OR ZONE B
E=ENVIRONMENTALLY SENSITIVE CHEMICALS
F=FLAMMABLE GAS 2.1
X=EXPLOSIVES 1.1 OR 1.2

Location of hazmat railcars from engine

INIT	NUMBER	L E	CONTAINER INIT	CONTAINER NUMBER	L E	COMMENT	ID NUMB	KEY	TRN POS	REVISION 1ST	2ND	3RD
CORX	4325	L	___	___	_	___	1824	__	050	___	___	___
DBEX	123456	L	___	___	_	___	2448	__	054	___	___	___
ABCX	002	L	___	___	_	___	1075	_F_	057	___	___	___
CDUX	543210	L	___	___	_	___	1075	_F_	058	___	___	___
ABCX	001	L	___	___	_	___	1830	__	059	___	___	___
___	___	_	___	___	_	___	___	__	___	___	___	___
___	___	_	___	___	_	___	___	__	___	___	___	___
___	___	_	___	___	_	___	___	__	___	___	___	___
___	___	_	___	___	_	___	___	__	___	___	___	___
___	___	_	___	___	_	___	___	__	___	___	___	___
___	___	_	___	___	_	___	___	__	___	___	___	___
___	___	_	___	___	_	___	___	__	___	___	___	___

Part C: Train Listing

TRAIN LISTING AND HAZARDOUS MATERIAL DESCRIPTIONS
==

TRAIN Q40629 CR TRN# ORIG: CFP 1 TIME: 0408290245 CONSIST# 991199

CARS IN THIS CONSIST COUNT FROM FRONT TO REAR

	CAR NUMBER			CAR TYPE	COMMODITY STCC	CODE ALPHA	DESTINATION	CONSIGNEE	YZCN
49	TTZX	84081	L	F483	2421184	LUMBER	LEXINGTKY	84LUMBER	DD002
50	CORX	4325	L	T105	4935240		ECHICAGIN	CSXTRANSFL	BI185

*********** 01 TC //0195900 LB //SODIUM HYDROXIDE SOLUTION//8//
* HAZMAT * UN1824//RQ (SODIUM HYDROXIDE)//HAZMAT STCC=
*********** 4935240

U.S.A. EMERGENCY CONTACT: 18004249300
FROM SHIPPER: PURECO AUGUSTA GA
TO CONSIGNEE:CS TRANSFER E CHICAGO IN

51	RBOX	400602	L	A403	2432158	PAPER	CHICAGOIL	CHICAGOTRI	BIH007
52	RBOX	400601	L	A403	2432158	PAPER	CHICAGOIL	CHICAGOTRI	BIH007
53	DBCX	11098	L	T105	2851915	LATEX	DANVILLIL	SHERWIWIL	0ZA001
54	DBEX	123456	L	T105	4961622		OTTAWAIL	BPAMOCO	BIF084

*********** 01 TC // 0187480 LB // SULFUR, MOLTEN// 9 //NA2448//
* HAZMAT * PGIII //HAZMAT STCC=4961622
****

U.S.A. EMERGENCY CONTACT: 18004249300
FROM SHIPPER: BRADCO GLENJUNCT CO
TO CONSIGNEE: PETROCO OTTAWA IL

55	GRAX	3333	L	C313	2041993	GRAIN	BATTLCRMI	KELLOGGS	CD025

Part D: Emergency Handling Instructions (for each hazardous material on the train)

HAZARDOUS SPECIAL HANDLING INSTRUCTIONS
==

IN CASE OF ACCIDENT PROVIDE THIS LIST TO RESPONSE TEAM

TRAIN Q40629 CR TRN# ORIG: CFP 1 TIME: 0408290245 CONSIST# 991199

EMERGENCY HANDLING INSTRUCTIONS

HAZARDOUS COMMODITY 4905752

ABCX 0002 CAR 057 FROM ENGINE
CDUX 543120 CAR 058 FROM ENGINE

CLASS 2.1 (FLAMMABLE GAS)
LIQUEFIED PETROLEUM GAS (BUTANE/PROPANE MIXTURE) UN1075

Butane/propane mixture is a colorless gas with a petroleum-like odor. It is shipped as a liquefied gas under pressure. For transportation it is odorized. Contact with the material can cause frostbite. It is easily ignited. Its vapors are heavier than air. Any leak can be either liquid or vapor. It can asphyxiate by the displacement of air. Under prolonged exposure to fire or intense heat the containers may rupture violently and rocket.

If material on fire or involved in fire
- Do not extinguish fire unless flow can be stopped
- Use water in flooding quantities as fog
- Cool all affected containers with flooding quantities of water
- Apply water from as far a distance as possible

If material not on fire and not involved in fire
- Keep sparks, flames, and other sources of ignition away
- Keep material out of water sources and sewers
- Attempt to stop leak if without undue personnel hazard
- Use water spray to knock-down vapors

Personnel protection
- Avoid breathing vapors
- Keep upwind
- Wear appropriate chemical protective gloves and goggles
- Do not handle broken packages unless wearing appropriate personal protective equipment
- Approach fire with caution

Evacuation
- If fire becomes uncontrollable or container is exposed to direct flame—consider evacuation of one-half (1/2) mile radius
- If material leaking (not on fire) consider evacuation from downwind area based on amount of material spilled, location and weather conditions

First aid responses
- Move victim to fresh air; call emergency medical care.
- If not breathing, give artificial respiration.

FIGURE 6-3 An example of a truncated train consist commonly used in rail transportation. *Courtesy of CSX Transportation.*

IMO DANGEROUS GOODS DECLARATION

1 Shipper/Consignor/Sender	2 Transport Document Number	
	3 Page of pages	4 Shipper's Reference
6 Consignee	5 Freight Forwarder's Reference	
	7 Carrier (to be declared by the Carrier)	

SHIPPER'S DECLARATION
I hereby declare that the contents of this consignment are fully and accurately described below by the Proper Shipping Name, and are classified, packaged, marked and labelled/placarded and are in all respects in proper condition for transport according to the applicable international and national governmental regulations.

10 Vessel/Flight & Date	11 Port/Place Handling	9 Additional Handling Information
12 Port/Place of Discharge	13 Destination	

14 Shipping Marks	*Number & Kind of Packages, Description of Goods	GW (kg)	CUBE (m^3)

15 CTU ID No.	16 Seal No.	17 CTU Size & Type	18 Tare Mass (kg)	19 Total Gross Mass (kg)

CONTAINER/VEHICLE PACKING CERTIFICATE
I hereby declare that the goods described above have been packed/loaded into the container/vehicle identified above in accordance with the applicable provisions of IMDG code 5.4.2.
MUST BE COMPLETED AND SIGNED FOR ALL CONTAINER/VEHICLE LOADS BY PERSON RESPONSIBLE FOR PACKING/LOADING

21 Receiving Organization Receipt
Received the above number of packages/containers/trailers in apparent good order and condition, unless stated hereon:
RECEIVING ORGANIZATION REMARKS:

20 Name of Company	Hauler's Name	22 Name of Company Preparing Note
Name/status of Declarant	Vehicle Registration No.	Name/Status of Declarant
Place and Date	Driver Name and Date	Place and Date
Signature of Declarant	Driver's Signature	Signature of Declarant

FIGURE 6-4 An example of a portion of a manifest used in marine transportation.

Print Reference Materials

Many print reference materials are available to the hazardous materials technician. These resources are generally not facility specific, and each has its own advantages and disadvantages. Any single reference book will contain very specific information. Some reference materials list chemical and physical properties, health effects of chemicals, response guidelines, and/or environmental effects of chemicals. The reference sources you choose will depend greatly on the challenges with which you are faced at any given hazardous materials incident. Let's take a look at some of the most common reference materials that are available to you and how they can benefit you at a hazardous materials release.

MATERIAL SAFETY DATA SHEETS (MSDSs)

Material Safety Data Sheets (MSDSs) are usually found at fixed-site facilities. However, MSDS may also be found in transportation attached to shipping papers in lieu of a copy of the current edition of the DOT *Emergency Response Guidebook*. Industrial facilities that use hazardous materials are required to keep an MSDS for each one of them according to the hazard communication standard for plant workers (29 CFR 1910.1200) and to the Emergency Planning & Community Right-to-Know Act (EPCRA). When responding to hazardous materials releases at fixed-site facilities always ask for the pertinent MSDSs.

Material Safety Data Sheet (MSDS) ■ A technical bulletin prepared by product manufacturers to provide workers with detailed information regarding the dangerous properties and handling recommendations of commercial products containing hazardous substances.

You should become familiar with the hazardous materials stored at the fixed site facilities in your response area. The facility plans should include the appropriate MSDSs. In addition, if you are called on to respond to one of these facilities, the MSDSs should be on location a safe distance away from where the hazardous material is stored or used.

MSDSs contain the following information organized into 16 sections that may be of use to technician level responders:

- manufacturer and emergency contact information
- product and hazard identification
- components and hazardous ingredients
- first-aid measures
- firefighting procedures
- accidental release procedures
- handling and storage procedures
- exposure control and personal protection
- chemical and physical properties
- stability and reactivity
- toxicological information
- ecological information
- disposal considerations
- transport information
- regulatory information
- other information

Although there is an ANSI standard for MSDSs, a wide variety of formats are used. Internationally, MSDSs are known as SDSs, which stands for Safety Data Sheets. You will see and hear the term SDS more often as global harmonization is implemented. SDSs contain virtually the identical information as MSDSs, but the order and format are standardized. Figure 6-5 shows an example of an MSDS for sulfuric acid.

At hazardous materials incidents, MSDSs can also be obtained directly from the manufacturer, from CHEMTREC, or from online databases in the event of an emergency. Organizations similar to CHEMTREC exist in Canada, where it is known as CANUTEC, and in Mexico, where it is known as SETIQ.

Material Safety Data Sheet

Revision Issued: 9/28/2011 Supercedes: 9/17/2008 First Issued: 1/02/1986

Section I - Chemical Product And Company Identification

Product Name: Sulfuric Acid

CAS Number: 7664-93-9 HBCC MSDS No. CS18100

HILL BROTHERS Chemical Co.

1675 NORTH MAIN STREET • ORANGE, CALIFORNIA 92867-3499
(714) 998-8800 • FAX: (714) 998-6310
http://hillbrothers.com

1675 No. Main Street, Orange, California 92867
Telephone No: 714-998-8800 | Outside Calif: 800-821-7234
Chemtrec: 800-424-9300

Section II - Composition/Information On Ingredients

			Exposure Limits (TWAs) in Air		
Chemical Name	CAS Number	%	ACGIH TLV	OSHA PEL	STEL
Sulfuric Acid (H_2SO_4)	7664-93-9	20-99	0.2 mg/m³	1 mg/m³	3 mg/m³
Sulfur Dioxide	7446-09-5	< 2	2 ppm	5 ppm	5 ppm

Section III - Hazard Identification

Routes of Exposure: Sulfuric acid can affect the body if it is inhaled or if it comes in contact with the eyes or skin. It can also affect the body if it is swallowed.
Points of Attack: Sulfuric acid attacks the respiratory system, eyes, skin, teeth, and lungs.
Summary of Acute Health Hazards: Concentrated sulfuric acid will effectively remove the elements of water from many organic materials with which it comes in contact. It is even more rapidly injurious to mucous membranes and exceedingly dangerous to the eyes.
Ingestion: Corrosive. Causes serious burns of the mouth or perforation of the esophagus or stomach. May be fatal if swallowed.
Inhalation: Corrosive and highly toxic. May be harmful or fatal if inhaled. May cause severe irritation and burns of the nose, throat and respiratory tract.
Skin: Corrosive. Splashes on the skin will cause severe skin burns. Burning and charring of the skin are a result of the great affinity for, and strong exothermic reaction with, water. Direct contact can be severely irritating to the skin and may result in redness, swelling, burns and severe skin damage.
Eyes: Corrosive. Direct contact with the liquid or exposure to vapors or mists may cause stinging, tearing, redness, swelling, corneal damage and irreversible eye damage. Splashes in the eyes will cause severe burns. Contact lenses should not be worn when working with this chemical.
Effects of Overexposure: May cause severe irritation and burns of the mouth, nose, throat, respiratory and digestive tract, coughing, nausea, vomiting, abdominal pain, chest pain, pneumonitis (inflammation of the fluid in the lungs), pulmonary edema (accumulation of the fluid in the lungs), and perforation of the stomach. Overexposure to acid mists has been reported to cause erosion to tooth enamel.

Product Name: Sulfuric Acid Page 1 of 6

Medical Conditions Generally Aggravated by Exposure: Persons with pre-existing skin disorders and/or respiratory disorders (e.g. Asthma-like conditions) may be more susceptible to the effects of this material, and may be aggravated by exposure to this material.
Note to Physicians: Sulfuric acid is reported to cause pulmonary function impairment. Periodic surveillance is indicated. Sulfuric acid may cause acute lung damage. Surveillance of the lungs is indicated. Ingestion may cause gastroesophageal perforation. Perforation may occur within 72 hours, but along with abscess formation, can occur weeks later. Long term complications may include esophageal, gastric or pyloric strictures or stenosis.

Section IV - First Aid Measures

Ingestion: If liquid sulfuric acid or solutions containing sulfuric acid have been swallowed and the person is conscious, give him one glass of water (1/2 glass of water to children under 5), immediately to dilute the sulfuric acid. Do NOT induce vomiting. Do not attempt to make the exposed person vomit. Do not leave victim unattended. GET MEDICAL ATTENTION IMMEDIATELY.
Inhalation: If a person breathes in large amounts of sulfuric acid, move the exposed person to fresh air at once. If breathing has stopped, perform artificial respiration. If breathing is difficult, give oxygen. Keep the affected person warm and at rest. GET MEDICAL ATTENTION AS SOON AS POSSIBLE.
Skin: If liquid sulfuric acid or solutions containing sulfuric acid get on the skin, immediately flush the contaminated skin with water for at least 15 minutes. If skin surface is damaged, apply a clean dressing. If liquid sulfuric acid or solutions containing sulfuric acid penetrate through the clothing, immediately remove the clothing, shoes and constrictive jewelry under a safety shower and continue to wash the skin for at least 15 minutes. GET MEDICAL ATTENTION IMMEDIATELY.
Eyes: If liquid sulfuric acid or solutions containing sulfuric acid get into the eyes, flush eyes immediately with a directed stream of water for at least 30 minutes while forcibly holding eyelids apart to ensure complete irrigation of all eye and lid tissue. GET MEDICAL ATTENTION IMMEDIATELY. Contact lenses should not be worn when working with this chemical.

Section V - Fire Fighting Measures

Flash Point: Non-flammable **Autoignition Temperature:** N/A
Lower Explosive Limit: N/A **Upper Explosive Limit:** N/A
Unusual Fire and Explosion Hazards: Not flammable but highly reactive and capable of igniting finely divided combustible materials on contact. Reacts violently with water and organic materials with evolution of heat. If involved in fire, may release hazardous oxides of sulfur. Vapors are heavier than air and may accumulate in low areas. Containers exposed to extreme heat may rupture due to pressure buildup. Contact with common metals may generate hydrogen, which can form flammable mixture with air. Fire may produce irritating, corrosive, and/or toxic gases.
Extinguishing Media: Fires involving small amount of combustibles may be smothered with suitable dry chemical, soda ash, lime, sand or CO2. Use water on combustibles burning in vicinity of this material but use care as water applied directly to this acid result in evolution of heat and causes splattering.
Special Firefighting Procedures: Causes severe, deep burns to tissue; very corrosive effect. Sulfuric Acid is extremely slippery. Emergency responders in the danger area should wear bunker gear and self contained breathing apparatus for fires beyond the incipient stage (29CFR 1910.156). In addition, wear other

Product Name: Sulfuric Acid Page 2 of 6

FIGURE 6-5 The material safety data sheet (MSDS) for sulfuric acid showing the detailed technical information that is provided. Useful information includes hazard identification information and first-aid measures, as well as firefighting and spill response advice.

appropriate protective equipment as conditions warrant (see Section 8). Water reactive. Contact with water may generate heat. Isolate damage area, keep unauthorized personnel out. If tank, railcar, or tank truck is involved in a fire, isolate for ½ mile in all directions. Consider initial evacuation for ½ mile in all directions. Stop spill/release if it can be done with minimal risk. Move undamaged containers from danger area if it can be done with minimal risk. Fires involving small amounts of combustibles may be smothered with suitable dry chemicals. Use water on combustibles burning but avoid using water directly on acid as it results in evolution of heat and causes splattering.

Section VI - Accidental Release Measures

If sulfuric acid is spilled or leaked, ventilate area. Wear appropriate protective equipment including respiratory protection as conditions warrant (see Section 8). Collect spilled or leaked material in the most convenient and safe manner for reclamation or for disposal in a secured sanitary landfill. Sulfuric acid should be absorbed in vermiculite, dry sand, earth, or a similar material. It may also be diluted and neutralized. Add slowly to solution of soda ash and slaked lime with stirring. Use Caution around spill area, Sulfuric Acid is extremely slippery. Stay upwind and away from spill release. Avoid discharge into drains, water courses or onto the ground.

Section VII - Handling and Storage

Protect against physical damage and water. Keep containers closed. Sulfuric Acid is extremely slippery. Do not enter confined spaces such as tanks or pits without following proper entry procedures such as ASTM D-4276. To prevent ignition of hydrogen gas generated in metal containers (from metal contact) smoking, open flames and sparks must not be permitted in storage areas. This product has a great affinity for water, abstracting it from the air and also from many organic substances; hence it will char wood, etc. When diluting, the acid should be added to the diluent. Separate from carbides, chlorates, fulminates, nitrates, picrates, powdered metals, and combustible materials. Keep away from strong oxidizing agents including oxygen and chlorine.
Other Precautions: Persons not wearing protective equipment and clothing should be restricted from areas of spills or leaks until cleanup has been completed.

Section VIII - Exposure Controls/Personal Protection

Respiratory Protection: Good industrial hygiene practices recommend that engineering controls be used to reduce environmental concentrations to the permissible exposure level. However, there are some exceptions where respirators may be used to control exposure. Respirators may be used when engineering and work practice controls are not technically feasible, when such controls are in the process of being installed, or when they fail and need to be supplemented. If the use of respirators is necessary, a NIOSH/MSHA approved air purifying respirator with N95 filter may be used under conditions where airborne concentrations are expected to exceed exposure limits (see Section II). Protection provided by air purifying respirators is limited (see manufacturers respirator selection guide). Use a positive pressure air supplied respirator if there is potential for an uncontrolled release, exposure levels are not known, or any other circumstances where air purifying respirators may not provide adequate protection. A respiratory protection program that meets OSHA'a 29 CFR 1910.134 and ANSI Z88.2 requirements must be followed whenever workplace conditions warrant a respirator's use.
Ventilation: General mechanical ventilation (typically 10 air changes per hour) may be sufficient to keep sulfuric acid vapor concentrations within specified time-weighted TLV range. If general ventilation proves inadequate to maintain safe vapor

concentrations, supplemental local exhaust may be required.
Protective Clothing: Employees should be provided with and required to use impervious clothing, gloves, face shields (eight-inch minimum), and other appropriate protective clothing necessary to prevent any possibility of skin contact with liquid sulfuric acid or solutions containing more than 1% sulfuric acid by weight.
Eye Protection: Employees should be provided with and required to use splash-proof safety goggles where there is any possibility of liquid sulfuric acid or solutions containing sulfuric acid contacting the eyes. Contact lenses should not be worn when working with this chemical.
Other Protective Clothing or Equipment: Rubber apron, rubber boots, eyewash stations and safety showers must be available in the immediate work area for emergency use.
Work/Hygienic Practices: Avoid contact with the skin and avoid breathing vapors. Do not eat, drink, or smoke in work area. Wash hands before eating, drinking, or using restroom. Do NOT place food, coffee or other drinks in the area where dusting or splashing of solutions is possible.

Section IX - Physical and Chemical Properties

Physical State: Liquid
pH: 0.03 (1N Solution)
Melting Point/Range: 11° C; 51.8°F
Boiling Point/Range: 105-325°C (221-616°F (20-100% H2SO4)
Appearance/Color/Odor: Colorless to dark brown; odorless unless hot, then choking irritant.
Solubility in Water: 100%
Vapor Pressure(mmHg): < .00120 mm

% Acid in Solution	20	30	35	36	40	50	72	75-99
Specific Gravity	1.09-1.10	1.2-1.3	1.3	1.3	1.3	1.4	1.6-1.7	1.6-1.84
Weight/Gallon in Lbs.	9.08-9.17	10-10.83	10.83	10.83	10.83	11.67	13.33-14.17	15.2

Evaporation Rate (N-Butyl Acetate=1): < 1
Molecular Weight: 98
Vapor Density(Air=1): 3.4
% Volatiles: Negligible
How to detect this compound : Sampling and analyses may be performed by collection of sulfuric acid on a cellulose membrane filter, followed by extraction with distilled water and isopropyl alcohol, treatment with perchloric acid, and titration with barium perchlorate. Also, detector tubes certified by NIOSH under 42 CFR Part 84 or other direct-reading devices calibrated to measure sulfuric acid may be used.

Section X - Stability and Reactivity

Stability: Sulfuric Acid reacts vigorously, violently or explosively with many organic and inorganic chemicals and with water.
Hazardous Polymerization: Will not occur
Conditions to Avoid: Temperatures above 150°F. Exposure to moist air or water.
Materials to Avoid: Contact of acid with organic materials (such as chlorates, carbides, fulminates, and picrates), alkaline materials and water may cause fires and explosions. Contact of acid with metals may form toxic sulfur dioxide fumes and flammable hydrogen gas. Contact with hypochlorites (e.g., chlorine bleach), sulfides, or cyanides will produce toxic gases.

FIGURE 6-5 *(Continued)*

(continued)

Hazardous Decomposition Products: Toxic gases and vapors (such as sulfuric acid fume, sulfur dioxide, and carbon monoxide) may be released when sulfuric acid decomposes. Decomposes to water and sulfur trioxide above 644°F.

Section XI - Toxicological Information

(Sulfuric acid) mist severely irritates the eyes, respiratory tract, and skin. Concentrated sulfuric acid destroys tissue due to its severe dehydrating action, whereas the dilute form acts as a mild irritant due to acid properties. The LC50 of mist of 1-micron particle size for an 8 hour exposure was 50 mg/m³ for adult guinea pigs and 18 mg/m³ for young animals. Continuous exposure of guinea pigs to 2 mg/m³ for 5 days caused pulmonary edema and thickening of the alveolar walls; exposure of guinea pigs to 2 mg/m³ for 1 hour caused an increase in pulmonary airway resistance from reflex bronchoconstriction. A worker sprayed in the face with liquid fuming sulfuric acid suffered skin burns of the face and body, as well as pulmonary edema from inhalation. Sequelae were pulmonary fibrosis, residual bronchitis, and pulmonary emphysema; in addition, necrosis of the skin resulted in marked scarring. In human subjects, concentrations of about 5 mg/m³ were objectionable, usually causing cough, an increase in respiratory rate, and impairment of ventilatory capacity. Workers exposed to concentrations of 12.6 to 35 mg/m³ had a markedly higher incidence of erosion and discoloration of teeth than was noted in unexposed individuals. Splashed in the eye, the concentrated acid causes extremely severe damage, often leading to blindness, whereas dilute acid produces more transient effects from which recovery may be complete. Repeated exposure of workers to the mist causes chronic conjunctivitis, tracheobronchitis, stomatitis, and dermatitis, as well as dental erosion. While ingestion of the liquid is unlikely in ordinary industrial use, the highly corrosive nature of the substance may be expected to produce serious mucous membrane burns of the mouth and esophagus.
Epidemiology: Workers exposed to industrial sulfuric acid mist showed a statistical increase in laryngeal cancer. This suggests a possible relationship between carcinogenesis and inhalation of sulfuric acid mist.

Section XII - Ecological Information

Ecotoxicity: Fish: Bluegill/Sunfish: 49 mg/L; 48 Hr; TLm (tap water @ 20°C)

Fish: Bluegill/Sunfish: 24.5 ppm; 48 Hr; TLm (fresh water)

Section XIII - Disposal Considerations

Sulfuric acid may be placed in sealed containers or absorbed in vermiculite, dry sand, earth, or a similar material and disposed of in a secured sanitary landfill. It may also be diluted and neutralized. Check with your Federal, State, and Local authorities as neutralized sulfuric acid may be allowed to be flushed down the drain. Empty containers must be handled with care due to material residue.

Section XIV - Transport Information

DOT Proper Shipping Name: Sulfuric Acid
DOT Hazard Class/ I.D. No.:
8, UN2796, II (with not more than 51% acid)
8, UN1830, II (with more than 51% acid)

Section XV - Regulatory Information

<u>**Sulfuric Acid**</u>
Section 302 Extremely Hazardous Substance (EHS): CAS # 7664-93-9
1000 Lbs. (454 Kilograms) (85 Gals.) Threshold Planning Quantity (TPQ)
Section 304 Extremely Hazardous Substance (EHS): CAS # 7664-93-9
1000 Lbs. (454 Kilograms) (85 Gals.) Reportable Quantity (RQ)
CERCLA Hazardous Substance: CAS #7664-93-9
1000 Lbs. (454 Kilograms) (85 Gals.) Reportable Quantity (RQ)
SARA 313: This material contains 20-99% Sulfuric Acid (CAS# 7664-93-9), which is subject to the reporting requirements of Section 313 of SARA Title III and 40 CFR Part 373. Sulfuric Acid (acid aerosols including mists, vapors, gas, fog, and other airborne forms of any particle size).
<u>**Sulfur Dioxide**</u>
Section 302 Extremely Hazardous Substance (EHS): CAS # 7446-09-5
500 Lbs. (227 Kilograms) (42.5 Gals.) Threshold Planning Quantity (TPQ)
Section 304 Extremely Hazardous Substance (EHS): CAS # 7446-09-5
500 Lbs. (227 Kilograms) (42.5 Gals.) Reportable Quantity (RQ)
Warning
This product contains Sulfuric Acid, listed as 'Strong inorganic acid mists contain', a chemical known to the State of California to cause cancer.
NFPA Rating: Health - 3. Flammability - 0; Instability – 2; Special Hazard: -W-
0=Insignificant 1=Slight 2=Moderate 3=High 4=Extreme
Carcinogenicity Lists:
ACGIH: A2 – Suspected Human Carcinogen (Sulfuric Acid contained in strong inorganic acid mists)
National Toxicology Program (NTP): Known carcinogen (listed as 'Strong inorganic acid mists containing Sulfuric Acid).
International Agency for Research on Cancer (IARC) Monograph: Group 1 carcinogen (Sulfuric Acid)
Occupational Safety & Health Administration (OSHA) Regulated: Yes

Section XVI - Other Information

Synonyms/Common Names: H_2SO_4; Oil of Vitriol; Spirit of Sulfur; Hydrogen Sulfate; Oleum
Chemical Family/Type: Inorganic Acid
Section changed since last revision: III, V, VI, VIII

IMPORTANT! <u>Read this MSDS before use or disposal of this product. Pass along the information to employees and any other persons who could be exposed to the product to be sure that they are aware of the information before use or other exposure.</u> This MSDS has been prepared according to the OSHA Hazard Communication Standard [29 CFR 1910.1200]. The MSDS information is based on sources believed to be reliable. However, since data, safety standards, and government regulations are subject to change and the conditions of handling and use, or misuse are beyond our control, **Hill Brothers Chemical Company** makes no warranty, either expressed or implied, with respect to the completeness or continuing accuracy of the information contained herein and disclaims all liability for reliance thereon. Also, additional information may be necessary or helpful for specific conditions and circumstances of use. It is the user's responsibility to determine the suitability of this product and to evaluate risks prior to use, and then to exercise appropriate precautions for protection of employees and others.

FIGURE 6-5 *(Continued)*

SOLVED EXERCISE 6-3

According to the MSDS for sulfuric acid shown in Figure 6-5, how should you handle a victim with chemical burns to his or her hands?

Solution: Skin: If liquid sulfuric acid or solutions containing sulfuric acid get on the skin, immediately flush the contaminated skin with water for at least 15 minutes. If skin surface is damaged, apply a clean dressing. If liquid sulfuric acid or solutions containing sulfuric acid penetrate through the clothing, immediately remove the clothing, shoes and constrictive jewelry under a safety shower and continue to wash the skin for at least 15 minutes. GET MEDICAL ATTENTION IMMEDIATELY.

NIOSH POCKET GUIDE TO CHEMICAL HAZARDS

The *NIOSH Pocket Guide to Chemical Hazards (NPG)* is a wonderful source of information about hazardous materials when their identity is known. NIOSH, which is part of the Centers for Disease Control and Prevention, developed this pocket guide containing detailed information for hazardous materials that pose an inhalation hazard. This information includes chemical and physical properties, chemical reactivity, exposure limits, personal protective actions, health effects, and first-aid measures. Almost 700 chemicals are listed in alphabetical order, with a synonym and trade name index in the back. The *NIOSH Pocket Guide* is updated every few years and can be ordered online or downloaded as a PDF file for free. Figure 6-6 shows an example of the *NIOSH Pocket Guide* entry for gasoline.

The *NIOSH Pocket Guide* contents include the following:

- Explanatory material (tables of abbreviations, definitions)
- Data on hazardous materials in alphabetical order
- Appendices (Carcinogens, Supplementary Exposure Limits, Respirator Requirements, and 1989 Vacated PELs)
- Chemical Abstracts Services (CAS) Number Index
- DOT UN/NA Identification Number Index
- Chemical, Synonym, and Trade Name Index

Gasoline	**Formula:**	**CAS#:** 8006-61-9	**RTECS#:** LX3300000	**IDLH:** Ca [N.D.]
Conversion: 1 ppm = 4.5 mg/m^3 (approx)	**DOT:** 1203 128			

Synonyms/Trade Names: Motor fuel, Motor spirits, Natural gasoline, Petrol [**Note:** A complex mixture of volatile hydrocarbons (paraffins, cycloparaffins & aromatics).]

Exposure Limits:	**Measurement Methods (see Table 1):**
NIOSH REL: Ca See Appendix A **OSHA PEL†:** none	**OSHA** PV2028
Physical Description: Clear liquid with a characteristic odor.	

Chemical & Physical Properties:	**Personal Protection/Sanitation (see Table 2):**	**Respirator Recommendations (see Tables 3 and 4):**
MW: 110 (approx) **BP:** 102°F **Sol:** Insoluble **Fl.P:** -45°F **IP:** ? **Sp.Gr(60°F):** 0.72-0.76 **VP:** 38-300 mmHg **FRZ:** ? **UEL:** 7.6% **LEL:** 1.4% Class IB Flammable Liquid	**Skin:** Prevent skin contact **Eyes:** Prevent eye contact **Wash skin:** When contam **Remove:** When wet (flamm) **Change:** N.R. **Provide:** Eyewash Quick drench	**NIOSH** **¥:** ScbaF:Pd,Pp/SaF:Pd,Pp:AScba **Escape:** GmFOv/ScbaE

Incompatibilities and Reactivities: Strong oxidizers such as peroxides, nitric acid & perchlorates

Exposure Routes, Symptoms, Target Organs (see Table 5):	**First Aid (see Table 6):**
ER: Inh, Abs, Ing, Con **SY:** Irrit eyes, skin, muc memb; derm; head, lass, blurred vision, dizz, slurred speech, conf, convuls; chemical pneu (aspir liquid); possible liver, kidney damage; [carc] **TO:** Eyes, skin, resp sys, CNS, liver, kidneys [in animals: liver & kidney cancer]	**Eye:** Irr immed **Skin:** Soap flush immed **Breath:** Resp support **Swallow:** Medical attention immed

FIGURE 6-6 The *NIOSH Pocket Guide* entry for gasoline. Useful information includes health effects and chemical and physical properties, as well as reactivity information. *Courtesy of National Institute for Occupational Safety and Health.*

It is sometimes difficult to determine hazards such as corrosivity and radioactivity from the description in the *NIOSH Pocket Guide* alone. The three-digit ERG guide number near the top in the DOT section can be used to help identify the hazards of chemicals. Corrosive materials have a guide number of 118 (bases – organic amines); or 123, 124, 125 (acidic and basic gases); or 167 (refrigerated liquid fluorine); or 136, 137, 153, 154, 155, 156, 157 (corrosive liquids and solids). Explosives have a guide number of 112, 113, or 114. Radioactive materials have a guide number between 161 and 166. Materials that have a guide number that ends in "P" polymerize.

CHRIS MANUAL

The CHRIS (Chemical Hazard Response Information System) manual is published by the U.S. Coast Guard and is especially useful for hazardous materials releases that pose an environmental hazard, especially to waterways. The CHRIS manual contains detailed temperature-specific chemical and physical property information as well as the solubility of chemicals in different solvents, typically including water, alcohols, and ether, for approximately 1300 different chemicals. The following sections are included on the typical CHRIS manual data sheet:

- Cautionary response information
- Corrective response actions
- Chemical designations
- Health hazards
- Fire hazards
- Chemical reactivity
- Water pollution
- Shipping information
- Hazard classifications
- Physical and chemical properties
- Temperature-dependent physical properties including saturated liquid density, liquid heat capacity, liquid thermal conductivity, liquid viscosity, solubility in water, saturated vapor pressure, saturated vapor density, and ideal gas heat capacity

Some of this information can be fairly difficult to find in other sources. Some of the more difficult physical properties to find are water pollution data and viscosity. Vapor pressure is reported in pounds per square inch (psi), where 1 psi is equivalent to 51.7 mmHg. Figure 6-7 shows an example of the CHRIS manual entry for toluene. The CHRIS manual entries have been incorporated into CAMEO (see further) and can be accessed as PDF files.

EMERGENCY ACTION GUIDE

The *Emergency Action Guide* published by the Association of American Railroads (AAR) provides detailed information—including spill response and cleanup information—for

SOLVED EXERCISE 6-4

According to the CHRIS manual entry for toluene shown in Figure 6-7, what are the implications of a toluene spill into a municipal reservoir?

Solution:

"Dangerous to aquatic life in high concentrations.
Fouling to shoreline.
May be dangerous if it enters water intakes.
Notify local health and wildlife officials.
Notify operators of nearby water intakes."
"6.1 Aquatic Toxicity: 1180 mg/l/96 hr/sunfish/TLm/fresh water"

TOLUENE

TOL

CAUTIONARY RESPONSE INFORMATION

Common Synonyms
Methylbenzene
Methylbenzol
Toluol

Watery liquid — Colorless — Pleasant odor

Floats on water. Flammable, irritating vapor is produced.

Keep people away.
Shut off ignition sources and call fire department.
Stay upwind and use water spray to ``knock down" vapor.
Avoid contact with liquid and vapor.
Notify local health and pollution control agencies.
Protect water intakes.

Fire

FLAMMABLE.
Flashback along vapor trail may occur.
Vapor may explode if ignited in an enclosed area.
Wear goggles and self-contained breathing apparatus.
Extinguish with dry chemical, foam, or carbon dioxide.
Water may be ineffective on fire.
Cool exposed containers with water.

Exposure

CALL FOR MEDICAL AID.

VAPOR
Irritating to eyes, nose and throat.
If inhaled, will cause nausea, vomiting, headache, dizziness, difficult breathing, or loss of consciousness.
Move to fresh air.
If breathing has stopped, give artificial respiration.
If breathing difficult, give oxygen.

LIQUID
Irritating to skin and eyes.
If swallowed, will cause nausea, vomiting or loss of consciousness.
Remove contaminated clothing and shoes.
Flush affected areas with plenty of water.
IF IN EYES, hold eyelids open and flush with plenty of water.
IF SWALLOWED and victim is CONSCIOUS, have victim drink water or milk.
DO NOT INDUCE VOMITING.

Water Pollution

Dangerous to aquatic life in high concentrations.
Fouling to shoreline.
May be dangerous if it enters water intakes.
Notify local health and wildlife officials.
Notify operators of nearby water intakes.

1. CORRECTIVE RESPONSE ACTIONS

Stop discharge
Contain
Collection Systems: Skim
Chemical and Physical Treatment: Burn
Clean shore line
Salvage waterfowl

2. CHEMICAL DESIGNATIONS

2.1 **CG Compatibility Group:** 32; Aromatic Hydrocarbon
2.2 **Formula:** $C_6H_5CH_3$
2.3 **IMO/UN Designation:** 3.2/1294
2.4 **DOT ID No.:** 1294
2.5 **CAS Registry No.:** 108-88-3
2.6 **NAERG Guide No.:** 130
2.7 **Standard Industrial Trade Classification:** 51123

3. HEALTH HAZARDS

3.1 **Personal Protective Equipment:** Air-supplied mask; goggles or face shield; plastic gloves.
3.2 **Symptoms Following Exposure:** Vapors irritate eyes and upper respiratory tract; cause dizziness, headache, anesthesia, respiratory arrest. Liquid irritates eyes and causes drying of skin. If aspirated, causes coughing, gagging, distress, and rapidly developing pulmonary edema. If ingested causes vomiting, griping, diarrhea, depressed respiration.
3.3 **Treatment of Exposure:** INHALATION: remove to fresh air, give artificial respiration and oxygen if needed; call a doctor. INGESTION: do NOT induce vomiting; call a doctor. EYES: flush with water for at least 15 min. SKIN: wipe off, wash with soap and water.
3.4 **TLV-TWA:** 50 ppm
3.5 **TLV-STEL:** Not listed.
3.6 **TLV-Ceiling:** Not listed.
3.7 **Toxicity by Ingestion:** Grade 2; LD_{50} = 0.5 to 5 g/kg
3.8 **Toxicity by Inhalation:** Currently not available.
3.9 **Chronic Toxicity:** Kidney and liver damage may follow ingestion.
3.10 **Vapor (Gas) Irritant Characteristics:** Vapors cause a slight smarting of the eyes or respiratory system if present in high concentrations. The effect is temporary.
3.11 **Liquid or Solid Characteristics:** Minimum hazard. If spilled on clothing and allowed to remain, may cause smarting and reddening of the skin.
3.12 **Odor Threshold:** 0.17 ppm
3.13 **IDLH Value:** 500 ppm
3.14 **OSHA PEL-TWA:** 200 ppm
3.15 **OSHA PEL-STEL:** 500 ppm, 10 minute peak once in 8 hour shift
3.16 **OSHA PEL-Ceiling:** 300 ppm
3.17 **EPA AEGL:** Not listed

4. FIRE HAZARDS

4.1 **Flash Point:** 55°F O.C. 40°F C.C.
4.2 **Flammable Limits in Air:** 1.27%-7%
4.3 **Fire Extinguishing Agents:** Carbon dioxide or dry chemical for small fires, ordinary foam for large fires.
4.4 **Fire Extinguishing Agents Not to Be Used:** Water may be ineffective
4.5 **Special Hazards of Combustion Products:** Not pertinent
4.6 **Behavior in Fire:** Vapor is heavier than air and may travel a considerable distance to a source of ignition and flash back.
4.7 **Auto Ignition Temperature:** 896°F
4.8 **Electrical Hazards:** Class I, Group D
4.9 **Burning Rate:** 5.7 mm/min.
4.10 **Adiabatic Flame Temperature:** Currently not available
4.11 **Stoichometric Air to Fuel Ratio:** 42.8 (calc.)
4.12 **Flame Temperature:** Currently not available
4.13 **Combustion Molar Ratio (Reactant to Product):** 11.0 (calc.)
4.14 **Minimum Oxygen Concentration for Combustion (MOCC):** N_2 diluent: 9.5%

5. CHEMICAL REACTIVITY

5.1 **Reactivity with Water:** No reaction
5.2 **Reactivity with Common Materials:** No reaction
5.3 **Stability During Transport:** Stable
5.4 **Neutralizing Agents for Acids and Caustics:** Not pertinent
5.5 **Polymerization:** Not pertinent
5.6 **Inhibitor of Polymerization:** Not pertinent

6. WATER POLLUTION

6.1 **Aquatic Toxicity:** 1180 mg/l/96 hr/sunfish/TL_m/fresh water
6.2 **Waterfowl Toxicity:** Currently not available
6.3 **Biological Oxygen Demand (BOD):** 0%, 5 days; 38% (theor), 8 days
6.4 **Food Chain Concentration Potential:** None
6.5 **GESAMP Hazard Profile:**
Bioaccumulation: 0
Damage to living resources: 2
Human Oral hazard: 1
Human Contact hazard: II
Reduction of amenities: XXX

7. SHIPPING INFORMATION

7.1 **Grades of Purity:** Research, reagent, nitration-all 99.8 + %; industrial: contains 94 + %, with 5% xylene and small amounts of benzene and nonaromatic hydrocarbons; 90/120: less pure than industrial.
7.2 **Storage Temperature:** Ambient
7.3 **Inert Atmosphere:** No requirement
7.4 **Venting:** Open (flame arrester) or pressure-vacuum
7.5 **IMO Pollution Category:** C
7.6 **Ship Type:** 3
7.7 **Barge Hull Type:** Currently not available

8. HAZARD CLASSIFICATIONS

8.1 **49 CFR Category:** Flammable liquid
8.2 **49 CFR Class:** 3
8.3 **49 CFR Package Group:** II
8.4 **Marine Pollutant:** No
8.5 **NFPA Hazard Classification:**

Category	Classification
Health Hazard (Blue)..........	2
Flammability (Red)............	3
Instability (Yellow)............	0

8.6 **EPA Reportable Quantity:** 1000 pounds
8.7 **EPA Pollution Category:** C
8.8 **RCRA Waste Number:** U220
8.9 **EPA FWPCA List:** Yes

9. PHYSICAL & CHEMICAL PROPERTIES

9.1 **Physical State at 15° C and 1 atm:** Liquid
9.2 **Molecular Weight:** 92.14
9.3 **Boiling Point at 1 atm:** 231.1°F = 110.6°C = 383.8°K
9.4 **Freezing Point:** –139°F = –95.0°C = 178.2°K
9.5 **Critical Temperature:** 605.5°F = 318.6°C = 591.8°K
9.6 **Critical Pressure:** 596.1 psia = 40.55 atm = 4.108 MN/m^2
9.7 **Specific Gravity:** 0.867 at 20°C (liquid)
9.8 **Liquid Surface Tension:** 29.0 dynes/cm = 0.0290 N/m at 20°C
9.9 **Liquid Water Interfacial Tension:** 36.1 dynes/cm = 0.0361 N/m at 25°C
9.10 **Vapor (Gas) Specific Gravity:** Not pertinent
9.11 **Ratio of Specific Heats of Vapor (Gas):** 1.089
9.12 **Latent Heat of Vaporization:** 155 Btu/lb = 86.1 cal/g = 3.61×10^5 J/kg
9.13 **Heat of Combustion:** –17,430 Btu/lb = –9686 cal/g = -405.5×10^5 J/kg
9.14 **Heat of Decomposition:** Not pertinent
9.15 **Heat of Solution:** Not pertinent
9.16 **Heat of Polymerization:** Not pertinent
9.17 **Heat of Fusion:** 17.17 cal/g
9.18 **Limiting Value:** Currently not available
9.19 **Reid Vapor Pressure:** 1.1 psia

NOTES

FIGURE 6-7 The CHRIS manual entry for toluene. Useful information includes response information, chemical and physical properties, and detailed environmental information. *Courtesy of the United States Coast Guard.*

(*continued*)

TOLUENE

TOL

9.20 SATURATED LIQUID DENSITY

Temperature (degrees F)	Pounds per cubic foot
-30	57.180
-20	56.870
-10	56.550
0	56.240
10	55.930
20	55.620
30	55.310
40	54.990
50	54.680
60	54.370
70	54.060
80	53.750
90	53.430
100	53.120
110	52.810
120	52.500

9.21 LIQUID HEAT CAPACITY

Temperature (degrees F)	British thermal unit per pound-F
0	0.396
5	0.397
10	0.399
15	0.400
20	0.402
25	0.403
30	0.404
35	0.406
40	0.407
45	0.409
50	0.410
55	0.411
60	0.413
65	0.414
70	0.415
75	0.417
80	0.418
85	0.420
90	0.421
95	0.422
100	0.424
105	0.425
110	0.427
115	0.428
120	0.429
125	0.431

9.22 LIQUID THERMAL CONDUCTIVITY

Temperature (degrees F)	British thermal unit inch per hour-square foot-F
0	1.026
10	1.015
20	1.005
30	0.994
40	0.983
50	0.072
60	0.962
70	0.951
80	0.940
90	0.929
100	0.919
110	0.908
120	0.897
130	0.886
140	0.876
150	0.865
160	0.854
170	0.843
180	0.833
190	0.822
200	0.811
210	0.800

9.23 LIQUID VISCOSITY

Temperature (degrees F)	Centipoise
0	1.024
5	0.978
10	0.935
15	0.894
20	0.857
25	0.821
30	0.788
35	0.757
40	0.727
45	0.700
50	0.673
55	0.649
60	0.625
65	0.603
70	0.582
75	0.562
80	0.544
85	0.526
90	0.509
95	0.493
100	0.477

9.24 SOLUBILITY IN WATER

Temperature (degrees F)	Pounds per 100 pounds of water
68	0.050

9.25 SATURATED VAPOR PRESSURE

Temperature (degrees F)	Pounds per square inch
0	0.038
10	0.057
20	0.084
30	0.121
40	0.172
50	0.241
60	0.331
70	0.449
80	0.600
90	0.792
100	1.033
110	1.332
120	1.700
130	2.148
140	2.690
150	3.338
160	4.109
170	5.018
180	6.083
190	7.323
200	8.758
210	10.410

9.26 SATURATED VAPOR DENSITY

Temperature (degrees F)	Pounds per cubic foot
0	0.00070
10	0.00103
20	0.00150
30	0.00212
40	0.00296
50	0.00405
60	0.00547
70	0.00727
80	0.00954
90	0.01237
100	0.01584
110	0.02007
120	0.02518
130	0.03127
140	0.03850
150	0.04700
160	0.05691
170	0.06840
180	0.08162
190	0.09675
200	0.11400
210	0.13340

9.27 IDEAL GAS HEAT CAPACITY

Temperature (degrees F)	British thermal unit per pound-F
0	0.228
25	0.241
50	0.255
75	0.268
100	0.281
125	0.294
150	0.306
175	0.319
200	0.331
225	0.343
250	0.355
275	0.367
300	0.378
325	0.389
350	0.400
375	0.411
400	0.422
425	0.432
450	0.443
475	0.453
500	0.462
525	0.472
550	0.482
575	0.491
600	0.500

FIGURE 6-7 *(Continued)*

Personnel protection
Avoid breathing dusts, and fumes from burning material
Keep upwind
Avoid bodily contact with the material
Wear appropriate chemical protective gloves, boots and goggles
Do not handle broken packages unless wearing appropriate personal protective equipment
Wash away any material which may have contacted the body with copious amounts of water or soap and water
Wear positive pressure self-contained breathing apparatus when fighting fires involving this material
If contact with the material anticipated, wear appropriate chemical protective clothing
Environmental considerations—land spill
Dig a pit, pond, lagoon, holding area to contain liquid or solid material
Cover solids with a plastic sheet to prevent dissolving in rain or fire fighting water
Environmental considerations—water spill
Use natural deep water pockets, excavated lagoons, or sand bag barriers to trap material at bottom
If dissolved, in region of 10ppm or greater concentration, apply activated carbon at ten times the spilled amount
Remove trapped material with suction hoses
Use mechanical dredges or lifts to remove immobilized masses of pollutants and precipitates
First aid responses
Move victim to fresh air; call emergency medical care.
Remove and isolate contaminated clothing and shoes at the site.
In case of contact with material, immediately flush skin or eyes with running water for at least 15 minutes.

STYPHNIC ACID **(SEE** TRINITRORESORCINOL)

STYRENE MONOMER, INHIBITED (FLAMMABLE LIQUIDS, N.O.S.) UN1993
CLASS 3 *(FLAMMABLE LIQUID)*
ENVIRONMENTALLY HAZARDOUS SUBSTANCE (RQ-1000/454)
Recycled styrene is a natural colored liquid with a sweet, aromatic odor. It contains styrene, toluene, and ethylbenzene. It has a flash point between 59 and 97 deg. F. Its vapors are irritating to the eyes and mucous membranes. If it becomes contaminated or is subjected to heat, it may polymerize. If the polymerization takes place inside a container, the container is subject to violent rupture. It is insoluble in water and its vapors are heavier than air.
If material on fire or involved in fire
Do not extinguish fire unless flow can be stopped
Use water in flooding quantities as fog
Solid streams of water may spread fire
Cool all affected containers with flooding quantities of water
Apply water from as far a distance as possible
Keep run-off water out of sewers and water sources
Use foam, dry chemical, or carbon dioxide
If material not on fire and not involved in fire
Keep sparks, flames, and other sources of ignition away
Keep material out of water sources and sewers
Build dikes to contain flow as necessary
Attempt to stop leak if without undue personnel hazard
Use water spray to knock-down vapors
Personnel protection
Avoid breathing vapors
Keep upwind
Wear appropriate chemical protective gloves, boots and goggles
Do not handle broken packages unless wearing appropriate personal protective equipment
Wash away any material which may have contacted the body with copious amounts of water or soap and water
Evacuation
If fire becomes uncontrollable or container is exposed to direct flame—consider evacuation of one-half (1/2) mile radius
If material leaking (not on fire) consider evacuation of one-half (1/2) mile radius based on amount of material spilled, location and weather conditions

Environmental considerations—land spill
Dig a pit, pond, lagoon, holding area to contain liquid or solid material
Dike surface flow using soil, sand bags, foamed polyurethane, or foamed concrete
Apply "universal" gelling agent to immobilize spill
Apply appropriate foam to diminish vapor and fire hazard
Environmental considerations—water spill
Use natural barriers or oil spill control booms to limit spill travel
Use surface active agent (e.g. detergent, soaps, alcohols), if approved by epa
Inject "universal" gelling agent to solidify encircled spill and increase effectiveness of booms
If dissolved, in region of 10ppm or greater concentration, apply activated carbon at ten times the spilled amount
Remove trapped material with suction hoses
Use mechanical dredges or lifts to remove immobilized masses of pollutants and precipitates
Environmental considerations—air spill
Apply water spray or mist to knock down vapors
First aid responses
Move victim to fresh air; call emergency medical care.
If not breathing, give artificial respiration.
If breathing, is difficult, give oxygen.
In case of contact with material, immediately flush skin or eyes with running water for at least 15 minutes.
Remove and isolate contaminated clothing and shoes at the site.
Compatible protective equipment construction materials include:
Polyethylene, polyvinyl alcohol, viton

STYRENE MONOMER, STABILIZED UN2055
CLASS 3 *(FLAMMABLE LIQUID)*
ENVIRONMENTALLY HAZARDOUS SUBSTANCE (RQ-1000/454)
Styrene monomer is a clear colorless to dark colored liquid with an aromatic odor. It has a flash point of 90 deg. F. Its vapors are irritating to the eyes and mucous membranes. If it becomes contaminated or is subjected to heat, it may polymerize. If the polymerization takes place inside a container, the container is subject to violent rupture. It is lighter than and insoluble in water. Its vapors are heavier than air. It weighs 7.6 lbs./gal. It is used to make plastics, paints, and synthetic rubber, and other chemicals.
If material on fire or involved in fire
Do not extinguish fire unless flow can be stopped
Use water in flooding quantities as fog
Solid streams of water may spread fire
Cool all affected containers with flooding quantities of water
Apply water from as far a distance as possible
Use foam, dry chemical, or carbon dioxide
If material not on fire and not involved in fire
Keep sparks, flames, and other sources of ignition away
Keep material out of water sources and sewers
Build dikes to contain flow as necessary
Attempt to stop leak if without undue personnel hazard
Use water spray to knock-down vapors
Personnel protection
Avoid breathing vapors
Keep upwind
Wear appropriate chemical protective gloves, boots and goggles
Do not handle broken packages unless wearing appropriate personal protective equipment
Wash away any material which may have contacted the body with copious amounts of water or soap and water
Environmental considerations—land spill
Dig a pit, pond, lagoon, holding area to contain liquid or solid material
Dike surface flow using soil, sand bags, foamed polyurethane, or foamed concrete
Absorb bulk liquid with fly ash, cement powder, or commercial sorbents
Apply "universal" gelling agent to immobilize spill
Apply appropriate foam to diminish vapor and fire hazard
Environmental considerations—water spill
Use natural barriers or oil spill control booms to limit spill travel

Continued on next page

842 **Important: Become familiar with front matter especially criteria used for sorting chemicals.** September 2005

FIGURE 6-8 The *Emergency Action Guide* entry for styrene. Useful information includes emergency response information. *Courtesy of the Association of American Railroads Bureau of Explosives.*

more than 260 chemicals commonly transported in bulk via the railroads (Figure 6-8). This guidebook is therefore especially helpful for hazardous materials incidents involving railcars and larger quantities of material. It contains the following information:

- general information
- chemical and physical properties
- identification information
- potential hazards
- personal protective clothing and equipment
- first-aid information
- fire response information
- spill response information
- DOT placard
- NFPA 704 diamond

ASSOCIATION OF AMERICAN RAILROADS (AAR) GUIDE

The AAR guide, officially known as *Emergency Handling of Hazardous Materials in Surface Transportation*, provides emergency response information for more than 3600 U.S. DOT–regulated chemicals. In fact, American and Canadian railroads use the AAR guide to provide emergency information that accompanies hazardous materials shipments to first responders. The guide contains the following information:

- chemical and physical properties
- recommended practices for responding to hazardous materials emergencies
- emergency environmental mitigation procedures
- first-aid information
- personal protective equipment suggestions

FIELD GUIDE TO TANK CAR IDENTIFICATION

The *Field Guide to Tank Car Identification* is also published by the Association of American Railroads and contains very detailed information on tank car construction and design features. The schematics of tank car valves and piping can be especially helpful.

GATX TANK CAR MANUAL

The *GATX Tank Car Manual* contains very detailed information on tank car construction and design features including schematics of tank car valves and piping. This manual is a comprehensive source for determining the weight of various chemicals in pounds per gallon. It is complementary to the AAR *Field Guide to Tank Car Identification* and contains the following information:

- emergency contact information
- DOT and AAR tank car information
- tank car arrangements and schematics including general service cars and pressure cars
- tank car components
- tank car anatomy
- loading and unloading of tank cars
- description of the Tanktrain system
- commodity weight in pounds per gallon

NFPA FIRE PROTECTION GUIDE TO HAZARDOUS MATERIALS

The *NFPA Guide* is an excellent source of information for a broad range of common industrial chemicals. It contains multiple sections that provide information on flammability, specific chemical hazards, and chemical reactivity (Figure 6-9). This is one of the few reference sources that provide information on chemical reactivity. The 2010 edition of the guide contains the following information:

- hazardous chemicals data (as contained in NFPA 49)
- fire hazard properties of flammable liquids, gases, and volatile solids (as contained in NFPA 325)
- hazardous locations for liquids, gases, and vapors (as contained in NFPA 497 and NFPA 499)
- hazardous chemical reactions (as contained in NFPA 491)
- NFPA 704 marking system ratings (as contained in NFPA 704)
- flashpoint, vapor pressure, and conductivity (from NFPA 77)
- explosive hazards of flammable metals such as tantalum, titanium, magnesium, and zirconium (from NFPA 484)

NAME: LITHIUM ALUMINUM HYDRIDE

SYNONYMS: aluminum lithium hydride: LAH; lithium tetrahydroaluminate

FORMULA: $LiAlH_4$

NFPA 30/OSHA CLASSIFICATION:

DOT CLASS: Class 4.3. Dangerous when wet material

SHIPPING LABEL: DANGEROUS WHEN WET

ID NO.: UN 1410

CAS NO.: 1302–30–3

MOL, WT: 37.9

STATEMENT OF HAZARDS: Corrosive and flammable solid. Water reactive.

EMERGENCY RESPONSE PERSONAL PROTECTIVE EQUIPMENT: Wear special protective clothing and positive pressure self-contained breathing apparatus.

SPILL OR LEAK PROCEDURES: Keep water away from release. Shovel into suitable dry container.

FIRE FIGHTING PROCEDURES: Use approved Class D extinguishers or smother with dry sand, dry clay, or dry ground limestone. DO NOT use carbon dioxide or halogenated extinguishing agents. DO NOT use water. Violent reaction may result.

HEALTH HAZARDS: Corrosive. Causes severe eye and skin burns. May be harmful if inhaled. Irritating to skin, eyes, and respiratory system. Symptoms of overexposure include spasms, inflammation and edema of larynx and bronchi, pulmonary edema, coughing, wheezing, laryngitis, nausea, and vomiting.

FIRE AND EXPLOSION HAZARDS: Flammable solid. Evolves hydrogen and ignites on contact with water. Decomposition begins at 257°F (125°C) and releases heat. Combustion may produce irritants and toxic gases.

INSTABILITY AND REACTIVITY HAZARDS: Reacts violently with a broad range of materials, including water.

STORAGE RECOMMENDATIONS: Store in a cool, dry well-ventilated location. Separate from ketones, aldehydes, nitrogenous organic compounds.

USUAL SHIPPING CONTAINERS: Glass bottles or metal cans inside wooden box. Packaged under nitrogen or argon gas.

PHYSICAL PROPERTIES: White crystalline powder.

MELTING POINT: 257°F (125°C) (decomposes)

BOILING POINT: decomposes

SPECIFIC GRAVITY: 0.92

SOLUBILITY IN WATER: decomposes

ELECTRICAL EQUIPMENT: Class II, Group Undesignated

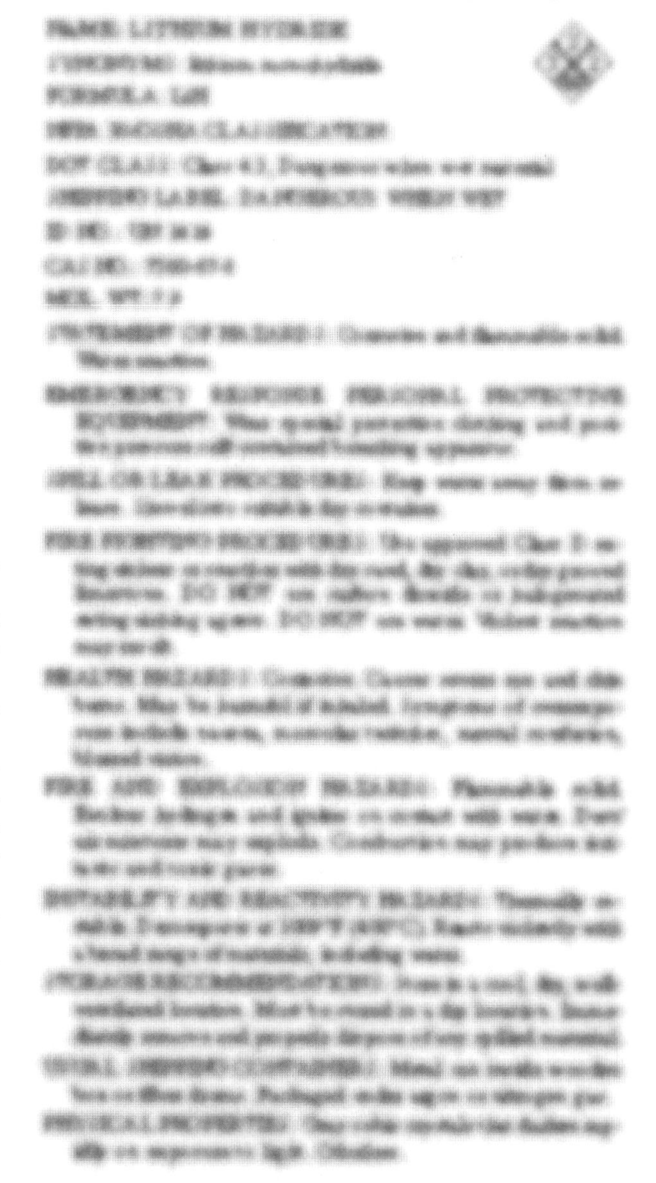

FIGURE 6-9 The *NFPA Guide* entry for lithium aluminum hydride. *Reproduced with permission from* Fire Guide to Hazardous Materials *(13th edition), Copyright © 2001, National Fire Protection Association.*

- parameters for determining degree and extent of hazardous locations for dusts (from NFPA 499)
- table for determining maximum allowable quantities of hazardous materials per industrial control area (from NFPA 400)
- chemical data and oxidizers (from NFPA 400)

CROP PROTECTION HANDBOOK

The *Crop Protection Handbook* is an excellent source of information for hazardous materials releases involving pesticides and herbicides. It is published annually and includes detailed information about common farm equipment and methods. There is no other agricultural reference that compares. This reference material is published for the agricultural industry and is not specifically designed for emergency response. It can therefore be difficult to use without some prior familiarity. Hazardous materials response teams that are responsible for rural areas should become especially familiar with this valuable reference source. It contains the following information:

- chemical and physical properties of pesticides
- agricultural equipment schematics and descriptions
- contact information for agricultural manufacturers

HAWLEY'S CONDENSED CHEMICAL DICTIONARY

This reference material contains information on thousands of common laboratory and industrial chemicals. It is laid out in a dictionary format and contains relatively general information regarding common uses and some esoteric facts, but it does not contain detailed chemical and physical properties for the vast majority of chemicals. This reference source can be used as a last resort and for general information.

BRETHRICK'S HANDBOOK OF REACTIVE CHEMICAL HAZARDS

This handbook is an excellent source of information when multiple chemicals are involved in a hazardous materials release. This reference source is a two-volume set that is also extensively cross-referenced. There are comparatively few resources available for determining chemical compatibility and reactivity.

CRC HANDBOOK OF CHEMISTRY AND PHYSICS

The *CRC Handbook* contains a wealth of information regarding the chemical and physical properties of laboratory and industrial chemicals. This guide also contains information about difficult-to-locate chemicals. In addition, recent editions of the *CRC Handbook* now contain ionization potential information, which is often difficult or impossible to find. The handbook contains the following information:

- physical constants of organic compounds
- physical constants of inorganic compounds
- ionization potentials
- table of radioactive isotopes
- physical and chemical properties of laboratory solvents
- handling and disposal of chemicals in laboratories

MERCK INDEX

The *Merck Index* is another reference source that contains many esoteric chemicals and is very useful for finding information about laboratory chemicals and pharmaceuticals (which are listed alphabetically). The listings include chemical and physical properties but do not contain practical response information, since the *Merck Index* was designed primarily for scientific use. The *Merck Index* is extremely well referenced and can be used as a starting point for further inquiry, especially for uncommon chemicals.

POCKET REFERENCE FOR HAZARDOUS MATERIALS RESPONSE

This reference material by the author contains a wide range of general response guidelines, as well as specific chemical and physical properties for radiological materials, chemical warfare agents, explosives, and biological agents.

FIRE FIGHTER'S HANDBOOK OF HAZARDOUS MATERIALS

This guide by Charles Baker provides chemical and physical property information for thousands of chemicals in alphabetical order. It is very encompassing and is a good primary reference source.

JANE'S CHEM BIO HANDBOOK

This is the definitive source on chemical and biological warfare agents. It contains detailed practical response guidelines and procedures for a wide variety of chemical agents as well as a wide variety of biological agents. It contains the following information:

- preincident planning for WMD
- on-scene procedures for WMD releases
- chemical agent information
- chemical agent treatment information
- biological agent information
- biological agent treatment information
- postincident procedures for WMD releases
- detection, identification, and monitoring equipment
- personal protective equipment
- decontamination procedures

EMERGENCY RESPONSE TO TERRORISM JOB AID

This guide is published by the FEMA and is a field operations guide particularly helpful for hazardous materials releases involving terrorism or WMDs. It contains the following information:

- operational considerations
- incident-specific information and actions for
  - biological agents
  - nuclear and/or radiological materials
  - incendiaries
  - chemical agents
  - explosives
- agency-related actions

SYMBOL SEEKER: HAZARD IDENTIFICATION MANUAL

The *Symbol Seeker* is an excellent reference material when you are faced with unfamiliar signage. This reference material contains information regarding military and ordnance signage as well as industry-specific signage. It also includes many international formats. This is a great reference to consult when you are faced with hazardous materials releases from containers or locations with uncommon markings.

HANDBOOK OF COMPRESSED GASES

The *Handbook of Compressed Gases* is published by the Compressed Gas Association (CGA) and provides detailed information on gases that are commonly produced and shipped in the compressed gas industry. An added benefit of this guide is that it contains excellent information about compressed gas cylinders.

SAX'S DANGEROUS PROPERTIES OF INDUSTRIAL MATERIALS

This reference contains detailed toxicological information and chemical and physical properties and is often useful at hazardous materials releases. It also contains information regarding the industrial uses of many chemicals.

EMERGENCY CARE FOR HAZARDOUS MATERIALS EXPOSURES

This reference written by Alvin Bronstein and Philip Currance contains extensive toxicological and emergency medical care information essential for any hazmat release involving victims or potential exposures. This guide also includes medical information in a format similar to that found in the *ERG*. For paramedics this guide includes treatment and drug protocols.

ACGIH GUIDE

The *ACGIH Guide* is updated annually and contains the latest detailed toxicological information on a large number of chemicals in the form of threshold limit values (TLVs)—including TWAs, STELs, and ceilings (C). The latest values for OSHA PELs, NIOSH RELs, and German MAKs are also included. There is a CAS number index in the back. The guide is often used to compare the TLVs with the PEL and REL to ensure that the most protective exposure limits are chosen.

Electronic Resources

Electronic resources are very popular and usually easy to use and easy to store. However, electronic resources are prone to failure owing to power supply issues such as dead batteries and generator problems. These resources will also fail if exposed to a powerful electromagnetic pulse such as from the detonation of a nuclear device or from an electronic device or weapons designed to disrupt electronic equipment. It is therefore prudent always to have paper backup reference materials.

INTERNET RESOURCES

The Internet contains a wealth of information that can be useful to the hazardous materials technician. When reliable Internet resources are consulted, information is usually very up-to-date and relatively easy to access and print out. Many online MSDS databases can be consulted 24 hours a day, 7 days a week. Hazardous materials response teams should therefore always have the ability to access the Internet, even when they are deployed in the field. With available cell phone technology and satellite telephones this should be relatively easy. As with other electronic resources, the Internet may become unavailable unexpectedly owing to disruptions, cyber attack, bandwidth problems, or Internet service provider (ISP) maintenance issues.

Gas Encyclopedia

The Gas Encyclopedia is a Web-based searchable encyclopedia compiled and maintained by Air Liquide and can be found at the following URL: http://encyclopedia.airliquide.com/encyclopedia.asp. The database is searchable by name, UN/NA identification number, and chemical formula. Each entry provides a wealth of information about compressed gases and compressed liquefied gases including the following:

- primary uses of the gas
- chemical and physical properties
- temperature-dependent vapor pressures

- liquid-to-gas conversion calculator
- Safety Data Sheets (SDSs)
- major hazards
- compatibility and reactivity data

At the bottom of the page the Gas Encyclopedia even lets you choose the units in which you would like the data presented. This is an excellent source of information for hazardous materials incidents involving industrial gases.

ChemSpider

ChemSpider is an online searchable chemical database owned and maintained by the Royal Society of Chemistry (RSC) and can be accessed at the following URL: http://www.chemspider.com/. This chemical database is primarily maintained by and for scientists, which has both advantages and disadvantages. The advantage is that the data are thoroughly vetted and very comprehensive in nature. The disadvantage is that there is potential for information overload. But if you know where to look it contains a wealth of information including the following general topics:

- Wikipedia article(s)
- associated data sources and commercial suppliers
- patents
- articles
- description
- names, synonyms, and database identifiers
- properties
- medical subject headings classification
- pharmacological links

Under the Properties tab you can find both predicted and experimentally determined chemical and physical property data. It is preferable to use the experimentally determined data whenever possible, but for some esoteric chemicals that may be encountered in academic and industrial research laboratories the Predicted Property tab may be the only readily accessible source of information during a hazardous materials incident.

Chemical Supply Companies

Chemical supply companies such as Sigma-Aldrich, Cole Parmer, Alfa Aesar, Fisher Scientific, and JT Baker, to name a few, are valuable sources of information, and most have online databases of MSDSs. One of the most helpful online chemical catalogs is the Sigma-Aldrich website, which can be accessed at the following URL: http://www.sigmaaldrich.com/united-states.html. It is an easy-to-use searchable chemical database that immediately gives valuable information such as chemical and physical properties as well as safety information. There is a tab to access the MSDS in PDF format for easy retrieval and access. Sigma-Aldrich sells well over 100,000 different chemicals. This is therefore also an excellent site for dealing with uncommon chemicals during a hazmat emergency.

Cole Parmer has an excellent chemical compatibility calculator that can be found at the following URL: http://www.coleparmer.com/TechInfo/ChemComp.asp. It is very easy to use, and can be used to find the chemical compatibility of different chemicals with different materials—such as chemical protective clothing and patching and plugging materials.

Emergency Response Guidebook

The DOT *Emergency Response Guidebook* in PDF format can be accessed and used as a searchable database at the Pipeline and Hazardous Materials Safety Administration (PHMSA) website at the following URL: http://www.phmsa.dot.gov/hazmat/library/erg. The 2012 *ERG* can be downloaded in PDF format at the following URL: http://www.phmsa.dot.gov/staticfiles/PHMSA/DownloadableFiles/Files/Hazmat/ERG2012.pdf.

NIOSH Pocket Guide to Chemical Hazards

The *NIOSH Pocket Guide* can be accessed and used as a searchable database online at the following URL: http://www.cdc.gov/niosh/npg/default.html.

WISER ELECTRONIC DATABASE

Electronic databases are becoming more widely used, especially with the advent of smartphones. WISER (Wireless Information System for Emergency Responders) is an electronic database developed by the National Library of Medicine (NLM) that contains detailed chemical and physical property information, spill and fire procedures, personal protective equipment suggestions, and health information for more than 440 unique chemicals (Figure 6-10). The database is intuitive and easy to use. It also contains radiological and biological information for more than 20 radioisotopes and six category A biological agents. It contains detailed medical information that cannot easily be found elsewhere and is therefore especially useful for EMS providers.

WISER also has additional tools, such as a searchable version of the *Emergency Response Guidebook,* a *WMD Response Guidebook,* the START adult triage and JumpSTART pediatric triage algorithms, and several radiological emergency response tools.

WISER is free of charge and available for download at the following URL: www.wiser.nlm.nih.gov. The WISER electronic database is also available for use online (WebWISER) at the following URL: http://webwiser.nlm.nih.gov/getHomeData.do.

CAMEO SUITE

The CAMEO suite is a family of programs developed by the EPA that consists of a chemical database, plume modeling program, and mapping software. This is an excellent source of information for hazardous materials technicians both for planning and for emergency response. CAMEO stands for computer-aided modeling and emergency operations. Some people have said the program is a bit cumbersome and difficult to use; however, the latest version is very easy and intuitive to use. The CAMEO suite is available free of charge and can be downloaded at the following URL: www.epa.gov/oem/content/cameo.

CAMEO

CAMEO is a database application designed to help hazardous materials technicians, firefighters, law enforcement officers, EMS workers, emergency managers, members of the LEPC, and employees in industry manage hazardous materials and facilities using and storing hazardous materials in the community. CAMEO consists of eight separate modules:

- Facilities
- Chemicals in Inventory
- Contacts
- Incidents

SOLVED EXERCISE 6-5

According to the WISER entry for hydrogen cyanide shown in Figure 6-10, would you expect hydrogen cyanide to be a flammability hazard?

Solution: Yes, the flashpoint of hydrogen cyanide is 0°F (−18°C) using the closed-cup laboratory measurement method. Notice that WISER has a typo in the flashpoint entry for hydrogen cyanide: someone omitted the negative sign when entering the conversion from Fahrenheit to Celsius temperature. Keep in mind that even the best reference materials will have errors! This is why it is important to consult at least three references as time permits.

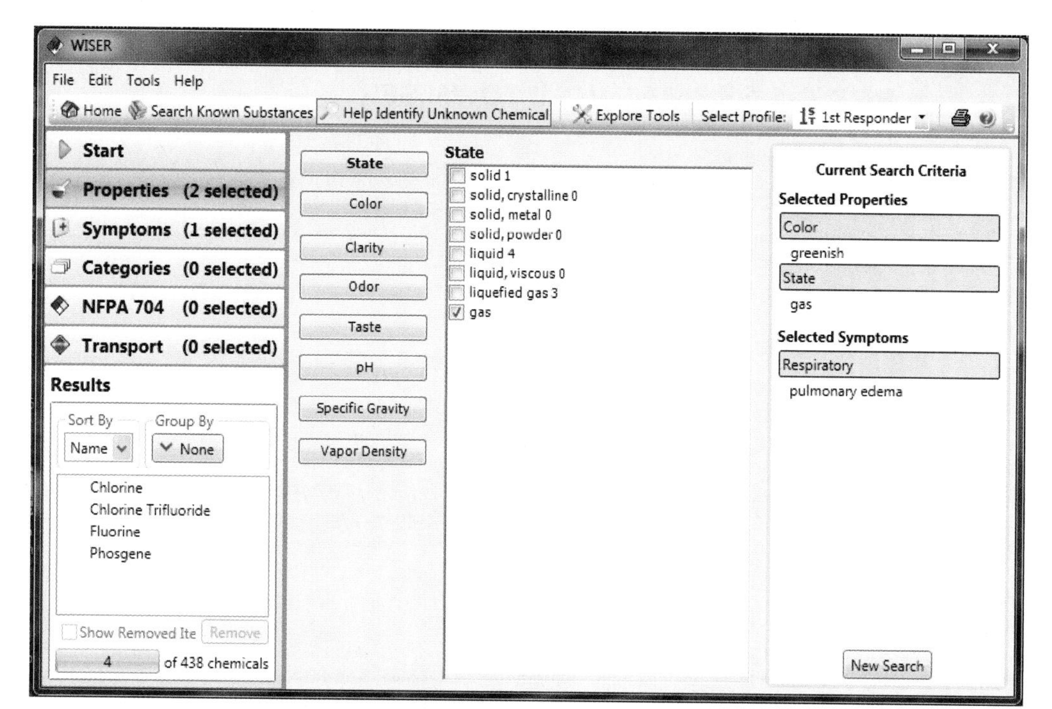

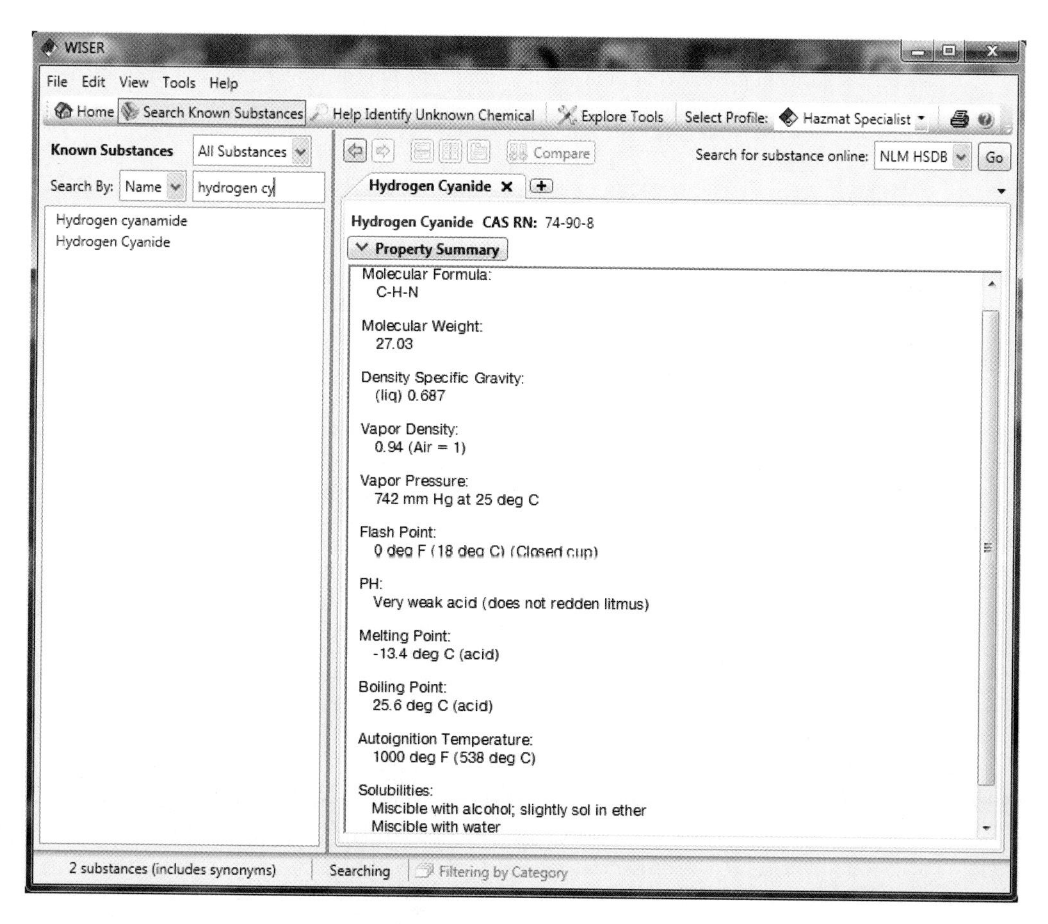

FIGURE 6-10 The WISER entry for hydrogen cyanide (bottom). Useful information includes detailed health information, chemical and physical properties, and a search feature based upon signs and symptoms and/or chemical appearance (top). *Courtesy of the National Institutes of Health (NIH).*

BOX 6-1 HAZMAT HOW TO: WISER

1. Open the WISER program.
2. Select the appropriate user profile from the drop-down list on the toolbar. The choices are 1st Responder, Hazmat Specialist, and EMS.
3. The application takes two paths, depending on whether the substance is known or not, as described further.
4. Help is available for each of the windows of the application via the Help menu in the toolbar.

SEARCH KNOWN SUBSTANCE

1. Select the Search Known Substance button.
2. Select the substance by scrolling through the list or entering the substance name or number. Use the Search by pull-down menu to indicate whether you are entering a name or one of the available identification numbers (UN/NA, CAS, or STCC).
3. Once you have located the desired substance, double-click on it to go to the Data Window. By default, the Data Window displays Key Info, a very brief summary of the most critical information about the substance.
4. Select the menu item or the drop down button on the Data Window to select more information about the substance. The top section of the pop-up menu represents quick, convenient links to the most pertinent information for the current user profile.
5. The bottom section of the menu provides access to all information available for the substance. Select the desired category to display a submenu that lists the data elements available in that category. Select one of the data elements to update the Data Window with the selected data.
6. Choose Protective Map at the bottom of the menu to calculate the threat zone area based on the chemical, spill parameters, and the site conditions.
7. Choose Reactivity at the bottom of the menu to determine the chemical reactivity of the chosen chemical with another chemical (which you choose from the Reactivity menu).
8. Select the Home tab in the toolbar to return to the main window.

HELP IDENTIFY UNKNOWN CHEMICAL

1. Press the Help Identify Unknown Chemical button on the main window to go to the search window, which will bring you to that window's Start tab.
2. If you can observe the physical properties of the substance, select the Properties tab. If you know symptoms of patients affected by the substance, select the Symptoms tab. In addition, you can search by Categories, NFPA 704, and Transport. If you know more than one characteristic, you can refine your search using multiple categories.
3. On the Properties tab, press one of the property category buttons, and select the appropriate value from the list.
4. On the Symptoms tab, click the body graphic in the affected areas: eyes, ears, nose, mouth/throat, neurological (brain), respiratory, cardiovascular, gastro/urinary (stomach or kidney), skin (arm), or body temperature (thermometer). Then, select the appropriate value from the list.
5. To unselect a property or symptom, click on the property or symptom from the selected list and uncheck the box. Alternatively, click on the category of the property or symptom and uncheck the box.
6. As you add properties, symptoms, or other data, the progress bar at the bottom left of the screen indicates the decreasing number of possible matching substances and their names.
7. To view a chemical, select it by double-clicking the name, or single-click the chemical name in the list and press the Details button. You are taken to the Data window for that substance. Close the Data window or select the back arrow on the toolbar to return to the Results tab.
8. Use the Group by menu to group the results by any of the symptom and property categories, or use the Sort by pull-down menu to order them by name or by one of the identification numbers.
9. To remove a substance from the list, right-click the substance, and a context menu appears. Select Remove.
10. To view removed substances, check the Show Removed Items box in the lower left-hand corner. Removed substances can be reinserted similarly to how they were removed.
11. Select the Home tab in the toolbar to return to the main window.

- Special Locations
- Routes
- Resources
- Screenings and Scenarios

The Facilities module can be used to store pre-incident planning information about facilities where chemicals are stored and used. Information that can be entered includes facility addresses, emergency contact information, site plans showing the building layout, and on-site chemicals. The records within this module can be linked to maps in MARPLOT, which can be helpful for pre-incident planning and plume modeling in ALOHA.

The Chemicals in Inventory module can be used to maintain inventory records for chemicals found within the entire community. Records include descriptions of each chemical's physical state, storage conditions, storage locations, and typical quantities on site. Chemicals that are entered into facility records and routes records are automatically transferred to the Chemicals in Inventory record.

The Contacts module can be used as a telephone-and-address book of emergency contacts, such as facility managers and engineers, chemical experts in the community, government agencies and organizations, outside response resources, and others who can assist with emergency response operations.

The Incidents module can be used to track information about hazardous chemical spills that have occurred in the community. Incidents that have occurred at either fixed-site facilities or along transport routes can be automatically linked to locations on the MARPLOT map.

The Special Locations module can be used to store pertinent data regarding high-risk occupancies such as hospitals, schools, nursing homes, and other facilities that may require special consideration during hazardous materials releases. These special occupancies can automatically be linked to locations on the MARPLOT map.

The Routes module can be used to compile information about transportation routes (such as roads, railroads, and waterways) commonly used to transport hazardous materials. Specific chemicals commonly transported along those routes can be entered into the database. These routes can also automatically be linked to locations on the MARPLOT map.

The Resources module can be used to maintain information about community resources that may be necessary during a hazardous materials emergency, such as environmental cleanup contractors, response equipment suppliers, subject-matter experts, and construction contractors. The locations of these resources can also automatically be linked to the MARPLOT map.

The Screening and Scenarios module can be used for hazards analysis within your community. Within this module you have the ability to enter a worst-case scenario spill size and conditions, and CAMEO will estimate the extent of plume migration around the modeled spill site and estimate the affected area. The affected zone can then be mapped on a MARPLOT map. Hazard analysis procedures are described in *Technical Guidance for Hazards Analysis,* also known as the Green Book, published by the EPA.

CAMEO is a first-rate database for comprehensive hazardous materials response pre-incident planning. The information in the database can quickly be accessed using any computer, and records can easily be printed as needed during an emergency.

CAMEO Chemicals

CAMEO Chemicals is a very versatile chemical database and analysis tool. When you start CAMEO Chemicals you are presented with three options:

- Search: Clicking here allows you to look up the properties of a particular chemical, including the CHRIS manual entry as well as the *ERG* entry.
- MyChemicals: Clicking here allows you to build a list of chemicals stored in a fixed-site facility or involved in a hazmat incident such as a train derailment.
- Reactivity: Clicking here allows you to determine how the chemicals you have listed under MyChemicals will react if they come in contact with one another.

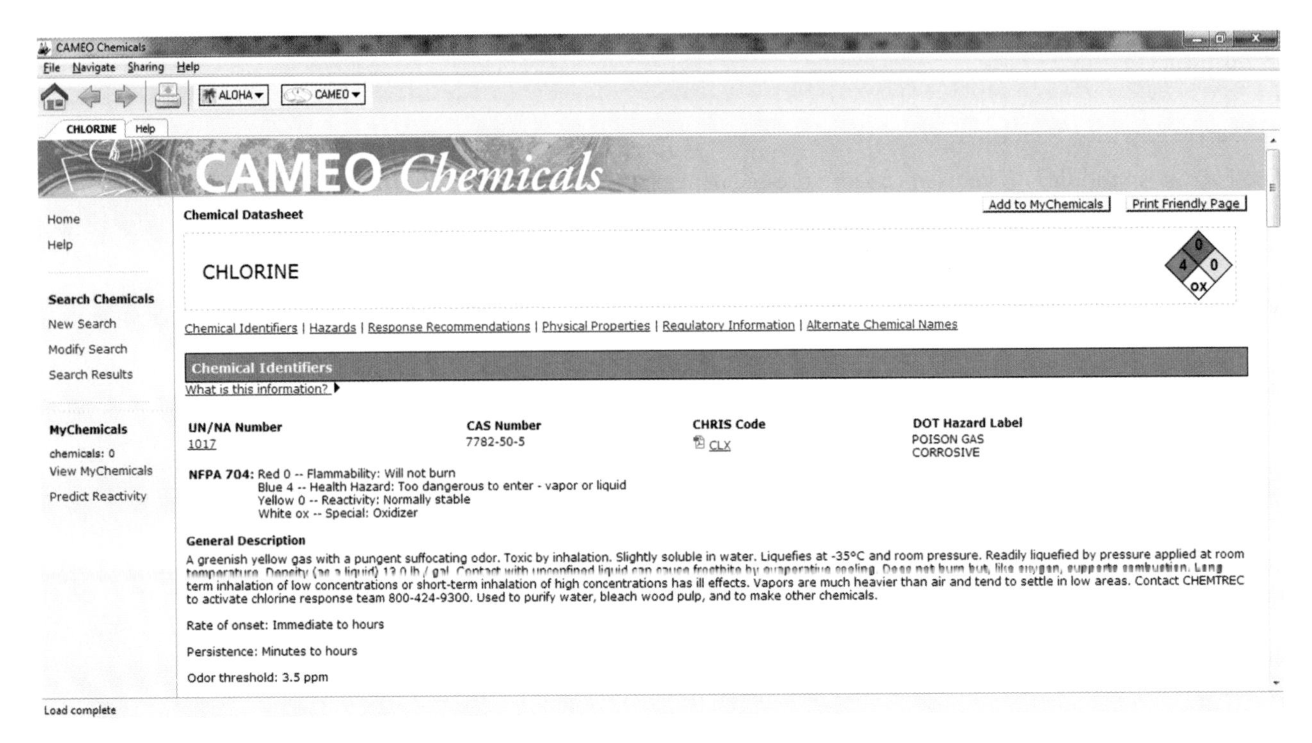

FIGURE 6-11 The CAMEO entry for chlorine. Useful information includes chemical and physical properties, personal protective equipment (PPE) suggestions, and easily accessible CHRIS manual pages. *Courtesy of the Environmental Protection Agency.*

CAMEO Chemicals contains technical data on almost 7000 chemicals (Figure 6-11). Information provided on chemical data sheets within the database includes the following:

- chemical and physical properties
- PPE recommendations
- health hazards
- recommendations for firefighting, first aid, and spill response
- regulatory information
- synonyms
- environmental hazards

The latest version of the *Emergency Response Guidebook* and CHRIS manual can also be accessed from within CAMEO Chemicals in PDF format. Chemicals that may be found in your community but are not in the CAMEO Chemicals database may be added to the MyChemicals collection within the database. Box 6-2 describes the stepwise use of CAMEO Chemicals.

Be cautious about using the Reactivity tool with your list in MyChemicals though. Other unforeseen reactions could very likely occur based on the chemical concentration, temperature, humidity, conditions of mixing, and any other impurities that may be present. Chemical reactivity data sheets can also be accessed by reactive group, such as alcohols, carboxylic acids, or organic peroxides.

This is an extremely versatile chemical database that can be downloaded and used locally on a laptop or computer, or the latest version of the program can be accessed online. The searchable online CAMEO Chemicals database can be accessed and queried at the following URL: http://cameochemicals.noaa.gov/.

ALOHA

The ALOHA plume modeling application is designed to predict the size of a gas or vapor cloud of an outdoor release based on a known chemical, the quantity of the release, and

SOLVED EXERCISE 6-6

According to the CAMEO entry for chlorine shown in Figure 6-11, what color would you expect the gas cloud emanating from a leaking chlorine railcar to be?

Solution: Under "General Description," CAMEO describes chlorine as "A greenish yellow gas with a pungent suffocating odor."

BOX 6-2 HAZMAT HOW TO: CAMEO CHEMICALS

LOOKING FOR CHEMICAL HAZARD AND RESPONSE INFORMATION FOR A KNOWN CHEMICAL

1. Click on Search from the home menu.
2. Type the information by which you would like to search (chemical name, CAS number, or UN/NA number). You also have the option of performing an advanced search using many other search parameters, such as general description, chemical and physical properties, and numerical values.
3. Choose the most appropriate entry from the list displayed. At this point you can view the data or add the chemical directly to MyChemicals (see next section).
4. Scroll down the page to view all the information. You can also click on the CHRIS code to get a PDF version of the CHRIS manual entry for that particular chemical (if one exists). Another option is to click on the UN/NA number of the chemical to find the *ERG* information and the orange guide number in PDF format for that particular chemical.

COMPILING A DATABASE OF CHEMICALS FOR A SARA TITLE III FACILITY OR A HAZARDOUS MATERIALS INCIDENT INVOLVING MULTIPLE CHEMICALS

1. You can add items to the MyChemicals collection three different ways:
   a. Run a search as described previously, select the chemical, and press the Add to MyChemicals button in the top right-hand corner of the screen.
   b. Add water or other reactive groups using either the Add Water button or the Add Reactive Group button near the top of the page.
   c. Import a previous list using the Import button at the bottom of the page.
2. You can export the list of chemicals using the Export button near the bottom of the page. This saves the list as an external file that you can later import using the Import button.

DETERMINING THE REACTIVITY OF TWO OR MORE CHEMICALS

1. In MyChemicals choose the Predict Reactivity toolbar in the lower left-hand corner of the menu.
2. Add water and generic functional groups in the MyChemicals menu.
3. A reactivity matrix will automatically be displayed for all the chemicals you've chosen (you must choose at least two chemicals, or one chemical and water).

the rate of the release (Figure 6-12). ALOHA can be very helpful in making evacuation versus in-place sheltering decisions. Using the Threat Point feature, you can determine the indoor versus outdoor atmospheric levels of contaminants. However, keep in mind that the model and estimated concentrations are only as good as the data! The chemical and physical properties necessary for plume modeling are accessed from CAMEO. ALOHA can be directly accessed from CAMEO using a hot button. Box 6-3 describes the stepwise use of ALOHA.

MARPLOT

MARPLOT is a mapping application designed to be used in conjunction with CAMEO and ALOHA. The plume that was modeled in ALOHA can be seamlessly overlaid onto a local map in MARPLOT. This feature is very helpful for medium- to long-term planning at hazardous materials releases. MARPLOT is accessed directly through ALOHA using a hot button.

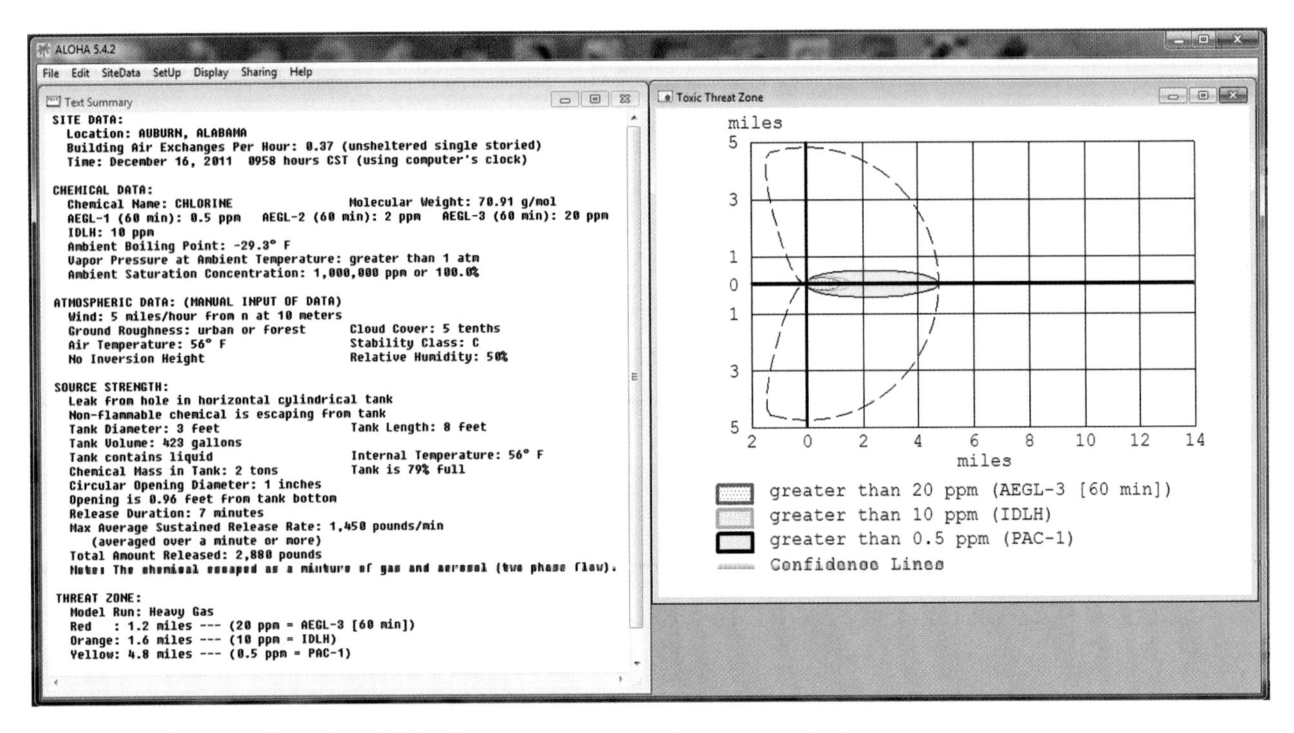

FIGURE 6-12 Plume modeling using ALOHA software. *Courtesy of the Environmental Protection Agency.*

BOX 6-3 HAZMAT HOW TO: ALOHA

1. After choosing a chemical in either CAMEO or CAMEO Chemicals, you can start the ALOHA program by pressing the ALOHA button in the toolbar at the top of the page in each program. Choose either Go to ALOHA to start ALOHA without choosing a chemical, or choose the Select this Chemical in ALOHA button to preselect the chemical in ALOHA.
2. ALOHA displays a warning that the program results may be inaccurate under certain conditions, such as very low wind speeds, very stable atmospheres, and terrain steering effects. The program also does not take into account chemical reactions, chemical mixtures, and particulates. Press the OK button to acknowledge these limitations.
3. Enter the site data by pulling down on the SiteData toolbar and first selecting Location. Choose your location from the options provided; then select Building Type and answer the questions; then select Date & Time and choose the appropriate option.
4. Next, choose the Set Up toolbar and select Chemical. Then, choose the chemical involved in the release (if you did not use the Select this Chemical in ALOHA button);
5. Next, choose the Set Up toolbar and select Atmospheric. Either manually input atmospheric conditions using the User Input tab, or use an attached weather station using the SAM Station tab.
6. Then, enter the release parameters by choosing the Set Up toolbar and selecting Source. Choose Direct, Puddle, Tank, or Gas Pipeline from the available options. Then, answer the applicable questions for each tab appropriately.
7. Go to the Display toolbar and choose Threat Zone. Choose the hazard to analyze (the choices are toxic area, flammable area, and blast area). Then, choose the measurement units to display for the zone boundary.
8. You may display this hazard zone boundary on a map or satellite image in MARPLOT.
   a. Start the MARPLOT program.
   b. To enter MARPLOT from ALOHA with this hazard zone, choose the Sharing toolbar, then the MARPLOT tab, and finally choose the Go to Map option. A map of the United States will open in MARPLOT.
   c. Follow the MARPLOT instructions in Box 6-4.

ALOHA has many additional options that can be used for more advanced hazardous materials release analysis. See the program manual for more information.

BOX 6-4 HAZMAT HOW TO: MARPLOT

1. Start the MARPLOT program.
2. Run the ALOHA program and set the site data, chemical information, weather information, and release information, and calculate a threat zone (see Box 6-3).
3. In MARPLOT, choose the map to display. The available options are Map, Satellite, and Topo. These maps will automatically be downloaded if you have Internet access.

PLOT THE THREAT ZONE ONTO THE MAP (DOWNWIND HAZARD ANALYSIS)

1. Point the cursor at the source point of the release using the cursor.
2. Choose the ALOHA tab from the toolbar menu.
3. Choose Set Source Point at Click Point. This will overlay the threat zone that was calculated in ALOHA over the map in MARPLOT.

DETERMINE THE HAZARD LEVEL AT A THREAT POINT (CAN BE USED FOR EVACUATION VERSUS IN-PLACE SHELTERING DECISIONS FOR SPECIFIC LOCATIONS)

1. Point the cursor at the threat location.
2. Choose the ALOHA tab from the toolbar menu.
3. Choose Set Threat Point at Click Point. This will open the ALOHA program and display the hazard analysis of the first 60 minutes at that location on the map. Both outdoor and indoor concentration levels will be displayed for a toxic threat zone.

ACCESS CURRENT WEATHER INFORMATION

1. Right-click the mouse at the desired location on the map.
2. Go to the CLICK POINT and choose Get Weather Conditions. If you are on the Internet, the National Oceanic and Atmospheric Administration (NOAA) weather Web page will automatically open for your location.

MARPLOT has many additional options that can be used for more advanced hazardous materials release analysis. See the program manual for more information.

TOMES SYSTEM

The TOMES system is an online database that is continuously updated and consists of the core components MEDITEXT Medical Managements, HAZARDTEXT Hazard Managements, and INFOTEXT documents. This is a very comprehensive database that includes detailed medical management guidelines, chemical and physical property information, chemical management information, as well as excerpts from other reference sources—such as *ERG*, *NIOSH Pocket Guide*, and MSDSs. TOMES can be accessed online, or the data can be installed locally using a CD-ROM.

DECISION SUPPORT SYSTEMS

Several excellent commercial electronic databases are also available. These databases can actually be best described as decision support systems, since they combine multiple databases and include response guidelines as well.

Adashi

The Adashi suite of programs is primarily an incident command platform, but the programs also contain hazardous materials–specific information. The software contains NIMS-compliant form-fillable ICS documents, street-level mapping features, the ability to integrate plume modeling data from ALOHA, and local pre-incident plans.

CoBRA Software

CoBRA is a multifunctional computer-based set of chemical databases, reference sources, and incident management resources. It is another commercial product with excellent support and includes the following capabilities:

- searchable databases such as the CHRIS manual data
- standard operating procedures and procedural checklists
- *NIOSH Pocket Guide and Jane's Chem Bio Handbook*
- incident-reporting and ICS forms
- ability to input local data including preplans

HazMasterG3 Software

HazMasterG3 is marketed as a comprehensive threat analysis and identification system for use in WMD, drug, and explosives response. The system also includes extensive pre-incident planning tools as well as chemical and physical property information on a number of chemicals and WMD agents. Its capabilities include the following:

- chemical agent identification using signs and symptoms and observable properties
- radiological isotope identification
- improvised explosive device (IED), homemade explosive (HME) threats, and detonators identification
- intermodal container information
- road and rail hazard identification
- identification of dual-use precursors and multiple-precursor product outcomes
- chemical reaction predictions
- database of detector capabilities and uses

PEAC-WMD Software

PEAC is conceptually similar to CAMEO in that it is a very good chemical database with a different user interface. PEAC software includes integrated mapping, GPS connectivity, multiple modeling tools, and more than 100,000 chemical names, synonyms, and trade names. It includes the following capabilities:

- chemical and WMD database
- more than 100 advanced and basic life support protocols
- database of methamphetamine chemicals
- calculates safe standoff distances from pool fires based on substance, pool size, and type of heat injury, and provides an estimate of pool burn duration
- Google Earth mapping
- data access for after-action reviews and generating incident reports
- access to NIMS forms that can be filled in, saved, and printed
- chemical and physical property search capabilities

Weather Information

Weather conditions are a critical factor at hazardous materials releases. Even indoor releases can be affected by weather conditions such as environmental heating or cooling of uninsulated buildings. It is important to know both the current as well as the future weather conditions. Otherwise, unexpected weather changes can have a disastrous effect on the outcome of the incident. Weather considerations include the following:

- temperature
- humidity
- precipitation
- wind
- weather fronts

WEATHER STATIONS

Most hazardous materials response teams have weather stations available on-site. These weather stations typically stand 3 meters (10 ft) tall and can be integrated into plume modeling applications such as ALOHA within the CAMEO suite. Weather stations are extremely helpful for determining real-time temperature, wind speed and direction, and even humidity with some models. This makes real-time plume modeling much more accurate. Plume patterns will vary significantly based on wind speed and variability. Figure 6-13 shows a deployed weather station at the scene of a hazardous materials release. A simple and cheap way to estimate wind speed is to tie a short length of barrier tape to a vehicle antenna, pole, or other handy unobstructed vertical item and observe.

FIGURE 6-13 A mobile weather station that can be used to gather current weather data at the scene of hazardous materials emergencies. Some weather stations can feed data directly into the ALOHA program, allowing for real-time plume modeling. *Courtesy of Chris Weber, Dr. Hazmat, Inc.*

INTERNET

When weather stations are not a viable option, the Internet can be very helpful. Many websites offer current weather conditions and are updated anywhere from hourly to every 5 minutes. Examples of weather forecasting websites include The Weather Channel (www.weather.com), Weather Underground (www.wunderground.com), and local news organizations.

AIRPORTS

Most airports have weather stations located at a height of 10 meters (33 ft). The nearest airport to a hazardous materials release can be contacted on a regular basis for current weather conditions. Even if you have your own weather station, a comparison of weather conditions at 3 meters and 10 meters—as long as they are located within a reasonable geographic distance—can often indicate significant weather patterns such as inversions.

Technical Specialists

Technical specialists are excellent resources at hazardous materials releases. Some hazmat teams are fortunate enough to have them on their teams directly. Technical specialists have very specific knowledge and typically know a great deal about their relatively small subject area. Therefore, even if you have a chemist on the team, it is always helpful to know whom to call in special circumstances. There are several national resources that we will discuss next. It can also be helpful to form ties with technical experts in your community, such as those in the academic research laboratories, chemical industry, petroleum industry, rail industry, shipping industry, and manufacturing industry.

CHEMTREC (UNITED STATES) AND CANUTEC (CANADA)

Often, shipping papers and MSDSs will list CHEMTREC as the emergency response contact number. CHEMTREC is a free service to emergency responders and can be reached by calling 1-800-424-9300. The CHEMTREC call center is staffed 24 hours a day, 7 days a week by trained personnel. Often, smaller companies and shippers will use CHEMTREC as a more economical alternative to providing their own technical experts around the clock as required by the U.S. DOT. CHEMTREC has the MSDSs and contact numbers for technical experts at its disposal. When you call CHEMTREC, the operator will expect you to have gathered certain information:

- your name and title
- your company or organization

- your location
- a callback number
- dispatch center number if available or appropriate
- fax number
- location of the incident
- current and predicted weather conditions
- time the incident occurred; estimates are OK
- shipping papers
- UN number or STCC number of product
- trade name of product
- carrier, shipper, and point of origin
- consignee and destination
- container or package information
- brief description of incidents and actions taken
- number and type of injuries and exposures
- amount of product involved and released
- whether industry representatives are on the scene
- the information you need ASAP

Although all this information is not absolutely necessary, having as much of it as possible will expedite the process. CHEMTREC will have specific information only for its registered industry customers; however, its MSDS database is very extensive, and other technical data may prove useful even for unregistered shipments.

CHLOREP

CHLOREP (CHLORine Emergency Plan) is a program that provides an organized and effective system for responding to chlorine emergencies in North America. It operates 24 hours a day, 7 days a week and is administered by The Chlorine Institute. Participating companies fund multiple CHLOREP groups, which consist of an emergency contact to whom the request for assistance is directed, a CHLOREP team made up of a team leader and assistants to handle the emergency, and a home coordinator to provide support at the home location. The personnel are employees at the various member company locations. These resources are supplemented by a network of emergency response contractors. CHLOREP teams can be requested through CHEMTREC and CANUTEC dispatchers. A CHLOREP team leader will contact the emergency caller at the scene and will determine what advice and aid is needed. The CHLOREP team leader decides whether or not to dispatch a team to the scene.

POISON CONTROL CENTER

The National Poison Control Hotline was established to respond to emergency calls from the public, first responders, and medical personnel in the event of exposure to poisons. The hotline is staffed 24 hours a day, 7 days a week by registered nurses or pharmacists with backgrounds in critical care who have passed a national certification exam in toxicology. The center can be reached at 1-800-222-1222 (emergency only).

NATIONAL PESTICIDE INFORMATION CENTER (NPIC)

The NPIC provides a wealth of information about a variety of pesticide-related subjects, including commercial pesticide formulations, recognition and management of pesticide poisonings, toxicology, and environmental chemistry. The website (http://npic.orst.edu) has excellent fact sheets describing dozens of common active ingredients in pesticide products. NPIC also lists state pesticide regulatory agencies, and provides links to their websites. The center can be contacted at 1-800-858-7378.

NATIONAL RESPONSE CENTER (NRC)

The NRC is the federal government's national communications center, which is staffed 24 hours a day by U.S. Coast Guard officers and marine science technicians. The NRC receives all reports of releases involving hazardous substances and oil that trigger the federal notification requirements under several laws. Reports to the NRC activate the National Contingency Plan and the federal government's response capabilities. The center can be reached at 1-800-424-8802.

MANUFACTURERS' TECHNICAL EXPERTS

If the number listed is not CHEMTREC (1-800-424-9300), you can call the listed number and speak directly with a manufacturer's representative. Often, this number is staffed by a call center that will transfer you to a technical expert. The technical experts may be chemists, chemical engineers, plant managers, or other qualified personnel. Technical experts will also require much of the same information listed under CHEMTREC.

Technical experts are also found at the fixed sites in your community. When responding to hazardous materials releases at fixed-site facilities it is a good idea to immediately ask facility personnel to contact those technical experts or subject-matter experts. It is always a good practice to be well prepared when speaking with technical experts, since they will need to know details about the hazardous materials release to be able to help you.

SCIENTISTS IN YOUR COMMUNITY

Scientists located in your community, especially chemists and toxicologists, can be a very helpful resource in hazardous materials incidents. The scientists may work in industry, for government agencies, or at universities or colleges. It is a good idea to approach scientists during pre-incident planning exercises and facility tours and ask if they would like to serve as an emergency contact in case of a complex hazardous materials release at their facility or elsewhere. Some scientists may even be willing to form a closer relationship with your hazmat response team and attend trainings and become a subject-matter expert (SME) for the team.

Standard Operating Procedures

The hazardous materials response team to which you are assigned is required to have standard operating procedures (SOP) or standard operating guidelines (SOG) that lay the groundwork for responding to hazardous materials releases, for training, and for planning, as well as describe the respiratory protection and personal protective equipment plans. Every new hazmat technician should examine these SOPs and thoroughly understand them.

CHAPTER **REVIEW**

Summary

Hazardous materials technician level responders should be able to gather information about one or more hazardous chemicals or weapons of mass destruction that have been released. This information can be gathered from such sources as the *ERG,* MSDSs, the *NIOSH Pocket Guide,* WISER, CAMEO, CHEMTREC, and technical experts. You should verify any information obtained from reference sources using a minimum of two additional sources owing to the possibility of typographical errors. It is also important to verify the spelling and chemical identity of the information relayed by the entry team. Do not fall into the trap of information overload. Determine only the necessary information and quickly pass that information along to the incident commander.

Review Questions

1. Why should you use multiple reference sources when determining chemical and physical properties?
2. What information can be found in the *NIOSH Pocket Guide?*
3. What is CHEMTREC?
4. What information is needed when contacting CHEMTREC or technical experts?
5. Why are technical experts useful at the scene of a hazardous materials release?

References

Alibek, Ken. et al. (2005) *Jane's Chem-Bio Handbook,* 3rd ed. Surry, UK: Jane's Information Group.

American Council of Governmental Industrial Hygienists (2011) Guide to Occupational Exposure Values-2011. Cincinnati, Ohio: ACGIH, Inc.

American Industrial Hygiene Association (2010) Emergency Response Planning Guidelines Series. Fairfax, Virginia: AIHA Press.

Arizona Emergency Response Commission. (2009) *CAMEO Companion.* Phoenix, AZ: author.

Baker, Charles. (2001) *Firefighters Handbook of Hazardous Materials,* 7th ed. Sunbury, MA: Jones and Bartlett.

Budavari, Susan, ed. (2006) *The Merck Index, 14th ed.* Rahway, NJ: Merck & Co., Inc.

Bureau of Explosives. (1990) *Emergency Action Guides.* Washington DC: Association of American Railroads.

Burns, Paul Peter. (2002) *Symbol Seeker: Global Edition.* Preston, UK: Symbol Seeker Ltd.

Centers for Disease Control and Prevention. (2007). *NIOSH Pocket Guide to Chemical Hazards.* Washington, DC: U.S. Government Printing Office.

Department of Transportation. (2012). *2012 Emergency Response Guidebook.* Washington, DC: Pipeline & Hazardous Materials Safety Administration.

Department of Transportation. (2011) 49 CFR 172.101. Hazardous Materials Table. Washington, DC: U.S. Department of Transportation.

EPA and NOAA. (2011) CAMEO Users Manual. Washington DC: Environmental Protection Agency.

Environmental Protection Agency. (1987) *Technical Guidance for Hazards Analysis: Emergency Planning for Extremely Hazardous Substances.* Washington DC: author.

Forsberg, Krister, and S. Z. Mansdorf. (2007). *Quick Selection Guide to Chemical Protective Clothing* (5th ed.). New York: John Wiley.

General American Transportation Corporation. (1994) *GATX Tank Car Manual* (6th Edition), Chicago, IL: General American Transportation Corporation.

Lewis, Richard (2007) *Hawley's Condensed Chemical Dictionary,* 15th ed. New York: John Wiley.

Lide, David R. (2008) *CRC Handbook of Chemistry and Physics, 89th edition,* CRC Press.

National Fire Protection Association. (1986) *Fire Protection Guide on Hazardous Materials.* Quincy, MA: author.

National Fire Protection Association. (2002) *Hazardous Materials Response Handbook,* 4th Ed. Boston, MA: author.

United States Coast Guard. (1984) *Chemical Hazards Response Information System (CHRIS) Manual.* Washington, D.C.: Department of Transportation.

Urben, Peter G. (2007) Bretherick's Handbook of Reactive Chemical Hazards, 7th ed. Elsevier.

Weber, Chris. (2007). *Pocket Reference for Hazardous Materials Response.* Upper Saddle River, NJ: Pearson/Brady.

CHAPTER 7 Personal Protective Equipment

Courtesy of Chris Weber, Dr. Hazmat, Inc.

KEY TERMS

breakthrough time, *p. 204*
doff, *p. 215*
don, *p. 215*
end of service life indicator (ESLI), *p. 198*
filter cartridge, *p. 196*
heat cramp, *p. 209*
heat exhaustion, *p. 209*
heat rash, *p. 209*
heat stress, *p. 208*
heatstroke, *p. 209*
hyperthermia, *p. 208*
hypothermia, *p. 209*
Level A PPE, *p. 190*
Level B PPE, *p. 190*
Level C PPE, *p. 190*
Level D PPE, *p. 190*
maximum use concentration (MUC), *p. 198*
protection factor, *p. 194*
respiratory protection, *p. 193*
self-contained breathing apparatus (SCBA), *p. 187*
splash protection, *p. 200*
supplied air respirator (SAR), *p. 199*

OBJECTIVES

After reading this chapter, the student should be able to:

- Select the appropriate personal protective equipment (PPE) for hot zone entry at a hazardous materials or WMD incident.
- Determine the respiratory protection necessary for incident conditions.
- Determine the chemical protective clothing (CPC) necessary for incident conditions.
- Don and doff Level C, Level B, and Level A personal protective equipment (PPE).
- Describe the types and dangers of thermal stress.
- Describe in-suit emergencies that may occur while wearing PPE.
- Describe how PPE should be stored, maintained, and tested.
- List the required components of a PPE program.

Resource Central For additional review and practice tests, visit **www.bradybooks.com** and click on Resource Central to access book-specific resources for this text! To access Resource Central, follow directions on the Student Access Card provided with this text. If there is no card, go to **www.bradybooks.com** and follow the Resource Central link to Buy Access from there.

You are dispatched to a local motel on a report of a clandestine methamphetamine laboratory. Guests in the motel complained about foul odors coming from a nearby room. When law enforcement officers arrived at the room they arrested two suspects and noticed a lot of chemical containers and what appeared to be some sort of chemistry contraption. As a new member of your agency's hazardous materials response team you have been asked to help take care of the situation. You have the following questions:

- How do I protect myself from all these chemicals?
- Is there one type of personal protective equipment (PPE) that can protect me from all the hazards—toxicity, flammability, and corrosivity?
- Are there ways of dealing with the hazards other than just using PPE?
- Because this looks like it's going to be a long operation, are there any consequences to wearing PPE for extended periods of time?

Let's see if we can get some answers to these questions and more.

Personal protective equipment, which consists of chemical protective clothing and respiratory protection, is a last line of defense against hazardous materials exposure at emergency incidents. You should not rely on PPE as your first line of defense. In the HAZWOPER regulation OSHA clearly states that there are three primary ways to protect yourself from hazardous materials:

1. Engineering controls
2. Administrative controls (workplace practices)
3. Personal protective equipment

Engineering controls are equipment and procedures—such as fume hoods, lockout–tag out, and positive pressure ventilation—that reduce or eliminate hazardous environments. Workplace practices are procedural changes that increase the safety of responders, such as using long-handled tools to gain standoff distance from hazardous materials. Personal protective equipment consists of chemical protective clothing and respiratory protection designed to minimize responder contact with a hazardous environment.

These safety measures should be implemented in the order listed—in other words, personal protective equipment should be a last resort. Like all things in life, PPE has its advantages and disadvantages. We discuss the features of PPE in detail in this chapter.

The Need for Personal Protective Equipment (PPE)

Let's take a look at some of the dangers you may face when responding to hazardous materials/WMD incidents. Among these are chemical, radiological, and mechanical hazards; you must be able to ensure that the chemical protective clothing and the level of respiratory protection you choose will protect you from all the hazardous properties of the environment. In Chapter 5 you learned to interpret some of the chemical and physical properties of hazardous materials so that you can predict their behavior. Biological and etiological hazards, radiological, and thermal and mechanical hazards may also be involved. There is PPE available to protect against these hazards as well; however, it is difficult to find PPE that protects against all hazards simultaneously.

CHEMICAL HAZARDS

Chemical hazards include corrosives, poisons, and a wide range of other hazard classes (Figure 7-1). Examples of corrosives include acids such as sulfuric acid, hydrochloric acid, and nitric acid; and bases such as ammonia, bleach, and sodium hydroxide. Examples of common poisonous substances include isocyanates, insecticides, and gases such as phosphine, hydrogen sulfide, arsine, and carbon monoxide. Chemical asphyxiants prevent the body from either transporting or using oxygen. Two common chemical asphyxiants are carbon monoxide and hydrogen cyanide. Oxidizers are a common type of chemical hazard that may increase the propensity of fuels to burn and may directly damage the human body. Examples of oxidizers include hypochlorite (bleach), hydrogen peroxide, and organic peroxides. It is very important to choose the appropriate respiratory protection and chemical protective clothing ensemble when responding to incidents involving chemical hazards.

FIGURE 7-1 Hazards within the hot zone include mechanical (top left), thermal (top right), chemical (bottom left), and biological (bottom right). *Photos courtesy of Chris Weber, Dr. Hazmat, Inc.*

Chemical hazards may be damaging by all four routes of entry, so choice of the appropriate level of respiratory protection and chemical protective clothing depends on the route of entry. If the chemical has a low vapor pressure—meaning that it does not evaporate readily—but is readily absorbed through the skin, you may be able to choose a lower level of respiratory protection with a high level of chemical protective clothing that offers splash protection to that particular chemical. However, if the chemical has a high vapor pressure and is readily absorbed through the skin, you must choose the highest level of respiratory protection and chemical protective clothing that is impermeable to that particular chemical.

SIMPLE ASPHYXIATING HAZARDS

The human body requires oxygen to survive. Simple asphyxiating hazards prevent the body from getting oxygen by displacing the ambient oxygen in the air. Such hazards are typically not chemically reactive and are otherwise not particularly harmful to the body. Thus, the most important personal protective equipment used to prevent asphyxiation is supplied air respiratory protection. Chemical protective clothing is a secondary consideration in this case. For example, if a tank of inert liquid argon leaks inside an enclosed space such as a laboratory, supplied air respiratory protection is essential for performing a rescue; chemical protective clothing is not. Therefore a firefighter in structural firefighter protective clothing using a **self-contained breathing apparatus (SCBA)** is

self-contained breathing apparatus (SCBA) ■ Supplied air respiratory protection consisting of a harness, compressed gas cylinder containing breathing grade air, mask, and one or more regulators.

well protected to perform this rescue. Examples of simple asphyxiants include nitrogen, helium, neon, argon, and carbon dioxide.

BIOLOGICAL/ETIOLOGICAL HAZARDS

Biological and etiological hazards can arise from typical sources such as contamination with human waste or human blood, or stem from properties of the hazardous materials themselves. For example, research laboratories use biological agents such as viruses, bacteria, and prions in their research. Medical and laboratory waste shipments may also contain etiological hazards. Universal precautions worn in the EMS field are an example of PPE that offers limited protection from biological and etiological hazards. Universal precautions consist of latex or nitrile gloves and eye and face protection. Typically, the most important type of PPE for protection from etiological hazards is respiratory protection. The skin is typically an excellent barrier to biological agents. Most chemical protective clothing and respiratory protection will adequately protect you from etiological hazards.

Wildlife is also a type of biological hazard and should be considered depending on the work environment (Figure 7-1). This can range from dangerous wild animals such as snakes to venomous insects such as bees and spiders. Africanized honey bees can be extremely dangerous, and swarms of them have been known to attack emergency responders.

RADIOLOGICAL HAZARDS

Radiological hazards are posed by radioactive materials and other nuclear processes that can emit radiation (see Chapter 3 for more detailed information on radiation). Examples of common radioactive materials include americium-241 in smoke detectors, tritium (hydrogen-3) in gun sights and exit signs, and medical and research radiopharmaceuticals. Examples of processes and equipment that generate radiation are the nuclear power industry and X-ray generators used for medical diagnostics. The most dangerous forms of radiation are gamma, X-ray, and neutron radiation because they travel long distances in air and are highly energetic. Chemical protective clothing, with the limited exception of Demron, is not effective at preventing exposure to these types of radiation; however, it will prevent exposure to the radioisotope itself. The radioisotope is the material that is emitting radiation. Chemical protective clothing will completely shield the wearer from alpha radiation and is partially effective at shielding the wearer from beta radiation.

THERMAL HAZARDS

Thermal hazards are associated with a wide range of hazardous materials, including flammable materials, combustible materials, cryogenics, and explosives (Figure 7-1). Flammable materials are a prime example of thermal hazards. Examples include gasoline, methane, propane, ethanol, and many other common industrial and household hazardous materials. Flammable hazardous materials are very commonly encountered—approximately 80° of hazmat incidents involve flammable materials.

It is extremely important not to overlook the thermal hazard potential of hazardous materials. In 1984 in Shreveport, Louisiana, a fire department hazmat team member was killed while responding to an anhydrous ammonia leak in a cold storage warehouse. He and his partner were wearing PPE to protect themselves from the corrosive properties of ammonia but neglected to protect themselves from the flammable properties of ammonia. While working in the hot zone they unwittingly found themselves in a flammable atmosphere in the presence of an ignition source. They both suffered severe burns, to which one ultimately tragically succumbed (Box 5-4).

Cryogenic materials also pose thermal hazards, although they are the direct opposite of the thermal hazards flammable materials pose. Cryogenic materials are defined as having a boiling point below -130°F by the U.S. DOT. Cryogenic materials are stored and transported as liquids. Often, the greatest danger with cryogenic materials is skin contact and the resulting cryogenic burn, which is a severe form of frostbite. Cold temperatures also cause chemical protective clothing to become brittle and may even cause the suit to crack or shatter on movement.

Thermal hazards require special protective clothing. One example of thermal hazard protection is structural firefighter protective gear. This type of PPE is designed to limit the amount of heat that reaches a firefighter's body during structural firefighting. Another example of thermal hazard protection is cryogenic gloves, which are designed to insulate the wearer from the extreme cold of the cryogenic liquid. They allow personnel to handle containers holding cryogenic materials without receiving cryogenic burns. Neither of these types of PPE offers any real chemical protection. Some chemical protective clothing is designed to be resistant to flash fires for a few seconds; however, this type of PPE offers little thermal protection. Aluminized suits are also available as overgarments and offer limited protection against radiant heat. This type of PPE is often used by airport and industrial firefighters who must face the intense heat of petroleum-based fires. It is important to note that a flame-resistant Nomex suit underneath plastic chemical protective clothing is not adequate flash fire protection. Several possible options are discussed later in this chapter.

Explosives pose thermal hazards as well as mechanical hazards, which are covered in the next section. Explosives are materials that detonate, and the detonation involves a very rapid release of energy. In contrast, *deflagration* is the very rapid burning of material. An example of a material that deflagrates is gunpowder. Unconfined, it burns rapidly but does not cause an explosion. In contrast, confined burning of gunpowder, may propel a bullet or cause a pipe bomb to explode. Explosives pose a thermal hazard, an overpressure hazard, and a mechanical hazard.

MECHANICAL HAZARDS

Mechanical hazards can be encountered in a wide range of typical hazardous materials incidents, especially in the hot zone (Figure 7-1). Mechanical hazards typically bring to mind explosions, in which fragmentation and shrapnel are deadly mechanical hazards. In the hot zone, more mundane mechanical hazards are encountered much more commonly. Examples of such hazards include jagged edges on metal drums, pneumatic and hydraulic equipment, sharp edges, and other sharp and jagged objects.

Personal protective equipment for mechanical hazards includes thicker chemical protective clothing, leather gloves, and bomb suits for explosive ordnance disposal (EOD) personnel. Bomb suits are designed to protect the wearer from limited shrapnel, thermal hazards, and overpressure waves.

Classification of Personal Protective Equipment

Personal protective equipment can be classified in many different ways. The NFPA uses ensembles that are based on a comprehensive system of chemical protective clothing, whereas the EPA assigns four distinct levels of protection based on respiratory protection and chemical protective clothing (a PPE ensemble).

SOLVED EXERCISE 7-1

What types of hazards would you expect to encounter in an illicit laboratory?

Solution: Illicit laboratories contain many chemicals, including flammable solvents, poisons, flammable solids, oxidizers, and corrosive materials. These laboratories also contain thermal hazards in the form of heated chemical reactions, and laboratory equipment such as hot plates and Bunsen burners. Many mechanical hazards also exist, such as propane cylinders that may be weakened through corrosion by anhydrous ammonia. Biological, etiological, radiological, and asphyxiating hazards cannot be ruled out.

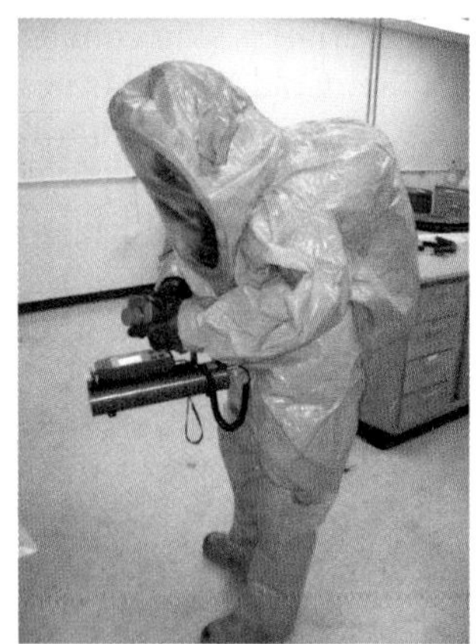
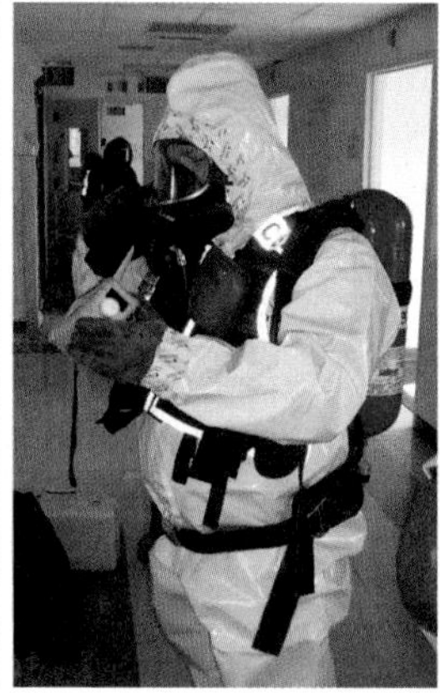
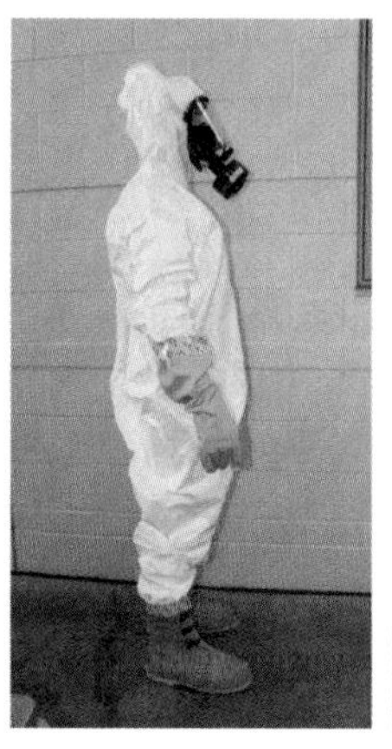
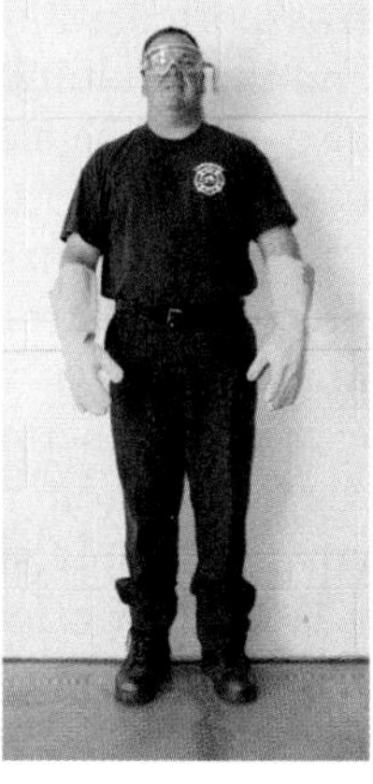

FIGURE 7-2 The four levels of PPE according to the EPA (from left to right): Level A, Level B, Level C, and Level D. *Photos courtesy of Chris Weber, Dr. Hazmat, Inc.*

EPA LEVELS OF PROTECTION

The lowest level of protection is designated Level D. Level D consists of nothing more than a uniform or other clothing; there is no respiratory protection, and there may be little or no chemical protective clothing. Level C consists of minimal respiratory protection, consisting of an air-purifying respirator (APR) and chemical splash protection. Level B consists of supplied air respiratory protection and chemical splash protection. Level A, the highest level of protection, consists of supplied air respiratory protection and a vapor-tight suit. Figure 7-2 shows examples of all four EPA levels of protection.

Level D

Level D PPE ■ Some chemical protective clothing combined with no or very limited respiratory protection (such as a dust mask).

Level D PPE essentially consists of very limited or no chemical protection and no respiratory protection at all. Street clothes or a daily work uniform may be considered Level D protection. However, even the universal precautions used by paramedics are also merely Level D protection. Even with the addition of chemical protective clothing, a dust mask, or a TB mask, it would still be considered Level D protection. Consistent with OSHA, Level D protection may only be used in environments with airborne contaminant levels below the permissible exposure limit (PEL).

Level C

Level C PPE ■ Liquid splash protective chemical protective clothing combined with an air-purifying respirator.

Level C PPE consists of chemical protective clothing that provides splash protection and limited respiratory protection consisting of an APR. Air-purifying respirators, also called *cartridge respirators,* filter the ambient air and thereby remove contaminants. This level of protection is typically used when some splash protection is required and some level of respiratory protection is required. Any level of respiratory protection below an APR is considered Level D by the EPA.

Level B

Level B PPE ■ Liquid splash protective chemical protective clothing combined with supplied air respiratory protection.

Level B PPE is used when some splash protection and the highest level of respiratory protection is needed. It consists of similar types of chemical protective clothing and offers approximately the same or slightly higher splash protection as Level C protection. However, instead of an APR, supplied air respiratory protection is included, which may consist of SCBA or a supplied air breathing apparatus (SABA). Because supplied air is used, Level B suits may also be fully encapsulating. Keep in mind, however, that taping does not make a Level B ensemble into a Level A ensemble! Tape should never be relied on to provide a vapor protective barrier.

Level A

Level A PPE ■ Vapor-tight chemical protective clothing combined with supplied air respiratory protection.

Level A PPE is used when the highest level of skin and respiratory protection is required. The chemical protective clothing used with Level A protection is vapor tight. Level A suits have sealed seams, integrated gloves and booties, an airtight zipper, and one-way exhalation valves (Figure 7-3). Level A suits tend to trap exhaled air, and if the one-way valves

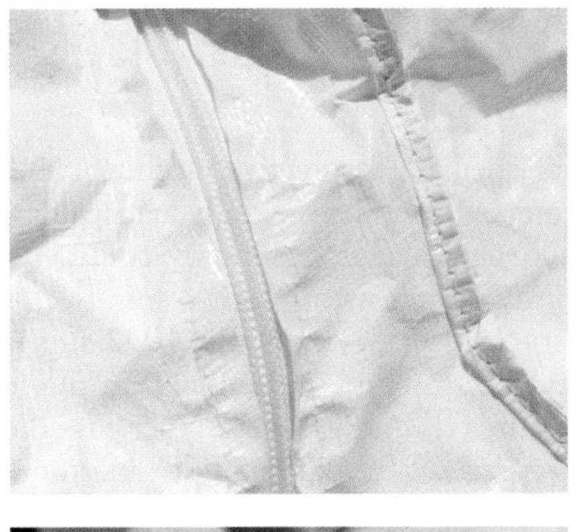

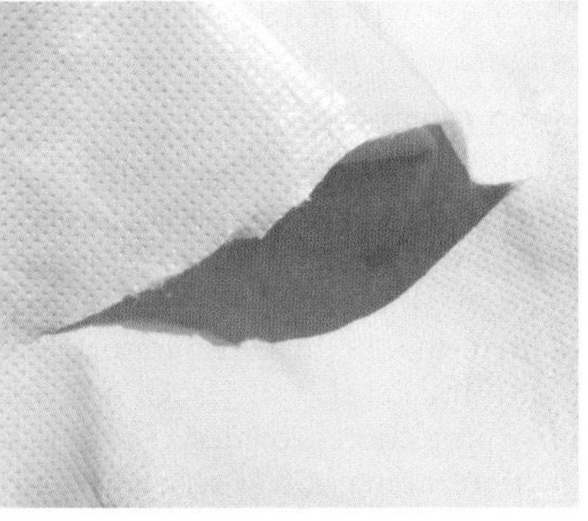

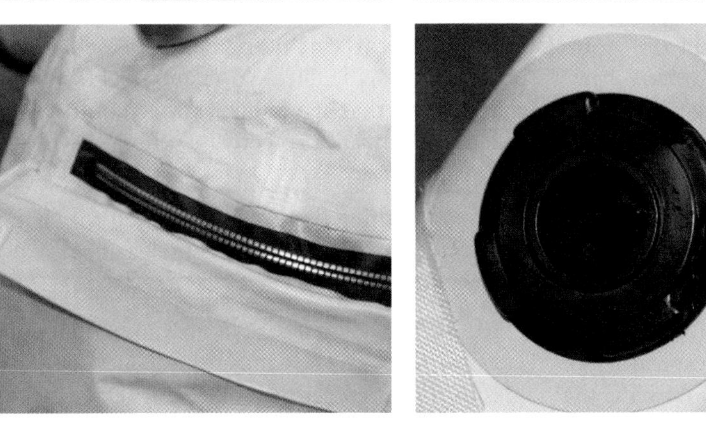

FIGURE 7-3 Level A suits have sealed zippers (bottom left) and one-way exhalation valves (bottom right), while Level B fully encapsulating suits have gas-permeable zippers (top left) and exhalation holes covered by a storm flap (top right). *Photos courtesy of Chris Weber, Dr. Hazmat, Inc.*

are not working perfectly, they tend to inflate. This means you may have to slowly and carefully bend over to burp the suit. Take care not to do this too fast, as you may damage the suit. In fact, you'll be able to float in water in a Level A suit owing to the added buoyancy of the trapped air.

Because these suits are vapor tight, they are also thicker and heavier than Level B or C suits, and wearers are more prone to heat stress. Level A is the highest level of personal protective equipment ensemble available.

Manufacturers also make Level A suits with flash protection—sometimes called *reflector suits*. These suits provide minimal protection against a flash fire—typically 5 seconds or less—and are not designed to be worn in flammable atmospheres.

NFPA CHEMICAL PROTECTIVE CLOTHING STANDARDS

The NFPA defines personal protective equipment in a different manner than the EPA does. NFPA 1991, *Standard on Vapor Protective Ensembles for Hazardous Materials Emergencies*, describes the chemical protective clothing that is analogous to the EPA's Level A protection. NFPA 1994, *Standard on Protective Ensembles for First Responders to CBRN Terrorism Incidents*, describes the specific chemical protective clothing for chemical, biological, and radiological WMD agents. Typically, these NFPA standards include the suit, gloves, and footwear but exclude respiratory protection equipment.

NFPA 1991 (2012 Edition)

The NFPA 1991 standard covers fully encapsulating chemical protective clothing that is gas and vapor tight. Thus, the NFPA 1991 standard covers the EPA Level A chemical protective clothing but makes no mention of respiratory protection. Specifically, the standard requires permeation testing against 21 different chemicals, including six gases, and two WMD agents. The CPC must also pass burst strength, seam strength, tear resistance, abrasion resistance, flammability resistance, and cold temperature performance tests. NFPA 1991 therefore quantifies the fully encapsulating Level A suit requirements and sets the bar for manufacturers to meet.

NFPA 1992 (2012 Edition)

Analogously, the NFPA 1992 standard covers chemical protective clothing that is designed for liquid splash protection. These suits may either be fully encapsulating or nonencapsulating. The NFPA 1992 standard covers what is commonly considered EPA Level B chemical protective clothing but again makes no mention of respiratory protection requirements. Because respiratory level is not defined, the NFPA 1992 suit may be used in an EPA Level B, Level C, or even Level D ensemble depending on the level of respiratory protection provided. The key is that the NFPA 1992 CPC must pass the burst strength, seam strength, tear resistance, abrasion resistance, flammability resistance, cold temperature performance, and flexural fatigue tests. The standard requires permeation testing against eight different challenge chemicals, including acetone, diethylamine, ethyl acetate, hexane, sodium hydroxide, sulfuric acid, tetrahydrofuran (THF), and toluene. The strength of the NFPA 1992 standard is that it quantifies the requirements for the complete system of liquid splash resistant chemical protective clothing.

NFPA 1994 (2012 Edition)

NFPA 1994 describes chemical protective clothing designed to be used for chemical, biological, radiological, or nuclear (CBRN) terrorism incidents. These garments are designed for one-time use. There are four classes within the NFPA 1994 standard:

- Class 1, which is covered by NFPA 1991, describes gas and vapor-tight suits with supplied air respiratory protection, otherwise known as a level A ensemble.
- Class 2 describes CPC that is designed to protect the user from IDLH atmospheres and approximately corresponds to the EPA Level B ensemble.
- Class 3 describes chemical protective clothing designed to protect the user from chemical hazards above the permissible exposure limit (PEL) but below the IDLH, corresponding roughly to Level C protection.
- Class 4 is essentially Level D protection (dust masks and HEPA filters).

The different NFPA 1991 and NFPA 1994 CPC classes are designed to be used in the following specific environments:

- Class 1 CPC is designed to be used in unknown environments, for unknown chemicals, when the chemical concentration is unknown, a little liquid contact can be expected, and skin contact cannot be permitted.
- Class 2 CPC is designed to be worn when liquid or aerosol contact is possible and when there are nonambulatory symptomatic victims.
- Class 3 CPC is designed to be worn when liquid or aerosol contact is possible and there are ambulatory symptomatic victims.

These descriptions have been generalized to give you a conceptual overview of the standard. The actual standard should always be consulted for applications requiring compliance.

NFPA 1999 (2008 edition)

NFPA 1999 sets forth the requirements for chemical protective clothing for emergency medical operations in the warm zone. The CPC must be liquid tight. Once again NFPA 1999 CPC must pass certain tests, including burst strength, seam strength, tear resistance, abrasion resistance, flammability resistance, cold temperature performance, and flexural fatigue.

Components of Personal Protective Equipment

As discussed in the last several sections, personal protective equipment consists of a number of different separate components, including both chemical protective clothing and respiratory protection.

OSHA PPE REGULATION (29 CFR 1910.132)

The OSHA PPE regulation, which was adopted in 2008, requires employers to provide workers with PPE at no cost when they must work in a hazardous location. Employers must

- provide PPE at no cost to the worker;
- ensure the PPE provides adequate protection for the hazard;
- monitor the work environment for hazards;
- provide employee training in PPE selection and use.

RESPIRATORY PROTECTION

Respiratory protection can be considered the most important aspect of personal protective equipment, since inhalation is considered the most efficient and dangerous route of entry for hazardous materials into the body (Table 7-1). Of course, there are exceptions to this rule of thumb, and different chemicals have different routes of entry that are most efficient. Therefore, it is very important to use the appropriate reference materials to determine the dangers of any particular chemical.

respiratory protection ■ PPE that prevents airborne contaminants from being inhaled.

OSHA Respiratory Protection Regulation (29 CFR 1910.134)

The OSHA respiratory protection regulation, which is referenced by HAZWOPER, is designed to protect employees from hazardous atmospheres. It requires the employer to

TABLE 7-1 Types of Respiratory Protection Devices

RESPIRATORY PROTECTION	ADVANTAGES	LIMITATIONS	USES	OPERATIONAL COMPONENTS
Self-contained breathing apparatus (SCBA)	Highest level of respiratory protection	Air bottle size (30 minutes to 1 hour of air)	Most common choice	Mask; air tank; harness; regulator
Air-line respirator	Highest level of respiratory protection with unlimited air supply	Hose line is vulnerable to damage; distance limitations	Longer-duration incidents; tight spaces	Mask; harness regulator; air hose; compressor or air tank in cold zone
Closed-circuit SCBA	Highest level of respiratory protection with longer entry times than SCBA	Expensive; chemical CO_2 scrubber creates heat	Longer-duration incidents	Mask; oxygen tank; chemical scrubber; harness; regulator
Powered air-purifying respirator (PAPR)	Less expensive than SCBA; longer entry times with positive pressure at seal	Cannot be used in IDLH atmospheres; must have appropriate cartridge	Decontamination line and situations requiring longer entry times	Mask; harness; blower motor; chemical absorptive cartridge
Air-purifying respirator (APR)	Comparatively inexpensive; lightweight; longer entry times; easy to operate tactically	Cannot be used in IDLH atmospheres; may leak significantly around seal; must have appropriate cartridge	Decontamination line and situations requiring longer entry times	Mask; chemical absorptive cartridge

HAZMAT HANDLE

Your respiratory protection is your last line of defense, keeping hazardous materials from entering your lungs. Make sure you understand how to use your respirator well!

provide workers with the appropriate respiratory protection, training in how to use the respirators, and a medical evaluation to ensure the employees are physically and mentally fit to wear the respirator. Depending on the type of respirator and whether the use is voluntary or mandated, a written respiratory protection plan must be in place that covers respirator selection, use, maintenance and care, fit testing, training, respiratory protection program evaluation, and record keeping. Those using a respirator should understand their employer's respiratory protection plan and be trained in the selection and use of the respiratory protection they will be wearing.

Respiratory Protection Program

According to 29 CFR 1910.134, employers must have a respiratory protection program if workers are required to wear respirators. This includes hazardous materials response teams. NIOSH-approved respirators must be provided to the employees at no cost. There must be a program administrator who implements and oversees a written respiratory protection program. The respiratory protection program must contain the following written records that must be made available to employees and OSHA on request:

- process for respirator selection
- standard operating procedures for respirator use (routine and emergency situations)
- employee training
- medical fitness to wear a respirator
- fit testing results
- cartridge filter change-out schedules
- method for evaluating hazardous atmospheres
- evaluation of program effectiveness
- medical evaluation
- respirator cleaning and maintenance
- breathing-air generation
- routine respirator inspection
- respirator inspection after use

Employees must be trained in the proper use, maintenance, inspection, and cleaning of respirators according to the manufacturer's instructions. In addition, procedures should be in place to ensure that supplied air is of breathing air quality and does not contain carbon monoxide or oil. Respirator selection should be based on documented objective data, such as quantitative air monitoring readings.

The respiratory protection program must be site-specific, since incident parameters affect respirator use and safety. Thus, there should be an addendum to the respiratory protection program for each hazardous materials incident that includes which respirators were selected and why. In contrast, voluntary respirator use has significantly less stringent requirements.

protection factor ▪ A unitless number denoting the effectiveness of respiratory protection. A protection factor of 100 indicates that up to 1 in 100 contaminant molecules can be expected to enter the respiratory tract.

Types of Respirators

Materials penetrate the respiratory passages at different efficiencies. Large particulates such as dust and soot are trapped by the nasal passages and upper respiratory tract (as any firefighter knows who has blown his or her nose after a structure fire). Dust masks and other particulate filters can protect against large particulates and dusts (Figure 7-4).

OSHA assigns **protection factors** to respirators based on how effectively they prevent contaminants from entering the body. The higher the protection factor, the more effective the respirator is at protecting the wearer. The following protection factors have been assigned by OSHA:

Half-face cartridge APR	10
Full-face cartridge APR	50
Self-contained breathing apparatus (SCBA)	10,000

The protection factor indicates the expected amount of contaminant that may enter the respiratory tract. For example, a full-face APR mask has a protection factor of 50. This means 1 in 50 contaminants can be expected to enter the breathing zone of the wearer. In other words, 98% of the contaminants are filtered out, while 2% of the contaminants enter the mask and lungs. In contrast, a positive-pressure SCBA has a protection factor of 10,000, indicating it filters out 9,999 contaminants and lets in only 1 for every 10,000 contaminant molecules it encounters. That's a much more comfortable feeling if you enter a dangerous atmosphere!

The following factors can lower the protection factor of respiratory protection:

- quality of the face mask seal (due to facial hair or poor fit)
- high breathing rate
- ambient temperature
- movement while working
- chemical effects on respirator parts

Half-mask APRs have the decided disadvantage that they leave the eyes unprotected. Although the respiratory tract is protected if the respirator is used properly, the eyes may be a ready route of entry or may be severely irritated, as by anhydrous ammonia. Therefore half-mask APRs are typically not suited for emergency response situations.

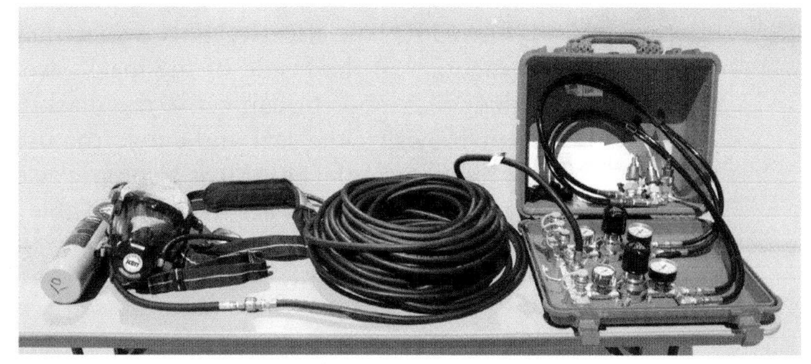

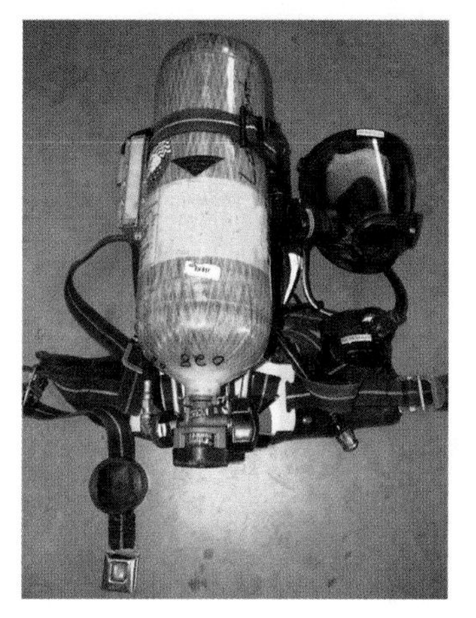

FIGURE 7-4 Respiratory protection is available in a variety of different forms. A dust mask (top left) provides minimal respiratory protection against particulates, an air-purifying respirator (top center) and a powered air-purifying respirator (top right) provide respiratory protection against specific gases and vapors. The highest level of respiratory protection consists of supplied air. The center image shows an in-line supplied air respirator. A self-contained breathing apparatus (SCBA) is shown in the bottom left-hand corner, and a rebreather is shown in the bottom right-hand corner. *Photos courtesy of Chris Weber, Dr. Hazmat, Inc.*

Particulate Respirator (Dust Mask) Particulate respirators, or dust masks, provide very limited respiratory protection. They are the simplest and least expensive type of respirator. This type of mask is what maintenance personnel may use to protect against dust, or EMS personnel may refer to as a tuberculosis (TB) mask (Figure 7-4). Officially, TB masks are known as N-95 filtering facepieces, which means they are rated to keep out 95% of the particulates. In other words, 5% of the contaminants make it into your respiratory tract. Particulate respirators are usually disposable and filter out dusts, fumes, and mists. The mask consists of a filtering type material and must be replaced when it becomes clogged (difficult to breathe through), discolored, or damaged. These respirators provide no protection against vapors and gases and are not considered acceptable respiratory protection against chemicals.

filter cartridge ■ The part of the respirator that removes contaminants from the ambient air. Cartridges must be chosen based on the specific contaminant and must be replaced at regular intervals.

Air-Purifying Respirator Air-purifying respirators (APRs), sometimes referred to colloquially as "gas masks," are tight-fitting respirators that filter out chemicals in the air using a filter cartridge system (Figure 7-4). **Filter cartridges** are designed to remove specific contaminants from the air, usually by binding the contaminant in an absorbent matrix or by neutralizing the chemical via a chemical reaction. Thus APRs clean the ambient air that passes through the cartridge. The APR mask forms a tight-fitting seal against the face and forehead, and straps that tighten against the back of the head maintain a tight seal. When the wearer inhales, the negative pressure (or vacuum) forces ambient air from the outside through the filtering cartridge. Depending on the type of cartridge and the physical fitness of the user, this can be straining on the lungs.

APRs are effective only when they are used properly. Because inhaling creates a vacuum on the inside of the mask, the mask's seal to the face must be air tight. Facial hair or a poor fit can cause the mask to leak, which means that contaminated ambient air blows by the seal and enters the user's respiratory system. Obviously not an acceptable situation! The APR is also only as effective as the selected cartridges. Many different types of cartridges are available (Table 7-2). If an organic vapor cartridge is selected for an acid gas exposure, there is essentially no protection. Again, this is obviously not an acceptable situation.

TABLE 7-2 Types of Cartridges Available for APRs

CARTRIDGE COLOR	CONTAMINANT FILTERED
Purple (magenta)	Particulates (P100) HE (HEPA) for PAPRs Radioactive materials (except tritium and noble gases such as radon)
Orange	Particulates (P95, P99, R95, R99, R100)
Teal	Oil-free particulates (N95, N99, N100)
White	Acid gases
White with 1/2-inch green stripe completely around the cartridge near the bottom	Hydrocyanic acid gas
White with 1/2-inch yellow stripe completely around the cartridge near the bottom	Chlorine gas
Black	Organic vapors

TABLE 7-2 Types of Cartridges Available for APRs (*Continued*)

CARTRIDGE COLOR	CONTAMINANT FILTERED
Black/purple (magenta)	Organic vapors and P100
Yellow	Acid gas and organic vapors
Yellow/purple (magenta)	Acid gas, organic vapor, and P100
Yellow with 1/2-inch blue stripe completely around the cartridge near the bottom	Hydrocyanic acid gas and chloropicrin vapor
Green	Ammonia gas and methylamine gas
Green with 1/2-inch white stripe completely around the canister near the bottom	Acid gases and ammonia gas
Blue	Carbon monoxide
Brown	Acid gases, organic vapors, and ammonia
Red	Acid gases, organic vapors, ammonia, and carbon monoxide
Olive	Multi-contaminant combinations not listed above CBRNE certified

Filter canisters have a limited shelf life, even when properly sealed in the original packaging. Always check the expiration date of the cartridge before using it in a potentially contaminated environment. As soon as the vacuum package containing the respirator cartridge is opened, the cartridge starts degrading. Previously opened cartridges, even if used only for training, should never be saved and used at another hazardous materials incident.

Because APRs filter the ambient air using a chemical specific process, several criteria must be met for their use:

1. Sufficient oxygen must be present (19.5%–23.5% per OSHA 29 CFR 1910.134(i)(1)(ii)(A)).
2. All dangerous contaminants have been identified (to be able to select the appropriate filtering cartridge).
3. Contaminant concentrations are known (to remain below IDLH levels and avoid oversaturating the cartridge filter).
4. Contaminant concentrations are below IDLH levels.
5. Appropriate cartridges are available that are able to filter out contaminants.
6. No other IDLH conditions exist (toxicity, flammability, oxygen deficiency, etc.).

Because APRs only filter the ambient air, you must first determine that ambient oxygen levels are above 19.5% and below 23.5%. Atmospheres are considered immediately dangerous to life and health (IDLH) when oxygen concentrations fall below 19.5% or rise above 23.5%. Other questions to consider are, Why has the oxygen level changed? What is displacing the oxygen? Is that material known and accounted for? Otherwise, an APR should not be used.

You must identify the contaminant to choose the appropriate air-purifying cartridge or canister. The absorbent or reactive material inside the canister must effectively exclude or neutralize the contaminant of interest. Furthermore, the cartridge or canister must have an **end of service life indicator (ESLI)**, or you must determine an appropriate change-out schedule based on the manufacturer's recommendations. You must know the contaminant level before using an APR to ensure ambient contaminant levels are, and remain, below IDLH levels. Contaminant levels can be determined using air monitoring instruments, the use of which will be covered in Chapter 8.

end of service life indicator (ESLI) ▪ A device on a filter cartridge that indicates to the wearer when contaminants are no longer being removed from the ambient air.

Finally, you must complete a site safety plan and determine that there are no other IDLH conditions present. Other IDLH conditions include flammable atmospheres, confined spaces, and other situations that pose an immediate danger to life and health.

Because cartridge respirators merely filter the contaminated ambient air, there is a **maximum use concentration (MUC)** based on the specifications of the APR that is calculated as follows:

maximum use concentration (MUC) ▪ The highest airborne concentration of a contaminant that a respirator may be used in based upon the permissible exposure limit (PEL) and the protection factor of the respirator.

$$\text{MUC} = \text{Protection factor} \times \text{Worker exposure limit}$$

For example, if you want to use an APR with a protection factor of 50 in an atmosphere of chlorine, which has a PEL of 1 ppm, then the MUC = 50 × 1 ppm = 50 ppm.

However, an APR may never be used in an IDLH atmosphere. Therefore, if the calculated MUC exceeds the IDLH, it should be lowered to the IDLH value for the chemical indicated in the *NIOSH Pocket Guide* or other reliable reference source. The IDLH for chlorine is only 10 ppm, so an APR should never be used to protect against chlorine concentrations at or above 10 ppm.

Some chemicals have very poor warning properties. A *warning property* alerts the wearer by the sense of smell to the presence of a dangerous gas or vapor below the PEL. When a chemical has poor warning properties there is no way to know when the filter becomes saturated, which could lead to serious injury or death from overexposure. Therefore, ESLIs are often built into these types of cartridges. However, many cartridges are not equipped with an ESLI. In these cases a qualified professional, such as a certified industrial hygienist (CIH), should develop an appropriate cartridge change-out schedule. Another option is to not rely on air-purifying respirators at all and use supplied air respiratory protection instead. The following are some chemicals which are known to have poor warning properties (adapted from MSA):

acrolein
aniline
arsine

bromine
carbon monoxide
diisocyanates
dimethyl sulfate
hydrogen cyanide
hydrogen selenide
methanol
methyl bromide
methyl chloride
methylene chloride
nickel carbonyl
nitric acid
nitrogen oxides
nitroglycerin
nitromethane
phosgene
phosphine
phosphorus trichloride
stibine
sulfur chloride
urethane
vinyl chloride

Powered Air-Purifying Respirator (PAPR) PAPRs are APRs with a blower motor that forces air through the filter cartridge. Thus, they essentially provide a positive-pressure environment (although it is not considered to be a true positive-pressure respirator). This makes them safer and easier to use. They are safer because a negative pressure situation never exists inside the mask (unless the blower motor malfunctions or the battery dies). For this reason a tight fitting mask is not necessary, and hooded PAPRs are very popular in agencies and industries where facial hair is common. Hospitals typically use hooded PAPRs to perform emergency decontamination and preliminary patient treatment on contaminated patients (Figure 7-4). They are easier to use because the lungs don't need to work as hard to force the ambient air through the filter cartridges. The drawback to PAPRs is that they are heavier (due to the blower motor and battery pack), and the motor may unexpectedly fail in the contaminated area.

Positive-Pressure Self-Contained Breathing Apparatus (SCBA) Self-contained breathing apparatus (SCBA) is a type of **supplied air respirator (SAR)** and offers the highest level of respiratory protection when used in positive-pressure mode (Figure 7-4). Positive-pressure mode refers to a constant air pressure delivery to the mask. If the seal of a positive-pressure SCBA is broken, air will rush out owing to the positive pressure inside the mask. SCBA consists of a compressed air bottle, a tight-fitting mask, and one or more regulators to reduce the air pressure. This is the respiratory protection that firefighters commonly use. As with APRs, the mask must be used properly and fit tightly against the skin. This is why the vast majority of fire departments have strict facial hair grooming standards.

supplied air respirator (SAR) ▪ A face mask and source of breathable air required to be used in IDLH atmospheres.

The main advantage of SCBA respiratory protection is that it does not rely on filtered ambient air but, rather, carries its own supply in the compressed air tank. The drawback is that SCBA bottles typically come in only 30- to 60-minute capacities. This limits the amount of time and the amount of work that can be accomplished while wearing SCBA, a distinct disadvantage over APRs. SCBAs are also significantly heavier than APRs; they can weigh 15–25 pounds depending on design and air capacity. A new generation of low-profile, lightweight SCBA is currently being developed and will be on the market soon. Nevertheless, SCBAs are the standard respiratory protection when unknown or IDLH atmospheres must be entered by hazmat personnel.

Positive-Pressure Air-Line Respirator with Escape Unit Air-line respirators are a type of supplied air respirator in which the air supply is not carried on the back but, rather, is supplied through an air hose from a remote location (Figure 7-4). By law, the maximum length of the air hose may be 300 feet with a minimum airflow of 115 L per minute for pressure-demand units, and 170 L per minute for continuous-flow units. The maximum airflow may be 435 L per minute. This remote location is usually one or more large compressed air tanks or an air compressor located outside the hot zone. This type of respirator is similar to an SCBA, since it supplies clean air, with the advantage that it is much lighter weight. The disadvantage is that the air line may become compromised by pinching, severing, or bursting. Therefore, a small escape bottle must be carried into the hot zone when using an air-line respirator. Supplied air respirators are popular when there is little room available, such as in confined spaces, and when longer operation times are required, such as in research laboratories, during cleanup operations, and at decontamination lines.

Closed-Circuit SCBA Closed-circuit SCBAs, also known as *rebreathers,* are a type of supplied air respirator designed to be worn on the back like an SCBA, but they have a longer work time. The work time can be 1–6 hours depending on the configuration. A closed-circuit SCBA consists of a tight-fitting mask, two or more regulators, a compressed air cylinder, a compressed oxygen cylinder, and a chemical scrubbing system (Figure 7-4). The chemical scrubbing system removes the exhaled carbon dioxide from the breathing air. This chemical reaction is exothermic, or heat releasing, which causes the breathing air to be heated as the unit is used. The unit also feels warm on the back. Both these factors lead to increased heat stress when using closed-circuit SCBA.

These respirators were developed many years ago for the mining industry and are used today in hazardous materials response when SCBA and air-line respirators are not practical. The advantage of the closed-circuit SCBA is an extended work period without the inconvenience and distance limitations of an air-line respirator. The disadvantage is that the breathing air is warm or even hot owing to the exothermic reaction of the scrubber, and that the units are even heavier than a standard SCBA unit. Today, rebreathers are often used in mines and tunnels, both during rescue operations and hazmat incidents. Examples of closed-circuit SCBAs are the Litpac and Biopack.

CHEMICAL PROTECTIVE CLOTHING (CPC)

Chemical protective clothing (CPC) is designed to keep chemicals from coming into contact with the wearer (Table 7-3). CPC includes chemical-resistant boots, suits, and gloves. The different types of CPC are designed to keep the chemical from contacting the skin. Some CPC protects from liquid exposures (**splash protection**); other CPC can protect from gaseous and vapor exposure as well (vapor protective suits).

splash protection ■ PPE designed to protect the wearer from liquid contact, but not from gases and vapors.

Chemical protective boots can be purchased in a variety of styles and materials. They are one of the most important components of CPC, since accidental or deliberate contact with chemicals is common. Entry team members often must walk through a spill to

SOLVED EXERCISE 7-2

What is the maximum use concentration (MUC) for a full-facepiece air-purifying respirator (APR) that will be used during cleanup of acetic acid? The permissible exposure limit (PEL) for acetic acid is 10 ppm, and the IDLH is 50 ppm.

Solution: The MUC is 50 ppm. Based solely on the protection factor of a full-facepiece APR, which is 50, the MUC would be 50 × 10 ppm, or 500 ppm. However, air-purifying respirators may never be used in IDLH atmospheres (per 29 CFR 1910.134(d)(2)). The IDLH for acetic acid is 50 ppm. Therefore, the MUC is the lower of the two values, or 50 ppm. If APRs are selected for the cleanup, continuous air monitoring for acetic acid must be performed to ensure that ambient acetic acid concentrations do not exceed 50 ppm.

TABLE 7-3 Types of Protective Clothing

PROTECTIVE CLOTHING	PURPOSE	ADVANTAGES	LIMITATIONS
Chemical protective clothing	Hazardous materials response	When appropriately selected, chemical resistant	No thermal protection; heat stress
Thermal protective clothing	High-temperature flammable liquid firefighting	Protects from radiant heat	Little, if any, chemical protection
Structural firefighter protective clothing	Structural firefighting	Protects from moderate radiant heat with some flash fire protection	Little, if any, chemical protection

perform emergency mitigation procedures. Sometimes, entry team members accidentally walk through a spill; however, this can usually be avoided through good work practices and proper training.

Gloves are also an important component of CPC. Upon exiting the hot zone, the boots and gloves will typically contain 85% or more of the contamination. Gloves can also be purchased in a variety of styles and materials, including many different polymers and laminates. When gloves and boots are not integrated into a suit ensemble, they are usually attached to the suit using chemical-resistant tape (Figure 7-2). It is important to use chemical-resistant tape rather than common duct tape to ensure that the chemical will not dissolve the adhesive, or that the adhesive will not damage other components of the CPC.

CPC comes in a variety of protective qualities, sizes, shapes, colors, thicknesses, flexibilities, shelf lives, and costs. Some CPC is disposable; other is reusable. Some is complicated to use and maintain, while other CPC is simple to use and maintain. You must become familiar with the type of CPC your agency uses. CPC is an extremely important component of your PPE that helps keep you safe during hazmat incidents.

Chemical protective suits come in a variety of configurations. The more effective—and expensive—CPC consists of several chemical-resistant polymers laminated together. Some chemical-resistant polymers are elastomers that return to their normal shape after being stretched and are desirable in CPC construction owing to their resiliency. However, many chemical-resistant polymers are nonelastomers, which are materials that do not return to their normal shape after stretching. Nonelastomers are typically constructed in multiple layers laminated to another substrate (such as Tyvek) to increase their resiliency. The different types of boots, gloves, suits, and chemical tape have different levels of resistance to chemicals and must be appropriately selected using manufacturer-supplied information (compatibility charts) (Figure 7-5).

Chemical protective clothing is primarily manufactured using three different types of seams (Figure 7-6). *Serged* seams are generally effective at keeping out solids, *bound* seams are generally effective at keeping out solids and the bulk of splashed liquids, and *sealed* seams are effective at keeping out solids, the bulk of splashed liquids, and gases.

There are three ways CPC may be breached by chemicals: penetration, degradation, and permeation. Each of these three processes can be accelerated by the unwitting actions of the hazmat technician wearing the suit. For example, it is very important not to kneel while wearing PPE to avoid stretching, tearing, or otherwise damaging the suit.

Penetration

Penetration refers to the passage of material through a macromolecular (relatively large) opening in CPC. Chemicals can enter chemical protective clothing via penetration of natural openings in the suit such as zippers, seams, and hand and foot openings (Figure 7-7). Splash

FIGURE 7-5 Manufacturer's compatibility chart for selection of chemical protective gloves. *Courtesy of Showa Best Glove, Inc.*

Key To Degradation Rating
E = Excellent
G = Good
F = Fair
P = Poor
NR = Not Recommended

Key To CPC Index Number
0
1
2
3
4
5

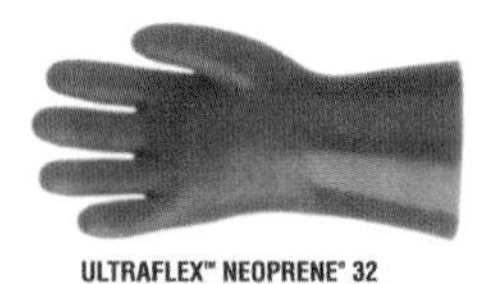

CHEMICAL	CAS NUMBER	EUROPEAN TOXICITY RISK CODE	NEOPRENE® 6780: NEOPRENE								ULTRAFLEX™ NEOPRENE® 32: NEOPRENE							
			DEGRADATION RATING				PERMEATION BREAKTHROUGH			CPC INDEX	DEGRADATION RATING				PERMEATION BREAKTHROUGH			CPC INDEX
			TIME IN MIN 5	30	60	240	MDL PPM	BDT MIN	RATE ug/cm²/min	RATING (0-5)	TIME IN MIN 5	30	60	240	MDL PPM	BDT MIN	RATE ug/cm²/min	RATING (0-5)
171. PROPYLENE GLYCOL MONOMETHYL ETHER	107-98-2	X	E	E	E	P	0.02	ND	ND	0	E	E	E	E	0.02	ND	ND	0
172. PROPYLENE GLYCOL MONOPROPYL ETHER	1569-01-3	X	E	E	E	E	0.02	ND	ND	0	E	E	E	E	0.02	177	52	3
173. PROPYLENE OXIDE	75-56-9	X, CANCER	G	G	F	F	0.02	11	204	5	G	G	G	G	0.02	43	47	3
174. PROPYL PROPASOL SOLVENT	1569-01-3	X	E	E	E	E	0.02	ND	ND	0	E	E	E	E	0.02	177	52	3
175. PYRIDINE	7291-22-7	X	E	F	P	P	0.02	NR	NR	5	G	P	NR	NR	0.02	NR	NR	5
176. REFRIGERANT 123A	306-83-2	V	E	E	E	E	0.02	73	62	3	E	E	G	G	0.02	85	180	4
177. REFRIGERANT 141B	1717-00-6	V	E	G	F	F	0.02	68	2688	5	E	G	F	F	0.02	21	602	4
178. RUBBER SOLVENT	8032-32-4	X	E	E	E	E	0.02	99	10	2	E	E	E	E	0.02	47	110	4
179. SAFROTIN (50% in ROH)	31218-83-4	X	E	E	E	E	0.02	ND	ND	0	E	E	E	E	0.02	ND	ND	0
180. *SODIUM HYDROXIDE* 50%	1310-73-2	Cx	E	E	E	E	0.02	ND	ND	0	E	E	E	E	0.02	ND	ND	0
181. SODIUM HYPOCHLORITE 4–6%	7681-52-9	C	E	E	E	E	45.0	ND	ND	0	E	E	E	E	45.0	ND	ND	0

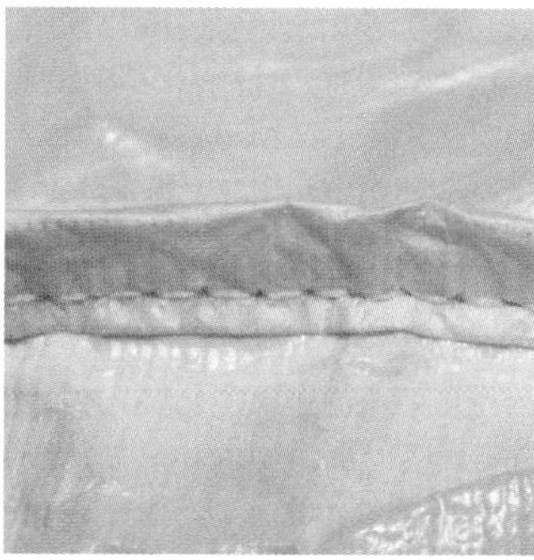
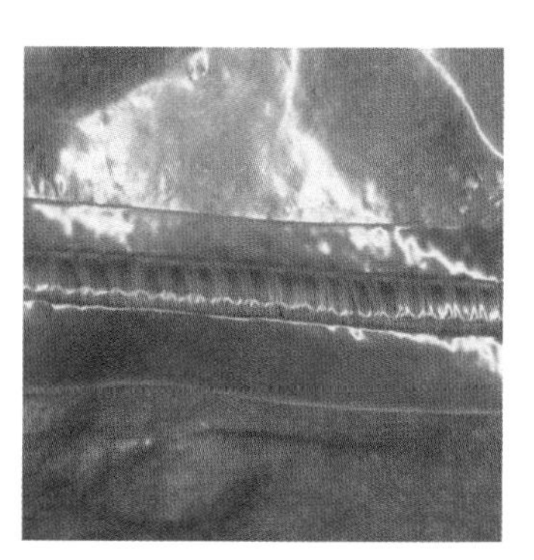

FIGURE 7-6 Three types of seams used in chemical protective clothing (CPC) include serged seams (left), bound seams (middle), and sealed seams (right). *Photos courtesy of Chris Weber, Dr. Hazmat, Inc.*

protective ensembles have several natural openings depending on their design criteria, and are not necessarily vapor- and liquid-tight. Penetration may also result from mechanical damage to the chemical protective clothing from tearing or from defective design openings in the suit. It is very important to safeguard your chemical protective clothing from mechanical damage while inside the hot zone, and to inspect the suit before entering the hot zone for any manufacturing defects that may be present. The ASTM F903 test method is used by manufacturers to test their suits for penetration.

Chemicals also may enter the CPC through mechanical damage to the suit, which may occur while navigating through the hot zone. Tears or rips in the suit fabric may be caused by sharp objects or shrapnel from an explosion, by structural features and equipment found normally at the facility, or by jagged edges at a transportation accident. Suits may also be punctured from the inside, if sharp objects such as rings are worn in the hot zone. Suit breaches may also occur if the chemical protective clothing is too small for the wearer. Zippers and seams, especially in the groin region, may split or tear when the wearer moves around during normal work-related tasks. Proper sizing of PPE is critical for entry team safety.

In addition, the equipment itself can fail owing to improper storage and use, or to a manufacturing defect. Many features of the CPC may fail, such as zippers, one-way valves, and seams; or the respiratory protection, such as the SCBA or APR, may fail. It is therefore very important to inspect PPE you will be wearing before every use.

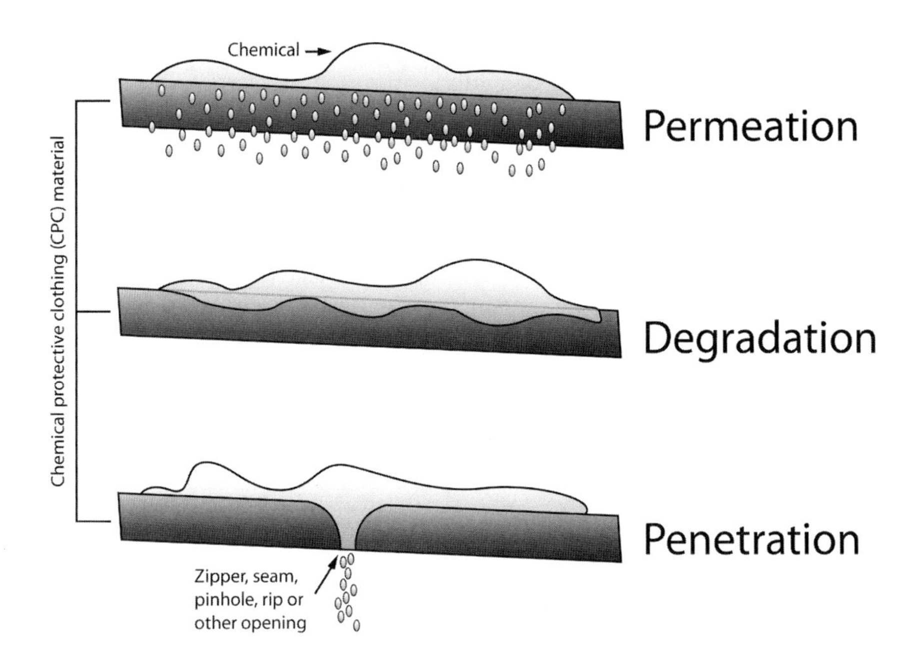

FIGURE 7-7 Chemicals can adversely affect personal protective equipment (PPE) in three ways: They can permeate the chemical protective clothing (CPC) by moving through the suit material at the molecular level (top), degrade the CPC (middle), or they can penetrate the CPC through openings such as zippers (bottom). *Art by David Heskett.*

Degradation

Degradation refers to the chemical breakdown of the CPC material. It is a process by which CPC is compromised by being dissolved by the chemical with which it comes in contact (Figure 7-7). Prolonged UV light exposure, as from sunlight, can also degrade CPC. Degradation reduces the integrity of one or more materials used in the CPC and renders it less effective or completely ineffective. A good example of degradation is the interaction between acetone and a polystyrene (Styrofoam) cup. The acetone dissolves the cup on contact.

Three signs of degradation are discoloration of the material, change in flexibility of the material, and thinning or melting of the suit material (see Box 7-1). Any one of these signs indicates that the chemical protective clothing may fail because of chemical degradation. It is very important to exit the contaminated area immediately and proceed to decontamination. Let the decon personnel know that you suspect suit failure due to degradation.

BOX 7-1 HAZMAT HISTORY: CHEMICAL PROTECTIVE CLOTHING FAILURE IN BENICIA, CALIFORNIA (1983)

In August 1983, two hazardous materials responders attempted to stop a leak on a defective sample well of a railcar carrying dimethylamine (DMA). DMA is a flammable, corrosive, and toxic gas. The sample well, which was approximately a 1/8- to 3/16-inch opening, started leaking as employees of the chemical plant were getting ready to offload the railcar.

The recommended PPE ensemble was chosen for the entry. During the second entry the polycarbonate facepiece of the suit started to fail. The two facepieces cracked and discolored owing to chemical incompatibility. The facepiece actually fell out of one entry team member's PPE as he descended the railcar. At the time, reference materials did not take into account all the components of PPE ensembles but merely the suit material itself. This incident was a key factor in the establishment of the initial NFPA chemical protective clothing standards (such as NFPA 1991).

Although the entry team members were not seriously injured during this response, given the hazards of DMA, a disaster was narrowly averted. This incident illustrates the importance of carefully selecting chemical protective clothing and PPE in general during hazardous materials response. It is vital that you consult the manufacturer's chemical compatibility chart before performing activities with the potential for direct contact with a hazardous chemical.

(From an interview with Jerry Grey, San Francisco Hazardous Materials Response Team [retired].)

Permeation

Permeation refers to the movement of material through a microscopic passage in the CPC. It is the process of chemical movement through the suit matrix at the molecular level (Figure 7-7). CPC is composed of materials, whether textiles or plastics, that have a weave of strands of material. There are various-sized spaces between the strands that chemicals can get through, or permeate. All chemicals permeate all materials at some rate. An example of permeation is the slow leakage of helium out of a rubber balloon over a day or two, so that it no longer rises. For a compatible piece of CPC the permeation rate will be several hours or more; for an incompatible material it will be minutes or seconds.

It is important to understand that visible suit integrity remains intact during this process. The suit material is not degraded, and chemical does not move across larger imperfections in the suit material. The permeation rate is a function of a combination of the properties of the chemical and of the suit material. A good analogy is an insect screen on your window. The effectiveness of the screen depends on both the mesh size and the size of the insect you're trying to keep out. If you decide to use chicken wire as a window screen, you won't be able to keep out flies and mosquitoes. However, if you use a smaller size mesh you'll keep those flies and mosquitoes out, but gnats might still make it through. You need to further reduce the mesh size of your screen to keep out those pesky gnats! In this analogy, the insects permeate the chicken wire, since the chicken wire itself has not been compromised (by penetration or degradation). Similarly, chemicals permeate chemical protective clothing if the weave of the material is essentially too loose.

Chemical properties that affect permeation rate include the molecular size of the contaminants, the state of matter of the contaminants, and the viscosity of the contaminants. Environmental factors also affect the permeation rate, such as temperature and pressure. Chemical protective clothing properties that affect the permeation rate include the type of polymer, the thickness of the material, and the density of the material.

breakthrough time ▪ The time it takes for a hazardous substance to pass through chemical protective clothing.

So how do you make the proper choices? The most reliable method for determining appropriate work times in chemical protective clothing is to consult the manufacturer's chemical compatibility charts that are based on breakthrough times using the ASTM F739 test method (Figure 7-8). **Breakthrough time** is the length of time it takes a chemical to pass through the chemical protective clothing material and desorb on the other side. A swatch of material to be tested is placed in a permeation chamber. One side of the chamber contains the challenge chemical, and the other side of the chamber is monitored using detection equipment. Breakthrough times are listed for up to 8 hours. Depending on how the chemical protective clothing ensemble is used, and how long it will be used, shorter breakthrough times may be acceptable under certain circumstances. Solved exercise 7-4 shows examples of breakthrough times for various types of chemical protective suits. The ASTM F1001 standard establishes a suggested set of challenge chemicals to be used to evaluate PPE.

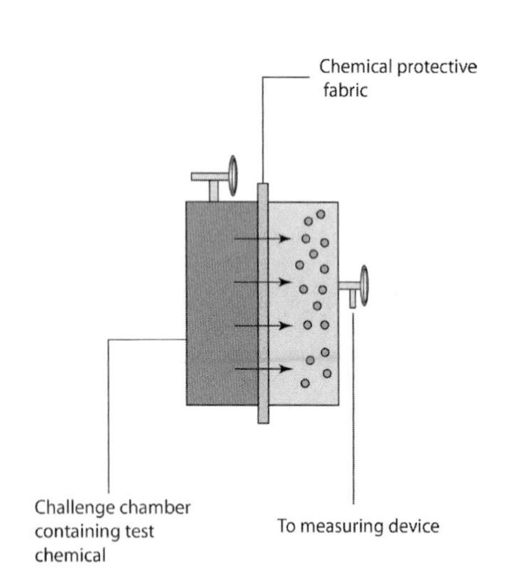

FIGURE 7-8 Schematic drawing of a permeation test cell used to measure the permeation rate of chemicals through chemical protective clothing (CPC). *Art by David Heskett.*

Flash Fire Protection

Chemical protective clothing is generally not effective for use in flammable atmospheres, since it does not provide either flash protection or thermal protection. There are currently some exceptions to this rule, since manufacturers have recognized that chemical and flammability hazards coexist. Several manufacturers have flash fire–resistant suits, such as the TyChem ThermoPro from DuPont. This particular suit was tested to NFPA 2112 (*Standard on Flame Resistant Garments for Protection of Industrial Personnel against Flash Fire*, 2007 Edition) criteria. Keep in mind that these suits offer only extremely limited flash fire protection on the order of 5 to 7 seconds and provide no thermal protection.

BOX 7-2 HAZMAT HOW TO: USING THE CPC SELECTION GUIDE

The *Quick Selection Guide to Chemical Protective Clothing* by Krister Forsberg and S. Z. Mansdorf is a compilation of compatibility data from a number of different manufacturers and literature sources. It is a very convenient single source for broad-spectrum chemical compatibility data to a wide range of materials and PPE ensembles. The three-step process used to determine chemical compatibility allows comparison of members of a chemical class of compounds.

1. Determine the class number to which the chemical of interest belongs to by consulting the chemical index near the beginning of the book.
2. Determine the selection recommendations by finding the chemical class in the compatibility tables that follow the chemical index.
3. Find the chemical of interest within the appropriate chemical class section in the selection recommendations table.

The following materials are listed: butyl rubber, natural rubber, neoprene rubber, nitrile rubber, polyethylene, polyvinyl alcohol, polyvinyl chloride, Viton, Viton/butyl rubber, Barrier, Silver Shield, various Trellchem ensembles, and various Tychem ensembles—including SL, F, BR, CPF 3, Responder, and TK.

Ionizing Radiation Protection

Chemical protective clothing has traditionally not been very effective at shielding against ionizing radiation, such as beta and gamma radiation and X-rays. Recently Radiation Shield Technologies (RST) has developed a heavy metal–impregnated nanotechnology fabric called Demron. This material contains dense ionizing radiation–absorbing metal atoms that effectively shield against 70% to 95% of low energy gamma and X-ray radiation (shielding efficiency is very dependent on the energy of the ionizing radiation). The class 2 suits they manufacture have been certified to NFPA 1994 (2007) as well as ISO 8194 (*Radiation Protective Clothing*) criteria. These suits can be considered the first true CBRN-capable CPC ensembles.

However, Demron - as used in the current class 2 suits - is not very effective at shielding against high energy gamma radiation and neutron radiation. Although Demron has approximately the same shielding effectiveness as lead, the relatively thin suit material (approximately 0.4 mm) severely limits its shielding capability at higher ionizing radiation energies. According to a report from Lawrence Livermore National Laboratory, Demron suits provide a factor of 3 protection against beta radiation, a factor of 10 protection against low energy gamma radiation, and relatively insignificant protection against higher energy gamma emissions (such as from ^{137}Cs and ^{60}Co which are most commonly encountered).

THERMAL PROTECTIVE CLOTHING

Thermal protective clothing is designed to protect the wearer from extremes in temperature. This may be a pair of cryogenic gloves used to handle liquid nitrogen, or structural firefighter protective gear used to fight fires, or aluminized high-temperature thermal protective clothing. High-temperature thermal clothing is often called a *proximity suit* and is used by refinery fire personnel and airport fire personnel owing to the extreme heat output of large quantities of burning hydrocarbons. Proximity suits provide excellent protection against radiant heat but not direct flame contact. In contrast, entry suits are designed to protect against limited direct flame contact.

Structural firefighter protective clothing, or *turnout gear,* is a type of thermal protective clothing. It is designed to protect the wearer from extremes in heat. But even this protection is limited. Turnout gear is very effective at providing thermal protection from the heat typically encountered during structural firefighting. This type of PPE provides little to no protection from chemicals, or from intense heat and flash fires. For this reason structural firefighter protective gear is not particularly effective for fires involving significant amounts of petroleum products. Turnout gear consists of a helmet, SCBA with a personal alert safety system (PASS) device, a thermal protective hood, thermal protective coat (bunker coat), thermal protective gloves, thermal protective pants (bunker pants), and steel-toed boots.

Structural firefighter protective clothing has very limited chemical protective qualities and should not routinely be used in the hot zone at hazmat incidents. However, some exceptions do apply. Turnout gear may be appropriate when the fire risk outweighs the other health and safety risks. An example would be a methane or propane leak in which a valve must be turned off. The primary hazard is flammability, the biggest health risk is inhalation and not skin absorption, and a positive-pressure SCBA is part of the turnout gear. Turnout gear is clearly much safer than chemical protective clothing in this case. Other examples include immediate lifesaving rescues in which the hot zone is entered solely for the purpose of rescuing known, live victims. Another example is a hazardous material release that requires thermal protection with minimal to no chemical protection (such as some cryogenic gases). Thermal protective clothing is *not* considered chemical protective clothing.

SPECIALTY SUITS

Other types of personal protective equipment are used under special circumstances, such as bomb suits, and LANX or MOPP suits (Figure 7-9). Bomb suits provide limited protection from the effects of an explosion, including shrapnel, thermal effects, and overpressure effects. These suits have special air-filtering systems to minimize the overpressure effect from an explosion. But as you can imagine, these suits are effective only for relatively small detonations. LANX and mission-oriented protective posture (MOPP) gear is primarily designed for protection from chemical warfare agents. MOPP gear is standard issue U.S. military chemical warfare agent protection on the battlefield. LANX suits are similar technology manufactured for emergency responders – especially law enforcement agencies. These suits have a lining containing activated charcoal, which absorbs most chemical warfare agents. The suits also offer limited protection against many toxic industrial chemicals (TICs).

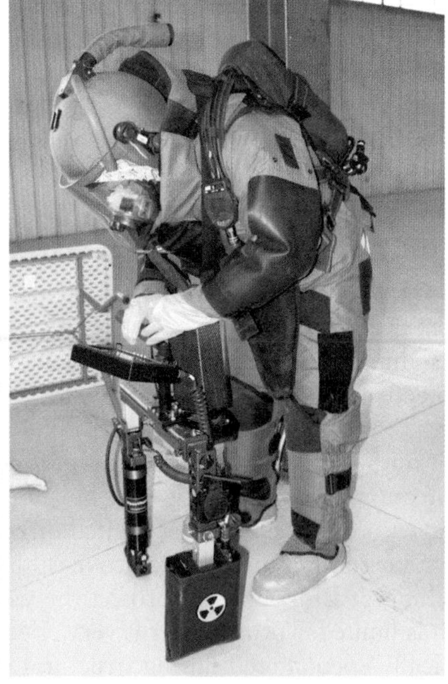

FIGURE 7-9 Specialty suits include aluminized proximity suits (left) and bomb suits (right). *Photo on left © Lakeland Industries, Inc. Photo on right courtesy of Chris Weber, Dr. Hazmat, Inc.*

COMMUNICATIONS

Communication between entry team members in the hot zone is essential, yet communication while wearing PPE can be a challenge. The most common type of communication between entry team members is verbal, while the most common type of communication between entry team members in the cold zone is via radio. Verbal communication can be challenging, especially when SCBA and Level A suits are used. Alternatives to verbal communication between entry team members are written communication or hand signals. Hand signals generally have the advantage of being visible over longer distances, but only relatively simple information can be transmitted. It is essential that the radio communications used in the hot zone not rely on fine motor skills. Typically, large press-to-talk buttons are used that can be strategically located and operated using an elbow or hand. For example, the button may be located on the SCBA belt or on an SCBA shoulder strap. Because speech is often difficult to recognize in PPE, throat microphones or bone microphones are often used to increase the accuracy of speech. Even when these technologies are used, radio communications are still a challenge at most hazardous materials incidents.

HEAT EXCHANGE UNITS

Heat exchange units are very effective at helping regulate body temperature and limiting the consequences of heat stress (Figure 7-10). When personnel work in protective clothing under extreme work conditions or in extreme climates, heat exchangers may prevent a heat stress injury. However, they will not prevent all heat injuries. Do not get a false sense of security—work cycles and hydration are very important!

Air-Cooled Heat Exchange Units

Air-cooled units are fairly uncommon owing to their cost and maintenance. The primary disadvantage of air-cooled units is that they can be heavy, bulky, and noisy owing to mechanical equipment such as fans and compressors, and the attached cooling line may be a trip hazard. The advantage of these units is that they cool in a consistent and predictable manner and can usually be precisely controlled and adjusted. These units are most effective during long-term operations.

Ice-Cooled Heat Exchange Units

Ice-cooled units are typically vests filled with water that are frozen and stored cold until they are needed. The benefits of ice-cooled vests are their simplicity and low cost. The disadvantages of ice vests are that they require a freezer, take a significant time to refreeze the vest after use, and produce an uneven cooling effect. Initially, the ice vest provides a significant amount of cooling—possibly too much—that then dissipates over a comparatively short amount of time.

Water-Cooled Heat Exchange Units

Water-cooled units are also vests filled with water; however, they have an inlet and an outlet that can be connected to a water chiller. This design has the advantage of providing a more consistent cooling effect and being able to be reused, or recharged, rapidly at the scene. The disadvantage of water-cooled vests is the

FIGURE 7-10 A heat exchange vest that uses ice packs for cooling. *Photos courtesy of Chris Weber, Dr. Hazmat, Inc.*

acquisition cost and the need for a water chiller. A more economical version of an ice-cooled vest uses a relatively inexpensive pump and ice water.

Phase-Change Coolers

Phase-change vests maintain a constant temperature, typically 59°F (15°C), which is a big advantage over the much lower temperatures of ice vests (32°F or 0°C). The phase-change material inside the vest changes from a liquid to semisolid gel that maintains a temperature of 59°F when electricity is applied, keeping the body cool by absorbing excess body heat. A disadvantage of this type of vest is higher cost and higher weight owing to the battery pack.

Limitations of Personal Protective Equipment

It is difficult to find one type of PPE that protects against all hazards. Therefore, it is essential to properly research the properties of the released hazardous material as well as the advantages and disadvantages, including chemical compatibility, of the available PPE ensembles as a whole. Hazardous materials and WMDs may exhibit many different hazards, so it is very important that you accurately assess the types of hazards the situation as a whole poses and choose PPE with the appropriate capabilities.

PHYSIOLOGICAL STRESSES

Working in chemical protective clothing and respiratory protection can be physically and mentally demanding. Encapsulating suits can cause feelings of claustrophobia in susceptible individuals because they limit sight and hearing. Chemical protective clothing, which is a barrier to chemicals, also prevents evaporative cooling by keeping perspiration and heat inside the suit. Anyone using PPE in a hazardous environment must be healthy and physically fit. It is a good idea for hazardous materials response teams to institute fitness programs for their members. NFPA 1500 describes the requirements for physical fitness programs.

Thermal Stress

heat stress ▪ Detrimental effects of elevated temperature on the body. Four signs of heat stress are heat rash, heat cramps, heat exhaustion, and heatstroke.

hyperthermia ▪ The excessive heating of the body above normal body temperature.

The biggest danger of using PPE is **heat stress**. Heat stress is the most common type of injury emergency responders wearing PPE incur. Because chemical protective clothing traps heat inside the suit and near your body, **hyperthermia** is a very real risk. The temperature inside the suit can be 25°F higher than the ambient temperature outside the suit. The humidity levels typically reach 100% inside the suit within 5 to 7 minutes, limiting evaporative cooling.

There are four medical conditions that indicate you are suffering from heat stress: heat rash, heat cramps, heat exhaustion, and heatstroke. The protective action that CPC provides must be carefully weighed against the thermal stress it may induce, especially in warmer climates and under strenuous working conditions. The best way to prevent heat-related injuries is to properly hydrate before donning PPE and to have appropriate length work cycles that depend on the workload and environmental conditions. In colder climates hypothermia and frostbite may also be issues while using personal protective equipment.

How real is heat stress? You can decide given the following statistics on the average number of deaths per year in the United States due to various causes (annual average 1979–2003):

Hurricanes	17
Avalanches	28
Lightning	55
Tornadoes	80
Heat stress	334

Let's take a look at some of the signs and symptoms of heat stress, the causes, and ways to avoid heat stress when wearing PPE.

Heat Rash and Heat Cramps The mildest form of heat stress is heat rash. **Heat rash** is a red, mottled appearance of the skin that may appear similar to a mild allergic reaction.

Some people may also suffer from heat cramps when they become overheated. **Heat cramps** occur when there is a rise in body temperature that induces uncontrollable muscle contractions. These contractions occur most commonly in the legs, stomach area, and arms and are quite painful. In fact, you've probably suffered from heat cramps on a hot summer day if you have become slightly dehydrated and overheated. Rest and fluid replenishment, including electrolytes, are the best treatments for muscle cramps and heat rash. It is extremely important to recognize the early signs of heat stress to avoid much more dire consequences!

heat rash ▪ A condition caused by exposure to excessive heat marked by discolored or mottled skin. Heat rash usually appears first on the legs and arms.

heat cramp ▪ A condition caused by exposure to excessive heat marked by severe involuntary muscle contractions.

Heat Exhaustion **Heat exhaustion** is a more serious reaction to heat stress than heat cramps. At this point the body is already significantly stressed and it has kicked into high gear to shed heat. Symptoms of heat exhaustion include weakness and fatigue, dizziness, nausea, headache, and—most visibly—cool and moist skin. A person suffering from heat exhaustion will have wet, clammy skin and feel cool to the touch because the body has overcompensated in its cooling attempts. Although the skin will feel cool to the touch, the core body temperature is still elevated. It is extremely important to quickly get the victim out of the hot environment, to rehydrate him or her, and to apply cold packs to the armpits, forehead, and groin region if necessary.

heat exhaustion ▪ A condition caused by exposure to excessive heat marked by pale and sweaty skin, weakness, and nausea.

Heatstroke **Heatstroke** is the most serious reaction to heat stress and is a life-threatening situation. At this point the body has failed to cope with the heat. Someone suffering from heatstroke will have dry, red skin and altered mental status, and can lose consciousness and/or have seizures. Each year many people die from heatstroke. Heatstroke victims must rapidly be cooled using cold packs, rehydrated, and quickly transported to the hospital for further medical care.

heatstroke ▪ A condition caused by exposure to excessive heat marked by dry and red skin. This is a true medical emergency and requires immediate advanced medical care. Many people die of heatstroke each year.

Cold Injuries Cold injuries can be caused by weather conditions or by the hazardous materials themselves—such as cryogenic materials and liquefied compressed gases. Across large parts of the United States and Canada the winter months bring intense cold, snow, and ice. These conditions make it much more difficult and dangerous to operate at hazmat and WMD incidents. Decontamination operations are of particular concern under these conditions, since wet decontamination can easily induce hypothermia. Temperatures as warm as 65°F can cause cold shock in susceptible individuals. Ambient weather conditions, such as humidity and wind chills, can contribute tremendously to cold injuries through heat loss.

The extremities, especially the fingers and toes, are most susceptible to frostbite, which is freezing of the skin and underlying tissue. The ice crystals that form in the cells of the tissue destroy the cells and cause tissue death. If frostbite is severe and not treated, gangrene may set in, and amputation may be necessary.

Hypothermia, a drop in core body temperature, is also a concern. Hypothermia sets in when the body cannot produce enough heat to maintain 37°C (98.7°F). Symptoms of hypothermia may include altered mental status, shivering, and cold-feeling and pale-looking extremities. Victims of hypothermia must be removed from the cold environment and gradually rewarmed. Hypothermic patients are more susceptible to heart attacks and must be handled gently.

hypothermia ▪ The excessive cooling of the body below normal body temperature.

PSYCHOLOGICAL STRESSES

The cramped and confined space inside chemical protective clothing may lead to psychological stress in some individuals. Claustrophobia is experienced by many first-time users of PPE. They may have a very intense desire—even panic—to get out of the CPC. If you are claustrophobic, or feel uncomfortable in small and confined spaces, please let your instructor or supervisor know before you don personal protective equipment. The good news is that psychological stress, especially due to claustrophobia, can often be managed with the appropriate training.

Use of Personal Protective Equipment

As we've already seen, using personal protective equipment has its risks. Heatstroke can be a life-threatening complication of using PPE. Some agencies have the philosophy "Level A all the way." However, sometimes using the highest level of protection is actually more dangerous owing to environmental or work conditions than the potential chemical exposure itself. For example, a 5-gallon spill of acetone in a small un-air-conditioned warehouse in Arizona, where 20 fully loaded 55-gallon drums need to be moved to get to the spill, is not a good candidate for Level A protection. Acetone is not particularly toxic, with a PEL of 1000 ppm, and it can pose a significant flammability hazard. In addition, the work conditions in this scenario would be brutal! A much safer alternative to a fully encapsulated Level A suit would be the use of continuous air monitoring for protection from fire, and minimal CPC with an air-purifying respirator if acetone concentrations are above the permissible exposure limit (PEL). This PPE ensemble would greatly reduce the heat stress to the wearer.

SELECTION

Personal protective equipment must be carefully selected based on the hazard the chemical poses, chemical compatibility, the quantity spilled, the airborne concentration, the environmental conditions, and the type of work that will be performed. Considerations include the following:

- Will you have a sufficient air supply to complete the assigned task?
- What is the potential for heat stress injuries owing to the design and weight of PPE, weather conditions such as temperature, and the exertion level required to complete the assigned duties?

The PPE selection process must be governed by a hazard- and risk-based approach that accurately assesses the entire situation, since all types of PPE have their advantages and disadvantages. Criteria for selection include the following:

1. Known versus unknown chemicals
2. Probability of exposure
3. Chemical hazards
4. Physical hazards

SOLVED EXERCISE 7-3

What types of physical and psychological stresses can responders expect to encounter when operating in personal protective equipment at a train derailment in Florida in August?

Solution: The weather conditions in Florida in August are very hot and humid. Heat stress will therefore be a primary concern. Train derailments are complex incidents involving large quantities of hazardous materials and many dangers. Owing to the large amounts of hazardous materials, the hot zone will be very large. This will require traveling a longer distance from the cold zone into the work area in the hot zone. PPE should be chosen that minimizes heat stress and work stress on the entry team members.

5. Chemical characteristics (such as state of matter and concentration)
6. Work function (such as reconnaissance versus overpacking of drums)
7. Work location (such as indoors versus outdoors)
8. Weather (such as temperature, precipitation, and humidity)
9. Training

PPE characteristics to consider include the following:

1. Chemical resistance
2. Flame resistance
3. Mobility
4. Durability
5. Ease of decontamination
6. Cost

For example, the selection of a Level A suit may involve determining the location of the zipper for ease of entry into and egress from the suit. Some Level A suits have their zippers in the front; others have their zippers in the back. The location may play a role in entry team safety in case rapid egress from the suit is required during an emergency or routine decontamination, and it may play a role in the number of decontamination personnel that are needed to help entry team members doff their Level A suits. The size of the visor may also play a role in your decision making. Increased visor size offers an advantage in visibility and may therefore increase safety; however, increased visor size may reduce the chemical protection owing to incompatibility and reduced breakthrough time. The visor is one of the weakest points in a Level A suit (see Box 7-1). Always consult the technical data package from the manufacturer when receiving and using PPE.

Choosing a Respirator

The airborne concentration of the contaminant plays the major role in respiratory protection selection (Figure 7-11). At airborne concentrations below the PEL no respiratory protection at all is required by OSHA. However, carefully consider long-term implications of low-level exposures, especially to suspected cancer-causing materials. Above the IDLH level supplied air respiratory protection is required by OSHA.

With all the options we have discussed, it may at first seem difficult to choose the appropriate respirator. Often, this choice is limited by the respirators provided by your employer. Most hazardous materials response teams will have more than one respirator available for use, typically including SCBA and APR. The Department of Health and Human Services (DHHS) has developed a respirator decision logic sequence that makes this decision much simpler (Figure 7-12). The logic sequence covers key concepts such as IDLH atmosphere, oxygen deficiency, whether the contaminant name and level are known, as well as whether the material is a carcinogen. By following the respirator decision logic you cover all the key bases. You may have noticed that flammability hazards are not covered in the metric. This is because OSHA has set the IDLH value to a maximum of 10% of the LEL for flammable materials (assuming the material is not toxic at a much lower level). Therefore, determining the IDLH value covers both toxicity and flammability.

Choosing Chemical Protective Clothing

Chemical compatibility is one of the most important factors to consider when choosing chemical protective clothing. As we've seen, chemicals may degrade chemical protective clothing and/or permeate the material. The breakthrough time of chemicals may be as short as seconds to minutes. If the chemical passes through the suit material quickly, you may as well not be wearing any chemical protective clothing

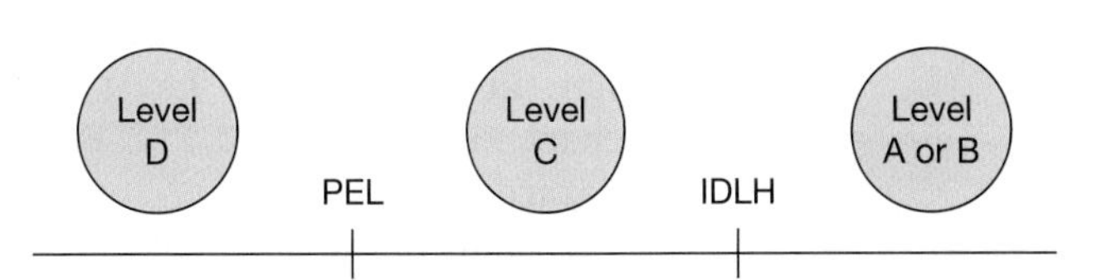

FIGURE 7-11 The permissible exposure limit (PEL) and immediately dangerous to life and health (IDLH) concentration should be used to select the correct level of personal protective equipment (PPE)

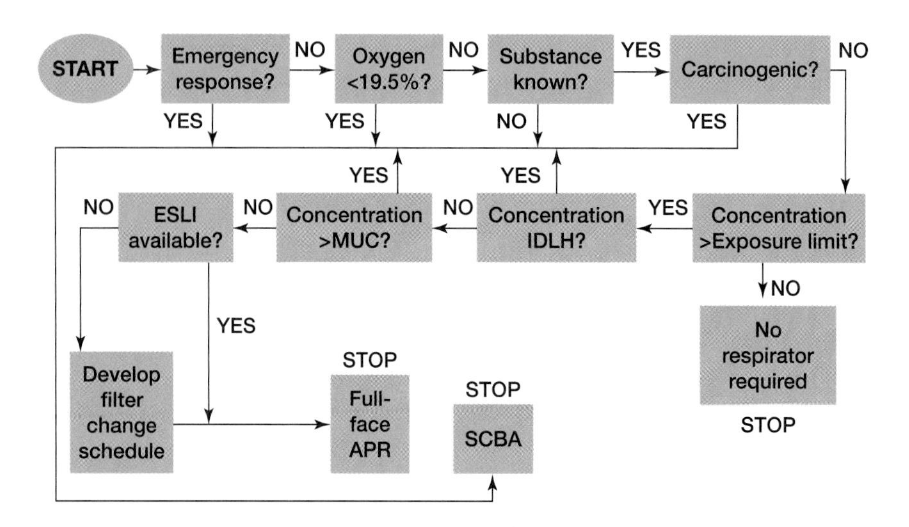

FIGURE 7-12 Respirator decision logic based on the EPA method. It is very important to know when to use supplied air respiratory protection versus air-purifying respirators (APR) or no respirator at all. *Courtesy of the National Institute for Occupational Safety and Health.*

at all. Manufacturers provide breakthrough timetables for their chemical protective ensembles, as illustrated in Figure 7-5. These tables may be color-coded, or they may list minutes to breakthrough. An appropriate PPE ensemble must be chosen based on the chemicals involved and the time that will be spent in proximity to them. The manufacturer may also be contacted for advice when confronted with unusual chemicals or complex hazmat incidents.

The quantity of the material spilled and the work function will play a major role in the type of CPC that is required. As discussed earlier, heat stress and other environmental conditions should play a major role in CPC selection as well. The risk of potentially deadly heatstroke should be weighed against the additional protection offered by more substantial chemical protective clothing.

Choosing a PPE Ensemble

The combination of respiratory protection and the level of chemical protective clothing will determine the level of personal protective equipment. If an APR is chosen, the EPA level of protection is automatically Level C. If an SCBA is chosen, the EPA level of protection may be Level A or Level B depending on the CPC chosen.

Above the IDLH level, supplied air respiratory protection is required by OSHA. This means that a minimum of Level B protection is required. Between the PEL and IDLH concentrations air-purifying respirators and powered air-purifying respirators—otherwise known as Level C protection—may be used. Under unknown conditions a minimum of Level B protection should be used. In confined areas such as indoors, Level A protection should be strongly considered under unknown conditions.

The type of the hazard or risk the chemical poses will indicate the category of PPE that will be required. For example, if a chemical is highly flammable, a Level A suit will not be appropriate in a confined area because of the risk of fire or explosion. In this case, PPE that offers thermal protection and flash protection will be required if no other engineering controls are available, such as ventilation or remote shutoff of the product. Similarly, a chemical that is highly skin absorptive will require a vapor-tight suit, in other words, Level A protection. Always consult the chemical compatibility chart for the selected PPE ensemble to ensure that it is compatible with the released hazardous material.

Many of the same factors will come into play when purchasing PPE. In this case, your agency should examine the SARA Title III sites in your jurisdiction and the transportation routes and determine what types of materials are commonly carried. Even with a good risk assessment based on the known hazards, there is always a level of uncertainty for

SOLVED EXERCISE 7-4

Based on the DuPont chemical protective clothing guideline shown, what is the breakthrough time of chlorine for the Tychem CPF 3 suit? Would this PPE be appropriate for use with chlorine?

Sub-Class	Chemical Name	CAS Number	Phase	Standardized Breakthrough Time (Minutes)										
				Tychem® CPF1	Tychem® QC	Tychem® CPF 2	Tychem® SL	Tychem® CPF 3	Tychem® F	Tychem® CPF 4	Tychem® BR/ Tychem® LV	Tychem® Responder®	Tychem® TK	Tychem® Reflector®
391	Acetone	67-64-1	L	imm.	imm.	12	12	>480	>480	>480	>480	>480	>480	>480
431	Acetonitrile	75-05-8	L	imm.	imm.	12	12	imm.	157	>480	>480	>480	>480	>480
350	Ammonia gas	7664-41-7	G		imm.	32	32	12	79	>480	46	>480	>480	>480
296	1,3-Butadiene	106-99-0	G		imm.	>480	>480	>480	>480	>480	>480	>480	>480	>480
502	Carbon disulfide	75-15-0	L	imm.	imm.	imm.	imm.	16	>480	>480	>480	>480	>480	>480
330/350	Chlorine gas	7782-50-5	G		imm.	>480	>480	>480	>480*	>480	>480	>480	>480	>480

Courtesy of DuPont.

Solution: The breakthrough time for chlorine is >480 minutes with the Tychem CPF 3. Work conditions notwithstanding, it would be appropriate to use a Tychem CPF 3 in a chlorine atmosphere.

public safety agencies that cover large geographical areas and large population densities. Other factors that should be considered when purchasing PPE are the resistance of the material to common insults such as abrasion, tears, and thermal stress (cold and heat), the ease with which the PPE can be cleaned and decontaminated, the acquisition and maintenance costs, the dexterity while in the suit, and even the color of the suit material. These factors should be evaluated in conjunction with the expected duties of the wearers. For example, cold resistance is probably not the most important consideration for responders in southern Florida.

PRE- AND POSTENTRY MEDICAL EVALUATION

Anyone entering the hot zone using personal protective equipment should receive a medical evaluation both before donning PPE and after exiting the hot zone. The pre-entry medical evaluation serves to ensure that personnel are physically fit at the time of the hazmat incident. When victim rescue is being performed in structural firefighter protective clothing in a time-critical fashion, the pre-entry medical evaluation may need to be reconsidered. The postentry medical evaluation serves to ensure that no significant ill effects were experienced during the hot zone entry.

Pre-entry medical monitoring should occur within 1 hour of donning the PPE. Medical monitoring should include heart rate, breathing rate, blood pressure, recent medical history, and mental status at a minimum. A more comprehensive medical evaluation may include the following:

- vital signs: heart rate, breathing rate, and blood pressure
- recent medical history: illnesses and pregnancy
- recent alcohol consumption and/or medication use

- mental status
- temperature
- body weight
- skin evaluation: rashes or open sores
- lung sounds
- electrocardiogram (EKG)

Medical exclusion criteria have not been formally established by law. Several different agencies and associations have their own recommended values, some of which are listed in Table 7-4. Exclusion criteria may include a diastolic blood pressure greater than 105 mmHg, a pulse greater than 70% of the maximum resting heart rate ([220 − age] × 0.7), an unusual heart rhythm, respirations greater than 24 per minute, a temperature above 100.4°F (38°C), abnormal lung sounds, any altered mental status, rashes or open sores that may lead to secondary infections, or an abnormal EKG. These exclusion criteria may be modified for an individual by the team physician based on careful consideration of the baseline medical evaluation and underlying medical history.

Postentry medical monitoring should be continued every 10 minutes until vital signs drop to within approximately 10% of the pre-entry values. Personnel should remain under medical supervision during this time. To avoid heat stress, it is especially important that entry personnel rehydrate properly after strenuous hot zone activity. It is not uncommon to literally pour sweat—water and electrolytes—out of boots and gloves during decontamination. It is critical that these fluids and electrolytes be replenished to prevent dehydration and electrolyte imbalance. A loss of 3% or more of body weight after exiting the hot zone indicates significant dehydration and warrants additional medical monitoring.

TABLE 7-4 Medical Hot Zone Entry Exclusion Criteria during Pre-Entry Check

CRITERION	NFPA 471 (2002) GUIDELINES	NOTES
Blood pressure	Diastolic >105 mmHg	
Pulse	> 70% maximum heart rate*	
Respiratory rate	>24 per minute	
Temperature	> 99.5°F or < 97.0°F (oral); > 100.5°F or < 98.0°F (core)	
Weight	None	Too large to comfortably fit in the suit
Skin	Open sores, large areas of rash, or significant sunburn	
Mental status	Altered mental status (e.g., slurred speech, clumsiness, weakness)	
Medical history	(a) Presence of nausea, vomiting, diarrhea, fever, upper respiratory infection, heat illness, or heavy alcohol intake within past 72 hours, all of which contribute to dehydration (b) New prescription medications taken within past 2 weeks or over-the-counter medications such as cold, flu, or allergy medicines, taken within past 72 hours (should be cleared through local medical control or hazardous materials medical director) (c) Any alcohol within past 6 hours (d) Pregnancy	
EKG	Dysrhythmia not previously detected (must be cleared by medical control)	

*Maximum heart rate = 220 − age.

According to NFPA 471 (2002) any of the following conditions during postentry medical monitoring require additional medical follow-up, possibly including transport to the hospital:

- Body weight loss of greater than 3%
- Positive orthostatic (pulse increase of 20 beats per minute or systolic blood pressure decrease of 20 mm of Hg at 2 minutes standing)
- Greater than 85% maximum pulse at 10 minutes
- Temperature greater than 38°C (101°F) (oral) or 39°C (102°F) (core)
- Nausea, vomiting, diarrhea, altered mental status, or respiratory, cardiac, or dermatologic complaints

Body harnesses capable of continuously monitoring vital signs in the hot zone and transmitting that data to the cold zone are now available. For example, the Rae Systems BioHarness has the capability to measure heart rate (including an electrocardiogram), breathing rate, body temperature, and posture and physical activity. This allows medical personnel to quickly remove entry team members from the hot zone when their vital signs deviate from an acceptable baseline.

DONNING AND DOFFING PERSONAL PROTECTIVE EQUIPMENT

Donning and **doffing** PPE should be an organized process. This is important for several reasons: Each step is very important, and following a step-by-step checklist ensures that no essential step is forgotten. OSHA requires that the buddy system be used on any hazardous materials incident. When all members of both the entry team and the backup team start dressing at the same time, it ensures that all members will be completely dressed at the same time and ready for their duties at the appropriate time. It is also important that team members spend the least amount of time waiting for their job assignments in chemical protective clothing to minimize the effects of thermal stress.

don ▪ To put on (as in chemical protective clothing).

doff ▪ To take off (as in chemical protective clothing).

Several key procedures should always be performed irrespective of the type of personal protective equipment used:

1. A medical evaluation should be performed prior to donning personal protective equipment to determine if the user is physically able to safely work in PPE. A medical evaluation should also be performed after the PPE is doffed to ensure that the user has not suffered injuries, thermal stress, or overexertion. (See previous section for more details.)
2. Personal protective equipment should be examined thoroughly for damage and other problems that may interfere with its proper function.
3. All personal effects should be removed before donning chemical protective clothing for two reasons: First, sharp personal effects such as jewelry may puncture or otherwise damage the personal protective equipment. As an example, rings with

BOX 7-3 HAZMAT HOW TO: DONNING AND DOFFING FULLY ENCAPSULATING SUITS

DONNING

1. Receive medical check.
2. Remove watches, jewelry, leather, shoes, and other personal items, and place in a plastic bag in a secure location.
3. Inspect suit (seams and zippers) and self-contained breathing apparatus (SCBA). Don fire-resistant suit scrubs (if required).
4. Don encapsulating suit to waist.
5. Don chemical-resistant boots with boot covers. For suits with splash guard, position splash guard over the outer boot.
6. Don a cooling vest.

View an illustrated step-by-step guide on Donning and Doffing Fully Encapsulating Suits

7. Don SCBA and record the bottle pressure.
8. Don communications gear (radio and connections should be secured with tape).
9. Don innermost cotton gloves (if desired) and nitrile gloves.
10. Continue donning suit over arms.
11. Don outermost chemical-resistant gloves.
12. Review hand signals:
    *Thumbs-up signal—OK
    *Hands clutching throat—SCBA malfunction
    *Tapping top of head—Emergency! Get me out of suit
    *Audible high–low siren—Leave Hot Zone Now!
13. Don SCBA face mask: Check seal, perform function test (don't stay on air), and test communications.
14. Don head gear (if required).
15. Assign each person a suit number. Advise entry team leader when ready to go on air. Ensure that the entry, backup, and decon teams are properly briefed. Connect regulator to face mask and record time. Ensure wearer is breathing air and ready with thumbs-up signal. Close suit. Record on-air time, entry time, egress time, and off-air time.

DOFFING

1. **NOTE: Remain on air until the last step!** Set out a large decontamination bag. This procedure may be performed as a stand-alone method (dry decontamination), or it may follow wet decontamination.
2. Step into a large bag (dry decon bag).
3. Unzip the suit and carefully peel the suit down to the boots. The outside (dirty) gloves should only touch the outside of the suit, and the inner (clean) gloves should only touch the inside of the suit.
4. Step out of the boots toward the cold zone. You may need the assistance of a tender or a long-handled tool for balance.
5. While remaining on air, remove head gear and communications equipment, loosen the backpack straps, and take off the pack and have a tender hold it or lay it on a chair. Remain on air! Do NOT remove the face piece at this time. Move towards the cold zone.
6. Using the inner gloves, remove the face piece and go off air (when in a clean area of the decon zone).
7. Remove the inner gloves.
8. Verify that decontamination was successful.
9. Perform post-entry medical monitoring and rehabilitation.

View an illustrated step-by-step guide on Donning and Doffing Nonencapsulating Suits

BOX 7-4 HAZMAT HOW TO: DONNING AND DOFFING NONENCAPSULATING SUITS

DONNING

1. Receive medical check.
2. Remove watches, jewelry, leather, shoes, and other personal items, and place in a plastic bag in a secure location.
3. Inspect suit (seams and zippers) and the self-contained breathing apparatus (SCBA) or the airpurifying respirator (APR).
4. Don fire-resistant suit/scrubs (if required). Don nonencapsulating suit to waist.
5. Don chemical-resistant boots with boot covers (if required). For suits with splash guard, position splash guard over the outer boot. For suits without splash guard, fold excess pant leg over outer boot, and tape seam between outer boot and suit (blousing). Leave tab on tape for easy removal.
6. Don a cooling vest (if required). Don suit and zip-up front zipper (if there is no splash flap for zipper, tape leaving tab for easy removal).
7. Don innermost nitrile gloves.
8. Don the appropriate chemical-resistant gloves. Tape gloves using appropriate method. Tape should be applied over widest part of hand and leave tab for easy removal. If using an outer chemical- or abrasion-resistant glove, don and secure with tape and tab.

9. Don SCBA and record tank pressure. Don communications gear (radio and connections should be secured with tape).
10. Review hand signals:
    *Thumbs-up signal—OK
    *Hands clutching throat—SCBA malfunction
    *Tapping top of head—Emergency! Get me out of suit
    *Audible high–low siren—Leave Hot Zone Now!
11. Don respirator face mask: Check seal, perform function test (don't stay on air), and test communications. Don attached hood of suit (if taping seems necessary, consider going to level A protection). Don head gear (if required).
12. Assign each person a suit number. Advise entry team leader when ready to go on air. Ensure that the entry, backup, and decon teams are properly briefed. Connect regulator to face mask and record time. Ensure wearer is breathing air and ready with thumbs-up signal. Close suit. Record on-air time, entry time, egress time, and off-air time.

DOFFING

1. **NOTE: Remain on air until Step 6!** Set out a large decontamination bag. This procedure may be performed as a stand-alone method (dry decontamination), or it may follow wet decontamination.
2. Step into the large bag (dry decon bag).
3. While remaining on air, remove head gear and communications equipment. If using an SCBA: While remaining on air, loosen the backpack straps, and take off the pack and have a tender hold it or lay it on a chair. Remain on air! Do NOT remove the face piece at this time.
4. Unzip the suit and carefully peel the suit down to the boots. The outside (dirty) gloves should only touch the outside of the suit, and the inner (clean) gloves should only touch the inside of the suit. The gloves should have been taped loosely enough to be able to remove the hands without removing the tape at the wrists.
5. Step out of the boots toward the cold zone. You may need the assistance of a tender or a long-handled tool for balance.
6. Using the inner gloves, remove the face piece and go off air (when in a clean area of the decon zone).
7. Remove the inner gloves.
8. Verify that decontamination was successful. Check the hands and feet first because typically 85% of the contamination is found there.
9. Perform post-entry medical monitoring and rehabilitation.

BOX 7-5 HAZMAT HOW TO: DOUBLE SEAM GLOVE TAPING METHOD

View an illustrated step-by-step guide on the Double Seam Glove Taping Method

1. Don glove and fold the glove back at the wrist, leaving approximately 1 to 2 inches folded over. Place the center of the folded-over flap at the widest part of the hand. Splay the thumb and fingers out to create maximum width.
2. Pull the suit cuff to the center of the folded-over flap. Make sure the seam is located at the widest part of the hand. This minimizes the number of wrinkles and allows the hand to be easily removed during doffing.
3. Tape the seam using a chemical-resistant tape (rated). Minimize the number of wrinkles (less liquid penetration), and make sure the hand can be removed from the glove without removing the tape (during doffing).
4. One or two wraps of tape are sufficient.
5. Fold the flap back. Make sure the seam has been taped well, at the widest part of the hand, and with few wrinkles.
6. Tape the newly created seam in the same way: over the widest part of the hand, with as few wrinkles as possible.
7. Two seams are now taped. The double taping method with few wrinkles gives maximum protection against liquid intrusion. The glove–suit ensemble is taped loosely enough to allow removal of the hand during dry decontamination.
8. Pull the glove back over the hand and repeat for the other side.

View an illustrated step-by-step guide on Using Air Purifying Respirators

BOX 7-6 HAZMAT HOW TO: USING AIR PURIFYING RESPIRATORS

1. Before wearing an APR in the hot zone, ensure that a fit test (qualitative or quantitative) has been performed, passed, and recorded with the exact same type and size of mask.
2. Carefully inspect the mask: Inspect the face piece head straps. Look for abnormal wear, degradation, or tears. Make sure all of the straps and fasteners are in place. Check face piece for dirt, scratches, cracks, discoloration, tears, or holes. If present, inspect the voice emitter or other electronics for signs of damage.
3. Check the filter canister to ensure that it is appropriate for use with the contaminant in the hot zone.
4. Check the end of service life indicator on the canister(s) (or determine the appropriate service life time), and make sure the canister(s) is not damaged.
5. Thread the canister(s) into the appropriate face piece port. Ensure that all ports have a canister! Some masks require up to three canisters. Remove both the tab covering the air inlet and the outlet holes of the canister.
6. Loosen the harness head straps. Place your chin in the face piece first, and pull the harness over your head.
7. Tighten the head straps according to the manufacturer, but typically the order is from the bottom (neck area) to the top (temple region).
8. Perform a seal check on the mask. Cover the air inlet with the palm of your hand. Breathe in and hold for 10 seconds. The mask should collapse into your face and remain there. If it leaks or does not stay collapsed, readjust or tighten the straps. **Do NOT use the APR if you cannot get a good seal!**

View an illustrated step-by-step guide on Donning and Doffing Level C with PAPR'S

BOX 7-7 HAZMAT HOW TO: DONNING AND DOFFING LEVEL C WITH PAPRS

DONNING

1. Organize and inspect all of the personal protective equipment. Is the equipment damaged or worn? Is it the correct level of PPE for the situation? Is it the proper size? Examine the face piece, blower, canister, seams, zippers, and materials for damage.
2. Perform pre-entry medical monitoring. At a minimum, check vital signs. Ensure personnel are hydrated.
3. Remove watches, jewelry, leather, shoes, and other personal items (place in plastic bag and store in a secure location).
4. Check the filter canister to ensure that it is appropriate for use with the contaminant in the hot zone.
5. Check the end of service life indicator on the canister(s) (or determine the appropriate service life time), and make sure the canister(s) is not damaged.
6. Thread all three canisters into the appropriate blower ports. Ensure that all ports have a canister and that the protective caps and tabs are removed (two per canister).
7. Check to make sure the blower flow rate is adequate (6 cfm) using the manufacturer-supplied tester.
8. If not already completed, attach the breathing hose to the blower by tightening the thumb screw.
9. Don innermost nitrile gloves.
10. Don nonencapsulating suit.
11. Don chemical-resistant boots with boot covers (if required). For suits with splash guard, position splash guard over the outer boot. For suits without splash guard, fold excess pant leg over outer boot and tape seam between outer boot and suit (blousing). Leave tab on tape for easy removal.
12. The hood of the suit is not used with a hooded PAPR. Tuck it under so it is under the neck on your back.
13. Don the outer pair of gloves and tape in an appropriate fashion (ensure that your hands can easily come out of the gloves for dry decon).
14. Don blower unit, and tuck the straps away to prevent trip and snag hazards.
15. Turn on blower unit battery.
16. Don the hood. Tuck the inner shroud into the suit, and let the outer shroud hang over the suit.

DOFFING

1. **NOTE: Remain on air until the last step!** Set out a large decontamination bag. This procedure may be performed as a stand-alone method (dry decontamination), or it may follow wet decontamination.
2. Step into the large bag (dry decon bag).
3. Remove tape at the boots and zipper seam, if they were used.
4. If wet decontamination was performed, wipe any excess liquid from the shroud using a towel.
5. Roll the outer shroud up and tape. Do not obscure the vision!
6. Unzip the suit and carefully undress down to the boots. The outside (dirty) gloves should only touch the outside of the suit, and the inner (clean) gloves should only touch the inside of the suit. The gloves should be taped loosely enough to be able to remove the hands without removing the tape at the wrists.
7. Step out of the boots toward the cold zone. You may need the assistance of a tender or a long-handled tool for balance.
8. Using the inner gloves, remove the hood touching only the inner shroud (when in a clean area of the decon zone). Go off air.
9. Remove the inner gloves.
10. **Verify that decontamination was successful.**
11. Perform post-entry medical monitoring and rehabilitation.

View an illustrated step-by-step guide on Donning and Doffing LANX Suits and MOPP Gear

BOX 7-8 HAZMAT HOW TO: DONNING AND DOFFING LANX SUITS AND MOPP GEAR

DONNING

1. Receive medical check.
2. Remove your duty belt, notepad, pens, uniform shirt and tie, and jewelry. Place in a plastic bag in a secure location.
3. Open the sealed pouches, and inspect the suit pants and suit coat (seams and zippers) and air purifying respirator (APR). Check the filter canister to ensure that it is appropriate for use with the contaminant in the hot zone. Check the end of service life indicator on the canister(s) (or determine the appropriate service life time) and make sure the canister(s) is not damaged.
4. Place a shoe/boot cover over each shoe or combat boot.
5. Don the LANX/MOPP trousers. Adjust the suspenders by sliding the adjustment clip up or down. Fasten the pants by snapping the front inner and outer snaps and zipping up the trousers. Adjust the waistband for a snug fit (do not fasten the bottom of the trousers at this time).
6. Don the boots. Secure the trouser legs over the boots by folding excess suit material over, making a snug fit, and attaching the Velcro straps.
7. Don the LANX/MOPP jacket and hood: Open the sealed bag and put on the jacket. Zip up the jacket. Reach between your legs and grab the elastic loop attached to the back of the jacket. Pull the loop under your legs and secure it with the front snap closure. Adjust the drawstrings at the bottom of the jacket by pulling snug and tying in a bow knot. Secure the jacket by attaching the Velcro in the front and by pulling the drawstrings snug using the adjustment clips to secure them. Lifting your chin up will help you in securing the Velcro and creating a good APR to jacket seal.
8. Don the air-purifying respirator (APR) and check for a good seal.
9. Pull the hood up and over the head, making sure the hood overlaps the APR.
10. Reattach duty belt.
11. Don nitrile innermost gloves.
12. Don chemical protective outer gloves. Secure the jacket sleeve over the gloves by folding excess suit material over, making a snug fit, and attaching the Velcro straps.
13. Check all zippers, ties, and Velcro, particularly around the APR and under the chin, and make sure they are all secured.

DOFFING

1. If you have a firearm, secure the weapon and place in the designated secure weapon-drop area or hand directly to the weapons officer.
2. Drop your duty belt and any tools into the designated tool drop area.

3. Loosen all Velcro closures, ties, zippers, and the boot straps.
4. Pull the hood off your head.
5. Remove the outer gloves and place them into an overpack drum (keep the inner gloves on).
6. Drop the coat off your back and place it into the overpack drum.
7. Unbuckle the suspenders, and unsnap and unzip the trousers.
8. Roll the trousers down, touching only the inside of the trousers, and pull your feet out of the boots and place the boots into the overpack drum.
9. Take a large step away from the potentially contaminated PPE toward the cold zone.
10. In one motion, remove the APR.
11. Remove the inner gloves and discard them.
12. Verify that decontamination was successful using monitoring equipment.
13. Perform post-entry medical monitoring and rehabilitation.

sharp edges may puncture gloves in the hot zone and cause penetration of the hazardous material. Second, wallets, cell phones, and jewelry may become contaminated and become hazardous waste if they cannot be properly decontaminated. For example, mercury will amalgamate with gold and silver jewelry and cannot be decontaminated. Leather and paper are porous materials and are difficult to impossible to decontaminate. A good rule of thumb is, don't take anything into the hot zone that you are not willing to give up during decontamination!

4. A user seal check should be performed to verify that there is a good seal between the mask and the user's face. Appendix B-1 of the respiratory protection regulation (29 CFR 1910.134) describes both a positive-pressure and a negative-pressure seal test.
5. The on-air and entry times should be recorded for all personnel entering the hot zone. Hot zone entry should be timed, and entry team members should be advised of time remaining at predetermined safety milestones. Remember to take into account the time it takes to walk to the decontamination line and be processed through the decontamination stations.
6. After doffing and decontaminating the personal protective equipment it is essential that decontamination be verified using appropriate monitoring equipment, such as a photoionization detector, pH paper, Geiger counter, or other appropriate detection equipment (see Chapter 8).

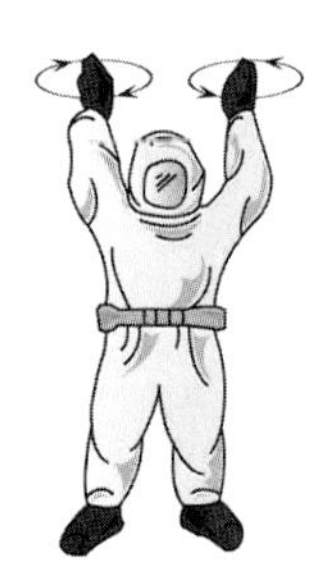

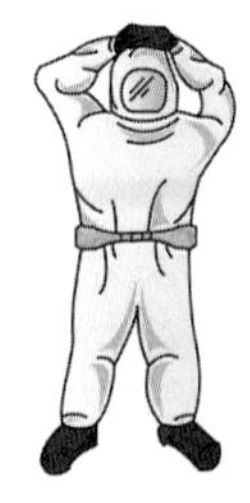

FIGURE 7-13 Hand signals are an effective emergency backup to radio communication systems. Shown are several examples of hand signals that can be used if radio communications fail. *Art by David Heskett.*

The accompanying boxes list examples of donning and doffing procedures for common personal protective equipment. These procedures are suggestions only; your agency or the manufacturer of your PPE may have variations.

Although entry team members typically will have radio communication systems, they should be aware of and know a system of hand signals for emergency communication. One possible set of hand signals is illustrated in Figure 7-13. It is important that all team members be aware of the meaning of all hand signals your team uses.

For ease of visual identification, entry team members should have their bottles or suits marked with numbers or letters. Ideally, place the identification on the SCBA bottle rather than the suit material itself. This will avoid damaging the chemical protective polymers of the suit with the adhesive on the tape or the solvents found in markers.

HOT ZONE ENTRY USING THE BUDDY SYSTEM

A minimum of two persons should always enter the hot zone together. This is referred to as the *buddy system*. Personnel should never enter the hot zone alone. In fact, 29 CFR 1910.120 (HAZWOPER) mandates the use of the buddy system inside the hot zone. The incident commander and others may be held criminally liable if injury or death occurs because this rule was violated. The buddy system increases the safety of hot zone entry personnel. If one person suffers a medical emergency, has equipment failure, or is overcome by the chemical, the other person may be able to effect a rescue and/or summon outside help in the form of a backup team. Immediately notify the entry team leader in the event of an emergency. Whenever possible, attempt to remove the incapacitated team member to a safe location (when immediate hazards are present), or remain with the incapacitated person until help arrives.

BACKUP TEAM

The "two in-two out" rule stipulates that for every two-member entry team at least one two-person backup team must remain in the cold zone at the ready. The backup team, or rapid intervention team (RIT), should be prepared to immediately effect a rescue in case the entry team experiences an emergency in the hot zone. This means that the backup team must have their PPE donned, tools at the ready, and be able to go on air immediately. Typically the backup team wears the same level of PPE as the entry team.

Much like the fire department rapid intervention team, backup team members should have appropriate tools at their disposal. This includes air monitoring equipment, nonambulatory victim retrieval devices, and any other tools the situation may dictate. Victim retrieval devices may include Sked boards, stretchers, Stokes baskets, or gurneys. Keep in mind that it is very difficult to remove a nonambulatory person from the hot zone dressed in full PPE. This will require practice and proficiency with your agency's rescue equipment.

Before entering the hot zone the backup team should try to determine what caused the entry team to go down, to avoid suffering the same fate. For example, the entry team may have come into contact with downed power lines or facility electrical equipment. Or the entry team did not have the appropriate chemical protective clothing and/or respiratory protection for the hazard. Information about unexpected hazards may come from the entry team's last communication, from the location the entry team went down, from a live video feed from the entry team (such as with a search and rescue video camera), or from other incident information such as technical experts or the obvious consequences of unforeseen events.

Both 29 CFR 1910.120 (HAZWOPER) and 29 CFR 1910.134 (*Respiratory Protection Regulation*) require that a backup team be used at hazardous materials incidents and in IDLH atmospheres in general.

IN-SUIT EMERGENCIES

Emergency procedures should be practiced for in-suit emergencies such as an air emergency or failure of an exhalation valve on Level A suits. Types of in-suit emergencies include the following:

- physical injury
- heat stress
- medical emergencies
- air loss

- suit breach
- chemical exposure

Low-air emergencies can occur when breathing apparatus malfunctions. It is a good idea to train on low-air emergencies under controlled situations so you know what to do. Low-air emergencies can often be fixed relatively quickly using the following procedures:

- Use the emergency bypass valve of the SCBA.
- Turn the main tank valve on and off to clear debris that may be stuck in the valve.
- Use buddy breathing.

When a one-way valve becomes inoperable on a Level A suit, it will continue to inflate until you look like the Michelin man. In this case you must attempt to reset the valve by burping the suit. You do this by slowly and methodically bending over with your arms crossed in front of you. This will gently force the air out of one of the other valves without puncturing the suit elsewhere. After burping the suit you should not continue entry operations but, rather, exit the hot zone to determine how and why it malfunctioned.

Believe it or not, there is still breathable air inside a Level A suit, even if your SCBA is completely empty or not functioning properly. Typically, there will be a few minutes of breathable air, but the air will quickly become severely oxygen deficient if emergency procedures are not initiated promptly. The key is to stay calm and move to the decontamination line as quickly as possible and let the decon personnel know of your emergency. It is also a good idea to practice doffing your PPE by yourself in case decon personnel become overwhelmed.

Preventive measures include good hydration before donning PPE, a thorough pre-entry medical evaluation, and well-fitting CPC suits. Visibility can be increased in PPE by using antifogging solutions on visors and facepieces, or taping a towel to the helmet when using fully encapsulating suits. The suit visor can then be cleaned of fog and moisture by rubbing the towel-containing helmet up and down against the visor.

Personal Protective Equipment Program

HAZWOPER requires that a written PPE plan be in place prior to using personal protective equipment. The PPE program must be designed to safely isolate the wearer from the chemical, physical, radiological, and biological hazards that may be found in the hot zone. This plan should cover the following:

- Hazard assessment
- Selection of PPE
- Inspection of PPE
- Use of PPE including duration of work mission
- Medical monitoring prior to use and immediately after use
- Decontamination of PPE
- Maintenance of PPE
- Testing of PPE
- Storage of the PPE
- Training program

The donning and doffing procedures discussed earlier should be part of this written PPE plan as well. The PPE program should cover the training requirements hazardous materials technicians must undergo before being allowed to use PPE in the warm zone or in the hot zone. Remember that there are inherent hazards to wearing PPE. A good hazard and risk assessment must be performed to ensure that the risks of wearing a particular level of PPE do not outweigh the hazards being protected against.

BREATHING-AIR QUALIFICATION

Breathing air that will be used in supplied air respirators must be grade D as defined by the Compressed Gas Association standard CGA G-7.1 "Commodity Specification for Air". Breathing-air compressors must be regularly inspected and tested. Common contaminants of breathing air include carbon monoxide and oil. The compressor filters should be regularly changed to prevent breathing-air contamination. Additionally, the moisture content of the breathing air must not be too high, or the regulators may freeze up during cold-weather operations. Compressed gas cylinders are regulated by the U.S. DOT and must be inspected according to their construction material and design. This information can be found stamped on the neck of the cylinder.

FIT TESTING

Fit testing is required for tight-fitting respirators to ensure that the facepiece of the mask makes a good seal with the wearer's facial features. Fit testing is not required for loose-fitting respiratory protection such as hooded powered air-purifying respirators. Fit testing is designed to ensure that a respirator will provide adequate protection under actual conditions of use. The fit testing conditions should correspond to those expected during use. During respirator use the wearer should have no facial hair, should not be wearing eyeglasses, and should not have foreign objects in their mouth such as gum or chewing tobacco. A written record of the fit test results must be maintained. Fit testing recommendations can be found in 29 CFR 1910.134, Appendix A.

Qualitative Fit Testing

In a qualitative fit test the respirator wearer is exposed to an irritating material such as stannous chloride or Bitrex solution. The fit test is deemed passed if the user does not smell the test chemical, and failed if he or she can smell it. Qualitative fit tests do not provide a numerical value for the protection factor. Qualitative fit tests may be used for all air-supplied respirators, and for APRs that require a fit factor of 100 or more. When a qualitative fit test is performed on an APR with an assigned protection factor of 100 or less, the APR may be used only to an exposure level of 10 times the chosen exposure limit (PEL, REL, or TWA).

Quantitative Fit Testing

Quantitative fit testing is performed using a machine that can measure contaminant levels inside and outside the mask to calculate the actual protection factor. Quantitative fit tests are performed while the users are performing certain tasks, such as speaking, grimacing, and moving the head—tasks that they may be expected to perform during their typical work functions. Quantitative fit testing is the most accurate type of fit testing and is the preferred method for hazardous materials technicians.

RESPIRATOR CLEANING AND DISINFECTION

Respirators must be cleaned and disinfected after each use. Many fire departments and hazardous materials response teams assign masks to individuals. In this case, it is still very important to clean and disinfect the mask regularly to prevent contaminants from permeating the surface, and bacteria from colonizing the mask. Respirator cleaning recommendations can be found in 29 CFR 1910.134, Appendix B-2.

MEDICAL SURVEILLANCE PROGRAM

Hazardous materials incidents place first responders in uncontrolled environments with the potential for chemical exposure. HAZWOPER therefore requires that a medical surveillance program be established by the employer, including public safety agencies, when they expect their employees to use PPE. The medical surveillance program must be designed to determine whether personnel are healthy and fit enough to safely and effectively use PPE.

Therefore the medical surveillance program should include a medical exam and work history with an emphasis on hazardous chemical exposure and the wearer's fitness to use PPE. The medical evaluation should include a medical history and a physical exam, and be documented in written form.

The medical surveillance program should be administered by, or under the supervision of, a licensed physician. The physician must be informed of the expected job duties and the types of PPE personnel will be wearing. Medical exams must be performed at specified times and at no cost to the employee:

- Prior to assignment to an organized hazardous materials response team
- Annually while a member (or biannually with approval of the physician)
- On chemical exposure
- On workplace changes, PPE changes, or health status changes
- After leaving the hazardous materials response team (exit physical)

The medical surveillance program is extremely important not only to determine fitness for duty but also to detect any chemical exposures early so that treatment can be started as soon as possible.

MAINTAINING PERSONAL PROTECTIVE EQUIPMENT

PPE must be maintained properly to ensure its effectiveness. Proper maintenance includes periodically inspecting the PPE visually, refolding the chemical protective clothing to limit damage from deep creases, and inspecting valves, zippers, and seams. During visual inspections look for the following:

- rips and tears
- failing seams
- fabric delamination
- visor damage
- zipper failure
- functioning exhaust valves

In addition, vapor- and gas-tight suits, or Level A suits, need to be pressure tested per the manufacturer's recommendations, which is usually on arrival and once a year thereafter. Pressurized cylinders that are used with supplied air respirators must be hydrostatically tested at intervals determined by the manufacturer based on cylinder construction. Level A suits often have a 5-year shelf life according to the manufacturer, even if the suit passes the pressure test.

PPE should be stored per the manufacturers' recommendations. This typically means storing the PPE in a cool, dry place away from sunlight, dust, and thermal stresses. Ideally, chemical protective suits should be stored flat and as wrinkle free as possible. PPE should also be kept away from solvents, abrasive materials, and sharp materials that may damage them. Periodic inspections should be documented in written form. If reusable PPE is used, any use of the PPE should be documented as well. This includes any training use of the PPE and any use at hazmat or WMD incidents. Most PPE has an expiration date. PPE should not be used after the expiration date, except for use in training that does not involve live chemical agents.

Compressed breathing-air cylinders that are part of your PPE must also be inspected and tested at regular intervals. SCBA cylinders should not be refilled if the hydrostatic test date is not in compliance, or if you notice any other cylinder damage or paint discoloration (from excessive heat or chemical damage). The hydrostatic test date is stamped in the shoulder of the cylinder.

Summary

Personal protective equipment is the last defense between you and a chemical released at an emergency incident. It is essential that you understand how to use PPE properly. This includes selecting chemical protective clothing based on the chemical hazards that are present and the chemical compatibility with the suit material using the breakthrough time. Heat stress and entry team work function must also be considered when making PPE decisions. Training is essential for learning how to don and doff PPE quickly and effectively. Between incidents the PPE must be maintained, stored, inspected, and tested per the manufacturer's recommendations. However, remember that PPE is only the last line of defense against chemical exposure after engineering controls and workplace practice changes have been determined to be inadequate.

Review Questions

1. What are the levels of protection according to the NFPA, and what are their key features?
2. How does the EPA classify PPE?
3. What are some specialty suits that special response teams may wear?
4. What is a common simple asphyxiant?
5. What is a common chemical asphyxiant?
6. What are the three ways PPE may be compromised?
7. What is breakthrough time?
8. What types of psychological stress can PPE use cause?
9. What is the difference between heat exhaustion and heatstroke?

References

Forsberg, Krister, and S. Z. Mansdorf. (2007). *Quick Selection Guide to Chemical Protective Clothing* (5th ed.). New York: John Wiley.

Forsberg, Krister and Keith, Lawrence H. (1999) *Chemical Protective Clothing Performance Index*, New York: Wiley-Interscience.

Klem, Thomas J. (1984, September 17). *Fire Investigations: Cold Storage Warehouse, Shreveport, Louisiana.* Quincy, MA: National Fire Protection Association.

National Fire Protection Association. (2008). NFPA 1999, *Standard on Protective Clothing for Emergency Medical Operations.* Quincy, MA: Author.

National Fire Protection Association. (2002). NFPA 471, Recommended Practice For Responding To Hazardous Materials Incidents. Quincy, Massachusetts: Author.

National Fire Protection Association. (2008). NFPA 472, *Standard for Competence of Responders to Hazardous Materials/Weapons of Mass Destruction Incidents.* Quincy, MA: Author.

National Fire Protection Association. (2008). NFPA 473, *Standard for Competencies for EMS Personnel Responding to Hazardous Materials/Weapons of Mass Destruction Incidents.* Quincy, MA: Author.

National Fire Protection Association. (2007). NFPA 1994, *Standard on Protective Ensembles for First Responders to CBRN Terrorism Incidents.* Quincy, MA: Author.

National Fire Protection Association. (2012). NFPA 1991, *Standard on Vapor-Protective Ensembles for Hazardous Materials Emergencies.* Quincy, MA: Author.

National Fire Protection Association. (2005). NFPA 1992, *Standard on Liquid Splash-Protective Ensembles and Clothing for Hazardous Materials Emergencies.* Quincy, MA: Author.

National Fire Protection Association. (2007). NFPA 1500, Standard On Fire Department Occupational Safety And Health Program. Quincy, MA: Author.

Occupational Safety and Health Administration. (1990). 29 CFR 1910.120, *Hazardous Waste Operations and Emergency Response (HAZWOPER).* Washington, DC: U.S. Department of Labor.

Occupational Safety and Health Administration. (1984). 29 CFR 1910.134, *Respiratory Protection.* Washington, DC: U.S. Department of Labor.

Occupational Safety and Health Administration. (1970). 29 CFR 1910.132, *Personal Protective Equipment.* Washington DC: US Department of Labor.

CHAPTER 8

Detection and Air Monitoring

Courtesy of Chris Weber, Dr. Hazmat, Inc.

KEY TERMS

action level, *p. 233*
atomic absorption spectroscopy (AAS), *p. 259*
background reading, *p. 232*
bump check, *p. 231*
calibration, *p. 231*
colorimetric tube, *p. 253*
combustible gas indicator (CGI), *p. 231*
correction factor, *p. 233*
cross-sensitivity, *p. 242*
data quality objective (DQO), *p. 230*
dosimeter, *p. 239*
electrochemical sensor, *p. 242*
field strength, *p. 236*
field zeroing, *p. 232*
flame ionization detector (FID), *p. 252*
flame spectrophotometry, *p. 262*
Geiger counter, *p. 238*
intrinsic safety, *p. 264*
ionization detector, *p. 238*
ionization potential (IP), *p. 247*
ion mobility spectroscopy (IMS), *p. 250*
photoionization detector (PID), *p. 246*
pH paper, *p. 240*
reaction time, *p. 232*
recovery time, *p. 232*
scintillation counter, *p. 238*
surface acoustic wave (SAW) detector, *p. 262*

OBJECTIVES

After reading this chapter, the student should be able to:

- State the optimal order of air monitoring and the reasons for the order.
- Describe the advantages and disadvantages of Geiger counters.
- Describe the advantages and disadvantages of pH paper.
- Describe the advantages and disadvantages of combustible gas indicators.
- Describe the advantages and disadvantages of electrochemical sensors.
- Describe the advantages and disadvantages of photoionization detectors.
- Describe the advantages and disadvantages of colorimetric tubes.
- Perform effective air-monitoring operations.

Resource**Central** For additional review and practice tests, visit **www.bradybooks.com** and click on Resource Central to access book-specific resources for this text! To access Resource Central, follow directions on the Student Access Card provided with this text. If there is no card, go to **www.bradybooks.com** and follow the Resource Central link to Buy Access from there.

You are dispatched to a local high school for a report of more than a dozen sick students. When you arrive you meet the paramedics who are actively treating patients. They all have relatively mild and generic signs and symptoms such as headaches, upset stomach, dizziness, and nausea. All the students were in the same two adjoining classrooms in the wing of the building that contains the science labs. These could very well be signs of chemical exposure, but they could also be psychosomatic symptoms—signs and symptoms caused by psychological stress. A few questions run through your mind:

- How can I determine if a chemical exposure took place?
- What chemicals are stored in the science labs?
- What chemicals were actively being used?
- What type of detection equipment do I have that will detect these materials?
- How sensitive is my detection equipment?
- Does my detection equipment return any false positives?
- Does my detection equipment report any false negatives?

Let's see if we can get some answers to these questions and more.

Detection and air-monitoring activities are among the most important first response and public safety measures at any hazardous materials incident. In fact, air-monitoring and detection equipment could be considered part of your personal protective equipment. Any emergency response should start with a hazard and risk assessment. How can you use a risk-based response if you don't know what you are dealing with? You must monitor the air to determine whether the hot zone boundaries have been chosen appropriately. Many hazardous gases and vapors are colorless and odorless above harmful concentrations. You will therefore be in danger if you do not use proper air-monitoring procedures. Furthermore, hazardous materials releases are dynamic incidents affected by changing release rates and weather conditions.

In this chapter we discuss specific detection, air-monitoring and sample-identification technology. For each technology we will cover the detection method, the limitations of the technology, and the specific advantages and disadvantages of the equipment currently in use. Before we do that let's set the stage for air-monitoring equipment selection.

Atmospheric Hazards

HAZWOPER (29 CFR 1910.120) requires that the incident commander identify hazards at hazardous materials incidents. Atmospheric hazards can take many forms, such as the following:

- radiation
- corrosive atmospheres
- oxygen-deficient atmospheres
- oxidizing atmospheres
- flammable and explosive atmospheres
- acutely toxic atmospheres

- chronically toxic atmospheres
- carcinogenic atmospheres
- biological/etiological agent containing atmospheres

Some of these hazards let you know you're in trouble right away—corrosive gases, flammable gases and vapors, oxygen-deficient atmospheres—if you make a mistake. Others don't—such as radiation and cancer-causing substances—and you may never know you made a mistake, since the effects may not manifest themselves for years. Proper monitoring is critical to keeping yourself healthy and alive.

Types of Monitoring Activities

There are several different types of radiation- and air-monitoring functions, including the following (Figure 8-1):

1. Perimeter air monitoring around the hazardous materials release site to determine isolation zones
2. Initial entry and reconnaissance duties
3. Periodic monitoring during mitigation activities
4. Continuous hot zone or downwind monitoring activities
5. Verification of decontamination
6. Verification of safe atmospheric levels to allow reoccupancy
7. Quantification of any employee or public exposures

Perimeter monitoring is used to ensure that the public is safe and you are and remain safe in the cold zone. The atmosphere is monitored in the hot zone to determine the nature of the release and diagnose the problem. If there are victims, monitoring can identify and give a concentration estimate of the exposure. After the release has been dealt with and the occupancy has been ventilated, air monitoring is used to verify that the indoor air quality is satisfactory. Radiation monitoring and air monitoring are vital for the safety of response personnel, by ensuring that the proper response zones have been chosen, and also for the protection of downwind populations.

There are several components to the atmospheric-monitoring process. The first is detection, which means determining the location of a hazardous atmosphere. This process often requires multiple detectors unless detailed site-specific, or incident-specific, information is known. Once a hazardous atmosphere has been detected and identified, the second component of the monitoring process is continuous monitoring during the incident, especially mitigation efforts, to ensure the atmosphere remains safe and stable.

A third component of the air-monitoring process may be sampling, which refers to collecting samples of the atmosphere, typically for off-site laboratory testing. Sampling equipment may be as simple as a pump and a plastic bag, or stainless steel vacuum canisters, or as complicated as continuous air sampling equipment.

The Monitoring Process

It is critical that you have the ability to accurately and quickly identify the hazards that radiation and airborne gases, vapors, and dusts pose to the responders on scene and to the general public. This means knowing how to use and interpret the results of your monitoring equipment accurately.

Air monitoring and radiation monitoring must be conducted in a systematic fashion to ensure accurate and reliable results. Training and equipment familiarity are essential for your safety as well as that of other first responders and of the public. Monitoring equipment must be maintained on a regular basis, prepared for use at the site of the hazardous materials release, and operated according to the manufacturer's instructions, and the results must

FIGURE 8-1 Air-monitoring and radiation-monitoring activities should be routinely performed at suspected hazardous materials releases and WMD incidents. Monitoring devices can be integrated with bomb squad robots (top left), can be used to help identify chemicals (top right), can be used to determine contamination levels on entry personnel (bottom right), and should be used to determine the air quality before entering buildings and rooms (bottom left). *Photos courtesy of Chris Weber, Dr. Hazmat, Inc.*

be interpreted correctly. Monitoring equipment should be stored in a cool, dry location. Batteries, including backup batteries, should be maintained in a ready state. Monitoring equipment typically does not give reliable results if the battery is almost depleted; using detection equipment with a low battery can be extremely dangerous. If monitoring activities are not carried out properly, the results could be disastrous.

Monitoring should be conducted systematically as the hot zone is approached and throughout the hot zone as the primary hazard area is assessed. Spills and leaks should be approached from upwind and uphill during the initial assessment. Sometimes, downwind monitoring activities must be conducted to determine downwind hot zone boundaries and to initiate public protective actions. When a building is entered, air monitoring should be performed both high and low as the door is carefully opened an inch or two. This technique should be repeated as doors are accessed inside the building.

Standard operating guidelines (SOGs) consistent with the manufacturer's recommendations should be written for each specific piece of monitoring equipment your agency uses. Become very familiar with these SOGs. They should cover performance and scheduling of routine maintenance such as calibration, pre-entry preparations such as bump checking and field zeroing, performance of measurements, as well as interpretation of results and application of correction factors. Personnel should be trained often on the use of monitoring equipment according to the SOGs. Any deficiencies in the SOGs noted during training should be promptly corrected, and personnel should be retrained on the new and improved SOG. Improper and inaccurate monitoring results can lead to unnecessary exposures and even death.

We will now explore in detail the nuts and bolts of the monitoring process and explain critical monitoring procedures.

DATA QUALITY OBJECTIVES

data quality objective (DQO) ■ The degree of accuracy needed from a measurement. It may be quantitative, such as determining whether a toxic material concentration is above or below the IDLH for a PPE level decision; or qualitative, such as discerning the presence or absence of a flammable solvent to determine whether there is a container leak.

It is important to consider the data quality objectives when considering the monitoring process. **Data quality objectives (DQO)** refer to the ultimate goal of the monitoring activities. Do you just want to detect the presence of a hazard in the atmosphere? Or do you need to identify the hazardous material? Or do you also want to quantify the amount of hazardous material that is present? Or do you need to do a combination of all three of these? Some air-monitoring equipment can identify the type of hazardous material present, other equipment can detect hazardous materials at low concentrations, and yet other air-monitoring equipment can identify the presence of hazardous materials only at higher concentrations. Thus, the selection of monitoring equipment and methods is very dependent on the type of data needed.

For example, if the incident commander wants to downgrade the level of PPE to an air purifying respirator (APR), he or she must know the identity of the contaminant and its concentration—and ambient oxygen levels. Thus, the air-monitoring equipment chosen must be able to identify the contaminant as well as measure its concentration at the appropriate level. Determining which equipment can detect the contaminant at the appropriate level can be challenging. We will discuss how to achieve data quality objectives with the available monitoring equipment later in the chapter.

UNITS OF MEASUREMENT

First, let's look at units of measurement. As mentioned in Chapter 3, airborne concentrations of gases and vapors are often reported in percentage (%) or parts per million (ppm). One percent (1%), or one part per hundred, refers to one molecule of gas or vapor for every 100 molecules of air. This value can be converted to parts per million:

$$1\% = 10{,}000 \text{ ppm}$$

The most common measurement units you will see are %, ppm, and ppb. Most air-monitoring instrumentation used in the field is not sensitive beyond parts per billion measurements. However, as usual, there are some exceptions, which we will discuss later.

Airborne concentrations of dusts and mists are usually measured in units of milligrams per cubic meter (mg/m^3). Because dust particles and mist droplets are not individual molecules, concentration values are reported as a weight of material in a given volume. Parts per million and milligrams per cubic meter can be interconverted using the following formulas, which take into account the molecular weight of the substance:

$$\text{ppm} = \text{mg/m}^3 \times 24.45 \text{ / molecular weight}$$

$$\text{mg/m}^3 = \text{ppm} \times \text{molecular weight / } 24.45$$

This formula is temperature and pressure dependent, and is valid at 25°C and 1 atmosphere of pressure (STP). The molecular weight of chemicals can be found in many of the reference materials we discussed in Chapter 6—including the *NIOSH Pocket Guide*.

BOX 8-1 THE METRIC SYSTEM

The metric system is a system of units used worldwide by many nations and is used almost exclusively by scientists. It is a system based on multiples of 10. Several common prefixes are used:

- mega (M) – 1,000,000
- kilo (k) – 1000
- deci (d) – 1/10 (one tenth)
- centi (c) – 1/100 (one hundredth)
- milli (m) – 1/1000 (one thousandth)
- micro (μ) – 1/1,000,000 (one millionth)
- nano (n) – 1/1,000,000,000 (one billionth)
- femto (f) – 1/1,000,000,000,000 (one trillionth)

Thus, the metric unit of distance—1 meter (m)—comprises 10 decimeters (dm), 100 centimeters (cm), 1000 millimeters (mm), 1,000,000 micrometers (μm), and so on. One thousand meters (1000 m) make up one kilometer (1 km). The same prefixes apply to the other metric units such as grams (mass or weight), and liters (volume). The metric system is widely used in hazardous materials response, especially in air monitoring.

calibration ■ The procedure of resetting an air-monitoring instrument to read correctly using a known concentration of sample. The known sample is referred to as calibration gas or cal gas, and is manufactured to rigorous technical standards.

combustible gas indicator (CGI) ■ A type of air monitor designed to detect flammable atmospheres and read in percent of the lower explosive limit (% LEL). The CGI has a non-catalytic element and a catalytic element (which allows flammable gases to burn below their LEL). The CGI is one type of sensor typically included in multi-gas detectors.

CALIBRATION

Calibration is the process of resetting electronic monitoring and sample-identification equipment to read a known value. The calibration process begins with a sample gas at an accurately known concentration—the calibration gas (Figure 8-2). The monitoring equipment is exposed to this known concentration of gas and the instrument is reset to this known value. For example, to calibrate a **combustible gas indicator (CGI)** an analytical grade sample of methane at 2.5% volumetric concentration may be used. This is 50% of the lower explosive limit (LEL) of methane (since the LEL of methane is 5%). If the instrument is reading 43%, the electronics are reprogrammed to read 50%. Calibration is typically performed on a set schedule per the manufacturer's recommendation during routine maintenance. However, if the instrument is found to be out of calibration at a hazardous materials incident, it must be field calibrated at the emergency scene before use. Field calibration refers to performing equipment calibration at an unscheduled time, usually in the field during a hazmat incident.

Calibration must be performed by trained personnel using high-quality calibration gases. As the preceding example shows, an air-monitoring instrument can easily be programmed to read incorrectly if an old or out-of-specification calibration gas is used or inappropriately trained personnel perform the calibration. For example, if someone mistypes or misreads the calibration gas and reprograms the instrument to 25% rather than 50% of the LEL, the CGI will read dramatically lower than the actual atmospheric concentration. Therefore, calibration must be performed with the utmost care!

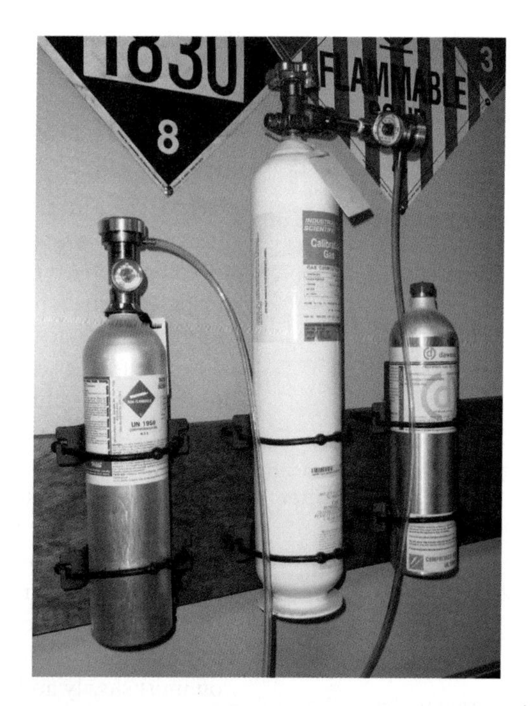

FIGURE 8-2 Bottles of calibration gas at a calibration station for multiple air-monitoring devices. *Photo courtesy of Chris Weber, Dr. Hazmat, Inc.*

BUMP CHECKING

Monitoring and sample-identification equipment must be set up and prepared at the scene of a hazardous material incident before use. Two common procedures used are bump checking and fresh air calibration (or normalizing) of the equipment.

A **bump check** is simply a field test to verify that the instrument is working properly, since sensors may fail to 0% or 0 ppm readings. For example, the electrolyte of an electrochemical sensor may be depleted causing the sensor to read 0 ppm regardless of the

HAZMAT HANDLE

Perform a functional check (bump check) before entering the hot zone every time you use a monitoring instrument!

bump check ▪ The procedure of verifying the ability of a piece of air-monitoring equipment to function properly. A bump check is performed by challenging the equipment with calibration gas or another reliable sample to which the equipment will alarm. Also known as bump test or functional check.

background reading ▪ The ambient level of contaminants that can be detected in a particular location using detection equipment. Background readings may refer to VOC levels when using a PID, or background radiation when using a Geiger counter.

field zeroing ▪ The procedure of resetting an air-monitoring instrument to read background levels when sensing normal levels of contaminant. This must be done carefully in a known clean environment. Also known as fresh air zeroing.

reaction time ▪ The time it takes a sensor to register an alarm after exposure to a contaminant. This includes travel time down the sample train, sensor operation, electronic conversion, and display.

recovery time ▪ The time it takes a sensor to return to background readings after an alarm.

actual concentration of the chemical. You can perform either a qualitative or a quantitative bump check. In a qualitative bump check you expose instrumentation to an unknown concentration of a material that the instrument is known to detect—such as by opening a permanent marker containing an organic solvent in front of a PID—and watch for a response. This tells you that the instrument is registering change, but you do not know how accurate it is. To determine that the instrument is functioning accurately, you must perform a quantitative bump check. You do this by passing the instrument's calibration gas over the sensor and comparing the instrument reading with the known concentration. For your own safety it is imperative that you perform a bump check before every use. Otherwise, how do you know that the instrument you are carrying into the hot zone even works?

FIELD ZEROING

Many air-monitoring instruments must be set to the clean, ambient **background reading** due to sensor drift. This is a simple procedure variously called **field zeroing**, fresh air calibration, or normalization. Oxygen sensors, CGIs, carbon monoxide sensors, hydrogen sulfide sensors, and photoionization detectors (PIDs) all require field zeroing before use. Zeroing involves setting the sensor readings to the known ambient conditions, usually 20.9% oxygen, 0% LEL on a CGI, and 0 ppm on a PID. When zeroing air-monitoring instruments you must ensure you are in a clean environment—for example, not next to the exhaust pipe of your vehicle! If you field zero a multi-gas meter next to a vehicle exhaust (which contains carbon monoxide), you will set a possibly dangerous carbon monoxide level as 0 ppm and consequently underreport carbon monoxide that may be in the residence you are investigating—with predictably dire results.

REACTION TIME AND RECOVERY TIME

Air-monitoring instrumentation readings are not instantaneous. There are two primary reasons for this: first, it takes time for the gas sample to travel through the instrument, called the *sample train*, to the sensor. The longer the sample train is, the longer it will take the sensor to be exposed to the sample and register a reading. This situation is extremely common during confined-space operations, in which a 20-foot or longer hose may be attached to the monitor. It is a good idea to factor in 1 to 2 seconds per foot of hose that is attached to the instrument. The second reason is sensor response time. Some sensors, such as PIDs are very rapid and react almost instantaneously. Other sensors, such as electrochemical sensors (for example, oxygen sensors), can take up to 20 to 40 seconds to register. These types of sensors operate using chemical reactions that take time to complete and register electronically.

You should know the **reaction time** of the air-monitoring instrumentation you are using because you'll have to wait until the instrument has had time to respond before moving on. The T_{90} value, which indicates the time it takes to reach 90% of the full response, can be found in the instrument's manual. For example, if your instrument has a 45-second response time, and it takes you 30 seconds to cross a 50-foot room, you may be in a dangerous atmosphere before your instrument can even alert you to that fact. Owing to limited response times, you must slowly and methodically carry out monitoring activities and not proceed too hastily.

Recovery time refers to the time it takes the sensor to be able to react to additional contaminant after it has previously sounded an alarm. There are three distinct components that affect recovery time: (1) the sensor recovery time, (2) the pump flow rate, and

HAZMAT HANDLE

Take your time when performing an air-monitoring survey!

(3) the sample train desorption process. All sensors use some sort of detection process that takes time. Some processes, such as ionization, are almost instantaneous and therefore have a very short sensor recovery time. Other processes, such as electrochemical detection or surface acoustic wave (SAW) detection, involve a chemical reaction or a binding process that takes many minutes to complete, register electronically, and revert to baseline. Some instruments have adjustable pump speeds. The more fresh air that passes through the sample train and over the sensors, the faster the recovery time will be.

Chemicals tend to deposit, or adsorb, onto the surface of various components of the sample train. The amount of chemical deposition will vary owing to the chemical and physical properties of the chemical contaminant, the chemical concentration, the chemical and physical properties of the sample train components, and atmospheric temperature in relation to instrument temperature (likelihood of condensation). For example, when warm atmospheric gases are pulled into a cold air-monitoring instrument, chemicals will condense on the walls of the sample train. Under these circumstances recovery time will be prolonged as the condensed liquid slowly evaporates. Some sample train construction materials can preferentially bind certain chemicals, or they may be porous. Under these circumstances recovery times may be extended, as desorption processes can be extremely slow. Some detectors, such as PIDs have very short recovery times. Other detectors, such as oxygen sensors and SAW sensors, tend to have longer recovery times. Recovery times can be a short as milliseconds or as long as several minutes to hours.

CORRECTION FACTORS

As discussed earlier, monitoring instruments must be calibrated to ensure their accuracy. Each instrument is typically calibrated to a specific gas. This does not pose a problem for sensors that are specific to one gas, such as oxygen sensors. However, nonspecific sensors such as CGIs and PIDs respond differently to different chemicals. A sensor may respond more vigorously to one gas than to another. If it responds more vigorously to the gas you are investigating at a hazmat incident than it does to the calibration gas, the instrument will overreport the concentration. However, if the air monitor responds less vigorously, it will underreport the concentration. If your instrument underreports, you might be lulled into a false sense of security and overexpose response personnel or the public. Conversely, if it overreports, you may take costly or unnecessary actions such as unwarranted evacuations or the use of excessive PPE.

Correction factors (sometimes called *relative response factors*) are designed to account for different instrument response efficiencies to different chemicals. The manufacturer supplies a table of correction factors (Table 8-1). Typically, the instrument reading is multiplied by the correction factor to obtain the actual concentration. For example, if a chemical has a correction factor of 2.0, and the instrument reading is 100 ppm, then the actual ambient concentration of the chemical is 2.0 × 100 ppm, or 200 ppm. However, it is important to read the manufacturer-supplied manual carefully, since a few manufacturers provide correction factors that are used to *divide* the instrument reading rather than to multiply it.

correction factor ▪ The value that air-monitoring readings must be multiplied by in order to get the true reading for any particular chemical. Instruments that require the use of correction factors include combustible gas indicators and photoionization detectors. Also known as relative response factor.

ACTION LEVELS

Once you obtain an air-monitoring or radiation monitoring reading what does it mean? You must compare the monitoring reading with an **action level** that will guide your hazardous materials response. Action levels are essential to monitoring. For example, what

action level ▪ The value an air-monitoring instrument reading is compared to in order to make a decision at a hazardous materials incident. An example of an action level is 19.5% oxygen, the IDLH level for oxygen, and the result of reaching this action level may be to leave the area immediately and don supplied air respiratory protection.

TABLE 8-1 Correction Factors for a Combustible Gas Indicator (CGI) Calibrated to Methane and Various Photoionization Detectors (PIDs) Calibrated to Isobutylene

CHEMICAL	AreaRAE CGI	9.8 eV RAE PID	10.6 eV RAE PID	11.7 eV RAE PID
Acetic acid	3.4	NR	22	2.6
Acetone	2.2	1.2	1.1	1.4
Ammonia	0.8	NR	9.7	5.7
Benzene	2.2	0.55	0.53	0.6
Chlorine	NR	NR	NR	1.0
Diesel fuel #2	NR	1.3	0.7	0.4
Ethanol	1.7	NR	10	8
Ethyl ether	2.3		1.1	
Formaldehyde		NR	NR	1.6
Gasoline #2	2.1	1.3	1.0	0.5
Hydrogen	1.1	NR	NR	NR
Hydrogen sulfide	Interferes	NR	3.3	1.5
Iodine	NR	0.1	0.1	0.1
Isobutylene		1	1	1
Methane	1.0 (cal gas)	NR	NR	NR
Methyl mercaptan	1.6	0.65	0.54	0.66
Mustard (HD)			0.6	
Nitromethane	2.1	NR	NR	4
Perchloroethene	NR	0.69	0.57	0.31
Phosphine	0.3	28	3.9	1.1
Propane	1.6	NR	NR	1.8
Vinyl chloride	1.8	NR	2.0	0.6

NR = no response.

does a reading of 5 ppm on a PID mean? The answer depends on how toxic the chemical is. An appropriate action level for toxic materials may be the permissible exposure limit (PEL), the immediately dangerous to life and health (IDLH) level, an acute exposure guideline level (AEGL) or emergency response planning guideline (ERPG) value (these values are explained in Chapter 3). The PEL may be the level that allows the building to be reoccupied. If a reading is below the IDLH yet above the PEL, you may be able to use Level C PPE. An AEGL or ERPG value may be chosen for public protective action decisions. These are examples of how action levels are used with real-time air-monitoring results. Table 8-2 shows common action levels for the monitoring equipment you may use at hazmat incidents.

ORDER OF MONITORING

There is a recommended order for monitoring for hazardous conditions at a typical incident:

1. Gamma and neutron radiation (from a distance, before approaching the suspected hot zone)
2. Corrosive atmospheres

SOLVED EXERCISE 8-1

You are using a multi-gas detector containing a combustible gas indicator that has been calibrated to methane at a vinyl chloride railcar release. You and your entry team are approaching the location of the release and have a reading of 7% on the CGI. What is the actual percentage of the LEL of vinyl chloride? *Hint:* The information found in Table 8-1 will be helpful.

Solution: You are at 12.6% of the LEL of vinyl chloride. The correction factor for vinyl chloride is 1.8. Therefore, you should multiply 7% by 1.8, which equals 12.6%. If your action level was 10% of the LEL, you have exceeded it.

TABLE 8-2 Action Levels

MONITORING ORDER	DETECTION EQUIPMENT	LEVEL OF CONCERN	RATIONALE FOR MONITORING ORDER
1. Radiation	Geiger counter Ionization detector (high level) Scintillation counter	>1 mR/hr above background (EPA) >2 mR/hr above background (NRC)	Gamma radiation will travel hundreds of feet through air!
2. Corrosives	pH paper	Varies: <4 or >10 as a rule of thumb	Prolonged exposure to corrosives will destroy most of the electronic detectors.
3. Oxygen	Oxygen sensor	<19.5% or >23.5% (from 29 CFR 1910.146 OSHA Confined Space Regulation)	Oxygen must be present at sufficient levels (>10% as a rule of thumb) for the CGI to work.
4. Flammables	Combustible gas indicator (CGI) Combustible dust meter	>10% of LEL or >25% of LEL	Detect explosive atmospheres.
5. Toxic	Photoionization detector (PID) Colorimetric tubes Ion mobility spectroscopy Surface acoustic wave (SAW) detector	PEL or IDLH	Identify and determine the level of toxic materials and CBRNE agents.

3. Oxygen
4. Flammable atmospheres
5. Toxic atmospheres

Why do you need to monitor in this particular order? Some hazards travel faster and farther than others, such as gamma and neutron radiation. In an unknown environment any of the aforementioned hazards could be present, so you must systematically determine which ones are present and which ones are not. In many cases a combination of chemicals and hazards will be present. These should be assessed in a priority that reflects the risk or danger they pose to hazardous materials responders and to the public. This is the foundation of risk-based response.

The first hazard you should investigate from a distance is strong ionizing radiation, specifically gamma and neutron radiation, because neutron and gamma radiation can travel hundreds of feet in air and cannot be detected by any of the senses until it is too late. In this age of terrorism and radiological dispersal devices, it may be prudent to routinely monitor for gamma radiation at any incident on arrival. As you pull up to what you assume to be a safe distance away, uphill and upwind, check for gamma radiation by holding the detector out of your response vehicle. If you detect levels greater than background, you're in the hot zone! Keep in mind, most radiation detectors are not intrinsically safe—that is, they may cause the ignition of a flammable atmosphere, so you must monitor for a flammable atmosphere before you bring radiation detectors into the hot zone.

As you approach the hot zone, the second hazard you should investigate is corrosiveness, because it will damage or destroy most monitoring equipment. Corrosive gases can disable sensors on exposure. These can be difficult or impossible to detect in a timely fashion while in PPE. The third hazard you should investigate is oxygen level, because most CGIs require oxygen to function properly. After you have determined that an appropriate oxygen level is present, the fourth hazard you should investigate is flammability. Finally, you should measure toxicity as well as alpha and beta radiation after you have determined there are no dangerous levels of gamma radiation, the atmosphere is not corrosive, and you are not entering an explosive atmosphere.

If you do not cover all your bases in a systematic fashion, it is easy to overlook a dangerous hazard and become exposed or injured. In practice, you will monitor for multiple hazards simultaneously, since integrated multi-gas detectors expose the sensors to the atmosphere at the same time. Remember to monitor for corrosive atmospheres before performing any monitoring activities involving electronic equipment, and monitor for a flammable atmosphere before performing any monitoring that uses equipment that is not intrinsically safe.

You don't always have to check for all these hazards at every hazmat incident. For example, if a leaking highway cargo tanker is placarded as acetone—a flammable liquid—it is probably unnecessary to monitor for radiation and corrosive materials. You can start monitoring for oxygen and flammability. However, if there is any question that radioactive materials or corrosive materials could be present, you should start air monitoring for either hazard using the appropriate monitoring equipment and standoff distance.

Monitoring Equipment

Fortunately, a wide range of monitoring equipment is available that is capable of detecting most common hazards and chemicals at useful concentrations. Let's explore the various types of available monitoring instrumentation in the appropriate monitoring order. Pay special attention not only to the capabilities but also to the limitations of each type of detection equipment!

RADIATION DETECTION

Most radiation is completely harmless to us, which is fortunate, since radiation is all around us. Types of radiation include visible light, radio waves, microwaves, and sunlight. When we discuss radiation in the hazardous materials response context, we're referring to ionizing radiation, which can be extremely dangerous at high levels. Ionizing radiation cannot be seen, heard, or smelled. Ionizing radiation is capable of knocking electrons out of molecules and causes damage by making these molecules much more reactive. In humans, this ionization causes cellular damage, including cell death, skin damage, organ failure, and cancer.

Ionizing radiation includes alpha, beta, gamma, X-ray, and neutron radiation. (Radiation is discussed in greater detail in Chapter 3.) Each type of radiation is unique, and different detectors are specific for one or more of these. You should always know which types of radiation your particular instrument will detect.

Radiation detectors report their results in one of two units: field strength or counts per minute (cpm). **Field strength**—typically reported in milliroentgens per hour (mR/hr) or in millisieverts per hour (mSv/hr)—refers to a dose of radiation received. Field strength readings from radiation detectors that are properly calibrated should be identical from instrument to instrument. Counts per minute readings, however, vary significantly from one instrument to another. The unit 'counts per minute', as the name describes, is the number

field strength ■ A measure of the amount of gamma radiation at a particular location.

of ionizing events that a given radiation detector records per minute, so this value will vary by detector type, detector configuration, and detector surface area (size). Field strength measurements are typically used to make risk-based assessments of a hazmat incident, such as determining exclusion zones, establishing the length of hot zone entry times, and calculating evacuation distances. Counts per minute measurements are typically used for

BOX 8-2 HAZMAT HOW TO: USE OF RADIATION DETECTION EQUIPMENT

1. Inspect the instrument. Look for visible damage to the case, inspect the detector and make sure the Mylar window is intact, and ensure that batteries are inserted.
2. Check the last calibration date of the instrument; if it is not within the manufacturer's recommended time period, use another instrument that is.
3. Bump check the instrument with a check source. The check source should come with the instrument. Ensure that the reading is approximately what is expected (based on the strength of the check source).
4. Obtain a background reading. Determine the normal background reading in the area. The normal background reading will depend on your location, including the region of the country, the altitude, and the type of building you are in or near. To obtain a background reading with an instrument that has a fast/slow switch use it in the slow mode to average out normal fluctuations in the background. You should observe the readings for at least 30 seconds before recording the results. Note the background reading on the monitoring log sheet (Figure 8-3). Normal background readings vary from 5 to 50 μR/hr (microroentgens per hour) in the United States.
5. Perform the measurement using the appropriate personal protective equipment (PPE) and record your readings. If there is a fast/slow switch initially put it in the fast mode with the audible chirp on to detect the radioisotope and then put it in slow mode to confirm the reading. The slow mode averages out natural fluctuations, but increases the response time of the meter. For example, a change from fast mode to slow mode for a Ludlum Model 3 survey meter changes the T_{90} response time from 4 seconds to 22 seconds.

AIR MONITORING LOG SHEET

DETECTOR NAME:				
DETECTOR TYPE:				
SERIAL NUMBER:				
LAST CALIBRATION:				
EXPIRATION DATE:				
FRESH AIR ZERO?				
BACKGROUND READING:				
DATE/TIME/LOCATION	READING	READING	READING	READING

FIGURE 8-3 An air-monitoring log sheet should be used to record monitoring readings at hazardous materials incidents. The serial number and preparation of each instrument should be recorded, along with the date, time, and location of each reading. *Courtesy of Chris Weber, Dr. Hazmat, Inc.*

contamination surveys and verification of decontamination. Instrument readings must be compared with the normal background radiation level for a particular instrument in a particular area.

Geiger counter ■ Radiation detection equipment that can detect alpha, beta, and gamma radiation when properly configured.

ionization detector ■ Radiation detection equipment that can detect gamma and neutron radiation and measures field strength.

Geiger-Mueller Tubes

Geiger-Mueller tubes are the detection technology used in **Geiger counters** and ionization detectors (Figure 8-4). A Geiger counter is probably the most useful broad-spectrum radiation detector for hazardous materials technicians. Geiger counters can be configured to detect alpha, beta, and gamma radiation. They report counts per minute or field strength depending on which probe has been attached.

Ionization detectors are typically designed to detect gamma radiation and report only field strength values. Field strength depends on the energy level of the ionizing radiation, which differs for alpha, beta, gamma, and neutron radiation. Within any individual radiation class (such as alpha), the various radionuclides, or materials that emit radiation, also differ in their energy emissions. This makes it very difficult to measure field strength across multiple radiation classes. Because gamma radiation is among the most penetrating ionizing radiation, detectors are typically configured to measure gamma radiation field strength—and therefore report accurate field strength values only for gamma radiation.

scintillation counter ■ Equipment that can be used to detect radiation. Scintillation counters can be very sensitive and radioisotope specific.

Scintillation Counters

Scintillation counters are designed to measure specific types of radiation and/or specific radiation energies. Scintillation counters can be configured for specificity or for sensitivity. Some common scintillating materials (phosphors) used in scintillation counters are zinc sulfide, cesium iodide, and sodium iodide. Scintillation counters are becoming much

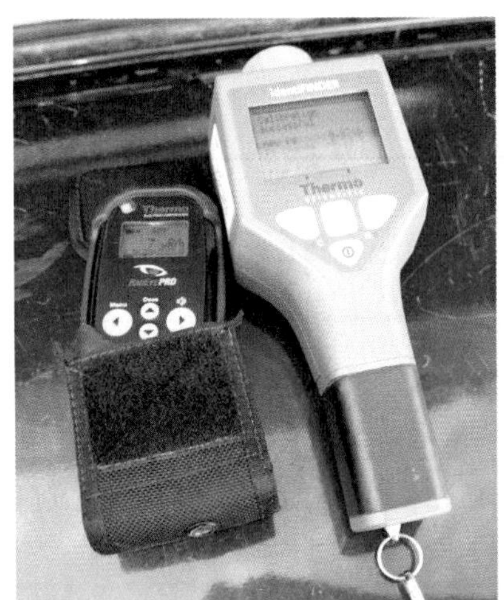

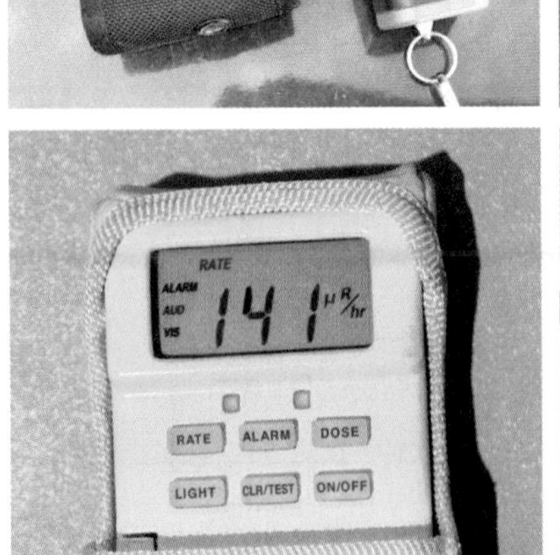

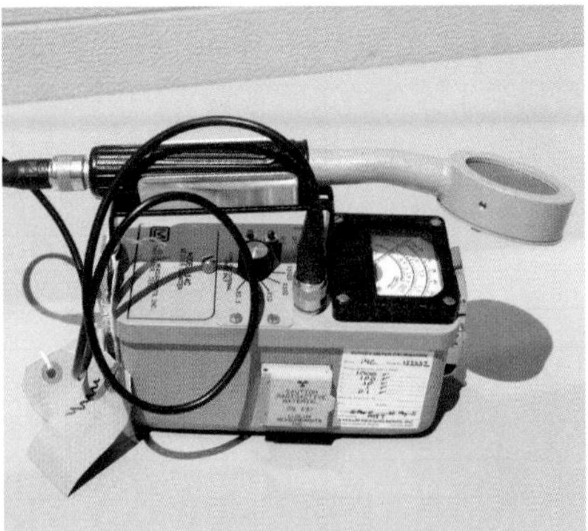

FIGURE 8-4 Several monitoring devices that can be used at radiological incidents. Radioisotope identifiers (top left) can identify the radioisotope by its unique energy signal. Geiger counters are able to detect high-energy radiation (top right) as well as alpha, beta, and gamma radiation using a pancake probe (bottom right). Dosimeters (bottom left) can measure dose rates and therefore are useful for determining personnel exposure levels. *Photos courtesy of Chris Weber, Dr. HazMat, Inc. Top left photo used with permission from Thermo Fisher Scientific.*

more commonly used by first responders. Radioisotope identifiers use gamma scintillation counter technology (Figure 8-4).

If radiation is detected, the first step is to classify it as alpha, beta, or gamma radiation—which can be accomplished using shielding: If a few sheets of paper placed between the source and the detector completely block the radiation, the radioisotope is most likely an alpha emitter. If a thin piece of metal (approximately 1/16th to 1/4th inch) shields the radiation a beta emitter is likely involved. Strong radiation which penetrates the metal is most likely due to a gamma emitter, although a select few radioisotopes emit neutron radiation as well. The second step is to identify the radioisotope, which is most easily accomplished using a radioisotope identifier. Identifying the radioisotope and consequently the precise nature of the emitted radiation allows for the implementation of better protective actions, including PPE selection.

Dosimeters

Dosimeters are used to monitor the amount of radiation individuals receive. At hazardous materials incidents involving radiation it is important to keep track of how much radiation each individual receives, to compare those levels with the action levels recommended by OSHA, EPA, and the Department of Energy (Table 8-3). The turn-back values listed in Table 8-3 are general guidelines, not absolute values, as each radiological emergency is unique.

dosimeter ■ An instrument used to detect and measure individual radiation exposure.

A dosimeter can be as simple as a thermoluminescent detector (TLD) or a film badge in which cumulative radiation dose is converted to a chemical signal that can later be analyzed in a laboratory. The major disadvantage of these traditional dosimeters is the absence of immediate feedback. These detectors are typically used in industries where low-level radiation exposures are relatively common and can be adequately anticipated. This is, of course, typically not the case at hazmat incidents. Therefore, electronic pocket dosimeters (EPD) are much more commonly used by first responders since they are a type of direct reading dosimeter (DRD). Electronic pocket dosimeters, also known as electronic personal dosimeters, are more versatile than traditional dosimeters such as TLDs or film badges. Pocket dosimeters can measure not only cumulative dose but also field

TABLE 8-3 Recommended Action Levels at Incidents Involving Radioactivity

DOSE	WORK FUNCTION	CONDITIONS
5 rem (0.05 Sv)	Maximum annual dose for all occupational exposures for industry workers	Dose minimization through ALARA principles.
10 rem (0.1 Sv)	Maximum dose for emergency responders engaged in valuable property protection	All appropriate safety actions and engineering controls have been implemented.
25 rem (0.25 Sv)	Maximum dose for emergency responders engaged in lifesaving operations or public protection	Exceeding 5 rem is unavoidable. Responders are fully aware of the risks of exposure. Dose over 5 rem must be on a voluntary basis. Respiratory protection and other PPE is provided and used. Radiological monitoring is available to directly measure dose or to calculate the projected dose.

Source: Adapted from the National Library of Medicine (REMM).

strength and/or counts per minute and provide immediate feedback of real-time radiation levels, which increases the safety factor tremendously.

CORROSIVE MATERIALS DETECTION

Corrosive materials include acids and bases that damage human tissue or corrode steel at a specified rate. It is important to identify the presence of a corrosive atmosphere early to prevent personal injury and damage to detection equipment and other electronic equipment you may be carrying (such as a radio).

The concept of pH is based on the dissociation of water molecules into protons (also called hydrogen ions or hydronium ions) and hydroxide ions. The reaction arrows show that the process is in equilibrium; that is, the ions recombine to form water:

$$H_2O \rightleftharpoons H^+ + OH^-$$

Thus, pH has meaning only in aqueous solutions, that is, solutions that contain water. Taking a pH reading of anhydrous solutions such as pure (neat) solvents will give misleading readings. Typically, these solutions will have pH values in the 4–6 range, giving the false impression of being acidic (which they generally are not).

The pH scale extends from 0 to 14 (Figure 8-5). A pH of 7 is considered neutral, a pH below 7 is considered acidic, and a pH above 7 is considered basic. The extremes of the pH scale (0 or 14) are also the extremes of corrosiveness (Figure 5-2).

The pH scale is logarithmic. This means that every change of one pH unit means a 10-fold change in corrosiveness. For example, a solution with a pH of 4 is 1000 times as acidic as a solution with a pH of 7, since there is a difference of three pH units ($10 \times 10 \times 10 = 1000$). This also means that a pH change from 7 to 6 is not nearly as dangerous as a pH change from 5 to 4. Each pH unit change of 1 is a 10-fold change in corrosivity (or hydrogen ion concentration), but because the 5 to 4 pH change is two pH units lower than the 7 to 6 pH change, it represents a 100-fold greater increase in hydrogen ion concentration. Practically, this means that the solution is much more acidic and dangerous, as well as significantly more difficult to neutralize.

pH Paper

pH paper ■ A colorimetric indicator used to determine the corrosiveness of an aqueous liquid. Red indicates an acid; blue indicates a base. Also referred to as litmus paper.

pH paper, also known as *litmus paper*, is used to detect corrosive materials (Figure 8-5). The paper turns red in the presence of an acid, and blue in the presence of a base. When pH paper is used for air-monitoring purposes, half should be wetted prior to use, and the other half should remain dry. Most corrosives such as ammonia are extremely water soluble and respond better to wet pH paper. Other corrosives such as hydrogen chloride and chlorine are much less water soluble and respond more quickly to dry pH paper. Under normal humidity conditions there is sufficient water in the paper to register a pH reading at low corrosive vapor concentrations.

FIGURE 8-5 Detection papers are very useful at hazardous materials incidents for air monitoring. pH paper is shown on the left, and fluoride paper is shown on the right. *Photos courtesy of Chris Weber, Dr. Hazmat, Inc.*

Typically, wetting the pH paper increases the sensitivity of the response, speeds up the response time of the paper, and makes the color change more permanent for most corrosives. This is because virtually all acids and bases are water soluble. For example, if anhydrous ammonia is released in an atmosphere with low humidity, dry pH paper will turn blue only temporarily. If you are not looking at the pH paper as you are passing through the anhydrous ammonia cloud, you may miss the response completely! In contrast, if the pH paper is wetted, anhydrous ammonia will react with the water to form ammonium hydroxide, and the pH paper will turn blue for a much longer period of time.

Electronic pH Meters

Electronic pH meters are essentially voltmeters that measure the hydrogen ion concentration in a solution electronically by measuring the voltage difference that develops across the bulb of a thin-walled glass electrode immersed in the solution. The electrode contains an electrolyte solution with a known concentration of protons. The meter is calibrated in pH units that correspond to the voltage differences. These pH meters are easily damaged, and residual contamination is not easily removed from the glass bulb and electrolyte solution. Most hazardous materials response teams therefore do not use electronic pH meters on a regular basis, especially since pH paper is much cheaper, has no maintenance, and is readily available.

Fluoride Test Paper

Fluoride test paper is extremely useful for detecting the presence of the highly toxic fluoride ion. Fluoride paper turns from pink to yellow on fluoride ion exposure (Figure 8-5). Fluoride paper can be used for air-monitoring purposes and works better when it is first wetted with 3N HCl or water before use. The fluorine atom is found in hydrogen fluoride, hydrofluoric acid, and many salts such as sodium fluoride.

When fluoride enters the body it irreversibly binds calcium and magnesium, both of which are essential in the human body. High doses of fluoride will bind calcium in the bones as well as the calcium and magnesium used in the cells. Chronic exposures to lower doses of fluoride can cause heart attacks. See Box 3-2 for more information about a train derailment where hydrogen fluoride generated from fluorosulfonic acid killed a first responder.

OXYGEN (O_2) SENSORS

Oxygen sensors operate using electrochemistry. Electrochemistry refers to the process of converting a signal from a chemical reaction into an electrical signal. You are very familiar with some common electrochemistry—a battery. Batteries convert a chemical reaction into electricity. Similarly, oxygen sensors convert a chemical reaction between oxygen and a metal into electricity. The oxygen oxidizes the metal, which produces ions that create an electrical current in the sensor. Measurement of the electricity generated is correlated with the amount of oxygen present. Oxygen detectors consist of a sensor composed of two electrodes and an electrolyte solution, a battery, and electronics that can quantify the amount of signal generated by the oxidation reaction produced by the ambient atmospheric oxygen (Figure 8-6).

HAZMAT HANDLE

For air-monitoring applications, use a combination of wet and dry pH paper by wetting half the pH paper with tap water before use. This increases the sensitivity and longevity of the reading.

FIGURE 8-6 Examples of electrochemical sensors: hydrogen sulfide (left), carbon monoxide (middle), and hydrogen cyanide (right). *Courtesy of Chris Weber, Dr. Hazmat, Inc. Used with permission from RAE Systems.*

electrochemical sensor ■ An air-monitoring device that senses the presence of a contaminant using a chemical reduction–oxidation reaction that is converted to electronic output.

cross sensitivity ■ The ability of a detection instrument to detect other chemicals for which it was not designed. Cross sensitivity usually results in a false positive; in other words, the chemical to which the detection equipment is cross sensitive is incorrectly assumed to be the chemical of interest.

Electrochemical sensors such as oxygen sensors have their advantages and limitations just like any other detection device. Advantages of electrochemical sensors include their reliability, specificity, relatively wide operational temperature range (roughly 20°F to 120°F), and comparatively low cost. Some limitations of electrochemical sensors include sensor readings that are affected by temperature, altitude effects due to barometric pressure, a relatively short lifespan (usually 6 months to 2 years), sensitivity to corrosive materials, and **cross-sensitivity** to chemically similar compounds. For example, oxygen sensors have a marked altitude effect, are sensitive to acids because their electrolyte solution is typically alkaline, and are cross-reactive to many oxidizers such as ozone, chlorine, and other halogen gases.

A drop in oxygen level can give you much more information than you may at first think. What caused the drop in oxygen level? There are two possibilities. The first possibility is that the oxygen has been used up in a chemical reaction, such as rusting inside a closed metal tank. The second possibility is that another gas has displaced the air. In the second case, you can determine the concentration of the oxygen-displacing gas from the drop in oxygen concentration. First, you need to recognize that air is not 100% oxygen. Rather, the atmosphere comprises roughly 20% oxygen and roughly 80% nitrogen. Thus, oxygen constitutes roughly 1/5 of the displaced air. This means that if oxygen levels drop by 1%, the concentration of the displacing gas is actually 5% (5 × 1%), because the gas displaces not only the oxygen but also the nitrogen that constitutes 4/5 of the air.

BOX 8-3 HAZMAT HOW TO: USE OF OXYGEN SENSORS

1. Inspect the instrument. Look for visible damage to the case, examine the sample train (the path the air takes inside the instrument), and ensure that batteries are inserted.
2. Check the last calibration date of the instrument. If it is not within the manufacturer's recommended time period, perform a field calibration:
   a. Access the calibration menu of the instrument.
   b. Ensure that the calibration gas (cal gas) values in the instrument correspond to the available cal gas. If they do not, change the values to match the available cal gas.
   c. Attach the cal gas to the instrument and perform the calibration per the manufacturer's directions.
3. Zero the instrument with fresh air. Follow the manufacturer's directions to normalize the instrument to the ambient air and temperature. Make sure you are in a clean environment!
4. Bump check the instrument. Apply cal gas to the instrument for a functionality check. Ensure that the instrument reads close to the cal gas specifications. "Close" depends on your SOGs and the data quality objectives, but within 10% of the value you expect is a good rule of thumb for reconnaissance and basic detection. A smaller percentage may be necessary for exposure level decisions and PPE level decisions.
5. Perform the measurement using the appropriate PPE, and record your results.

SOLVED EXERCISE 8-2

You are investigating a gas release in the warehouse of a local manufacturing plant using a multi-gas meter. As you perform air monitoring at the door, you notice that oxygen levels are at 18.7%, but there is no change in the other sensors (including a CGI and PID). What is the concentration of the escaped gas in the warehouse based on your oxygen reading?

Solution: The concentration of the gas is 11%. Assuming your oxygen sensor was zeroed in fresh air to 20.9%, 2.2% of the oxygen has been displaced (20.9% − 18.7% = 2.2%). Oxygen constitutes only approximately 1/5 of the atmosphere, and the escaped gas has also displaced the remaining 4/5 of the atmosphere made up of nitrogen and other gases. Therefore, 11% of the atmosphere has been displaced (2.2% × 5 = 11%).

COMBUSTIBLE GAS INDICATORS

Combustible gas indicators (CGIs) are designed to detect flammable and combustible gases and vapors below the lower explosive limit (LEL) (Figure 8-7). If CGIs could detect the level of flammable and combustible gases and vapors only at or above the LEL, it would literally be too little too late. Therefore, CGIs detect the presence of flammable gases using a catalytic sensor that can burn the gases or vapors below their LEL. The chemistry involved is very similar to that used in the catalytic converter in cars. Automotive catalytic converters are designed to burn off common by-products of incomplete combustion that would otherwise become components of smog. Combustible gas indicators have two elements: One element is the catalytic sensor coated with platinum or palladium that allows the flammable gas to burn below its LEL (Figure 8-7). The second element is the noncatalytic reference sensor. This element is incapable of burning the flammable gas at concentrations below the LEL and does not increase in temperature when the concentration of flammable vapors is below the LEL. Therefore the catalytic sensor increases in temperature while the reference sensor does not. The increased temperature leads to an increase in resistance in only the catalytic sensor. The difference in resistance between the two sensors is compared and quantified by an electronic circuit called a *Wheatstone bridge.*

A very important fact to remember is that CGIs report their values as a percentage of a percentage. The LEL is a percentage—the percentage of a flammable vapor or gas in air. For example, the LEL of methane is 5%. Methane is rich enough to burn when it constitutes 5% of ambient air. Your CGI reports a percentage of the LEL (% LEL). For example, a 1%

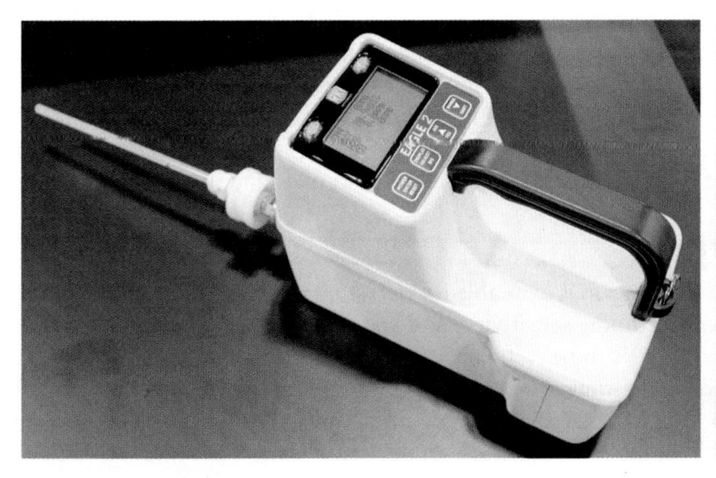

FIGURE 8-7 The RKI Eagle 2 multi-gas detector is shown on the left. On the right is an internal view of the sensing components of a combustible gas indicator (CGI). The catalytic and non-catalytic cells that form the basis of the Wheatstone bridge circuit are visible. The flame arrester is visible to the right in front of the sensor, and the permeable membrane is visible in front of the sensor to the left. *Photos courtesy of Chris Weber, Dr. Hazmat, Inc. Photo on left used with permission of RKI Instruments, Inc. Eagle 2-1-6 Gas Monitor.*

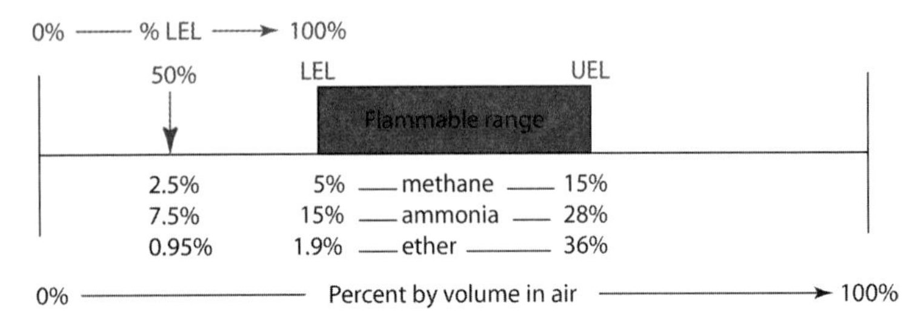

FIGURE 8-8 Comparison of percent lower explosive limit (LEL) versus percent by volume. It is very important to know the difference between these two when using a combustible gas indicator (CGI) that reads in percent LEL.

reading on a CGI calibrated for, and detecting methane (which has a LEL of 5%), is actually detecting 0.05% by volume of methane in air (0.01 × 5% = 0.05%). The conversion of 1% to a decimal is 0.01 (1 divided by 100). Figure 8-8 graphically illustrates this concept.

The sensitivity of CGIs varies from one flammable gas to another, based primarily on the size of the gas molecule but also on the configuration of the CGI. Typically, there is a diffusion barrier that limits gas flow to the catalytic sensor, so smaller molecules reach the sensor in greater numbers. Larger hydrocarbons are therefore detected less readily, since it is more difficult for them to reach the catalytic sensor. This means that CGIs calibrated to methane tend to underreport the concentration of most gases (which are typically larger). However, this is accounted for in the correction factor.

CGIs have advantages and limitations just like any other detection devices. Advantages of CGIs include reliability and low cost. Limitations include the need for oxygen. Depending on the gas or vapor that is being measured, CGIs need at a minimum 10%–16% oxygen to function adequately. However, at oxygen levels much below 20.9% CGIs tend to underreport the level of flammable gas. Below 10% oxygen levels CGIs should be considered unreliable.

The catalytic sensor in CGIs can be poisoned by various chemicals including compounds containing heavy metals such as tetraethyl lead and lead acetate, phosphorus-containing compounds such as red phosphorus and phosphorus trichloride, and sulfides such as hydrogen sulfide and mercaptans. Other materials, such as hydrogen chloride, ammonia, phosphine, silicones and halogenated hydrocarbons such as Freon, can coat or otherwise inhibit the sensor and cause it to underreport the level of flammable gases. One advantage of using methane as the calibration gas is that inhibited or poisoned sensors lose their sensitivity to methane first, before larger molecules such as pentane. Therefore a poisoned sensor should become evident more readily during calibration with methane gas.

Sometimes, coating compounds can be burned off by operating the CGI for an extended period of time; however heavy metals irreversibly poison the sensor by forming an alloy with the palladium or platinum catalyst. In addition, corrosive gases are able to corrode sensor components.

Some CGIs can also be operated in a supersensitive mode. In that mode the CGI can detect as low as 20 ppm of flammable gases, reporting the value in parts per million (ppm). Supersensitive CGIs often use a solid-state sensor in supersensitive mode, but this is not necessary. For example, a 0.1% reading on a CGI calibrated for, and detecting, methane (which has a LEL of 5%) is already detecting 50 ppm methane (0.001 × 5 × 10,000).

TOXICS DETECTION

A wide range of instrumentation is available for detecting toxic chemicals. These instruments vary in their detection methods, sensitivity, and specificity. Therefore, choosing the appropriate toxic detection device can be difficult. We describe several air-monitoring instruments for detecting toxic materials that are commonly available to hazardous materials responders at the technician level. We do not cover manufacturer-specific procedures owing to the large number of different instruments in use in the field. Always become intimately familiar with the manufacturer supplied instruction manual for your particular detection device!

Electrochemical Sensors

Electrochemical sensors can be used to detect not only oxygen, as was discussed earlier, but also toxic gases such as carbon monoxide and hydrogen sulfide. When these sensors are

BOX 8-4 HAZMAT HOW TO: USE OF COMBUSTIBLE GAS INDICATORS (CGIs)

1. Inspect the instrument. Look for visible damage to the case, examine the sample train (the path the air takes inside the instrument), and ensure that batteries are inserted.
2. Check the last calibration date of the instrument. If it is not within the manufacturer's recommended time period, perform a field calibration:
   a. Access the calibration menu of the instrument.
   b. Ensure that the calibration gas (cal gas) values in the instrument correspond to the available cal gas.
   c. Attach the cal gas to the instrument and perform the calibration according to the manufacturer's directions.
3. Zero the instrument with fresh air. Follow the manufacturer's directions to normalize the instrument to the ambient air and temperature. Make sure you are in a clean environment!
4. Bump check the instrument. Apply cal gas to the instrument for a functionality check. Ensure that the instrument reads close to the cal gas specifications. "Close" depends on your SOGs and the data quality objectives, but within 10% of the value you expect is a good rule of thumb for reconnaissance and basic detection. A smaller percentage may be necessary for exposure level decisions and PPE level decisions.
5. Perform the measurement using the appropriate PPE, and record the results.
6. Apply the appropriate correction factor when quantitative results are needed.

bundled together—the oxygen sensor, the combustible gas indicator, the carbon monoxide sensor, and the hydrogen sulfide sensor—they constitute what is often called a "four-gas meter" or multi-gas meter. Multi-gas meters are very common in the fire department and hazardous materials communities owing to their broad utility (Figure 8-7).

Carbon monoxide (CO) and hydrogen sulfide (H_2S) electrochemical sensors operate using the same type of chemical reaction at the sensing electrode: Carbon monoxide (or hydrogen sulfide) is reduced on the sensing electrode and generates a current. Oxygen is needed at the counting electrode to complete the circuit. Thus, oxygen is also required for the operation of carbon monoxide and hydrogen sulfide sensors. The primary difference between a hydrogen sulfide and a carbon monoxide sensor is an activated carbon filter in front of the carbon monoxide sensor, which is absent from the hydrogen sulfide sensor. This explains some cross-reactivity problems these sensors experience (Table 8-4). Owing to the very small size of hydrogen, even a very low concentration of this gas is cross-indicated on these sensors.

There are several dozen electrochemical sensors available for the detection of specific toxic gases and vapors. These sensors may be swapped in and out of multi-gas detectors—even in the field when necessary. Other available electrochemical sensors include ammonia, arsine, phosphine, chlorine, Freon, hydrogen chloride, hydrogen cyanide, hydrogen fluoride, phosgene, and sulfur dioxide. Any of these sensors can typically be substituted or added into a four-gas meter. In fact, many so-called four-gas meters actually have five or six sensor ports. Therefore, if there are facilities in your jurisdiction that use some of these gases, it may be a good idea to add the appropriate sensor to your instrument. However, be aware that some of the listed gases are corrosive and may adversely affect other sensors in the instrument. Nevertheless, four-gas meters are comparatively cheap, especially when cost-recovery ordinances are in place, and may need to be sacrificed in the name of safety.

Metal-Oxide Semiconductor Sensors

Metal-oxide semiconductor (MOS) sensors can be found in some combustible gas detectors as well as in toxic gas and vapor detectors. MOS detectors can be very sensitive, with a lower detection limit down to 1 ppm, or they can be somewhat less sensitive than that depending on the specific instrument. Metal-oxide sensors are a subtype of solid-state sensors. Metal-oxide sensors use semiconductors coated with metal oxide to detect the presence of gases and vapors. The conductivity of semiconductors changes in the presence of certain gases and vapors.

TABLE 8-4 Cross-Sensitivity of Selected Electrochemical Sensors			
RAE SYSTEMS SENSOR TYPE	CROSS-SENSITIVITIES		
	LARGE EFFECT (>50%)	MODERATE EFFECT	SMALL EFFECT (<10%)
Hydrogen sulfide	Phosphine	Methyl mercaptan Sulfur dioxide Nitrous oxide	Carbon monoxide Nitric oxide Methyl sulfide Ethyl sulfide Turpentine Ethylene
Carbon monoxide	Hydrogen Hydrogen sulfide (only with a used CO sensor or used charcoal filter) Ethylene oxide (without filter) Isobutene (with a used CO sensor)	Trichloroethylene (with a new sensor) Ethylene (especially with a used sensor)	Isobutene Isobutylene
Ammonia	Hydrazine (3:1, extremely high) Triethylamine Hydrogen sulfide Sulfur dioxide Chlorine	Nitric oxide	Hydrogen cyanide
Chlorine	Bromine Chlorine dioxide	Sulfur dioxide	Hydrogen sulfide Nitric oxide Nitrous oxide
Phosphine	Silane Diborane Germane	Sulfur dioxide Hydrogen sulfide	Hydrogen Hydrogen cyanide Ethylene

Source: Adapted from RAE Systems Technical Note TN-114.

FIGURE 8-9 Photoionization detectors (PIDs) are able to detect chemicals in the parts per million (ppm) or even the parts per billion (ppb) range. *Courtesy of Chris Weber, Dr. Hazmat, Inc. Used with permission from RAE Systems.*

Some metal-oxide coatings—such as aluminum oxide and zinc oxide—are very nonspecific and detect a wide range of chemicals and are incorporated into broadly selective instruments such as CGIs. Other semiconductors are more specific and detect only a few chemicals, or a chemical family, and are incorporated into toxic gas detectors.

MOS sensors need oxygen to operate, typically at least 10%. They are also sensitive to humidity and dust. Allowing for an extended (5- to 10-minute) warm-up time will help burn off any moisture or dust on the sensor and stabilize readings. As with all detectors it is important to read the manufacturer-supplied manual to determine the capabilities and limitations of your equipment.

Photoionization Detectors

Photoionization detectors (PIDs) are designed to detect organic and inorganic gases and vapors in the low parts per million range (Figure 8-9). They were developed in the early 1970s primarily to detect volatile organic compounds (VOCs) such as benzene, so they are sometimes referred to as VOC detectors. The great advantage of the PID is that it is a reliable, low-cost, low-maintenance, extremely sensitive broad-spectrum detector. This makes it an ideal tool for hazmat incident decision making. A PID is often used to establish the hot zone perimeter for toxic substances, to make the initial PPE assessment for entry, to delineate spills, to detect leaks, and to monitor decontamination procedures.

BOX 8-5 TACTICS: RESPONDING TO CHEMICAL SUICIDES

Chemical suicide—the use of chemicals to commit suicide—has unfortunately been around for years. Historically, suicidal subjects have ingested pesticides such as cyanide salts, aluminum phosphide, or organophosphates. These incidents could pose a danger to first responders because patients may vomit the deadly concoction, because cyanide salts produce deadly hydrogen cyanide gas on contact with stomach acid, and because aluminum phosphide produces deadly phosphine gas on contact with body fluids. According to the Centers for Disease Control and Prevention (CDC) these types of suicides should be treated as hazardous materials incidents.

Recently, a disturbing phenomenon known as "detergent suicide" has emerged. Detergent suicide was first documented in Japan in 2007. Sadly, the detergent suicide recipes are spreading rapidly across the Internet. These types of suicides involve mixing a sulfur-containing compound—such as a detergent—together with an acid to form the deadly gas hydrogen sulfide. Suicidal subjects typically choose small enclosed spaces such as vehicles and small rooms in buildings—such as closets and bathrooms—that they can seal up to concentrate the hydrogen sulfide gas. Often, suicidal subjects will post a note on a window or door warning others, and first responders in particular.

When vehicles are parked outdoors the interior can typically be ventilated with little problem or danger to the surrounding area. However, indoors the hydrogen sulfide gas can migrate to other areas of the building. There have been incidents in which other building occupants have been severely sickened and even died (in Japan). Some locations have seen a variation of the detergent suicide in which suicidal subjects use a cyanide salt and an acid to generate hydrogen cyanide gas in an enclosed space. Both hydrogen sulfide and hydrogen cyanide are toxic and flammable gases. Although unlikely, at high concentrations both gases could pose explosion and flammability hazards at the scene of a suicide.

When approaching any chemical suicide you should take the following precautions:

- Wear full personal protective equipment, including SCBA.
- Investigate the area using a multi-gas meter that contains a hydrogen sulfide sensor and preferably a hydrogen cyanide sensor, as well as an oxygen sensor and combustible gas indicator (CGI).
- Quickly determine whether the patient is still alive, and start medical treatment accordingly. The patient may need to be decontaminated depending on his or her contact with the liquid or solid chemicals. A patient who has only inhaled hydrogen sulfide or hydrogen cyanide will not need to be decontaminated. However, it is important to verify that the patient does not need decontamination using monitoring and detection equipment before transporting him or her to the hospital.
- Treat the suicide location as a crime scene, especially if the subject has died.
- Do not move containers unless absolutely necessary. This can be dangerous and disturbs potential evidence. If you move a container, any unreacted compounds may contact the acid solution and reinitiate the production of hydrogen sulfide or hydrogen cyanide gas.

PIDs operate by ionizing the target gas using an ultraviolet (UV) lamp. A PID can detect any chemical with an **ionization potential (IP)** at or below the bulb strength. The IPs of chemicals can be found in numerous reference materials, including the *NIOSH Pocket Guide* and the WISER program. The IP values of some common gases are listed in Table 8-5. The most common PID bulb used in emergency response applications is 10.6 eV. Other available bulb strengths include 8.3, 9.5, 10.0, 10.2, and 11.7 eV. Different bulb strengths will detect different chemicals. For example, benzene, with an ionization potential of 9.24 eV, can be detected with a PID that has a 10.6 eV bulb. However, methanol, with an ionization potential of 10.84 eV, cannot be detected, since its IP is greater than 10.6 eV. An 11.7 eV bulb is necessary to detect methanol using a PID. Methane has an ionization potential of 13.0 eV and cannot be detected by any PID.

photoionization detector (PID) ■ A highly sensitive volatile organic compound detector. The detection capability of the PID depends on the ionization potential (IP) of the chemical and the bulb strength of the detector.

ionization potential (IP) ■ The amount of energy needed to knock an electron off a chemical.

Table 8-6 lists the ionization potentials of hydrocarbons and hydrocarbon derivatives from one to six carbons in size. As a general rule, as the size of the molecule increases, it becomes easier to ionize and detect. Smaller alkanes—including methane and propane—are impossible to detect with a PID equipped with a 10.6 eV bulb. Some of the most difficult hydrocarbon derivatives to detect are the nitriles—including hydrogen cyanide and acetonitrile; the chloroalkanes—including methyl chloride; the nitroalkanes—including nitromethane; and the carboxylic acids—including formic acid and acetic acid.

TABLE 8-5 Ionization Potentials of Selected Hazardous Materials That Are Difficult or Impossible to Detect

Red indicates chemicals that cannot be detected using a PID, while yellow indicates an 11.7 eV bulb is necessary to detect the chemical (cannot be detected with a 10.6 eV bulb).

CHEMICAL	IONIZATION POTENTIAL
Acetic acid	10.66 eV
Acetylene	11.40 eV
Carbon dioxide	13.8 eV
Carbon monoxide	14.0 eV
Chlorine	11.48 eV
Formaldehyde	10.87 eV
Hydrogen	15.43 eV
Hydrogen cyanide	13.6 eV
Methane	12.61 eV
Nitrogen	15.8 eV
Nitromethane	11.02 eV
Oxygen	12.1 eV
Propane	10.95 eV
Water	12.6 eV

Like combustible gas indicators, photoionization detectors will detect some gases more readily than others. The sensitivity of a PID is determined by the chemistry of the gas or vapor being detected, the flow rate of the PID, and, most important, the surface area of the UV bulb. As with a CGI, correction factors must be applied to PID readings when gases other than the calibration gas are measured. The most common calibration gas for PIDs is 100 ppm isobutylene. The correction factor is supplied by the PID manufacturer and must be consulted when quantitative airborne contaminant concentrations are needed.

Photoionization detectors are very robust, and there are very few sensor poisons, although moisture can easily damage 11.7 eV bulbs owing to their construction. The 11.7 eV bulb window is constructed of lithium fluoride which is very hygroscopic, or moisture absorbing, and is degraded by UV light. These lamps therefore typically last only a few months. Phosphine gas and its UV breakdown products can coat PID lamp windows and should therefore be measured only briefly or at low concentrations.

The biggest problem with PIDs is loss of sensitivity, owing to the presence of dust, smoke, humidity, and other gases such as carbon monoxide and methane that cannot be detected. Loss of sensitivity also occurs when high concentrations of gases and vapors are measured. In each case, the interfering gas or particulate absorbs part of the UV energy without causing any ionization. This keeps that part of the UV energy from ionizing detectable gases and causes underreporting. When the concentration of a detectable gas gets too high, typically over 2000 ppm, there is insufficient UV energy to ionize all the gas molecules passing the sensor, and the detector becomes saturated. The linear range of most PIDs is 0 to 1000 ppm.

Some manufacturers make supersensitive PIDs, such as the ppbRAE made by RAE Systems, that can detect down to the low parts per billion range. Yet other PIDs are able to detect up to 10,000 ppm. Consult the instruction manual of your instrument for your exact parameters. As mentioned previously, PID sensors are being bundled with four-gas meters, and the package is commonly called a multi-gas meter.

TABLE 8-6 Ionization Potential by Compound Class

Red indicates chemicals that cannot be detected using a PID, while yellow indicates an 11.7 eV bulb is necessary to detect the chemical (cannot be detected with a 10.6 eV bulb).

FUNCTIONAL GROUP	SYMBOL	1 C (METHYL-)	2 C (ETHYL-)	3 C (PROPYL-)	4 C (BUTYL-)	5 C (PENTYL-)	6 C (HEXYL-)
Hydrocarbons							
Aromatics	R—C_6H_5	9.24 eV Benzene	8.82 eV Toluene	8.76 eV Ethyl benzene	8.72 eV *n*-Propyl benzene	8.69 eV *n*-Butyl benzene	
Alkenes	R═R	N/A	**10.51 eV** Ethene	**9.73 eV** Propene	**9.58 eV** 1-Butene	**9.50 eV** 1-Pentene	**9.46 eV** 1-Hexene
Alkynes	R≡R	N/A	11.40 eV Acetylene	**10.36 eV** 1-Propyne	**10.18 eV** 1-Butyne	**10.10 eV** 1-Pentyne	**10.03 eV** 1-Hexyne
Alkanes	R—R	**12.61 eV** Methane	11.56 eV Ethane	11.07 eV Propane	10.63 eV Butane	**10.34 eV** Pentane	**10.18 eV** Hexane
Hydrocarbon Derivatives							
Primary amines	R—N—H_2	**8.97 eV** Methylamine	**8.86 eV** Ethylamine	**8.78 eV** *n*-Propylamine	**8.71 eV** *n*-Butylamine		**8.63 eV** Hexylamine
Thiols	R—S—H	**9.44 eV** Methyl mercaptan	**9.29 eV** Ethyl mercaptan	**9.20 eV** 1-propanethiol	**9.15 eV** *n*-Butyl mercaptan		
Ethers	R—O—R	**10.03 eV** Dimethyl ether	**9.53 eV** Diethyl ether	**9.30 eV** Dipropyl ether	**9.28 eV** Dibutyl ether		
Ketones	R—(C═O)—R	N/A	N/A	**9.69 eV** Acetone	**9.54 eV** 2-Butanone	**9.39 eV** 2-Pentanone	**9.34 eV** 2-Hexanone
Aldehydes	R—(C═O)H	10.88 eV Formaldehyde	**10.22 eV** Acetaldehyde	**9.96 eV** 1-Propanal	**9.83 eV** 1-Butanal	**9.74 eV** 1-Pentanal	**9.72 eV** Hexanal
Esters	R—(C═O)O—R	10.82 eV Methyl formate	**10.27 eV** Methyl acetate	**10.11 eV** Ethyl acetate	**10.04 eV** *n*-Propyl acetate	**10.00 eV** *n*-Butyl acetate	
Alcohols	R—OH	10.84 eV Methanol	**10.47 eV** Ethanol	**10.15 eV** *n*-Propanol	**10.04 eV** *n*-Butanol	**10.00 eV** Amyl alcohol	**9.89 eV** *n*-Hexanol
Carboxylic acids	R—(C═O)OH	11.05 eV Formic acid	10.66 eV Acetic acid	**10.24 eV** Propionic acid	**10.17 eV** Butanoic acid	**10.12 eV** Pentanoic acid	
Nitroalkanes	R—NO_2	11.08 eV Nitromethane	10.88 eV Nitroethane	10.78 eV 1-Nitropropane	10.71 eV 1-Nitrobutane		
Chloroalkanes	R—Cl	11.28 eV Methyl chloride	10.97 eV Ethyl Chloride	10.82 eV 1-Chloropropane	10.64 eV 1-Chlorobutane		
Nitriles	R—C≡N	**13.60 eV** Hydrogen cyanide	**12.20 eV** Acetonitrile	**11.84 eV** Propionitrile	11.67 eV *n*-Butyronitrile		

Source: CHRIS, CAMEO, *NIOSH Guide*, WISER, *CRC Handbook*, MSDS, ChemSpider.

BOX 8-6 HAZMAT HOW TO: USE OF PHOTOIONIZATION DETECTORS (PIDS)

1. Inspect the instrument. Look for visible damage to the case, examine the sample train (the path the air takes inside the instrument), and ensure that the batteries are inserted properly.
2. Check the last calibration date of the instruments, if it is not within the manufacturer's recommended time period, perform a field calibration:
   a. Access the calibration menu of the instrument.
   b. Ensure that the calibration gas (cal gas) values in the instrument correspond to the available cal gas.
   c. Attach the cal gas to the instrument and perform the calibration according to the manufacturer's directions.
3. Zero the instrument with fresh air. Follow the manufacturer's directions to normalize the instrument to the ambient air and temperature. Make sure you are in a clean environment!
4. Bump check the instrument. Apply cal gas to the instrument for a functionality check. Ensure that the instrument reads close to the cal gas specifications. "Close" depends on your SOGs and the data quality objectives, but within 10% of the value you expect is a good rule of thumb for reconnaissance and basic detection. A smaller percentage may be necessary for exposure level decisions and PPE level decisions.
5. Perform the measurement using the appropriate PPE, and record the results.
6. Apply the appropriate correction factor when quantitative results are needed.

ion mobility spectroscopy (IMS) ■ A detection device that ionizes contaminants and determines the identity of a material by measuring the drift time of the ions.

Ion Mobility Spectroscopy

Ion mobility spectroscopy (IMS) is commonly used to detect chemical warfare agents, explosives, and illegal drugs (Figure 8-10). It can also be used to detect toxic industrial chemicals (TICs) and toxic industrial materials (TIMs). The most common IMS

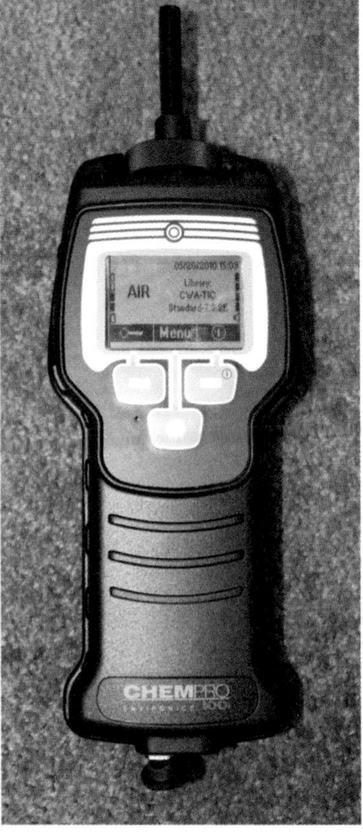

FIGURE 8-10 A selection of ionization mobility spectroscopy (IMS) instruments—the Smiths APD 2000 (top left), the Environics ChemPro (right), and the Smiths Sabre 4000 (bottom left). *Photos courtesy of Chris Weber, Dr. Hazmat, Inc. Top and bottom left photos used with permission from Smiths Detection, Inc. Right photo used with permission from Environics USA.*

instrument used by first responders for WMD agents is the APD 2000 made by Smiths Detection. The Sabre 2000 and Sabre 4000 are broad-spectrum IMS detectors that are able to detect chemical warfare agents, explosives, drugs, and a number of toxic industrial chemicals. The ChemPro 100, manufactured by Environics, has several unique survey capabilities and is an excellent broad-spectrum IMS instrument.

IMS instruments typically use a radioactive isotope (radioactive source) to ionize the gases and vapors to be identified. The ions are then separated by their size and mobility in a drift tube. At the end of the drift tube is a collector electrode where the ions are collected and produce a signal. The type and number of ions, combined with the time it takes them to move down the drift tube, provides a unique pattern or signature that identifies the unknown material. Some IMS detectors have large databases and good sensor technology, while others do not. IMS detectors have been notorious in the past for large numbers of false positives, especially from such common materials as diesel fuel, vehicle exhaust, household cleaners, solvents, and even wintergreen oil. This cross-sensitivity has been vastly reduced using ion filters, longer drift tubes, and larger ion-fragment databases. IMS technology is another good tool to have in the toolbox.

BOX 8-7 TACTICS: RESPONDING TO BUILDING OCCUPANT CHEMICAL EXPOSURES

Building-occupant chemical exposures, often called *sick building* responses, are a bread-and-butter call for the typical hazardous materials response team. A sick building call can range from reports of strange smells in the building to sickened people inside the building. It is the hazmat team's responsibility to identify the problem. Sometimes a chemical or a mixture of chemicals is the root cause, such as roofing solvents or an incompatible mixture of cleaning chemicals. At yet other times there may not even be a legitimate cause, for example, on a Friday afternoon when employees want the afternoon off, or a person feels ill and suddenly his or her coworkers also feel ill—classic psychosomatic symptoms. As a hazmat technician who is part of a hazmat response team, it will be your responsibility to determine what is happening. These types of calls require excellent training, good air-monitoring skills, and good discipline. These calls can quickly escalate out of control for an unprepared hazmat team.

PREPARATION & PLANNING

There are four primary causes of building-occupant chemical exposures: exposure to chemicals off-gassing from building contents such as carpet, foam, furniture, or paint; exposure to chemicals that are used in the building, such as cleaning chemicals, construction chemicals, production facility chemicals, and consumer items such as perfumes, colognes, and self-defense sprays; exposure to chemicals from exterior sources that are brought in through the HVAC system such as vehicle exhaust, roofing compounds, or overspray from nearby pesticide application; or exposure to chemicals from poor building design and maintenance such as carbon dioxide or mold.

Safety Considerations

Some chemicals involved in these types of responses can be quite dangerous, so it is important always to wear respiratory protection while conducting air-monitoring operations in sick buildings. It is easy to become complacent when responding to these types of calls because in some instances the chemical dissipates before the hazmat team arrives, or it is a false alarm owing to the Friday afternoon effect or psychosomatic effects.

Another reason not to become complacent is the possibility of a terrorist attack. This may be especially likely in high-profile buildings or in buildings with a large number of occupants. The HVAC system of these buildings can be very effective at dispersing both biological and chemical agents.

Response Considerations

Common building-occupant chemical exposures include elevated levels of carbon dioxide; volatile organic compounds that off-gas from recently installed building fixtures such as carpeting, furniture, and paint; allergies due to mold; chlorine or ammonia fumes from cleaning supplies; self-defense sprays; and even overexposure of hypersensitive individuals to colognes and perfumes. Because such a wide array of chemicals and biological agents can be involved, it takes very good detective work—including

(continued)

excellent interview skills and excellent air-monitoring skills— to identify the problem. The following are some questions to consider and attempt to answer:

1. Are occupants truly ill, or are they possibly exhibiting psychosomatic symptoms?
2. What chemical is involved?
3. What is the airborne concentration of the chemical?
4. Where is the chemical?
5. Do the patient's signs and symptoms correlate to the suspected chemical?

Even when you perform the investigation flawlessly, you may still come up empty-handed. This may be a result of early building ventilation or because the building occupants were not exposed to chemicals at all but, rather, were exhibiting psychosomatic effects instead. Psychosomatic effects are physiological signs and symptoms brought about by fear. They include stomach ache, nausea, rapid pulse, headache, and nervousness. It is very difficult, if not impossible, to distinguish psychosomatic effects from actual chemical exposure.

RESPONSE

A good sick building response starts when dispatch is contacted by occupants of the affected building. The dispatcher should ask the occupants to close all the windows and doors, shut down the HVAC system, and then immediately evacuate the building. Keeping the building sealed will provide the hazmat team with the highest probability of finding the offending chemical. If the building is ventilated prior to the team's arrival, there will be very little chance of finding the chemical unless large amounts are present. On arrival, the hazmat team should immediately begin interviewing affected occupants, the building manager, and maintenance personnel. The following are key questions to ask:

1. Can you describe exactly how you feel?
2. What part of the building were you in when you started feeling signs and symptoms?
3. How long have you been feeling signs and symptoms?
4. What was the time of onset of signs and symptoms?
5. Have any unusual activities take place in the building, such as renovations, cleaning, or changes in production?
6. Has anything unusual happened to you personally, on or off premises?
7. What types of chemicals are routinely used on the premises?
8. What types of cleaning chemicals are used and how often are they used?
9. How many air exchanges occur per hour?
10. Is there anything else you would like to tell me?

While interviews are being conducted the air-monitoring equipment should be warmed up, bump checked, and calibrated (if necessary).

Initial air-monitoring activities should focus on the areas of the building that were indicated by the occupants during the interview. For a completely unknown situation, monitoring activities should be performed systematically starting with gamma-radiation monitors from the outside, followed by wet and dry pH paper, oxygen levels, combustible gases, then a sensitive, broad-range toxic chemical detector such as a PID. Keep in mind that PIDs do not detect all chemicals. If there are reliable indications that a specific chemical or class of chemical is involved, targeted air monitoring should be performed—such as with colorimetric tubes.

Often, it is difficult to find significant levels of contaminants. When contaminants are found, air monitoring should be performed in the direction of higher contaminant readings. This technique is called *following the gradient.* Following the gradient will typically lead you to the source of contamination, whether that is a leaking container, the location where incompatible chemicals were mixed, or a bucket of roofing compound near an air intake on the roof.

Flame Ionization Detectors

flame ionization detector (FID) ■ An air-monitoring device that senses the presence of a contaminant by ionizing the gas or vapor using a hydrogen flame in a destructive process.

Flame ionization detectors (FIDs) have capabilities similar to those of PIDs. They both operate by ionizing the target molecule. In contrast with photoionization detectors, flame ionization detectors ionize gases and vapors using a hydrogen flame, which is a destructive process (Figure 8-11). Essentially, FIDs are limited less by the ionization potential of the chemical and more by the ability of the hydrogen flame to produce detectable levels of ions. FIDs tend to detect any materials that contain carbon and hydrogen in the range of 0.1 to 1000 ppm. They have an effective ionization energy of 15.4 eV. Therefore, one advantage of FIDs is that they will detect gases like hydrogen cyanide and methane that have high ionization potentials, which a conventional PID would not be able to detect. Disadvantages of FIDs are that they require a source of highly pure hydrogen gas, are more expensive than conventional PIDs, and require oxygen for the hydrogen flame to remain lit. It is sometimes difficult to keep the hydrogen flame of an FID lit, especially in cold weather. FIDs typically use 100 ppm methane as a calibration gas. They are an excellent complement to conventional PIDs.

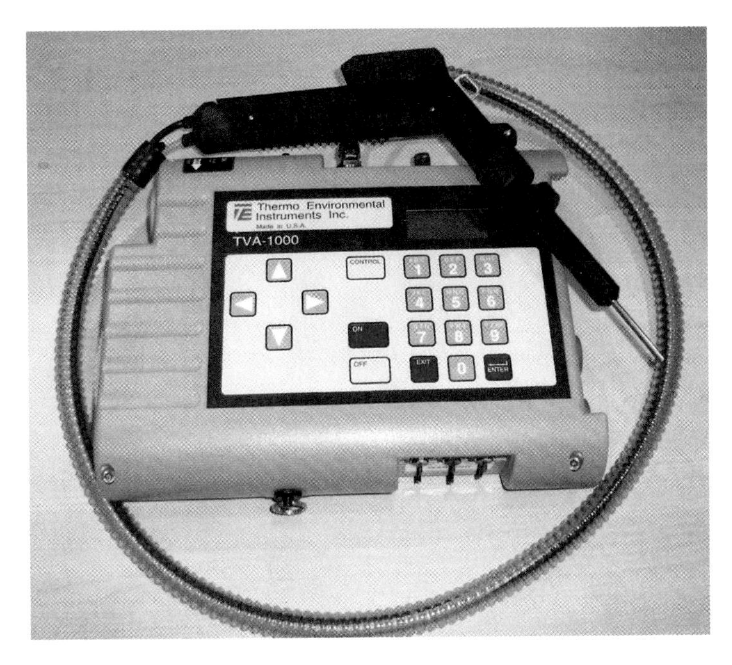

FIGURE 8-11 A flame ionization detector (FID) that is capable of detecting chemicals in the parts per million (ppm) range. *Photo courtesy of Chris Weber, Dr. Hazmat, Inc. Used with permission from Thermo Fisher Scientific.*

Colorimetric Tubes and Chips

Colorimetric tubes are low-cost, relatively specific, highly sensitive air-monitoring devices. Colorimetric tubes can be very useful in hazardous materials emergency response, especially when it involves the release of a known material, or when information must be verified. Colorimetric tubes can also be helpful in the identification process. Several manufacturers make colorimetric tubes (Gastec Polytec tubes) or tube sets (Dräger Simultest kits), that can be used to characterize and place an unknown contaminant into a chemical family.

Colorimetric tubes are small glass tubes filled with a chemical indicator and sealed at both ends (Figure 8-12). The chemical indicator inside the tube changes color in response to certain chemicals or chemical families. The length of the stain indicates approximate concentration. The concentration is only approximate because colorimetric tube accuracy can be as low as ±35%. Sensitivity ranges from parts per million to percent. Various manufacturers have more than 100 different colorimetric tubes on the market for different applications in different industries.

colorimetric tube ■ A type of air-monitoring device that uses a chemical reaction to detect airborne contaminants. A pump is used to move a known quantity of air through a glass tube containing the detection medium.

The first step when using colorimetric tubes is to check their expiration date, appropriate operation range (sensitivity), and tube integrity (including the tube itself and the colorimetric indicator inside). Using expired tubes can be dangerous because the colorimetric indicating reagents lose their effectiveness over time. The second step is to check the accuracy of the pump (leak check). During normal operation after each end of the tube has been opened, a known quantity of air, usually 100 cubic centimeters (cc), is pumped through the glass detector tube using either a piston pump or a bellows pump (Figure 8-12). A bellows pump can be checked by depressing the bellows and placing an unopened tube on the end. After 5 minutes, the pump should remain almost completely depressed, which means that there is no leak. Once the equipment has passed muster, a measurement can be taken.

When using colorimetric tubes it is extremely important to read and understand completely the directions supplied with the tubes. At a minimum, the following information should be noted:

- The preparation of the tube (if any).
- The number of pump strokes needed per measurement, and how long this process is expected to take. (For example, sulfuric acid tubes require 100 pump strokes, and the overall measurement takes approximately 75 minutes. This is obviously a very important piece of information to know before entering the hot zone in full PPE! It should be noted that tubes requiring 100 pump strokes are extremely rare. Most tubes require 10 pump strokes or fewer.)

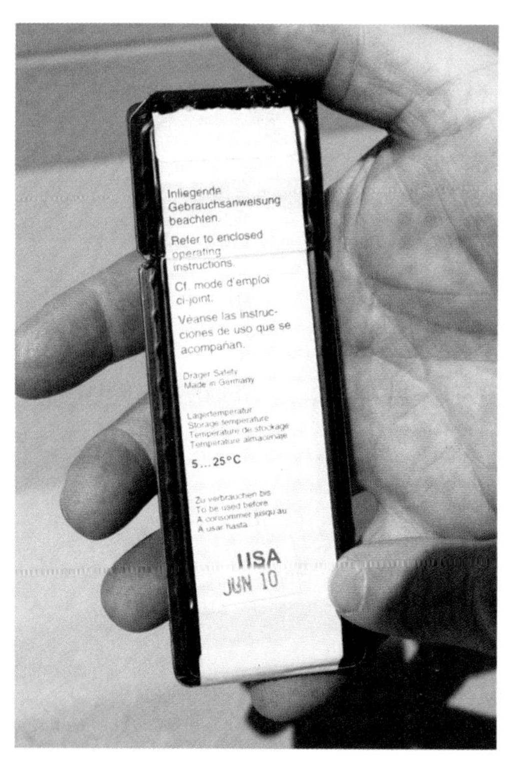

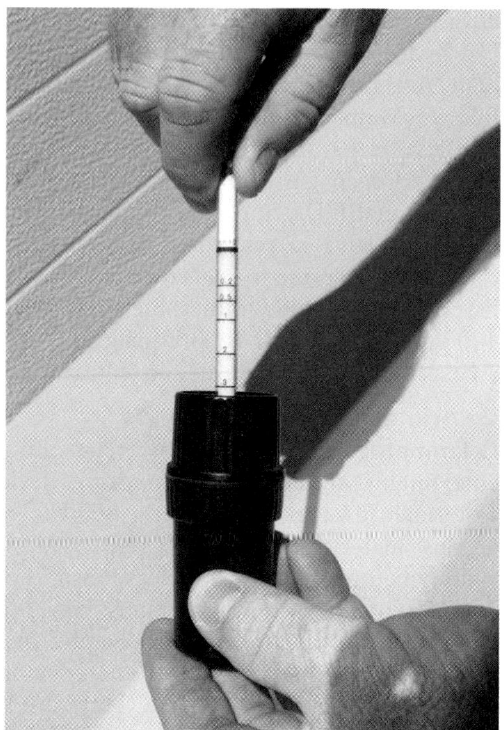

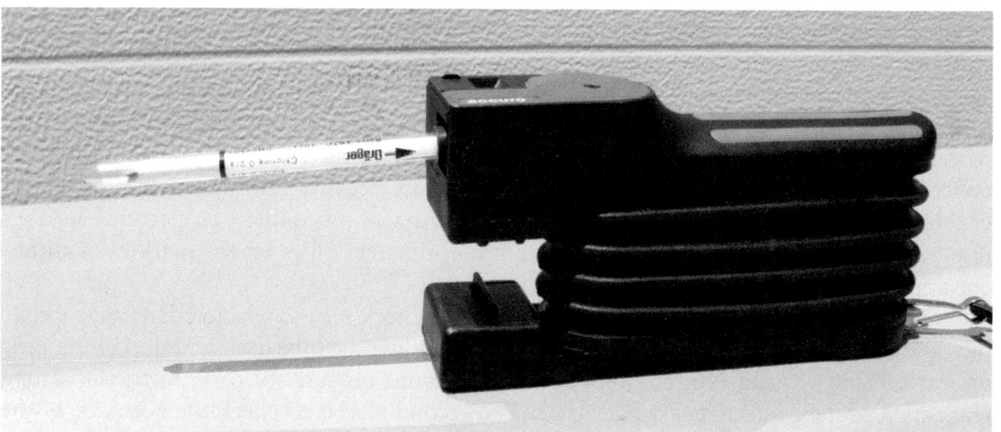

FIGURE 8-12 Colorimetric tubes are very useful air-monitoring tools. It is important to check the expiration date of the indicator tubes before use (top left). The tubes must be opened before use by scoring the glass and breaking off the tip on both ends (top right). Ensure that the indicator tube is inserted correctly into the pump (bottom). *Photos courtesy of Chris Weber, Dr. Hazmat, Inc. © Drägerwerk AG & Co. KGaA, Lubeck. All rights reserved. No portion hereof may be reproduced, saved or stored in a data processing system, electronically, or mechanically copied or otherwise recorded by any other means without our express written permission.*

- The interval between pump strokes (if any).
- Cross sensitivities the tube may exhibit.
- The color change to be expected with the chemical of interest and any chemicals to which the tube is cross-sensitive.
- Tube detection range and accuracy.
- Any special instructions (such as taking 10 pump strokes, then breaking an ampoule and taking 10 more pump strokes—such as with phosphoester tubes)

The advantages of colorimetric tubes include comparatively low cost, ease of use, and minimal maintenance and upkeep. The limitations of colorimetric tubes include a limited shelf life, the possible need for temperature controlled tube storage, comparatively poor accuracy, and multiple cross-sensitivities. The shelf life of most colorimetric tubes is 1 to 3 years, depending on the reagent used in the colorimetric reaction. Most colorimetric tubes will last longer if stored in a refrigerator, because the reagents break down more slowly at lower temperatures. Some tubes may even be reused immediately on scene if no color changes occurred during the first use. However, be careful to read the directions carefully

BOX 8-8 HAZMAT HOW TO: USE OF COLORIMETRIC TUBES

1. Inspect the pump and perform a leak test. Look for visible damage to the pump, such as cracking or visible leaks.
2. Check the expiration date of the colorimetric tubes.
3. Ensure that the sensitivity range is appropriate (for example, ppm versus %).
4. Read the instructions carefully and note the number of pump strokes, the color change to be expected, and any cross-sensitivities.
5. Score and break open both ends of the colorimetric tube and insert with the arrow pointing toward the pump (in the direction of air flow). Ensure that the lowest numbers are farthest away from the pump!
6. Take at least three tubes into the hot zone: One tube will be required for the measurement, a second unopened tube will serve as a reference, and a third tube should serve as a backup.
7. Carefully take a pump stroke and wait for the pump to completely reinflate. Repeat this procedure for the required number of pump strokes.
8. Read the length of stain along the tube. If the staining is uneven, take the most conservative reading (highest number).
9. Compare the reading to an unopened tube as reference. Responders have in the past misinterpreted a rapidly saturated tube as a zero reading because they did not remember the original color of the tube.

and ensure that this is a practice sanctioned by the manufacturer and that the tube has not already been saturated. This is one reason it is necessary to take an unopened tube into the hot zone for reference purposes.

Colorimetric tubes are generally not chemical specific. Rather, they react to a class of chemicals (Figure 8-13). This leads to cross-sensitivities, or cross-reactivities, otherwise known as *false positives* (Figure 8-14). For example, MSA makes an aromatic hydrocarbon colorimetric tube that detects benzene, toluene, and xylene. Dräger makes a benzene-specific colorimetric tube with a prelayer that filters out the other aromatics (such as toluene and xylene). However, if the prelayer becomes saturated, the readings will be inaccurate.

Colorimetric chips are much smaller versions of colorimetric tubes built into a small, credit card–sized plastic chip that can be exposed and automatically read by a chip-reading device. The chip reader determines the colorimetric reading and quantifies the result. Each chip contains up to 10 identical individual colorimetric tubes and can therefore be used for 10 separate readings for any particular gas or vapor. Colorimetric chips have the advantage that the chip reader not only reads the colorimetric stain but also pumps the air through the tube on the chip. This reduces user error greatly by standardizing the pumping process as well as standardizing colorimetric stain interpretation. The disadvantage of the chip system is that it is more expensive than traditional glass colorimetric tubes.

Some manufacturers make screening tubes or sets of tubes. These systems can be used to screen multiple hazards simultaneously. For example, Gastec manufactures several Polytec tubes that identify chemical classes by varied color changes. The Polytec I tube indicates the following chemicals at varying concentrations by the following color changes:

carbon disulfide or hydrogen sulfide	green
carbon monoxide	green or brown
acetone, acetylene, or ethylene	brown or green
gasoline, benzene, propane, or propylene	brown
styrene	yellow or brown
trichloroethylene	pale brown
toluene or xylene	purple

GASTEC No.81 Instructions for Acetic Acid Detector Tube

FOR SAFE OPERATION :

Read this manual and the instruction manual of your Gastec Gas Sampling Pump carefully.

⚠ **WARNING:**

1. Use only Gastec detector tubes in a Gastec Pump.
2. Do not interchange or use non-Gastec parts or components in Gastec's detector tube and pump system.
3. The use of non-Gastec parts or components in Gastec's detector tube and pump system or use of a non-Gastec detector tube with a Gastec pump or use of a Gastec detector tube with a non-Gastec pump may result in property damage, serious bodily injury, and death; voids all warranties; and voids all performance and data accuracy guaranties

⚠ **CAUTION : If not observed, injuries to the operator or damage to the product may result.**

1. When breaking the tube ends, keep away from eyes.
2. Do not touch the broken glass tubes, pieces and reagent with bare hand(s).
3. The sampling time represents the time necessary to draw the air sample through the tube. The tube must be positioned in the desired sampling area for the entire sampling time or until the flow finish indicator indicates the end of the sample.

△ **NOTES : For maintaining performance and reliability of the test results**

1. Use Gastec Gas Sampling Pump together with Gastec Detector Tubes only for the purposes specified in the instruction manual of the detector tube.
2. Use this tube within the temperature range of 0 - 40℃ (32 - 104℉).
3. Use this tube within the relative humidity range of 0 - 80%.
4. This tube may be interfered with by the coexisting gases. Please refer to the "INTERFERENCES".

APPLICATION OF THE TUBE : Use of this tube for the detection of Acid gases in air or the industrial areas and environmental atmospheric condition.

SPECIFICATION : (As a result of Gastec's commitment to continued improvement, specifi-cations are subject to change without notice.)

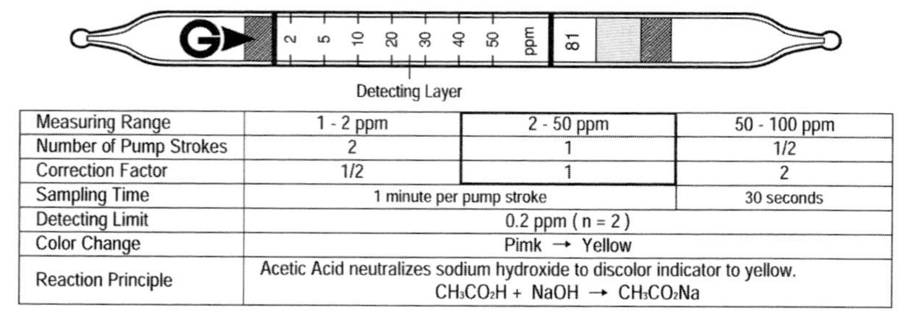

Measuring Range	1 - 2 ppm	2 - 50 ppm	50 - 100 ppm
Number of Pump Strokes	2	1	1/2
Correction Factor	1/2	1	2
Sampling Time	1 minute per pump stroke		30 seconds
Detecting Limit	0.2 ppm (n = 2)		
Color Change	Pimk → Yellow		
Reaction Principle	Acetic Acid neutralizes sodium hydroxide to discolor indicator to yellow. $CH_3CO_2H + NaOH \rightarrow CH_3CO_2Na$		

Coefficient of Variance : 10% (for 2 to 10 ppm), 5% (for 10 to 50 ppm)
**** Shelf Life : Please refer to the Validity Date printed on the box of tube.**
**** Store the tubes in dark and cool place.**

CORRECTION FOR TEMPERATURE, HUMIDITY AND PRESSURE :

Calibration of the Gastec detector Tube No.81 is based on a tube temperature of 20℃ (68℉) and not the temperature of the gas being sampled, appoximately 50% relative humidity and normal atmospheric pressure.

Temperature : No Correction is required.
Humidity : Correct for humidity by the table below.

Relative Humidity (%)	0	20	40	50	60	80
Correction Factor	0.5	0.7	0.9	1.0	1.2	1.5

Pressure : To correct for pressure, multiply the tube reading by

$$\frac{\text{Tube Reading (ppm)} \times 1013\ (\text{hPa})}{\text{Atmospheric Pressure (hPa)}}$$

MEASUREMENT PROCEDURE :

1. For leak tight check of the pump insert a fresh sealed detector tube into pump. Follow instructions provided with the pump operating manual.
2. Break tips off a fresh detector tube in the tube tip breaker of the pump.
3. Insert the tube securely into pump inlet with arrow G➤ on the tube pointing toward pump.
4. Make certain pump handle is all the way in. Align guide marks on pump body and handle.
5. Pull the handle all the way out until it locks on 1 pump stroke (100ml). Wait 1 minute and confirm the completion of the sampling.
6. For lower than 2 ppm measurement, repeat the above sampling procedure one more times untill the stain attains to the first calibration mark. For higher than 50 ppm measurement, prepare fresh tube and take 1/2 pump strokes.
7. Read concentration at the interface of the stained-to-unstained reagent.
8. If correction is needed, multipty the correction factors of temperature, pump strokes and pressure.

INTERFERENCES :

Substance	Concentration	Interference	Changes color by itself to
Hydrogen chloride, Hydrogen cyanide, Nitric acid	≧ 3 times	Plus error	Yellow
chloride, Sulfur dioxide, Nitrogen dioxide	≧ 1/2	Plus error	Yellow

The table of this interference gases primarily expresses the interference of each coexisting gas in the gas concentration range, equivalent to the gas concentration. Therefore, the test result may be given positive result by the other substances not listed in the table. For more information is needed, please contact us or our distributors in your territory.

APPLICATION FOR OTHER SUBSTANCES :

Substance	Correction	No. of Pump Strokes	Measuring Range
Acetic anhydride	0.3	1	0.6 - 15 ppm
Acrylic acid	1.0	1	2 - 50 ppm
Formic acid	2.6	1	5.2 - 130 ppm
Isovaleric acid	1.0	1	2 - 50 ppm
Maleic anhydride	0.4	1	0.8 - 20 ppm
Methacrylic acid	0.9	1	1.8 - 45 ppm
Propionic acid	1.5	1	3 - 75 ppm

CORRECTION FACTOR : Detector tubes are primarily designed to measure specific gases. But it is also possible to measure other substances of similar chemical properties with the aid of a correction factor or chart. Therefore, please make use of the correction factor/chart measuring ranges as a reference. For a more precise factor please contact your Gastec distributor.

DANGEROUS AND HAZARDOUS PROPERTIES :

Threshold Limit Value-Time Weighted Average by ACGIH (2004) : 10 ppm (7-8 hours)
Threshold Limit Value-Short Term Exposure Limit by ACGIH (2004) : 15 ppm (15 min.)
Explosive Range: 4 -19.9 %

DISPOSAL INSTRUCTION : Reagent of the tube does not use any hazardous substances. When disposing the tube regardless of used or unused, follow the rules and regulations of the local government.

WARRANTY : If you have any questions regarding gas detection and quality of the tubes, please feel free to contact your Gastec representatives.

Manufacturer : Gastec Corporation
8-8-6 Fukayanaka, Ayase-City, 252-1195, Japan
http://www.gastec.co.jp/
Telephone +81-467-79-3910 Facsimile +81-467-79-3979

IM0081E1
Printed in Japan
05K1Z

FIGURE 8-13 Colorimetric indicator tubes come with detailed instructions that indicate the number of pump strokes, method of indication, and any cross-indicating chemicals.
Courtesy of Gastec Corporation.

116 | Dräger-Tubes for short-term measurements

Benzene 0.5/a

Order No. 67 28 561

Application Range

Standard Measuring Range:	0.5 to 10 ppm
Number of Strokes n:	40 to 2
Time for Measurement:	max. 15 min
Standard Deviation:	± 30 %
Color Change:	white → pale brown

Ambient Operating Conditions

Temperature:	10 to 40 °C
Absolute Humidity:	3 to 15 mg H_2O / L

Reaction Principle

$2\ C_6H_6 + HCHO \rightarrow C_6H_5\text{-}CH_2\text{-}C_6H_5 + H_2O$

$C_6H_5\text{-}CH_2\text{-}C_6H_5 + H_2SO_4 \rightarrow$ p-quinoid compound

Cross Sensitivity

Other aromatics (toluene, xylene, ethyl benzene) are indicated as well. It is impossible to measure benzene in the presence of these aromatics. Petroleum hydrocarbons, alcohols and esters do not affect the indication.

Additional Information

Before performing the measurement the ampoule must be broken and the liquid transferred onto the indicating layer so that it is completely saturated.

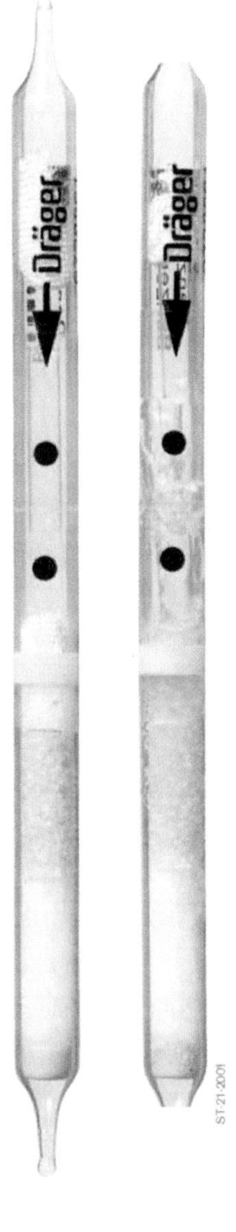

FIGURE 8-14 Some colorimetric indicator tubes—such as this Dräger benzene tube—require more extensive preparation—such as breaking an ampoule—before they are ready for use. *© Drägerwerk AG & Co. KGaA, Lubeck. All rights reserved. No portion hereof may be reproduced, saved or stored in a data processing system, electronically or mechanically copied or otherwise recorded by any other means without our express prior written permission.*

Dräger manufactures a number of Simultest kits for several different toxic industrial chemicals, as well as different classes of chemical warfare agents. These kits consist of five tubes bundled together in a manifold system designed to simultaneously draw air through all five tubes. These Simultest kits require 50 pump strokes, and some tubes require preparation both before and during sampling. Nevertheless, these can be very useful for screening unknown atmospheres as well as for providing additional detection technology to complement those already discussed and others to be discussed later.

Mercury Detection

Elemental mercury, also known as *inorganic mercury*, is an unusually volatile liquid metal at room temperature which gives off toxic vapors. Mercury is contained in consumer products including older thermometers, regular and compact fluorescent lightbulbs,

BOX 8-9 HAZMAT HISTORY: LIQUID CARBON DIOXIDE LEAKS IN RESTAURANTS (2011)

Restaurants, especially fast food restaurants, are increasingly using liquid carbon dioxide storage units inside their buildings to carbonate the soft drinks in the soda fountains. Traditionally, soda fountains have been supplied by compressed carbon dioxide gas rather than liquefied carbon dioxide. Liquefied carbon dioxide storage tanks pose a much greater danger than gas cylinders of carbon dioxide. The large danger with liquefied carbon dioxide is that the expansion ratio from liquid to gas is 845:1. This means that 1 cubic foot (ft^3) of liquid carbon dioxide will expand to 845 ft^3 of carbon dioxide gas, thereby displacing the oxygen in the immediate area.

Carbon dioxide is a simple asphyxiant. However, it also has physiological effects similar to those of a narcotic in that it causes breathing suppression and can lead to unconsciousness, coma, and even death. On mixing with water in saliva, carbon dioxide forms the weak acid carbonic acid. This reaction often creates a bitter smell and taste, since acids taste bitter.

The liquid carbon dioxide dewar, which can hold up to 100 gallons, is commonly stored in either the basement or the main floor of the establishment. The dewar is typically refilled from a tanker truck approximately once a month through an exterior fill line. The dewar is constructed of stainless steel, and much of the internal piping is made of plastic. It is not uncommon for these fittings and hoses to leak, as is illustrated by the following two cases—one of which caused a fatality.

PHOENIX, ARIZONA

On May 31, 2011, the Phoenix Fire Department received a call that a pregnant employee had fainted and collapsed at a fast food establishment. The first arriving units treated the patient in the kitchen area, at the top of the stairs to the basement (where they found her). At this time, the restaurant was still fully operational, and employees offered little information except that she fell as she was coming up the stairs. Two firefighters went downstairs to investigate. They found paint cans, miscellaneous cleaning chemicals, and a liquefied carbon dioxide storage unit in the basement. After approximately 15 seconds one firefighter noticed a slight burning sensation in his throat, and the other firefighter noticed a strange odor. They therefore quickly exited the area, and one of them subsequently collapsed at the top of the stairs. The fast food restaurant was quickly evacuated, and the hazmat unit was called.

The hazmat technicians investigated the area in full PPE. When they entered the basement, their air-monitoring instrumentation—which included an oxygen sensor as well as a natural gas detector—sounded an alarm. They briefly investigated the area, since the CGI was not indicating an alarm, and noticed that the facility carbon dioxide sensor located immediately next to the carbon dioxide dewar was not issuing an alarm. In addition, there were no emergency shutoff signs or instructions near the dewar. Owing to this conflicting information the entry team exited the basement and formed a plan.

The basement was therefore ventilated using a confined-space fan. After monitoring the air and ensuring there were no other hazardous readings, the entry team members shut off all visible valves on the dewar. On investigation, they noticed that there was tape on the facility carbon dioxide sensor, which explained why this sensor did not sound an alarm. In a conversation with the manufacturer of the natural gas detector it was determined that carbon dioxide mimics methane and is therefore a cross-indicating false positive. This incident illustrates the importance of air monitoring and correctly interpreting air-monitoring results—including false positives due to cross-sensitivities.

SAVANNAH, GEORGIA

On September 12, 2011, firefighters in Pooler, Georgia, were called to a local fast food establishment for a report of several ill customers. On arrival, firefighters noticed a strange smell, quickly exited the building, donned PPE including SCBA, and reentered the building. The firefighters pulled two unconscious victims from the bathroom and evacuated the building. One of the victims subsequently died at the hospital from asphyxiation. It was later determined that a carbon dioxide line in the wall between the kitchen and the bathroom was leaking and had caused the bathroom to fill with carbon dioxide.

Therefore, when investigating unknown odors or medical emergencies at fast food establishments or other occupancies that may have liquefied carbon dioxide on-site, it is important to ask the right questions, bring a multi-gas meter to determine oxygen levels, and have SCBA at the ready.

batteries, and mercury switches. When elemental mercury is released in poorly ventilated areas of the home or workplace, toxic levels of mercury vapor can accumulate.

A very good video showing mercury evaporating has been created at Bowling Green State University in collaboration with the EPA and can be accessed online. In the video, mercury vapors are detected using a fluorescent screen and UV light (however, this visualization technique is not practical in the field). Approximately 70% to 75% of inhaled inorganic mercury vapor is absorbed by the body. In contrast, virtually no inorganic mercury is absorbed through the skin or through the digestive tract.

Do not confuse inorganic mercury with the much more dangerous organic mercury compounds, such as methylmercury. Methylmercury is extremely deadly and readily absorbed through the skin and gastrointestinal tract. Methylmercury killed a Dartmouth chemist several months after she spilled one drop of it on her latex glove. The organic mercury quickly permeated the latex and was rapidly absorbed through the skin. This is how toxic and skin absorptive organic mercury compounds can be!

Hazardous materials technicians are often called on to clean up and determine whether mercury levels are safe for reoccupancy. Two types of detectors can be used to determine the initial mercury vapor levels, to find areas of spot contamination, and to determine whether a building is safe to reoccupy after a cleanup operation.

Atomic Absorption Spectroscopy **Atomic absorption spectroscopy (AAS)** is a very effective technique for detecting elemental materials such as mercury. The most common mercury detector that uses AAS is the Lumex mercury detector (Figure 8-15). Atomic absorption spectroscopy detects atoms of heavy elements by exciting specific electrons of the element in the gaseous state using electromagnetic energy. The Lumex 915+ meter uses a mercury bulb to specifically detect mercury atoms through electronic excitation. The Lumex 915+ has a detection limit of 2 ng/m^3 mercury vapor, and the Lumex 915 has a detection limit of 100 ng/m^3. The dynamic range of AAS detectors is 2 to 50,000 ng/m^3 (0.002 to 50 μg/m^3). Gasoline may interfere with AAS detection of mercury. Temperature swings also may affect the accuracy of the instrument. For this reason atomic absorption instruments should be allowed to warm up at least 10 to 15 minutes before use.

atomic absorption spectroscopy (AAS) ■ An analytical technique for determining the concentration of gaseous chemical elements through the absorption of light by atoms.

Gold-Film Detectors Gold-film detectors, which are often also called *Jerome meters,* are able to detect mercury by forming a gold–mercury amalgam, which changes the electrical resistance of a thin gold detection film (Figure 8-15). These detectors are less sensitive than the atomic absorption detectors and typically have minimum detectable levels of between 0.5 and 5 μg/m^3 mercury vapor (which equals approximately 500 to 5000 ng/m^3). The dynamic range of gold film detectors is from 0.5 to 999 μg/m^3.

The gold film needs to be regenerated prior to storage after contact with mercury, by heating the gold film and volatilizing the amalgamated mercury. If the instrument is not regenerated prior to storage after use, the gold film must eventually be replaced. These meters typically require annual factory maintenance and calibration. They are typically accurate to between ±5% and ±15% depending on the manufacturer. Gold-film detectors should be used cautiously after sulfur has been applied to absorb mercury in the spill area, since sulfur irreversibly amalgamates with gold film and may ruin the detector. However, some manufacturers such as the makers of Jerome units utilize sulfur exclusion filters that allow gold-film detectors to operate in the presence of sulfur.

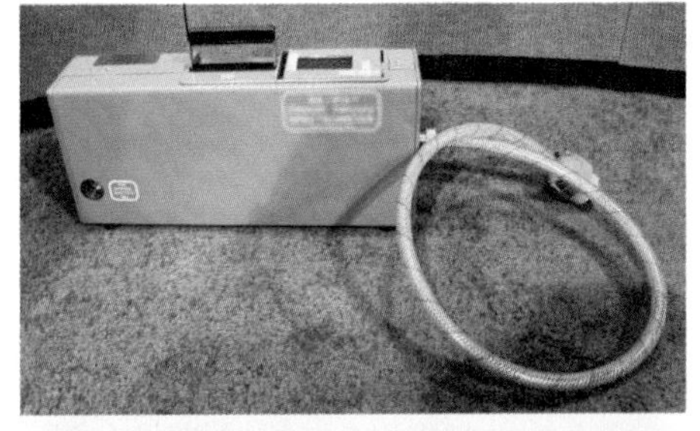

FIGURE 8-15 Mercury detection devices include atomic absorption instruments (top) and gold-film detectors (bottom). *Top photo courtesy of Chris Weber, Dr. Hazmat, Inc. Bottom photo courtesy of Arizona Instruments, LLC.*

Chemical Test Strips

There are a wide range of chemical test strips on the market, including pH paper, fluoride test paper, oxidizer paper, M8 and M9 papers, water detection paper, organic solvent detection paper, hydrogen sulfide detection paper, nitrate and nitrite detection papers, and halogen detection paper, among many others. Manufacturers have created multitest strips such as the Spilfyter chemical strip, the Spilfyter wastewater strip, and the HazMat Smart Strip. Chemical

BOX 8-10 TACTICS: MERCURY SPILL RESPONSE

Courtesy of Chris Weber, Dr. Hazmat, Inc.

Elemental mercury is a surprisingly common metal found in industry and in many consumer commodities such as thermometers, mercury switches, fluorescent bulbs, and medications. When elemental mercury is released into the home or workplace, it can generate vapors that are toxic over weeks and months (chronic exposures). Elemental mercury is poorly absorbed through the skin and digestive tract, but the vapors are readily absorbed in the lungs. It is therefore very important to find and clean up any spilled mercury, especially in an enclosed environment such as a home or office setting. When a visible amount of mercury is released from a thermometer or mercury switch, it appears as small shiny beads. However, these beads can easily be overlooked if they fall into cracks, carpeting, and carpet backing, or if they disperse into very small beads that cannot be seen with the naked eye. For this reason it is important to rely on detection equipment to locate spilled mercury and to verify that the cleanup operation has been successful.

Elemental mercury should not be confused with organic mercury compounds. Organic mercury compounds are extremely toxic and usually readily absorbed through the skin. Responses involving organic mercury compounds should be treated as extremely dangerous, and the guidelines described in this box do not apply.

PREPARATION & PLANNING

The location of an elemental mercury spill is an important consideration. Residential mercury spills must be treated with a great deal of caution owing to the possibility of extended exposure times to residents and the possibility of exposure to young children and pregnant women. The Agency for Toxic Substances and Disease Registry (ATSDR) has taken this difference between residential occupancies and commercial occupancies into account. Residential mercury spills should be cleaned to a mercury level below 1 µg/m^3, while commercial structures should be cleaned to a mercury level of 1–3 µg/m^3. These guidelines require that the mercury source be completely removed.

Safety Considerations

The IDLH for elemental mercury is 10 mg/m^3, the OSHA PEL is 100 µg/m^3, the NIOSH REL is 50 µg/m^3, and the ACGIH TLV is 25 µg/m^3. It is recommended that chemical protective clothing be used for all mercury cleanups—including boots, boot covers, Tyvek or Saranex coveralls, a double layer of nitrile gloves, and respiratory protection. The primary reason for the chemical protective clothing is to limit cross-contamination, not necessarily skin exposure, since dermal absorption is not a primary route of entry. A double layer of gloves allows removal and exchange of the outermost gloves in the hot zone if they become contaminated during cleanup activities thereby minimizing cross contamination.

For mercury vapor concentrations at or above 100 µg/m^3 respiratory protection must be worn. Between 50 µg/m^3 and 10,000 µg/m^3 (10 mg/m^3) an APR may be worn; however, in the rare cases where mercury vapor concentration exceeds 10 mg/m^3, supplied air respiratory protection must be worn. APR cartridges with end of service life indicators (ESLIs), such as Mersorb-P100 or P100 mercury vapor/chlorine cartridges, have been used successfully. Remember, inhalation is the primary route of exposure for elemental mercury.

Response Considerations

It is important to have air-monitoring capabilities when responding to mercury incidents. This is crucial not only to ensure first-responder safety during cleanup operations but also to confirm that the cleanup has been adequately completed and it is safe for occupants to return to the building. If you do not have these capabilities, you should call the local health department, the EPA, and/or a qualified cleanup contractor. Typically, wet decontamination is not needed at mercury spills, since there will only be vapor exposure; dry decontamination is usually sufficient (see Chapter 12).

RESPONSE

Many hazardous materials response teams are starting to clean up residential mercury spills owing to the high cost to homeowners for cleanup contractors. This trend may expose hazmat teams to liability in the future if the cleanup was not properly completed, air-monitoring equipment was not used to verify complete cleanup, or the cleanup procedures and air-monitoring results were not properly documented. An airborne mercury vapor analyzer must always be used during any mercury spill response.

Equipment

Several pieces of equipment are vital for proper elemental mercury cleanup. One of the most important pieces of equipment is an air monitor capable of reading airborne mercury levels in the micrograms per cubic meter range or nanograms per cubic meter range. The most sensitive mercury detectors are atomic absorption meters such as the Lumex. Tools are needed to pick up visible mercury beads. These tools include a flashlight to visualize small mercury beads, a small brush and paper to collect the beads, shaving cream to trap the beads, and a syringe to pull the beads off surfaces. In addition, mercury spill kits are available that contain either sulfur or zinc powders, which form an amalgam with elemental mercury that can be used to remove trace mercury from surfaces. An amalgam is a mixture of different metals.

Cleanup contractors often use mercury vacuum cleaners owing to their efficiency and ease of use. These devices are extremely expensive, and most hazmat teams do not own one. Mercury should absolutely never be vacuumed up using a household vacuum or a shop vacuum, since neither of these traps the mercury, and, furthermore, both are very efficient at volatilizing the mercury and aerosolizing it. Mercury should also not be swept using a broom, since this action tends to disperse larger beads into smaller ones that are harder to clean up.

In cases where hair or pets have been contaminated with mercury, dandruff shampoos that contain selenium sulfide are effective at chelating (or binding) elemental mercury. An example of such a shampoo is Selsun Blue.

Techniques

Depending on the size of the spill and its location, cleanup techniques will vary. According to the EPA, mercury spills should be handled as follows by first responders:

1. Dispatch should have occupants isolate the immediate spill area, turn off building ventilation systems, and exit the area. Occupants should leave all possibly contaminated possessions behind to avoid cross-contaminating other areas. Have the responsible party wait outside to meet first responders.
2. Using proper PPE, locate the visible mercury and remove it. Large mercury beads can be removed by scooping them into cones, using duct tape, or using shaving cream on a paintbrush. Never use a household vacuum cleaner or shop vacuum! Keep in mind that large mercury spills will typically require the services of a competent mercury cleanup contractor. Your local public health department can typically recommend mercury cleanup contractors.
3. Using a mercury vapor analyzer, locate microbeads and mercury that may have fallen into cracks, underneath baseboards, between cushions on furniture, or into and underneath carpeting.
4. Sulfur compounds (fine powder, not pellets) can be used to absorb residual mercury in these areas. Liquid amalgamation can be used on tiling grout by mixing the sulfur with water into a slurry. In either case, the sulfur should be left in place overnight, or at least for several hours. When sulfur complexes with elemental mercury it will turn from yellow to brown. This color change can be used as an indicator for the presence of mercury.
5. Porous items that have been contaminated should be disposed of properly. It is difficult, if not impossible, to remove elemental mercury from porous items. That having been said, mercury can be "cooked off" valuable or sentimental items by placing them in a black trash bag, then placing the trash bag in the sun, thereby heating it. The headspace of the bag should be analyzed using a mercury vapor detector to determine the level of contamination. The bag can be heated and ventilated outdoors to volatilize the mercury. Sometimes, valuable items can be decontaminated to safe levels using this method, but not always. It is important to verify that decontamination has been successful using a mercury vapor analyzer. This process may take days.
6. To remove the last residual mercury from a building, it should be heated and ventilated. The building should be heated to at least 80°F.
7. After stabilization, use a mercury vapor analyzer to determine that the building is safe to reoccupy. The building should never be reoccupied after a mercury spill without analyzing for mercury vapors. Ultimately, the local or state health department should make the final decision regarding reoccupancy of the building.
8. All actions should be documented—including proper waste disposal and a list of items that have been disposed of. It can be helpful to offer the homeowner or building manager a mercury fact sheet. These are available online from the EPA and many state public health and/or environmental agencies.

(continued)

For mercury spills in schools additional precautions should be taken. Owing to the high foot traffic in schools, any spill that is discovered more than 24 hours after the fact indicates that the entire school should be surveyed for mercury using a mercury vapor detector. In addition, children should be surveyed prior to leaving school. Any contaminated items, such as shoes, should remain at school. If it is found that multiple classrooms are contaminated, school closure should be considered. Sometimes, homes will have to be surveyed as well—especially when extensive school contamination is found. The EPA is a great resource to contact for any mercury spill, but especially those involving schools.

test strips are an economical and versatile tool in the hazmat technician's chemical and air-monitoring detection toolbox. Many of these papers can't be used effectively for air monitoring and will be discussed in the next chapter.

M9 Paper M9 paper detects the presence of liquid or aerosol chemical warfare agents. It cannot be used to identify the specific chemical warfare agent. It can, however, be used in all types of weather and temperatures; it sticks to clothing, vehicles, and other equipment; and it conveniently comes on a dispenser role with an adhesive backing. When M9 paper reacts, it turns from an olive green color to a reddish pink color.

M256A Kits

The M256A chemical agent detector kit was designed many years ago for the military and is now being widely used by hazardous materials response teams for the detection of airborne chemical warfare agents, including blood agents, blister agents, and nerve agents (Figure 8-16). These kits are very sensitive and are excellent tools for air monitoring of suspected chemical warfare agents. The directions on the back of the package are quite difficult to read in PPE. It is therefore advantageous to copy and enlarge the instructions on the back of the package or have the directions radioed one step at a time from the cold zone to the entry team.

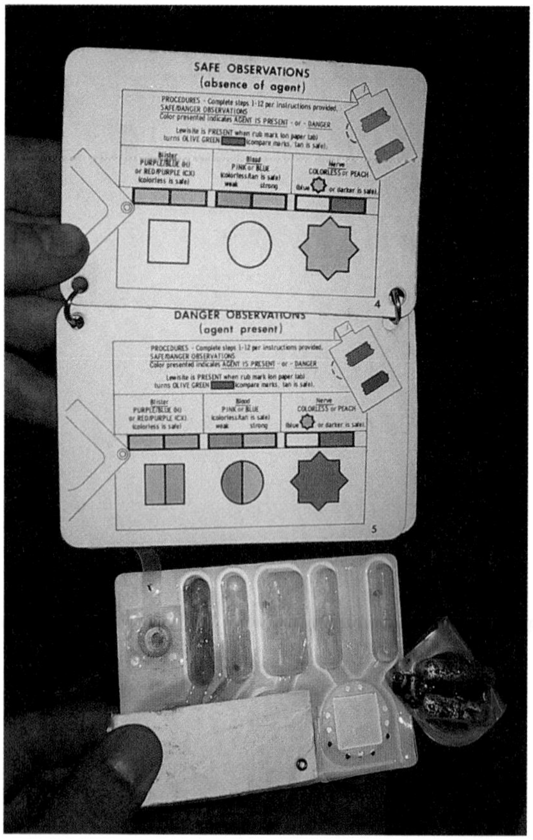

FIGURE 8-16 The M256A1 chemical warfare agent detectors are very effective colorimetric air monitors.
Courtesy of Chris Weber, Dr. Hazmat, Inc.

Flame Spectrophotometry

Flame spectrophotometry, also known as *flame photometric detection* (FPD), uses a hydrogen flame to ionize a gas or vapor sample, which causes the emission of unique wavelengths of light depending on the nuclei present (Figure 8-17). These unique wavelengths are detected and indicate the presence of different atoms such as phosphorus, sulfur, nitrogen, or arsenic that may be unique to chemicals of concern. These instruments can detect vapors, aerosols, and dusts.

The AP2C and AP4C are examples of chemical warfare agent detectors that use FPD technology. The AP2C has two channels, sulfur and phosphorus, that can indicate the difference between nerve agents and blister agents. The AP4C has four channels, sulfur, phosphorus, HNO, and arsenic, and can tell the difference among a wider range of chemical warfare agents—such as sulfur and nitrogen mustards, organophosphates, and arsenicals like lewisite. Flame spectrophotometers can detect GA and GB nerve agents down to 2 ppb, GD and VX nerve agents down to 1 ppb, and HD vesicant down to 60 ppm.

Surface Acoustic Wave Detectors

Surface acoustic wave (SAW) detectors operate on a two-part system. The chemical of concern binds to a target-specific coating on a piezoelectric crystal, which changes the acoustic properties of the crystal. *Piezoelectric* refers to materials that can transform an acoustic, or sound, wave into an electrical signal. The acoustic changes are registered, and the unit sounds an alarm for the indicated chemical.

SAW detectors are only as good as the target-specific coating on the piezoelectric crystal and the number of unique crystals in the detector. In the past, SAW detectors were prone to false positives and failure owing to poor specificity of the coating material. Recent SAW detectors have been vastly improved in both coating specificity and in reversibility of agent binding. Modern SAW detectors also have multiple uniquely coated piezoelectric crystals, which increase the reliability of these detectors by reducing the number of false positives. These detectors are still prone to oversaturation, and many industrial and household chemicals can irreversibly bind to the coating, thereby destroying it. SAW detectors do add a unique technology that can be used as independent confirmation for results obtained by other, more traditional chemical warfare agent detectors.

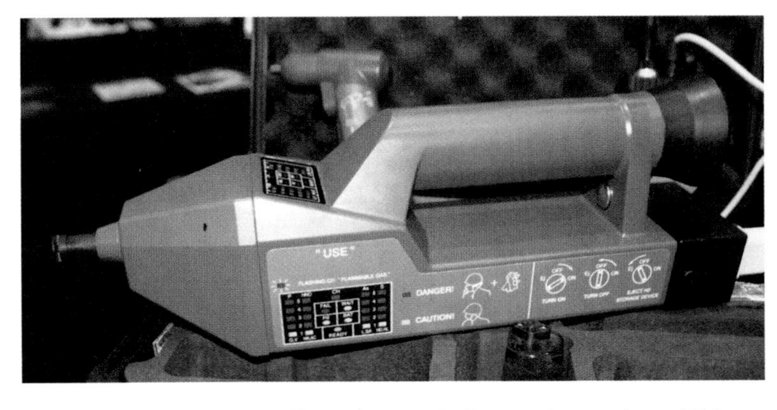

FIGURE 8-17 The AP4C is a flame photometric detector that can detect CHO, nitrogen, sulfur, and arsenic channels. *Courtesy of Chris Weber, Dr. Hazmat, Inc. Used with permission from Proengin, Inc.*

flame spectrophotometry ■ An analytical technique that uses a hydrogen flame to ionize a gas or vapor sample, which causes the emission of unique wavelengths of light depending on the nuclei present. Also known as flame photometric detection.

surface acoustic wave (SAW) detector ■ A detection device that senses chemicals through a change in acoustic properties of a piezoelectric crystal after the contaminant reversibly binds to a selective matrix attached to the crystal.

Selection Criteria for Detection Equipment

Monitoring equipment can be expensive to acquire, requires advanced training, and has continuing annual maintenance costs, which can be substantial. However, monitoring is an essential component of hazardous materials response and must be performed to respond safely and effectively.

The first step is to choose the monitoring equipment you will own and operate. It is important to select the appropriate monitoring equipment to keep the public and first responders safe at hazmat incidents. First and foremost, detection equipment should be chosen based on the chemicals you're likely to find in your jurisdiction. For example, if you do not have a nuclear power plant in your jurisdiction, a neutron detector should probably not be one of your first detector acquisitions. Additional criteria for instrumentation selection include the following:

- portability
- ruggedness
- ease of use
- reliability
- accuracy
- response time
- selectivity
- intrinsic safety

As you acquire detection instrumentation, you should diversify the detection technology in your toolbox. It makes little sense to own three PIDs with essentially the same capabilities and limitations. It is much more prudent to own three different technologies, such as a PID, IMS, and colorimetric tubes. It is also wise to verify the presence of airborne hazards using at least two or three different detection technologies. This will increase the chance of discovering false positives and false negatives early in any hazardous materials incident.

Once the monitoring equipment is acquired, personnel should be trained on its use per the manufacturer's recommendations, and the equipment must be maintained

HAZMAT HANDLE

Whenever air-monitoring instrumentation is used to make a presumptive identification, at least three different detectors should give consistent results to avoid false positives.

according to the manufacturer's recommendations. Standard operating guidelines (SOGs) should be written and made available to ensure all personnel have the ability to safely and effectively operate and interpret the monitoring equipment. Regular hands-on training with the equipment is an essential part of training to competency, which is required by OSHA under the HAZWOPER regulation.

INTRINSIC SAFETY

intrinsic safety ■ A certification process that ensures equipment is safe to operate in a potentially flammable atmosphere.

Operation around potentially flammable or combustible materials requires the use of intrinsically safe equipment, among other safety practices. **Intrinsic safety** is a certification process that ensures equipment is safe to operate in a potentially flammable atmosphere. There are intrinsic safety ratings for different types of flammable and combustible materials as well as for different types of locations. The most common intrinsic safety standard is that from the National Electrical Code (NEC), which itself is based on NFPA standards. Classes and groups define the flammable/combustible material, while divisions define the location in which the equipment will be used.

Classes and groups are defined in the following way:

Class I: combustible gases and vapors (Group A, B, C, and D)
Class II: combustible dusts (Group E, F, and G)
Class III: combustible fibers and/or flyings

Groups are assigned the following chemicals per Article 500 of the NEC:

Group A: acetylene
Group B: 1,3-butadiene; ethylene oxide; formaldehyde; hydrogen; propylene oxide; propyl nitrate; allyl glycidyl ether; and *n*-butyl glycidyl ether
Group C: such chemicals as acetaldehyde; carbon monoxide; diethyl ether; diethylamine; dimethylamine; epichlorohydrin; ethylene; ethyl mercaptan; hydrogen cyanide; hydrogen selenide; hydrogen sulfide; nitropropane; tetrahydrofuran; triethylamine, among others
Group D: such chemicals as acetone; acetonitrile; acrylonitrile; ammonia; benzene; butane; chlorobenzene; methane; methanol; methyl ethyl ketone; naphtha; propane; styrene; and vinyl chloride, among others
Group E: conductive metallic dusts
Group F: semivolatile carbonaceous dusts
Group G: other dusts such as from flour, starch, grain, and chemical thermoplastic compounds

Divisions are defined as follows:

Division I: locations in which combustible atmospheres exist continuously, intermittently, or periodically under normal working conditions
Division II: locations in which combustible atmospheres do not typically exist under normal working conditions

It is important to ensure that any equipment taken into the hot zone where a flammable atmosphere may be present is certified intrinsically safe for the appropriate class, group, and division. Typically, air-monitoring instrumentation is rated as Class I, Division I intrinsically safe, however there are many exceptions to this rule.

Summary

Every piece of monitoring equipment has its advantages and disadvantages. No one technology or manufacturer can do everything. Every piece of equipment will give false positives and false negatives. As the user of air-monitoring and radiation monitoring instrumentation you must learn to recognize the limitations of your equipment in the field. Ensure that you know the minimum detection limits of each piece of equipment you own and operate. It is good practice to confirm instrument readings with at least one other technology—preferably two—to avoid false positives. Each instrument reading is a piece of the puzzle, and the sum of the parts will give you a whole picture of the atmospheric hazards at the hazardous materials incident.

Review Questions

1. What is a false positive?
2. What is a false negative?
3. What is the recommended order of monitoring and why is this order necessary?
4. How do data quality objectives affect monitoring activities?
5. What is a correction factor and why is it necessary?
6. What units does a combustible gas indicator (CGI) display? Why is this significant?
7. How are photoionization detectors (PIDs) useful in the hazmat decision-making process?
8. What information should you know about colorimetric tubes before using them?

References

Centers for Disease Control and Prevention. (2007). *NIOSH Pocket Guide to Chemical Hazards*. Washington, DC: U.S. Government Printing Office.

Draeger. (2011) *Draeger-Tubes & CMS Handbook*, 16th ed. Luebeck, Germany: Draeger Sicherheitstechnik, GmbH.

Gastec (2009) *Environmental Analysis Technology Handbook*, 9th ed. Kanagawa, Japan: Gastec Corporation.

Hawley, Christopher. (2007). *Hazardous Materials Air Monitoring & Detection Devices* (2nd ed.). Florence, KY: Delmar Cengage Learning.

International Safety Equipment Association (2010) *ISEA Statement on Validation of Operation For Direct Reading Portable Gas Monitors*. Washington DC: ISEA.

Maslansky, Carol J., and Steven P. Maslansky. (1993). *Air Monitoring Instrumentation*. New York: John Wiley & Sons.

Rae Systems. (2005). *Application & Technical Notes Guide* (3rd ed.) San Jose, CA: Author.

CHAPTER

9

Sample Collection and Identification

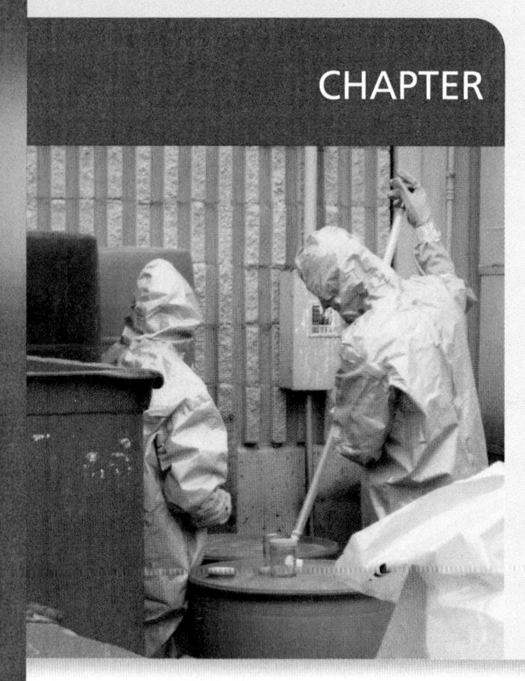

Courtesy of Mike Becker, Longmont (CO) Fire Department.

KEY TERMS

aseptic technique, *p. 290*
chain of custody, *p. 268*
evidence, *p. 268*
field screening, *p. 268*
fluorescence, *p. 278*
fluoride test paper, *p. 275*
Fourier transform infrared (FTIR) spectroscopy, *p. 282*
gas chromatography (GC), *p. 285*
immunoassay, *p. 292*
M8 paper, *p. 275*
mass spectrometry, *p. 286*
oxidizer paper, *p. 270*
polymerase chain reaction (PCR) assay, *p. 293*
protein assay, *p. 292*
Raman spectroscopy, *p. 277*
spectrum, *p. 277*

OBJECTIVES

After reading this chapter, the student should be able to:

- Name the components of a sampling plan.
- Describe the importance of field screening samples.
- Collect chemical samples.
- Collect biological agent samples.
- Perform effective sampling activities at a hazardous materials or WMD release.
- Describe the advantages and disadvantages of FTIR spectroscopy.
- Describe the advantages and disadvantages of Raman spectroscopy.
- Perform effective sample identification operations at a hazardous materials or WMD release.

ResourceCentral For additional review and practice tests, visit **www.bradybooks.com** and click on Resource Central to access book-specific resources for this text! To access Resource Central, follow directions on the Student Access Card provided with this text. If there is no card, go to **www.bradybooks.com** and follow the Resource Central link to Buy Access from there.

Several dozen abandoned drums have just been discovered by the County Road Commission in a remote area. You've been called to the site to assess the situation. Specifically, the incident commander needs to know if any of the drums are leaking, what the hazards of the drum contents are, and if possible, the identity of the materials within the drum. As you survey the site as part of the initial entry team, several questions enter your mind:

- How do I access these abandoned drums safely?
- What type of sampling equipment will I need?
- What type of sampling protocols should I use?
- What type of chemical identification equipment do I have?
- What are the capabilities of the chemical identification equipment that I do have?
- What information do I need to obtain?
- Will knowledge of the hazard class be enough?

Let's see if we can get some answers to these questions and more.

Sample identification of liquids and solids is a very important part of hazmat response. It has many similarities to air-monitoring activities at the technical level, including similarities and overlap in the detection technology. Important differences include the priority of sample identification during a response. While air monitoring should be one of the first activities performed at a hazardous materials release to help protect first responders from dangerous gases and ensure public safety, sample identification is usually performed later in a response to prepare for mitigation activities such as chemical segregation, overpacking, and other remediation activities. Recent technical advances have made sample identification a much faster and accurate process than in the past. In this chapter we discuss effective and safe methods for identifying unknown chemicals at hazardous materials incidents.

Samples must be obtained for a number of different reasons: At hazardous materials releases unknown chemicals must often be identified, necessitating sample collection and sample identification using one or more techniques. At possible WMD incidents chemical samples must be obtained for field identification purposes, laboratory analysis, and evidentiary purposes. Obtaining chemical samples can be dangerous and tricky. Some explosive materials may be sensitive to friction, heat, or pressure. If samples must be obtained from within a container, opening the container may cause a friction-sensitive explosive to detonate, or a pyrophoric material to react with the oxygen or moisture in the air. To minimize the hazards of sample collection and identification all available information must first be gathered.

Sampling Plan

The first step in sample collection and sample identification is to create a sampling plan. The sampling plan should take into account entry team safety—including air monitoring.

evidence ■ Anything that may be linked to a suspect and used to convict the suspect in a court of law.

chain of custody ■ A chronological written record of individuals who have had physical possession of evidence.

It should also take into account **evidence** preservation and sample **chain of custody** when necessary. Key components of the sampling plan should include the following:

- initial air monitoring for hazardous atmospheres
- continuous air monitoring of the sampling area as containers are accessed and samples are collected
- safe access of containers (closed versus open; clear versus opaque)
- field screening of samples for hazards
- biological versus chemical sample collection protocols
- documentation of field screening, sample collection, and chain of custody
- map indicating sample collection locations
- decontamination plan
- evidence preservation
- evidence collection under direct supervision of law enforcement
- chain of custody of samples
- shipping of samples to an analytical laboratory

Many different agencies at the federal and state levels have their own sampling procedures and protocols. For example, the EPA is concerned with waste site sampling procedures and testing methods, and the CDC is concerned with biological agent samples for epidemiological studies during disease outbreaks. Depending on the agencies with which your hazardous materials response team works, you may want to consult their respective sampling plans and guidelines.

DOCUMENTATION AND DATA SHARING OF SAMPLING ACTIVITIES

Documentation is a critical aspect of field screening and sample identification. Good documentation of sample identification results will allow others to use and interpret them to protect first responders and the public. Your documentation also verifies the accuracy of your instrument readings and procedures. Figure 9-1 shows an example of a field screening log sheet that can be used for data recording and documentation.

Your sample identification results will be of interest to other agencies such as law enforcement, public health, EMS, hospitals, and the EPA. It is very important to share your results with the appropriate agencies to maximize the safety of other first responders and the public. These presumptive field identifications will also be of interest to the laboratory performing a definitive analysis. You and your agency may also be interested in comparing the laboratory results with your field identification results. Any discrepancies between your results and the laboratory results may indicate that your standard operating guidelines or training program needs to be improved.

SAMPLE FIELD SCREENING

field screening ■ The process of investigating a sample that will be sent to an analytical lab for hazardous properties such as corrosiveness, radioactivity, explosiveness, flammability, and oxidizing potential.

Prior to sample collection, the material of interest should be field screened for hazards. **Field screening** is performed to ensure that sample handling will not pose a danger to sample collection personnel, to ensure that the tools and containers chosen are compatible with the material, and to comply with shipping regulations and laboratory safety protocols prior to shipping the sample to a forensics or public health laboratory.

Each sample should be field screened for the following hazards before being removed from the site:

1. Corrosive vapor
2. Flammability
3. Oxidizing potential
4. Explosivity
5. Corrosive liquid/solid
6. Radioactivity

FIELD SAMPLE SCREENING LOG SHEET

INCIDENT NAME:	
LOCATION:	
DATE/TIME:	
AGENCY NAME:	

FIELD SCREENING TEAM MEMBERS	

AT MINIMUM SCREEN FOR THE FOLLOWING HAZARDS:

1. EXPLOSIVENESS
2. FLAMMABILITY
3. RADIOACTIVITY
4. WATER REACTIVITY
5. CORROSIVENESS
6. OXIDIZING POTENTIAL

TEST TYPE:								
DETECTOR TYPE:	EXPLOSIVE	WATER REACTIVITY	OXIDIZING POTENTIAL	CORROSIVE	FLAMMABILTY	RADIATION	TOXICITY	TOXICITY
INSTRUMENT NAME:								
SERIAL NUMBER:								
LAST CALIBRATION:								
EXPIRATION DATE:								
FRESH AIR ZERO?								
BACKGROUND READING:								
SAMPLE NAME & ID	RESULT	RESULT	RESULT	RESULT	RESULT	RESULT	RESULT	RESULT

FIGURE 9-1 All samples should be field screened in a systematic fashion before they are sent to a diagnostic laboratory.
Courtesy of Chris Weber, Dr. Hazmat, Inc.

Samples and evidence must be field screened because the U.S. Department of Transportation (DOT) requires that hazardous materials be properly packaged, labeled, and placarded for transport. Therefore, at a minimum, the hazard class must be determined. Second, most analytical laboratories will not accept samples unless they have been field

BOX 9-1 PUBLIC SAFETY SAMPLES OR EVIDENCE?

Samples of suspected hazardous materials must be collected for varying reasons, such as presumptive field identification, definitive laboratory verification, or evidentiary purposes. This raises the question, What is the difference between a public safety sample and an evidentiary sample?

Evidence may be collected only by sworn law enforcement officers and fire investigators, or by other first responders under their direct supervision. Even more importantly, evidence may be collected only after probable cause has been established or a warrant has been issued. Therefore, if a fire department–based hazmat team collects a sample as "evidence" at an emergency incident, that sample may subsequently be thrown out of court owing to improper search and seizure. Because the hazmat team was called to an emergency, probable cause may not exist for a law enforcement operation, which evidence collection is by definition. Non-law-enforcement agencies should therefore always call their sampling activities "public safety sampling" instead of "evidentiary sampling."

Nevertheless, agencies' public safety sampling protocols should adhere to high standards of documentation, certified clean tools and containers, and good sampling protocols that minimize any chance of cross-contamination. Such guidelines not only will give more reliable field test results and lab analysis but will also allow law enforcement to seize these public safety samples into evidence at a later time, after a warrant has been obtained. Thus, in practice, public safety sampling and evidentiary sampling look very similar; the difference is intent. The intent of a public safety sample is to identify a substance to mitigate an emergency, whereas the intent of an evidentiary sample is to solve and prosecute a crime. It is extremely important not to confuse the two!

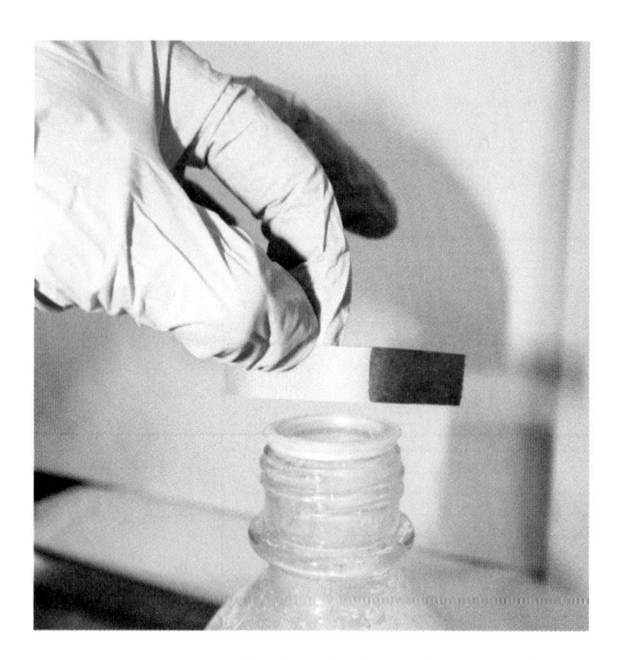

FIGURE 9-2 A piece of half wet/half dry pH paper can be used to screen unknown samples for corrosive vapors. In this case, the wet portion of the pH paper (right half) very readily indicates the ammonia vapors, while the dry portion of the pH paper (left half) only weakly indicates the presence of the caustic ammonia vapors. *Courtesy of Chris Weber, Dr. Hazmat, Inc.*

screened for potential hazards that may harm laboratory personnel. Third, do you really want an uncharacterized piece of evidence sitting in your station or evidence locker for days, months, or years?

Each sample should generally be screened in the indicated order. It is important to determine the pH of the material before exposing sensitive electronic equipment to potentially corrosive atmospheres and/or materials. For air monitoring, the pH paper should be half wet and half dry and should be wafted in the vapor space of the material to determine the presence of volatile corrosive vapors; for determining the pH of liquids, a drop of the sample should be placed on the pH paper (Figure 9-2). Never dip the pH paper into the sample itself! The pH paper could react violently with the sample, or the dyes in the pH paper could leach into the sample, thereby contaminating it.

Flammability hazards should be identified next. Flammability may be determined with a combustible gas indicator (CGI) or by placing a small amount of the material on a watch glass and attempting to ignite it. For safety reasons, this should be done with only a small amount of material (a pea-sized amount or less) and well away from the original sample, its container, or other uncharacterized samples. Destructive testing is less desirable than nondestructive testing (such as air monitoring) and should be avoided whenever possible. If destructive testing is absolutely necessary, use the smallest amount practical and ensure that enough of the sample is left for laboratory analysis and evidence collection.

oxidizer paper ▪ An indicator paper impregnated with potassium iodide (KI) and starch that turns black in the presence of many oxidizers.

Next, the evidence must be screened for oxidizing potential and explosives. **Oxidizer paper** is used to screen for oxidizing potential. The oxidizer paper is used much like pH paper and should be wetted before use with solids. It may also be activated by application of 3N HCl. To use oxidizer paper with liquids, place a drop of the unknown material onto the oxidizer paper. Once again, never dip the oxidizer paper into the sample or container itself to prevent contamination of the evidence. Several manufacturers produce explosives test kits with the capability of identifying nitro-based explosives, peroxide-based explosives, and oxidizers such as perchlorates and nitrates.

Finally, radioactive hazards should be identified, primarily alpha and beta radiation and weak gamma radiation that may have been missed earlier during the primary monitoring activities. Remember, the initial radiation-monitoring activities should already have screened for strong gamma radiation.

Chemical Samples

The majority of the sampling most hazardous materials response teams conduct is of chemicals, consisting of solids, liquids, and gases found at hazardous materials incidents ranging from chemical dump sites, to overturned tanker trucks, to derailments, to releases at fixed-site facilities.

SAMPLE COLLECTION TECHNIQUES

Samples of suspected hazardous materials often must be collected for presumptive field identification, definitive laboratory verification, or evidentiary purposes. How do you collect and process gas, liquid, and solid samples of possible hazardous materials?

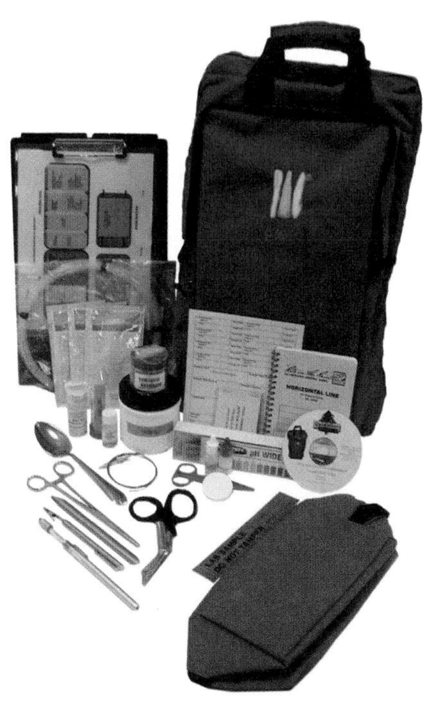

FIGURE 9-3 Sampling kits can be homemade (left) or can be purchased as a kit with certified containers and equipment (right). *Photo on the left courtesy of Chris Weber, Dr. Hazmat, Inc. Photo on the right provided courtesy of the manufacturer, QuickSilver Analytics, Inc. FAC is a registered trademark.*

Sample collection protocols are designed to ensure the consistency and reproducibility of sampling protocols, to ensure that the acquired sample actually contains the material of interest, to minimize the likelihood of cross-contamination of samples, and to prevent external contamination of the samples.

Sampling equipment includes compatible receiving containers; tools for acquiring the samples such as spatulas, pipettes, and swabs; and equipment for preventing cross-contamination such as towels, wipes, and surface covers. Whenever possible, sampling equipment should be certified clean. Sampling equipment must be certified clean for lab testing and evidentiary sampling. It is important to maximize the compatibility of the sampling equipment with the sample, especially the receiving container. Chemicals are generally collected in glass containers, while biological samples are generally collected in plastic containers. However, some exceptions do apply. For example, hydrofluoric acid is incompatible with glass and must be collected in a plastic container. Several manufacturers make and sell sampling kits, many of which contain certified clean containers and equipment (Figure 9-3).

Two primary types of samples are typically collected: the unknown sample and a control sample. Sometimes an equipment sample, or equipment blank, is also collected. The unknown sample contains the material of interest. A control sample is taken from a clean area near the sampling point using the identical sampling method. The control sample is designed to ensure that the sampling technique itself does not introduce contamination. The control sample should be taken after the unknown samples have been collected. An equipment blank is a complete, duplicate unused set of all the equipment used to obtain the unknown sample. The equipment sample is designed to ensure that the equipment being used for sampling has not been contaminated prior to use. Typically one equipment blank is taken per lot of sampling equipment, or every 20 to 30 samples.

The way the sampling points are determined can affect the outcome of the sampling mission. There are several ways to collect samples: a systematic method, a judgmental method, or a random method. The systematic sampling method involves setting up a grid and collecting samples at predetermined intervals along that grid. The advantages of the systematic method are that samples are taken at regular intervals, and it is the most thorough and unbiased sampling method. The disadvantages of the systematic method are that it is time- and labor-intensive and requires a substantial amount of equipment to collect the required number of samples.

The judgmental sampling method involves determining the sampling points based on the observations and expertise of the sampling team. The advantages of the judgmental sample method are that a large number of samples are not required, and there is a higher likelihood of getting the sample of interest with an experienced sampling team. The disadvantage of the judgmental sampling method is that the material of interest may not be obtained if the judgment of the sampling team is faulty.

The random sampling method involves taking samples at random within a predetermined area. The random sampling method is used when resources are limited and there is no immediate indication of which location to sample. It is the least desirable of the three sampling methods. Whichever sampling method you choose, always be prepared to justify your actions!

SAMPLING PROCEDURES: TWO-PERSON TECHNIQUE

The two-person sampling technique is designed to minimize cross-contamination of samples and was specifically designed for evidentiary sample collection (Figure 9-4). However, this technique also ensures that public safety samples are as pristine as possible, which leads to better sample identification results and laboratory test results. It also helps minimize problems with samples in court proceedings if they are seized as evidence by law enforcement at a later time (see Box 9-1).

The two-person sampling technique uses a minimum of two people per team. The assistant handles only clean sampling tools and equipment as well as the double-bagged sample of interest. The sampler collects the sample of interest using the opened tools handed to him or her by the assistant. All sampling tools are disposable and are used for one sample only. The sampler wears a minimum of two layers of gloves, the outer layer of which is always changed between sampling points to eliminate any chance of cross- contaminating the sample.

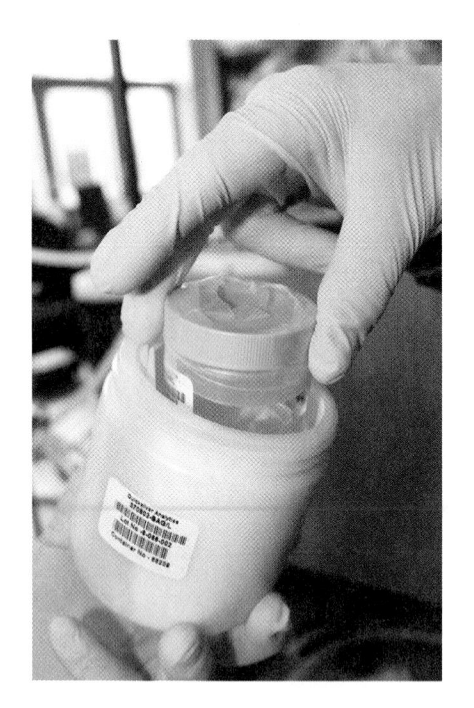

FIGURE 9-4 Good sampling technique is very important to ensure that samples do not become cross- contaminated. *Courtesy of Chris Weber, Dr. Hazmat, Inc.*

Documentation is very important during sampling activities. The names of the assistant and sampler, the time of sampling, the sample location, and the sample tracking number should all be noted for each individual sample. Sometimes public safety samples unexpectedly become evidence. When the sampling techniques and sampling personnel are well documented, the samples will be much more likely to stand up in court.

Gases

Gas samples may include the contents of abandoned cylinders, air samples within buildings, headspace samples of containers, atmospheric samples, gaseous contents of bags, and the like. Sampling gases can be a difficult process. Hazardous materials technicians or environmental contractors may use pumps or vacuum canisters (such as SUMMA canisters).

A neat trick is to use a photoionization detector (PID) as a pump: PIDs typically have a threaded gas outlet, and most manufacturers supply a nipple that can be screwed into the outlet and connected to a gas bag. Recall that PIDs use a nondestructive form of ionization in the detection process, and the sample train is usually made of stainless steel and glass, which minimally absorb chemicals (Figure 9-5). Check with the manufacturer of your PID to determine if these assumptions are valid. This is not the ideal method of gas sampling, but it can be useful as an improvised method.

BOX 9-2 HAZMAT HOW TO: TWO-PERSON SAMPLE COLLECTION TECHNIQUE

1. The assistant or scribe photographs the area before anyone enters the area or disturbs it.
2. The assistant prepares tools.
3. Sampler retrieves sterile tool.
4. Sampler collects sample.
5. Sampler places sample into container and seals it.
6. The assistant overpacks the container.
7. The assistant or scribe documents the sample.
8. Sampler removes outermost gloves and replaces them before moving to next sample.

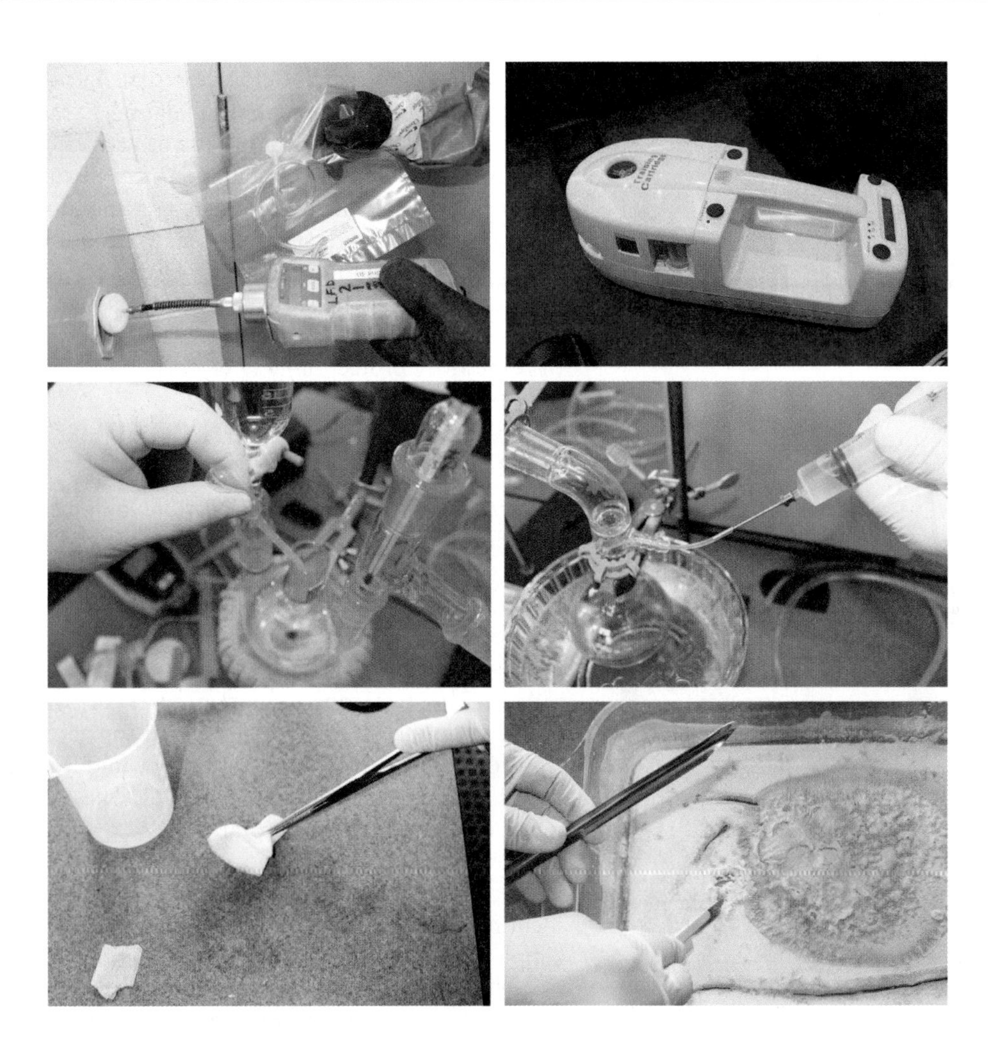

FIGURE 9-5 Various sample collection techniques. The top row illustrates gas and vapor sampling techniques. Photoionization detectors (PIDs) are nondestructive, so a sample hose and Tedlar bag can be connected to the exhaust to effectively collect a sample (top left). Air samplers can also be used to collect dust or particulates (top right). The middle row illustrates liquid sampling techniques using a pipette (middle left) and a syringe with tubing (middle right) to access a difficult-to-reach laboratory process. The bottom row illustrates solid sampling techniques using a wet or dry swab (bottom left) and a knife or spatula (bottom right). *Photos courtesy of Chris Weber, Dr. Hazmat, Inc.*

Liquids

Liquid samples may include contents of abandoned containers, freestanding liquids, surface waters, liquid runoff, sewage, liquids found in clandestine laboratories, pastes, and so forth. Liquids are typically sampled using plastic or glass pipettes (Figure 9-5). A cotton or Dacron swab can be used to sample small amounts of liquid, or liquids in hard-to-reach places (Figure 9-5). Pastes and gels can be scraped up using a spatula or other device. Tubing and a syringe may be used to access difficult-to-reach liquids, such as those inside laboratory glassware setups.

SOLVED EXERCISE 9-1

How could you collect a vapor sample from the headspace of a drum at the incident described in the opening scenario?

Solution: A sampling pump or the outlet of a stand-alone PID could be used to collect the vapor in a suitable container (such as a Tedlar bag).

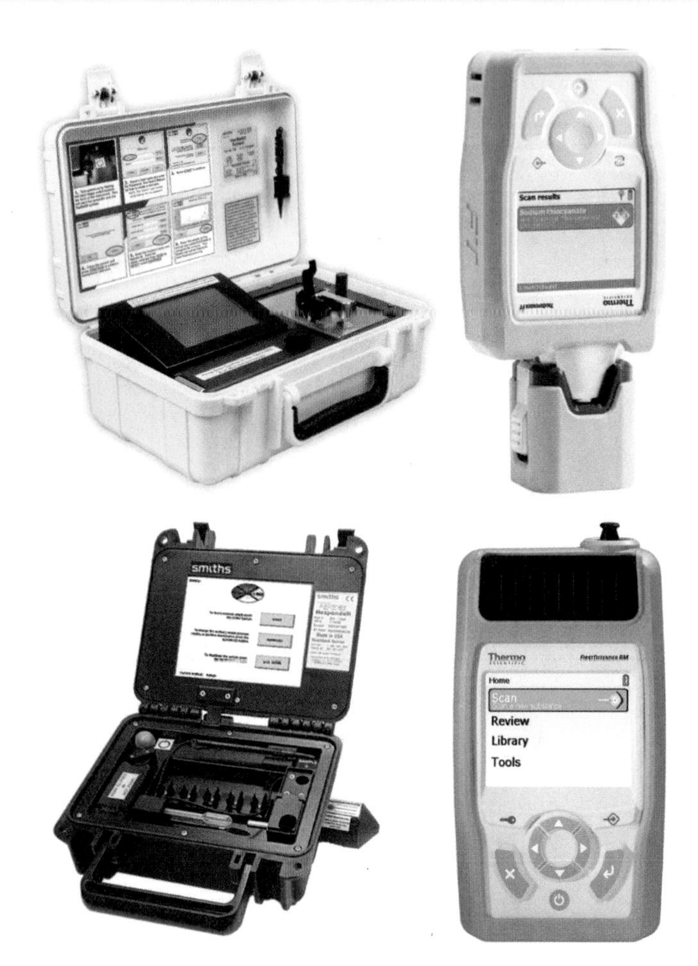

FIGURE 9-6 Infrared and Raman spectrometers can be used to identify liquids and solids. The top row shows two FTIR spectrometers, the HazMatID (top left) and the TruDefender FT (top right). The bottom row shows two Raman spectrometers, the ResponderR RCI (bottom left) and the FirstDefender RM (bottom right). *Top and bottom left photos courtesy of Smiths Detection, Inc. Top and bottom right photos courtesy of Thermo Fisher Scientific.*

Solids

Solid samples may include pellets, powders, dusts, granules, sediments, soils, plants, residues, clothing, and many other solid materials. Solids are typically sampled using a spatula (Figure 9-5). Some solids may be fine powders that aerosolize easily and may be a potential inhalation hazard if proper respiratory protection has not been chosen, or may cause downwind populations to be exposed or property to be damaged. Solids typically do not have a high vapor pressure and usually do not pose a significant respiratory hazard. However, under certain circumstances, such as at incidents involving white powder or fine particulates, the highest level of respiratory protection may be warranted.

SAMPLE IDENTIFICATION OF SOLID AND LIQUID CHEMICALS

Solid and liquid sample identification equipment—including both high-tech instrumentation and low-tech papers—has become much more common in recent years owing to the increased funding provided by the federal government, most significantly the U.S. Department of Homeland Security (DHS). The two most common types of sample identification instruments are based on Fourier transform infrared (FTIR) spectroscopy and Raman spectroscopy technologies (Figure 9-6). These technologies are complementary and are very effective for presumptive identification of unknown liquid and solid materials. However, many other sample identification methods are available. These include simple and cheap devices such as detection papers as well as more complicated chemical identification systems based on chemical reactivity, such as the HazCat Kit.

Detection Papers

Detection papers, or test papers, are a cheap and effective way to identify a wide range of chemicals including acids, bases, oxidizers, fluoride containing materials, heavy metals, and even water. These detection papers are comparatively cheap and are readily available from chemical supply houses and environmental response company suppliers.

pH Paper pH paper, which is sometimes referred to as litmus paper, is capable of detecting corrosive materials in a wide range of sensitivities and over a wide pH range. The

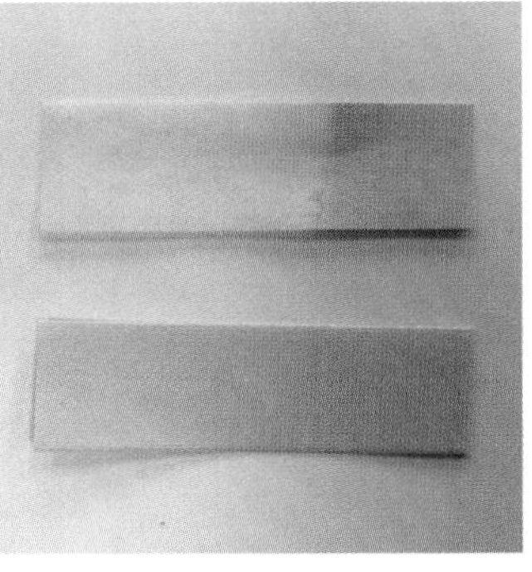

FIGURE 9-7 Detection papers are an economical and reliable source of information for hazardous materials technicians. pH paper turns from yellow to red in the presence of acids, and yellow to blue in the presence of bases (left). Fluoride detection paper turns from pink to yellow in the presence of the fluoride ion (middle). Oxidizer paper—also known as potassium iodide–starch paper—turns from white to dark blue or black in the presence of strong oxidizers (right). Not all oxidizers are detected by oxidizer paper, including ammonium nitrate and many organic peroxides. *Courtesy of Chris Weber, Dr. Hazmat, Inc.*

most common pH paper detects both acids and bases in the pH range 0–13. This is the best broad-range corrosive detection method for most hazmat teams. Wide pH paper is the most practical for use while wearing personal protective equipment.

pH paper contains dyes that change color based on the pH of the solution (Figure 9-7). The hydrogen ion concentration affects the wavelengths of light the indicating dye will absorb; therefore, it changes color on a change in pH. Figure 8-5 shows the pH range of a common brand of pH paper.

Oxidizer Paper Oxidizer paper is very useful for determining whether a material will react with fuels. Oxidizer paper turns black in the presence of most oxidizers (Figure 9-7). There are several common oxidizers that oxidizer paper will not detect, such as ammonium nitrate. Oxidizers will react violently with fuels and release significant amounts of energy in the form of combustion.

Oxidizer paper is impregnated with potassium iodide (KI) and starch. Potassium iodide is a white salt that is oxidized to elemental iodine (I_2) by a strong enough oxidizing agent. Be aware that some oxidizers will bleach the black indicator back to white. Therefore, any color change—no matter how brief—with oxidizer paper should be considered a positive result.

Although oxidizer paper turns black, it does not indicate the strength of the oxidizer. In fact there is a continuum between oxidizing and reducing agents. Another name for reducing agents is fuel. If a strong oxidizer is mixed with a weak oxidizer, the weak oxidizer will act as a reducing agent, and the reaction may produce large amounts of energy. It is therefore important to prevent different oxidizers from coming into contact with each other, even though they are in the same chemical family.

Fluoride Test Paper **Fluoride test paper** is extremely useful for detecting the presence of the highly toxic fluoride ion. When fluoride enters the body it irreversibly binds calcium and magnesium, both of which are essential in the human body. High doses of fluoride will bind calcium in the bones as well as the calcium and magnesium used in the cells. Chronic exposures to lower doses of fluoride can even cause heart attacks. See Box 3-2 for more information about a train derailment where hydrogen fluoride generated from fluorosulfonic acid killed a first responder. The fluoride ion is found in hydrogen fluoride, hydrofluoric acid, and many salts such as sodium fluoride. Fluoride paper turns from pink to yellow on exposure to fluoride ion (Figure 9-7).

fluoride test paper ▪ An indicator paper that turns from pink to yellow on exposure to fluoride ion.

M8 Paper **M8 paper** is a chemical test strip originally designed and used by the military to detect liquid chemical warfare agents on the battlefield. M8 paper is now commercially available and widely used by civilian first responders. It cannot be used as an air-monitoring tool, since it works on the basis of solubility: The M8 paper is impregnated with three dyes: yellow, red, and blue. Different chemical warfare agents, and many other chemicals that induce false positives, dissolve one or more of these three dyes (Figure 9-8).

M8 paper ▪ A chemical indicator paper originally designed and used by the military to detect liquid chemical warfare agents including nerve agents and blister agents on the battlefield.

FIGURE 9-8 M8 paper was developed by the military to detect chemical warfare agents on the battlefield. It is based on the solubility of three different dyes—yellow, blue, and red. On the battlefield, a yellow dye indicates G-type nerve agents, a red color indicates blister agents, and a green color indicates the presence of V-type nerve agents (bottom right corner). However, many common household and industrial chemicals also dissolve one or more of the dyes and therefore produce false positives. Chlorinated hydrocarbons, ketones, furans, esters, and organic peroxides turn red. Amines and brake fluid turn green. Many organic solvents wet the M8 paper and give a yellow appearance. One chemical, arsenic trichloride, turns M8 paper blue. *Courtesy of Chris Weber, Dr. Hazmat, Inc.*

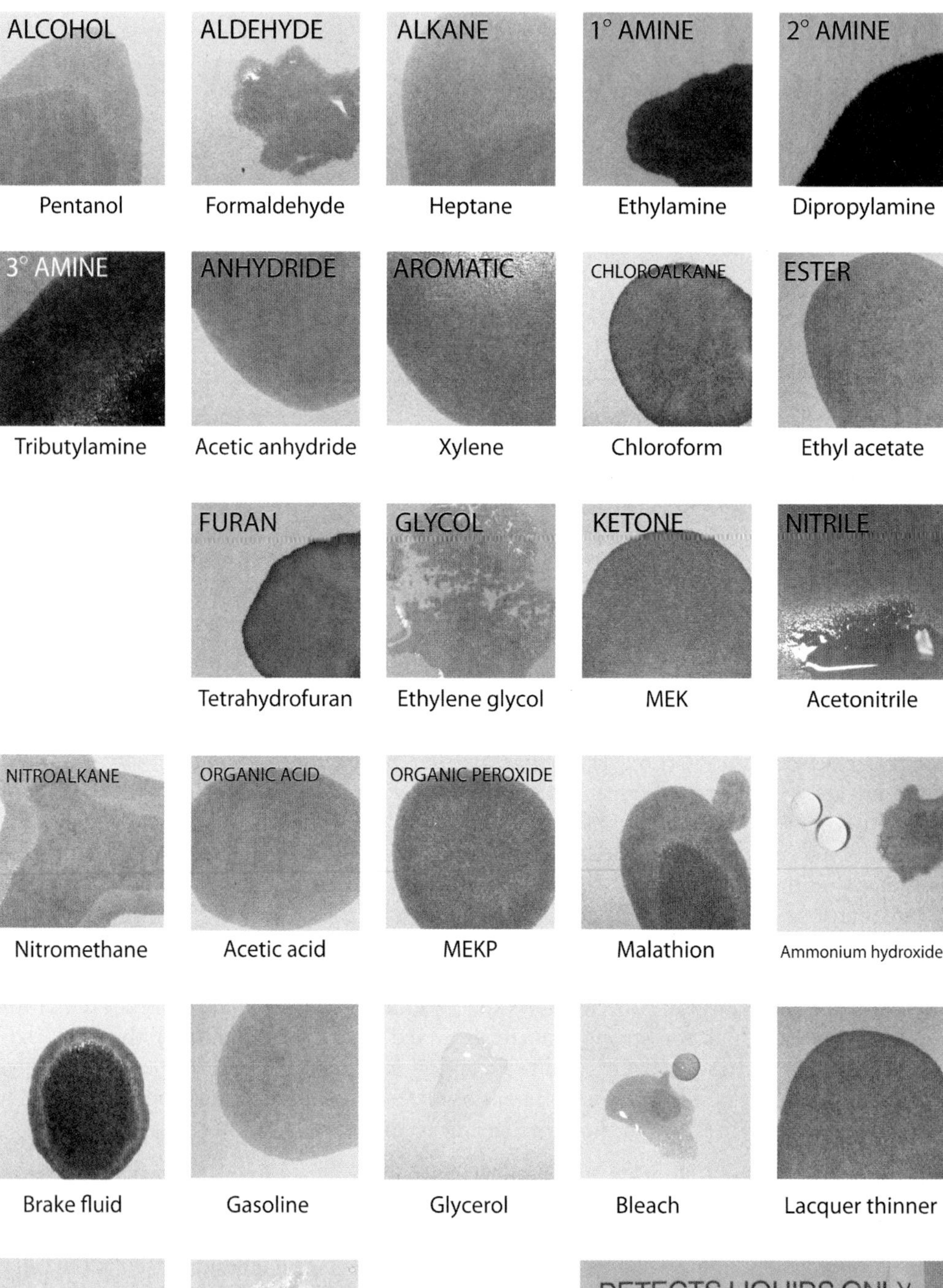

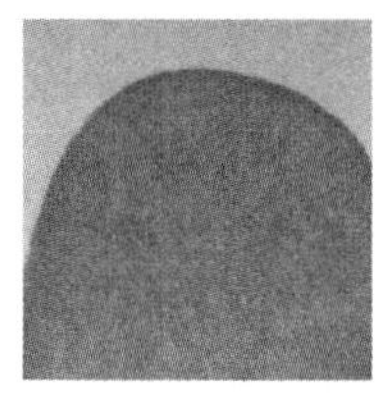

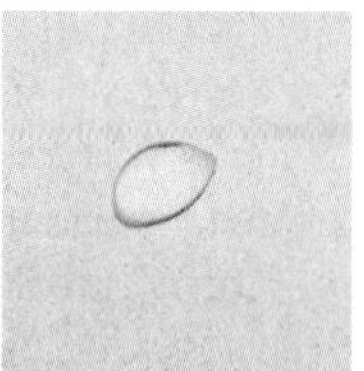

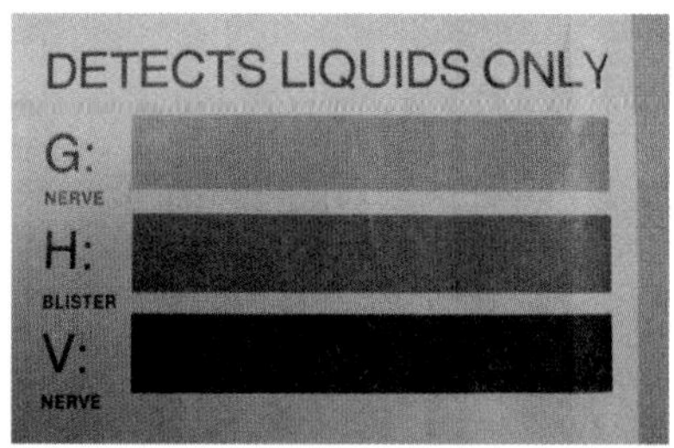

G-series nerve agent is indicated by a yellow color, V-series nerve agents are indicated by a dark green color, and blister agents are indicated by a pink/red color. Arsenic trichloride, a precursor to lewisite and other arsenicals, is indicated by a deep blue color. The green color generated by V-series nerve agents results from dissolution of both the yellow and the blue dyes in the M8 paper. Many industrial and household chemicals generate false positives with M8 paper. M8 paper was initially designed for battlefield use only, where cross-indicating chemicals are much more unlikely to be encountered.

The routine use of M8 paper can be beneficial to hazmat technicians as an additional confirmatory tool for the identification of unknown liquids using a unique technology—solubility. As Figure 9-8 indicates, certain classes of hydrocarbon derivatives dissolve particular dyes. For example, amines tend to dissolve the yellow and blue dyes while ketones and aromatic hydrocarbons tend to dissolve the red dye. This can help confirm or rule out the presence of certain chemicals at hazmat releases. Organic solvents (nonpolar liquids) tend to wet the M8 paper while aqueous solutions (polar liquids) tend to bead up on the surface of the M8 paper (as seen with hydrogen peroxide in the lower left hand corner of Figure 9-8).

FIGURE 9-9 Spilfyter chemical classifier strips contain pH paper, oxidizer paper, fluoride paper, halogen detection paper, and petroleum product detection paper. These are a convenient rapid testing product. *Courtesy of Chris Weber, Dr. Hazmat, Inc. © 2012 NPS Corporation; Spilfyter is a trademark of NPS Corporation.*

Spilfyter Strips Spilfyter strips are multiple test papers in one product and are therefore an excellent initial screening tool (Figure 9-9). There are two types of Spilfyter strips: chemical classifier and wastewater classifier. The chemical classifier strips are commonly used by hazardous materials response teams for broad-spectrum unknown liquid classification. The chemical classifier strips contain a swatch of pH paper, oxidizer paper, fluoride paper, hydrocarbon paper, and halogen paper—with the ability to identify iodide, bromine, and chlorine. The limits of sensitivity for the chemical classifier strip are as follows:

Test #1: pH paper – 0–13 pH units
Test #2: oxidizer paper – 1 mg/L (hydrogen peroxide)
Test #3: fluoride – 20 mg/L
Test #4: organic solvents and hydrocarbons – 10mg/L (gasoline)
Test #5: iodine, bromine, and chlorine – 1 mg/L (chlorine)

The wastewater classifier can be helpful under certain circumstances. The wastewater strips contain pH paper, hydrocarbon paper, sulfide paper, nitrite paper, nitrate paper, and fluoride paper. The limits of sensitivity for the wastewater classifier strip are as follows:

Test #1: pH paper – 0–13 pH units
Test #2: organic solvents and hydrocarbons – 10 ppm (gasoline)
Test #3: hydrogen sulfide – 10 ppm
Test #4: nitrite – 1 ppm
Test #5: nitrate – 10 ppm
Test #6: fluoride – 20 ppm

These are very convenient and comparatively cheap test kits for hazmat technicians, although each test strip costs upward of $10.

Raman Spectroscopy

Raman spectroscopy uses a near-infrared laser to excite covalent bonds within an unknown material to produce a pattern of scattered light. The scattered light is measured and generates a **spectrum** that can be compared with a known database of chemical spectra, similarly to the way a fingerprint is compared with a database of known fingerprints.

Raman spectroscopy ▪ A method of sample identification that uses near-infrared single-wavelength scattering and can identify most covalently bonded compounds when they are present in sufficient concentration.

spectrum ▪ A visual representation of the intensity of electromagnetic energy over a range of frequencies or wavelengths.

SOLVED EXERCISE 9-2

You place a drop of an unknown liquid on a piece of pH paper and it turns dark blue for less than a second and then rapidly turns white. What type of material is the unknown solution?

Solution: The material is most likely a corrosive and an oxidizer—an oxidizing base. The pH paper briefly turned blue, indicating that the material has caustic properties and is therefore corrosive. Since the material turned the paper white, it bleached the pH-indicating dyes in the paper. This could indicate the presence of an oxidizer such as bleach—also known as hypochlorite solution, but there are several other alkaline oxidizers that behave this way.

BOX 9-3 HAZMAT HISTORY: ESCONDIDO, CALIFORNIA, BOMB FACTORY (2010)

On November 18, 2010, an explosion occurred when a gardener stepped on a gravel pathway in the backyard of a residence in Escondido, California. The Escondido Fire Department and San Diego Sheriff's Office Bomb Squad responded to the 911 call.

Multiple bomb-making materials were found in the residence, including 50-pound bags of hexamine, sulfuric acid, nitric acid, hydrogen peroxide, and hydrochloric acid. Six mason jars of the peroxide-based explosive HMTD were identified with Raman spectroscopy using the scan delay function of the FirstDefender RMX. The mason jars of HMTD were detonated in-place using bomb robots.

Ultimately the house was burned down to destroy the explosives and minimize the dangers of possible improvised explosive devices (IEDs) planted by the resident. During the burn on December 9th, a major interstate was closed, 40 air monitoring devices were used for perimeter air monitoring, weather conditions were closely monitored, and the fire temperature was kept intentionally hot to ensure thermal destruction of all potential explosive residues. In fact, thermal strips that are typically used in restaurants to ensure sufficient heat in dishwashers for disinfection were placed in the backyard to ensure that any explosive residue was completely destroyed (which is 167°F or 75°C for HMTD).

This incident illustrated the importance of good sampling procedures and the value of sample identification equipment. In addition, more than 60 local, state, and federal agencies were needed for a safe and effective response to this WMD incident. In Chapters 13 and 14 we will discuss incident command considerations and response considerations, respectively.

Raman spectroscopy can be used to identify a material through translucent containers. Most glass containers, plastic containers, and thinner envelopes are translucent enough to make field identification possible without even sampling the contents. The color of the material does not play a significant role—clear, brown, green, blue, and most other colors normally work just fine. However, if the container is not translucent, then the contents must be sampled.

Raman spectroscopy has advantages and limitations, just like any other technique. It can be used to identify only materials that contain covalent bonds. Fortunately, most chemicals do contain covalent bonds and can be readily identified using Raman spectroscopy. However, materials that contain only ionic bonds, such as table salt or sodium chloride, cannot be identified with this tool. A big advantage of Raman spectroscopy is that it cannot identify water. Water is completely Raman silent. Raman spectrometers are therefore more sensitive (about 10-fold more sensitive) to hazardous materials dissolved in water than other liquid and solid sample identification technologies such as Fourier transform infrared spectroscopy. The minimal detection limit of Raman spectroscopy is approximately 1% in water.

fluorescence ▪ The intense production of a broad range of electromagnetic radiation by a substance in response to the input of energy. Infrared fluorescence sometimes interferes with Raman spectroscopy.

Fluorescence can interfere with Raman spectrometers. Fluorescence occurs when a material emits a broad range of electromagnetic radiation, often in the infrared region, on laser excitation (Figure 9-10). Fluorescence doesn't give any specific information about the material; it just interferes with the Raman signal needed for identification. Infrared

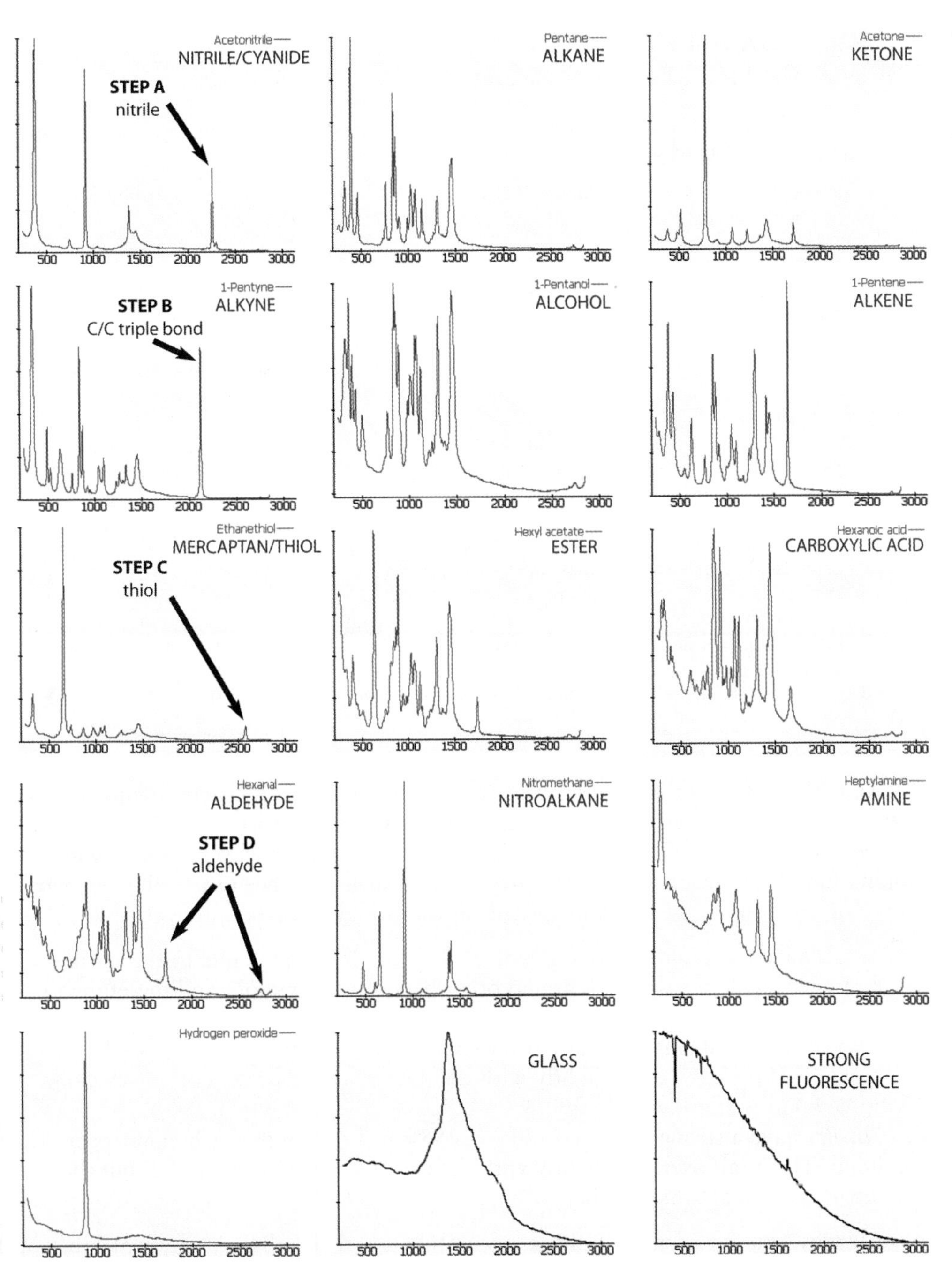

FIGURE 9-10 Raman spectra contain peaks at specific locations based on the molecular structure of the sample. When spectrometers are unable to identify the material, it can be helpful to look at these unique peaks while waiting for the manufacturer's reachback support.
Raman spectra courtesy of Thermo Fisher Scientific.

fluorescence interferes with a Raman spectrometer in the same way that an oncoming vehicle with its high beams on interferes with your vision while driving.

Raman Spectral Signatures All the major Raman spectrometer manufacturers offer some sort of reachback capabilities. *Reachback* refers to technical support when a spectrum cannot be automatically identified using the onboard software. By calling the technical support number you can reach a professional spectroscopist who will attempt to interpret the spectrum and identify the chemical. This process can take time though, commonly up to 1 to 2 hours. It can therefore be helpful to have a rudimentary understanding of the location of key peaks in the spectrum to bridge the time until reachback support is received. Always initiate the reachback support process before attempting to interpret spectra.

BOX 9-4 HAZMAT HOW TO: RAMAN SPECTROSCOPY

The following procedure can be used to identify an unknown solid or liquid material using Raman spectroscopy in point-and-shoot and vial modes:

1. Inspect the instrument. Look for visible damage to the case, make sure the laser pathway is not heavily contaminated or obstructed, and ensure that batteries are inserted.
2. Turn the instrument on.
   SAFETY WARNING: Most Raman spectrometers use a powerful laser, which can be an eye hazard at close quarters, commonly within 20 inches from the laser aperture. Protect your eyes!
   SAFETY WARNING: Never scan dark materials or suspected explosives in their original containers, using a larger than pea-sized amount, or in capped sample vials! An explosion causing serious injury or death may otherwise occur. If scanning is absolutely necessary, take a tiny sample (the size of a grain of sand) and scan it on a noncombustible surface away from the rest of the suspect material to avoid sympathetic detonation.
3. In point-and-shoot mode: Position the sample at the focal point of the laser. It is critical that the sample to be identified is at the focal point of the laser to ensure accurate and effective identification! This is the most common reason for sample identification failure. Also, be aware of the eye hazard that may be posed by the external laser.
4. In vial mode: Ensure that the vial is at least one-quarter full. (With the exception of dark and/or potentially energetic materials noted previously.)
5. The instrument will automatically perform a library search. Matches with a 0.95 confidence factor or above are considered reliable by most manufacturers; however, the spectrum of the unknown sample should always be compared with that of the library entry to make sure that the peaks match well.
6. If there are any doubts, access the manufacturer reachback service.

Raman spectra are a bit more difficult to interpret than FTIR spectra owing to greater overlap of key spectral regions. Generally, Raman spectra consist of many narrow peaks between 250 cm^{-1} and 2000 cm^{-1} (reciprocal centimeters, or wavenumbers) with overlapping functional group regions (Figure 9-10). That having been said, alkynes, nitriles, and thiols can effectively be distinguished from other hydrocarbon derivatives:

- *Thiols* have a medium to strong peak at 2550 to 2600 cm^{-1} due to the sulfur–hydrogen single bond. There are no other peaks in this region, and therefore a peak at this position is a good indication of a thiol.
- *Alkynes* have a strong peak at 2100 to 2250 cm^{-1} due to the carbon–carbon triple bond. This peak overlaps slightly with the nitrile bond, but the carbon–carbon triple bond is at a lower energy.
- *Nitriles* have a strong peak at 2220 to 2255 cm^{-1} due to the carbon–nitrogen triple bond. This peak overlaps slightly with the carbon–carbon triple bond, but the nitrile bond is at a higher energy.

Alkanes can be identified more accurately by their absence of peaks between 1500 cm^{-1} and 2800 cm^{-1}, but this is not a definitive measure. Alkenes and dienes have peaks in the alkane region and an additional peak between 1630 and 1660 cm^{-1}. This region has other overlapping peaks including the carbonyl group and carboxylic acids, making it difficult to interpret.

Alcohols, amines, ethers, nitroalkanes, organic peroxides, and halogenated hydrocarbons do not scatter in a unique region and overlap the alkanes completely. Therefore these functional groups cannot readily be distinguished from one another by the hazmat technician using Raman spectra without advanced spectroscopy training.

It is difficult to distinguish aldehydes, ketones, carboxylic acids, and esters from one another, since they all have the characteristic carbonyl peaks between 1680 cm^{-1} and 1820 cm^{-1}. Table 9-1 summarizes the characteristic spectral signatures for selected covalent bonds in both Raman and FTIR spectra.

TABLE 9-1 Spectral Signatures of Selected Covalent Bonds

FUNCTIONAL GROUP	EXAMPLE(S)	RAMAN PEAKS	FTIR PEAKS
Alkane	Pentane	250–500 cm^{-1} 600–1300 cm^{-1} 1380 cm^{-1} 1400–1500 cm^{-1}	Strong: 3000-2800 cm^{-1} 1500–1350 cm^{-1} Weak: 1300–600 cm^{-1}
Alkene	Pentene	Alkane peaks plus 1630–1660 cm^{-1}	Weak: 3080-–3020 cm^{-1} 1680–1640 cm^{-1} 1000–675 cm^{-1}
Alkyne	Pentyne	Alkane peaks plus 2100–2250 cm^{-1}	3333–3267 cm^{-1} Weak: 2260–2100 cm^{-1} Broad: 700–610 cm^{-1}
Aromatic	Benzene	Alkane peak region Two medium peaks at roughly 1580 cm^{-1} and 1620 cm^{-1}	3100–3000 cm^{-1} 1600 cm^{-1} 1500 cm^{-1}
Thiol	Ethanethiol	Alkane peaks plus 2550–2600 cm^{-1}	Weak: 2550–2600 cm^{-1} *Aliphatic*: 630–790 cm^{-1} *Aromatic*: 1080–1100 cm^{-1}
Nitrile	Valeronitrile	Alkane peaks plus 2220– 2255 cm^{-1}	2260–2220 cm^{-1}
Alcohol	Methanol	Alkane peak region only	Broad: 3600–3100 cm^{-1} Weak: 1300–1000 cm^{-1}
Water	Water	No peaks	Broad: 3100-–3650 cm^{-1} 1640 cm^{-1}
Amine	Ethylamine	Alkane peak region only Possibly a weak peak at 1600 cm^{-1}	3500–3300 cm^{-1} 1650–1580 cm^{-1} 1340–1020 cm^{-1}
Aldehyde	Hexanal	Alkane peaks plus 1700–1750 cm^{-1}	Sharp: 1760-–1670 cm^{-1} Weak: 2840–2720 cm^{-1}
Organic acid	Acetic acid	Alkane peaks plus wide peak or two peaks in 1650–1780 cm^{-1} region	Broad: 3000–2500 cm^{-1} Sharp: 1760–1670 cm^{-1} 1300–1000 cm^{-1}
Ketone	Acetone	Alkane peaks plus 1670–1730 cm^{-1}	Sharp: 1760–1670 cm^{-1}
Ester	Amyl acetate	Alkane peaks plus weak 1680–1820 cm^{-1}	Sharp: 1760–1670 cm^{-1} 1300–1000 cm^{-1}
Nitroalkane	Nitromethane	Complete overlap with alkane peak region (1340–1380 cm^{-1} 1530–1590 cm^{-1})	1660–1500 cm^{-1} 1390–1260 cm^{-1}
Ether	Diethyl ether	Complete overlap with alkane peak region (800–970 cm^{-1})	1300–1000 cm^{-1} Weak: 800–970 cm^{-1}
Peroxide	*t*-Butyl peroxide Hydrogen peroxide	Complete overlap with alkane peak region (840–900 cm^{-1})	Weak: 840–900 cm^{-1}
Chloroalkane	Chloropropane	Complete overlap with alkane peak region (550–800 cm^{-1})	550–800 cm^{-1}
Bromoalkane	Bromopropane	Complete overlap with alkane peak region (500–700 cm^{-1})	500–700 cm^{-1}
Iodoalkane	Iodopropane	Complete overlap with alkane peak region (480–660 cm^{-1})	480–660 cm^{-1}

SOLVED EXERCISE 9-3

According to the Raman spectrum shown, what material(s) may be present?

Solution: The peaks below 1500 cm^{-1} indicate that this material has hydrocarbon characteristics. The strong peak at approximately 2250 cm^{-1} indicates the presence of a carbon–carbon triple bond or a carbon–nitrogen triple bond, strongly suggesting this is either an alkyne or a nitrile. Both of these hydrocarbon derivatives are toxic, flammable, and reactive. The Raman spectrometer identified this material as acetonitrile using the onboard library. The black spectrum is the unknown material, and the red spectrum is the library entry for the match found.

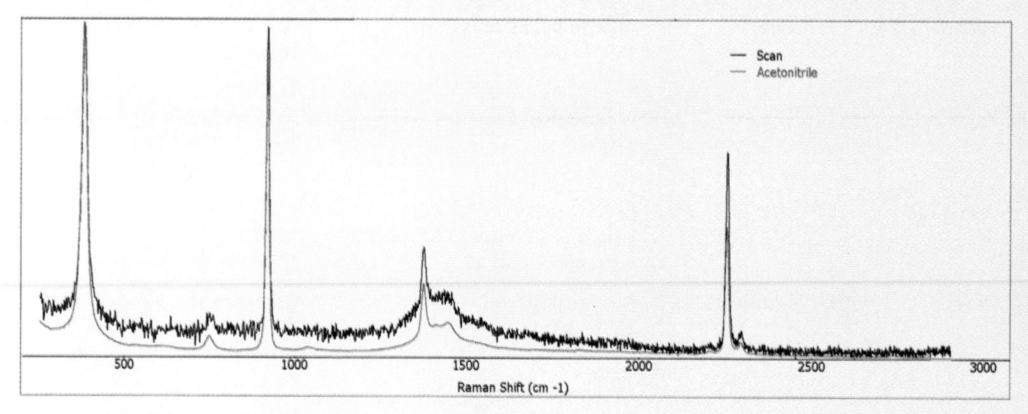

Image courtesy of Thermo Fisher Scientific.

Fourier Transform Infrared Spectroscopy

Fourier transform infrared (FTIR) spectroscopy ■ A method of sample identification that uses infrared absorption and can identify most covalently bonded compounds when found in a sufficient concentration.

Fourier transform infrared (FTIR) spectroscopy is a spectroscopic technique capable of identifying solids, liquids, and gases depending on instrument configuration. FTIR identifies chemicals by exposing them to a broadband infrared source and measuring which wavelengths of infrared light are absorbed by the covalent bonds in the sample. The absorbed wavelengths are measured and generate a spectrum that can be compared with a database of known chemical spectra, similarly to the way a fingerprint is compared with a database of known fingerprints. In this way, FTIR spectroscopy is conceptually similar to Raman spectroscopy.

Unlike Raman spectroscopy, FTIR spectroscopy requires direct contact with the unknown material being identified, because it is an absorbance-based technique. Field-ready FTIR equipment uses the ATR illumination mode. ATR stands for attenuated total reflectance. Like Raman spectroscopy, FTIR spectroscopy can be used to identify materials that have covalent bonds. FTIR spectroscopy cannot identify chemicals that contain only ionic bonds.

FTIR and Raman spectroscopy differ in the types of covalent bonds that each technique is able to identify. Polar covalent bonds absorb infrared energy particularly well, while nonpolar covalent bonds scatter infrared energy particularly well. Thus, FTIR spectrometers see polar covalent bonds very well, while Raman spectrometers see nonpolar covalent bonds particularly well, which is why water is Raman silent, even though it contains covalent bonds. The covalent bonds in water are extremely polar. Nevertheless, both types of spectrometers see the vast majority of chemicals very well. The identification capabilities of Raman and FTIR spectrometers are summarized in Table 9-2.

FTIR Spectral Signatures Although the spectral databases included in FTIR instruments are progressively becoming more complete, you may encounter a chemical or mixture of chemicals that is not in the database. Spectral signal interpretation should not be attempted by everyone, and the manufacturer's reachback process should be initiated first in any case. Nevertheless, having a little bit of knowledge about where key functional group peaks can be found can be very helpful when confronted with such a case.

TABLE 9-2	Identification Capabilities and Limitations of Raman and FTIR Spectrometers
RAMAN SPECTROSCOPY CAPABILITIES	**FTIR SPECTROSCOPY CAPABILITIES**
Compounds containing covalent bonds can be identified (most chemicals)	Compounds containing covalent bonds can be identified (most chemicals)
Minimum detection threshold in water is approximately 1%	Minimum detection threshold in water is approximately 10%
Hazardous material does not contact Raman spectrometer (less decon needed)	Water is readily detected by FTIR
Unknown material can be identified through clear containers and standard-thickness envelopes	
Some elemental substances, such as sulfur and red phosphorus, can be detected	
LIMITATIONS	**LIMITATIONS**
Purely ionic compounds are not seen (NaCl, HF, etc.)	Purely ionic compounds are not seen (NaCl, HF, etc.)
Dark-colored materials may ignite or explode	Unknown material must be sampled and must have direct contact with instrument
Highly fluorescent materials are difficult or impossible to identify	Water has a strong FTIR signal and makes it more difficult to detect material of interest
Water is not detected	Cannot detect elemental sulfur or phosphorus (such as red phosphorus)

BOX 9-5 HAZMAT HOW TO: FTIR SPECTROSCOPY

The following procedure can be used to identify an unknown solid or liquid material using FTIR spectroscopy:

1. Inspect the instrument. Look for visible damage to the case, make sure the diamond surface is not contaminated, and ensure that batteries are inserted.
2. Turn the instrument on.
3. Clean the diamond surface with an isopropanol wipe. Ensure that the diamond is dry and no residual water or isopropanol remains.
4. Collect a background reading.
5. Place a small amount of the sample on the diamond. Solid materials need to be compressed onto the diamond surface. Most instrument manufacturers include a tool to facilitate this procedure. Liquids need only a thin coating on the diamond surface. Identification occurs within 5 μm of the diamond surface, which is why solid materials must be pressed against the diamond with moderate force.
6. Most FTIR spectrometers will automatically subtract the background reading from the spectrum of the unknown sample and perform the library search. Matches with a 0.95 confidence factor or above are considered reliable by most manufacturers; however, the spectrum of the unknown should always be compared with that of the library entry to make sure that the peaks match well.
7. If there are any doubts, access the manufacturer reachback service.

Let's take a look at unique spectral features as well as spectral features of functional groups that are particularly dangerous to hazardous materials technicians, namely, nitro groups, nitrile groups, alcohols, and ethers (Figure 9-11). The following algorithm—which starts with higher hazard chemicals and progresses to lower hazard chemicals—can be used to interpret FTIR spectra:

STEP 1: The first functional group to consider is the nitro group owing to the high risk of explosion. Fortunately, the nitro group has a very distinct signature–typically

FIGURE 9-11 FTIR spectra contain peaks at specific locations based on the molecular structure of the sample. When spectrometers are unable to identify the material, it can be helpful to look at these unique peaks while waiting for the manufacturer's reachback support. *FTIR spectra courtesy of Thermo Fisher Scientific.*

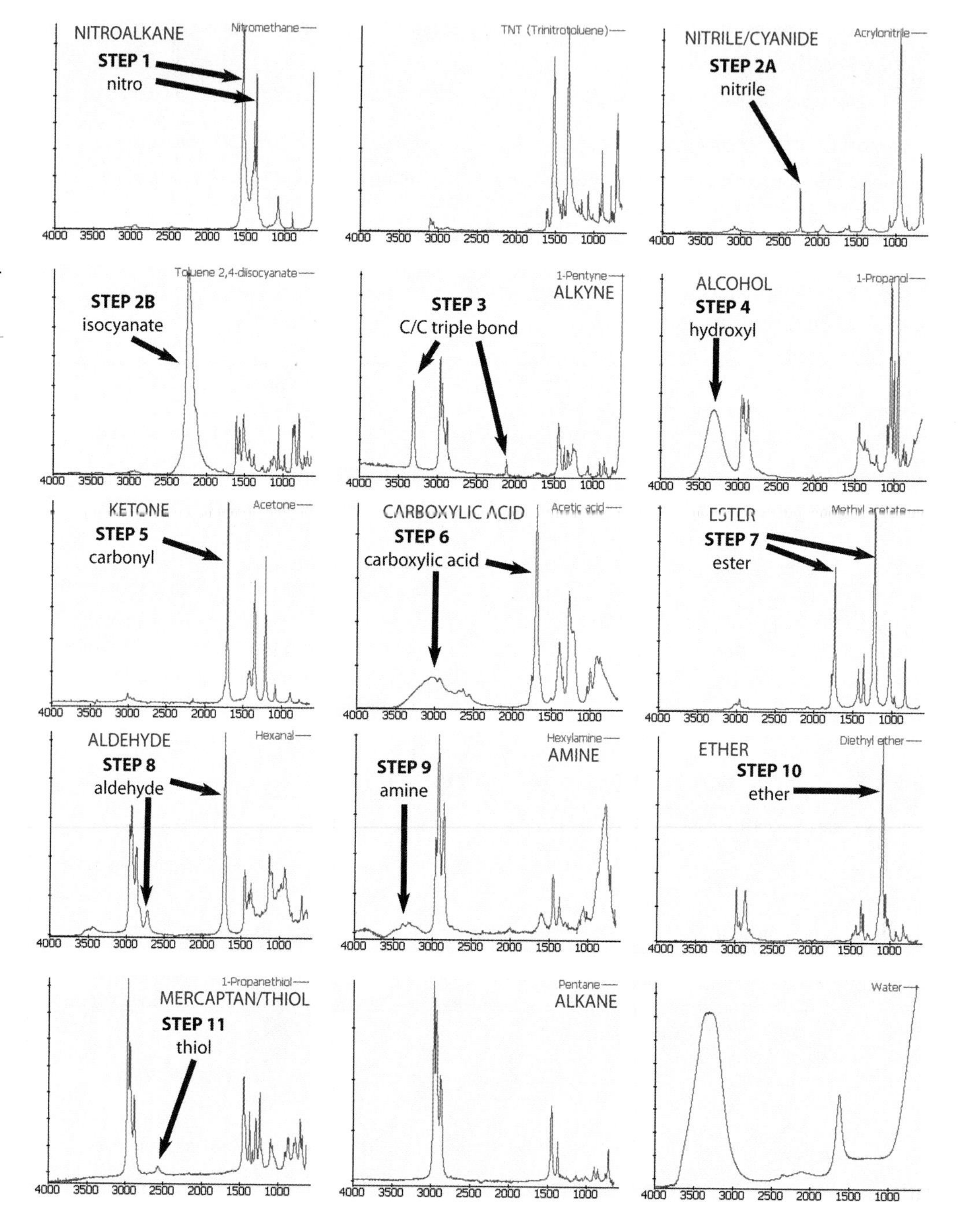

two strong and narrow peaks to the left and right of 1500 cm^{-1}, which are often called the twin towers. If you see these peaks, the chemical may contain one or more nitro groups. The larger these peaks are, the more dominant the nitro group is in the chemical and the more dangerous the material is. This may be a small chemical, such as nitromethane, with a single nitro group, or it may be a larger chemical with several nitro groups, such as trinitrotoluene (TNT).

STEP 2: The next functional group to consider is the nitrile group owing to its toxicity. Nitriles (such as acetonitrile) have a single narrow peak at approximately 2250 cm^{-1}, whereas isocyanates have a wide peak in this region (such as toluene diisocyanate).

STEP 3: Carbon-carbon triple bonds weakly absorb in the same region as the nitriles, but also contain a strong peak near 3300 cm-1.

STEP 4: The next functional group to consider is the hydroxyl group. Alcohols have a large, broad peak ranging from 3200 to 3600 cm^{-1}.

STEP 5: The next functional group to consider is the carbonyl group, which is a strong, medium-sharp peak centered near 1750 cm^{-1} with a range of ±100 cm^{-1} depending on the hydrocarbon derivative. If no other characteristic peaks are present, you may be dealing with a ketone, such as acetone.

STEP 6: If in addition to the carbonyl group there is a large, broad peak ranging from 2500 to 3600 cm^{-1}, which is due to a hydroxyl group, you are likely dealing with a carboxylic acid. The carbonyl group of carboxylic acids is typically located near 1700 to 1725 cm^{-1}. This combination of peaks is characteristic of organic acids such as formic acid, acetic acid, and citric acid.

STEP 7: If in addition to the carbonyl group there is a medium-intensity peak at 1000 to 1300 cm^{-1}, there may be a carbon–oxygen single bond in the molecule, and you are likely dealing with an ester (such as amyl acetate).

STEP 8: If in addition to the carbonyl group there is a weak peak between 2850 cm^{-1} and 2750 cm^{-1}, there may be an aldehyde carbon–hydrogen single bond present, and you are likely dealing with an aldehyde (such as formaldehyde or glutaraldehyde).

STEP 9: If there are one or two narrow, medium peaks near 3500 cm^{-1}, you may be dealing with a primary or secondary amine (such as methylamine).

STEP 10: If there is a strong, narrow peak in the range of 1000 to 1300 cm^{-1}, there may be a carbon–oxygen single bond in the molecule. In the absence of a strong, wide hydroxyl peak it is likely you are dealing with an ether (such as diethyl ether).

STEP 11: If there is a weak peak centered near 2575 cm^{-1} there may be a thiol (also known as a mercaptan) present. However, thiols are very difficult for the hazmat technician to distinguish by eye in an FTIR spectrum owing to overlap and weak peaks.

Gas Chromatography

Gas chromatography (GC) is a technique that is capable of separating and tentatively identifying chemicals in a mixture. These instruments are fairly expensive to acquire and maintain. A gas chromatograph consists of a long thin tube filled with a packing material such as hydroxyapatite that is able to separate the different chemicals found in the mixture based on their

gas chromatography (GC) ■ A technique that is capable of separating and tentatively identifying chemicals in a mixture based on their affinities for the column packing material.

SOLVED EXERCISE 9-4

According to the FTIR spectrum shown, what material(s) may be present?

Solution: This is water. The two signature water peaks are present: the very broad characteristic hydroxyl peak in the region 3100–3500 cm^{-1}, as well as the characteristic peak at 1640 cm^{-1}. It is well worth remembering this spectrum due to the abundance of water in the environment!

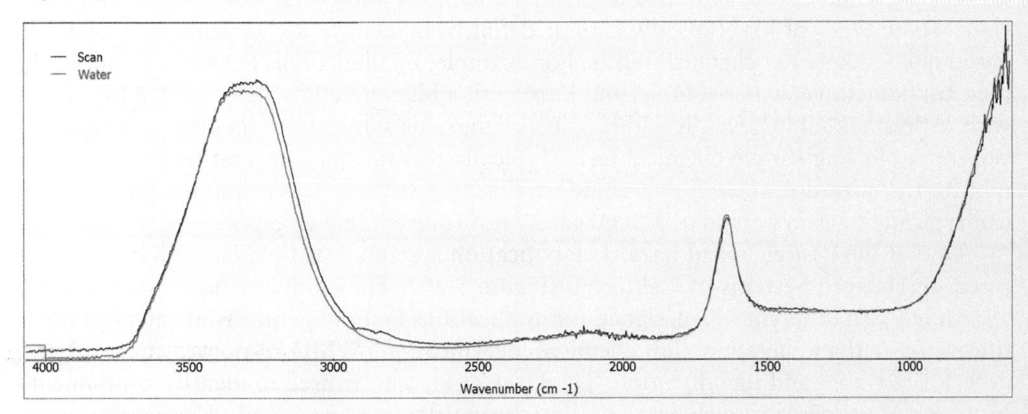

Image courtesy of Thermo Fisher Scientific.

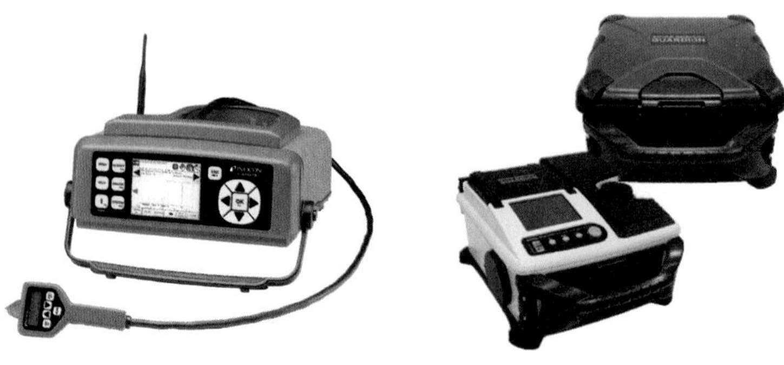

FIGURE 9-12 Gas chromatography/mass spectroscopy (GC/MS) instruments can be used to separate mixtures of gases and volatile liquid vapors and identify individual components. The Inficon Hapsite (left) and Smiths GUARDION (right) GC/MS are pictured. *Photo on left courtesy of INFICON. Photo on right courtesy of Smiths Detection, Inc.*

chemical characteristics and their different affinities for the column packing material. This means that a chemical that binds to hydroxyapatite will spend more time in the column than a material that does not interact with hydroxyapatite. The amount of time a chemical spends in the column is called the *retention time.* The retention time is a unique value for a given chemical and a given column configuration, although many chemicals have similar or identical retention times. Gas chromatographs are commonly used in conjunction with more definitive identification technology such as a mass spectrometer (like the Hapsite and GUARDION, seen in Figure 9-12).

Mass Spectrometry

mass spectrometry ▪ A technique that is capable of uniquely identifying chemicals by ionizing a chemical into a large number of different fragments of varying size and charge. These fragments are then separated and quantified by their mass-to-charge ratio using a strong magnetic field under vacuum.

Mass spectrometry is a technique that is capable of uniquely identifying chemicals. Mass spectrometers are quite expensive and are not yet routinely used by most hazardous materials response teams. Mass spectrometers work by ionizing a chemical into a large number of different fragments of varying size and charge. These fragments are then separated and quantified using a strong magnetic field under vacuum. The fragment pattern is unique for any given chemical under similar ionization conditions and can be considered similar to a fingerprint for the chemical.

The most commonly used field portable mass spectrometer is the Hapsite (Figure 9-12). The Hapsite consists of a gas chromatograph and a mass spectrometer and is an excellent tool for both air monitoring and chemical identification. The GUARDION uses solid-phase microextraction (SPME) fiber technology and can be used for both air monitoring and liquid trace sample analysis. The advantages of GC/MS technology is that it can both separate a mixture (using the gas chromatograph) and definitively identify the mixture components (using both the retention time in the GC column and the ion fragment fingerprint of the mass spectrometer) using large spectral databases. The disadvantages of GC/MS technology are the high cost—both acquisition and maintenance costs—and the extensive training that is required to maintain and use the equipment.

Chemical Hazard Classification Systems

Chemical hazard classification systems are designed to identify materials based on their chemical reactivity. Several manufacturers make kits that contain test reagents and detailed instructions to safely experiment with small aliquots of the unknown material to learn more about how it interacts with other chemicals and thereby determine its hazard class. These types of kits typically cannot definitively identify a chemical, since they are based only on how the chemical reacts. For example, small alcohols behave a certain way: they are flammable, water soluble, and burn with a blue or yellow flame with a blue base, with little smoke. However, it is difficult to determine whether an alcohol is ethanol or isopropanol using simple chemical tests. Typically this further information is not important for the hazardous materials technician. Once the hazard class is known, the chemical can typically be safely mitigated, segregated, transported, and disposed of.

One of the first chemical hazard classification systems was the Hazcat Kit manufactured by Haztech Systems of California (Figure 9-13). The company now manufactures almost a dozen different kits designed for applications including industrial chemical identification, methamphetamine lab chemical identification, WMD response, and lead and asbestos response and identification. The HazCat kit can be used to identify components of mixtures, to perform elemental analysis (using flame colors), and to determine basic chemical and physical properties of unknown materials.

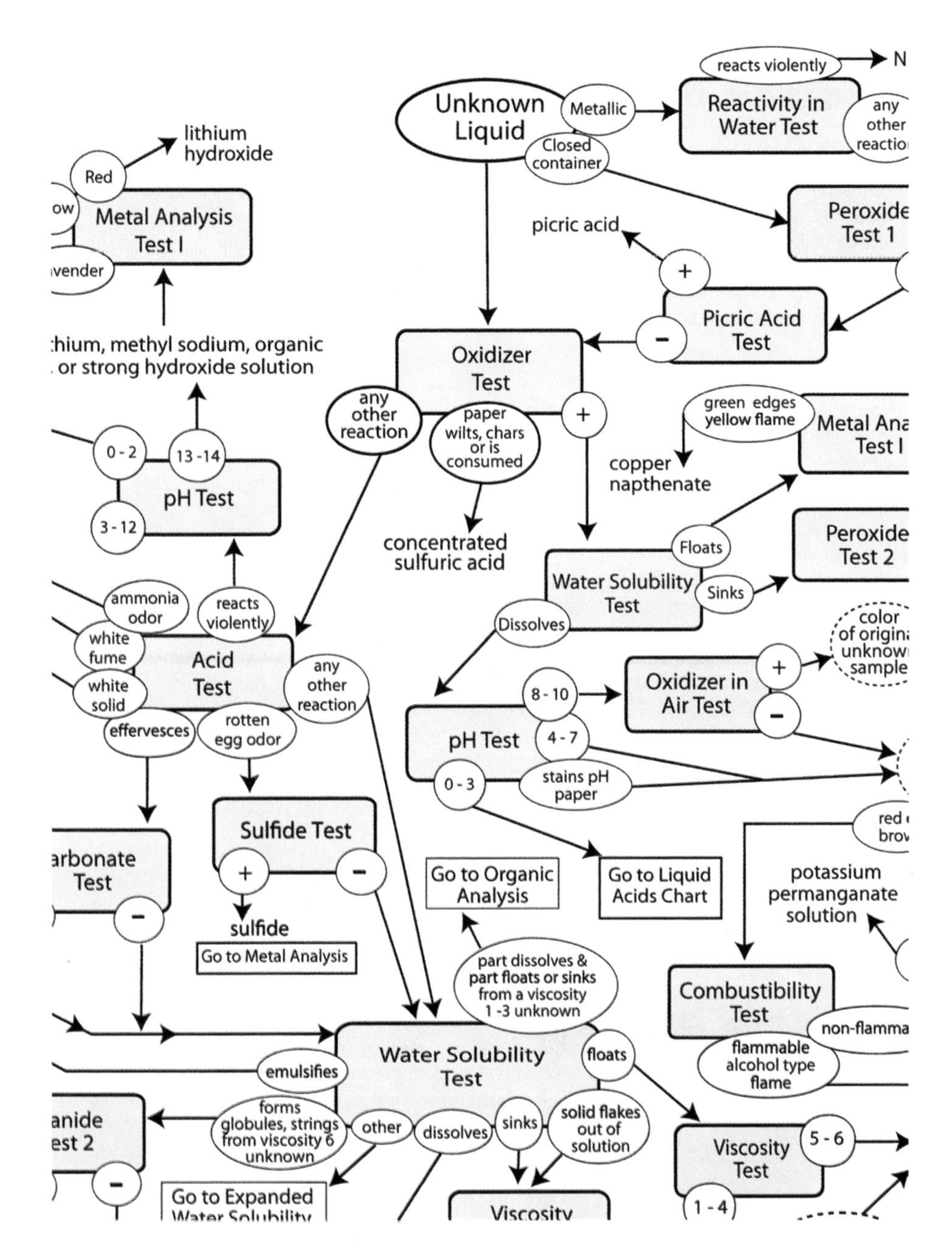

FIGURE 9-13 Chemical hazard classification systems are very effective at identifying the hazard classes of chemicals and mixtures using a step-by-step process to determine chemical and physical properties of the sample as well as to identify key functional groups present. *HazTech Systems, Inc. Established 1986.*

Another excellent chemical hazard classification system is the Heinz 5 Step Kit. HazChem LLC manufactures a simpler chemical identification kit as well. In addition, some agencies and hazmat teams, including the EPA, have developed their own chemical hazard classification systems that are typically simplified versions of commercially available kits designed for the responses common in their jurisdictions. These kits are much less costly than spectrometers but require more training and may have higher maintenance and long-term consumables and training sustainment costs.

Biological Agents

Responding to releases of biological agents can be difficult. One scenario is that a number of people are sickened, and epidemiologists have determined it is not a natural outbreak. Then, law enforcement, with the aid of the public health department and hazardous

BOX 9-6 TACTICS: ABANDONED CHEMICAL DUMPSITE RESPONSE

Abandoned chemical dumpsites can arise from illegal dumping of industrial hazardous waste, chemical waste from illegal drug production such as methamphetamine labs, abandoned industrial chemicals, hazardous wastes at shuttered industrial facilities, and unnoticed accidental releases such as when a valve is left partially open on a tanker. In all these cases responding hazardous materials technicians must identify the risks these chemicals pose to themselves, the public, and the environment. One of the first steps that must be taken is to monitor the air in the immediate area. Then, typically, a sample of the material must be obtained and sample identification must be started.

PREPARATION AND PLANNING

In most cases responses to abandoned chemical dumpsites are not as time critical as some other incidents to which hazmat teams respond. An exception to this rule is a case in which the dumpsite is discovered owing to a chemical exposure to a person. In this case time is of the essence to permit proper treatment of the victim. However, if there are no life-safety issues, the area can methodically be assessed and systematically analyzed. This is the safest approach for first responders.

Safety Considerations

The most significant safety considerations at chemical dumpsites are unwanted and unanticipated chemical reactions. These can take many forms. While chemicals have been exposed to the environment they may have reacted with ambient oxygen and moisture to form more dangerous compounds or slowly reacted inside sealed containers, thereby pressurizing them. Incompatible chemicals at the dumpsite can also be inadvertently mixed during reconnaissance work, sampling activities, or during mitigation. In addition, other chemicals may be partially or completely buried underground, and it will be difficult to locate these containers without ground-penetrating radar. Therefore it is very important to take a slow and systematic approach when dealing with chemical dumpsites. Most hazmat response teams will call in a cleanup contractor to deal with chemical dumpsites after verifying that no emergency conditions exist.

Response Considerations

The EPA and your state department of the environment should be notified immediately when an illegal chemical dumpsite is found. These agencies are mandated by law to oversee hazardous waste cleanup operations, they have mechanisms for funding remediation activities, and they have standing contracts with trusted cleanup contractors. The NRC must be notified as well when chemicals above the reportable quantity (RQ) are released.

Courtesy of Chris Weber, Dr. Hazmat, Inc.

RESPONSE

The first step in responding to an abandoned chemical dumpsite is to perform perimeter air monitoring and a thorough site assessment to determine whether volatile chemicals are present and if they threaten downwind exposures. If perimeter air-monitoring readings are positive, the hot zone must be expanded, and threatened downwind populations must be evacuated or sheltered in place.

The second step should be reconnaissance and up-close air monitoring to determine the contents of the various containers. It is important to develop a systematic approach to analyze, label, and document the different containers and leaks at the dumpsite. Documentation is critical for first responder safety. It is wise to photograph the scene before entry and to photograph each container encountered. This will minimize the need for extra reconnaissance entries as questions arise.

The initial reconnaissance should ideally be performed from a distance using binoculars. As more information is obtained from the site an entry team can perform reconnaissance to determine any drum markings and get up-close air-monitoring readings. At this stage the containers should be minimally handled or not at all. Depending on how long the containers have been abandoned, weathering, internal chemical reactions, and overpressurization may make them prone to fail unexpectedly.

After the initial reconnaissance a sampling and identification plan should be drafted. This plan should take into account information taken from drum labels, keeping in mind that hazardous waste may have been placed in the drum with the original labels and markings left in place. In this case air-monitoring readings may be more reliable than container markings. Containers showing overpressurization, such as drums with distended ends (as shown above), should be opened remotely. Cleanup contractors should have these devices at their disposal.

When sampling the containers keep in mind that they may contain multiple chemical layers. These may be as simple and innocuous as rainwater lying on top of a chlorinated solvent, or water underneath a hydrocarbon such as diesel fuel. However, the drum labeled "motor oil" may contain hazardous waste underneath a thin layer of motor oil designed to camouflage the waste. There may be multiple immiscible chemical layers in any given container, including an aqueous or water-soluble layer, a nonaqueous layer, and a solid sludge at the bottom. Therefore, all containers at the dumpsite should be sampled using a drum thief or COLIWASA tube. These devices are essentially straws that can be pushed through all layers in a container when the top is uncapped. When the tube reaches the bottom the top is capped and all the layers in the container can be pulled up. These tubes can be made of glass or plastic for a wide range of chemical compatibilities. Usually, each layer is placed in a separate sample container so that each can be identified with FTIR spectroscopy, Raman spectroscopy, GC/MS, or chemical identification kits.

Photos courtesy of Mike Becker, Longmont (CO) Fire Department.

materials response teams, must determine the likely dissemination location, collect samples, and attempt field identification of the infectious substance. Another scenario is that an envelope with or without visible powder has been received by a credible target and is accompanied by a credible threat. Law enforcement officers and hazmat technicians must attempt to retrieve the substance, take samples, and get them to a laboratory within the Laboratory Response Network (LRN) that has the capability of definitively identifying the pathogen. In almost all cases of biological agent sample collection, biological agent field identification will need to be performed as well.

BIOLOGICAL AGENT SAMPLE COLLECTION

Biological agent sample collection is a more difficult task than chemical sample collection. Biological agents, such as bacteria and viruses, are sensitive to ultraviolet (UV) light, temperature extremes (especially heat), extremes of pH, and many other chemicals and environmental factors. Therefore, biological agent samples must be carefully collected and stored to avoid damaging the sample, which would then preclude definitive laboratory analysis. The gold standard for biological agent identification is culturing and growing the biological agent in a controlled laboratory environment. If the sample is not properly collected, transported, and maintained, the biological agent may not be alive when it arrives at the testing laboratory.

ASTM International has formulated two standards that describe operational guidelines for the initial response to suspected biological agent releases, and specific guidance for bulk sample collection and swab sample collection of visible powders (ASTM E2770-10 and ASTM E2458-10, respectively). These standards were implemented in cooperation with first responders, public health laboratories, the military, and the Federal Bureau of Investigation (FBI) and therefore represent best practices for biological agent sample collection.

Before biological agent sampling activities commence in the hot zone, a risk assessment should be performed, a decontamination line should be established, and appropriate PPE must be donned. Additionally, based on the risk assessment and reconnaissance activities, a sampling plan should be developed based on a two-person sampling team at a minimum, the sampler and the assistant sampler (or facilitator).

The sampler, as the name implies, collects the sample. The assistant sampler communicates with the cold zone, ensures that the scene is properly documented through photography and sketches, fills out sampling documentation, and hands the sampler the necessary containers and tools. The assistant sampler should never touch the sample or contaminated surfaces. The sampler is the only person to touch the sample, surface areas, and the primary sample container. The sampler must also sign the chain-of-custody form in the cold zone on transfer of the sample.

aseptic technique ▪ The systematic minimization of cross-contamination using proper sampling procedures, sterilized and certified clean tools, and carefully controlled sampling conditions.

All sampling should be conducted using **aseptic technique**, which is the systematic minimization of cross-contamination using proper sampling procedures, sterilized and certified clean tools, and carefully controlled sampling conditions. Both the sampler and the assistant sampler should change their outermost layer of gloves between samples. Sampling procedures for bulk sampling and swab sampling are discussed in Boxes 9-7 and 9-8, respectively.

All tools used for biological sample collection should be individually packaged and sterile. Sampling conditions should be controlled to the extent reasonably possible, such as limiting wind currents, direct sunlight, and moisture until the sample is collected. Negative controls should be included with the sample. The negative control—also called a *media blank*—should include unopened sampling tools and sample containers that can be tested to verify that positive results were not originally present in the sampling tools and equipment.

All containers that will be used to hold samples should be labeled prior to hot zone entry, including the primary and secondary containers. For the sampling procedures described in Boxes 9-7 and 9-8, the following containers should be labeled:

1. Two sample collection containers and two transparent 1-gallon resealable bags for the bulk powder sample should be labeled "Powder Sample" and have a unique sample identification number.
2. The transparent 1-gallon resealable bag for the dry swab sample should be labeled "Dry Swab" and have a unique sample identification number.
3. The container into which the source or dissemination device will be placed should be labeled "Primary Source" and have a unique sample identification number.

Bulk sampling of biological agents should be carried out when there is visible powder on a nonporous surface. The primary reason for bulk sampling is to collect a sample for definitive testing in an LRN laboratory or for evidentiary purposes. If a sample is to be collected for an on-site biological assessment—a *presumptive test*—the swab sampling procedure should be used after the majority of visible powder has been collected. Swab sampling should be conducted only after bulk sampling is complete. The reason for this is that the first priority for sample allocation must be the LRN reference laboratory and/or a forensics laboratory. The swab sampling for field identification purposes should be a second priority, since confirmatory analysis must be carried out in a licensed laboratory.

BOX 9-7 HAZMAT HOW TO: PROCEDURES FOR BULK SAMPLING OF SUSPECTED BIOLOGICAL AGENTS

The following procedure is consistent with standards ASTM E2458-10 and ASTM E2770-10 and is designed to preserve the integrity of the sample for testing purposes and for evidentiary purposes.

1. Field screen the sample for explosives, corrosiveness, flammability, radioactivity, and oxidizing potential (if not already completed during reconnaissance activities).
2. Assistant sampler: Lay down a clean drop cloth or laboratory diaper to serve as a clean work area. Locate the work area close enough for convenience, yet not directly on the powder itself or at the likely release point.
3. Assistant sampler: For each sample, ensure that the following information is accurately recorded either through radio communication with the cold zone, and/or by making the appropriate entries on the paperwork:
   a. sample identification number
   b. name of sampler and assistant sampler
   c. time and date of sample
   d. sample location and address
   e. size of the sampled area
   f. map and/or photographs of the sample area (before and after)
   g. type of sample
4. Collect the primary source first, if present. Primary sources include letters and dissemination devices.
   a. Assistant sampler: Open the 1-gallon resealable bag labeled "Primary Source" and hold it open next to the sample.
   b. Sampler: Place the primary source gently into the open bag. Handle the primary source as little as possible to preserve forensic evidence such as fingerprints and DNA.
   c. Assistant sampler: Seal the bag and place the sealed bag into the second 1-gallon resealable bag, seal it as well, then place it in the transport container, which will be decontaminated in the warm zone.
5. Collect the bulk sample next.
   a. Assistant sampler: Give the sampler a sterile 1.5" × 2.5" (4 × 6.5 cm) plastic laminated card to scrape and move the bulk powder into the sampling container. Also, loosen the cap to the tube containing the swab.
   b. Sampler: Remove the swab from the tube that is being held by the assistant sampler. Hold the laminated card at an angle on the surface next to the bulk powder sample. If necessary, use the card to push the powder into a pile. Use the sterile swab to slowly push the powder onto the laminated card. Be careful not to resuspend fine powders and avoid causing clumps of solid to flick off in random directions when scraping with the card.
   c. Assistant sampler: Hold the sterile tube while the sampler carefully places the swab into the tube.
   d. Assistant sampler: Close the tube, place it into the resealable bag labeled "Dry Swab," seal it, then place it in the transport container.
   e. Assistant sampler: Open the container labeled "Powder Sample" and hand the open container to the sampler.
   f. Sampler: Gently place the laminated card containing the bulk powder into the container, being careful to minimize powder resuspension, then take the lid from the assistant sampler and cap the container.
   g. Assistant sampler: Open the resealable bag labeled "Powder Sample"
   h. Sampler: Carefully place the container holding the powder sample into the bag.
   i. Assistant sampler: Seal the bag and place it in the transport container.
6. If all the powder cannot fit into a single sample collection container, repeat the process until all the powder is collected. If an inordinate amount of material is present, coordinate disposal of the material with the FBI and the LRN laboratory.
7. Perform field identification on any residual material left according to the swab sampling procedure described in Box 9-8.
8. In the hot zone, decontaminate each of the resealable bags using a 10% household bleach solution that has been adjusted to a pH of 6.8–8.0 with vinegar. The pH-adjusted bleach solution should be made fresh immediately prior to decontamination.
9. In the warm zone, place each individual wetted bag into another resealable bag. Do not completely dry the decontaminated resealable bags, to ensure sufficient contact time for sterilization.
10. In the cold zone, place the samples into the final, hard-sided shipping container, complete all the sampling forms, and attach a copy of the sampling form to the outside of the shipping container. All of the documentation should be checked for accuracy, including sample identification numbers, sample team members, and chain-of-custody forms and signatures.

Before performing this procedure in the field, carefully read ASTM E2458-10 and ASTM E2770-10 standard practices.

BOX 9-8 HAZMAT HOW TO: PROCEDURES FOR WET SWAB SAMPLING OF RESIDUAL SUSPICIOUS POWDERS FOR FIELD IDENTIFICATION

The following procedure is consistent with standards ASTM E2458-10 and ASTM E2770-10:

1. Field screen the sample for explosives, corrosiveness, flammability, radioactivity, and oxidizing potential (if not already completed during reconnaissance activities or the bulk sampling procedure).
2. Follow the instructions provided by the manufacturer of the field identification kit or continue with this procedure for wet swab sampling.
3. Assistant sampler: Record pertinent sample information and the results of any testing.
4. Collect wet swab samples using the following methodology:
   a. Assistant sampler: Remove the lid from the buffer solution and hand the buffer solution to the sampler. Also, loosen the lid containing the swab, and hand the tube and swab to the sampler without removing the swab from the container.
   b. Sampler: Remove the swab from the tube and wet it in the buffer solution. Remove excess buffer from the swab tip by pressing it against the inside (sterile) surface of the buffer solution container and squeezing out excess buffer.
   c. Sampler: Wipe the swab vertically in an S- or Z-shaped pattern across an area not to exceed 8" × 8" (20 cm × 20 cm).
   d. Sampler: Rotate the swab to expose the clean side of the swab and wipe the swab horizontally over the same area using an S- or Z-shaped pattern.
   e. Sampler: Place the swab into the original sterile tube and seal it.
   f. Assistant sampler: Hold open a resealable bag labeled "Wet Swab."
   g. Sampler: Carefully place the tube into the bag.
   h. Assistant sampler: Seal the bag and place it in the transport container.
   i. Decontaminate these samples and ship them in the same manner as described for the bulk sample collection procedure (if necessary).

Before performing this procedure, carefully read ASTM E2458-10 and ASTM E2770-10 standard practices.

BIOLOGICAL AGENT IDENTIFICATION

Biological agent identification technology has vastly improved over the last decade. Three primary field identification technologies are commonly used: protein and pH assays, immunoassays, and DNA-based assays. However, the gold standard for biological identification remains culturing (growing) the organism in a public health laboratory.

Protein Assays

protein assay ▪ A colorimetric test to detect amino acids, which are the building blocks of proteins.

All biological organisms contain proteins and have an approximately neutral pH. **Protein assays** consist of colorimetric tests to detect amino acids, which are the building blocks of proteins. The test reagent turns from clear to purple in the presence of amino acids (Figure 9-14). Since many chemical formulations contain protein, a pH test is usually coupled with these kits. Biological organisms—especially pathogens—invariably require a neutral pH to function and survive. Protein and pH assays are therefore good tests for ruling out the possibility of biological agents. The colorimetric tests are quite sensitive and can therefore detect small amounts of protein. However, some biological agents have minimal effective doses of a few organisms and may therefore not be detected by these assays, resulting in false negatives. The advantage of protein assays is their low cost. The disadvantage of protein assays is their high rate of false positives, since many other materials contain proteins and are neutral, including flour and most food products. Some test kits therefore include a starch test to further eliminate false positives.

immunoassay ▪ An antibody-based technique that is specific for certain proteins found in virulent strains of common biological agents such as anthrax. Commonly known as a *handheld assay*.

Immunoassays

Immunoassays, commonly known as *handheld assays*, are an antibody-based technology (Figure 9-14). Antibodies are proteins made by living organisms to fend off pathogens inside the body. The immune system relies on antibodies against common disease organisms

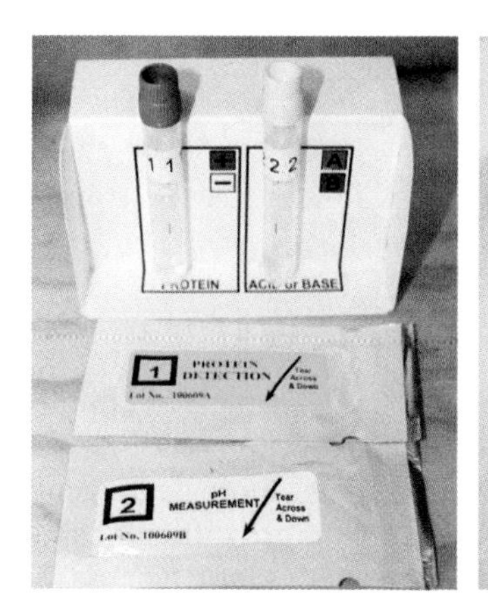

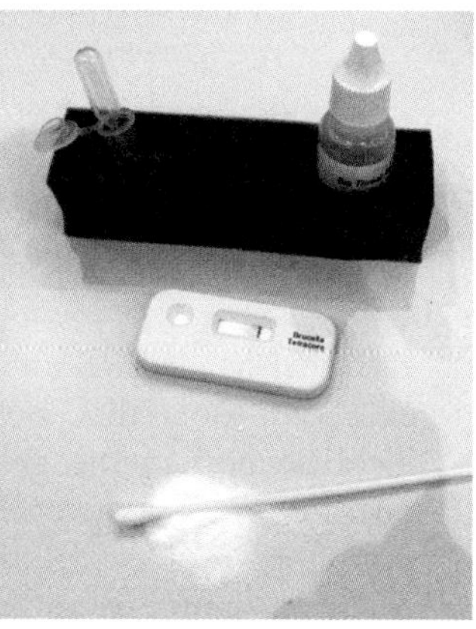

FIGURE 9-14 Biological agent field detection and presumptive identification is usually a three-step process. In the first step, the pH of the sample, the protein content, and starch and/or sugar content of the sample can be analyzed (left). The second step usually involves immunoassay (middle). The final step involves a DNA-based assay (right). *Photos courtesy of Chris Weber, Dr. Hazmat, Inc. Photo on right used with permission from Smiths Detection, Inc.*

to recognize the invaders and produces antibodies each time the body is exposed to foreign invaders. Vaccines are designed to stimulate the body into producing antibodies. When horses, rabbits, or mice are vaccinated, these antibodies can be extracted and purified in a laboratory and then incorporated into simple handheld testing devices. These testing devices work on the same principle as a pregnancy test (which uses antibodies specific for the primary hormone produced by pregnant women).

Immunoassays are only as good as the antibody that is used as the test reagent. This is why in 2001, handheld assays acquired a really bad reputation owing to their many false positives and false negatives. In the intervening years manufacturers have greatly improved the antibodies used as test reagents, including designing and purifying antibodies specific for certain proteins found in virulent strains of common biological agents such as anthrax. Immunoassays can be used to identify biological agents, such as anthrax and plague, as well as many common toxins including ricin and botulinum toxin.

Polymerase Chain Reaction Assays

The definitive field screening technology for biological agents is DNA-based assays such as the **polymerase chain reaction (PCR) assay** (Figure 9-14). PCR theoretically has the ability to detect a single microorganism, as long as the DNA of that microorganism is intact. Because this is not always the case, the detection limit for PCR is commonly stated as 5 to 10 organisms. This is extremely sensitive. PCR can even detect nonviable organisms that have been autoclaved. This can be a disadvantage in the field, where the concern is primarily about viable organisms for public health reasons. However, this can be an advantage in the case of law enforcement activities such as evidence collection and prosecution.

polymerase chain reaction (PCR) assay ■ An analytical technique capable of uniquely identifying biological agents based on the amplification of deoxyribonucleic acid (DNA).

DNA-based assays detect deoxyribonucleic acid (DNA), which is the genetic material of the majority of pathogenic microorganisms. The genetic material is the roadmap to reproducing that organism and encodes all the proteins that make up that living organism, including toxins such as anthrax lethal factors, botulinum toxin, and ricin. PCR technology works by repeatedly copying specific areas of the target genome using DNA polymerase until a detectable quantity of DNA is obtained. PCR is an excellent tool for biological agent identification because the portion of the DNA that is amplified can be chosen to incorporate those parts of the genome that are critical for virulence. In this way, PCR assays are extremely specific. Not only can they tell related organisms such as *Bacillus anthracis* from *Bacillus subtlis* from each other, but they also can tell the difference between a virulent strain of anthrax and a veterinary strain of anthrax. Veterinary strains have been genetically engineered to remove the deadly anthrax lethal factor toxins that lead to mortality. Although PCR assays are unable to detect toxins themselves, they can detect genetic material from the host organism that contaminates the toxin preparation. For this reason you will see PCR assays for ricin and botulinum toxins that rely on minute amounts of DNA impurities that remain in the toxin preparation.

CHAPTER **REVIEW**

Summary

It is often necessary to collect samples and perform field identification. It is important to systematically sample and identify the contents of unmarked or suspicious containers. It is also very important to understand the difference between public safety sampling and evidentiary sampling. Test papers are a cheap and efficient way to analyze a wide range of chemical hazards. Raman and FTIR spectroscopy are excellent tools that provide semiautomated sample identification in the field and are simple enough to operate under emergency conditions. Biological agent sampling and identification can be more difficult than chemical sampling, and using proper aseptic technique is imperative.

Review Questions

1. Name three reasons for collecting samples at hazardous materials incidents.
2. Explain how the two-person sample collection technique works.
3. Why is it important to minimize cross-contamination?
4. What techniques can be used to identify solid and liquid samples?
5. Compare and contrast the advantages and limitations of Raman spectroscopy and FTIR spectroscopy.
6. Why is documentation critical during sampling and sample identification?

References

E2458 Practices for Bulk Sample Collection and Swab Sample Collection of Visible Powders Suspected of Being Biological Agents from. Nonporous Surfaces. (2010). Washington DC: ASTM International.

E2770 Guide for Operational Guidelines for Initial Response to a Suspected Biothreat Agent. (2010). Washington DC: ASTM International.

Houghton, Rick. *Emergency Characterization Of Unknown Materials* (2007), CRC Press.

Houghton, Rick. *Field Confirmation Testing for Suspicious Substances* (2009), CRC Press.

CHAPTER

10 Containers and Damage Assessment

Courtesy of Eugene Ngai, Chemically Speaking LLC.

KEY TERMS

annealing, *p. 297*
cold work, *p. 328*
distillation, *p. 301*
ductility, *p. 297*
heat-affected zone, *p. 299*
intermodal container, *p. 303*
label, *p. 345*
placard, *p. 345*
quenching, *p. 297*
reflux, *p. 299*
specific activity, *p. 342*
stiffening ring, *p. 336*
tempering, *p. 297*
tensile strength, *p. 297*
transition temperature, *p. 299*
transport index, *p. 343*

OBJECTIVES

After reading this chapter, the student should be able to:

- Describe the types of containers used to store solids, liquids, and gases and their hazards.
- List the features and uses of 10 container construction materials.
- List common safety features of pressurized vessels.
- Describe the difference between QT and NQT steel.
- Describe the damage assessment process.
- Describe common types of container damage.
- Identify critical damage to hazardous materials containers.
- Define the heat-affected zone of a weld.
- Interpret a radioactive label.
- Determine if a package containing radioactive material is damaged and leaking.

ResourceCentral For additional review and practice tests, visit **www.bradybooks.com** and click on Resource Central to access book-specific resources for this text! To access Resource Central, follow directions on the Student Access Card provided with this text. If there is no card, go to **www.bradybooks.com** and follow the Resource Central link to Buy Access from there.

There has been a major storage tank fire at a regional refinery caused by a lightning strike. You've been dispatched to the scene as part of the regional hazmat team to do a damage assessment on the other oil storage tanks, gasoline tanks, and volatiles tanks. Although the fire suppression system and the fire brigade held the fire to the involved tank, the incident commander needs to know whether neighboring tanks were damaged by impinging heat or through other mechanisms. The following are some questions running through your mind:

- How bad could the damage from impinging heat really be?
- How dangerous will the damage assessment process be to me and my crew?
- How do I determine the severity of the damage?
- What resources are available to help me assess the damage?
- How much internal pressure do these oil storage tanks have?
- How much internal pressure do these gasoline tanks have?
- How much internal pressure do these volatiles tanks have? They certainly do look different.
- How do I recognize significant damage?

Let's see if we can get some answers to these questions and more.

Containers are used to store, transport, use, and produce chemicals. These substances may be harmless or potentially harmful materials and are found in a wide variety of quantities. The quantity of the material, the state of matter of the material, and the type of container can have a profound impact on the outcome of a hazardous materials incident. Generally, the larger the quantity of the hazardous material, the bigger the problem it is. In other words, larger containers tend to lead to larger problems. Pressurized containers, such as those storing compressed gases or compressed liquefied gases, pose a much greater hazard than unpressurized containers. The type of container any given hazardous material is in reveals a lot of information about the potential danger we will face at a hazmat incident.

Containers that carry hazardous materials are usually required to be constructed to specifications required by such agencies or organizations as the U.S. Department of Transportation (DOT), the Association of American Railroads (AAR), the American Society of Mechanical Engineers (ASME), or the Compressed Gas Association (CGA)—especially when they contain large quantities of dangerous goods or extremely hazardous materials, such as poison inhalation hazards or radioactive materials with high specific activity. When a container is exposed to stresses—such as fire or mechanical damage—it may show obvious signs indicating the level of stress, but others may remain hidden—such as internal cracks or damage located in inaccessible areas.

Damage assessment is the estimation of the harm a container has suffered and the impact of this harm on its integrity. Damage assessment is one of the most important functions at hazmat incidents involving containers, whether they are highway cargo tankers, rail tank cars, marine vessels, non-bulk packaging, or fixed-site facility tanks. It is important to understand container construction and safety features to respond effectively to hazardous materials releases from them. The container construction material, container stresses, and internal pressure are crucial pieces of information. It is important to reduce or stabilize the stress on a damaged container as quickly and as safely as possible. Ideally, damage will be assessed by a technical specialist with extensive experience with the damaged container. Let's explore some of these containers and how they can affect emergency operations when they fail.

Container Construction

Container construction materials are strongly dependent on the chemical and physical properties of the hazardous material being transported. Chemical reactivity precludes some materials altogether, such as the use of many metals for corrosive materials, and the use of glass for hydrofluoric acid. The vapor pressure of the material will dictate the internal pressure to which the container will be subjected. Therefore, compressed and liquefied gases can't be stored in glass containers, and liquids and gases can't be stored in cardboard containers.

Glass, paper, and plastic containers are cheap to produce, and plastic containers are quite light compared with their strength. Aluminum is a comparatively expensive metal, but it is extremely light relative to its strength. This is why aircraft are made of aluminum. Aluminum is a very pliable metal and dents and gouges easily. The most common steel used in the United States is carbon steel. Depending on the carbon content, other added doping elements, and metal treatments, this type of steel can have a wide range of properties. Increased carbon content leads to increased hardness and strength, but also to brittleness. High-strength low-alloy (HSLA) steel, which generally has low carbon content, is formulated to specific mechanical properties including strength, hardness, corrosion resistance, and abrasion resistance. Table 10-1 lists common container construction materials and their primary characteristics as well as commodities they often contain.

Most bulk containers, especially those used for shipping, are constructed of metal owing to its increased strength and durability characteristics. Two mutually exclusive characteristics are ductility and tensile strength. **Ductility**, also referred to as *malleability*, describes how easily a metal can be shaped or drawn without cracking or breaking. In practical terms, ductility describes how easily a metal can be bent or elongated when subjected to mechanical forces, such as a train wreck or rollover accident. *Elongation* is the expansion of a metal without breaking. Extremely ductile metals are not very strong because they bend or stretch easily when subjected to force. **Tensile strength** describes how resistant a metal is to pulling. Often, metals with high tensile strength are also brittle and crack easily—they are not very malleable.

Metal treatments are used to optimize desired characteristics in the final container. Metals are microcrystalline structures. In untreated metals these crystals are randomly oriented, which reduces metal strength. Metal treatments are designed to realign these crystals parallel to one another, which increases tensile strength. These treatments include annealing, quenching, and tempering. **Annealing** is the process of heating the metal above its recrystallization temperature and then letting it slowly cool over a period of time.

Tempering is the process of heating a metal rapidly to a high temperature. **Quenching** is the process of quickly cooling a metal after it has been heated.

Two types of steel are commonly used, QT and NQT. QT stands for quenched and tempered, which means it was heated rapidly and then cooled rapidly. QT steel is more susceptible to corrosion. NQT stands for nonquenched and tempered, which means the steel was heated rapidly and then cooled slowly. Each of these treatments produces a different effect depending on the type of metal, the temperature to which it is heated, and how rapidly it is heated and cooled. These nuances can be quite important. For example, anhydrous ammonia with less than 0.2% moisture content may not be stored and

ductility ▪ A measure of how easily a metal can be shaped or drawn without cracking or breaking. Also referred to as malleability.

tensile strength ▪ A measure of the resistance of a metal to elongation.

annealing ▪ The process of heating a metal to a predetermined temperature and then letting it slowly cool over a period of time.

tempering ▪ The process of heating a metal rapidly to a high temperature.

quenching ▪ The process of quickly cooling a metal after it has been heated.

TABLE 10-1	Container Construction Materials			
MATERIAL	**CONTAINER EXAMPLES**	**CHARACTERISTICS**	**STATE OF MATTER**	**SELECTED PRODUCTS**
Fiberboard	Fiberboard drums, cardboard boxes	Light weight, easily damaged by water, easily punctured, combustible	Solids	Citric acid, sodium bicarbonate
Fiberboard with plastic liner	Lined fiberboard drums, lined boxes	Light weight, easily damaged by water, easily punctured, combustible	Solids, pastes	Sodium hydroxide pellets
Wood/plywood	Gaylord boxes	Combustible, stronger than fiberboard	Solids	Solid hazardous waste
Glass	Jugs, laboratory glassware, dewars	Brittle, resistant to corrosives and solvents	Solids, liquids	Liquid laboratory solvents and corrosives, solid materials such as sodium hydroxide
Plastic	Jugs, drums, IBCs, storage tanks	Less brittle than glass, easily punctured, resistant to corrosives and solvents	Solids, liquids	Liquid laboratory solvents and corrosives, solid materials such as sodium hydroxide
Aluminum-40 series	Bulk tanks	Light weight, ductile, weak	Solids, liquids, gases	
Aluminum-50 series	Bulk tanks	Light weight, medium ductility, medium strength	Solids, liquids, gases	
Aluminum-60 series	Bulk tanks, cylinders	Lightweight, strong, not ductile	Solids, liquids, gases	SCBA tanks, oxygen
Mild steel	Bulk tanks, cylinders	Ductile with a low tensile strength	Solids, liquids, gases	
HSLA	Bulk tanks, cylinders	Very strong, hard, and abrasion resistant	Solids, liquids, gases	Propane, anhydrous ammonia
Carbon steel	Bulk tanks, cylinders	As carbon content increases the metal becomes harder and stronger but more brittle	Solids, liquids, gases	Commodities carried in pressure rail tank cars
Quenched and tempered (QT) steel	MC 330, fixed- site storage tanks	Very strong but brittle, tensile strengths can approach 80,000 psi, prone to corrosion	Solids, liquids, gases	Propane
Nonquenched and tempered (NQT) steel	MC 331, bulk storage tanks, cylinders	Strong and less brittle than QT steel	Solids, liquids, gases	Propane, anhydrous ammonia
Stainless steel	Non-bulk containers, bulk containers, cylinders	Resistant to corrosion, strong, ductile, and abrasion resistant	Solids, liquids, gases	Food-grade chemicals, pharmaceuticals, corrosives
Nickel	Cylinders	Resistant to corrosion	Liquids, gases	Tungsten hexafluoride

transported in QT steel containers per DOT regulations owing to the heightened risk of corrosion.

Almost all containers must be welded at some point (with the exception of gas cylinders and tubes). Most containers are made up of sheets of metal that must be curved during the construction process, and two heads. Large tanks are made from several plates.

When multiple plates are used to form the core of the tank, the welds are offset from one another to increase the strength of the tank. Welds pose a particular problem at hazardous materials incidents involving container damage. A weld weakens the structural integrity of the container along its edge—an area referred to as the **heat-affected zone** (Figure 10-1). The heat-affected zone is an area directly next to the weld bead, approximately the width of the weld bead, which is less ductile—and therefore more brittle and weaker—than the rest of the metal. Any damage to the weld area is of particular concern during any damage assessment.

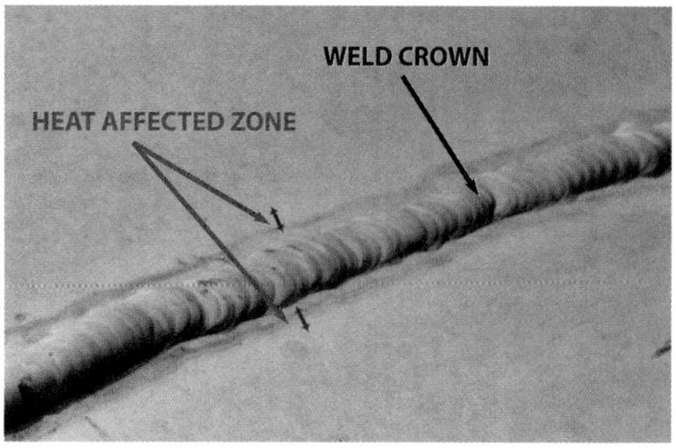

FIGURE 10-1 Weld seams are among the weakest points of containers. The weld seam is composed of the weld crown and the heat-affected zone. The heat-affected zone is located on both sides of the weld crown and is approximately the width of the weld crown. Any damage affecting the weld seam is considered critical damage. *Courtesy of Chris Weber, Dr. Hazmat, Inc.*

Extremes of temperature, both high and low, affect the elasticity of metals. If tank temperature is greater than 100°F, it will be entirely ductile regardless of the steel used in its construction. Cold temperatures make metals brittle. Stainless steel and aluminum cylinders and dewars are used for cryogenic service (materials with boiling points less than or equal to −130°F). Below the **transition temperature**, which is approximately −50°F for commonly used TC-128 steel, metals are brittle; above the transition temperature they are ductile. Temperature guns can therefore be very helpful in assessing the ductility of metals. A temperature gun is essentially an infrared thermometer that can be used to measure the temperature of an object from a distance by reading the infrared emissions of the object.

heat-affected zone ▪ An area directly next to each side of a weld bead, approximately the width of the weld bead, that is less ductile—and therefore more brittle and weaker—than the rest of the metal.

transition temperature ▪ The temperature below which metals are brittle; above the transition temperature they are ductile.

Container design is driven by practical applications such as ease of loading and unloading. Shipping containers containing solids often have one or more funnel-shaped openings near the bottom of the container to facilitate complete gravity unloading. But this is not always the case, since fine powders and solids can be pneumatically loaded and unloaded. Compressed gas containers usually have rounded ends to evenly distribute high internal pressure.

Shipping container parameters are described in great detail in 49 CFR parts 100 through 199. The Hazardous Materials Regulations (HMR) describe the packaging and container requirements for listed hazardous materials. Container construction requirements are formulated in 49 CFR parts 178 to 199. These requirements include minimum container thickness, construction materials, installation requirements, safety device requirements, maximum allowable working pressure (MAWP), and minimum burst pressure.

Types of Transportation and Storage Containers

Chemical storage containers come in a wide variety of shapes and sizes. They are also constructed of a wide variety of materials, including paper, wood, plastic, fabrics, glass, and a number of metals. One critical component to consider is the size of the container. Generally, the greater the amount of product that can escape, the bigger the problem will be.

LABORATORY GLASSWARE

Hazardous materials releases can occur in laboratories, whether they are academic or industrial research laboratories, or illicit drug or WMD laboratories. Labs usually contain exotic glassware and equipment (Figure 10-2A). Small- and large-scale laboratory glassware and equipment is commercially designed for use in industrial or research labs. Clandestine drug and WMD labs may have improvised homemade equipment constructed of normal household items. All this equipment is designed to facilitate one or more chemical reactions.

Let's explore some of the chemical reactions that you might run across in chemical laboratories. **Reflux** reactors are used to heat a chemical reaction over a period of time,

reflux ▪ The process of carrying out a chemical reaction with the addition of heat.

FIGURE 10-2A Laboratory glassware comes in a variety of shapes, sizes, and purposes—including (clockwise from top left) glass beaker and hot plate, round-bottom flask and heating mantle, desiccator, condenser, Erlenmeyer flask, addition funnel, and tabletop centrifuge. *Figures A and B courtesy of Chris Weber, Dr. Hazmat, Inc.*

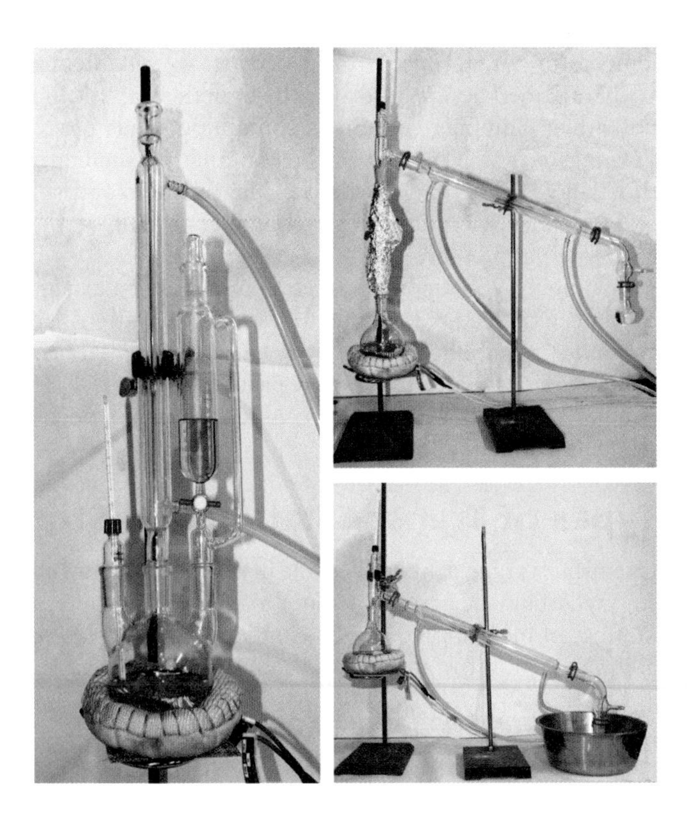

FIGURE 10-2B Laboratory glassware is assembled to perform chemical functions. A reflux reactor (left) is used to heat reagents and to produce a final product. Components of the pictured reflux reactor include a heating mantle, a three-necked round-bottom flask, a thermometer, a pressure-equalizing addition funnel, a condenser, and tubing for cooling water. Fractional distillation (top right) is used to separate materials with similar boiling points. Components of the pictured fractional distillation set up include a heating mantle, a single-necked round-bottom flask, a fractionating column insulated with aluminum foil, a thermometer and distillation head, a condenser and tubing, a vacuum adapter, and a receiving single-necked round-bottom flask. Simple distillation is used to separate materials with dissimilar boiling points. A simple distillation (bottom right) differs from fractional distillation by the absence of the fractionating column.

to increase the speed of the reaction (Figure 10-2B). The reflux reactor is designed to limit evaporation of the solvent or other volatile chemicals used in the reaction. The reflux reaction used in illicit laboratories is similar in concept to simmering a stew on a stove using a covered pot. The steam condenses on the lid and drips back into the stew to continue the cycle. The pot and lid are a simple reflux reactor.

Distillation reactions are used to separate liquids. Typically they are used to separate the final product or intermediate from the reagents, solvents, and precursors. The reaction vessel at the bottom is heated to volatilize the components creating a vapor. A fractional distillation apparatus is used to separate materials that have similar boiling points (Figure 10-2B). The distillation column is used to remove less volatile components from the vapor, which allows the more volatile components to continue to rise and move into the condensation column, or condenser. A simple distillation apparatus does not have a vertical distillation column and is used to separate materials that have markedly different boiling points (Figure 10-2B).

FIGURE 10-3 Pictured clockwise from the top left, non-bulk containers include cryogenic dewars, compressed gas cylinders, 5-gallon metal drums, gallon plastic containers, plastic and metal 55-gallon drums, and cardboard drums. *Courtesy of Chris Weber, Dr. Hazmat, Inc.*

NON-BULK CONTAINERS

Non-bulk containers are ubiquitous (Figure 10-3). They are found in the home in the form of bottles, bags, aerosol cans, and many other small containers. These containers come from a variety of sources, including grocery stores, home improvement stores, auto parts stores, and pool supply companies. Larger-sized non-bulk containers are found in industry in the form of 55-gallon drums, bags, and reagent bottles. You may see these types of containers in almost any mode of transport and in almost any type of vehicle, placarded or not. Some of the properties of non-bulk containers are listed in Table 10-2.

distillation ■ The process of separating a mixture of liquids by heating the mixture to the boiling point of the individual substances, starting with the lowest boiling point liquid, and subsequently condensing the purified substance in a condenser. Each substance is purified in turn by primarily converting the most volatile liquid in the mixture into a vapor at any one time, and not the other liquids in the mixtures.

INTERMEDIATE BULK CONTAINERS

Intermediate bulk containers hold larger quantities than non-bulk packaging; however, they are still mobile and used in transport (Figure 10-4). Some of the most common intermediate bulk containers, or IBCs as they are known, are 1-ton containers carrying materials such as chlorine or sulfur dioxide, and totes that typically carry liquids such as nitric acid or caustic soda. This type of packaging is very popular in industries using intermediate amounts of materials. IBCs are cheaper to acquire, transport, and store than multiple 55-gallon drums, for example, yet the facilities do not need to install expensive fixed storage tanks. The product can generally be used directly from the intermediate bulk container. Once again, the container shape and size give clues to the contents, but the label or shipping papers are more reliable. Table 10-3 lists examples of intermediate bulk containers.

INTERMODAL CONTAINERS

An intermodal tank is a tank mounted inside a metal frame of standardized dimensions. These tanks were developed to facilitate transport of goods, including hazardous materials, in many modes of transportation, as their name implies (Figure 10-5). Let's follow a shipment of one of the most heavily produced and used hazardous materials in the world: sulfuric acid. The chemical plant that produces the sulfuric acid places it directly

TABLE 10-2 Non-bulk Packaging

CONTAINER	CHARACTERISTICS	TYPICAL CONTENTS
Carboy	Glass or plastic 1- to 10-gallon capacity	Liquids of all hazard classes
Bottle	Glass, plastic, or metal May be coated on the interior and/or exterior	Liquids or solids of all hazard classes
Bag	Paper, fiber, or plastic Possibly lined Supersacks or tote sized available	Solids of all hazard classes
Cardboard box	Should be DOT specification for most hazardous materials Often contains the primary hazmat container (such as glass, plastic, or metal bottles)	Solids, liquids, and gases of all hazard classes
Drum	Steel, plastic, fiberboard, stainless steel, or aluminum Up to 95 gallon capacity	Liquids and solids of all hazard classes
Cylinder	200–6000 psi (13.8-415 bar) Have pressure-relief devices such as fusible links or frangible disks Steel or aluminum	Compressed gases, or liquefied compressed gases of any hazard type
Dewar	Stainless steel Low pressure Insulated	Cryogenic gases such as liquid nitrogen

FIGURE 10-4 Intermediate bulk containers include metal totes (top left), plastic totes (top right), super sacks (bottom right), and 1-ton containers (bottom left). *Courtesy of Chris Weber, Dr. Hazmat, Inc.*

TABLE 10-3 Intermediate Bulk Containers

CONTAINER	CHARACTERISTICS	TYPICAL CONTENTS
Totes	Steel, plastic, stainless steel, or aluminum May be lined May be collapsible Up to 500 gallon (1900 L) capacity	Liquids of all hazard classes
One-ton containers	Steel Pressurized	Liquefied compressed gases such as chlorine and sulfur dioxide

FIGURE 10-5 Intermodal containers come in many shapes and sizes and have a specification plate detailing the construction features. *Courtesy of Chris Weber, Dr. Hazmat, Inc.*

in an intermodal tank. This intermodal tank is loaded on a semi truck, which transports it to a railroad terminal, where the intermodal tank is moved onto a railcar. The railcar then transports the intermodal tank to a shipyard, where it is loaded onto a container ship. This process continues in reverse through various modes of transportation until the container reaches its destination, at which point the sulfuric acid is used directly from the intermodal tank by the end user without ever having been unloaded at any stage of transportation (Figure 10-6).

intermodal container ■ Bulk packaging designed to be used in multiple modes of transportation.

Intermodal containers are gaining wide popularity in industry owing to their safety and low transportation costs. Often, accidents occur when products are transferred from one container to another, and incompatible materials are inadvertently mixed. For example, in 1999 at a tannery in Whitehall, Michigan, a truck driver was killed by hydrogen sulfide gas that was generated when he accidentally unloaded sodium hydrosulfide solution into a "pickle acid" (ferrous sulfate) tank. One of the ways this accident could have been prevented is if the sodium hydrosulfide had been delivered and used in an intermodal tank.

The width, height, and length of intermodal containers are standardized so they can easily be moved and stacked during transportation and storage. They are usually 8' wide and 8' to 8½' tall. Domestic intermodal containers, which are used only in rail and road transportation, are often 8½' wide. The 8½' wide intermodal containers do not conform to international standards. Lengths range from 20' to 53'. In addition, the corner castings, by which the intermodal containers are lifted and moved, are also standardized internationally (Figure 10-7). Intermodal containers must always be lifted and moved by their corner castings using the appropriate heavy machinery. If loaded intermodal containers

FIGURE 10-6 The transportation cycle of an intermodal container. The use of intermodal containers is convenient for industry and minimizes hazmat incidents caused by transferring the wrong product into the wrong tank. *Art by David Heskett.*

FIGURE 10-7 Intermodal containers should be lifted only by specifically designed equipment (bottom and top right) using the corner castings (top left). *Photos courtesy of Chris Weber, Dr. Hazmat, Inc.*

are lifted by a conventional industrial forklift along their midsection, they will almost certainly fail catastrophically.

The maximum capacity of intermodal tanks is typically less than 24,000 L (6340 gallons). Four types of intermodal tanks are commonly encountered in the United States: the IMO type 1 (also known as IM-101), the IMO type 2 (also known as IM-102), the IMO type 7, and the IMO type 5 (also known as DOT Spec 51). Table 10-4 lists the characteristics of these four types of intermodal containers. The main difference among them is the MAWP and, consequently, the products each is capable of carrying. The tank type and capacity, as well as other construction features of the tank, may be found on the

TABLE 10-4	Intermodal Containers
INTERMODAL CONTAINER TYPE	**SPECIFICATIONS**
Nonpressure or pressure tank Corner casting Beam type Box type	Non-pressure: 90% of tank containers in use Capacity: 5000–6300 gallons IM 101 or IMO Type 1 (25.4–100 psi) Pressure-relief valves Vacuum-relief valves Flammables with flashpoint less than 32°F IM 102 or IMO Type 2 (14.5–25.4 psi) Flammables with flashpoint between 32 and 140°F. Pressure: (LPG, anhydrous ammonia) Capacity: 4500–5500 gallons DOT Specification 51 or IMO Type 5 (100–600 psi) Enclosed fittings
Cryogenic tank	IMO Type 7 Liquefied compressed gases (cryogenic liquids) such as liquid nitrogen, liquid oxygen, liquid helium, liquid hydrogen, liquid carbon dioxide, and liquid argon These containers will vent periodically under normal operating conditions, especially warm weather
Tube module	3000–5000 psi T3 cylinders permanently mounted in an IMO frame Compressed gases such as nitrogen, oxygen, helium, and hydrogen

specification plate (Figure 10-5). The specification plate is found on the front of intermodal tanks, typically near the valves. The valves on some intermodal containers are not on the front of the tank. For example, on cryogenic containers they may be on the side, and on trichlorosilane containers they will be in a manway on top of the tank.

In addition to the specification plate, the tank will also have container markings that typically include the following information:

- reporting marks
- country code, tank size, type code
- approvals, compliance, and conformity information
- DOT or AAR specification markings
- DOT special permit (SP) marking (if applicable)
- hazardous materials markings and placards

The meaning of the codes for containers manufactured after 1996 can be found in Table 10-5, and those for containers manufactured before 1996 can be found in Table 10-6. Intermodal containers have become more prevalent in North America with the increase in international commerce owing to rapid globalization.

HIGHWAY CARGO TANKERS

Several different highway cargo tank configurations have been U.S. DOT approved for over-the-road transportation and are currently in widespread use (Figure 10-8). The Motor Carrier (MC) 300 series, which includes the MC 306, MC 307, MC 312, MC 331, and MC 338 cargo tankers, was the standard highway cargo tank design for several decades. In August 1995 the highway cargo tanker specifications were modernized into

TABLE 10-5 New Intermodal Container Codes (Alphanumeric Code for Containers Manufactured Since 1996)

1st Char	1	2	3	4	A	B	C	D	E
LENGTH	**2.991 m (10 feet)**	6.068 m (20 feet)	9.125 m (30 feet)	12.192 m (40 feet)	7.150 m	7.315 m (24 feet)	7.430 m (24.5 feet)	7.450 m	7.820 m
		2nd Character	**0**	**2**	**4**	**5**	**6**	**8**	**9**
		WIDTH	2.438 m (8 feet)	2.438 m (8 feet)	2.438 m (8 feet)	2.438 m (8 feet)	2.438 m (8 feet)	2.438 m (8 feet)	2.438 m (8 feet)
		HEIGHT	2.438 m (8 feet)	2.591 m (8.5 feet)	2.743 m (9 feet)	2.895 m (9.5 ft)	> 2.895 m (> 9.5 ft)	1.295 m (4.25 ft)	≤ 1.219 m (≤ 4 ft)
3rd Character		**A (AS-Air/ surface)**	**B (BU/BK-Dry bulk)**	**G (GP-General purpose w/o ventilation)**		**H (HI/HR-Thermal)**		**P (PC/PF/PL/ PS-Platform)**	
4th Char (Digit)	**0**		Closed, non-pressure	Openings at one or both ends		Refrigeration and/or heating with removable, external equip. (k = 0.4 W/m² K)		Plain platform	
	1		Airtight, non-pressure	Passive vents near top		Refrigeration and/or heating with removable, internal equip.		Two complete and fixed ends	
	2		-	Openings at one or both ends and full openings one or both sides		Refrigeration and/or heating with removable, external equip (k = 0.7 W/m² K)		Fixed posts, either free-standing or with removable top member	
	3		Horizontal discharge (1.5 bar test pressure)	Openings at one or both ends and partial openings one or both sides		-		Folding complete structure	
	4		Horizontal discharge (2.65 bar test pressure)			-		Folding posts, either freestanding or with removable top member	
	5		Tipping discharge (1.5 bar test press)			Insulated (k = 0.4 W/m² K)		Open top, open ends	
	6		Tipping discharge (2.65 bar test press)			Insulated (k = 0.7 W/m² K)			
	7		-						
	8		-						
	9		-						

(m = meters; ft = feet; k = heat transfer coefficient)

F	G	H	K	L	M	N	P
8.100 m	12.500 m (41 feet)	13.106 m (43 feet)	13.600 m	13.716 m (45 feet)	14.630 m (48 feet)	14.935 (49 feet)	16.154 m
C	**D**	**E**	**F**	**L**	**M**	**N**	**P**
> 2.438 m & < 2.5 m	>2.438 m & < 2.5 m	> 2.438 m & < 2.5 m	> 2.438 m & < 2.5 m	> 2.5 m	> 2.5 m	> 2.5 m	> 2.5 m
2.591 m (8.5 feet)	2.743 m (9 feet)	2.895 m (9.5 ft)	> 2.895 m (> 9.5 ft)	2.591 m (8.5 feet)	2.743 m (9 feet)	2.895 (9.5 ft)	> 2.895 m (> 9.5 ft)
R (RE/RS/RT-Thermal)	**S (SN-Named)**	**T (TD/TG/TN-Tanks)**		**U (UT-Open top)**			**V (VH-General purpose with ventilation)**
Mechanical refrigeration	Livestock carrier	Non-dangerous liquids (minimum pressure 0.45 bar)		Openings at one or both ends			Non-mechanical vents in lower and upper space
Mechanical refrigera-tion and heating	Automo-bile carrier	Non-dangerous liquids (minimum pressure 1.50 bar)		Openings at one or both ends with removable top members in end frames			-
Self-powered me-chanical refrigeration	Live fish carrier	Non-dangerous liquids (minimum pressure 2.65 bar)		Openings at one or both ends and openings one or both sides			Internal mechanical ventilation system
Self-powered me-chanical refrigeration and heating		Dangerous liquids (mini-mum pressure 1.50 bar)		Openings at one or both ends, openings one or both sides, and removable top members in end frames			-
		Dangerous liquids (mini-mum pressure 2.65 bar)		Openings at one or both ends and partial openings one side and full opening other side			External mechanical ventilation system
		Dangerous liquids (min press 4.00 bar)		No doors			-
		Dangerous liquids (min press 6.00 bar)					-
		Gas (min press 9.10 bar)					-
		Gas (min press 22.0 bar)					-
		Gas (min press TBD)					-

TABLE 10-6 Old Intermodal Container Codes (Completely Numeric Code for Containers Manufactured Prior to 1996)

1st Char		1	2	3		4
LENGTH		2.991 m (10 feet)	6.068 m (20 feet)	9.125 m (30 feet)		12.192 m (40 feet)
2nd Character		1	2	3		4
HEIGHT		2.438 m (8 feet)	2.438 m (8 feet) TUNNEL	2.591 m (8.5 feet)		2.591 m (8.5 feet) TUNNEL
3rd Char		0 General purpose container	1 Ventilated closed container	2 Dry bulk (0–4) and named (5-9) containers	3 Thermal containers	4 Thermal containers
4th Char (Digit)	0	Openings at one or both ends	Passive vents near top (< 25 cm^2/m vent cross section)	Closed, non-pressure box type	Refrigeration, expendable refrigerant	Refrigeration and/ or heating with removable, external equip.
	1	Openings at one or both ends and full openings one or both sides	Passive vents near top (≥ 25 cm^2/m vent cross section)	Vented, non-pressure box type	Mechanical refrigeration	Refrigeration and/ or heating with removable, internal equip.
	2	Openings at one or both ends and partial openings one or both sides	-	Ventilated, non-pressure box type	Refrigeration and heating	Refrigeration and/ or heating with removable, external equip.
	3	Openings at one or both ends and opening roof	Non-mechanical vents in lower and upper space	Airtight, non-pressure box type	Heated	Refrigeration and/ or heating with removable equip., spare
	4	Openings at one or both ends, partial openings one or both sides and openings roof	-	-	Heated	Refrigeration and/ or heating with removable equip., spare
	5	-	Internal mechanical ventilation system	Livestock carrier	Heated	Insulated
	6	-	-	Automobile carrier	Self-powered mechanical refrigeration	Insulated
	7	-	External mechanical ventilation system	-	Self-powered refrigeration and heating	Insulated, spare
	8	-	-	-	Self-powered heating	Insulated, spare
	9	-	-	-	-	Insulated, spare

5	6	7	8	9
-	-	-	10.668 m (35 feet)	13.716 m (45 feet)
5	**6**	**7**	**8**	**9**
>2.591 m(>8.5 ft)	>2.591 m (>8.5 ft) TUNNEL	>4′ and <4.25′	>4′ and <4.25′ TUNNEL	>4.24′ and <8′
5 Open top containers	**6** Platform containers	**7** Tank containers	**8** Dry bulk containers	**9** Air/surface containers
Openings at one or both ends	Plain platform	Non-dangerous liquids (test pressure 0.45 bar)	Closed, non-pressure hopper type	
Openings at one or both ends and removable top members in end frames	Incomplete superstructure with complete and fixed ends	Non-dangerous liquids (test pressure 1.50 bar)	Vented, non-pressure hopper type	
Openings at one or both ends and openings one or both sides	Incomplete superstructure with fixed free-standing posts	Non-dangerous liquids (test pressure 2.65 bar)	Ventilated, non-pressure hopper type	
Openings at one or both ends, openings one or both sides, and removable top members in end frames	Incomplete superstructure with folding complete and structure	Dangerous liquids (test pressure 1.50 bar)	Airtight, non-pressure hopper type	
Openings at one or both ends and partial openings one side and full opening other side	Incomplete superstructure with folding free-standing posts	Dangerous liquids (test pressure 2.65 bar)	Non-pressure hopper type, spare	
-	Complete superstructure with roof	Dangerous liquids (test pressure 4.00 bar)	Horizontal discharge (1.5 bar test pressure)	
-	Complete superstructure with open top	Dangerous liquids (test pressure 6.00 bar)	Horizontal discharge (2.65 bar test pressure)	
-	Complete superstructure with open top and open ends	Dangerous gases (test pressure 10.5 bar)	Tipping discharge (1.5 bar test press)	
-	-	Dangerous gases (test pressure 22.0 bar)	Tipping discharge (2.65 bar test press)	
HT, spare	-	Dangerous gases (test pressure unassigned)	Pressurized, spare	CO, spare

SOLVED EXERCISE 10-1

According to the intermodal container marking shown, what are the specifications of this container?

Courtesy of Chris Weber, Dr. Hazmat, Inc.

Solution: The intermodal code visible in the picture is "22T6." Since the four-digit code contains the letter *T*, this is a new code, with the following meaning:

2: The intermodal container is 20 feet long.
2: The intermodal container is 8 feet wide and 8½ feet tall.
T6: The intermodal container is able to hold dangerous liquids with a minimum pressure of 6 bar (87 psi).

the DOT 400 series, which created the DOT 406, DOT 407, and DOT 412 tankers. The primary difference between the 300 series and the 400 series are thicker tank shells, improved rollover protection, and improved manhole assemblies that can withstand higher static pressures. MC 306, MC 307, and MC 312 cargo tankers may no longer be manufactured, but existing cargo tankers may continue to be used in transport, and you will likely continue to encounter them on a regular basis for many years to come.

Highway cargo tanker properties are summarized in Table 10-7. As with intermodal tanks, the main difference among these tanks is their construction material and the MAWP, and, consequently, the types of products each tank can carry. Once again, highway cargo tankers have a specification plate, typically located on the frame rail on either side of the vehicle (Figure 10-8). The specification plate lists the important features of the tank, including tank type, construction material, number of compartments, and

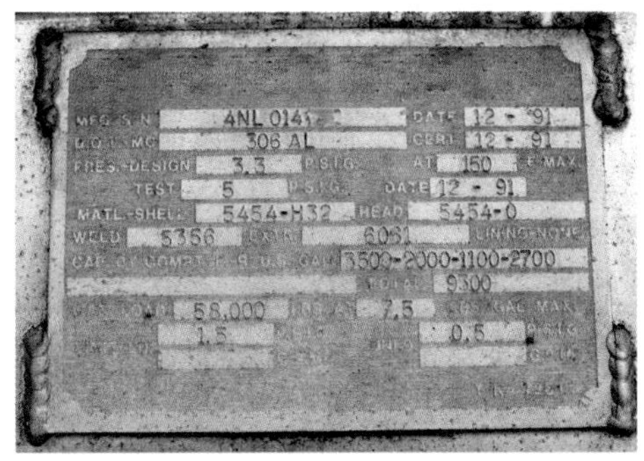

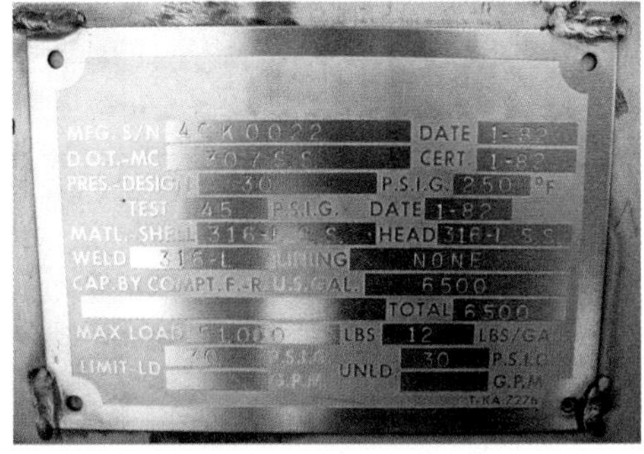

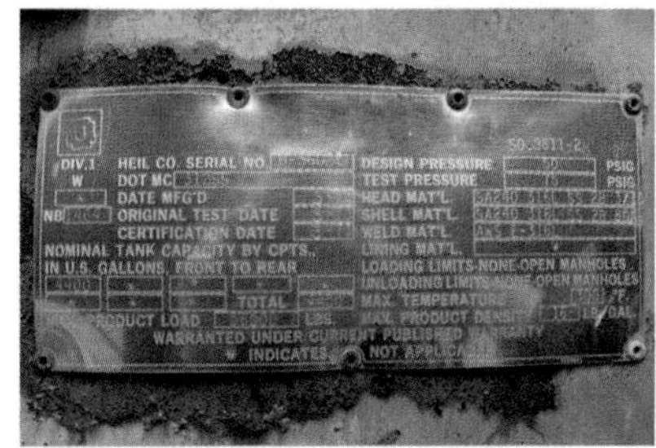

FIGURE 10-8 Highway cargo tankers come in many different shapes and design criteria (left column). The design criteria are listed on the specification plate (right column) that can be found on the frame rail of the trailer. (Top row): DOT 406/MC 306 highway cargo tanker; (second row): DOT 407/MC 307 highway cargo tanker; (third row): DOT 412/MC 312 highway cargo tanker; (fourth row): MC 331 high-pressure highway cargo tanker; (bottom left): MC 338 highway cargo tanker. The specification plate shown at the bottom right is for an MC 331 highway cargo tanker. *Photos courtesy of Chris Weber, Dr. Hazmat, Inc.*

TABLE 10-7 Highway Cargo Tankers

DOT 406/MC 306-non-pressure liquid tank

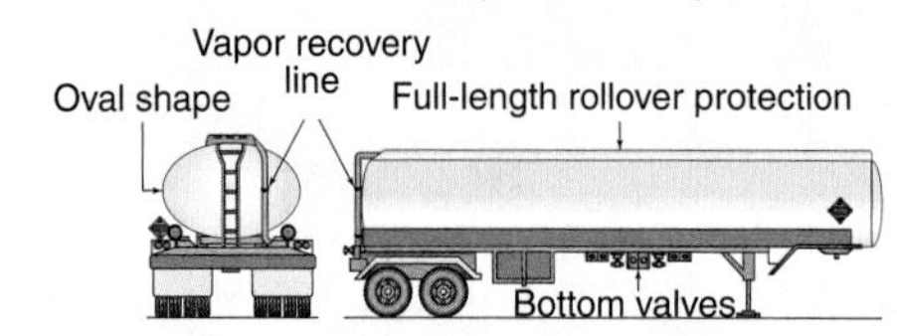

Tank pressure:	Less than 5 psi
Maximum capacity:	9000 gallons
Tank construction:	Aluminum (new); steel (old) Multiple compartments Recessed manholes/rollover protection Bottom valves Vapor recovery system (newer)
Typical contents:	Flammable liquids Combustible liquids Gasoline Diesel fuel Fuel oils Alcohols Solvents

DOT 407/MC 307-low pressure chemical tank

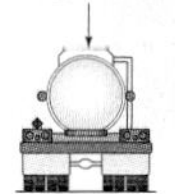
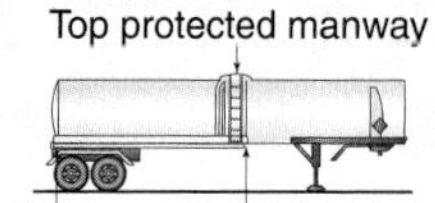

Tank pressure:	25–40 psi
Maximum capacity:	6000 gallons
Tank construction:	Steel with stiffening rings (typically double shell) May be rubber lined Single or double top manhole Single outlet discharge for each compartment at bottom (midship or rear) Discharges are sometimes customized by owner Often pressure unloaded (pneumatic) Rollover protection May have multiple compartments Horseshoe or round profile as viewed from rear
Typical contents:	Flammable liquids Combustible liquids Corrosives Poisons Hazardous waste Food-grade materials/products

TABLE 10-7 Highway Cargo Tankers (*Continued*)

DOT 412/MC 312-corrosive liquid tank

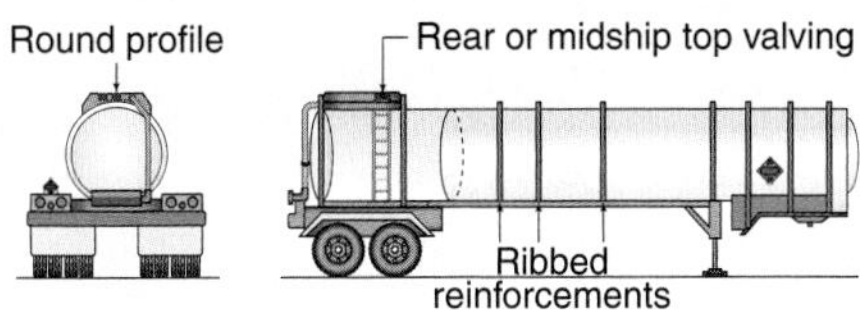

Tank pressure:	Less than 75 psi
Maximum capacity:	6000 gallons
Tank construction:	May be lined (butyl rubber or polyethylene) Recessed manhole Rollover protection around valving Corrosive-resistant paint band below valving Steel/stainless steel/aluminum with stiffening rings Top loading at center and/or rear Usually single compartment Baffled compartment to slow product surges during transport
Typical contents:	Corrosive liquids Strong acids Strong bases

MC 331: high-pressure tanker

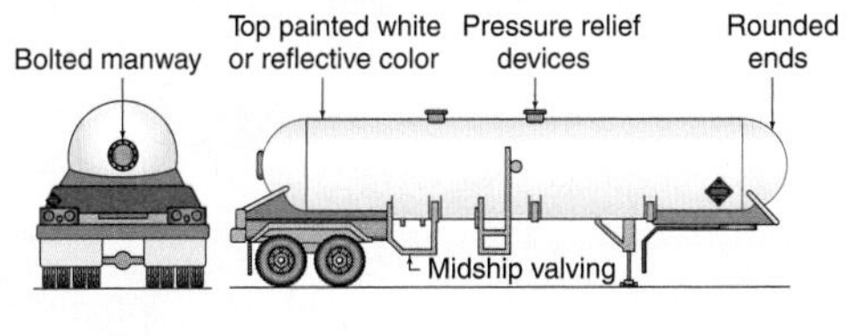

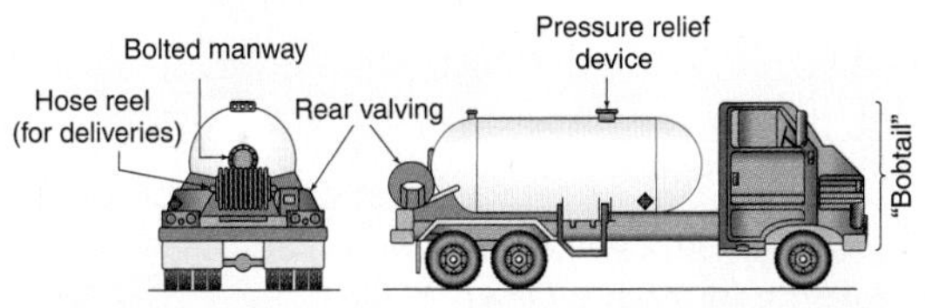

Tank pressure:	300 psi
Maximum capacity:	11,500 gallons
Tank construction:	Steel single compartment/noninsulated Bolted manhole at front or rear Internal and rear-outlet valves Excess flow shutoff valve Typically painted white or other reflective color May be marked "Flammable Gas" and "Compressed Gas" Round, dome-shaped ends
Typical contents:	Propane Anhydrous ammonia Butane Pressurized gases and liquids

(*continued*)

TABLE 10-7 Highway Cargo Tankers (*Continued*)

MC 338—cryogenic liquid tanker

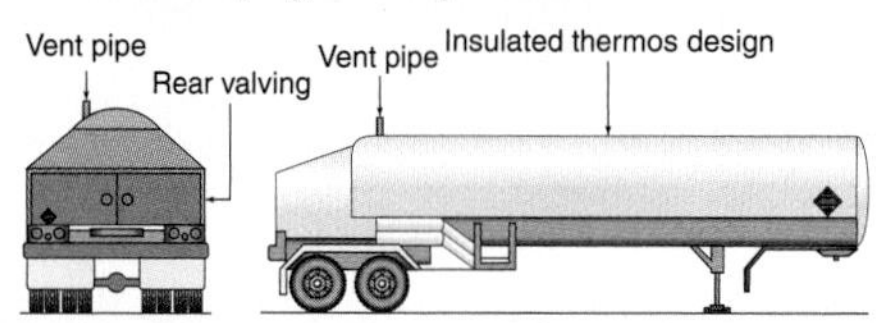

Tank pressure:	Less than 22 psi
Maximum capacity:	9000 gallons
Tank construction:	Well-insulated steel tank May have vapor discharging from vent pipe under normal operation Loading/unloading valves enclosed at rear Pressure relief valve May be marked "Refrigerated Liquid" Round tank with cabinet at rear
Typical contents:	Liquid nitrogen Liquid carbon dioxide Liquid hydrogen Liquid oxygen

Compressed gas/tube trailer

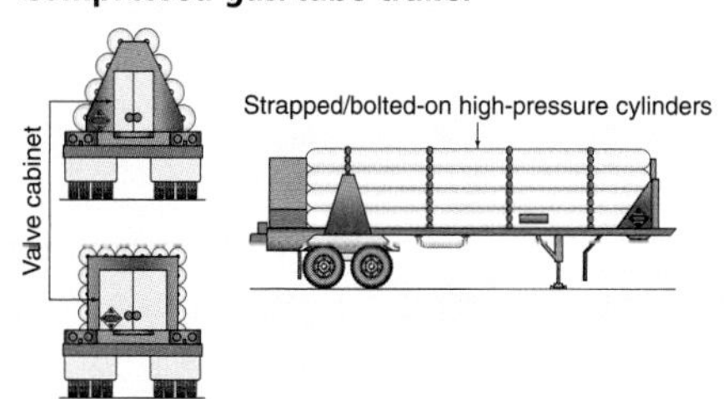

Tank pressure:	3000–6000 psi
Maximum capacity:	1500 cubic feet per cylinder
Tank construction:	Steel cylinders stacked and banded together Overpressure device for each cylinder Protected valving at rear Flat truck with multiple cylinders stacked in modular or nested shape
Typical contents:	Helium Hydrogen Methane Oxygen Argon Nitrogen

Dry bulk cargo tanker

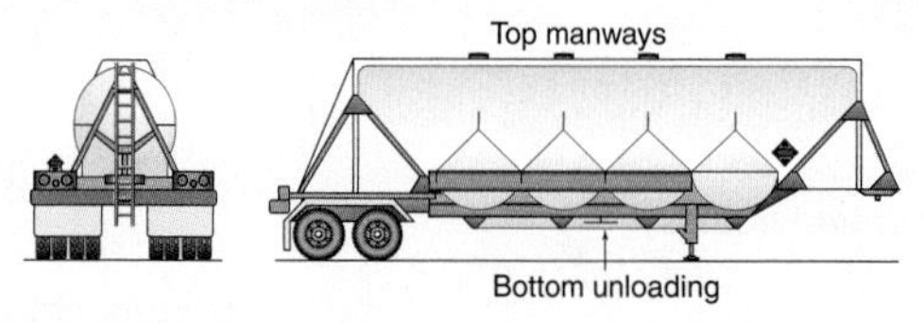

TABLE 10-7	Highway Cargo Tankers (*Continued*)
Tank pressure:	Not under pressure
Maximum capacity:	2500 cubic feet
Tank construction:	Top-side manholes Bottom valves Air-assisted loading/unloading Shapes vary but will have hoppers Aluminum shell Multiple compartments
Typical contents:	Calcium carbide Oxidizers Corrosive solids Cement Plastic pellets Fertilizers Cyanides

tank compartment capacity. The following are some common abbreviations found on specification plates:

AL: aluminum
CS: carbon steel
SS: stainless steel
MS: mild steel
HSLA: high-strength low-alloy steel

Some highway cargo tankers are termed "non-specification" cargo tankers. This means they are not required to comply with DOT specifications. The following are examples of non-specification cargo vessels:

- dry bulk
- pneumatically offloaded hoppers
- tube trailers
- molten sulfur haulers
- asphalt haulers

RAILROAD TANK CARS

With the exception of oil tankers and liquefied natural gas tankers, railroad tank cars are the largest tanks found in transportation. The extremely large quantities they carry makes them very dangerous vehicles. There is a world of difference between the pint of acetone you may have in the house being used as nail polish remover, and 40,000 gallons of acetone in a railcar. As with highway cargo tankers, railroad tank cars have a number of different railcar specifications, which are listed in Table 10-8. Railroad tank cars can be divided into two general classes: low-pressure tanks and high-pressure tanks.

Railcar Markings

Railcars have markings that indicate their contents, tank construction, and ownership. This information is crucial to a safe and effective risk-based response to a train derailment involving railcars containing hazardous materials. Figure 10-9 shows typical stenciling that can be found on the sides and ends of railcars. The stenciling indicates the company name, reporting mark, load limits, and tare weights. The reporting mark indicates the

TABLE 10-8 Railcar Tanker Specifications and Shapes

LOW-PRESSURE TANK CARS (NON-PRESSURE)	PRESSURE TANK CARS	CRYOGENIC
DOT **103**, 104, **111**, 113, and 115 AAR 201, 203, 204, 206, 207, 208, and 211	DOT **105**, 107, 109, **112**, and **114** DOT 106 and 110 (ton container or 1-ton cylinders)	DOT 113 AAR 204

Boldface indicates most commonly used railcars.

TANK FEATURES	SAFETY FEATURES	TYPICAL CONTENTS
Class 103 General Service Car 35–60 psi safety valves 10,000–40,000 gal. capacity Low pressure Bottom outlets allowed **Class 104** Low pressure		Benzene Corrosives Phosphorus trichloride Hydrogen peroxide Acrylonitrile
Class 105 High pressure	DOT 105A500W insulated carbon dioxide tank car	
75–450 PSI safety valve	Insulated top and bottom shelf couplers Thermal protection Head protection	Ammonium nitrate solution Anhydrous ammonia Anhydrous hydrofluoric acid Carbon dioxide Chlorine Ethylene oxide Hydrogen cyanide Liquefied hydrocarbon gas LPG Metallic sodium Sulfur dioxide Vinyl chloride
Hydrogen cyanide 1⅛-inch inner shell	DOT 105A500W or DOT 105J500W or DOT 105A600W or DOT 105J600W	
4-inch cork insulation ¼-inch outer shell Enclosed valves		

TABLE 10-8 Railcar Tanker Specifications and Shapes (*Continued*)

TANK FEATURES	SAFETY FEATURES	TYPICAL CONTENTS
Chlorine 90-ton capacity 375 psi safety valve	DOT 105A500W insulated	
Class 106A High pressure DOT 106A500X 375-600 psi safety valve		Chlorine Anhydrous ammonia Sulfur dioxide Butadiene Refrigerant gases
Class 107A High pressure		
Class 109A High pressure		
Class 111 Low pressure		Hydrochloric acid Phosphoric acid Aluminum sulfate Sulfuric acid Vegetable oil Phosphorus Gasoline Ethyl ether Ammonia solution Acetic acid Ethanol; Whiskey
Class 112 High pressure Bottom outlets	Thermal protection Head protection 10,000–40,000 gal. capacity 75–450 psi safety valve	LPG Sulfur dioxide Vinyl chloride
Class 114 High pressure No bottom outlets		
Class 115 Low pressure		Methyl methacrylate
Tube car (high pressure)	3000–5000 psi	Gases Helium, hydrogen, methane, oxygen

(*continued*)

TABLE 10-8 Railcar Tanker Specifications and Shapes (*Continued*)

TANK FEATURES	SAFETY FEATURES	TYPICAL CONTENTS
Class 113 Cryogenic tank car	16 psi outer shell 75 psi tank 30,000–40,000 gal. capacity	Liquefied gases Liquid hydrogen, liquid oxygen, liquid helium, liquid argon, liquid nitrogen, liquid carbon dioxide
Flat bed car with intermodal tanks		
Covered hopper car		Calcium carbide Cement Grain
Open-top hopper car		Materials are exposed to weather Coal, rock, sand
Pneumatic hopper car	15 psi 2600-5900 cubic feet	Fine powdered materials Plastic pellets, flour
Gondola car (open roof)		Materials are exposed to weather Sand, rolled steel
Box car		Varied May contain tank inside!

owner of the railcar as well as a five-digit numerical serial number unique to that tank car. The reporting mark is a crucial piece of information that can be used to gather other information such as tank contents, load level, origination, and destination by contacting the owner. From a safe distance, every effort should be made to locate the reporting marks to relay this information to the incident commander or the railroad representative.

Specification markings indicate information about tank construction. Figure 10-9 illustrates typical specification markings for a railcar. These markings indicate the authorizing agency, tank car class, safety features such as jackets, tank test pressure, tank construction material, type of welds, and even fittings and tank liners. This information is important to hazmat technicians performing damage assessments and offloading of the tank contents. Figure 10-10 shows the standard railcar layout and terminology for damage assessment and reporting. Table 10-9 lists tank shell thicknesses for common commodities.

FIXED STORAGE TANKS

Fixed-site storage tanks come in many sizes and shapes (Figure 10-11). The storage tanks may be above or below ground—the latter are commonly found at gasoline stations. At industrial facilities, fixed-site storage tanks are usually found above ground with secondary containment walls around them. The secondary containment system should be able to hold the larger of 100% of the largest container or at least 10% of the aggregate container volume when there are multiple storage tanks in the area. Aboveground storage tanks may be vertical or horizontal depending on the facility. Storage tanks used for chemical processes may be funnel shaped to facilitate product movement. Spherical storage tanks are used for high-pressure gas or liquefied gas storage (such as methane). In the petroleum

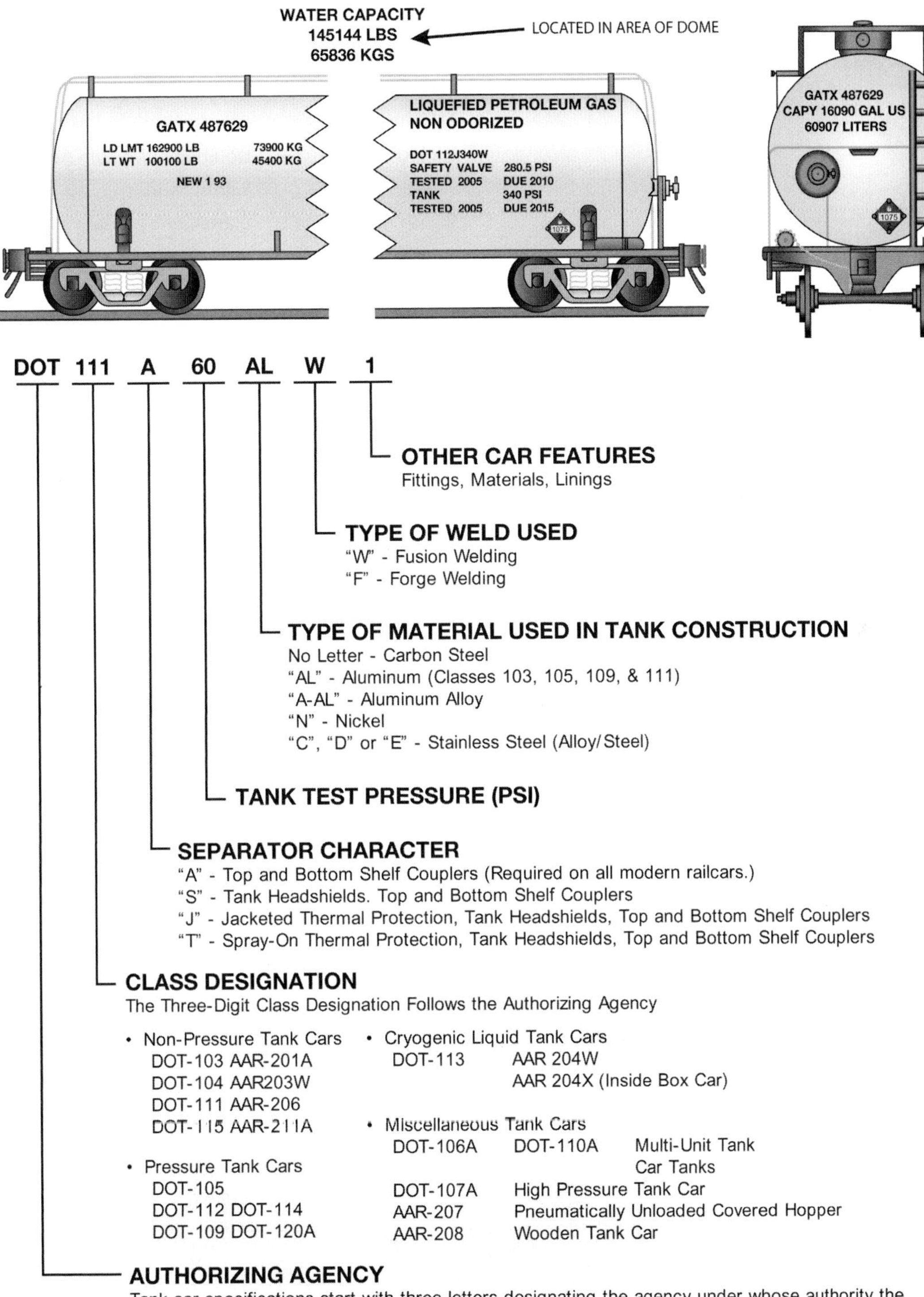

FIGURE 10-9 Railcars are required to be stenciled with important information concerning their contents and design criteria. *First Hazardous Materials Guide for First Responders, FEMA, USFA.*

TABLE 10-9	Tank Shell Thicknesses for Common Commodities	
TANK THICKNESS:	1⅛ inch	Hydrogen sulfide
	¾ inch	Chlorine
	11/16 inch	Liquified petroleum gas (LPG)
	7/16 inch	Sulfuric acid
	¼ inch	Container material
	⅛ inch	Jacket material

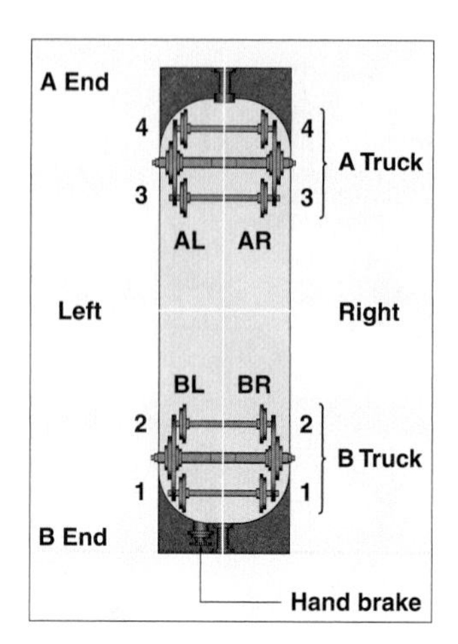

FIGURE 10-10 Common terminology for describing railcar construction for damage assessments. The end of the railcar with the brake is always referred to as the "B" end, regardless of its orientation with respect to direction of travel. *First Hazardous Materials Guide for First Responders, FEMA, USFA*

industry extremely large storage tanks may be used for crude oil, gasoline, diesel fuel, and other petroleum products. These tanks may be able to hold tens of millions of gallons of product. As you can imagine, catastrophic failure of one of these tanks could spell disaster. It is extremely important to plan and routinely visit industrial facilities in your response area to be adequately prepared in the event of an emergency. Fixed-site storage tank properties are summarized in Table 10-10.

The container shape and size often indicates the state of matter of the material it contains. For example, fiberboard drums typically carry solids, plastic 55-gallon drums often carry corrosives, and steel 55-gallon drums often carry flammable liquids. Cylinders typically carry compressed gases, and dewars usually contain cryogenic liquids. However these are not hard-and-fast rules; there are many exceptions! For example, a cylinder carrying acetylene gas actually contains a solid, porous material (such as calcium silicate) and acetone in which the gas is dissolved. The most reliable way to identify the contents of non-bulk containers is to look at the label, and/or the shipping papers if the container is involved in transport.

FIGURE 10-11 Fixed-site storage tanks come in a variety of shapes and sizes depending on their function. (Top row): a cryogenic tank (top left), a high-pressure storage vessel (middle), and a hydrocarbon fuel low-pressure storage tank (top right); (middle row): large-volume petroleum storage tanks; (bottom row): a horizontal high-pressure storage tank typically used to hold propane or anhydrous ammonia. *Courtesy of Chris Weber, Dr. Hazmat, Inc.*

TABLE 10-10 Fixed-Site Storage Tanks

TANK TYPE	TYPICAL HAZARDOUS SPECIFICATIONS	CONTENTS
Cone roof tank	Atmospheric pressure Pressure-vacuum valve for breathing Welded steel tank (*Caution*: older tanks may still be riveted and present a greater risk of failing under fire.) Diameter: 20–300 feet Capacity: 40,000–4,000,000 gallons Vapor space between liquid level and roof Weak roof-to-shell seam for emergency venting Possibly lined	Flammable liquids Combustible liquids Solvents Oxidizers Corrosives
Open floating roof tank Note the wind girder around the top.	Atmospheric pressure Steel tank Roof floats on the surface of product using pontoon or honeycomb system Seal area between tank shell and roof No vapor space (unless nearly empty) Limited drainage system to carry water off roof Diameter: 40–400 feet Capacity: 40,000–20,000,000 gallons Issues: seal fires from lightning strikes, roof sinking from incorrect water application or excessive precipitation	Flammable liquids
Covered floating roof tank Note the vent system at the top of the tank. Vent	Atmospheric pressure Welded steel tank Floating roof on liquid Fixed flat, geodesic, or conical roof supported by cylindrical tank walls No vapor space (unless nearly empty) Diameter: 40–400 feet Capacity: 40,000–20,000,000 gallons Issues: Internal tank fires can be difficult to extinguish owing to inaccessible void spaces.	Flammable liquids Combustible liquids
Vertical Storage tank Pressure relief device Tank supports Valving and piping	Low pressure (2.5–15 psi) Pressure-vacuum valve Welded steel tank (*Caution*: Older tanks may still be riveted and present a greater risk of failing under fire.) Some tanks may be fiberglass Weak roof-to-shell seam for emergency venting Diameter: up to 25 feet Capacity: 4000–400,000 gallons Possibly lined	Flammable liquids Combustible liquids Solvents Oxidizers Poisonous liquids Corrosives

(continued)

TABLE 10-10 Fixed-Site Storage Tanks (*Continued*)

TANK TYPE	TYPICAL HAZARDOUS SPECIFICATIONS	CONTENTS
Horizontal storage tank	Low pressure (up to 0.5 psi) Pressure-vacuum valve Wide range of construction: welded/riveted steel; often, tank within-a-tank design (double walled) Diameter: up to 20 feet Capacity: 300–100,000 gallons Issues: Tank supports must be protected during fires.	Flammable liquids Combustible liquids Solvents Oxidizers Poisonous liquids Corrosives
Pressure vessels Pressure relief valve	High pressure	Flammable gases

PIPELINES

Pipeline emergencies can be difficult to respond to, since they are often discovered and reported late, the actual location of the leak may be difficult to locate, and the exact product being transported may be difficult to identify. Most pipelines in the United States are buried below ground and are difficult to locate. Buried pipelines are required to be marked using pipeline markers (Figure 10-12). These markers are required to have the owner's name, emergency telephone number, and product the pipeline transports listed. However, pipelines often carry multiple materials. For example, a pipeline that carries gasoline and diesel fuel, among other materials, may just be labeled "petroleum products."

Long-distance pipelines typically do not transport alcohols such as ethanol because, owing to their high water solubility, alcohols cause much more galvanic corrosion than do the nonpolar hydrocarbons. Therefore, the majority of ethanol is currently transported by rail and road to the terminals where it is blended with hydrocarbon fuels.

Pipelines have aboveground valving stations positioned periodically. If you have one of these stations in your jurisdiction, it is very important to remember not to open or close valves unless directed to do so by an authorized representative of the pipeline company. Given the large diameter and pressures of pipelines, unauthorized valve movement (opening or closing) may lead to overpressurization and subsequent rupture or explosion of the pipeline. These effects may be felt many miles away.

FIGURE 10-12 An example of a pipeline marker. The pipeline company name and telephone number will be located prominently on the pipeline marker. *Courtesy of Chris Weber, Dr. Hazmat, Inc.*

Performance Packaging Requirements

Packaging for non-bulk containers is regulated in 49 CFR 178. Dangerous goods listed in the Hazardous Materials Regulations typically require specific containers and combination packages (multiple individual containers shipped at once in the same package) to be tested and certified. Table 10-11 describes the required tests for non-bulk packaging. If the package is certified, the UN symbol will be embossed for certified packaging (Figure 10-13).
Additional information useful to the hazmat technician will follow the UN symbol, such as container construction material, packing group designation,

SOLVED EXERCISE 10-2

According to the photo of a railroad tank car shown, what are the specifications of this container?

Courtesy of Chris Weber, Dr. Hazmat, Inc.

Solution: The railcar specification markings indicate this is a DOT 105 J 600 W tank car. This means it is a high-pressure tank car (DOT 105) with a test pressure of 600 psi, it has a jacket with insulation underneath for thermal protection (J), and fusion welded seams (W). The stenciling indicates it contains chlorine. The pressure relief device (PRD) will activate at 375 psi (based on the information in the table stenciled to the side of the railcar).

TABLE 10-11 Selected Required Tests for Non-bulk Packaging

TEST	REQUIREMENTS	SUBJECTED CONTAINERS
Drop test	49 CFR 178.603	All packaging
Leakproofness test	49 CFR 178.604	All liquid packaging
Hydrostatic test	49 CFR 178.605	All metal, plastic, and composite liquid packaging
Stacking test	49 CFR 178.606	All packaging with the exception of bags
Cooperage test	49 CFR 178.607	All bung-type wooden barrels
Vibration test	49 CFR 178.608	All packaging
Pressure differential	49 CFR 173.27	Packaging intended for air transport

capacity, container manufacturer, and certain testing criteria. The specific information included will depend on whether it is combination packaging, packaging designed for solids, packaging designed for liquids, or packaging designed for intermediate bulk containers (IBC) (Figure 10-14).

Packaging code designates the type of packaging and its construction material (Table 10-12). A packaging code followed by the letter *W* indicates associate administrator approval. A code followed by the letter *V* indicates special variation packaging.

FIGURE 10-13 The United Nations symbol found on performance packaging. *Courtesy of Chris Weber, Dr. Hazmat, Inc.*

Performance level identifies which type of packing group materials may be shipped in this packaging:

X: packing groups I, II, and III (strongest packaging)
Y: packing groups II and III
Z: packing group III only

The *specific gravity* field for liquid packaging indicates the maximum specific gravity of the material that may be placed in the container. When the specific gravity is less than 1.2, this field may be omitted.

Gross mass is the maximum gross quantity permitted to be contained in the packaging in kilograms (kg). An *S* following the gross mass field indicates that the packaging is designated to hold solids or inner packaging for combination packages.

Hydrostatic test pressure gives the internal hydrostatic test pressure in kilopascals (kPa). This field is not required for inner packaging of combination packaging.

Manufacturing information is required information. The *year of manufacture* field indicates the last two digits of the year of manufacture. Selected packaging, such as plastic drums and jericans, must also be marked with the month of manufacture. The *country*

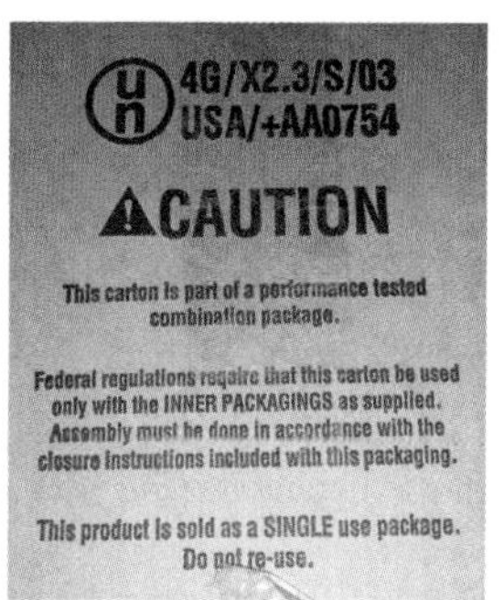

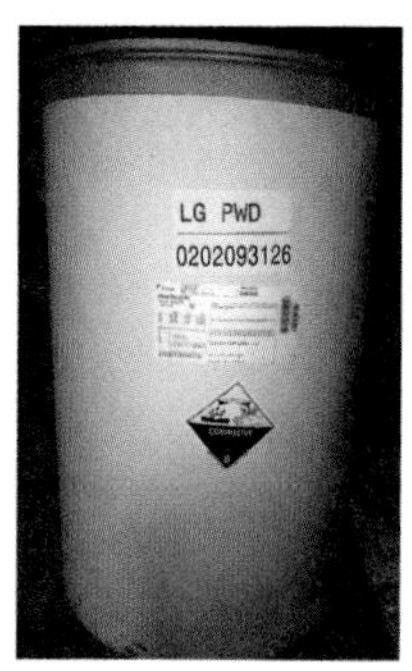

FIGURE 10-14 A variety of commonly encountered performance packaging that often contains hazardous materials, and their specification markings. *Photos courtesy of Chris Weber, Dr. Hazmat, Inc.*

TABLE 10-12 Selected Non-bulk Packaging Codes

§§178.504–178.521	
1A1	Steel drum, nonremovable head
1A2	Steel drum, removable head
1B1	Aluminum drum, nonremovable head
1B2	Aluminum drum, removable head
1D	Plywood drum
1G	Fiber drum
1H1	Plastic drum, nonremovable head
1H2	Plastic drum, removable head
1N1	Metal drum, nonremovable head
1N2	Metal drum, removable head
2C1	Wooden barrel, bung type
2C2	Wooden barrel, slack type, removable head
3A1	Steel jerican, nonremovable head
3A2	Steel jerican, removable head
3B1	Aluminum jerican, nonremovable head
3B2	Aluminum jerrican, removable head
3H1	Plastic jerican, nonremovable head
3H2	Plastic jerican, removable head
4A	Steel box
4B	Aluminum box
4C1	Wood box, ordinary
4C2	Wood box, sift-proof walls
4D	Plywood box
4F	Reconstituted wood box
4G	Fiberboard box
4H1	Plastic box, expanded
4H2	Plastic box, solid
5H1	Woven plastic bag, unlined or noncoated
5H2	Woven plastic bag, sift-proof
5H3	Woven plastic bag, water-resistant
5H4	Plastic film bag
5L1	Textile bag, unlined or noncoated
5L2	Textile bag, sift-proof
5L3	Textile bag, water-resistant
5M1	Paper bag, multiwall
5M2	Paper bag, multiwall water-resistant

§§178.522 AND 178.523	
6HA1	Plastic receptacle within a protective steel drum
6HA2	Plastic receptacle within a protective steel crate or box
6HB1	Plastic receptacle within a protective aluminum drum
6HB2	Plastic receptacle within a protective aluminum crate or box
6HC	Plastic receptacle within a protective wooden box
6HD1	Plastic receptacle within a protective plywood drum
6HD2	Plastic receptacle within a protective plywood box
6HG1	Plastic receptacle within a protective fiber drum
6HG2	Plastic receptacle within a protective fiberboard box
6HH1	Plastic receptacle within a protective plastic drum
6HH2	Plastic receptacle within a protective plastic box
6PA1	Glass, porcelain, or stoneware receptacles within a protective steel drum
6PA2	Glass, porcelain, or stoneware receptacles within a protective steel crate or box
6PB1	Glass, porcelain, or stoneware receptacles within a protective aluminum drum
6PB2	Glass, porcelain, or stoneware receptacles within a protective aluminum crate or box
6PC	Glass, porcelain, or stoneware receptacles within a protective wooden box
6PD1	Glass, porcelain, or stoneware receptacles within a protective plywood drum
6PD2	Glass, porcelain, or stoneware receptacles within a protective wickerwork hamper
6PG1	Glass, porcelain, or stoneware receptacles within a protective fiber drum
6PG2	Glass, porcelain, or stoneware receptacles within a protective fiberboard box
6PH1	Glass, porcelain, or stoneware receptacles within a protective expanded plastic packaging
6PH2	Glass, porcelain, or stoneware receptacles within a protective solid plastic packaging

of authorization field designates the country in which the packaging was manufactured. Country codes include the following:

CA: Canada
CN (PRC): China
FR: France
DE: Germany
JP: Japan
KR: Korea
MX: Mexico
US (USA): United States of America
UK: United Kingdom

The *manufacturer identification* field indicates the name and address or authorized symbol of the manufacturer or certifying agency.

The *minimum thickness* field is included on metal drums, plastic drums, jericans, or composite packaging intended for reuse. The letter *R* indicates reconditioned packaging, and the letter *L* indicates reconditioned packaging that has successfully passed a leakproofness test.

Intermediate bulk containers (IBC) will also show the *month and year of manufacture* and *stacking test load* in kilograms (kg) (Figure 10-14).

The performance packaging code fields give the hazmat technician valuable information regarding container contents, manufacturing materials, and container toughness. More information on container parameters can be found in 49 CFR.

Mechanisms of Damage

Container failure is caused by damage, and damage is caused by stress on the container. Let's examine all three of these phenomena—container failure, container damage and container stress—in reverse order.

SOLVED EXERCISE 10-3

According to the performance packaging UN marking shown, what are the specifications of this container?

Courtesy of Chris Weber, Dr. Hazmat, Inc.

Solution: Since this marking contains the UN symbol on the left, this is performance packaging code and has the following meaning:

1H1: This is a plastic drum with a nonremovable head.
Y1.8: This container is authorized for packing groups II and III and may contain materials with specific gravity up to 1.8.
100: The internal hydrostatic test pressure of the package is 100 kPa.
07: The container was manufactured in 2007.
USA: The container was manufactured in the United States.
M4723: This is the authorized symbol of the manufacturer.
RL: This is reconditioned packaging that has successfully passed a leakproofness test.

BOX 10-1 HAZMAT HISTORY: CHEMICAL OXYGEN GENERATORS: VALUJET FLIGHT 592 CRASH (1996)

On May 11, 1996, ValuJet flight 592 crashed in a remote portion of the Florida Everglades. The cause of the accident was determined to be a chemical fire caused by improperly packaged oxygen generators that were shipped by an airline maintenance contractor. Oxygen generators are used aboard aircraft to generate the oxygen for the oxygen masks that drop from the ceiling in the event of an emergency. When firing pins are activated, an exothermic chemical reaction is initiated that generates oxygen gas. Apparently, the firing pins were accidentally engaged during aircraft taxiing at Miami International Airport prior to takeoff because they were not covered with the required safety caps. In all, 110 people lost their lives including 105 passengers and five crew members. This crash, along with other near-misses involving lithium-ion batteries caused the FAA to strengthen regulations and enforcement of hazardous material shipments in passenger aircraft.

CONTAINER STRESS

Three primary types of stress can be placed on a container—and two or more are often applied at the same time (Figure 10-15):

- mechanical
- thermal
- chemical

Mechanical stress applies a force to the container. When the mechanical stress is within the design parameters of the container, no damage occurs—only wear and tear. Most mechanical stress is due to impact and pressure. Impact can cause compression in one portion of the container, and elongation in other portions. Unequal impact can cause torsion or twisting of the container, which may cause the container to tear. Repeated mild to moderate mechanical stress can cause metal fatigue and other chronic damage. Severe mechanical stress causes immediate visible damage.

Thermal stress applies heat or cold to the container. The container may be heated by flame impingement, compression, infrared or microwave irradiation, or an exothermic chemical reaction. Fifty-five-gallon drums of flammable liquids can explode and/or rocket when placed under fire conditions. Always ensure the isolation zone is large enough in such conditions. Filling a compressed gas cylinder too rapidly can significantly heat the container to the point of rupture. High temperatures typically weaken metals as they approach their melting temperature. Cooling may occur owing to extreme weather conditions or cryogenic liquids. Cooling typically makes metals more brittle as they approach and fall below their transition temperature.

Chemical stress is caused by a chemical reaction that typically degrades the container. Corrosive materials can cause chemical stress when they come into contact with an incompatible material. This often happens when lined metal containers corrode owing to a damaged or punctured liner. This type of damage is

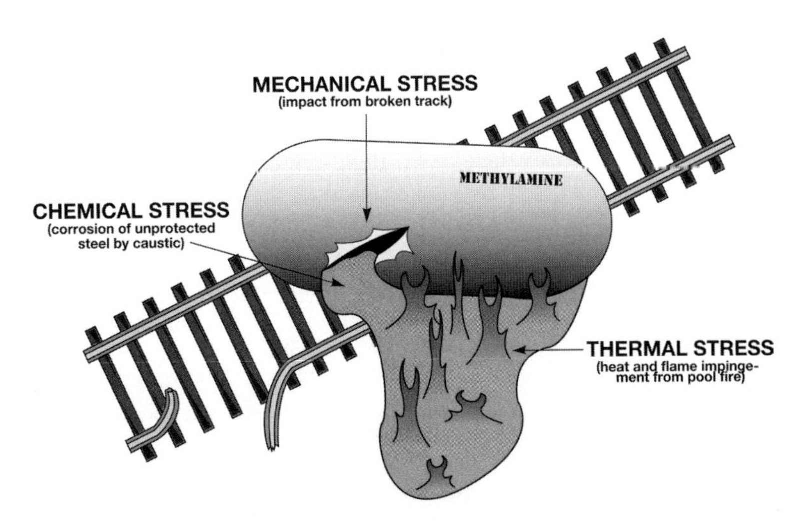

FIGURE 10-15 Containers can be stressed mechanically, chemically, and thermally. If the stress exceeds the design parameters, the container will leak, break, or violently rupture. *Art by David Heskett.*

particularly difficult to detect, since it starts on the inside of the container. Some chemical reactions, such as polymerization, may also produce heat and pressure.

Increased stress can occur if a chemical reaction produces a gas, as when a sodium hydroxide solution reacts with aluminum metal to produce hydrogen gas. In fact, in 2011 an unlined aluminum highway cargo tanker exploded when it was mistakenly loaded with a sodium hydroxide solution. Under atmospheric conditions the surface of the aluminum metal oxidizes to form an aluminum oxide coating that protects the underlying metal. Bases, such as sodium hydroxide, are able to destroy the aluminum oxide coating and attack the metal. Acids also can react with metals and produce hydrogen gas, for example, when liners inadvertently get punctured in metal containers.

CONTAINER DAMAGE

When container stress exceeds the container design parameters, visible and invisible damage ensues. Mechanical stress may cause dents, scores, gouges, tears, cracks, fractures, or punctures. Chemical stress causes corrosion.

A *dent* is a deformation of the container caused by an impact (Figure 10-16). If the impact is forceful enough, the container may be punctured, gouged, or scored. A *puncture* occurs when a relatively sharp object penetrates the container completely (Figure 10-17). Punctures typically occur in softer metals such as aluminum and mild steel. A *gouge* occurs when a relatively sharp object scrapes the container surface and removes container material, thereby thinning the container wall. A *score* occurs when a relatively blunt object scrapes the container surface and displaces container material, thereby thinning the container wall. The depth of the gouge or score is important in the damage assessment process (Figure 10-18).

cold work ■ Damage or deformation of steel at comparatively low temperatures—usually ambient temperature but always below the recrystallization temperature of the metal. Examples include rail burns, road burns, and wheel burns.

Cold work occurs when an object moves across a container forcefully for some period of time and the container is damaged and deformed at comparatively low temperatures (typically ambient temperature). Rail burns, road burns, and wheel burns are all examples of cold work usually accompanied by scoring. Cold work is caused by intense mechanical stress at temperatures below the recrystallization point of the metal. The *recrystallization point* is the temperature to which a metal must heated in order for the fine structure of the metal to form a more highly ordered structure thereby making it more ductile—an essential component of the annealing process. In contrast, cold work makes the container more brittle—therefore less ductile—at the damage location. This dramatically increases the likelihood of catastrophic container failure at the damage location due to cracking.

A *crack* or *fracture* is a narrow split in the container that may or may not penetrate the full thickness of the container. Cracks are often a result of *denting*, which causes both compression and tension on opposite sides of the dent (Figure 10-16). The concave surface of the dent is the area of compression and is not prone to cracking. The convex surface of the dent, typically located on the inside of the container, is the area of tension and is prone to cracking. This can be a major problem, since small cracks on the inside of the container will not be noticed during an external inspection. These cracks can then grow as temperature and pressure changes occur or when the container is moved and can lead to catastrophic container failure. Containers made of brittle materials, such as QT steel or glass, are much more prone to cracking than ductile materials such as aluminum and mild steel. Visible cracks in the base metal of the tank, especially the weld areas, are extremely dangerous in pressurized containers, and therefore extreme

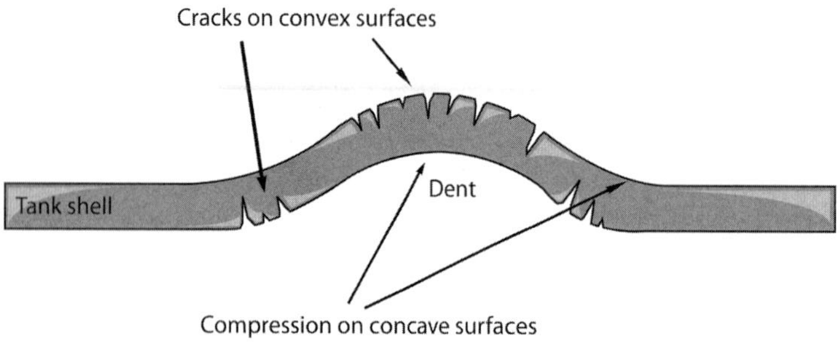

FIGURE 10-16 Dents cause unseen damage on the inside of the container. Container intrusion causes compression on the outside concave surface and causes cracks to form on the inside convex surface. *Art by David Heskett.*

caution should be exercised—this includes reducing the internal pressure and unloading the contents—before moving the container (see Chapter 11).

CONTAINER FAILURE

Container failure occurs when it suffers severe damage (Figure 10-17). Container failure may occur instantaneously, such as when a glass jug disintegrates on impact. Rapid container failure also occurs under fire conditions with plastic containers, which tend to melt. Plastic 55-gallon drums and larger IBCs will melt under fire conditions and fail catastrophically. If they contain flammable liquids, a burning chemical flow will ensue. Ensure that all exposures, especially downhill, are protected or moved—such as fire apparatus. In October 2011 a ladder truck was lost in exactly this way in Waxahachie, Texas, at a large chemical plant fire when burning flammable liquids flowed out of the plant and overwhelmed the truck company. Fortunately, no one was seriously injured or killed in that incident.

Container failure may also occur slowly over time, such as when cracks propagate slowly or when thermal stress builds up slowly—as in the February 1978 BLEVE that occurred in Waverly, Tennessee, several days after the initial train derailment (see Box 5-5). Runaway cracking typically occurs in pressurized containers, since the higher internal pressure applies a greater stress on the container. Containers with low design test pressures typically tear or burst, as when an aluminum soda can is frozen or dropped.

Container failure often occurs at openings such as valves, manway covers, gauging devices, sampling ports, and pressure relief devices. Container openings tend to get sheared off during accidents. Modern manufacturing standards typically mandate that large-bore openings have automatically closing internal valves and sheer grooves on external piping.

Damage Assessment

Damage assessment is the process of identifying the type of container—including design pressure and contents—and evaluating the structural integrity of the container. Damage assessment is the first step in formulating offensive product control strategies and tactics.

FIGURE 10-17 Containers can be punctured by sharp objects—such as railcar couplings. *Courtesy of Chris Weber, Dr. Hazmat, Inc.*

For safety reasons the first inspection should be done at a distance using binoculars. This is especially important with large containers (such as rail tank cars) and when weather conditions are rapidly changing. For example, if a derailment occurs at night when temperatures are cooler and solar radiation is absent, and then emergency response operations commence around dawn, rapid temperature and pressure changes in damaged containers could lead to unexpected runaway cracking and catastrophic failure of a container.

There will likely be portions of the container you cannot see and evaluate visually. This can be extremely dangerous, since the worst damage may be underneath. You'll have to take into account the mechanism of damage—in other words, how the damage occurred. How much energy was involved? What path did the container take? What sharp objects or edges were in the likely path of travel? Answering these questions correctly may save your life.

The damage assessment process should take into account the following key components:

- container type and features
- container construction material
- container contents
- visible container damage
- inaccessible container areas

It is important to determine the container type, as this will provide information about its construction material, its safety features, and the type of safety relief devices present. Container type and construction can be determined from the specification plate on highway cargo tankers, intermodal containers, and unjacketed rail tank cars, or the stenciling on jacketed and unjacketed rail tank cars. The built date can also be helpful in ascertaining tank car construction, because different metals were used during different time periods. As a last resort, the certificate of construction can be obtained from the tank car owner.

The container contents will affect container stability during an emergency. Contents may be solids, liquids, cryogenic liquids, compressed liquefied gases, or compressed gases. The identity of the product will tell you the vapor pressure of the material and consequently the internal pressure of the tank at any given temperature. The appropriate reference materials should be consulted that contain charts of commodity vapor pressure versus temperature. Two good sources of this information are the Gas Encyclopedia maintained online by Air Liquide and the *Handbook of Compressed Gases* (see Chapter 6 for a description of these references).

Critical damage is any damage that can lead to catastrophic failure of the container at any time during the emergency response or cleanup phases of the hazmat incident. If critical damage is present, the tank car should be unloaded without moving it. Critical damage includes all the following:

- any cracks in the tank shell metal
- any damage across a weld or touching the heat-affected zone—including dents, gouges, scores, and cold work such as rail burns
- any score or gouge deeper than ⅛ inch (until evaluated by a technical expert)
- dents that include any gouges, scores, or cracks
- dents with a radius of less than 2 inches or showing evidence of cold work if the tank car was built in 1967 or after
- dents with a radius of less than 4 inches if the tank car was built before 1967
- impinging flame

Crack growth should be closely watched. One way to monitor cracks is to mark each end of the crack with paint, marker, or chalk and observe the marks over time. Cracks tend to grow rapidly in brittle steel, and more slowly in ductile steels.

The heat-affected zone within a weld's width on either side of the weld is a major area of concern. This area does not have the same properties as the rest of the tank and is less ductile and more brittle than the rest of the container. Catastrophic tank shell cracks are more likely to propagate from the heat-affected zone than anywhere else on the container. However, damage merely to the crown of a weld is not considered significant.

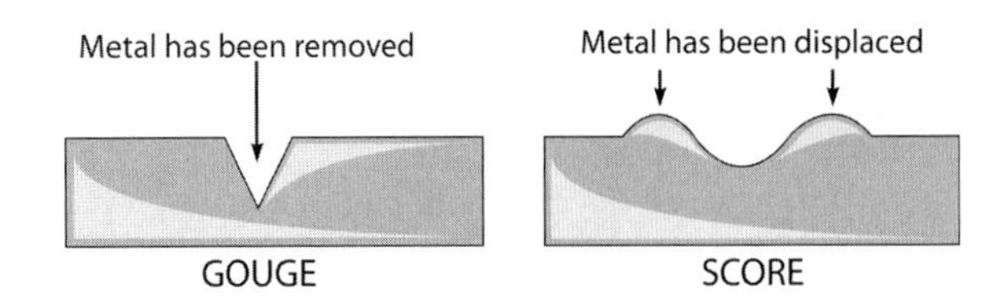

FIGURE 10-18 Gouging removes metal, while scoring displaces metal. In either case, the depth of the damage should be measured from the surface of the container (not the lip in the case of a score). *Art by David Heskett.*

Gouge and score depth are measured from the surface of the tank shell to the bottom of the gouge or score. Any displaced metal forming a lip at the edge of the score need not be measured as part of the score depth (Figure 10-18). Longitudinal gouges and scores—those that run parallel to the axis of the container—are generally of greatest concern.

Dents are not necessarily a problem for containers. Dents running parallel to the axis of the container are of greater concern than those running along the circumference. Red flags concerning dents are those with a small radius, and long dents, especially when they cross weld seams. Dents are usually evaluated using a dent gauge designed for that container type. Dents in the tank shell head are of concern only if they are accompanied by cracking. Rail burn is often of concern because it is usually a long dent accompanied by cold work. Remember, cold work weakens steel by reducing its ductility.

Insulated tanks may be difficult to inspect. The outer jacket that covers the insulation is not structural, so any damage to the jacket is irrelevant to the damage assessment. Of concern is the tank shell, which is underneath the outer jacket and 2 to 4 inches of insulation. Sometimes, portions of the outer jacket and underlying insulation may need to be removed. This must be done very carefully using proper air-monitoring techniques, grounding and bonding when necessary, and nonsparking tools when necessary.

When many different chemicals are transported together, there is no guarantee that one container will be compatible with the chemical stored in another container in close proximity. Therefore, when one or more chemicals are released from their container it is important to assess the condition of the other containers as well as their construction material.

Any containers that have critical damage should be gauged as soon as possible. *Gauging* refers to using or installing a pressure gauge on the tank to monitor the internal pressure. When this is not feasible, the internal pressure can be estimated from the vapor pressure of the material at the ambient temperature. The two values are approximately equal.

SOLID CARGO

Solid materials typically are not very volatile, and containers carrying solids typically operate at atmospheric pressure. The exception to this rule is the pneumatic loading and unloading of fine solids and powders. Most other solids are top loaded and bottom unloaded by gravity. Containers that carry solid materials typically have a funnel, or hopper, shape.

Container Hazards

Containers carrying solid cargo may have an engulfment hazard depending on their size. Pneumatic containers may have a pressure hazard as well during part of their work cycle.

Types of Containers

Non-bulk Containers Non-bulk containers are defined by a maximum capacity of less than 882 pounds (401 kg) for solids. Non-bulk containers carrying solids will typically be fiberboard drums, boxes, bags, and plastic and glass bottles.

Highway Dry Bulk Commodity Trailers Dry bulk commodity trailers, also known as *pneumatic trailers* or *hopper trailers,* fall under the category of non-specification cargo tanks (Figure 10-19). These trailers are constructed of aluminum or steel and typically

SOLVED EXERCISE 10-4

According to the container damage shown, how critical is the damage to this container?

Courtesy of Chris Weber, Dr. Hazmat, Inc.

Solution: This is a jacketed MC 307 highway cargo tanker (based on the specification plate, which is not pictured). The damage appears to be a dent in the outer jacket. These cargo tankers typically have several inches of insulation underneath a thin stainless steel jacket. Based on the radius of the dent, the damage appears to be superficial and has not affected the tank shell. A transportation specialist should be consulted to verify this assumption.

transport dry, flowable products. These trailers may be unloaded through gravity or pneumatic means and are designed to operate up to a pressure of 15 psi. These trailers may carry such diverse products as foodstuffs like grains, water-reactive metals such as magnesium powder, or even explosives such as ammonium nitrate/fuel oil (ANFO) mixtures.

Dry Bulk Railcars Dry bulk rail cars are larger versions of dry bulk highway cargo tankers. They have an internal capacity of up to 6000 ft^3 depending on the weight of the product they are hauling. Heavier products, such as cement, may occupy only half this volume.

Container Construction

Dry bulk railcars are typically constructed of steel owing to the large amount of weight they are able to carry. They may be covered or uncovered depending on the nature of the cargo. Many dry bulk railcars are pneumatically loaded and unloaded and are designed for a minimum internal pressure of 15 psig. Some of these railcars may be mechanically unloaded or gravity unloaded.

Damage Assessment

Damage assessment is not as critical with dry bulk containers, since they are not high-pressure vessels. When these containers fail they tend to spill and do not detonate or cause large chemical plumes.

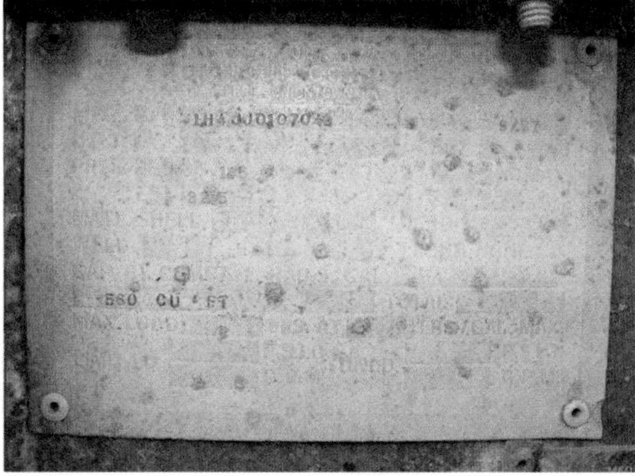

FIGURE 10-19 There are many non-specification highway cargo tankers carrying hazardous materials legally on the roads today. Some non-specification cargo tankers were once certified but are now decertified and can contain such materials as hot asphalt (top row). Dry bulk cargo tankers (middle row) and compressed gas tube trailers (bottom row) are both non-specification cargo tankers that often carry hazardous materials.
Courtesy of Chris Weber, Dr. Hazmat, Inc.

LIQUID CARGO

Liquids are generally more volatile than solids and when released tend to migrate in two dimensions—they flow. Therefore, concerns include direction of flow—such as toward people, occupancies, bodies of water, sewer openings, or other reactive materials that may be nearby.

Container Hazards

Liquids may have higher vapor pressures, and heating of the liquid can cause a boiling-liquid, expanding-vapor explosion (BLEVE) in susceptible containers. Typically, containers that BLEVE are made of steel and are able to hold pressure but have insufficient pressure relief valves (PRVs). Aluminum containers typically fail by melting or tearing rather than exploding owing to their lower melting temperature than steel.

FIGURE 10-20 A scrubber used to neutralize or remove dangerous vapors emitted from a chemical process. The scrubber may contain an acid or base solution (inside the glass beaker) to neutralize a corrosive vapor, or it may contain a reactive chemical such as bleach that breaks down a toxic vapor. *Courtesy of Chris Weber, Dr. Hazmat, Inc.*

Types of Containers

Laboratory Glassware It is important to understand the purpose of different glassware setups to be able to assess their hazards and determine the most effective sampling points for evidence collection and sample identification. Both reflux apparatus and distillation apparatus must remain open to the atmosphere so that temperature changes will not cause an explosion or implosion. Typically, there will be a tube that leads to a scrubber, which removes any harmful vapors (Figure 10-20). A scrubber usually contains a corrosive solution, either an acid or a base, that reacts with the harmful vapors and inactivates or neutralizes them. When investigating clandestine labs be extremely careful not to disconnect any scrubbers, because you may inadvertently create an extremely hazardous—even deadly—atmosphere. Also be aware that even if there is a scrubber, it may be set up incorrectly and thus not effectively removing dangerous contaminants. Proper air monitoring is essential to alert you to a dangerous atmosphere.

Non-bulk Containers Non-bulk containers are defined by a maximum capacity of less than 119 gallons (450 L) for liquids. Non-bulk liquid containers may carry flammable liquids, oxidizers, poisonous liquids, radioactive liquids, and corrosive liquids. Examples of flammable liquids include gasoline and acetone; examples of oxidizers include strong hydrogen peroxide solutions; examples of poisonous liquids include pesticides; examples of radioactive liquids include radioisotopes used in biomedical research; and examples of corrosives include muriatic acid and sodium hydroxide solutions.

IM-101 Intermodal Tanks IM-101 intermodal tanks, or IMO type 1 when used internationally, are designed to accommodate pressures between 25.4 psig and 100 psig (Figure 10-21). Their maximum capacity is typically 6300 gallons. These tanks have a housing on top that contains the product piping. Typical contents of IM-101 tanks include food-grade products, nonflammable liquids, and mild corrosives.

IM-102 Intermodal Tanks IM-102 intermodal tanks, or IMO type 2 tanks when used internationally, are designed to accommodate lower pressures than the IMO-101 tanks (Figure 10-21). Tank pressures will be between 14.5 psig and 25.4 psig, and the maximum capacity is 6300 gallons. An indication that you are dealing with an IMO type 2 tank is that product piping will be visible on the outside of the tank. All valves must be contained within the structural framework. Typical contents of IM-102 tanks include food-grade commodities, alcohols, pesticides and insecticides, industrial solvents and flammable liquids, and strong corrosives.

DOT 406/MC 306 Highway Cargo Tankers Some tankers with which you may be very familiar, at least from seeing them on the road, are gasoline tankers (UN 1203), which are DOT 406 or MC 306 specification tanks (Figure 10-8). Visually, their most striking feature is typically an elliptical (oval) cross section when viewed from either end (although there are exceptions to this rule). They are typically constructed of aluminum or carbon steel and transport petroleum products such as gasoline, diesel fuel, fuel oil, motor oil, jet fuel (JP 8), kerosene, E85, ethanol and a variety of hydrocarbon-based solvents.

Often, these tankers have multiple compartments. When these tankers are involved in accidents, one or more of their compartments may be damaged or leaking. Compartments are separated by bulkheads, and baffles are located periodically inside the tank to increase

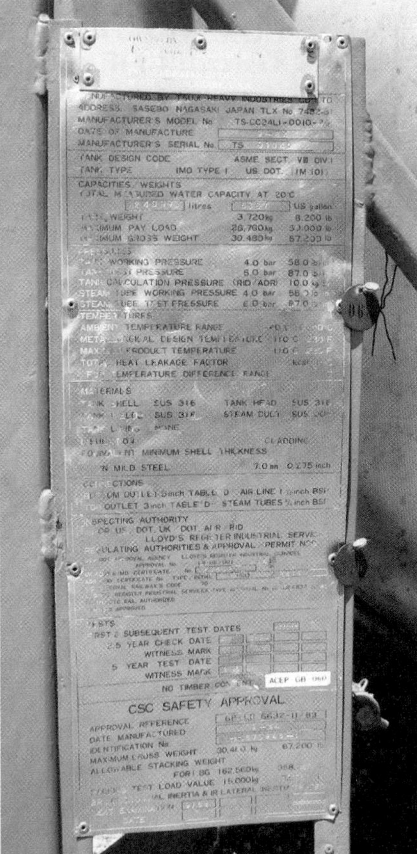

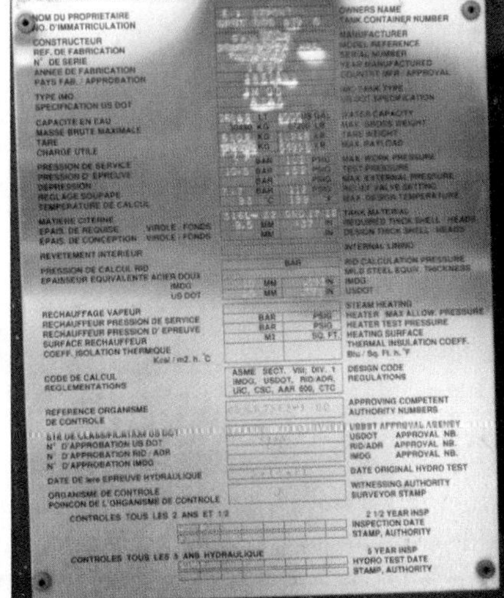

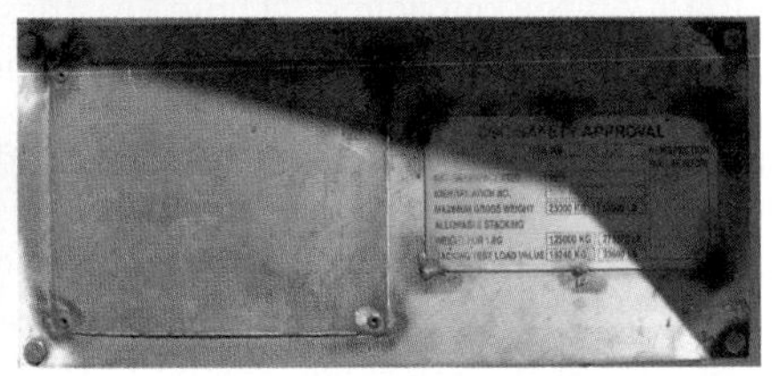

FIGURE 10-21 Intermodal containers come in a variety of sizes and shapes depending on their intended function (left column). Intermodal containers also have a specification plate, typically near the valves, that lists the design parameters of the container (right column). Not pictured is the IMO Type 7 intermodal container which carries cryogenic liquids. *Bottom left photo courtesy of Dean DeMark, HAZMAT/CBRNE Specialist (N.C.). All other photos courtesy of Chris Weber, Dr. Hazmat, Inc.*

BOX 10-2 TACTICS: GASOLINE TANKER FIRES

When a DOT 406 or an MC 306 tanker carrying hydrocarbons such as gasoline, gasohol, E85, or ethanol catches fire, it presents a special problem. A decision must be made whether to let it burn or extinguish it. Trying to extinguish these tankers using water can exacerbate the problem tremendously. Hydrocarbons such as gasoline are lighter than water. The water will sink to the bottom of the tanker and cause the burning gasoline to overflow and creating a river of burning gasoline seeking the lowest point in the immediate area. The lowest points are usually sanitary and storm sewers as well as basements of dwellings. Then, instead of having just a gasoline tanker fire, you may very well have multiple structure fires and grass fires as well.

Foam may be used to fight hydrocarbon fires, but is imperative that you have enough foam on hand to extinguish the fire and maintain a foam blanket to suppress vapors before you begin operations. Mixtures of gasoline and ethanol, such as gasohol and E85, can pose problems as well. Gasohol usually contains 10%–20% ethanol and behaves more like a hydrocarbon. E85 (which contains 85% ethanol, 15% gasoline) behaves more like an alcohol and dissolves completely in water. The higher the ethanol content, the more soluble the fuel will be in water, so alcohol-resistant foam is required to fight the fire. As with hydrocarbon fires, it is imperative that you have enough foam on hand to extinguish the fire and maintain a foam blanket to suppress vapors before you begin operations. If you have the appropriate resources—primarily the correct foam and a steady water supply, it is best to extinguish the fire. This must be done systematically (see Box 11-8).

Courtesy of Gary D. Brandt.

Sometimes a better way to solve the problem is to let the tanker burn. The risk of a BLEVE is minimal owing to DOT 406 and MC 306 tanker construction safety features. The aluminum shell of the tanker will burn down as the gasoline is consumed. Environmental agencies usually prefer this option rather than risk having the gasoline overflow and enter the soil and migrate into the groundwater table. This tactic should be avoided with tankers constructed of steel or when the fire is burning near critical infrastructure such as a bridge or overpass. As with all decisions at hazardous materials incidents, it is up to the incident commander to make an informed decision based on training and experience.

structural stability. The baffles have four holes to allow movement of product within a given tank compartment. Tank capacities range from 1500 gallons to 13,000 gallons (in some states). Do not allow a truck driver to disconnect the tank trailer from the rig and set it down on the outriggers. Typically, these tanks are not designed to have loaded weight supported by the outriggers. In fact, in some cases the outriggers have punctured the cargo tank.

stiffening ring ■ A component of a cargo tank that adds structural strength to the tank. Stiffening rings can be observed welded onto the outside of MC 312 and DOT 412 cargo tankers and uninsulated MC307 and DOT 407 cargo tankers.

DOT 407/MC 307 Highway Cargo Tankers Another common tank is the DOT 407 or MC 307 chemical tank, which is the workhorse of the chemical industry (Figure 10-8). These cargo tankers are often insulated, which gives them a horseshoe-shaped appearance from behind. The tanks themselves are circular cylinders and have **stiffening rings** on the outside. Stiffening rings are included because these tankers often carry heavier products, such as corrosives, water-based solutions, and halogenated hydrocarbons. These tankers may transport just about any hazard class material, including flammable and combustible liquids, corrosives, poisons, and miscellaneous dangerous goods. The tanks are typically constructed of stainless steel to withstand corrosion and provide greater structural stability.

Both the 406/306 series and 407/307 series of tankers are considered low-pressure cargo tankers, although the DOT 406 tank can accommodate up to 3 psi residual pressure and the DOT 407 tanker, up to 40 psi residual pressure.

DOT 412/ MC 312 Highway Cargo Tankers Corrosives are carried in specialized cargo tankers owing to their reactivity with most metals. These tankers have a circular cross section, a relatively small diameter, and many stiffening rings (Figure 10-8). Stiffening rings are necessary because most corrosives are quite heavy. Corrosives generally weigh significantly more than water. For example, water weighs 8.3 pounds per gallon at room temperature, while pure sulfuric acid weighs 15.35 pounds per gallon. In comparison, gasoline weighs only 6.2 to 6.3 pounds per gallon. Table 10-13 lists the weight per pound of material for some common commodities transported in cargo tankers and railcars. The 412/312 series tankers transport acids and bases such as sodium hydroxide, sulfuric acid, hydrochloric acid, phosphoric acid, and nitric acid. These tanks are typically constructed

TABLE 10-13 Density in Pounds per Gallon of Common Liquid Commodities

COMMODITY	WEIGHT (POUNDS/GALLON)	COMMODITY	WEIGHT (POUNDS/GALLON)
Acetic acid (glacial)	8.75	Methanol	6.59
Acetone	6.59	Methylamine	5.77 (at 20°F)
Acrylonitrile	6.72	Methyl bromide	14.00
Allyl chloride	7.81	Methyl ethyl ketone	6.71
Aluminum chloride (32°Be)	10.69	Methyl mercaptan	7.42 (at 43°F)
Anhydrous ammonia	5.15	Muriatic acid	7.29
Bromine	26.00	Nitric acid (67%)	11.66
Butane	4.87	Nitromethane	9.49
Carbon disulfide	10.50	Phosphorus trichloride	13.13
Chlorine (liquid)	11.74	Propane (LPG)	4.42
Crude oil	7.00	Propylene glycol	8.66
Dimethylamine	5.59	Sodium hydroxide solution (50%)	12.70
Ethanol	6.58	Sugar (liquid)	11.6
Ethyl ether	5.95	Sulfur dioxide	11.94
Ferric chloride solution (42°Be)	11.73	Sulfur (molten)	15.0 (at 255°F)
Fluorosulfonic acid	14.4	Sulfuric acid (98%)	15.35
Formaldehyde (37%)	9.16	Tetrachloroethane	13.28
Fuel oil #2	6.96–7.49	Thionyl chloride	13.64
Gasoline (average)	6.10	Titanium tetrachloride	14.38
Hexane	5.50	Toluene	7.22
Hydrocyanic acid	5.74	Trichlorosilane	11.20
Hydrogen peroxide (50%)	10.0	Triethylamine	6.07
Liquid argon	11.66 (at −303°F)	Trimethylamine (anhydrous)	5.27
Liquid hydrogen	0.59 (at −423°F)	Vinyl chloride	7.68 (at 56°F)
Liquid nitrogen	6.72 (at −320°F)	Water	8.33
Liquid oxygen	9.50 (at −297°F)	Xylene (isomers)	7.17–7.33

Data from GATX Tank and Freight Car Manual.

of mild steel and are usually lined with a polymer coating that resists the corrosive properties of the tank contents.

Low-Pressure Rail Tank Cars General-service tank cars are commonly referred to as "non-pressure" tank cars in the industry, although this is an extremely dangerous misnomer (Figure 10-22). There are no "non-pressure" tank cars being used in the railroad industry today. Low-pressure tank cars can have test pressures as high as 100 psi and can transport materials with vapor pressures of up to 25 psi. Tank capacities range from 4000 to 45,000 gallons. Low-pressure tank cars make up approximately 75% of tank cars in service. To prevent overpressurization, these tank cars have rupture disks (for hazardous materials with a low vapor pressure) or spring-activated reseating safety relief valves (for flammable liquids and poisonous materials with a high vapor pressure). Safety relief valves are typically set to discharge at 75% of the tank test pressure. Examples of common low-pressure tank cars include the DOT 103, the DOT 111, and the AAR 211 tank cars with typical tank test pressures of 60 psi or 100 psi. Table 10-8 lists common low-pressure and high- pressure rail tank car configurations.

Ships Many maritime vessels carry hazardous liquid cargo. On the open seas there are oil tankers, liquefied propane tankers, and liquefied natural gas tankers. Large rivers such as the Mississippi have significant flat-bottom barge traffic. Shipboard emergencies involving hazardous materials can be quite challenging.

Pipelines Liquid pipelines may be up to 42 inches in diameter and may carry everything from petroleum products to liquefied gases. Most pipelines are steel. The pressure within the pipeline depends on the viscosity of the product being transported. Thus, pressures within pipelines may cycle from high to low and vice versa.

The petroleum pipeline system is composed of gathering mains—which transport product from the source to the refinery; transmission mains—which transport product from the refinery to storage facilities; and smaller distribution lines—which transport product from the storage facilities to the end user.

Container Construction

Containers that carry liquids may be constructed from a wide variety of materials, including glass, plastic, and metal. Smaller containers are often glass or plastic, while bulk liquid containers are typically constructed of aluminum or steel. The containers are designed to hold a variety of pressures, up to approximately 25 psi.

FIGURE 10-22 Railcars come in a variety of sizes and shapes depending on their intended function. Since 2006, railcars also have a specification plate that lists the design parameters of the container (not pictured). *Second photo from bottom courtesy of Dean DeMark, HAZMAT/CBRNE Specialist (N.C.). All other photos courtesy of Chris Weber, Dr. Hazmat, Inc.*

Owing to temperature variations the MAWP may be significantly higher, as with railcars that have test pressures up to 100 psi. Railcars are constructed of steel and may be prone to BLEVE if exposed to heat or fire. Pipelines are constructed primarily from metal, but smaller lower-pressure pipelines may also be made of plastic.

Damage Assessment

It is crucial to perform a thorough damage assessment on liquid containers even though they have comparatively low pressures. When exposed to heat these containers may still fail catastrophically. Fixed-site crude oil tanks are designed to fail at the top seam. However, these tanks usually contain dirty, salty water at the bottom, which leads to corrosion there.

Liquid levels in containers can be assessed using a thermal imaging camera (TIC), temperature gun, or an acoustic wave identifier. Low-tech options include observing the frost line on containers that have undergone autorefrigeration (Figure 10-23) or gently sounding the container.

It is impossible for the hazmat technician to perform a damage assessment on a pipeline in the field. The pipeline industry inspects pipelines using remote sensing devices called "pigs" that travel inside the pipeline. By law, the industry is required to periodically inspect pipes for corrosion. Corrosion may occur for a variety of reasons, including chemical corrosion from the products or impurities in the products they carry, and electrochemical corrosion due to inadequate cathodic protection.

FIGURE 10-23 A frost line will form on uninsulated liquefied compressed gas cylinders when gas is released at a rapid rate or moderately fast over a longer period of time. The frost line will form at the liquid level as the evaporation of the liquefied compressed gas cools the remaining liquid to its boiling point. In this photograph, approximately 20 pounds of anhydrous ammonia was released in approximately 8 minutes from a 100-pound tank before the leak was stopped. *Courtesy of Eugene Ngai, Chemically Speaking LLC.*

GASEOUS CARGO

Gases migrate in three dimensions and are very difficult to control once released. By definition, gases have a vapor pressure above atmospheric pressure (14.7 psi) at room temperature. This makes containers carrying gases very dangerous.

Container Hazards

Gases may be transported as gases in cylinders with pressures up to 10,000 psi, as liquefied compressed gases, or as cryogenic liquids. Gas cylinders are constructed of steel or aluminum and pose a rocketing hazard if the valve is sheared off, and an explosion hazard when exposed to intense heat or fire. Gas cylinders therefore typically have pressure-relief devices (PRDs). Low-pressure liquefied compressed gas cylinders typically have a fusible-plug type of PRD. Nonliquefied compressed gases have rupture disks. Flammable and high-pressure liquefied gases usually have a combination of a frangible disk and fusible metal. Some gas cylinders, such as LPG, have PRVs. Some gas cylinders are prohibited from having any pressure-relief devices owing to their extreme toxicity, such as zone A poisons. The following markings refer to pressure-relief devices:

CG-1: Rupture disk
CG-2: Fusible metal 165°F
CG-3: Fusible metal 212°F
CG-4: Fusible metal 165°F/rupture disk
CG-5: Fusible metal 212°F/rupture disk
CG-7: Spring-loaded valve

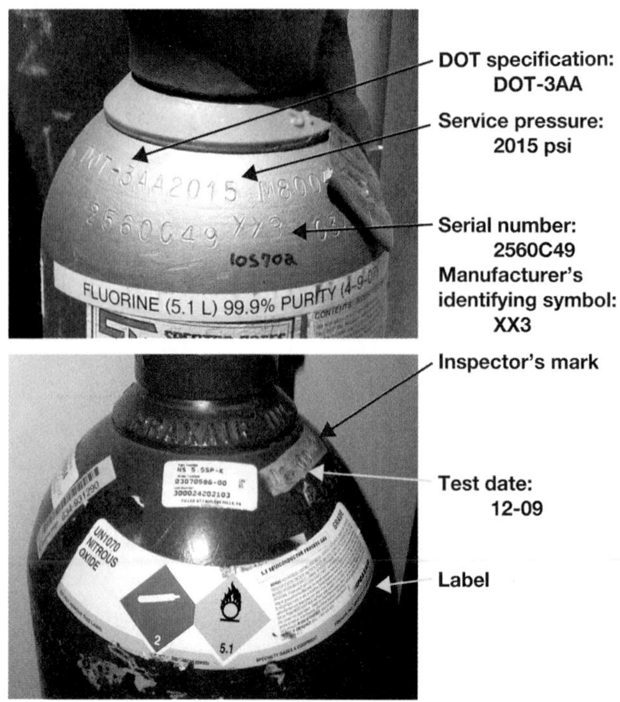

FIGURE 10-24 Compressed gas cylinders are stamped with container design specifications on the neck and also typically contain a label specifying their contents in that general area. *Courtesy of Chris Weber, Dr. Hazmat, Inc.*

Liquefied compressed gases are shipped under pressure in steel containers such as 1-ton cylinders used for chlorine and sulfur dioxide. Liquefied compressed gases are shipped at ambient temperature under pressure and will expand several hundredfold when exposed to atmospheric pressure (see expansion ratios in Chapter 5). Liquefied compressed gas containers may contain up to 600 psi of pressure and have been involved in many deadly BLEVEs, as when carrying LPG.

Cryogenic liquids are shipped at low pressure, or even atmospheric pressure, which means the liquefied gas is extremely cold. Cryogenic liquid containers are heavily insulated and often have a thermos bottle–type design. This means there is an inside tank that holds the cryogenic liquid and an outside tank, and between the two tanks is a vacuum or insulation that helps keeps the liquid cold. Typically, the biggest hazard from cryogenic containers is cryogenic burns from their contents. Cryogenic tanks must continuously vaporize some liquid to maintain their extremely cold temperature (autorefrigeration). If a tank is not in use, the spring-loaded vent valve will periodically vent the gas. This can cause an asphyxiation hazard in an enclosed room.

Types of Containers

Non-bulk Containers Non-bulk containers are defined by a maximum capacity of 1001 pounds (454 kg) water for gas cylinders. Compressed gas cylinders will have the following information permanently stamped into the cylinder, typically at the neck or shoulder (Figure 10-24):

- DOT specification
- service pressure
- manufacturer's serial number
- tare weight (for liquefied gases)
- test date
- inspector's mark

IMO Type 7 Intermodal Tanks IMO type 7 intermodal tanks carry cryogenic liquids and are comparatively low-pressure vessels. They are designed to periodically vent their contents to the atmosphere during autorefrigeration. These tanks carry comparatively nontoxic cryogenic liquid gases such as nitrogen, oxygen, helium, ethylene, hydrogen, and argon.

DOT Spec 51 Intermodal Tanks DOT Spec 51 intermodal tanks, or IMO type 5 tanks when used internationally, are designed to accommodate high pressures from 100 psig to 500 psig (Figure 10-21). These tanks carry materials similar to those transported by the MC 331 highway cargo tanker, such as the liquefied compressed gases propane, ammonia, and chlorine. Their maximum capacity is typically 6300 gallons.

Intermodal Tube Containers Intermodal tube containers consist of high-pressure cylinders mounted within an 8' × 8' or an 8' × 4' ISO frame. The high-pressure cylinders are typically tested to 3000 to 5000 psi and are permanently mounted within the framework. The cylinders are typically 12" to 24" in diameter and constructed of steel. The cylinder valves are contained in a compartment at one or both ends of the frame (tubes containing liquefied compressed gases have valves on both ends, while those holding compressed gases have valves on one end). Typical contents include inert, toxic, corrosive, and/or flammable compressed gases such as nitrogen, oxygen, helium, hydrogen, and argon. Intermodal tube containers may also carry some liquefied compressed gases such as hydrogen chloride and carbon dioxide.

MC331 Highway Cargo Tankers Another cargo tank you may be familiar with is the high-pressure MC 331 cargo tanker, which often carries LPG and anhydrous ammonia (Figure 10-8). The MC 331 may accommodate pressures up to 300 psi and is made of hardened steel. These tanks have rounded ends, which distribute the high internal pressure more evenly across the tank shell than the low-pressure vessels, which have almost 90° bends in their tank construction. These highway cargo tankers usually transport propane, LPG, butane, anhydrous ammonia, and chlorine.

Resource Central

View a Hazmat History on Liquid Petroleum Gas (LPG) and Propane: The Deadly Years

MC 338 and CGA 341 Highway Cargo Tankers The MC 338 series of cryogenic liquid cargo tankers have circular cross sections with slightly rounded ends (Figure 10-8). These cargo tankers have a thermos bottle–design, which means they contain a tank inside a tank. In addition, they are also heavily insulated on the outside. This insulation keeps the very cold liquefied gases from evaporating too quickly. These tankers have pressures below 22 psi, and may be seen releasing small amounts of vapor owing to autorefrigeration as they travel, which sometimes generates calls to 911 reporting a leaking tanker. However, this is normal operation for the MC 338 tank, and there is usually no leak to be concerned about. For this reason MC 338 tankers may carry only nontoxic gases such as oxygen, nitrogen, carbon dioxide, argon, helium, hydrogen, and methane. These tankers typically have an enclosed valve compartment at the rear.

Heat exchangers are used to convert the liquefied material into a gas for final use or to build pressure for offloading the container. When the boiling point of the cryogenic material is less than that of oxygen, the heat exchanger can condense oxygen out of the atmosphere, which can drip down on surfaces. Remember, oxygen—and liquefied oxygen (LOX) in particular—is a strong oxidizer. When liquid oxygen drips onto asphalt it forms a contact explosive. This is why LOX tanks at facilities are always mounted on concrete pads.

When cryogenic materials are spilled, a white cloud is often visible. This is condensed water vapor from the air and can be very difficult to see through. A TIC can effectively be used to see through these white clouds and gain perspective on the incident.

There is also a class of non-DOT specification cryogenic trailers (such as the CGA 341 specification) that carry atmospheric gases such as nitrogen, oxygen, carbon dioxide, helium, and argon. Non-specification cryogenic trailers may transport only nonflammable atmospheric gases, and the internal tank pressures may not exceed 25.3 psig during transport. These tanks are not required to have any emergency discharge controls. They also are not required by law to display a placard for the product, but often do. Flammable cryogenic gases must be transported in MC 338 cargo tanks.

Highway Tube Trailers Tube trailers fall under the category of non-specification cargo tanks and trailers, since DOT does not classify them as cargo tanks (Figure 10-19). However, the components of the tube trailer are DOT specification cylinders (which are regulated) that are connected together and mounted to the vehicle. All cylinders on the tube trailer carry the same product. Typically, the rear end of the tube trailer contains the valves and manifold assembly. The cylinders carry compressed gases at up to 6000 psi, and each will have a PRD. For flammable compressed gases, PRDs must discharge upward and not impinge on the other cylinders in the tube trailer assembly. Compressed gases such as nitrogen, helium, argon, oxygen, hydrogen, and methane are typically carried in tube trailers. Tube trailers may also carry toxic, corrosive, and/or some liquefied compressed gases such as hydrogen chloride and carbon dioxide.

High-Pressure Rail Tank Cars High-pressure railcars have tank test pressures from 100 to 600 psi and typically carry hazardous materials such as flammable liquids and poison gases (Figure 10-22). Approximately 25% of the tank cars in service today are high-pressure tank cars. Tank capacities range from 4000 to 45,000 gallons. These tank cars have a single compartment, are typically constructed of steel, and have rounded ends to help contain the pressure. Unlike low-pressure tank cars, high-pressure tank cars have all their fittings (such as valves, thermometer wells, sampling ports, and safety relief devices) located inside a protective housing that provides rollover protection. But there are exceptions to every rule, such as the DOT 111A100W5 low-pressure tank car, which also has all its fittings located within a protective housing, as shown in Figure 10-25. Therefore, it is always best to locate the specification markings and determine exactly the type of tank car with which you are dealing!

High-pressure tank cars typically have safety relief valves set to 75% of tank test pressure. One exception to this rule is the 340 psi high-pressure tank car, which has a safety relief valve set at 280.5 psi (82.5% of tank test pressure). Examples of common high-pressure tank cars include the DOT 105, the DOT 112, and the DOT 114.

Pipelines Gas pipelines may be up to 48 inches in diameter and commonly carry natural gas. However, they may also carry many other products such as propane, ethylene, and even chlorine and ammonia.

Container Construction

Containers carrying gases or liquefied compressed gases are typically constructed of thick metal—typically steel—to withstand high internal pressures. Some small-diameter low-pressure gas pipelines may be constructed of plastic—such as residential natural gas pipelines.

View a Hazmat History on Critical Damage to an Ethylene Oxide Railcar in Conroe, TX

Damage Assessment

These containers can cause extreme damage when they fail catastrophically and must be thoroughly inspected on arrival, and at regular intervals during mitigation operations—especially during changing weather conditions and when containers are moved.

RADIOACTIVE MATERIALS

Radioactive materials containers come in a wide variety of shapes, sizes, and strengths. These containers must be recognized early, and radiation monitoring must be instituted quickly to protect first responders and the public, since gamma and neutron radiation travel long distances in the absence of shielding.

Interpreting Radioactive Placards and Labels

specific activity ■ The decay rate of a radioactive material. A measure of the strength of a radioactive material in terms of the quantity of radiation emitted.

The three types of radioactive package labels are based on the **specific activity** of the radioactive material contained (Figure 10-26). Specific activity refers to the strength of the radioactive material—in other words, the quantity of radiation it emits. The higher the activity, the more danger it poses to first responders. We discussed the various types of radiation and their hazards in Chapter 3.

FIGURE 10-25
Although recognizing high-pressure railcars by their single valve dome is usually fairly reliable, some general-service cars look like high-pressure railcars. The best way to determine the design parameters of the railcar is from the stenciling located on the right-hand side of the railcar. *Courtesy of Dean DeMark, HAZMAT/CBRNE Specialist (N.C.).*

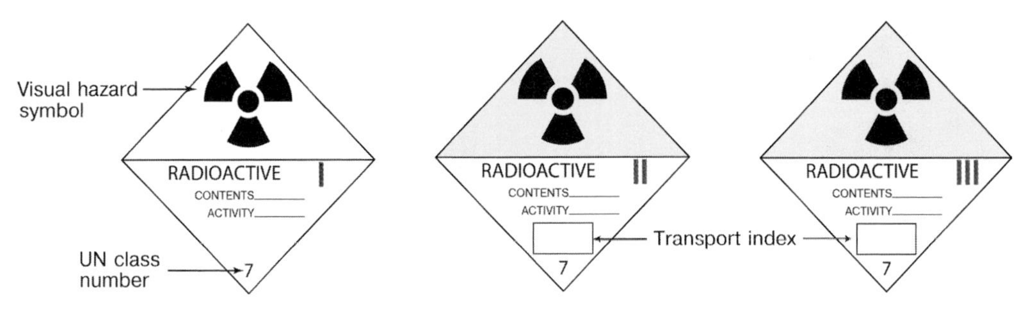

FIGURE 10-26 The U.S. DOT Radioactive I, II, and III labels. The transport index indicates the maximum amount of radiation permitted at 1 meter (3.3 ft) from the package in millirem per hour. *Art by David Heskett.*

Radioactive I

- Label has a completely white background with black and red writing.
- Label indicates the radioactive isotope and its activity (for example ^{137}Cs and 2 Bq).
- Radiation levels up to 0.5 mrem/hour may be present on the outside surface of the package.

Radioactive II

- Label has a yellow and white background with black and red writing.
- Label indicates the radioactive isotope and its activity (for example ^{137}Cs and 2 Bq) as well as the **transport index (TI)**.
- Radiation levels may be between 0.5 mrem/hr and 50 mrem/hr on the outside surface of the package.
- *Note:* Radiation levels may also be up to 1 mrem/hr at 1 meter (3.3 ft) from the package (maximum TI = 1).

transport index ■ A dimensionless number used to designate the degree of control to be exercised during transportation; indicates the maximum amount of radiation permitted at 1 meter (3.3 ft) from the package in millirem per hour.

Radioactive III

- Label has a yellow and white background with black and red writing.
- Label indicates the radioactive isotope and its activity as well as the TI.
- Radiation levels may be between 50 mrem/hr and 200 mrem/hr on the outside surface of the package.
- *Note:* Radiation levels may be up to 10 mrem/hr at 1 meter (3.3 ft) from the package (maximum TI = 10).

Fissile

- Label is a white rectangle with black lettering.
- Label indicates the package contains a radioisotope that has the ability to undergo fission under the right conditions.

EMPTY

- Label is a white rectangle with black lettering.
- Label indicates a dedicated type 2 radioactive material container is empty.

It may come as a surprise to you that radioactive packaging is legally permitted to emit radiation, and under certain circumstances this can be dangerous. Needless to say, it would not be wise to pick a Radioactive III package and use it as a makeshift seat! This is why it is very important that you understand the meanings of the various radioactive labels.

Container Hazards

Radioactive materials containers in and of themselves usually do not present a hazard. Most radioactive materials are solids or low vapor pressure liquids. Some exceptions do exist, and containers holding high vapor pressure liquids or gases pose the same hazards as mentioned previously.

Types of Containers

Owing to the danger of radioactive materials, five types of packaging are used: excepted, industrial, Type A, Type B, and Type C (Figure 10-27). Table 10-14 lists the characteristics of each of these regulated package types. Excepted packaging is used to ship very low level

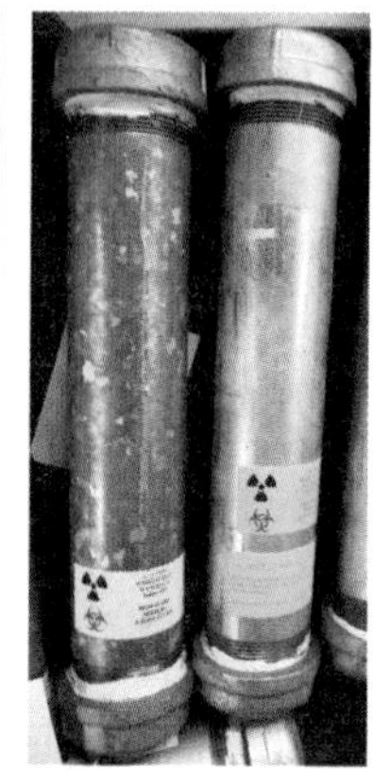

FIGURE 10-27 Radioactive materials are contained in a wide variety of packages. Industrial packaging is often used to ship radioisotopes used in biomedical research and patient treatment (top left). At fixed-site facilities non-specification containers may be used to store radioactive materials (top middle)—including improvised sharps containers (top right). A Type B package is designed to withstand significant damage (bottom left). A Type A package is shown in the bottom right-hand photograph.
Photos courtesy of Chris Weber, Dr. Hazmat, Inc.

radioactive materials such as smoke detectors, which pose a minimal hazard if released. Industrial packaging is used to ship low specific activity materials such as concrete-solidified radioactive waste, which pose little hazard if released (non-life-threatening amounts).

Type A packaging is used to ship higher-activity materials than industrial packaging but still non-life-threatening amounts of radioactive materials. Type A packages are constructed of steel, wood, or fiberboard with an inner containment vessel for the radioactive material made of glass, plastic, or metal. Radiopharmaceuticals and research isotopes are commonly shipped in Type A packages. Certified packaging materials are marked with "US DOT 7A—Type A." This type of packaging should withstand normal shipping and handling, but not necessarily vehicle accidents or fires.

Type B packaging is used to ship high-activity, potentially life-threatening amounts of radioactive materials. Type B packaging consists of heavily shielded steel casks—such as those carrying 55-gallon drums of highly radioactive waste inside—that weigh over 100 tons. This type of packaging is tested to withstand severe accidents and fires. Spent nuclear fuel and high-level radioactive waste is shipped in Type B packages.

Type C packaging is used for international shipments of radioactive materials on aircraft. Type C packaging requirements are even more stringent than Type B packaging. Both Type B and Type C packaging must be permanently marked with the trefoil symbol to ensure that they can be identified even after a severe accident. Any type of radioactive materials packages should be approached only after proper screening with radiation detection equipment by qualified personnel.

Container Construction

Container construction varies widely depending on the type of radioactive material involved. Excepted and industrial packaging is generally quite weak and designed to

TABLE 10-14 Radioactive Packaging

RADIOACTIVE PACKAGING TYPE	CHARACTERISTICS
Excepted packaging	Limited quantity of radioactive material Maximum external radiation 0.5 mR/hr
Industrial packaging	Higher radioactivity than excepted packaging Low specific activity (LSA) material or surface-contaminated object (SCO) May be boxes, freight containers, tanks, etc. There are Type 1 (IP-1), Type 2 (IP-2) and Type 3 (IP-3) industrial packages
Type A	Low-level radioactive material shipments Designed to retain integrity and shielding under normal transport conditions Must have tamper-evident security seal (or the vehicle or cargo compartment has it)
Type B	High-level radioactive material shipments (within limits) Designed to retain integrity and shielding under extreme transport conditions and accident conditions
Type C	Air transport of radioactive materials

withstand only normal stresses associated with shipping and handling. Type B and Type C radioactive material containers are constructed to very rigorous standards—including flame impingement, impact, and multistory drop tests.

placard ▪ The written warnings and pictograms affixed to vehicles transporting hazardous materials designed to rapidly communicate hazard information.

label ▪ The written warnings and pictograms affixed to containers of hazardous materials designed to rapidly communicate hazard information.

Damage Assessment

Damage assessment for radioactive materials containers should initially mirror those described previously for general containers. Radioactive II and Radioactive III packages show the transport index in the middle of the **placard** and **label**. The TI indicates the measured radioactivity in millirem per hour (mrem/hr) at 1 meter (3.3 ft) from the package when it was shipped. Radioactivity levels should be measured as part of the damage assessment. If they exceed the TI value at 1 meter, the package has been damaged and is likely leaking.

SOLVED EXERCISE 10-5

According to the radioactive package label shown, what is the transport index and radioisotope contained in this package? What radiation level would you expect at 3.3 feet from the package surface?

Solution: This is a Radioactive III label with a transport index of 1.2. This means that the radiation level at 1 meter (3.3 ft) from the surface of the package can be expected to be 1.2 mrem/hr. If radiation levels are higher than this value, the package has most likely been damaged.

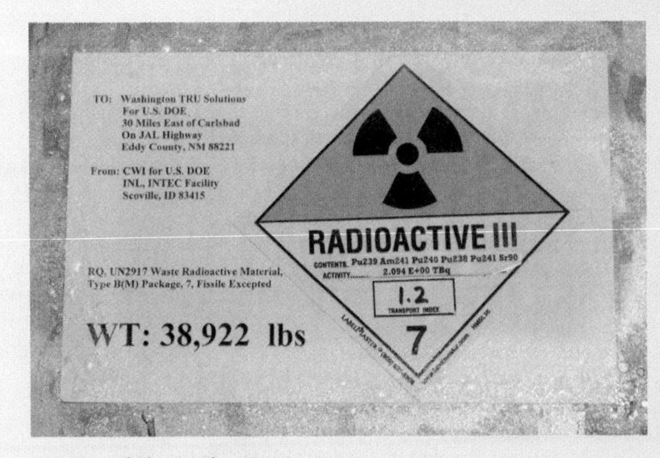

Courtesy of Chris Weber, Dr. Hazmat, Inc.

CHAPTER REVIEW

Summary

Containers can be very dangerous at hazardous materials releases. Although the focus at hazmat releases tends to be on the chemical, the chemical and its container act in concert. An intact and undamaged container holding even the most dangerous chemical is of minimal risk. However, when containers are damaged, the hazard and risk rise exponentially. Whenever a container has possibly been damaged in a hazardous materials incident, a damage assessment must be conducted as quickly and safely as possible to conduct a risk-based response. The damage assessment must be thorough and accurate to keep responders and the public safe. Damage assessments should be conducted at regular intervals during mitigation activities to detect damage caused by incident activities, changes in internal pressure due to changing weather conditions or pressure release device failure, and crack propagation. In most cases damage should be assessed by specially trained technical specialists.

Review Questions

1. What information will compressed gas cylinders have permanently stamped into the cylinder?
2. Name two materials that are commonly carried in 1-ton cylinders.
3. What type of materials does the IMO 101 intermodal tank carry?
4. What type of materials does the IMO 102 intermodal tank carry?
5. What type of materials does the IMO Type 7 intermodal tank carry?
6. What type of materials does the DOT Spec 51 intermodal tank carry?
7. What information does the specification plate on intermodal tanks carry?
8. What type of materials does the MC 306/DOT 406 highway cargo tanker carry?
9. What is the maximum allowable working pressure of DOT 406 tanks?
10. What are the locations of the emergency shutoff valves on MC 306/DOT 406 highway cargo tankers?
11. What type of materials does the MC 307/DOT 407 highway cargo tanker carry?
12. What is the maximum allowable working pressure of DOT 407 tanks?
13. What type of materials does the MC 312/ DOT 412 highway cargo tanker carry?
14. How are MC 312/DOT 412 tanks typically loaded?
15. What type of materials does the MC 331 highway cargo tanker carry?
16. What is the maximum allowable working pressure of MC 331 tanks?
17. What type of materials does the MC 338 highway cargo tanker carry?
18. What type of materials do tube trailers carry?
19. What dangerous materials may dry bulk commodity trailers carry?

References

Association of American Railroads, Emergency Response Training Center (2003) Highway Emergency Response Specialist Course Manual. Pueblo, Colorado: author.

Association of American Railroads, Emergency Response Training Center (2003) Tank Car Specialist Course Manual. Pueblo, Colorado: author.

Association of American Railroads (1992) Field Guide to Tank Car Identification. Pueblo, Colorado: author.

Compressed Gas Association, Inc. (1999) Handbook of Compressed Gases, 4th ed. Kluwer Academic Publications, New York.

Department of Transportation. (2012). *2012 Emergency Response Guidebook*. Washington, DC: Pipeline & Hazardous Materials Safety Administration.

Department of Transportation. (1998) 49 CFR. Hazardous Materials Regulations. Washington, DC: U.S. Department of Transportation.

GATX Tank and Freight Car Manual (1994) General American Transportation Corporation.

Hildebrand, Michael and Noll, Greg (1997) *Storage Tank Emergencies: Guidelines and Procedures*. Annapolis, Maryland: Red Hat Publishing.

National Fire Protection Association. (2000) Standard on Static Electricity, NFPA 77. Boston, MA: author.

Noll, Hildebrand, and Donahue. (1996) *Gasoline Tank Truck Emergencies: Guidelines and Procedures*, 2nd ed. Stillwater, Oklahoma: Fire Protection Publications.

Noll, Hildebrand, and Donahue. (1995) *Hazardous Materials Emergencies Involving Intermodal Containers: Guidelines and Procedures*. Stillwater, Oklahoma: Fire Protection Publications.

Noll, Hildebrand and Yvorra. (2005) *Hazardous Materials Managing the Incident*, 3rd ed., Annapolis, Maryland: Red Hat Publishing.

Union Pacific Railroad (2006) *Tank Car Safety Course*. Omaha, Nebraska: Union Pacific Railroad Company.

CHAPTER 11

Product Control Techniques

Courtesy of Chris Weber, Dr. Hazmat, Inc.

KEY TERMS

alcohol-resistant foam (ARF), *p. 358*
aqueous film-forming foam (AFFF), *p. 357*
booming, *p. 355*
charge relaxation, *p. 381*
conduction, *p. 380*
damming, *p. 353*
diking, *p. 352*
diversion, *p. 355*
fluoroprotein foam, *p. 358*
foam eductor, *p. 357*
high-expansion foam, *p. 358*
ignition energy, *p. 380*
remote valve shutoff, *p. 363*
static electricity, *p. 380*
unit train, *p. 362*
vapor dispersion, *p. 359*
vapor suppression, *p. 357*

OBJECTIVES

After reading this chapter, the student should be able to:

- Describe the planning requirements for successfully applying product control techniques at hazardous materials incidents.
- Explain the importance of remote valve shutoff in product control.
- Explain three techniques for controlling moving liquids.
- Explain three techniques for controlling hazardous materials in bodies of water.
- Explain two techniques for controlling vapors.
- Describe the types of foam and their uses.
- Explain the need for grounding and bonding.
- Explain the cable connection order for grounding and bonding operations.
- Name five product transfer methods.
- Describe the flaring process.
- Describe the cold tapping process.

- Explain the use of the three chlorine kits.
- Describe three methods of drum overpacking.
- Describe how to apply a dome clamp to a gasoline tanker.
- Describe how to perform patching and plugging on three different containers.

You've just been dispatched to the wastewater treatment plant, which is located next to a river adjacent to a large metropolitan area. The wastewater treatment plant has at least one chlorine rail tank car on the premises at all times for the water treatment process. The caller indicated there was a large chlorine leak somewhere between the railcar and the treatment plant. As the rookie member of the first-due hazmat company you have the following questions running through your mind:

- How much chlorine does this railcar hold?
- How is product transferred from the railcar to the water treatment process?
- Does the treatment plant use liquid chlorine or gaseous chlorine?
- What is the diameter of the transfer line?
- How will we be able to stop the leak?
- Do we have the equipment to control a leak from the railcar valves?
- Do we have the equipment to control a leak from the transfer line?

Let's see if we can get some answers to these questions and more.

Product control techniques are required when actively leaking containers are the cause of hazardous materials releases. Often, the leak will have run its course by the time first responders arrive or are ready to act and begin mitigation; therefore, cleanup operations become the responsibility of the cleanup contractor. However, when large containers leak, or response teams have short response times, they may encounter actively leaking containers and must be prepared to intervene in the emergency.

Defensive product control techniques aim to minimize contact with spilled hazardous materials or deployed weapons of mass destruction. There are two broad categories of intervention: defensive and offensive product control. You should have learned a number of defensive product control techniques in your hazmat operations class.

Offensive product control techniques require the hazmat technician to come in close proximity to the hazardous material and, often, to contact it directly. This makes these types of operations significantly more dangerous than the primarily defensive product control techniques learned in the hazmat operations class. It is imperative that the air be continuously monitored during product control operations, that the appropriate personal protective equipment be chosen, and as at all hazardous materials incidents, that a well-trained and well-prepared backup team be standing by and monitoring the situation.

The first step in planning a product control operation is to review the fixed-site facility plans or the vehicle specification plate and shipping papers to gather as much product and container information as possible. Fixed-site facility plans may include the SARA title III paperwork the facility provides to the LEPC and the local fire chief annually, facility maps, emergency response plans, and any specific facility plans that were formulated by local agencies. These facility plans are important for identifying any safety hazards posed by other chemicals, industrial equipment, or building layout. Remember to find a technical expert from the facility to help you. Product information can be obtained from material safety data sheets (MSDSs); shipping papers; facility inventories, which can be found in the SARA title III paperwork or Tier 2 forms; and research materials. Once you understand the hazards of the chemical(s), the container properties, and the dangers the facility may pose, you can plan a safe and effective product control operation.

General Work Practices

First and foremost, minimize contact with the released product by not walking through contaminants. This minimizes cross-contamination. Contaminants include such obvious materials as spilled liquids or powders, or less obvious materials such as gases and vapor clouds. Although the spilled liquids and solids will typically be visible—as long as you're looking for them—gases and vapor clouds often are not. For this reason the right monitoring equipment is essential for any personnel entering the hot zone.

Whenever possible, do not come into contact with the hazardous material. For technician level personnel this is often not possible, since you are performing offensive product control techniques. Good work practices dictate that you remain as far away as practical from the hazardous material while accomplishing your objectives. One way to avoid contacting the hazardous material directly is to use remote sampling and handling tools such as drum grapplers whenever possible (Figure 11-1). For example, when spreading solidification materials, you should use a long-handled shovel to put distance between you and the spilled material.

Before attempting confinement or containment operations in the vicinity of damaged containers, ensure that the containers will not move unexpectedly. If necessary, you may need to brace and block the container using cribbing or other stabilization equipment. On vehicles, ensure that all brakes are set, and chock the wheels when feasible. This is especially important to remember when bringing vehicles back to an upright position.

Control options should be carefully chosen based first and foremost on safety considerations, such as available equipment and training, whether the chemical is known or unknown, the chemical and physical properties of the product, and the nature of the release. When you select a product control option you should be able to identify the purpose for implementing it, have defined operating procedures in which you have been trained, have access to the appropriate product control equipment, know the risks associated with the operation, and know the appropriate safety precautions. Product control options should

FIGURE 11-1 Drum dollies (left) and drum grapplers (right) make lifting and moving heavy drums safer and easier. *Photos courtesy of Chris Weber, Dr. Hazmat, Inc.*

be carefully selected based on scene considerations to maximize the safety of the personnel that will be engaging in the product control procedures, other emergency responders on the scene, and the general public. There are many product control options at your disposal.

Confinement Control Options

Confinement operations limit the spread of leaking product to a defined area such as a secondary containment pool, or limit vapor generation through vapor suppression; or mitigate the spill or leak with other tactics. Let's review some of the confinement techniques with which you should already be familiar from your hazardous materials operations training. These techniques are still very commonly used by hazardous materials technicians.

DILUTION

When large amounts of water are available, dilution is a very simple method of lowering the concentration of a hazardous material—ideally, to less-than-dangerous levels. Dilution typically is not a practical solution owing to the large amounts of water required to make a hazardous material nonhazardous. In addition, many hazardous materials cannot be rendered nonhazardous by dilution with water, since they may not be soluble in water. However, alcohols, for example—which are water soluble—may be diluted to the point at which they are no longer flammable.

Acids and bases may be diluted to the point that the pH is no longer considered corrosive. However, as a rule of thumb, each change of one pH unit will require 10 times the amount of water. Therefore, to dilute 10 gallons of sulfuric acid at a pH of 2—which is already a fairly dilute sulfuric acid solution—to pH 7 (a change of five pH units), would require 1,000,000 gallons of water. This volume was arrived at by multiplying 10 gallons by 100,000, which is $10 \times 10 \times 10 \times 10 \times 10$ (for each of the five pH unit changes). In this example a large amount of water would be needed, which is typically not practical. Keep in mind that the addition of water to concentrated acids or bases may initially cause a violent reaction through the generation of heat and will almost always create a large vapor cloud so the entry team must remain a safe distance upwind.

ABSORPTION, ADSORPTION, AND SOLIDIFICATION

Absorption, adsorption, and solidification are three similar processes by which a liquid can be more easily handled and disposed of. *Absorption* refers to trapping a liquid in the pores of a solid material, as in trapping water in a sponge or a paper towel. *Adsorption* is the process in which a liquid adheres to the surface of the adsorbent material, such as when oil coats the surface of solid plastic beads. *Solidification* is a chemical process in which the liquid being disposed of reacts with the solidifying agent to form a less hazardous solid product. For example, the hazardous liquid hydrofluoric acid may be treated with the sorbent calcium chloride to form the solid calcium fluoride, which is significantly less hazardous. Sorbents can also be used to catch slowly dripping liquids: the top of a floor dry bag is cut open and placed beneath the leak. Figure 11-2 shows the use of sorbent products.

Each of these processes has its advantages and disadvantages. Absorption and adsorption merely contain the hazardous material within the fibers or on the surface of the sorbent material. The disadvantage is that if the hazardous material is released, it is still dangerous. The used sorbent material must be disposed of as hazardous waste. Solidification, in contrast, may render the hazardous material significantly less hazardous or even nonhazardous, possibly allowing for normal solid waste disposal. You should always check with the manufacturer for proper disposal guidelines before disposing of any spent solidification matrix. Improper disposal of spent sorbent may be a violation of RCRA

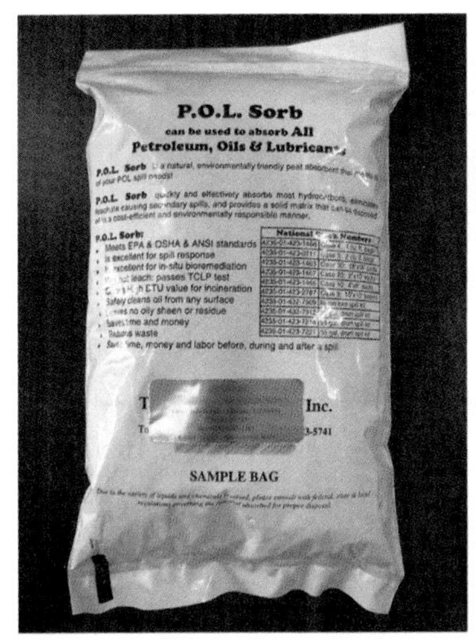

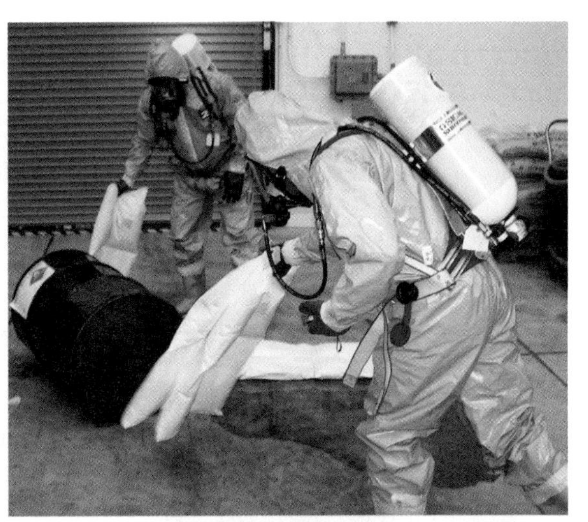

FIGURE 11-2 Absorbents—in the form of powders, pillows, or socks—can be effectively used to clean up liquid spills and help reduce vapor production. *Photos courtesy of Chris Weber, Dr. Hazmat, Inc.*

and may result in large civil and possibly criminal penalties. Specialists should always be consulted before disposal of hazardous waste or potentially hazardous waste.

Selective Sorbents

Selective sorbents are able to absorb a certain chemical or class of chemicals preferentially over other materials. One class of selective sorbents is oil-absorbing booms, which preferentially absorb hydrocarbons over water. This allows them to be effectively employed for oil spills in waterways.

Nonselective Sorbents

Nonselective sorbents will absorb almost any material with which they come into contact. For example, if a nonselective boom is placed in a waterway, it will quickly soak up not only spilled oil but also the surrounding water. The water quickly saturates the boom, and it becomes useless. It is therefore very important to choose the correct sorbent for the job at hand.

View an illustrated step-by-step guide on Absorption, Adsorption, and Solidification

BOX 11-1 HAZMAT HOW TO: ABSORPTION, ADSORPTION, AND SOLIDIFICATION

1. Choose the appropriate material.
2. Test a small amount to ensure compatibility.
3. Apply material from the outside in.
4. Continue to apply absorbent toward the middle or source of the leak.
5. Rake absorbent back and forth to maximize absorption.
6. Shovel absorbent into appropriate waste container.

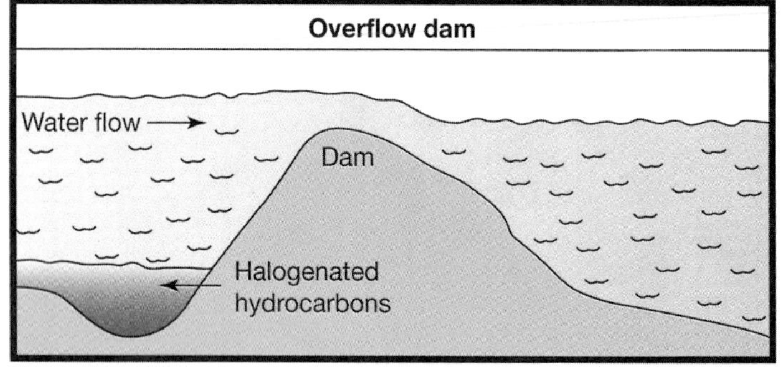

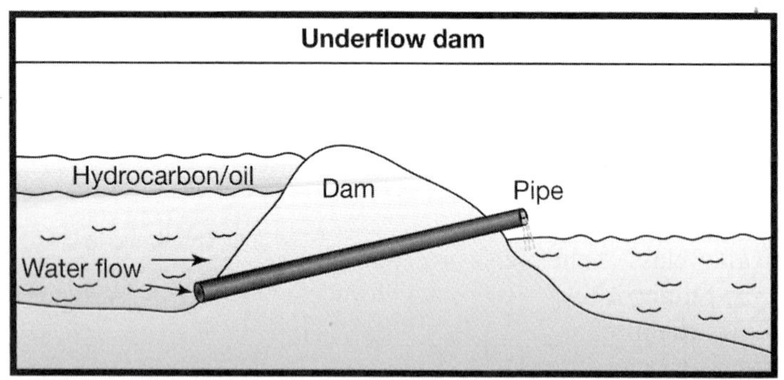

FIGURE 11-3 Diking, damming, and retention can be used to confine a liquid spill to a defined area. Slow leaks from vehicles or tanks can be effectively confined using retention basins (top left). Drains and other sensitive areas can be protected by diking (top right). Overflow and underflow dams can be used to confine liquids in moving bodies of water (bottom). *Photos courtesy of Chris Weber, Dr. Hazmat, Inc.*

Sorbents come in many shapes and sizes: square and rectangular pads, cylindrical socks of varying lengths and thicknesses, and pillows. Pads, socks, and pillows should be placed using long-handled tools to minimize contact with the spilled material. Sorbents also come as loose powders or granules. Loose sorbent is easily spread over liquid puddles and pools using shovels. Information on how much product a given quantity of sorbent is able to absorb to determine how much is needed for the job at hand can be found in the literature and directions for use supplied by the manufacturer.

It also is very important to ensure that the sorbent material is compatible with the spilled material. Otherwise, a spilled oxidizer could ignite a combustible sorbent, or a corrosive material could react with an incompatible sorbent and generate toxic vapors. Once the spilled liquid has been solidified, the sorbent material is typically placed in a salvage drum for disposal. The drum, too, must be compatible with the used sorbent.

DIKING

Diking is the process of protecting sensitive drainage openings such as sewer openings from hazardous runoff, as shown in Figure 11-3. Diking materials can be purchased commercially and are very effective for a wide range of hazardous materials. As with solidification materials, there are selective and nonselective products. Some diking materials are designed for water-soluble materials, while others are designed for oily or water-insoluble materials. It is important to use the appropriate diking material on the appropriate hazardous material. Otherwise, it will be ineffective at stopping the material or may lead to an unexpected chemical reaction. Diking materials will often have embedded sorbent materials. The simplest

BOX 11-2 HAZMAT HOW TO: DIKING

1. Choose the appropriate diking material based on chemical compatibility and functional considerations.
2. Test a small amount of the diking material to ensure compatibility.
3. Place the diking material around the area to be protected or directly around the container that is leaking. When protecting manhole covers or sewer openings, it can be helpful to place plastic sheeting over the opening and then the diking material around that.
4. In soft ground it is helpful to dig a ditch or depression on the product side of the dike. The ditch can be lined with chemically compatible plastic sheeting to make it more effective.
5. Monitor the dike and reinforce as necessary.
6. Repeat dike construction once or twice downstream of the first dike to ensure redundancy in case of failure or leakage.
7. Dispose of used diking material as hazardous waste.

Resource Central

View an illustrated step-by-step guide on Diking

diking materials are sand, clay, and soil. These are often available in quantity from your local public works department, road commission, or even on-site with a little digging.

diking ▪ The procedure of keeping a spilled hazardous material in a defined location using barriers such as sand, berms, or absorbent material.

RETENTION, DAMMING, AND DIVERSION

If a hazardous material is leaking out of its container, the simplest way to retain it is to use catch basins, tubs, buckets, pools, or other readily available containers (Figure 11-3). However, before choosing a container, ensure that the container material is compatible with the hazardous material and does not contain incompatible residual chemicals. Otherwise, the hazardous material may eat through the container or, worse yet, may react violently with the container or its residue. If a flammable material is involved, static electricity is generated when droplets of flammable liquid fall through the air. The drier the air, the more static electricity is generated. Flammable materials should be retained in metal containers that are grounded and bonded to the leaking container whenever possible. When approaching the area of the leak, always be very careful not to contact a hazardous material directly.

Damming is used to retain liquid in a specific area on land, or to retain a heavier-than-water liquid at the bottom of a stream or river, as shown in Figure 11-3. Materials used in dam construction are usually obtained locally, often at the hazardous materials incident itself. Very common damming materials are soil and sand. When the soil or sand is very porous, the dam may be lined with an impermeable material such as plastic sheeting or clay to increase its effectiveness. Dams work best on slow-moving waterways. The hydraulic energy of moving bodies of water can be enormous; therefore, always have a backup plan in place downstream.

damming ▪ The process of restricting the flow of a spilled hazardous material by building a barrier to redirect or contain it.

Underflow Dam (Siphon Dam)

Siphon dams are used to trap materials that float on water. Figure 11-3 illustrates the construction of a siphon dam. The siphon dam requires 4-inch or larger piping material as well as water-impermeable diking material, such as dirt or sand. It may be helpful to line the dike with plastic sheeting to make it more effective. A sufficient number of pipes are placed at an angle through the dike to allow underlying water movement through the dike. Because of the large amount of time and construction materials that would be needed on large bodies of water, siphon dams work only on smaller bodies of water such as creeks or streams.

Filter Fence

Filtering is a quick and easy method for catching lighter-than-water materials such as gasoline and diesel fuel in a moving body of water. This technique consists of stringing a relatively rigid, permeable barrier such as chicken wire fencing across the stream or river.

BOX 11-3 HAZMAT HOW TO: UNDERFLOW DAM

1. Determine that an underflow dam is appropriate for the spilled product: the specific gravity of the material must be less than 1—meaning it floats on water—and the solubility of the material in water must be negligible.
2. Determine that an underflow dam is appropriate for the moving body of water in question. For example, this technique will not work on a large river or cascading rapids.
3. Choose the appropriate damming material. This can be sand, stone, or earth found in the area. Plastic sheeting is usually necessary to minimize damage from water erosion.
4. Choose the appropriate damming location.
5. Place the dam at a slight angle to the flow of the river. This will force the product toward one side, where it can be recovered using a vacuum truck, other pumping equipment, or absorbents.
6. Place the plastic sheeting at the bottom of the moving body of water and weigh it down with rocks. This works best with a roll of plastic sheeting that has approximately 1 to 2 feet of plastic unrolled. Place the rocks on this lip.
7. Place the bulk of the damming material over the rock weights and unfurl the plastic as the height of the dam increases.
8. Place several 3- to 6-inch rigid plastic or metal pipes at an upward angle into the dam as it is constructed. Punch holes through the plastic to accommodate pipe insertion. The height and angle of the pipes will depend on the depth of the water. The number of pipes will depend on the volume of water flowing in the stream. For higher-volume streams larger-diameter corrugated metal drainage pipe may be used.
9. Build the dam to a height greater than the level of the water. The product will be trapped behind the dam.
10. Use appropriate air-monitoring equipment to monitor product vapor levels at the collection area.
11. Periodically monitor the integrity of the dam.
12. Repeat dam construction once or twice downstream of the first underflow dam to ensure redundancy in case of failure or leakage.

Straw or other sorbent material that floats on water is placed upstream (behind) the barrier. This sorbent material catches the floating contaminant and can be augmented or replaced as it becomes saturated. This is a cost-effective method, and the needed materials are available in most communities at the local hardware store.

Overflow Dam

An overflow dam is used to trap materials that are heavier than water. The dam is essentially an impermeable barrier at the bottom of a moving body of water that traps the

BOX 11-4 HAZMAT HOW TO: FILTER FENCE

1. Determine that a filter fence is appropriate for the spilled product: the specific gravity of the material must be less than 1—meaning it floats on water—and the solubility of the material in water must be negligible.
2. Choose the appropriate absorbent material. This can be straw or a commercially available absorbent.
3. Choose the appropriate damming location.
4. Gather the necessary supplies: pickets, chicken wire, and absorbent material.
5. String the chicken wire across the moving body of water.
6. Stake the chicken wire securely into the ground, or tie it to available trees.
7. Place the absorbent material upstream of the chicken wire.
8. Use appropriate air-monitoring equipment to monitor product vapor levels at the collection area.
9. Monitor the filter fence and add additional absorbent material as necessary to prevent leakage.
10. Dispose of used absorbent material as hazardous waste.
11. Construct one or two additional filter fences downstream of the first one to ensure redundancy in case of failure or leakage.

heavier-than-water hazardous liquid behind it. If the barrier is substantial enough, or a suitable location with a preexisting depression or pit is chosen, significant amounts of contaminant can be trapped with minimal effort. Keep in mind that this technique works only with materials that are insoluble in water and have a specific gravity greater than 1. The denser liquid can be pumped out as it accumulates using pumps or vacuum trucks, which must be compatible with the material being recovered. Most vacuum trucks are either MC 307/DOT 407 or MC 312/DOT 412 specification.

Diversion

Diversion involves digging temporary ditches and using pipes or other channels to redirect the flow of liquids to a safer or more practical location. Diversion may be as simple as placing a few shovel loads of sand at a strategic location to redirect the flow of a hazardous material into a low-lying area that can be used as a retention pond, or it may be as complex as digging a deep channel with a backhoe to redirect the flow of a large amount of hazardous materials into a retrieval station.

diversion ▪ The procedure of redirecting the flow of a spilled hazardous material away from a sensitive area.

BOOMING

Floating containment booms are used on a body of water to retain a hazardous material in a specific area, as shown in Figure 11-4. **Booming** materials can be purchased commercially and are very effective for a wide range of hazardous materials. Because booms will be deployed in water, they are typically designed for water-insoluble liquids with a specific gravity of less than 1 (which means they float on water). By far the most commonly boomed hazardous materials are petroleum products such as crude oil, diesel fuel, and gasoline.

booming ▪ The process of using floating barriers to contain nonpolar, lighter-than-water hydrocarbons on bodies of water.

Depending on the affected body of water, booming operations can be logistically very difficult. A small pond may be effectively boomed using a few personnel from shore, while a large lake or river may require watercraft and a large number of personnel. It is extremely important to avoid becoming exposed to the hazardous material while performing booming operations. This means that booms should be erected well downstream of the hazmat release, especially on faster moving bodies of water. Be aware of the fact that booms don't work well in rough water or on fast-moving rivers, although booms are available that have variable length polymer "skirts" hanging below the floating boom to prevent material from being washed under it by rough water. These are typically used with spills on large bodies of water and the ocean.

Product Redirection

It is often helpful to direct released products on bodies of waters into collection points at which skimming operations can be conducted. On moving bodies of water, such as rivers, the direction of flow and angle of boom placement can be used to direct the product to one side of the river and the collection point. Still bodies of water may require hydraulic or pneumatic assistance in directing the product to the desired location. On moderate-sized bodies of water powerful fans or fire department hose lines can be used to blow or push the product to the collection point. These techniques should be used carefully to avoid mixing the product and water, thereby creating an emulsion that will be more difficult to deal with.

FIGURE 11-4 Booming operations can be used to confine non-water-soluble liquids on top of bodies of water. *Courtesy of Chris Weber, Dr. Hazmat, Inc.*

BOX 11-5 HAZMAT HOW TO: BOOMING

1. Determine that booming is the appropriate technique for the spilled product: the specific gravity of the material must be less than 1—meaning it floats on water—and the solubility of the material in water must be negligible.
2. Determine that booming is appropriate for the moving body of water in question. For example, this technique will not work on an extremely fast moving river or cascading rapids.
3. Choose the appropriate booming material. This may be an absorbent boom or a containment boom.
4. Choose the appropriate booming location.
5. Place the dam at a slight angle to the flow of the body of water. This will force the product toward one side (the collection area), where it can be recovered using a vacuum truck, other pumping equipment, or absorbents. When possible, choose a side and location with good access for heavy equipment.
6. Use appropriate air-monitoring equipment to monitor product vapor levels at the collection area.
7. Monitor the collection area and periodically remove accumulated product.
8. Repeat booming once or twice downstream of the first boom to ensure redundancy in case of failure or leakage.

Skimming

Hydrocarbons and other lighter-than-water products can be collected and removed from the body of water using vacuum trucks or marine vessel–based skimmers. Vacuum trucks are typically able to hold approximately 4700 gallons of liquid. Depending on the size of the release, it is very important to plan for the correct number of vacuum trucks so that the booming operation is not overwhelmed and swamped by product.

Siphoning and Dredging

Halogenated hydrocarbons and other heavier-than-water products must be siphoned from the bottom of the body of water, or in the case of solids, may need to be dredged from

BOX 11-6 HAZMAT HISTORY: DEEPWATER HORIZON OIL SPILL (2010)

On April 20, 2010, the deep-sea oil rig Deepwater Horizon located 41 miles off the Louisiana coast exploded, killing 11 men and injuring 17 others. The explosion is believed to have been caused by a methane gas explosion. The Deepwater Horizon sank after burning for 36 hours. On the seafloor, the oil well was gushing approximately 60,000 barrels of oil per day despite safety equipment—such as a blowout preventer—that should have stopped the leak immediately. Ultimately, it was estimated that 4.9 billion barrels of crude oil flowed into the Gulf of Mexico, damaging hundreds of miles of coastline, valuable fisheries, and the tourism industry.

A multitude of remediation methods were used including the following:

- skimmer ships to collect surface oil
- floating containment booms
- burning the oil in place on the surface of the ocean
- oil dispersants, such as Corexit EC9500A and EC9527A

Floating containment booms were used to protect sensitive areas along the coast, including ports, beaches, and wildlife refuges. Skimmer ships were used to collect as much oil as possible, including using ships and booms to corral bodies of oil that were then burned in place on the ocean surface out at sea. Owing to the extremely large quantity of oil, dispersants were also used to break down huge oil slicks. This was a controversial tactic in that, it is believed, dispersants may have resulted in large underwater oil plumes that may affect fisheries for years to come. Despite an extensive effort by BP to clean up visible oil, large underwater oil plumes reportedly continued to migrate and sink to the bottom of the ocean. This incident illustrates the importance of planning and emergency response—both of which were severely lacking during the initial days of the Deepwater Horizon oil spill. It took several weeks for the responsible parties to obtain the necessary resources to begin to mitigate this incident.

the bottom. Dredging requires specialized equipment and specially trained personnel. Vacuum trucks can often be used to siphon heavier-than-water products from the bottom of shallow bodies of water.

VAPOR SUPPRESSION

Vapor suppression is a technique for limiting the amount of vapors that enter the atmosphere after a hazardous materials release. Vapor suppression may take the form of covering the released material with either plastic sheeting or vapor suppressing foam.

Plastic sheeting, such as Visqueen, can be placed over a released hazardous material to reduce the evaporation rate of the material (Figure 11-5). It is always necessary to make sure that the spilled product is compatible with the plastic sheeting.

FIGURE 11-5 Vapors can be suppressed by placing plastic sheeting over the spill location. In this case mercury vapors are being suppressed. *Courtesy of Chris Weber, Dr. Hazmat, Inc.*

Foam is used to suppress vapors when volatile materials such as gasoline are spilled. Depending on the type of chemical that has been released, several types of foam may be used. The most important chemical and physical property to consider when choosing foam is the water solubility of the spilled product. Water-insoluble materials, such as the common hydrocarbons gasoline and diesel fuel, typically require the use of aqueous film-forming foam (AFFF). Water-soluble materials, such as alcohols, require the use of alcohol-resistant foam (ARF), which are usually used in higher concentrations because water-soluble materials tend to attack and degrade the foam much more rapidly than do water-insoluble materials.

vapor suppression ▪ The process of limiting the evaporation of a pooled liquid.

Foam is applied using standard fire hoses with a **foam eductor**, which applies a consistent and measured amount of foam concentrate into the water stream using the venturi effect. Standard nozzles or air-aspirating nozzles may be used, although the air-aspirating nozzle creates much higher quality foam. The foam layer tends to degrade over time no matter which foam product is used or which hazardous material has been released. Before starting any foam application, you must be certain that you have enough foam on hand to maintain the foam blanket. Otherwise, dangerous vapors will quickly start to build up and may find an ignition source and ignite unexpectedly—possibly leading to an explosion. Many hydrocarbon fires and releases have turned tragic because the protective foam blanket could not be maintained.

foam eductor ▪ A device used to mix foam concentrate and water in the correct proportion.

Aqueous Film-Forming Foam

Aqueous film-forming foam (AFFF) is used to suppress the vapors generated by hydrocarbon liquids. Hydrocarbons are the most common type of flammable liquids transported in the United States and Canada. Fuels such as gasoline, diesel fuel, kerosene, and Jet A are all liquid hydrocarbons. Hydrocarbons are water insoluble and are the easiest types of liquid hazardous material on which to keep a foam blanket intact. AFFF is typically used in either a 3% or 6% formulation and is designed for optimal flow properties. AFFF foam forms a thin, impermeable film layer on top of the hydrocarbon surface that spreads and polymerizes quickly and thereby suppresses the flammable vapors. AFFF foam concentrate can be used with freshwater or seawater, and it can be created without a foam eductor (although the results will be suboptimal). Because a solid film is formed by the foam on the hydrocarbon surface, it is important to avoid disturbing it, for example, by walking through the foam blanket. If you have an airport in the area, you likely have

aqueous film-forming foam (AFFF) ▪ A frothy liquid-air mixture prepared by mixing a concentrate with water, intended to extinguish fires involving water insoluble, flammable nonpolar hydrocarbon liquids. Examples of nonpolar hydrocarbon liquids are gasoline, crude oil, kerosene, diesel fuel, and heating oil.

large amounts of AFFF available in the event of an emergency in the form of airport crash trucks. By FAA regulations these units must carry large amounts of AFFF.

Alcohol Resistant Foam (ARF)

alcohol resistant foam (ARF) ■ A frothy liquid-air mixture prepared by mixing a concentrate with water, intended to extinguish fires involving water-soluble, flammable polar hydrocarbon liquids. Examples of polar hydrocarbon liquids are E85, ethanol, methanol, and acetone.

Alcohol-resistant foam (ARF) is designed to be used on polar, water-soluble solvents such as alcohols. ARF is becoming much more important with the advent of E85 fuels. E85 fuel is a mixture of 85% ethyl alcohol (ethanol) and 15% gasoline. Standard AFFF foam is quite ineffective in fighting E85 fuel fires. Alcohol-resistant-AFFF are made of a combination of synthetic detergents, fluorochemical surfactants, and polysaccharide polymers. They can be used on both hydrocarbon fuels and polar fuels, making this foam a good choice owing to its flexibility of use. ARF is typically used in either a 3% or 6% formulation, depending on the type of fuel. When AR-AFFF is used on polar solvents, the polysaccharide polymer forms a durable membrane that keeps the fuel and the foam separate, extending the lifetime of the foam blanket. Because a solid film is formed by the foam on the surface of the polar liquid it is important to avoid disturbing it, for example, by walking through the foam blanket.

Fluoroprotein Foam

fluoroprotein foam ■ A frothy liquid–air mixture consisting of a fluorinated protein polymer, intended to extinguish fires involving flammable liquids.

Fluoroprotein foam is a protein foam with a fluorochemical surfactant additive mixed in. This makes the foam fairly heat resistant, improves resistance to fuel pickup, and makes the foam compatible with dry-chemical extinguishing agent. Fluoroprotein foam is most commonly used for hydrocarbon fuels, the concentrate can be mixed with freshwater or seawater, and a foam eductor must be used.

High-Expansion Foam

high-expansion foam ■ A type of foam that expands at least 200:1 when mixed in a foam eductor.

High-expansion foams have an expansion ratio above approximately 200:1. These foams are designed to be used primarily on class A fuels (common combustible materials such as paper, wood, and plastics) in small, well-defined spaces. The foam is made from synthetic foaming agents and is designed to be used with a high-expansion foam generator. High-expansion foam is often used in confined spaces, such as warehouses, mines, and aboard ships. Although it is designed to be used on class A fuels, it has been used on small flammable liquid fires (class B fuels) as well. High-expansion foam is much less effective on large class B fires such as tanks, highway cargo tankers, and railcars and should generally not be used in that application.

BOX 11-7 HAZMAT HOW TO: FOAM APPLICATION

1. Remove all ignition sources from the area of flammable product pooling.
2. Before starting to apply foam, ensure you have enough foam on hand to continue to maintain the foam blanket until product control operations are completed.
3. Choose the appropriate foam for the appropriate product: hydrocarbons will require aqueous film-forming foam (AFFF), while water-soluble products, including gasoline/ethanol blends, will require alcohol-resistant foam (AR-AFFF).
4. Continuously monitor the area using an appropriate air-monitoring instrument— typically a combustible gas indicator (CGI) for flammable liquids. Air-monitoring instruments should be deployed on all four sides of the product pool/foam blanket.
5. Open the nozzle away from the pool and test the foam consistency. If it is not well aerated, adjust the foam eductor, the flow rate, or the nozzle.
6. From a safe distance, apply the foam gently to the pool of product:

RAIN-DOWN METHOD

1. Direct the nozzle up into the air so that the foam gently falls through the air and blankets the pool of product. Keep in mind that this technique will not work well with an intense pool fire that creates thermal updrafts!
2. Make sure the foam flows smoothly across the surface of the pool.

3. Once the foam covers the entire surface area of the pool, stop the flow.
4. Continuously monitor the integrity of the foam blanket.
5. Add more foam as necessary to maintain the integrity of the foam blanket.

ROLL-IN METHOD

1. Direct the nozzle onto the ground a few feet in front of the flammable liquid pool. This technique works best when there is a relatively long, smooth surface directly in front of the pool.
2. Make sure the foam flows smoothly across the surface of the pool.
3. Once the foam covers the entire surface area of the pool, stop the flow.
4. Continuously monitor the integrity of the foam blanket.
5. Add more foam as necessary to maintain the integrity of the foam blanket.

BOUNCE-OFF METHOD

1. Direct the nozzle against a wall or solid object behind the product pool, and allow the foam to gently flow down onto the product. This technique works well for storage tanks and vehicle tanks when the burning liquid is deeper down and enclosed by walls on all sides.
2. Move the nozzle back and forth to maximize coverage. Different angles may be necessary to reach more pockets of fire and pooled liquid. Under no circumstances allow the nozzle to directly inject foam into the pooled liquid!
3. Make sure the foam flows smoothly across the surface of the pool.
4. Once the foam covers the entire surface area of the pool, stop the flow.
5. Continuously monitor the integrity of the foam blanket.
6. Add more foam as necessary to maintain the integrity of the foam blanket.

VAPOR DISPERSION

When vapor suppression operations have failed or cannot be safely and effectively performed, **vapor dispersion** is used to reduce the airborne concentration of gases or vapors. There are several different methods of vapor dispersion: the simplest method is to use a fan or other mechanical device to create airflow that pushes the vapors out of the immediate area and dilutes them in the process. This process is also called *ventilation* and can be used indoors and outdoors.

vapor dispersion ▪ The process of moving and diluting a gas or vapor.

When larger volumes of air need to be moved, fire department hose lines using a fog nozzle set on wide can be effective at dispersing vapors (Figure 11-6). This technique relies on the venturi effect, which pulls air along with the moving water. This technique has long been used successfully during ventilation operations at structure fires and is referred to as *hydraulic ventilation* in that application. Hydraulic vapor dispersion can be used for both water-soluble and water-insoluble gases or vapors. When this technique is used with water-insoluble gases and vapors, it will remove them from the immediate area and push them downwind. When this technique is used with water-soluble gases and vapors, the chemical will dissolve in the water, and the resulting runoff will likely be hazardous and must be contained (corrosive vapors—such as anhydrous ammonia—are one example).

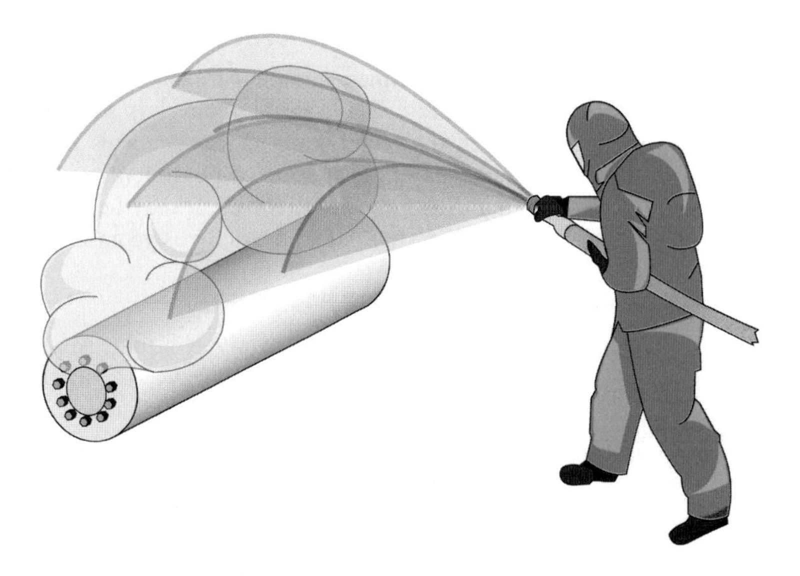

FIGURE 11-6 Water-soluble gases and vapors can be suppressed using a hose line. It is important to contain the runoff from these operations to avoid spreading the problem. *Art by David Heskett.*

Any of these techniques may be used indoors or outdoors, keeping in mind that the smaller the volume of vapors and the smaller the space that needs to be ventilated, the more effective the operation will be. Generally, when these techniques are used outdoors, they are designed to protect the entry team in the hot zone or exposed populations that are immediately downwind and cannot be effectively evacuated or sheltered in place. When moving indoor gases or vapors to the outside, keep in mind where the vapor or gas will go and what populations are downwind. It is wise to have downwind air monitoring in place before starting vapor dispersion or ventilation activities to avoid creating a bigger problem.

REPOSITIONING CONTAINERS

Repositioning containers is a quick and easy solution for leaks from small containers. During the initial reconnaissance the entry team can solve many problems by repositioning a leaking container so that the container breach is above the liquid level line or positioning a liquefied compressed gas cylinder so that the breach is above the liquid phase. This effectively turns a liquid leak into a gas or vapor leak, reducing the volume of escaping material several hundredfold owing to the expansion ratio of gases and the much slower evaporation rates of liquids when they are contained and expose less of their surface area. In addition, autorefrigeration will lower the container pressure dramatically, thereby slowing, if not almost completely stopping, the leak.

CATCHING A LEAK

When a container is too large to reposition or cannot effectively be accessed, the leaking material can often be contained by placing a catch basin beneath the leak. The catch basin may be a 5-gallon bucket, a decon pool, an overpack drum, or a specialty container. As long as the container fits under the leak and is compatible with the leaking chemical, it should work just fine. As the catch basin fills, it can either be pumped out or be replaced with another suitable receptacle. This technique is often used for diesel fuel spilling from ruptured saddle tanks from a semitrailer truck and for liquids leaking from the back of semitrailers or other elevated containers. Make sure that the construction material of the catch basin is compatible with the spilled material and that the catch basin is clean. Otherwise, unexpected chemical reactions could occur.

SEPARATING MATERIALS

When multiple containers are leaking or involved in a hazmat release, it is very important to separate incompatible materials such as acids and bases and/or fuels and oxidizers. When the recon team spots chemicals that may mix, it is important that they separate those materials as soon as possible if it is safe to do so. Research takes time, and if multiple chemicals are involved, it may be too late to prevent explosive reactions or reactions that liberate toxic and/or corrosive gases.

COOLING CONTAINERS

When a material, especially a liquid, is heated it will expand. Liquefied gases are especially dangerous, since they have already been cooled and compressed, they are typically stored in a pressurized container, and it takes comparatively little extra heat to cause container failure. When a fire involves liquefied compressed gas containers or flammable liquid containers, these containers must be cooled to avoid a boiling-liquid expanding-vapor explosion (BLEVE).

A BLEVE occurs when a container is heated to a temperature above the boiling point of the liquid contained in it, and the container then fails catastrophically, sending shrapnel and vaporized liquid in all directions. If the gas or vapors are flammable, a sizable fireball and shock wave will ensue (such as with LPG). If the liquid or liquefied gas is toxic (such as with chlorine), a large toxic vapor cloud will be created that can be deadly to downwind populations (see Box 11-13).

Large vapor clouds created from nonflammable liquids retain their shape and high chemical concentrations for extended distances depending on weather and terrain. Containers that are being impinged on with flame should be cooled if it is safe to do so. These containers should be cooled with unmanned monitors and master streams. If there are no prepositioned and piped master streams, which petroleum storage and refining facilities usually have, it may not be safe even to approach and set up an unmanned monitor. In these cases the immediate area must rapidly be isolated, and downwind populations must be evacuated as quickly as possible. Containers impinged on by direct flame contact can fail within minutes of first responder arrival, so conditions must be accurately analyzed before offensive, close-contact operations are initiated. Many first responders have needlessly been killed or injured during LPG emergencies. Historically, propane and LPG tanks have been especially dangerous in this regard (see Box 5-5).

EXTINGUISHING FLAMMABLE LIQUID FIRES

Gasoline tanker fires are an impressive sight. When 9000 gallons of volatile fuel burns, large amounts of smoke, fire, and heat are released. The first instinct of most firefighters is to rapidly extinguish the fire. This can be a more difficult proposition than it at first appears. Gasoline is not soluble in water and is lighter than water and will therefore float on top of the water and flow along the path of least resistance, usually into the sewer system, storm drains, and basements. If water is applied into a container, the burning fuel will float on top and swamp the tank. This will tend to spread the fire, causing extensive property and environmental damage, and worse.

Often, it is better to let a burning gasoline tanker burn for a time while all the necessary equipment and supplies are obtained (such as foam concentrate and a consistent water supply). Although this fire creates air pollution, the resulting environmental damage is significantly less than when hydrocarbons get into the groundwater or other water sources. DOT 406 and MC 306 type fuel tankers usually are constructed of aluminum and burn down like candle wicks as the aluminum melts or burns away. The gasoline burns down at approximately 6 to 12 inches per hour. Because these tankers are not pressurized vessels (like LPG tankers), the risk of a BLEVE is minimal.

In some instances gasoline tanker fires should be extinguished as quickly as possible with available resource, as when life safety or significant property damage is at stake. Examples of life safety include occupants trapped in vehicles after a crash, or occupants of nearby dwellings who cannot be safely evacuated, such as in a hospital. An example of significant property damage would be infrastructure such as a bridge overpass, which would cost months of time and millions of dollars to replace. In such cases the fire can be extinguished using the appropriate foam. It is important to have a sufficient amount of foam concentrate on hand before extinguishment operations begin so that the fire can be completely extinguished, and a foam blanket can be maintained for an extended period of time over the remaining gasoline. Difficulties include accessing the burning gasoline when fire has entered the tank after it has been torn open, extinguishing a moving fire, and maintaining a foam blanket.

It is also vitally important that the appropriate foam be selected. Gasoline is a nonpolar hydrocarbon and can be extinguished with standard AFFF. Fuel mixtures containing ethanol, which is a polar alcohol, will require ARF. When AFFF is applied to fuels such as E85 (85% ethanol, 15% gasoline) the foam blanket is quickly degraded, and the problem becomes worse. After the fire is extinguished, the remaining fuel must be transferred to another container as quickly as possible to prevent reignition. Transferring heated, volatile fuels is very risky, so the fuel should be allowed to cool to a safe temperature (which depends on the selected transfer operation and equipment). A thermal imaging camera (TIC) or temperature gun can be used to assess the fuel temperature.

Carefully consider the cost versus benefit when attempting to extinguish gasoline tanker fires. Sometimes it may be a better idea to let it burn temporarily or even let it burn itself out.

BOX 11-8 HAZMAT HOW TO: EXTINGUISHMENT OF HYDROCARBON TANKER FIRES

This technique uses both foam (either AFFF or AR-AFFF) and dry-chemical fire extinguishing agent for knock down and is therefore often called the dual-agent approach.

1. Cool the thermal column with three 1-3/4" hand lines flowing water. This will reduce the intensity and heat of the fire. Make absolutely sure you do not get any water into the cargo tanker.
2. Obtain at least 100 to 150 gallons of the appropriate foam concentrate. You will need AFFF for hydrocarbon fires like gasoline, and you will need AR-AFFF for E85 and ethanol fires. Ensure that you have the correct type of foam! Do not start extinguishment operations until you have enough foam concentrate on hand.
3. To extinguish the tanker fire you'll need two 1-3/4" hand lines that can flow approximately 100 gallons per minute (gpm) of foam solution each. After this point apply only foam or dry chemical to the tanker. Flowing water now will only degrade the foam blanket.
4. Extinguish spot fires in hard-to-reach places with dry-chemical fire extinguishers.
5. Maintain the foam blanket as necessary. It is extremely important to maintain the foam blanket while the fuel is still hot! Ensure that sufficient foam concentrate is available to maintain the foam blanket until the remaining fuel can be off-loaded.
6. It is absolutely imperative that continuous air monitoring using several CGIs be used to ensure that vapor suppression is effective. Otherwise, accumulated vapors may spontaneously ignite and explode.

BOX 11-9 HAZMAT HISTORY: ETHANOL UNIT TRAIN DERAILMENT (2011)

unit train ■ A train that carries a single commodity.

On February 6, 2011, a Norfolk Southern 65-car ethanol unit train derailed in Arcadia, Ohio. A **unit train** is a train that carries a single commodity—in this case E95 ethanol. The train was going approximately 64 mph when the last 32 cars derailed. Cars 28 and 29 breached, and more than 800,000 gallons of ethanol was released and started burning. Owing to a large pool fire multiple other cars ruptured, some violently. Fortunately, the train derailment occurred in a rural area and no one was injured or killed.

The derailment occurred several hundred feet away from an agricultural co-op that stored large quantities of anhydrous ammonia. In fact, one car was blown in half and half the railcar flew over the 100-foot-high anhydrous ammonia storage tanks. Clearly, even low-pressure tanks can BLEVE. Ethanol tankcars are general-service tankcars that are constructed from 7/16" steel. Typically, ethanol tankcars stretch and burst rather than explode under fire conditions. The large pool fire may have overwhelmed the pressure relief devices (PRDs) and may be the reason several of the tank cars exploded. There were reports that one tank car had liquid ethanol shooting out of its PRD and was burning like a blowtorch.

The firefighting operations were carried out almost exclusively by railroad contractors. The water supply operation was carried out by the local fire department. Norfolk Southern Railroad supplied their 550-gallon foam concentrate trailer. Because the derailment occurred in the winter, the foam concentrate arrived frozen. During wreck-clearing operations individual spot fires were extinguished, which led to at least one flash fire from the pooled ethanol on the ground. The ignition source appears to have been smoldering solids that adhered to the inside roof of the tank car. This incident happened after 11 railcars were already cleared. It is important not to let complacency set in during dangerous product control operations.

SOLVED EXERCISE 11-1

You are dispatched to a semi truck, placarded as environmentally sensitive material, leaking liquid from the rear doors. How could you initially quickly and effectively handle this incident?

Solution: Place a catch basin beneath the leak.

Containment Control Options

Plugging and patching are offensive product control techniques designed to stop a leak directly at the container. Properly overpacking the container in a certified salvage drum or overpack drum will allow the container to be moved and/or shipped. Patching and plugging minimize the amount of released product, so they are very effective at hazard mitigation for public safety as well as property and environmental conservation.

Containment operations can be very difficult depending on a number of factors including the following:

- size of the container
- type of container (high internal pressure versus low internal pressure)
- chemical contents of the container
- damage to the container
- position of the container
- geographic conditions (down a steep incline, for example)
- weather conditions
- available mitigation equipment

The size and type of container play a significant role in how difficult containment operations will be. Larger containers typically contain more product and create a larger hazard area as well as a greater internal head pressure. The type of container plays an arguably larger role, since it determines the internal pressure and limits the types of plugging and patching operations that can be performed. MC 306/DOT 406 highway cargo tankers always need to be unloaded before they are moved owing to their less substantial construction features. High-pressure vessels typically only need to be vented, flared, or off-loaded if they have sustained critical damage. Their high internal pressure makes plugging and patching generally ineffective.

The chemical contents of the container are also very important in determining both the type of container and the mitigation equipment that will be used in confinement operations. Chemical compatibility is of primary concern. If plugs or patches react with the chemical, they may fail or, worse yet, cause an exothermic reaction with the contents of the container. For example, plugging and patching material could cause runaway polymerization in certain reactive materials such as organic peroxides and unsaturated hydrocarbons (such as styrene).

During transfer operations chemical incompatibility can cause leaks, fires, or even explosions as transfer lines fail or flammable gases are generated through chemical reactivity. As discussed in the last chapter, damage to the container—especially high-pressure containers—can be extremely dangerous and may warrant a defensive operation. The position of the container, especially if it is upside down and valves are inaccessible, often limits available containment options. Some of these options may not be readily feasible if the necessary mitigation equipment is not available or may cause long response delays as the necessary equipment and trained personnel are transported to the scene.

REMOTE VALVE SHUTOFF

Closing a remote shutoff valve is one of the most effective ways to quickly stop the release of a hazardous material from its container. **Remote valve shutoff** switches can be found at fixed-site facilities and in the transportation industry (Figure 11-7). Often, facility workers or drivers will have used the remote shutoff before first responders arrive at the scene of a hazmat incident. This action is often the critical step in preventing a small release from spiraling into a community-wide disaster. Sometimes, however, the workers or drivers do not have the time or are incapacitated and cannot work the remote shutoff valves.

remote valve shutoff ▪ A method to safely stop a leak issuing from a pipe by closing an outlet at a distance from the leak.

FIGURE 11-7 Emergency shutoff valves are located on many vehicles transporting hazardous materials and at many fixed-site facilities that use hazardous materials. Gasoline tankers (top left), DOT 407 cargo tankers (top right), and fixed-site facilities (bottom row) should have marked emergency shutoff valves located away from expected problem areas. *Photos courtesy of Chris Weber, Dr. Hazmat, Inc.*

Transportation

Hazardous materials incidents commonly occur during transport. Road and rail incidents are the most common, but hazmat incidents involving marine transport and air transport also occur with regular frequency. Highway cargo tankers are equipped with remote valve shutoff switches. These switches are typically located on the left front side of the tank, and a second switch may be located on the right rear side of the tank, diagonally opposite each other. The switches operate an internal emergency shutoff valve completely contained within the tank. This means that even if external piping is sheared away, the internal valve should stay intact and operate. The remote shutoff devices are typically mechanically, pneumatically, or hydraulically operated and contain an additional fusible link that fails at a predetermined temperature in case of fire.

DOT 406/MC 306 highway cargo tankers typically carry flammable liquids such as gasoline, diesel fuel, and E85. Historically, approximately 50% of hazmat releases have been hydrocarbon fuels such as gasoline. These cargo tankers are typically constructed of aluminum, and the remote shutoff valves are usually operated mechanically or pneumatically. The MC 306/DOT 406 emergency shutoff switches are located on the driver side at the front of the tank.

DOT 407/MC 307 highway cargo tankers are known as the workhorse of the chemical industry and are typically constructed of stainless steel. Their remote shutoff valves are typically hydraulically operated and are typically located on the left front side of the tank. A second switch may be located on the right rear side of the tank, diagonally opposite the first.

DOT 412/MC 312 highway cargo tankers are designed to carry corrosive liquids. These tankers are not required to have any emergency remote shutoff valves.

MC 331 highway cargo tankers carry compressed liquefied gases such as propane and anhydrous ammonia. These tankers are constructed of steel. MC 331 tankers with a capacity of over 3500 gallons must have two remotely operated emergency shutoff switches, while those with a capacity of less than 3500 gallons must have at least one emergency shutoff switch. The emergency shutoff switch is located on the left front side of the tank, and a second switch will be located on the right rear side of the tank, diagonally opposite the other one. In addition, while off-loading an MC 331 tanker the driver must have a remote shutoff switch that can be operated 100 feet away from the vehicle and must be able to shut off the engine and close the internal shutoff valve.

MC 338 and CGA-341 highway cargo tankers are designed to carry relatively non-toxic liquefied compressed gases. Typically these gases are cryogenic, which means they have a boiling point below −130°F. These tankers are designed to vent small amounts of their contents under normal operating conditions, in the process known as autorefrigeration. The extremely cold product is designed to cool the tank by evaporation if the internal pressure rises too high. Thus, it is not uncommon to see a white cloud coming off these tankers. This white cloud is water vapor that is condensed from the air by the cold gaseous product. Not surprisingly, more humid atmospheres tend to produce larger white clouds. These cargo tankers are not required to have remote shutoff valves.

Intermodal tanks typically have an emergency remote shutoff valve located on the bottom right-hand corner.

Fixed-Site Facilities

Remote shutoff valves are often difficult to locate at fixed-site facilities and may have poor signage (Figure 11-7). Therefore, facility plans and routine training tours are essential for gaining familiarity with fixed-site facilities in your jurisdiction. The use of any fixed-site facility remote shutoff valves should be discussed with the facility's technical experts and covered in the appropriate facility emergency response procedures for your agency. The remote shutoff valves should be indicated on all facility maps.

It is important not to confuse remote shutoff valves with other valves at the facility. It can be extremely dangerous to turn valves on or off that are not designed to be emergency remote shutoff valves. Doing so may make chemical reactions unstable or cause overpressurization of pipes or vessels and lead to further and more severe hazardous materials releases.

DIRECT VALVE SHUTOFF

Direct valve shut off is an offensive product control technique that puts the hazmat technician in close proximity to the hazardous chemical (Figure 11-8). This technique must be used for stand-alone containers that do not have piping coming from or feeding into the container that can be closed remotely. Whenever any container is approached directly—such as to shut off a valve—the following safety precautions must be observed:

- All indications are that the container will not fail catastrophically.
- Appropriate PPE is available that adequately protects against all hazards (including chemical, thermal, and cryogenic hazards).
- Appropriate air-monitoring equipment is available.
- The necessary tools to close the valve are available, and personnel are trained in their use.

Because entry team personnel will be in very close contact with the released product, direct valve shut off may require the highest level of PPE, Level A protection.

SOLVED EXERCISE 11-2

You are dispatched to an anhydrous ammonia leak at the local ice rink. On arrival you find an MC 331 tank truck delivering anhydrous ammonia and a large white cloud emanating from the area. What control measures could be implemented?

Solution: The vehicle remote shutoff valve could be used to stop transfer of the product from the tank truck. The facility should have a remote shutoff valve that will contain the product to the facility tank. This would limit the leak to the product contained in the piping.

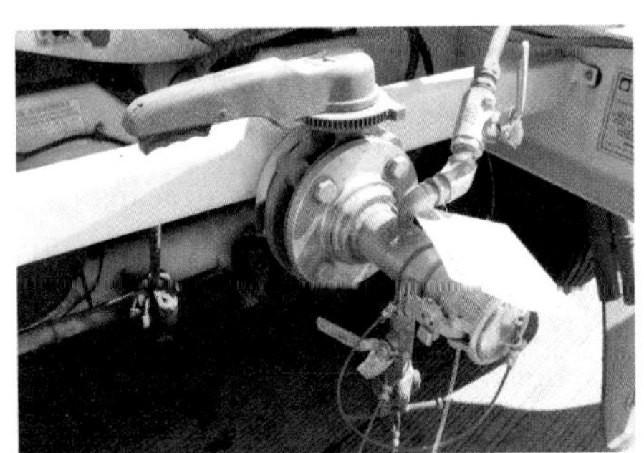

FIGURE 11-8 Direct valve shutoff is an option when there is no emergency shutoff valve or if it has become inoperable. *Photos courtesy of Chris Weber, Dr. Hazmat, Inc*

CYLINDER PATCH KITS

When container valves are inoperable—or the container has been damaged—plugging, patching, or capping product control techniques must be used. As with direct valve shutoff, this places the hazmat technician in close proximity to and almost certain contact with the hazardous material. These activities—especially when dealing with liquefied compressed gases—can therefore be very dangerous and must be attempted only with the appropriate tools, training, and experience. Be cognizant of cold injuries when dealing with liquefied compressed gas leaks. Sometimes it may be best to let the container vent itself, such as when there are no life safety issues or significant exposures nearby.

Chlorine Kits

Chlorine is used by many public and private entities, including municipal drinking water and wastewater treatment plants, the chemical manufacturing industry, and research laboratories. The following are the most common causes of chlorine container leaks:

- incompletely closed valve
- failure of the fusible plug or relief valve
- leaking valves (loose valve, worn valve seat, or damaged packing)
- valves sheared off
- cylinder shell damage (puncture or corrosion)

By far the most common cause of a chlorine release is an incompletely closed valve. This type of emergency can be mitigated by quickly deploying an entry team to manually close the valve. Sometimes a piece of debris can lodge in the valve seat, causing it to remain partially open. The debris can often be dislodged by opening and then closing the valve, thereby reseating it. This procedure allows the piece of debris to flow through the valve, as long as the valve has been opened far enough to permit the debris to flow through.

When the simple fixes do not work, or the problem is obviously more serious, a chlorine cylinder patch kit must be applied (Figure 11-9). Three types of chlorine kits are manufactured. The red A kit is designed for use with 100- and 150-pound cylinders, the yellow B kit is designed for use with 1-ton cylinders, and the green C kit is designed for use with highway cargo tankers and rail tank cars. It is worth noting that the A and B Kits are authorized for use in transportation of leaking chlorine cylinders under DOT regulations.

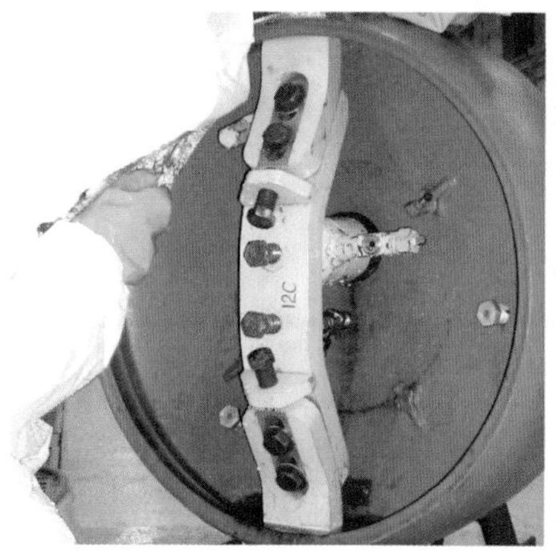

FIGURE 11-9 There are three types of chlorine kits: the chlorine A kit is used for 100-pound and 150-pound cylinders (right), the B kit is used for 1-ton containers (bottom left), and the C kit is used for railcars and highway cargo tankers (top left). *Photos courtesy of Chris Weber, Dr. Hazmat, Inc.*

Chlorine cylinders are normally constructed of carbon steel, with aluminum–silicon–bronze alloy fittings, and Viton gaskets. A spray bottle of aqueous ammonia can be used to visualize chlorine leaks, since chlorine reacts with ammonia to form ammonium chloride salt, which appears as a white haze when aqueous ammonia is sprayed on a chlorine release.

A Kit DOT 3A480 cylinders, which have 100- and 150-pound compressed liquefied chlorine capacity, have a fusible plug located on the neck of the valve that melts between 158°F and 165°F.

TABLE 11-1 Temperature-Dependent Internal Pressure of Containers Containing Anhydrous Ammonia, Chlorine, and Propane

AMBIENT TEMPERATURE	INTERNAL CONTAINER PRESSURE (VAPOR PRESSURE)		
	ANHYDROUS AMMONIA	CHLORINE	PROPANE
32°F	46 psig	39 psig	50 psig
68°F	115 psig	82 psig	105 psig
95°F	180 psig	139 psig	165 psig
165°F	510 psig	325 psig	375 psig

BOX 11-10 HAZMAT HOW TO: CHLORINE A KIT

1. Verify that the appropriate container (a standard DOT 3A480 or DOT 3AA480 cylinder) is indeed leaking. These are 100- and 150-pound chlorine cylinders.
2. Do not apply the chlorine A kit to a cylinder completely full of liquid (liquid full).
3. Familiarize yourself with the contents of the chlorine A kit in the cold zone before donning PPE.
4. Prepare the equipment by inserting the correct gaskets onto the base of the hood (kit part 1A2), removing the outlet cap on the vent valve of the hood, opening the outlet valve on the hood, and loosening the set screws on the yokes that will be used.
5. Don appropriate PPE: Level A ensembles are recommended for indoor chlorine leaks, while a Level B ensemble may be appropriate for mitigating certain outdoor leaks.
6. Position the cylinder so that the leak is in the vapor phase and not the liquid phase. This will reduce the severity of the leak several hundredfold.
7. Remove the valve protective housing if it is in place. This will facilitate finding the leak if it is located in the valve assembly area.
8. Use a solution of household ammonia (dilute ammonium hydroxide solution) in a spray bottle to locate the chlorine leak. The interaction of chlorine and ammonia will produce a dense white cloud.

VALVE STEM LEAK

1. If the leak is coming from the valve area, attempt to close the valve using the small square end of the supplied wrench (kit part 200A).
2. If this does not stop the leak, open and close the valve briefly to dislodge any foreign material that may have seated at the gasket.
3. If this does not stop the leak, tighten the valve stem packing with the wrench (200A).

HOOD ASSEMBLY APPLICATION

1. If the leak appears to be coming from the valve inlet threads, the threads may be corroded. It is possible to tighten the valve into the cylinder, but this can dislodge the valve completely if the threads are corroded. It is therefore generally safer to apply the hood assembly (1A2) instead.
2. Set up the base assembly (1EFP) next to the damaged cylinder. Do not forget to insert the stabilizer into the center. This will ensure that the three legs are equidistant and stable.
3. Set up the ramp (1R) between the damaged cylinder and the base assembly.
4. Carefully roll the cylinder onto the base assembly.
5. If necessary, clean the shoulder of the cylinder using the scraper (A8). This will help the hood gasket make a better seal.
6. Place the yoke (1C1) on top of the hood assembly. The set screws should be over the dimples in the top of the hood.
7. Straighten the chains and hook the tightest link possible onto the yoke. Each of the three chains should have the same link hooked onto the yoke. It is very important to ensure that the chain links are not twisted and that they are all fully extended and not sideways. Otherwise it will be difficult to get an even seal.
8. Tighten the set screws on the yoke until the leak stops. Do not over tighten the set screws, as this may damage the gasket!
9. Close the vent valve (1V) on the hood. You may need to tighten the set screws again at this point as the pressure increases underneath the hood.
10. Test for leaks as the cylinder pressure comes to equilibrium.

FUSIBLE PLUG PATCH APPLICATION

1. Attempt to tighten the fusible plug slowly with the wrench (203).
2. If this does not stop the leak, the fusible plug can be patched using the clamp assembly (Device 2) or the entire hood assembly (1A2) can be applied (see previous section).
3. Saw the fusible plug off flush with the valve body using the supplied hacksaw.
4. Then, file the surface more using the supplied file.
5. Insert the appropriate gasket material into the clamp (2C).
6. Place the clamp over the valve and tighten the set screw until the leak stops.
7. Test for leaks as the cylinder pressure comes to equilibrium.

SIDEWALL PATCH APPLICATION

1. Leaks in the sidewall of the cylinder can be patched using the sidewall patch (Device 8). If the sidewall leak was caused by corrosion, areas immediately adjacent to the leak may be weakened. Therefore, use extreme caution when tightening the set screw!
2. Wrap the chain around the cylinder.
3. Place the gasket and patch (8D) over the leak.
4. Place the cap screw and yoke on top of the patch.
5. Straighten the chain and hook the tightest link possible onto the yoke.
6. Tighten the set screw until the leak stops. You may need to tighten the set screw again at this point as the cylinder pressure increases. Use extreme caution when tightening the set screw to avoid further damaging the container! If the cylinder show signs of weakening, immediately stop tightening the set screw.
7. Test for leaks as the cylinder pressure comes to equilibrium.

DRIFT PIN INSERTION

Drift pins (A3 and A4) can be used to stop leaks from the valve assembly. The drift pin is driven into the opening with the supplied hammer. However, the drift pin may become a projectile if cylinder pressure increases. It is therefore generally safer to apply the hood assembly instead.

BOX 11-11 HAZMAT HOW TO: CHLORINE B KIT

1. Verify that the appropriate container (a DOT 106A500X chlorine 1-ton cylinder) is indeed leaking.
2. Do not apply the chlorine B kit to a cylinder completely full of liquid (liquid full).
3. Familiarize yourself with the contents of the chlorine B kit in the cold zone before donning PPE.
4. Prepare the equipment by inserting the correct gaskets onto the base of the hoods, opening the diversion valve on the hood, then loosening the clamp screws on the yoke.
5. Don appropriate PPE: Level A ensembles are recommended for indoor chlorine leaks, while a Level B ensemble may be appropriate for mitigating certain outdoor leaks.
6. Position the cylinder so that the leak is in the vapor phase and not the liquid phase. This will reduce the severity of the leak several hundredfold.
7. Remove the valve protective housing if it is in place. This will facilitate finding the leak if it is located in the valve assembly area.
8. Use a solution of household ammonia in a spray bottle to locate the chlorine leak. The interaction of chlorine and ammonia will produce a dense white cloud.

VALVE STEM LEAK

1. If the leak is coming from the valve area, attempt to close the valve using the small square end of the supplied wrench (kit part 200).
2. If this does not stop the leak, open and close the valve briefly to dislodge any foreign material that may have seated at the gasket.
3. If this does not stop the leak, apply the outlet cap with a fresh gasket.
4. If this does not stop the leak, tighten the valve stem packing with the wrench (200).

BONNET APPLICATION

1. Prepare the hood (12A) by inserting the proper gaskets into the base of the hood and by opening the vent valve (12V) on the hood. Prepare the yoke assembly by loosening all the set screws and the jack screw. This is most easily accomplished in the cold zone.
2. If the leak coming out of the valve is liquid, attempt to roll the cylinder until the valve is in the vapor space. This will reduce the leak intensity by several hundredfold.
3. Chock the container to prevent unexpected movement.
4. If the leak is at a valve, attempt to close the valve. If this does not stop the leak, open the valve a quarter turn and quickly close the valve again. This should stop 90% of the leaks you will encounter!
5. Clean the area around the valve assembly using the scraper (B5) if necessary.

(continued)

6. If there is a ridge between the valves, you'll need to use the indented gasket, which allows the gasket to fit over the ridge.
7. The bonnet is most easily applied by two people: One person holds the bonnet over the leaking valve while the second person installs the yoke underneath the lip of the 1-ton container.
8. Ensure that the jack screw is aligned vertically over the bonnet using the positioning screws on the yoke. This is the most important aspect to getting a good seal.
9. Tighten the set screw over the bonnet until the leak stops. Be careful not to overtighten the set screw, or you may damage the gasket.
10. If the leak is not stopped, and chlorine seems to be escaping from one side, the set screw is not centered over the bonnet. Loosen all set screws and reapply the bonnet.
11. Test for leaks as the cylinder pressure comes to equilibrium.
12. Monitor the chlorine container to ensure that the leak is completely stopped.

SIDEWALL PATCH APPLICATION

1. If the leak is liquid, attempt to roll the 1-ton cylinder until the leak is in the vapor space. Rolling the cylinder will change the liquid leak into a vapor leak, reducing the leak intensity by several hundredfold.
2. Chock the container to prevent unexpected movement.
3. Leaks in the sidewall of the cylinder can be patched using the sidewall patch (Device 9). If the sidewall leak was caused by corrosion, areas immediately adjacent to the leak may be weakened. Therefore, use extreme caution when tightening the set screw!
4. Wrap the chain underneath the cylinder.
5. Place the gasket and patch (9D) over the leak.
6. Place the cap screw and yoke on top of the patch.
7. Straighten the chain and hook the tightest link possible onto the yoke.
8. Tighten the set screw until the leak stops. You may need to tighten the set screw again at this point as the cylinder pressure increases. Use extreme caution when tightening the set screw to avoid further damaging the container! If the cylinder shows signs of weakening, immediately stop tightening the set screw.
9. Test for leaks as the cylinder pressure comes to equilibrium.

FUSIBLE PLUG PATCH AND HOOD APPLICATION

1. Attempt to tighten the fusible plug slowly with the wrench (104).
2. If the fusible plug is blown out, file the surface smooth.
3. Place a gasket underneath the stud (4E).
4. Fit the yoke (4C) and stud over the head of the fusible plug and tighten the stud.
5. If tightening the fusible plug does not stop the leak and the fusible plug is not blown out, the fusible plug can be capped using the fusible plug hood assembly (4A plus appropriate gasket).
6. Clean the area around the fusible plug using the scraper (B5) if necessary.
7. Fit the yoke (4C) and stud (4E) over the head of the fusible plug and tighten the stud.
8. Place the hood over the fusible plug/yoke/stud assembly.
9. Place the nut onto the stud and very carefully tighten the nut until the leak stops. Be extremely careful when tightening the nut. This assembly will pull the fusible plug out by the threads if it is overtightened.
10. Test for leaks as the cylinder pressure comes to equilibrium.

DRIFT PIN INSERTION

Drift pins (B1, B2, B3, and B4) can be used to stop leaks from the fusible plug and/or the valve assembly. The drift pin is driven into the opening with the supplied hammer (B6). However, the drift pin may become a projectile if cylinder pressure increases. It is therefore generally safer to apply the hood assembly (when this is an option) instead.

B Kit The DOT 106A500X 1-ton chlorine cylinders containing compressed liquefied chlorine have six fusible plugs; three fusible plugs are located at each end. These fusible plugs also melt between 158°F and 165°F.

C Kit MC 331 highway cargo tankers and DOT 105A500W rail tank cars typically carry compressed liquefied chlorine. Railcars have a spring-loaded safety relief device with a rupture disk underneath located on top of the railcar inside the dome cover. The safety relief valve operates at 375 psi. When climbing onto highway cargo tankers

BOX 11-12 HAZMAT HOW TO: CHLORINE C KIT APPLICATION

1. Do not apply the chlorine C kit to a railcar or highway cargo tanker completely full of liquid (liquid full).
2. Familiarize yourself with the contents of the chlorine C kit in the cold zone before donning PPE.
3. Prepare the equipment by inserting the correct gaskets onto the base of the hoods, opening the diversion valve on the hood, then loosening the clamp screws on the cross bar.
4. Don appropriate PPE.
5. Use a roof ladder to ascend the rail car or highway cargo tanker. Do not use the ladder on the vehicle itself, since it may be damaged or otherwise unsafe.
6. Use a solution of household ammonia in a spray bottle to locate the chlorine leak. The interaction of chlorine and ammonia will produce a dense white cloud.
7. If the leak is at a vapor or liquid valve, attempt to close the valve. If this does not stop the leak, open the valve a quarter turn and close the valve again. This should stop 90% of the leaks you will encounter!
8. The angle valve packing gland, angle valve seat, angle valve gasket, pressure relief device gasket, and manway cover gasket may be loose. These can all be tightened with the appropriate wrench (see the chlorine C kit instructions).
9. If the leak is at the PRD, place a pressure gauge on one of the vapor lines. If the railcar is on its side be careful: the vapor line may now be a liquid line! Never place a capping device on a PRD without monitoring the container's pressure. Remember, PRDs are designed to vent excess pressure in the container to keep it from failing catastrophically!
10. Carefully open the vapor line and observe the pressure. If the pressure is at or above 375 psi, the PRD is operating normally: DO **NOT** CAP THE PRESSURE RELIEF DEVICE, AND LEAVE THE AREA IMMEDIATELY!
11. If capping a valve or PRD is necessary, clean the area around the leaking fitting with the scraper.
12. Place the hood and gasket over the leaking fitting. The valve outlet plug may need to be removed if it is too big.
13. Assemble the brace and place the set screw over the hood and latch the hooks into the eyelets of the dome. Be sure to vertically align the set screw with the hood. Otherwise, the hood will leak as the gasket experiences uneven pressure.
14. Tighten the set screw carefully until the leak stops. Be careful not to overtighten the set screw, as you may damage the gasket!
15. Close the vent valve on the hood and replace the outlet cap.
16. Observe the leak for approximately 10 to 15 minutes as the tank pressure rises to ensure that the set screw is tight enough.
17. Test for leaks as the cylinder pressure comes to equilibrium.
18. Monitor the chlorine container periodically to ensure that the leak is completely stopped.

or railcars, be sure to use a ladder—preferably a roof ladder with hooks—instead of the ladder found on the vehicle (Figure 11-10). It could have been damaged or otherwise compromised.

The following considerations apply to responses to any leaking gas, including chlorine:

- Monitor vapor concentrations at the release location and downwind.
- Evacuate and/or use in-place sheltering for at-risk populations if necessary.
- Select PPE based on the release location and the airborne concentration.
- Move the container so that a liquid leak becomes a gas leak (the container breach should be in the vapor space).
- Cool cylinders using ice, dry ice, or liquid nitrogen to reduce the leak rate when appropriate and under the guidance of a technical expert.
- Use water streams to control vapor clouds when appropriate.
- Be careful about applying water directly to the leaking container, as it will typically warm the container and increase the rate of release.
- Apply a patch kit when appropriate.
- Neutralize the material when appropriate.
- Use proper decontamination methods and procedures.

FIGURE 11-10 When climbing onto a rail car or highway cargo tanker, it is important to use your own ladder, preferably a roof ladder with hooks. Do not rely on the permanently affixed ladders on the vehicle, as they may have been damaged during the accident or derailment. *Photos courtesy of Chris Weber, Dr. Hazmat, Inc.*

BOX 11-13 HAZMAT HISTORY: GRANITEVILLE, SOUTH CAROLINA (2005)

On January 6, 2005 two Norfolk Southern trains collided in the middle of Graniteville, South Carolina. The 42 car train was carrying hazardous materials including chlorine, sodium hydroxide, and cresol. A rail car loaded with 90 tons of chlorine ruptured and released about 60 tons of its load, killing nine people and injuring at least 250 people. The accident was caused by an improperly positioned mainline switch that allowed a through bound freight train to collide with a freight train parked on a railroad siding located next to the Avondale Mills sugar plant (where 5 employees died). Over 5000 residents within a mile of the derailment were evacuated for almost 2 weeks during the immediate aftermath and the cleanup operations. Area hospitals treated over 200 patients.

A chlorine scrubber can be created by dissolving 125 pounds of sodium hydroxide (caustic soda or lye) in approximately 50 gallons of water, or dissolving 300 pounds of soda ash in approximately 100 gallons of water. This scrubber should have a capacity of approximately 100 pounds of chlorine.

Sulfur Dioxide Kits

Sulfur dioxide—like chlorine—is used in water treatment plants and is commonly used in 150-pound cylinders and 1-ton containers. Patch kits are available that are similar to the chlorine kits but with some slight design as well as gasket material differences. The gasket material most commonly used for sulfur dioxide is ethylene propylene diene monomer (EPDM). As with chlorine, DOT authorizes the shipment of leaking sulfur dioxide cylinders sealed with the kits.

SOLVED EXERCISE 11-3

You have been dispatched to a chlorine leak at the local water treatment plant, which uses 1-ton containers. What is the appropriate leak control equipment you must retrieve from the response vehicle?

Solution: The chlorine B Kit.

PLUGGING

Plugging a container involves stopping or slowing the leak by placing a mechanical or pneumatic device into the breach. Examples of plugs include wooden and rubber wedges; metal, plastic, wood, or rubber plugs attached to a butterfly nut; pneumatic rubber plugs; T-bolts; and sheet metal screws with a compatible gasket (Figure 11-11). Plugs that are held in by friction can be used only on low-pressure containers, while butterfly nuts, T-bolts, and sheet metal screws may be used on slightly higher pressure containers or on low-pressure containers that have a high back pressure owing to a large lading. Plugs are very rarely used on high-pressure containers because of the high probability of failure with the potential to injure responders in the immediate area.

FIGURE 11-11 Plugging and patching tools are limited only by your imagination. Wooden plugs (top left), steel wool and epoxy (top right), toggle bolts (bottom right), and pipe clamps (bottom left) can all be used successfully to stop leaks. Keep in mind that chemical compatibility of the plug or patch is crucial. *Photos courtesy of Chris Weber, Dr. Hazmat, Inc.*

BOX 11-14 HAZMAT HOW TO: DRUM PLUGGING AND PATCHING

1. Wear necessary PPE.
2. Use appropriate air-monitoring equipment to ensure safety.
3. Gather compatible plugging and patching tools. Nonsparking tools will be necessary for drums containing flammable liquids.
4. If the drum is leaking liquid, position the drum so that the damaged area is in the vapor space.
5. IF SAFE TO DO: Clean the area to be patched with a wire brush for nonflammable products.
6. Place patching material into the hole. Patching material may consist of such items as wooden, plastic, or metal plugs; sealing putties; or toggle bolts and T-bolts.
7. IF SAFE TO DO SO: Cut off the protruding end of the plug (if present).
8. Place chemically compatible tape over the patch.
9. Place epoxy over the taped patch.
10. Let the epoxy dry per manufacturer's suggestions.
11. The drum is not legal for transportation per DOT until it is overpacked in a certified salvage drum.

Containers also can be sealed using an epoxy compound, rubber compound, polymer foams, or commercial mixtures such as plumber's putty and Plug N Dike. Often, sealants are used in combination with plugs to completely stop the leak and make the plug more durable, as in plugging and patching damaged 55-gallon drums to facilitate overpacking and transportation for disposal. Keep in mind that the plug must be able to fit inside an overpack drum. Wooden plugs and some metal screws can be sawed off when necessary. Lead wool has been used to help plug leaks in the past, but it should be used sparingly—if at all—because of environmental concerns.

PATCHING

Patching of containers involves stopping or slowing a leak by placing a mechanical, magnetic, hydraulic, or pneumatic device over the breach (Figure 11-11). Patches should conform to the container and typically use a gasket that must be compatible with the contents of the container. Patches can be used on low-pressure and moderately high pressure containers. The higher the container pressure, the greater the strength of the patch must be. Patches can be held on using magnets, pneumatic devices such as airbags, chains, or straps. Also available are chemical pipe repair/patch systems that use an epoxy and gauze system to wrap the pipe. Once the epoxy hardens, the damaged pipe is sealed, much like a cast that helps stabilize a broken arm or leg.

Patch and plug kits are commercially available or can be assembled by a trip to the hardware store. Typically, a good plug and patch kit will contain a variety of hand tools (such as screwdrivers, hammers, and wrenches); a wide variety of different sizes of closures (such as self-tapping screws, butterfly nuts, T-bolts, and toggle bolts); a wide variety of plugging materials (such as rubber stoppers, wooden and neoprene wedges and dowels, pipe plugs and caps, and polymer balls); a variety of pipe clamp sizes; a variety of pneumatic rubber pipe stops; and a wide variety of gasket material (such as neoprene, Viton, rubber, and Teflon). Be careful not to overtighten clamps and screws, as this may cause container failure or gasket failure.

Pipe Patch Kits

Pipe patch kits contain various clamps and gaskets designed to fit over piping of various diameters. Pipe clamps are very effective at stopping leaks caused by holes or small cracks in a wide variety of pipes, including both plastic and metal. The clamps are tightened only until the leak stops. Overtightening, especially when dealing with more fragile plastic piping, may cause the pipe to fail. Pipe patch kits are commercially available, or they can be put together using components found in most hardware stores.

FIGURE 11-12 Dome clamps can be used on DOT 406/MC 306 cargo tankers to stop leaks from dome covers. *Photos courtesy of Chris Weber, Dr. Hazmat, Inc.*

Dome Clamps

Dome clamps are specialty pieces of equipment designed to fit over the manway dome covers of DOT 406/MC 306 highway cargo tankers (Figure 11-12). Several manufacturers produce different types of dome clamps that vary in their effectiveness and ease of application in specific situations. It is best to have several different types of dome clamps on hand, and since these cargo tankers often have multiple compartments, it is wise to have at least six dome clamps available at any given time. The dome clamps are tightened until the leak stops. Expect leaks from dome covers

BOX 11-15 HAZMAT HOW TO: DOME CLAMP APPLICATION

1. Gather enough dome clamps so there is one for each manway cover.
2. Prepare dome clamps by moving the set screw to its most retracted position.
3. Suppress flammable liquid vapors using foam.
4. Enter the hot zone using a CGI to monitor for flammable vapors.
5. Enter the hot zone wearing appropriate PPE—typically, structural firefighter's protective equipment.
6. Secure the damaged tanker from movement using blocking, bracing, cribbing, and/or the vehicle brake.
7. Apply a dome clamp to the leaking manway cover and tighten the dome clamp set screw. Tighten the set screw until the leak stops. Overtightening the set screw may cause the leak to intensify.

that have been clamped to get worse owing to internal pressure changes as tanks are drilled and pumped off.

Pneumatic Patches

Pneumatic patches, also sometimes referred to as Vetter bags, are essentially airbags that can be strapped onto containers over the leak using webbing or rope attached to a ratchet system. The uninflated pneumatic patch is securely strapped on top of an appropriate gasket material over the damaged area of the container and slowly inflated until the leak stops. The airbag should be inflated only until the leak stops to avoid damaging the container further, or crushing it altogether. Products from different manufacturers vary in their airbag, gasket, strap type and length, and ratchet configurations. Two manufacturers of pneumatic patches are Vetter and Maxi-Force.

Pneumatic plugs of various sizes are also available for inserting into pipe openings. These are similar to the pneumatic plugs available in hardware stores for pressure testing building plumbing. Those used for product control and hazardous materials response are typically a heavier gauge material and are available in larger sizes. In a pinch, pneumatic plugs from a hardware store may be useful on smaller-diameter pipes that are flowing at relatively low pressures (typically 30 psi or less).

Magnetic Patches

Magnetic patches, as their name implies, can be used with ferrous metals such as steel to stop container leaks (Figure 11-13). A gasket is placed beneath a curved magnet that is attached to the container at the damaged location. Commercial magnetic patches typically have a way to release the magnet using a fulcrum-type system to pry it from the container surface. These patches can be very effective at stopping leaks quickly. One problem with magnetic patches is that they can inadvertently be attached at the wrong location, and they are fairly difficult to pry loose, especially while wearing personal protective

FIGURE 11-13 Magnetic patches can be used on containers constructed of ferrous metal to stop leaks. *Photos courtesy of Chris Weber, Dr. Hazmat, Inc.*

SOLVED EXERCISE 11-4

An ethanol rail tank car is leaking from a puncture hole in its side. What are your leak control options?

Solution: Magnetic or pneumatic patches would work very well. Other patches, such as plugs, could also work well if they could be secured in place. The hydrostatic pressure of the product will tend to force plugs back out if not held in place securely.

equipment. These patches are commonly used on railroad tank car leaks, highway cargo tanker leaks, and fixed-site storage tank leaks.

OVERPACKING

Overpacking is the process of placing a larger undamaged container over a smaller damaged container (Figure 11-14). For example, a 65-gallon salvage drum can be used to overpack a damaged 55-gallon drum. It is essential to always choose a compatible overpack container. Generally, metal containers should be overpacked with the same type of metal container (e.g., steel versus aluminum); plastic drums should be overpacked using plastic salvage drums constructed of the same type of polymer; compressed gas cylinders should be overpacked into a compatible emergency response containment vessel (ERCV).

Typically, the damaged container should be plugged and patched before being placed into the salvage drum. In addition, the salvage drum should contain compatible absorbent material to soak up any additional leakage as well as to stabilize the damaged drum within the overpack container. When a damaged drum is overpacked, the openings of the drum—called *bungs*—should be facing up whenever possible. Under time-critical circumstances this may not always be possible. However, if the bungs are facing up, it will be much easier to sample the damaged drum as well as transfer its contents without having to drill it first.

Laboratory packs, or *lab packs* as they are commonly known, are salvage drums full of smaller containers ranging in size from ounces to approximately 5 gallons. Only compatible materials should be placed inside a common salvage drum. Packaging

FIGURE 11-14 There are three common methods used to overpack damaged and patched 55-gallon drums. The damaged drum can be placed on the salvage drum lid, and the salvage drum can be lifted over the damaged drum (left); the damaged drum and the salvage drum can be rolled into one another (top right); or the damaged drum can be slid into the salvage drum directly (bottom right). *Photos courtesy of Chris Weber, Dr. Hazmat, Inc.*

BOX 11-16 HAZMAT HOW TO: DRUM OVERPACKING

View an illustrated step-by-step guide on Drum Overpacking

1. Place plastic sheeting on the ground next to the damaged drum. Plug and/or patch the damaged drum before overpacking it (see Box 11-14).
2. Prepare the salvage drum by removing the lid and ensuring chemical compatibility of the seals, gaskets, and salvage drum material.
3. Always attempt to overpack the damaged drum with the bungs up. This will facilitate later sampling and transfer efforts when necessary.

SLIDE-IN METHOD

1. Gently place the drum to be overpacked on its side using a drum grappler to lower it. Remember, a full drum can weigh up to 1000 pounds!
2. Place the open salvage drum on its side with the open end next to the bottom of the damaged drum.
3. Lift the bottom of the damaged drum over the lip of the salvage drum.
4. Using a drum grappler, slide the damaged drum into the salvage drum.
5. Once the damaged drum is completely inside the salvage drum, upright the salvage drum using a drum grappler.
6. Pack the space between the overpacked drum and the salvage drum with compatible absorbent material.
7. Place the lid on the salvage drum. Do not place a lid on the salvage drum if the contents are undergoing a chemical reaction! This could lead to catastrophic failure of the salvage drum.

ROLLING-V METHOD

1. Gently place the drum to be overpacked on its side using a drum grappler to lower it. Remember, a full drum can weigh up to 1000 pounds!
2. Place the open salvage drum on its side with the open end next to the bottom of the damaged drum.
3. Lift the bottom of the damaged drum over the lip of the salvage drum.
4. Place the damaged drum at an angle with respect to the salvage drum.
5. Start rolling both drums together in the direction of the arms of the V. The damaged drum should start moving into the salvage drum.
6. Once the damaged drum is completely inside the salvage drum, upright the salvage drum using a drum grappler.
7. Pack the space between the overpacked drum and the salvage drum with compatible absorbent material.
8. Place the lid on the salvage drum. Do not place a lid on the salvage drum if the contents are undergoing a chemical reaction! This could lead to catastrophic failure of the salvage drum.

SLIP-OVER METHOD

1. Place the lid of the salvage drum upside down on the ground next to the drum to be overpacked.
2. If the damaged drum is upright, gently lower the damaged drum onto its side with the chime of the damaged drum resting on the outside edge of the salvage drum lid (if it is safe to do so).
3. If the damaged drum is upside down, carefully slide the damaged drum onto the salvage drum lid.
4. Invert the salvage drum and place it over the damaged drum.
5. Screw or clamp the salvage drum lid to the salvage drum body.
6. Upright the salvage drum using a drum grappler.

incompatible materials together—such as acids and bases or fuels and oxidizers—could create an extremely dangerous situation. Reactions between such combinations of materials will create heat and possibly generate gases that will overpressurize the container and cause it to rupture violently. Each smaller container should be carefully packaged and surrounded by absorbent to minimize breakage during transport.

Forklift Assist

There are several different methods of overpacking. Using a forklift with a drum attachment is the easiest and safest way to overpack a drum, by lifting the damaged (but patched) drum by the chime and vertically placing it into the salvage drum. This minimizes the danger to the entry team. Drums can be very heavy, averaging several hundred pounds and weighing up to 1000 pounds with denser materials. Using a rated forklift to do the heavy lifting will save your back, and possibly feet and legs, if the drum decides to tip over or roll unexpectedly. Before placing the lid onto the salvage drum, ensure that no exothermic chemical reactions or reactions producing gases are occurring inside. Otherwise, the salvage drum may burst or explode.

Slip-Over Method

The slip-over method is the next best way of overpacking, and it has two options: the single inversion method and the double inversion method. The single inversion method is the faster and easier of the two and consists of inverting the salvage drum and placing it over the damaged drum. If the damaged drum is sitting upright, the bungs will end up at the bottom of the salvage drum, which is not an ideal situation. It is thus imperative to thoroughly tighten the bungs before overpacking a drum by this method. After the salvage drum has been placed over the damaged drum, the drum combination is gently laid on its side and subsequently righted. It is important to use either mechanical advantage tools or an adequate number of personnel to perform the drum manipulations. As mentioned earlier, full drums can be very heavy, and partially full drums can unexpectedly shift their center of equilibrium and topple onto unsuspecting response personnel.

The double inversion method avoids the problem of having the bungs at the bottom of the salvage drum. In this method the salvage drum lid is placed upside down on the ground a few feet away from the damaged drum. The damaged drum is slowly tipped over so that the chime of the damaged drum rests at the edge of the salvage drum lid. The damaged drum is then carefully flipped completely onto the salvage drum lid. The salvage drum is then inverted and placed onto the lid, which is then clamped onto the salvage drum, or the salvage drum is rotated to screw the lid on. The final step involves inverting the salvage drum with the damaged drum inside. If the salvage drum will be shipped off-site, the salvage drum should be reopened, and absorbent packing material should be placed inside.

Slide-In Method

The slide-in method of overpacking can be used effectively with relatively light drums or with heavier drums when drum handlers are available. In this method both the damaged drum and the salvage drum are placed on their side. The drums are oriented so that the bungs of the damaged drum will be at the top of the salvage drum. Then, the bottom of the damaged drum is slid onto the lip of the salvage drum and pushed all the way into the salvage drum. A drum handler makes this process much easier!

Rolling-V Method

The rolling-V method of overpacking is similar to the slide-in method, except that the two drums are placed at an angle to each other and both drums are rolled together, which forces the damaged drum into the salvage drum (Figure 11-14). Because the rolling process forces the damaged drum into the salvage drum, this method can be used with heavier drums than the slide-in method. This operation is more easily completed when two entry team members work in tandem.

Cylinder Containment Vessels

Emergency response containment vessels (ERCVs), sometimes referred to as *cylinder coffins,* are much more expensive than a patching kit but may be the only option available for a high-pressure toxic gas cylinder. Containment vessels designed to meet the DOT requirements under 49 CFR 173.3(d) are DOT approved to transport the leaking cylinder to a final destination (Figure 11-15). Before moving the loaded ERCV ensure that the damaged cylinders internal pressure is equal to the recovery vessels pressure to make sure the ERCV seals will hold during transport. Containment vessels are manufactured for certain types and sizes of cylinders as well as certain classes of chemical products, since the containment vessel construction material as well as the gaskets used to seal the vessel door must be

SOLVED EXERCISE 11-5

A metal 55-gallon drum has fallen off a flatbed truck and landed upside down. What is the most efficient overpacking method to use?

Solution: Since the bungs are facing down, a metal overpack drum should be placed over the damaged drum. Then, the overpack drum should be righted and the lid attached. This is referred to as the "slip-over" drum overpacking method. If the damaged drum is leaking, it should be patched first.

compatible with the leaking product. Gas supply companies often have appropriate containment vessels on-site, and the major gas manufacturers have mobile emergency response teams at their disposal. Many chemical waste disposal companies have them available as well.

FIGURE 11-15 Cylinder recovery vessels can be used to contain damaged and leaking compressed gas cylinders. Many of these recovery vessels are also certified to be transported full, as evidenced by the DOT special permit designation on the right-hand side of the cylinder. *Photo courtesy of Eugene Ngai, Chemically Speaking LLC.*

Venting and Transfer Operations

Plugging, patching, and overpacking aren't always the most feasible options: the internal pressure of the damaged container may be too high; the container may be too severely damaged to allow the manipulations necessary for plugging, patching, or overpacking; the available equipment may be inadequate; or there may be compatibility issues. Sometimes it is better to transfer the material to an undamaged certified container for transport and/or disposal.

Whenever one material moves against another, static electricity may be generated. If a spark occurs, it will ignite a flammable atmosphere if one exists. An explosion can be prevented by removing the flammable atmosphere, by preventing the generation of static electricity, by preventing the accumulation of static electricity, or by preventing the discharge of the static electricity. It is a good idea to use more than one of these preventative measures to ensure an explosion does not occur.

BOX 11-17 HAZMAT HOW TO: GROUNDING AND BONDING

1. Set up a grounding field using the necessary number of grounding rods to reach a resistance of at most 25 ohms (10 ohms in Canada).
2. Allow sufficient time between each grounding and bonding step to permit charge relaxation. The time needed will depend on the product in the container.
3. Ground the source container, typically the damaged container, by connecting it to the grounding field.
4. Ground the receiving container by connecting it to the grounding field.
5. Ground the transfer pump by connecting it to the grounding field.
6. Bond the damaged container and the receiving container by connecting them to each other.
7. Bond the damaged container and the transfer pump by connecting them to each other.
8. Bond the transfer pump and the receiving container by connecting them to each other.
9. Test each connection for good conductivity.
10. Allow the product sufficient charge relaxation time before beginning transfer operations.

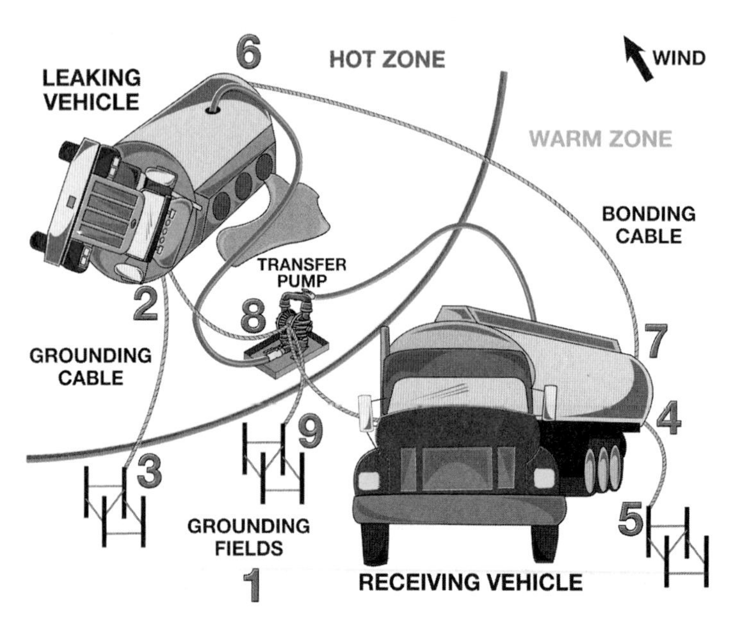

FIGURE 11-16 Grounding and bonding is important before moving flammable or combustible liquids or gases during transfer operations. This diagram illustrates the grounding and bonding of the damaged container, the receiving container, and the transfer pump. Connections should first be made in the hot zone, and final connections should be made in an uncontaminated area. It is important always to use a combustible gas indicator (CGI) to determine safe operating areas. *Art by David Heskett.*

GROUNDING AND BONDING

Grounding and bonding are necessary whenever flammable or combustible materials are moved (Figure 11-16). The **static electricity** buildup caused by the movement of solids, liquids, and gases against piping or the atmosphere may be enough to cause the material to ignite. Static electricity may build up via a number of different mechanisms, such as the following:

- liquid flowing through piping (especially at high rates)
- liquid free falling through air
- dumping powders
- pouring liquid from one container to another
- agitating or mixing liquids (movement)
- friction between different objects (conveyor belts)
- misting, spraying, and splashing of liquids
- impurities such as rust, dirt, and metal in containers
- gaseous generators (steam/gases flowing from opening in pipe)

static electricity ▪ The buildup of an electric charge on an insulator.

ignition energy ▪ The minimum amount of energy required to ignite a combustible vapor, gas, or dust cloud.

Sparks can discharge with a wide variety of energy levels, and flammable materials have a wide variety of **ignition energies** (Table 11-2).

Therefore, before any flammable liquids are transferred or flared, all containers and transfer lines must be properly grounded and bonded together to facilitate the safe discharge of any static charge that may build up during the transfer process. Otherwise, a catastrophic explosion may occur.

Four conditions must exist for a static discharge–caused explosion to occur:

1. Flammable atmosphere
2. Flammable material movement
3. Accumulation of static electricity on objects
4. Discharge of static electricity (spark) with enough energy to ignite the flammable atmosphere

conduction ▪ The movement of an electric charge within an object or between two objects.

Static electricity is the buildup of an electric charge on an insulator. Insulators include the air, plastic containers and piping, and even conducting objects like metals that are not grounded. A static charge forms when two materials have an unequal buildup of electrons. When this electron imbalance is great enough a static discharge—often in the form of a spark—can occur. The static buildup discharges through conduction. **Conduction** occurs when two objects touch and release the static buildup. When the static buildup is great enough the energy may be released even if the two objects are not in direct contact, through conduction across air. In this case the voltage difference between the two objects must be high enough to ionize the air causing it to become a conductor. An example of the buildup of static electricity is that caused by shuffling your feet along a carpet when the atmosphere is dry, for example, in winter. An example of static discharge is the spark that occurs when you then touch the doorknob or another person.

TABLE 11-2 Ignition Energies and Charge Relaxation Times of Selected Flammable Materials

FLAMMABLE MATERIAL	IGNITION ENERGY (MJ)	RELAXATION TIME (SEC)
Acetaldehyde	0.37	0.0000011
Acetone	1.15	0.00003
Acetylene	0.017	
Acrolein	0.137	
Alcohols	**0.14**	**Fast**
Alkanes	**0.28**	**Slow**
Ammonia	680	
Aromatics	**0.2**	**Slow**
Benzene	0.22	~100 (dissipates)
Carbon disulfide	0.009	~100 (dissipates)
Dichlorosilane	0.015	
Diesel	0.23	~100 (dissipates)
Diethyl ether	0.19	1.4
Ethane	0.24	
Ethanol		0.0016
Ethylene	0.07	
Gasoline	0.23–0.8	~100 (dissipates)
Heptane	0.24	~100 (dissipates)
Hydrocarbons	**0.3**	**Slow**
Hydrogen	0.011	
Hydrogen sulfide	0.068	
LPG	0.25	
Methane	0.28	
Methanol	0.14	0.0000066
Propane	0.29	
Propylene	0.28	
Toluene	0.24	21

The generation of static charge can be prevented by doing the following:

- Minimizing flow rates of solids, liquids, and gases in pipes.
- Preventing or minimizing the free fall of solids and liquids through air (such as dripping, splashing, and spraying).
- Allowing sufficient time for charge relaxation after grounding and bonding.

Charge relaxation refers to the time it takes for a material to transfer electrons and thereby dissipate static electricity and come to charge equilibrium (Table 11-2). Some materials, especially metals, have very fast relaxation times. In fact, metals are such good conductors that their relaxation times are effectively instantaneous. Other materials—such as hexane—have much longer relaxation times, on the order of minutes. In practice, relaxation times range from seconds to hours.

charge relaxation ■ The time it takes for a material to transfer electrons and thereby dissipate static electricity and come to charge equilibrium.

Grounding prevents the accumulation of static electricity, and bonding prevents the discharge of static electricity. Grounding allows electrons to dissipate from an object by flowing into the earth. Bonding allows electrons to flow between objects, thereby equalizing their electrical potential.

Whenever making grounding or bonding contacts make sure that metal-to-metal contact is made. Connecting a grounding wire or bonding wire to a painted surface will not lead to good electrical contact. Therefore, always ensure that you have a good connection and verify it using an ohmmeter to ensure that resistance is below 25 ohms in the United States (per the National Electric Code) and below 10 ohms in Canada (per Petro Canada). In addition, intermittently check the grounding and bonding system during transfer and flaring operations to ensure that conditions have not changed, such as by the accidental disconnection of a wire or changes in ground conditions (such as drying out, which can lead to reduced conductivity and increased resistance). Materials in nonconductive containers can be grounded and bonded by inserting a grounding rod into the container. Ionizers also can be used to remove static charge on nonconductive containers. This is a specialty piece of equipment that should be used only by qualified personnel.

Whenever possible, Underwriters Laboratories (UL)- or FM Global-approved grounding rods and cables should be used in the grounding and bonding system. However, it is much more important to measure the resistance of the system and ensure that it is within the appropriate parameters. When driving grounding rods you must be careful of buried power lines, gas lines, water and sewer lines, as well as communications cables. If these are breached, they can be extremely dangerous and/or costly to repair and leave many people without vital utilities. Per NFPA 77, grounding and bonding is not strictly required when transferring flammable materials through a closed-loop system. A closed-loop system is one that is not open to the atmosphere.

INERTING

Inerting the atmosphere of a flammable liquid removes oxygen—the oxidizer—from the fire triangle. If a mixture of flammable gas and oxygen is not within the flammable range, even an ignition source such as a spark will not cause a fire or explosion. Inerting can be accomplished by flooding a tank with nitrogen, argon, carbon dioxide, or another inert gas that is able to displace oxygen. Argon and carbon dioxide are good inerting gases because they are heavier than air and remain in low-lying areas such as inside tanks. Nitrogen, which is slightly lighter than oxygen, is not able to displace the oxygen in low areas since it tends to rise and disperse more quickly. The disadvantage of using argon is its high cost compared with that of nitrogen and carbon dioxide. Atmospheric

BOX 11-18 HAZMAT HOW TO: VENTING TOXIC AND CORROSIVE VAPORS USING A SCRUBBER

1. Secure the damaged container from movement using blocking, bracing, cribbing, and/or the vehicle brake.
2. Protect downwind populations in case of scrubber failure or too rapid of a release rate.
3. Determine the appropriate scrubbing solution. Always consult a technical expert during venting operations!
4. Determine the appropriate gas/vapor flow rate to minimize exothermic reactions and maximize product neutralization. Ensure that the ventilation rate can be controlled.
5. Determine the needed venting equipment. A sparging device may be needed.
6. Monitor the air in the hot zone and at the perimeter during venting operations to ensure the safety of the entry team and the surrounding public.
7. Test all hoses for leaks.
8. Slowly open the vapor valve on the damaged container.
9. Carefully monitor airborne concentrations of product.

SOLVED EXERCISE 11-6

According to the drawing shown, what are the appropriate connections, and what is the appropriate connection order for grounding and bonding?

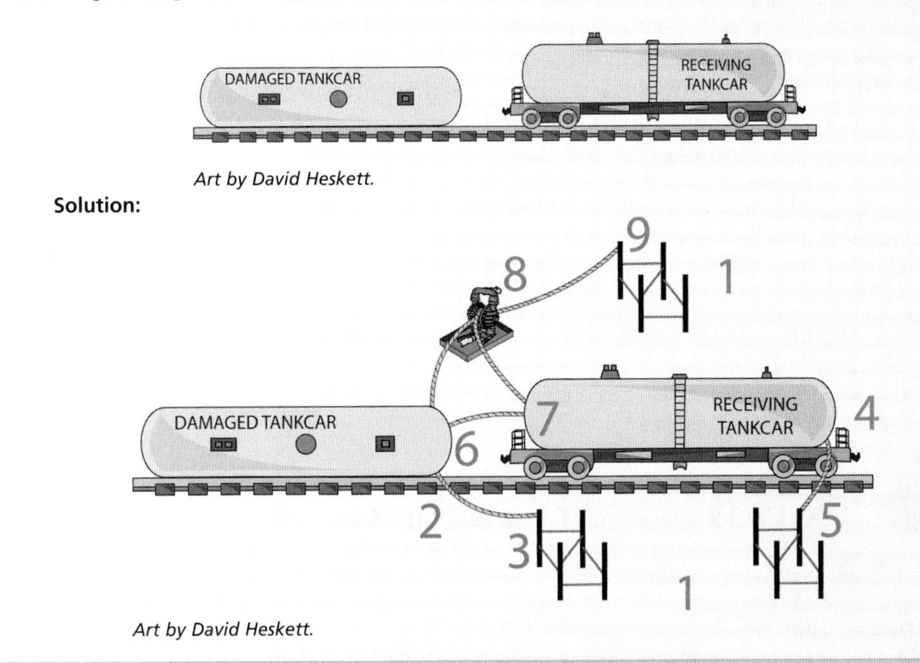

Art by David Heskett.

Solution:

Art by David Heskett.

monitoring, consisting of oxygen readings and combustible gas indicator readings, should be conducted after an atmosphere has been inerted to verify that oxygen deficiency has indeed been achieved. The atmosphere is considered inert if the oxygen concentration is less than 2%–3%. The air should be monitored continuously to ensure that the atmosphere remains inert.

VENTING

Venting is the process of reducing internal container pressure by the controlled release of a gas or liquefied compressed gas into the atmosphere, through a scrubber, or by flaring (Figure 11-17). A scrubber is designed to reduce or eliminate the toxicity or corrosiveness of the vented gas by reacting with and neutralizing or inactivating the hazardous gas. A container is usually vented through a vapor valve when one or more are operable. When the vapor valves are not operable, advanced techniques must be used to access the container contents (which are described later).

Venting to reduce internal container pressure is always done using the vapor phase, not the liquid phase. Releasing vapors from the damaged container reduces the internal pressure much more rapidly

FIGURE 11-17 When a container is overpressurized and in danger of violently rupturing, a viable option is to vent the container to the atmosphere. Keep in mind this tactic may cause problems downwind of your location and/or may be in violation of environmental regulations. *Courtesy of Dean DeMark, HAZMAT/CBRNE Specialist (N.C.).*

than releasing liquid, because more vapors are generated to replace the vented material, which cools the remaining liquid and lowers its vapor pressure. This process is known as autorefrigeration. Remember that vapor pressure is temperature dependent and decreases with decreasing temperature. The vapor pressure of the material at the container temperature determines the internal pressure of the container. Therefore evaporative cooling of the liquid contents very effectively reduces the internal pressure of the container. Anyone who has pulled propane too rapidly from a cylinder attached to a grill realizes how effective autorefrigeration is at lowering internal container pressure when the supply of fuel slows down dramatically and chokes out the flame.

Before starting a venting operation the following key issues must be considered:

- Are downwind populations and property adequately protected in the event of an accidental release?
- Is the container position amenable to venting (in other words, is there an accessible valve that will release the container contents in the vapor phase)?
- Can the venting rate be controlled? Is the valve operable or can it be made operable?
- Are the vapors nontoxic, or can they be made nontoxic by scrubbing?

BOX 11-19 HAZMAT HOW TO: FLARING

1. Determine that flaring is appropriate for this product. Some products, such as flammable halogenated hydrocarbons, will form toxic products of combustion such as hydrogen chloride, hydrogen fluoride, and phosgene.
2. Secure the damaged container from movement using blocking, bracing, cribbing, and/or the vehicle brake.
3. Place a pressure gauge on the damaged container and closely monitor the internal pressure for unsafe changes (increases in pressure).
4. Ground and bond the damaged container (see Box 11-17).
5. Control all ignition sources within a 300-foot radius.
6. Keep the flare stack at least 200 feet from all exposures.
7. Clear vegetation within 100 feet of the flare stack or flare pit (for liquid flares and 2-inch-diameter vapor flares) or within a 30-foot radius for 1-inch-diameter vapor flares.
8. Have fire suppression resources immediately available in the event spot fires or exposures are accidentally ignited. Wet the area using a fog nozzle if necessary.
9. Determine what type of flaring is necessary:
   a. Liquid flaring is used to dispose of product.
   b. Vapor/gas flaring is used to reduce internal tank pressure.
10. Gather all necessary equipment and verify that it functions. Use a back flow preventor.

LIQUID FLARING

1. Dig a large pit that is able to contain any unburned liquid upwind of the damaged container or at a 90° angle with respect to the wind and the damaged container. Monitor wind conditions carefully throughout the flaring operation.
2. Connect an inert gas line to the vapor valve of the container to be flared. This will be used to push the liquid out of the container. Keep the maximum pressure inside the damaged container at 10 psi or more below the pressure at which you found the container.
3. Place a 100-foot rigid flare pipe into the pit.
4. Connect a hose from the liquid valve of the container to a ground control valve at the rigid pipe.
5. Test all hoses, fittings, and pipes for leaks using inert gas.
6. Purge the hose and pipe with inert gas.
7. Place lit flares or another suitable ignition source into the containment pit.
8. Slowly open the liquid line on the container to be flared.
9. Ensure that the flammable compressed liquefied gas ignites.
10. Continuously monitor the area closely for optimal product flow, that the fire remains lit, and that the levels of heat output are safe.

GAS/VAPOR FLARING

1. Erect a flare stack at least 100 feet from the damaged container upwind of the damaged container or at a 90° angle with respect to the wind and the damaged container. Monitor wind conditions carefully throughout the flaring operation.
2. Place a ground control valve between the damaged containers vapor valve and the flare stack.
3. Connect a vapor line to the vapor valve of the damaged container and to the ground control valve, then connect a vapor line from the ground control valve to the flare stack. Protect the last 60 feet of the vapor line (for 2-inch flares) or the last 30 feet (for 1-inch flares) by burying it under 3 to 6 inches of dirt.
4. Test all hoses, fittings, and pipes for leaks using inert gas.
5. Purge the hose and pipe with inert gas.
6. Open the flare valve. There should be only inert gas in the line from the testing; however, monitor the vapors using a CGI. There should be no flammable vapors. If there are, purge the vapor lines with inert gas.
7. Tape a road flare to an 8-foot section of wood or PVC pipe.
8. Using the proper flame-resistant PPE, ignite the flare and hold it next to the flare stand outlet, a few inches downwind.
9. Have a second person slowly open the vapor valve.
10. Once the flare stack is ignited, the person with the road flare should move back to a safe distance.
11. Ensure that the flammable gas is burning.
12. Continuously monitor the area closely to ensure that product flow is optimal, that the flare stack remains lit, and that the levels of heat output are safe.

FLARING

Flaring is the process of reducing internal container pressure by venting and burning a flammable gas or compressed liquefied gas. In this case, the flammable gas is destroyed instead of being scrubbed (as with toxic or corrosive nonflammable gases). A properly designed flaring system will create water and carbon dioxide and some other minor combustion by-products. Flaring can also be used to dispose of flammable compressed liquefied gases. In this case, liquid flaring is used instead of vapor flaring.

Flaring of gases in the vapor space of the damaged container is very effective at reducing its internal pressure. Therefore, flaring may be used in an emergency when environmental conditions are causing a pressure buildup in the damaged container. Liquid flaring can be used to dispose of a flammable material more quickly than vapor flaring.

Flaring is performed using a flare stand (Figure 11-18). The flare stand should be positioned approximately 200 feet crosswind or upwind of the damaged container when flaring vapors, and approximately 300 feet crosswind or upwind when flaring liquids. The flare stand, piping, and hoses must be protected from radiant heat. Otherwise, an uncontrolled release may occur that can burn back to the damaged container. With liquid flares, beware that "ice balls"—frozen flammable liquid nodules—could cause grass fires downwind of the flaring location. Plan for this eventuality.

FIGURE 11-18 Flaring can be used to dispose of flammable vapors and gases. Keep in mind that burning halogenated gases will release an inorganic binary acid (such as hydrochloric acid). *Courtesy of Chris Weber, Dr. Hazmat, Inc.*

COLD TAPPING

Cold tapping is the process of gaining access to the contents of nonpressurized aluminum or nonpressurized

HAZMAT HANDLE

Vapor flaring is used to reduce internal container pressure; liquid flaring is used to dispose of product.

BOX 11-20 HAZMAT HOW TO: COLD TAPPING

1. Secure the damaged container from movement using blocking, bracing, cribbing, and/or the vehicle brake.
2. Ground and bond the damaged container (see Box 11-17).
3. Control all ignition sources within a 100-foot radius.
4. Suppress flammable vapors using the appropriate type of foam (see Box 11-7)
5. Gather the necessary equipment: This includes a hole saw with a diameter slightly greater than the diameter of the stinger that will be used to draw off the product, a pneumatic drill, and a soap solution in a spray bottle.
6. Use a ladder to access the damaged gasoline tanker.
7. If the top of the work area is slippery (owing to foam), it may be necessary to wash it off. It is advisable to use a nonslip mat as a stable work surface.
8. Have fire protection in place in the form of a foam hose line.
9. Don flame-resistant PPE (such as structural firefighter's PPE) with SCBA.
10. Using the hole saw, drill a hole through the aluminum shell of the container in-line with the manhole cover of the compartment that needs to be emptied. Drilling in-line with the manhole cover reduces the chances of drilling between compartments or over an internal baffle.
11. Use soap solution to cool the metal and keep the hole saw sharp.
12. After the hole is drilled, the product can be off-loaded using a stinger and transfer pump, as described in Box 11-21.
13. Once the transfer is complete, plug the hole before moving to the next compartment.

stainless steel containers through drilling (Figure 11-19). Never drill a container carrying corrosive liquids, since the tank compartment may contain explosive hydrogen gas. This technique is necessary when the normal loading and unloading valves have become inoperable owing to damage or are inaccessible owing to container position. Cold tapping must be done in the vapor space of the tank. Pressurized vessels should never be cold tapped except under the guidance of a technical expert. Otherwise, catastrophic failure of the container could occur.

Cold tapping involves using a 2- to 4-inch circular hole saw connected to a pneumatic drill to cut a hole into the damaged container into which a stinger tube can be placed to siphon off the container contents. A *stinger* is a long hollow tube that can be connected to a double-diaphragm pump through a flexible hose. It is very important to ensure that all components of the cold tapping operation are compatible with the contents of the container.

It is important to drill the holes into the appropriate container compartment and at the appropriate locations, so it is important to know the approximate locations of internal baffles and compartment boundaries. Double-head compartments will have a drainage hole at the bottom of the void space between compartments. The easiest way is to drill a hole in the same plane as the dome cover (Figure 11-19). This will ensure that you are not drilling on top of a baffle.

The actual drilling of the container should be the last step before the product is pumped off into the receiving container. This means that the grounding field should already be set up, all containers and equipment should have been grounded and bonded together, sufficient time for charge relaxation should have been allowed, and all transfer line connections should have been made. This preparation will ensure that the smallest amount of vapor is allowed to escape into the atmosphere after the holes have been

FIGURE 11-19 Cold tapping is used to access aluminum containers to transfer their contents when valves are inaccessible or inoperable (top). Pneumatically powered hole saws are used to drill a hole (bottom left). Once the transfer operation is complete, plugs are used to minimize vapor production and the residual product in the container (bottom right). *Top photo courtesy of Sue Blauser, Hazmat Solutions, Inc. Bottom photos courtesy of Chris Weber, Dr. Hazmat, Inc.*

drilled. Some areas require the use of vapor recovery systems when cold tapping; therefore, check your state and local regulations.

The drilling process itself takes only a few minutes with a sharp hole saw. Personnel assigned to the drilling operation should be protected with a hose stream for safety (when dealing with flammable liquids). One person should drill the hole while the other person cools the hole saw and tank shell with a spray bottle of water or soap and water.

As the product is removed, especially on multicompartment cargo tankers, watch for tank shifting and movement—even if the tank has been cribbed. As product is pumped off, the weight distribution of the tank will change and may cause rapid movement and shifting of the tank. This could lead to injuries or death if not properly addressed during both the planning and off-loading stages of the incident. After the transfer operation is complete, verify that the tank is indeed empty by visual inspection, sounding of the tank walls, or using a thermal imaging camera or temperature gun. If the tank or tank compartment is not completely empty, it may be necessary to drill additional holes on either side of the internal baffle.

HOT TAPPING

Hot tapping is the process of gaining access to the contents of a steel container through a combination of welding a fitting onto the damaged container, drilling a hole in the container, and then off-loading the product through the newly attached valve. This technique is necessary when the normal loading and unloading valves have become inoperable owing to damage or are inaccessible owing to container position. Hot tapping must occur in the liquid space of the tank to diffuse the heat of welding.

Hot tapping has many significant risks, including catastrophic failure of the tank due to welding and drilling, initiation of a violent chemical reaction by the addition of heat, and leakage of the transfer valve by metal shavings from the drilling process. In addition, hot tapping is a time-consuming process that takes approximately 2 hours to complete by qualified and experienced personnel and requires a specially sealed drilling apparatus. Hot tapping should be attempted only by appropriately trained hazmat specialists!

Hot tapping is significantly more dangerous than cold tapping, since the former process involves welding, which causes any metal to weaken. Welding is especially dangerous when the container has already been damaged and is under pressure. You must make certain that hot tapping will not cause further damage or catastrophic failure by first ensuring the following:

- The tank is not exposed to fire.
- Hot tapping will not occur in a flammable atmosphere.
- The additional heat from welding will not cause an unwanted chemical reaction.
- The hot tapping location is in contact with the liquid.
- A competent and certified welder is available.

Never attempt hot tapping with the following materials:

- bromine
- chlorine
- cryogenic liquids
- elemental sulfur
- ethylene
- ethylene oxide
- hydrocarbons in stainless steel tanks
- hydrochloric acid
- nitric acid
- propylene oxide
- sulfuric acid
- unsaturated hydrocarbons

Ensure that the required hot tapping equipment is present and functional. For jacketed containers the tank shell must be accessed by removing the jacket skin and underlying insulation. The hot tapping drill has seals that prevent product loss during the drilling operation. After the hole is drilled, hoses are attached to the valve, and the product is transferred by one of the methods described in the next section.

BOX 11-21 HAZMAT HOW TO: PRODUCT TRANSFER

1. Secure the damaged container from movement using blocking, bracing, cribbing, and/or the vehicle brake.
2. Place a pressure gauge on the damaged container and closely monitor the internal pressure for unsafe changes (increases in pressure).
3. Control all ignition sources within a 100-foot radius.
4. Determine the optimal transfer method. Transfer methods that increase the internal pressure of the damaged container are sometimes necessary when the damaged container and receiving

container are a greater distance apart. However, critically damaged containers should be transferred using only methods that reduce pressure on the damaged container!

a. Transfer methods that *reduce* internal pressure of the damaged container:
   i. Gas transfer using a liquid pump (Figure 11-21).
   ii. Gas transfer using product vapor pressure (Figure 11-21).
   iii. Liquid transfer using a liquid pump (Figure 11-23).
b. Transfer methods that *increase* internal pressure of the damaged container (potentially dangerous):
   i. Gas transfer using a liquid pump and a vapor compressor (Figure 11-22).
   ii. Gas transfer using a vapor compressor (Figure 11-22).
   iii. Gas transfer using compressed inert gas (Figure 11-22).
   iv. Liquid transfer using compressed inert gas (Figure 11-23).

5. Gather the necessary transfer equipment. Equipment and fittings must be chemically compatible and clean of other chemical residue.
6. Obtain the receiving container. The receiving container must be large enough to accommodate the product, and it must be clean of other chemical residue. Place the receiving container upwind of the damaged container whenever possible.
7. Monitor the air in the hot zone and at the perimeter during transfer operations to ensure the safety of the entry team and the surrounding public.
8. Allow only trained personnel to set up and perform transfer operations.
9. Position the receiving container as close as practical to the damaged container, and position the pumps and transfer hoses appropriately.
10. Wrap all joints in chemically compatible pipe joint tape (most commonly Teflon tape).
11. Connect all hoses and fittings.
12. Ground and bond the damaged container, receiving container, and transfer pump (see Box 11-17). Although grounding and bonding is not strictly required per NFPA 77 for closed-loop transfers (in which the vapors are recovered), it is nevertheless wise to do so in case a leak should occur.
13. Purge all hoses with inert gas and test for leaks.
14. Slowly open valves starting at the receiving container. If the receiving container has been cleaned and purged, it should have an inert atmosphere and a slight vacuum.
15. Slowly open the tank valve on the damaged container to prevent the excess flow valve from engaging.
16. Allow the pressures to equalize in the damaged container and the receiving container.
17. During transfer operations using a vapor recovery system, open the vapor valve on the damaged container halfway and then open the vapor valve on the receiving container fully.
18. During the transfer operation, keep the damaged tank internal pressure at least 10 psi below the tank pressure at which you found it whenever possible. This is especially important for a critically damaged container.
19. Monitor the transfer operation carefully! You should be able to shut down the process at a moment's notice—either automatically or manually. DOT regulations require continuous attendance during transfer operations.
20. When transfer is complete, close the product line on the damaged container and purge the transfer hoses with inert gas in a way that forces the remaining product into the receiving container.

TRANSFERRING

Transferring is the process of moving a product from a damaged or overloaded container to one or more undamaged receiving containers (Figure 11-20). For example, a rolled-over gasoline tanker that has been damaged must have its contents transferred to an undamaged gasoline tanker of sufficient size. Products must be removed from their containers and transferred or destroyed under the following circumstances:

- The container is too fragile to be moved full (such as an aluminum DOT 406/MC 306 highway cargo tanker).
- The container is too damaged to be safely moved.
- The container will not be accepted by the consignee or shipper.
- The container is no longer authorized to be used in transport.

The following are some of the risks associated with transfer operations:

- product release into the environment
- catastrophic failure of transfer equipment (such as pumps and hoses)

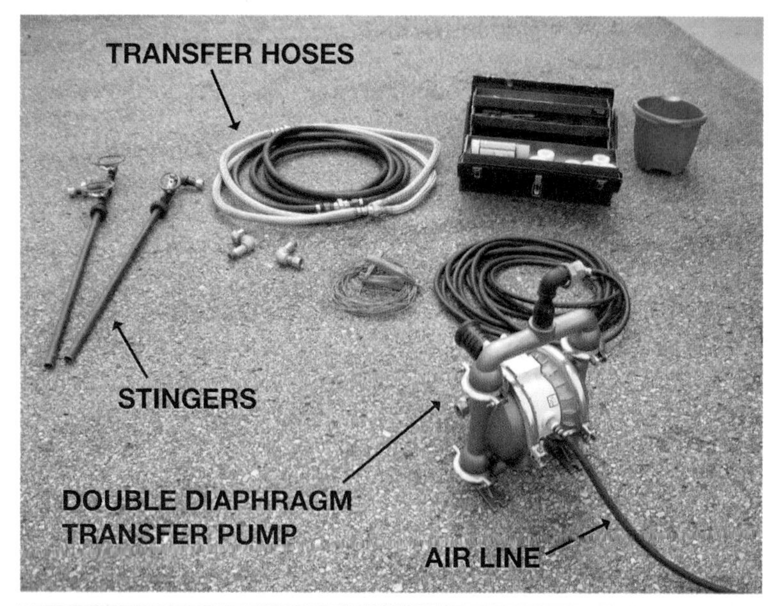

FIGURE 11-20 Transfer operations are equipment intensive. The top photograph shows much of the necessary equipment, and the bottom photograph shows a double-diaphragm transfer pump connected to product hoses, and the inert gas line that powers the pump. *Courtesy of Chris Weber, Dr. Hazmat, Inc.*

- cross-contamination of the product (if it will still be used)
- dangerous chemical reactions with residual contaminants in the transfer equipment
- overfilling of the receiving container (if the receiving container is too small)

Some of the hazardous material will likely be released into the environment during transfer operations. This is inevitable during normal operations owing to vapor release while opening the container, dripping of fittings, and residual material on stingers, pumps, hoses, and other equipment. Catch basins should be used to minimize product loss and environmental damage. Significant unanticipated larger releases may occur when incompatible equipment is used, such as when pump seals fail or hoses fail. It is therefore very important to check for chemical compatibility of the product with all equipment and transfer components.

Nonconductive plastic pails should never be used in transfer operations involving flammable liquids owing to the buildup of static charge and the difficulty of removing it. Conductive metal containers are the ideal choice. Always keep containers containing flammable materials closed until the container has been grounded, the container has been bonded to other equipment being used in the transfer operation, and a sufficient amount of time has been allowed for charge relaxation of all components in the system.

Products can be transferred through gravity or by pumping—that is, by means of a pressure differential. Gravity transfers are often not practical, since the receiving container must be a sufficient distance below the damaged container. This is an unsafe location in the case of catastrophic rupture or an unforeseen rapid release. One advantage of gravity transfers is their simplicity and lack of pumping equipment; however, safety considerations almost always outweigh simplicity, and gravity transfers should be used as a last resort.

There are three types of pressure-differential transfer operations: the material can either be pulled or pushed, or both. When a material is pumped using pressure, it is being pushed, whereas when a vacuum pump is employed, it is being pulled. Pulling the product reduces the internal pressure of the damaged container and is therefore typically significantly safer than pushing product. The following equipment combinations are commonly used to transfer product from a damaged container to an undamaged one:

- gas transfer using a liquid pump (Figure 11-21)
- gas transfer using the inherent vapor pressure of the product (Figure 11-21)
- gas transfer using a vapor compressor (Figure 11-22)
- gas transfer using a vapor compressor and a liquid pump (Figure 11-22)
- gas transfer using a compressed inert gas (Figure 11-22)
- liquid transfer using a liquid pump (Figure 11-23)
- liquid transfer using a compressed inert gas (Figure 11-23)

As in all operations, good preparation makes for a much safer and efficient transfer operation. Ensure that the container will not move—using bracing and cribbing when necessary, that all the transfer equipment is present and in good condition, that none of the equipment is contaminated with other chemicals, that all threads are in good condition and clean, that all pipe joints are wrapped with Teflon tape or other suitable material before being joined together and that those connections are clean, and that O-rings are present and in good condition. All nonessential personnel must be moved a safe distance away from the transfer operation in case a transfer line ruptures or blows a connection.

RIGHTING VEHICLES AND TANKS

The safety and ease with which a container can be uprighted depends on several factors, including the container design and construction, damage to the container, the accessibility of the container, and its weight (Figure 11-24). For example, relatively weak containers such as the aluminum DOT 406/MC 306 highway cargo tanker should never be uprighted while loaded with product—not even using airbags. These containers usually fail catastrophically when they are lifted while full. In contrast, undamaged MC 331 highway cargo tankers can often be uprighted full owing to their inherent strength. However, any

GAS TRANSFER USING A LIQUID PUMP

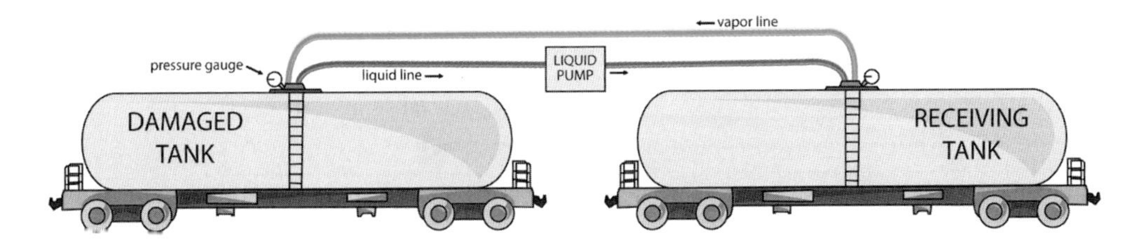

GAS TRANSFER USING PRODUCT VAPOR PRESSURE

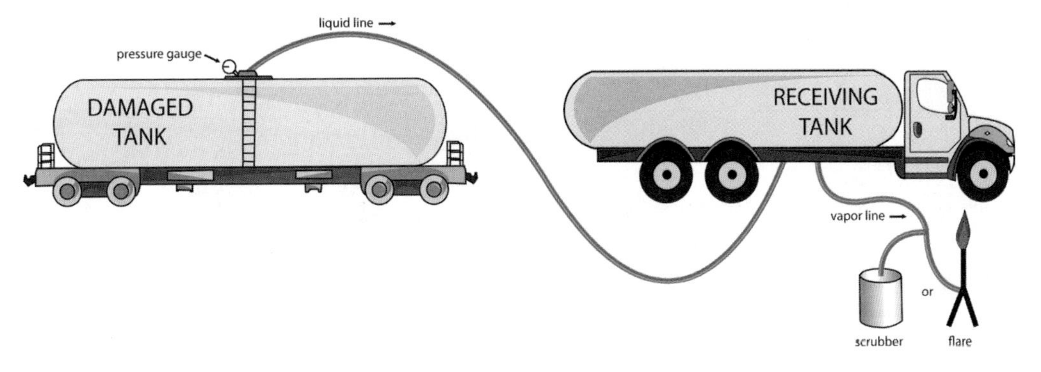

FIGURE 11-21 This diagram illustrates gas transfers that may be used with critically damaged containers, since these techniques will reduce container pressure in the damaged vessel. *Art by David Heskett.*

GAS TRANSFER USING A LIQUID PUMP AND A VAPOR COMPRESSOR

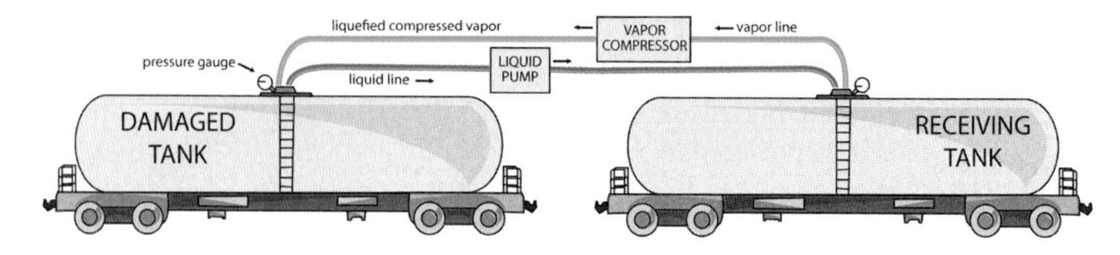

GAS TRANSFER USING A VAPOR COMPRESSOR

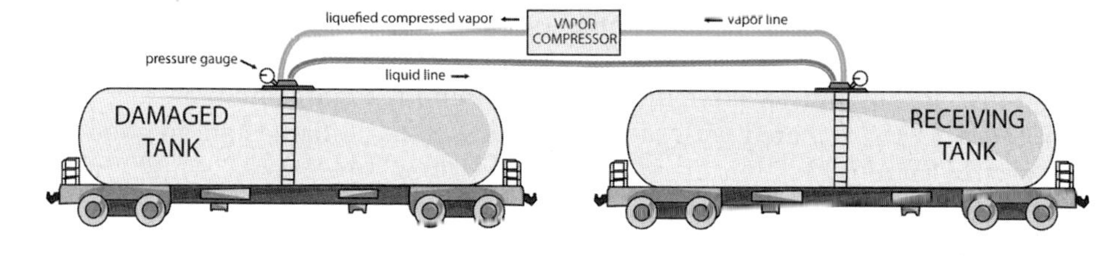

GAS TRANSFER USING COMPRESSED INERT GAS

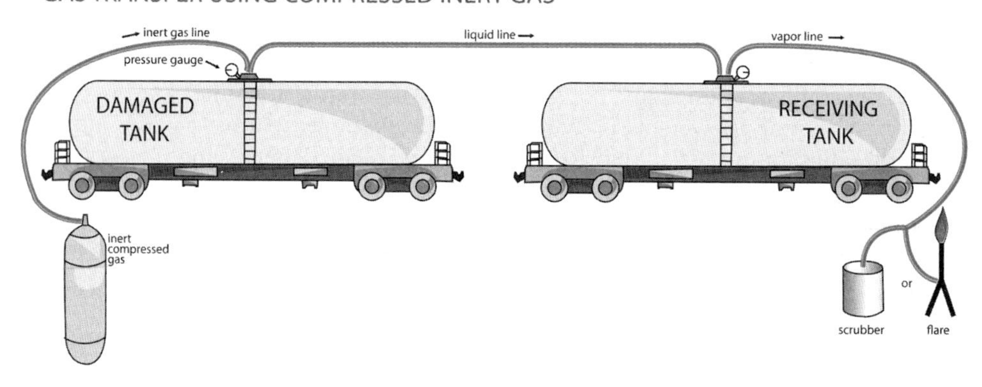

FIGURE 11-22 This diagram illustrates gas transfers that should *not* be used with critically damaged containers, since these techniques will increase container pressure in the damaged vessel. Vapor compressors are designed to return liquefied compressed gases into the damaged container, thereby increasing its internal pressure. Liquefied compressed gases can be pushed from the damaged container using an inert gas, but this likewise increases the internal pressure of the damaged container. *Art by David Heskett.*

LIQUID TRANSFER USING A LIQUID PUMP

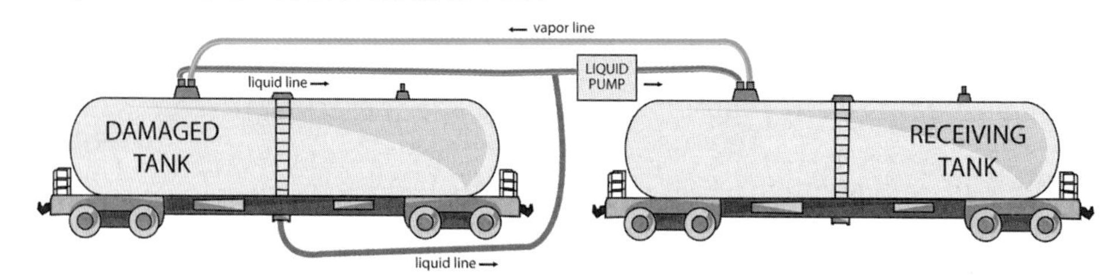

LIQUID TRANSFER USING COMPRESSED INERT GAS

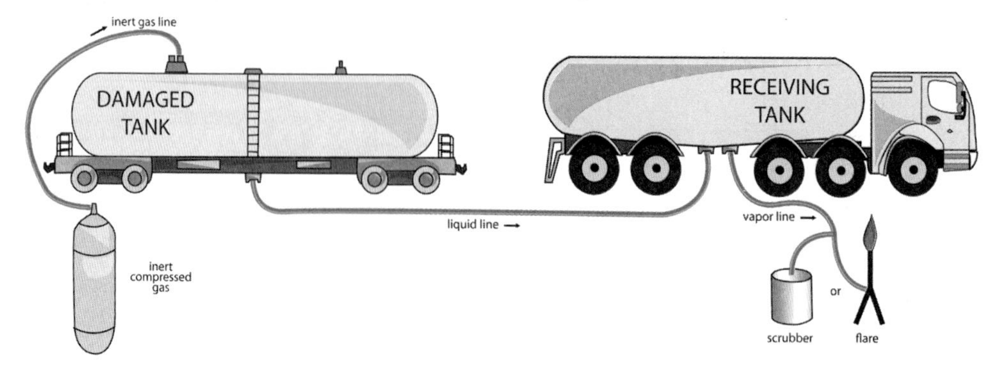

FIGURE 11-23 This diagram illustrates two techniques that can be used for liquid transfers. The top technique reduces internal pressure in the damaged container by using a liquid pump to remove liquid from the damaged container. The bottom technique increases pressure in the internal container by using an inert gas to push a liquid out of the tank and therefore should *not* be used with critically damaged containers. *Art by David Heskett.*

FIGURE 11-24 Vehicles will eventually need to be moved and uprighted. It is important to set all brakes on the overturned vehicle and to make sure that the equipment and operators are properly rated and trained to prevent dangerous and costly accidents. *Courtesy of Chris Weber, Dr. Hazmat, Inc.*

critical damage to the container will require the cargo to be unloaded before the tanker is uprighted (see Chapter 10).

It is very important to have a good relationship with your local wrecker service and/or crane operators in the event they are needed to upright containers. You should know what type of equipment they use and its capabilities, as well as the training received by the equipment operators. Is the wrecker or crane rated for the intended load? Or will it tip over? It can be very helpful to conduct joint training so that everyone is prepared when they are needed at an emergency incident. Always ensure that the parking brake and emergency brake are set on any vehicles that will be righted. In addition, other portions of the vehicle—such as the tank or trailer—may need to be braced and/or restrained at appropriate locations.

Segregating Materials for Transport

When multiple chemicals are involved in an incident, it is very important to properly identify and segregate these chemicals. DOT regulations require that chemicals be transported in certified containers and be segregated by chemical compatibility. The EPA has developed a very good compatibility chart that can help with the sorting and segregating process (Figure 11-25). Keep in mind that the oxidizing power of chemicals is relative. If you have one very strong oxidizer and a second moderately strong oxidizer, the weaker oxidizer (the moderately strong oxidizer) actually becomes the reducing agent—or fuel—to the stronger oxidizer (very strong oxidizer). It is therefore best to allow technical experts to sort and segregate chemicals, especially if they will be comingled in laboratory packs or if they will be transported together.

SOLVED EXERCISE 11-7

A DOT 406 gasoline tanker has overturned and is lying on its side. What are the appropriate mitigation methods?

Solution: Air-monitoring operations should be started immediately and continued throughout the operation. Dome clamps should be applied to any leaking manhole openings. Proper grounding and bonding should be set up and the product transferred to an undamaged DOT 406 tanker of sufficient size. Before transfer operations can commence the tanker will likely need to be cold tapped to access the compartments. After the gasoline has been offloaded, the tanker can be righted using a licensed and trained wrecking service.

EPA's Chemical Compatibility Chart

EPA-600/2-80-076 April 1980

A METHOD FOR DETERMINING THE COMPATIBILITY OF CHEMICAL MIXTURES

Please Note: This chart is intended as an indication of some of the hazards that can be expected on mixing chemical wastes. Because of the differing activities of the thousands of compounds that may be encountered, it is not possible to make any chart definitive and all inclusive. It cannot be assumed to ensure compatibility of wastes because wastes are not classified as hazardous on the chart, nor do any blanks necessarily mean that the mixture cannot result in a hazard occurring. Detailed instructions as to hazards involved in handling and disposing of any given waste should be obtained from the originator of the waste.

CODE	CONSEQUENCE
H	Heat Generation
F	Fire
G	Innocuous and non-flammable gas generation
GT	Toxic Gas formation
GF	Flammable Gas formation
E	Explosion
P	Violent Polymerization
S	Solubilization of toxic substance
U	May be hazardous, but Unknown

#	REACTIVITY GROUP NAME	1	2	3	4	5	6	7	8	9	10	11	12	13	14	15	16	17	18	19	20	21	22	23	24	25	26	27	28	29	30	31	32	33	34	101	102	103	104	105	106	107
1	Acids, Mineral, Non-oxidizing	1
2	Acids, Mineral, Oxidizing		2
3	Acids, Organic		G H	3
4	Alcohols and Glycols	H	H F	H P	4
5	Aldehydes	H P	H F	H P		5
6	Amides	H	H GT				6
7	Amines, Aliphatic and Aromatic	H	H GT	H		H		7
8	Azo Compounds, Diazo Compounds and Hydrazines	H G	H GT	H G	H G	H			8
9	Carbamates	H G	H GT						G H	9
10	Caustics	H	H	H		H				H G	10
11	Cyanides	GT GF	GT GF	GT GF					G			11
12	Dithiocarbamates	H,F GF	H,F GF	H,GT GF		GF GT		U	H G				12
13	Esters	H	H F						H G		H			13
14	Ethers	H	H F												14
15	Fluorides, Inorganic	GT	GT	GT												15
16	Hydrocarbons, Aromatic		H F														16
17	Halogenated Organics	H GT	H,F GT					H GT	H G		H GF	H						17
18	Isocyanates	H G	H,F GT	H G	H P			H P	H G		H,P G	H G	U						18
19	Ketones	H	H F						H G		H	H								19
20	Mercaptans and Other Organic Sulfides	GT GF	H,F GT						H G									H	H	H	20
21	Metals, Alkali and Alkaline Earth, Elemental	H,F GF	H,F GF	H,F GF	H,F GF	H,F GF	GF H	GF H	GF H	GF H	GF H	GF H	GF,H GT	GF H				H E	GF H	GF H	GF H	21
22	Metals, Other Elemental & Alloys as Powders, Vapors, or Sponges	H,F GF	H,F GF	G F					H,F GT	U	GF H							H E	GF H		H,F GF		22
23	Metals, Other Elemental & Alloys as Sheets, Rods, Drops, etc.	H,F GF	H,F GF						H,F G									H F						23
24	Metals and Metal Compounds, Toxic	S	S	S			S	S			S														24
25	Nitrides	GF HF	H,F E	H GF	H,E GF	GF H			U	H G	U	GF H	GF H	GF H				GF H	U	GF H	GF H	E				25
26	Nitriles	H,GT GF	H,F GT	H							U											H P			S	GF H	26
27	Nitro Compounds, Organic		H,F GT			H					H E											H,E GF				H,E GF		27
28	Hydrocarbons, Aliphatic, Unsaturated	H	H F			H																	H E						28
29	Hydrocarbons, Aliphatic, Saturated		H F																											29
30	Peroxides and Hydroperoxides, Organic	H G	H E		H F	H G		H GT	H,F E	H,F GT		H,E GT	H,F GT					H E	H	E	H,F GT	H E	H G		H G	H,E GF	H,P GT		H P		30
31	Phenols and Cresols	H	H F						H G										H P			GF H				GF H					H	31
32	Organophosphates, Phosphothioates, Phosphodithioates	H GT	H GT						U		H E											H									U		32
33	Sulfides, Inorganic	GT GF	HF GF	GT		H			E										H												H GT			33
34	Epoxides	H P	H P	H P	H P	U		H P	H P		H P	H P	U								H P	H P	H P		H P	H P					H P	H P	U	H P	34
101	Combustible and Flammable Materials, Miscellaneous	H G	H,F GT																			H,F G				H,F GF					H,F GT					101
102	Explosives	H E	H E	H E					H E		H E			H E								H E	H E	H E	E	E					H E	H E		H E	H E	H E	102
103	Polymerizable Compounds	P H	P H	P H					P H		P H	P H	U									P H	P H	P H	P H	P H					P H	P H		P H			H E	103
104	Oxidizing Agents, Strong	H GT		H GT	H F	H F	H,F GT	H,F GT	H E	H,F GT		H,E GT	H,F GT	H F	H F		H F	H GT	H,F GT	H F	H,F GT	H,F E	H,F E	H F		H,F E	H,F GT	H E	H F	H F	H G	H F	H,F GT	H,F GT	H,F G	H,F G	H E	H,F GT	104
105	Reducing Agents, Strong	H GF	H,F GT	H GF	H,F GF	H,F GF	H GF	H G				H GT	H F				H,F E	H E	H GF	H GF	H GF						H GF	[illegible] E			H E	H GF	H,GT GF		H	H GF	H E	H,P GF	H,F E	105
106	Water and Mixtures Containing Water	H	H						G										H G			H GF	H GF		S	H GF								GT GF						GF GT	106
107	Water Reactive Substances	<---EXTREMELY REACTIVE! DO NOT MIX WITH ANY CHEMICAL OR WASTE MATERIAL! EXTREMELY REACTIVE!--->																																								107

FIGURE 11-25 When segregating chemicals and disposing of them in comingled laboratory packs it is important to consult a compatibility chart and a chemist. *Courtesy of the Environmental Protection Agency.*

Summary

Product control is one of the core components of hazardous materials response at the technician level. When possible, the safest and most effective product control technique is to use a remote valve shutoff. Highway cargo tankers and fixed-site facilities typically have remote emergency valve shutoff switches located at a distance from likely hazardous materials release points. You should become familiar with the location and operation of these switches for hazards in your jurisdiction. Other effective techniques for product control are absorption and solidification, diking, damming, diversion, retention, vapor suppression, and vapor dispersion. When containers with high internal pressures are damaged, it is crucial to quickly and effectively control that pressure by using venting and/or flaring. Before any venting or flaring operations are initiated, the container must be properly grounded and bonded. After the pressure in the damaged container has been controlled, the contents of the container are normally transferred and the container is uprighted. These are all advanced offensive product control techniques requiring a significant amount of specialized equipment and training. Only well-equipped and well-trained hazmat response teams should attempt to carry out these operations.

Review Questions

1. Where are the remote shutoff valves located on a DOT 406/ MC 306 highway cargo tanker?
2. What type of remote shutoff valves do DOT 407/ MC 307 highway cargo tankers have?
3. How can you identify the location of emergency remote shutoff valves at fixed-site facilities?
4. What are the advantages and disadvantages of using dilution for product control?
5. Why must diking operations be performed well in advance of a hazardous material that is flowing?
6. What are the advantages and disadvantages of booming operations?
7. Why must the runoff from vapor dispersion activities be contained?
8. What is the importance of grounding and bonding?
9. What types of containers may be cold tapped?
10. Where should a flare stand be located?
11. What are the operational and safety issues to be considered before starting a transfer operation?
12. What are the risks of hot tapping?

References

Association of American Railroads, Emergency Response Training Center (2003) Highway Emergency Response Specialist Course Manual. Pueblo, Colorado.

Association of American Railroads, Emergency Response Training Center (2003) Tank Car Specialist Course Manual. Pueblo, Colorado.

Britton, LG (1992) Using Material Data in Static Hazard Assessment, Plant/Operations Progress, American Institute of Chemical Engineers, **11**:56-70

Compressed Gas Association, Inc. (1999) Handbook of Compressed Gases, 4th ed. Kluwer Academic Publications.

Department of Transportation. (1998) 49 CFR. Hazardous Materials Regulations. Washington, DC: U.S. Department of Transportation.

Ebadat, V (1996) Electrostatic Hazards in the Chemical Process Industry, Chemical Engineering

GATX Tank and Freight Car Manual (1994) General American Transportation Corporation.

Lesak, David (1999) Hazardous Materials Strategies and Tactics, Brady/Prentice Hall.

National Fire Protection Association. (2000). "*Standard on Static Electricity,*" NFPA 77, Boston, MA: National Fire Protection Assocation.

National Fire Protection Association. (2008). NFPA 472, *Standard for Competence of Responders to Hazardous Materials/Weapons of Mass Destruction Incidents.* Quincy, MA: Author.

Noll, Hildebrand and Yvorra (2005) Hazardous Materials Managing the Incident, 3rd ed., Red Hat Publishing.

Occupational Safety and Health Administration. (1990). 29 CFR 1910.120, *Hazardous Waste Site Operations and Emergency Response (HAZWOPER).* Washington, DC: U.S. Department of Labor.

Oldfield, Kenneth W. (2005) Emergency Responder Training Manual for the Hazardous Materials Technician. Hoboken, New Jersey: Wiley – Interscience.

The Chlorine Institute, Inc. (2009) Chlorine Institute Emergency Kit "C" For Chlorine Tank Cars & Tank Trucks, Instruction Booklet, Ed. 9

The Chlorine Institute, Inc. (2009) Chlorine Institute Emergency Kit "B" For Chlorine Ton Containers, Instruction Booklet, Ed. 10

The Chlorine Institute, Inc. (2009) Chlorine Institute Emergency Kit "A" For 100 LB & 150 LB Chlorine Cylinders, Instruction Booklet, Ed. 11

Union Pacific Railroad (2006) Tank Car Safety Course.

Weber, Chris. (2007). Pocket Reference for Hazardous Materials Response. Upper Saddle River, NJ: Pearson/Brady.

York, Kenneth J. and Gerald L. Grey (1989) Hazardous Materials/ Waste Handling For the Emergency Responder, Fire Engineering.

CHAPTER 12

Decontamination

Courtesy of Chris Weber, Dr. Hazmat, Inc.

KEY TERMS

adsorption, *p. 401*
ambulatory, *p. 409*
chemical degradation, *p. 401*
cross-contamination, *p. 397*
dilution, *p. 400*
disinfection, *p. 402*
evaporation, *p. 401*
evidence preservation, *p. 418*
fire engine alley, *p. 408*
mass casualty incident, *p. 417*
neutralization, *p. 402*
nonambulatory, *p. 413*
solidification, *p. 401*
triage, *p. 409*

OBJECTIVES

After reading this chapter, the student should be able to:

- Formulate a decontamination plan for a hazardous materials and a WMD incident.
- Perform emergency decontamination.
- Describe the equipment and methods needed to set up a mass decontamination line.
- Set up a mass decontamination line using available equipment.
- Describe crowd control procedures at a mass casualty incident.
- Explain why technical decontamination is necessary at hazardous materials incidents.
- Explain the differences between dry decontamination and wet decontamination.
- Describe the equipment and tools necessary to perform technical decontamination for both ambulatory and nonambulatory personnel.
- Set up a technical decontamination line.
- Perform technical decontamination on ambulatory personnel.
- Perform technical decontamination on nonambulatory personnel.
- Describe why proper documentation is crucial during decontamination operations.
- Explain how to properly dispose of contaminated PPE, tools, equipment, and decontamination runoff.

Resource Central For additional review and practice tests, visit **www.bradybooks.com** and click on Resource Central to access book-specific resources for this text! To access Resource Central, follow directions on the Student Access Card provided with this text. If there is no card, go to **www.bradybooks.com** and follow the Resource Central link to Buy Access from there.

Your hazardous materials response team has been dispatched to a report that several workers have been contaminated with a cleaning solution at a residence. At this point you're thinking, how bad can this really be? When you arrive the first two paramedics point you to two victims who have had their clothing removed and have dark patches of skin that is starting to slough off. When you ask the paramedic what chemical the victims have been exposed to, he replies that the workers don't know. You ask yourself the following questions:

- What do we need to do before we can safely treat the patients?
- What is the best decontamination method?
- How can we tell where the contamination is?
- What type of decontamination solution should be used?
- Do we need to control the runoff?
- How can we verify that decontamination has been completed?
- Are the patients internally contaminated as well?

Let's see if we can get some answers to these questions and more.

Decontamination is the process of removing or neutralizing unwanted substances from personnel and equipment. Hazardous materials technicians, because they perform offensive product control strategies, often become contaminated during the course of their normal job duties. To prevent **cross-contamination** or transfer of contamination to clean areas, these contaminants must be removed before entering the cold zone. This process is called *technical decontamination.* Sometimes, members of the public may become contaminated owing to an accident or a criminal event. Victims of hazmat incidents or WMD incidents must also be decontaminated. When there are multiple victims, a less comprehensive but faster form of decontamination, called *mass decontamination,* may be performed. We discuss the performance of emergency decontamination, technical decontamination and mass decontamination in this chapter.

cross-contamination ■ The transfer of contaminants from one person or location to another.

Contamination

Contamination is the transfer of unwanted substances from one location to another. Most commonly, spilled hazardous material contaminates either victims or hazmat personnel and equipment, so it must quickly be detected and removed through proper decontamination. When this does not happen, the result is cross-contamination between people, locations, and equipment during the following types of contact:

- victim to victim
- victim to rescuer
- victim to EMS staff
- hot zone entry personnel to decontamination and/or cold zone personnel
- decontamination personnel to cold zone personnel
- equipment to decontamination and/or other personnel

Entry personnel thus must remain vigilant and be aware of the location of contamination, avoid contacting the contamination, and be aware of potential cross-contamination pathways.

INTERNAL CONTAMINATION VERSUS EXTERNAL CONTAMINATION

Contamination can be located on the exterior of the body (*external contamination*), or it can be internalized (*internal contamination*). External contamination is typically easier to remove than internal contamination. In fact, depending on the route of entry, internal contamination can be almost impossible to remove in the field and may even require surgery. It is extremely important to advise medical personnel and the receiving hospital of suspected internal contamination. When hazardous materials are ingested, additional dangers to medical personnel exist in the form of emesis. Patients will often vomit hazardous stomach contents, thereby contaminating EMS personnel, the ambulance, or even the emergency room.

EXPOSURE VERSUS CONTAMINATION

Not all victims or people found in the hot zone are contaminated. Contamination refers to the presence of the hazardous material on or in a victim. *Exposure,* in contrast, means that patients may have been near the hazardous material and received a dose of radiation or been exposed to hazardous vapors or gases but do not have significant amounts of contamination on or in their bodies that can be removed through decontamination.

The topic of exposure versus contamination routinely comes up during discussions of radiation. A radioactive material emits radiation. The radiation dose, once received, cannot be reversed through decontamination. In other words, the exposed person is not contaminated. In fact, victims of radiation exposure may be given aggressive medical care for whatever injuries they have with no additional risk to EMS personnel from the radioactive material itself. This logic also applies to gas and vapor exposures. A victim exposed to chlorine gas has inhaled toxic material, and some of that chlorine gas has been absorbed into the body, but is very difficult to decontaminate the individual owing to the comparatively small amounts of hazardous material present. In addition, the victim will not off-gas chlorine and therefore poses virtually no risk to EMS personnel rendering care.

In contrast, a victim who handles a radioactive powder and gets it on or in his or her body should be decontaminated to avoid continuously exposing him- or herself and spreading contamination. This situation is starkly illustrated in Box 12-1, which recounts one of the world's worst incidents of radioactive contamination. However, it is important to emphasize that seriously injured victims should be medically stabilized and receive life saving treatments—including transport and surgical intervention when necessary—immediately, before decontamination measures are initiated according to the U.S. Department of Energy. The acute radiation doses received by EMS personnel during medical care of a viable patient will be negligible in the vast majority of cases.

Similarly, a victim that has become contaminated with a chemical on or in their body should be decontaminated to avoid prolonged physical damage to their body and spreading the contamination to other locations. In certain cases seriously injured victims should be medically stabilized and receive life saving treatments immediately—even in the hot zone, before definitive decontamination measures are initiated.

Decontamination Plan

HAZWOPER requires that a decontamination plan be put in place before hazardous materials technicians enter the hot zone (29 CFR 1910.120(q)(2)(vii)). Technician level hazardous materials response teams have generally interpreted this to mean that a

BOX 12-1 HAZMAT HISTORY: RADIOACTIVE CONTAMINATION FROM A MEDICAL SOURCE IN GOIÂNIA, BRAZIL (1987)

On September 13, 1987, an old radiotherapy source was stolen from an abandoned medical facility in Goiânia, Brazil, by scrap-metal scavengers. This theft resulted in four deaths and 245 contaminated people over approximately a 2- week period before the source of the widespread illness was discovered.

The 1375 Ci radiotherapy source in question consisted of approximately 95 g of cesium-137 (^{137}Cs) in the form of the water-soluble salt cesium chloride. The source was initially located in a shielded canister containing an iridium window that allowed gamma exposure to the patient for cancer treatment. The window could be opened and closed by rotating the internal radioactive source. When the window was open, the dose rate at 1 m was over 450 rad/hr (4.6 Gy/hr).

The thieves sold the radiotherapy unit to a scrap-metal dealer who started to dismantle the equipment. One of the scrap-metal employees took the billiard ball–sized radioactive cesium capsule out of its protective canister. Within a few days two of the employees were suffering from acute radiation sickness. They went to a clinic, which diagnosed them with a severe allergic reaction and sent them home, where they continued their dismantling effort.

Many nearby people were fascinated by the deep blue light emitted by the radioactive cesium. This included one six-year-old who painted the radioactive powder on her body and accidentally ingested some of the radioactive cesium while consuming a meal. Tragically, this young child eventually died from acute radiation sickness at the hospital. This little girl was buried in a lead-lined fiberglass coffin to contain the high levels of radiation remaining in her body.

Approximately 2 weeks after the initial extraction of the cesium-137, an alert family member postulated that the blue powder must be causing the sudden illnesses all around her. She proceeded to place the cesium source in a plastic bag and take it to a local hospital. The hospital staff immediately recognized the potential hazard of the radiotherapy source and placed it in an isolated courtyard and called in a radiation physicist, who confirmed their suspicions. As the news started leaking out approximately 130,000 people overwhelmed local hospitals and clinics. These people were screened using Geiger counters. Out of this enormous number approximately 250 people were found either to be suffering from radiation sickness (20 people) or to being internally or externally contamination. Unfortunately, the woman, who undoubtedly saved many lives, also died from her radiation exposure.

This incident graphically illustrates the importance of maintaining strict accountability of strong radioactive sources and extremely hazardous materials in general. The large number of people flooding area hospitals required a herculean effort to thoroughly evaluate their levels of contamination and subsequently to decontaminate them—both externally and internally. How would your agency and your jurisdiction in general handle an emergency of this magnitude? Do you have the detection equipment, the decontamination equipment, and the personnel to handle an incident like this safely, efficiently, and effectively?

decontamination line must be set up before anyone enters the hot zone. The decontamination plan should be written and cover the following key components of the technical decontamination process:

- Patient treatment location and assigned medical personnel
- The decontamination equipment
- The number and layout of decontamination stations
- The decontamination methods
- Procedures that minimize cross-contamination of clean areas
- Procedures that minimize cross-contamination during doffing of PPE
- Disposal procedures for hazardous waste, including disposable PPE and equipment

The decontamination plan must be revised whenever scene conditions change, the incident hazards change, or the type or level of PPE is changed.

Emergency decontamination, technical decontamination, and mass decontamination methods should be covered in your standard operating guidelines or procedures to ensure that all first responders understand the decontamination methods and equipment that will be used in a response to a hazmat or WMD incident. Written guidelines or procedures allow decontamination to be performed consistently during training, and the guidelines can be referenced in the decontamination plan during hazmat incidents.

Decontamination Methods

Many people, pieces of equipment, and occupancies may need to be decontaminated. People who may need to be decontaminated include victims located in the hot zone, entry team personnel, decontamination line personnel, and EMS personnel and law enforcement personnel if they become cross-contaminated. Equipment that may need to be decontaminated includes air monitors, sample identification equipment, tools used to mitigate the hazardous material spill, apparatus and vehicles, personal effects of victims, and furnishings and contents of contaminated buildings. In fact, often the buildings themselves and the surrounding environment—including streets, soil, and waterways—also need to be decontaminated.

dilution ▪ The process of reducing the concentration of a hazardous material by mixing it with a nonhazardous material (usually water) in the hope of rendering the entire mixture nonhazardous. Typically, large quantities of water are needed to dilute a hazardous material, especially corrosives, to safe levels.

Many different options are available for removing contaminants from people, equipment, buildings, and the environment. The decontamination method you select will depend on what you are decontaminating, the resources available, the urgency with which decontamination is needed, the weather conditions, and the chemical and physical properties of the hazardous material involved. Additionally, the work assignments of hot zone entry team members, the levels of contamination that can be expected on victims and entry team members, the toxicity of the contaminants, and the propensity for cross-contamination should all be considered before choosing a decontamination method.

Some decontamination methods will not be appropriate in all situations. For example, chemical neutralization, aggressive scraping, or pressurized air are not appropriate decontamination methods for people—whether victims or entry personnel. However, any of the aforementioned methods may generally be used on robust equipment such as metal mitigation tools. Carefully consider the decontamination method and try to maximize safety and efficiency while minimizing waste generation. Hazardous materials waste is expensive to dispose of and is regulated by the Resource Conservation and Recovery Act (RCRA).

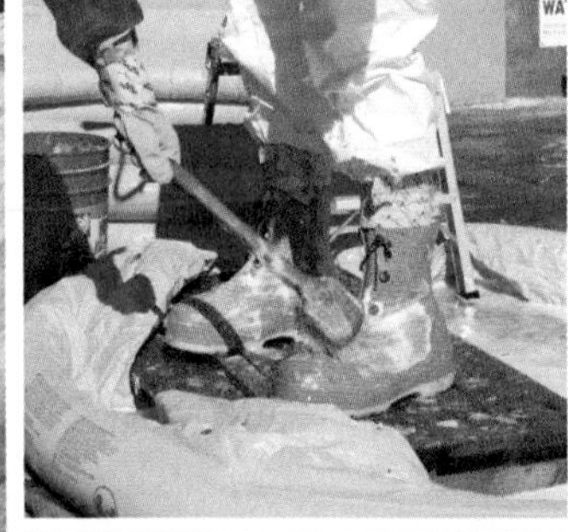
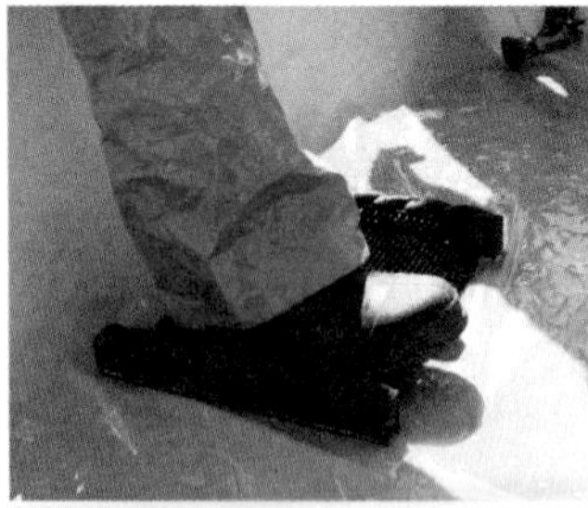

FIGURE 12-1 Decontamination can be accomplished in many different ways—including rinsing, washing, scrubbing, absorbing, and disposing of contaminated items (shown clockwise from the top left). *Photos courtesy of Chris Weber, Dr. Hazmat, Inc.*

DILUTION AND WASHING

Dilution is the reduction of the concentration of harmful contaminants below a dangerous level (Figure 12-1). Water-soluble contaminants are diluted with water, and water-insoluble contaminants, with soap and water. The removal of a corrosive material from the surface of the skin with copious amounts of water is an example of dilution.

Washing with brushes adds the physical or mechanical removal of contamination to the dilution process. Brushes or sponges are most commonly used in the washing process with or without soap. Pressure washing may be used on equipment that is durable enough to withstand the process—most commonly, heavy equipment or metal tools. Not all materials can be decontaminated by washing owing to permeation, which is the penetration of a chemical into a material at the molecular level. Clearly, washing is not going to remove a hazardous material that has permeated a material. The following factors affect the permeation rate:

- contact time
- concentration of contaminant
- temperature
- contaminant size
- molecular pore size of the material to be decontaminated
- physical state of the contaminant (solid, liquid, or gas)

EVAPORATION

Evaporation is the removal of hazardous materials by volatilization. Evaporation is a successful decontamination method for hazardous materials that have relatively high vapor pressures. Gases and cryogenic liquids evaporate extremely rapidly, while solvents with high vapor pressures such as acetone evaporate relatively quickly. Dry decontamination is a viable option when contaminants evaporate quickly.

evaporation ▪ (1) The process by which molecules in a liquid escape into the vapor state; the process by which a liquid becomes a vapor. (2) In decontamination procedures, the removal of hazardous materials by volatilization.

ISOLATION AND DISPOSAL

Isolation and disposal are the primary decontamination methods used in dry decontamination (Figure 12-1). Isolation and disposal is the placement of contaminated PPE or equipment in an impermeable container as hazardous waste. The contamination is not removed or treated in any other way. This is a very effective decontamination method for disposable equipment and PPE. Isolation and disposal are not typically practical for costly equipment such as air monitors and reusable PPE.

SOLIDIFICATION THROUGH ABSORPTION AND ADSORPTION

Solidification is accomplished by using the appropriate sorbent material to irreversibly bind an unwanted and dangerous contaminant. Absorption and adsorption are two solidification processes (Figure 12-1). In **adsorption** the adsorbent retains the contaminant through a wetting effect on its surface. In contrast, in the process of **absorption** the contaminant binds to the absorbent and becomes internalized, much like a sponge holds water.

Common sorbent materials include cat litter, diatomaceous earth, clay, powdered lime, shredded paper, and sand or soil. Depending on the nature of the contaminant, cornstarch or flour may be used on victims because of its low level of toxicity. Solidification can be used to rapidly and effectively remove the majority of contaminants adhering to footwear. It is important to check for chemical compatibility of the sorbent material prior to applying it! Be aware that many sorbents contain moisture and so must be carefully considered before being used to solidify water-reactive materials. Solidification can be used on PPE, equipment, and sometimes on patients when permitted by the sorbent manufacturer.

solidification ▪ The process of trapping a liquid in a solid form by absorption or concretion.

adsorption ▪ A physical phenomenon characterized by the adherence of a substance to the surface of another substance.

absorption ▪ (1) One of the four routes of entry of toxic substances into the body by passing through the skin, mucous membranes, or eyes. (2) The process of using a material or media to soak up a spilled hazardous material through incorporation.

CHEMICAL DEGRADATION

Chemical degradation renders contamination less hazardous by chemical reaction with a cleaning solution, thereby rendering the contaminant inert or destroying it. Because these cleaning chemicals are usually hazardous materials in their own right, great caution must be exercised. Examples of chemical degradation include the inactivation of toluene diisocyanate (TDI) using ammonia, the binding of mercury using zinc amalgams, and the inactivation of chemical warfare agents using bleach. Since the decontamination solutions, such as bleach and ammonia, are highly reactive and may generate heat, they should never be used on the skin during patient decontamination and should be carefully considered for use on PPE. Chemical degradation is often used to decontaminate equipment.

chemical degradation ▪ The breakdown of a material through the action of a hazardous material.

NEUTRALIZATION

neutralization ■ An exothermic reaction between an acid and a base, resulting in the production of heat, water, and a salt. The reaction between a concentrated acid and a concentrated base may be extremely violent.

Neutralization is a form of chemical degradation in which an acid reacts with a base to form a noncorrosive product, or vice versa—a base is neutralized by an acid. Neutralization reactions yield water and a salt:

$$\text{Acid} + \text{Base} \rightarrow \text{Water} + \text{Salt}$$

$$HCl + NaOH \rightarrow H_2O + NaCl$$

This process generates a great amount of heat and should *never* be used on patients or PPE, to prevent injury or damage. However, neutralization is a very effective decontamination method for equipment that is not sensitive to or damaged by corrosives or heat.

DISINFECTION AND STERILIZATION

disinfection ■ The process of killing a disease-causing microorganism. Methods of disinfection include autoclaving, irradiation, boiling, and chemical inactivation (such as with bleach).

Disinfecting agents are used to kill biological organisms. **Disinfection** can be accomplished using chemical disinfectants, high heat, or radiation. Chemical disinfectants include bleach, ammonia solutions, and quaternary ammonium–based commercial disinfectants. Heat disinfection includes using autoclaves, boiling water, and steaming. Disinfection using radiation is accomplished using ultraviolet (UV) radiation or a source of gamma radiation such as cobalt-60 (^{60}Co). Needless to say, strong disinfectants are not appropriate for use on patients. Some forms of chemical disinfection may be used with PPE if compatibility is not an issue. As a rule of thumb, disinfection is used only with equipment.

VACUUMING

Vacuuming may be used to remove liquids and solids from contaminated surfaces. Wet vacuum cleaners can be used to remove liquid contamination from PPE and equipment. Wet vacs are so cheap that they can be disposed of along with the hazardous waste, minimizing further personnel exposure and decontamination efforts. HEPA vacuum cleaners may be used to remove fine particulates and dusts from PPE and equipment. Specialized vacuum cleaners, such as those designed to remove mercury contamination, may have applications in certain situations.

Personal Protective Equipment

Personal protective equipment must be carefully selected for decontamination operations. The PPE must be compatible with the spilled hazardous material as well as the decontamination solution. Some equipment may be decontaminated with a strong decontamination solution. For example, TDI, a component of foam manufacturing, is often decontaminated using an ammonia solution. If not handled properly, the toxic and corrosive ammonia decontamination solution poses its own hazards. In this example PPE must be worn

SOLVED EXERCISE 12-1

What is the best decontamination method for the injured workers in the opening scenario?

Solution: Based on the limited amount of information, washing with flooding amounts of water for at least 20 minutes would be the first step (based on Guide 111 from the DOT *Emergency Response Guidebook*). Runoff should be contained if practical, but is not absolutely necessary if you are faced with limited resources. As soon as practical, all runoff should be contained, and all contaminated clothing, medical equipment, and PPE must be isolated, decontaminated, and/or disposed of as appropriate. After the patients have been treated (or simultaneously when resources permit) personnel should determine the exact identity of the cleaning solution, its hazards, and its chemical and physical properties. This information will dictate any further decontamination methods (if appropriate and necessary).

that is compatible with both TDI and ammonia. The ammonia solution should never be used to decontaminate victims directly.

PPE selection for mass decontamination operations depends on the situation. The PPE available will dictate the type of mass decontamination and follow-up medical care you can perform. For example, a fire department that has only structural firefighter protective gear may still perform mass decontamination but must remain upwind and uphill and a sufficient distance away from the contaminated victims. In this case, the firefighters will not be able to provide a comprehensive level of medical care, and crowd control is of utmost importance. In contrast, if chemical protective clothing (CPC) and respiratory protection are available, a greater level of decontamination and medical care may be performed. For mass decontamination, CPC and an air purifying respirator (APR) or powered air purifying respirator (PAPR) with a NIOSH CBRNE-rated cartridge or SCBA is generally acceptable personal protective equipment.

Types of Decontamination

There are three categories of decontamination depending on the urgency and number of victims: emergency decontamination, technical decontamination, and mass decontamination.

In the case of radioactive materials—whether from a nuclear accident, an industrial accident, or a radiological dispersal device (RDD)—screening should be done before decontamination is started to verify that contamination exists and to locate the areas that need decontamination. If decontamination, especially rinsing, is carried out before screening, the rinse water will shield weak radioactive sources such as alpha emitters and some beta emitters from detection.

EMERGENCY DECONTAMINATION

Emergency decontamination is performed on employees and even emergency response personnel who become injured or incapacitated early on at an emergency incident, or on members of the public who may have become victims of a hazmat release or WMD incident (Figure 12-2). Emergency decontamination is performed when technical decontamination is not available or when its capabilities are overwhelmed. Emergency decontamination is field expedient and may be as simple as a charged fire department hose line or an emergency shower at a fixed-site facility.

FIGURE 12-2 Emergency decontamination can be accomplished using a hose line. *Courtesy of Chris Weber, Dr. Hazmat, Inc.*

You may have concerns about water-reactive materials after learning about chemistry in Chapter 4 and the chemical and physical properties of materials in Chapter 5. However, the small amount of chemical contamination on persons will not lead to a large or dangerous chemical reaction when *flooding* amounts of water are used for decontamination. It is imperative to perform emergency decontamination quickly with large amounts of water even when chemical information is not available, or even if research indicates the material is water reactive. For example, the MSDS for sodium hydroxide states one should "avoid contact with water" because it produces intense heat when it is mixed with water (heat of solution). Copious amounts of water will quickly absorb that heat, thereby keeping the patient from suffering a thermal burn. And, of course, removing the caustic chemical quickly will keep the patient from experiencing a chemical burn as well.

Whenever an entry team enters the hot zone, arrangements should be made for emergency egress and emergency decontamination. Sometimes, the victim can rapidly be removed from the hot zone and can quickly be decontaminated. If egress times will be lengthy, limited emergency medical treatment should be administered in the hot zone. This

View an illustrated step-by-step guide on Emergency Decontamination

BOX 12-2 HAZMAT HOW TO: EMERGENCY DECONTAMINATION

1. Remove the victim's clothing.
2. Rinse with flooding amounts of water from an emergency shower or from a low-pressure fire department hose line for 2 to 5 minutes. *Note:* Beware of causing hypothermia!
3. When resources permit, follow the rinsing by brushing with soap and water.
4. Rinse again with flooding amounts of water.
5. Re-dress the patient in a Tyvek suit or a black trash bag and place him or her in a warm environment.

may include immediate lifesaving measures such as artificial respirations using a bag valve mask (BVM) and/or cardiopulmonary resuscitation (CPR). Not all medical emergencies in the hot zone will be due to a hazardous material. Heat stress may cause respiratory arrest and/or cardiac arrest. If it takes more than a few minutes to get the victim from the hot zone through decontamination to emergency medical care, he or she will likely not survive. Therefore, there should be arrangements for rudimentary hot zone medical treatment that includes artificial respirations, perfusion, and cervical spine precautions as indicated. Basic emergency medical care—including the administration of antidotes—should be performed in the warm zone at the latest, and should rarely be delayed further.

TECHNICAL DECONTAMINATION

Technical decontamination involves decontaminating hot zone entry personnel and their equipment in a slow and systematic fashion. Technical decontamination is also used to decontaminate small numbers of victims rapidly, including nonambulatory victims. In contrast, mass decontamination operations involve the expedient removal of hazardous materials from a large number of victims. The mass decontamination process is typically less thorough and takes less time per person than the technical decontamination process and is described in the next section.

The technical decontamination area is located in the warm zone, which is carved out of the cold zone before any personnel have entered the hot zone. The decontamination line is therefore typically set up without PPE and becomes the warm zone when the first entry team members return from the hot zone. The decontamination area should be located upwind, uphill, and upstream of the hazmat release. This location minimizes the chance of exposure to hazardous materials owing to the wind or gravity, which might allow a liquid to flow toward the support zone and decontamination line. This has been known to happen at hazmat incidents.

Decontamination is performed systematically, and decontamination stations should be arranged in a straight line going from the hot zone side to the cold zone side whenever possible. Entry and exit points should be conspicuously marked. The most heavily contaminated items, such as outer gloves and boots, are removed first, closest to the hot zone. By the time the last station is reached near the cold zone, the contamination should have been completely removed. Figure 12-3 shows an example of a layout of a technical decontamination line. To prevent cross-contamination between stations, workers should remain at their assigned station and not wander among them. There should be sufficient physical separation between the stations to avoid cross-contamination from overspray.

The optimal number of decontamination workers depends one of the following factors:

- incident resources
- number of entry team personnel
- level of PPE
- nature of the contaminant

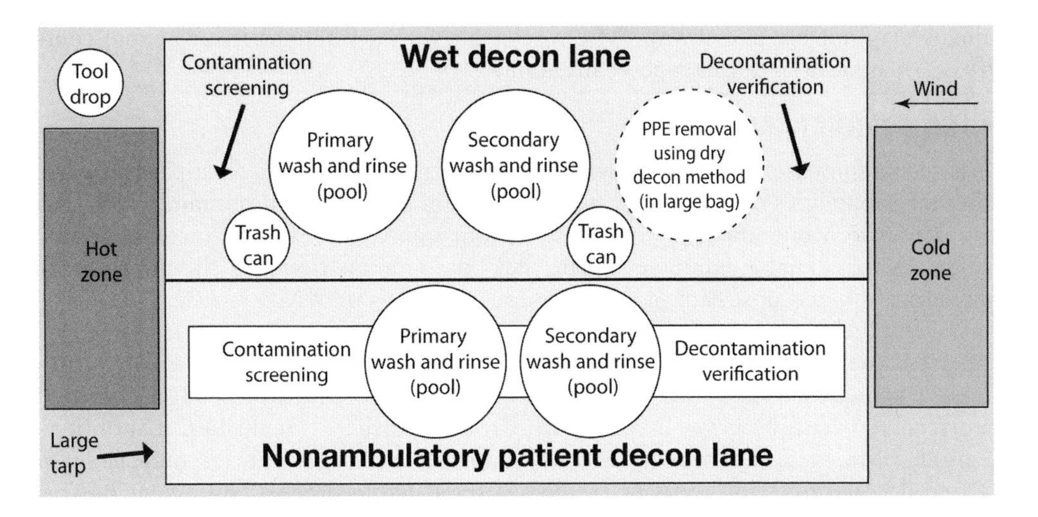

FIGURE 12-3 The layout of one possible technical decontamination corridor. Keep in mind that technical decontamination may be complex or simple depending on the circumstances of the incident. *Art by David Heskett.*

- decontamination method
- nature of the work function of the entry team
- number of victims
- medical condition of the victims (ambulatory versus nonambulatory)
- weather conditions

Each decontamination station should have a minimum of one person and typically will have two people. However, if the decontamination activities are labor- and time-intensive—such as with nonambulatory patients or extremely contaminated entry personnel—more decontamination personnel will be needed, and decon personnel will have to be rotated in and out more frequently.

Typically, the same level of protection, or at a minimum one level of protection lower than what the entry team is wearing, is chosen for decon workers performing technical decontamination nearest the hot zone. If the entry team is in Level A protection, the primary wash and rinse station personnel may be in Level B protection and the secondary wash and rinse station personnel may be in Level C protection. Thus, the level of protection may be downgraded as the decontamination personnel are stationed closer to the cold zone (farther away from the hot zone).

Once the entry team operations have ended and technical decontamination is no longer needed, the decontamination line is broken down, and the decontamination workers are decontaminated. This process also occurs in a systematic fashion. The decontamination workers at the dirtiest station, closest to the hot zone, are decontaminated first. They then step into the next closest station in the direction of the cold zone and proceed with decontamination. Then, the workers at this station (the second station in from the hot zone) proceed to the next closest station toward the cold zone and are decontaminated there. This process continues until the station with the last remaining decontamination workers is reached. These remaining decon workers should be uncontaminated, since they were using instrumentation to verify that decontamination operations were effective and should not have had any appreciable contact with the hazardous material. They can therefore doff their own PPE and exit the warm zone.

Occasionally, technical decontamination of animals must be performed, perhaps as a result of a hazardous materials spill in a rural area and the subsequent contamination of livestock such as horses, cows, and pigs; or as a result of a hazardous materials spill in a residential area where pets need to be decontaminated. A hazmat team recently responded to a mercury spill where they needed to decontaminate a dog. The team used dandruff shampoo, which contains selenium sulfide and binds mercury very well, to decontaminate the dog's fur. This was a relatively easy process, since the dog was docile, but other animals may not be. It is always wise to consult with a veterinarian or animal control officer when faced with this type of situation.

MASS DECONTAMINATION

Mass decontamination is necessary when the number of contaminated people overwhelms either the capabilities of emergency decontamination or technical decontamination. Mass decontamination operations must be carried out quickly, often with incomplete information, using available equipment. The mass decontamination line should be located upwind, uphill, and upstream of the hazardous materials or WMD release.

Several techniques can be used for mass decontamination operations (Figure 12-4). Fire departments for example, typically carry water on their fire engines and trucks. Water is primarily used during a mass decontamination incident, since it is considered a universal solvent; however if soap is available in sufficient quantities, it should also be used. EMS, fire department, and law enforcement organizations may have specialized decontamination tents or trailers at their disposal. This equipment has been designed to decontaminate large numbers of people. However, unless they are readily available, and personnel are well trained in their deployment, they may take longer to deploy.

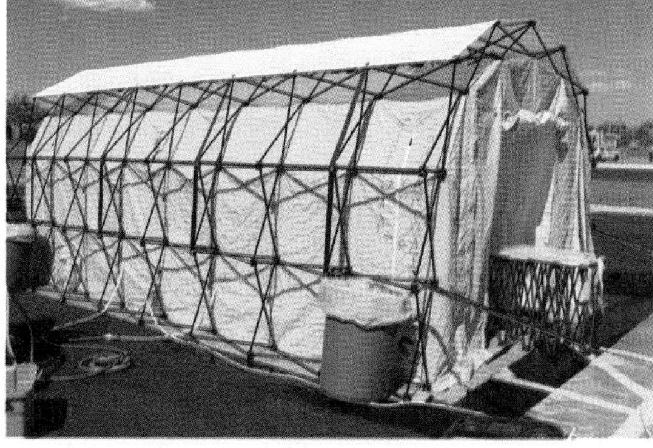

FIGURE 12-4 Mass decontamination can be accomplished for ambulatory and nonambulatory victims using tents (top left and right) or a fire engine alley (bottom left and right). In either case, the victims must disrobe completely and remove any jewelry such as rings that may trap residual chemical. After decontamination the victims must be given clean garments to put on, such as commercially available redress kits (middle). *Photos courtesy of Chris Weber, Dr. Hazmat, Inc.*

The easiest population to decontaminate is ambulatory victims who can follow verbal commands. Using a bullhorn, or another means of public address, ambulatory victims can be instructed to leave the hot zone, remove their contaminated clothing, move to the triage area, and walk through a mass decontamination line. Nonambulatory victims and those who cannot follow verbal commands must be rescued and removed from the hot zone.

The personal property of the victims that is removed during the mass decontamination operation must be carefully tracked, for two reasons. First, it must eventually be returned to the victims and it may include valuables such as wallets, diamond wedding rings, and other jewelry. Second, the clothing and other property may be valuable evidence during the investigation. The clothing of victims may contain residue that will aid in identifying the agent and prosecuting the case. When feasible, law enforcement should be present at the decontamination line to maintain and document the chain of custody of the victims' personal effects.

All clothing and other property should be removed prior to showering and placed in a plastic bag and then double bagged for safety. The contents must eventually be matched up with the victim. This may happen hours, days, weeks, or even months after the incident. The easiest way to track patient belongings is to use bar-coded triage tags with removable stickers or tags that can be placed directly on valuable property and into the bag containing the artifacts. The triage tag number should remain part of the permanent patient record. In addition, for stable, ambulatory patients a name and triage tag number should be recorded by decontamination personnel immediately after they exit the mass decontamination corridor.

Supplies and Equipment

The supplies and equipment needed to set up the decontamination line will vary based on the type of decontamination necessary and the nature of the released hazardous materials (Figure 12-5). Decontamination supplies and equipment may be off the shelf, homemade, or specifically designed for hazmat decontamination operations. Make sure to train with your supplies and equipment to ensure that your layout and decontamination method are practical. When the hazmat call comes in is not the time to find that you are missing equipment or that you can't get the job done the way you had planned.

TECHNICAL DECONTAMINATION

Tents can be very useful during technical and mass decontamination operations. They can be used during inclement weather to protect decon workers, entry team members, and victims from extremes in temperature as well as rain and snow. During the summer, tents can be used to shade personnel from the sun and heat, especially in warmer and sunnier climates. Often, tents come with heaters and air conditioning, as well as the capability to heat the water used during decontamination. This is especially helpful in colder climates and during the winter months, where wet decontamination would not otherwise be an option.

When life safety is not at stake, the EPA requires that the runoff from the decontamination line be contained. In practical terms, this means that a tarp will be placed underneath all the decontamination stations. The tarp should have a berm around the edges that can contain any overspray from the wash and rinse stations. The berm can be constructed using an uncharged hose line tucked under the tarp. In addition, each wash and rinse station should have a pool of sufficient size to contain the runoff. The pool should contain a grate at the bottom so that entry team members do not stand in the contaminated wash/rinse solution. At long and complex hazmat incidents, provisions should be made for pumping off the pools into suitable secondary containment vessels, such as bladders, totes, or 55-gallon drums.

FIGURE 12-5 Technical decontamination is an equipment-intensive operation. Showers may be designed from PVC pipes and fittings or bought commercially (top left). Pools, brushes, and handholds are useful during technical decontamination (top right and bottom left). A manifold can be quite helpful in supplying the many water needs during technical decontamination (middle right). Benches or sawhorses can be used to support nonambulatory patients (bottom right). *Photos courtesy of Chris Weber, Dr. Hazmat, Inc.*

MASS DECONTAMINATION

A wide variety of mass decontamination equipment and supplies is commercially available or can be assembled and arranged quickly. Although preconfigured tents and trailers, showers and pools, and specialized decontamination solutions are helpful and make the mass decontamination process faster and more efficient, all that is needed initially is a good water supply.

People who are decontaminated will require some sort of clothing or covering to put on once they are decontaminated. This may be a Tyvek suit, a dark-colored trash bag, or a commercially available re-dress kit. Whatever the covering, it should ensure that victim modesty is taken into account and that it provides sufficient warmth for the prevailing weather conditions.

fire engine alley ■ Emergency decontamination using two fire engines that are positioned side by side with their side discharges operating that are used to wash victims as they walk through the resulting corridor.

Fire Engine Alley

With a little practice every fire department can set up mass decontamination relatively quickly and effectively. The simplest way to perform mass decontamination is to set up a **fire engine alley.** Figures 12-4 and 12-6 illustrate several common configurations of fire engine alleys. In its simplest form, two fire engines are parked side by side approximately 10 to 15 feet apart, and a fog nozzle is applied to each side discharge. **Ambulatory** victims are directed to disrobe at one end, walk through the shower produced by the fog nozzles, and re-dress in clean clothing at the other end, where **triage** can occur. This system can be enhanced by providing a soap solution using a foam eductor, increasing privacy by adding tarps above and beside the shower corridor, and adding a rinse station by providing another fire engine alley immediately behind the first. This technique is especially useful in urban areas, since reliable water supplies are common.

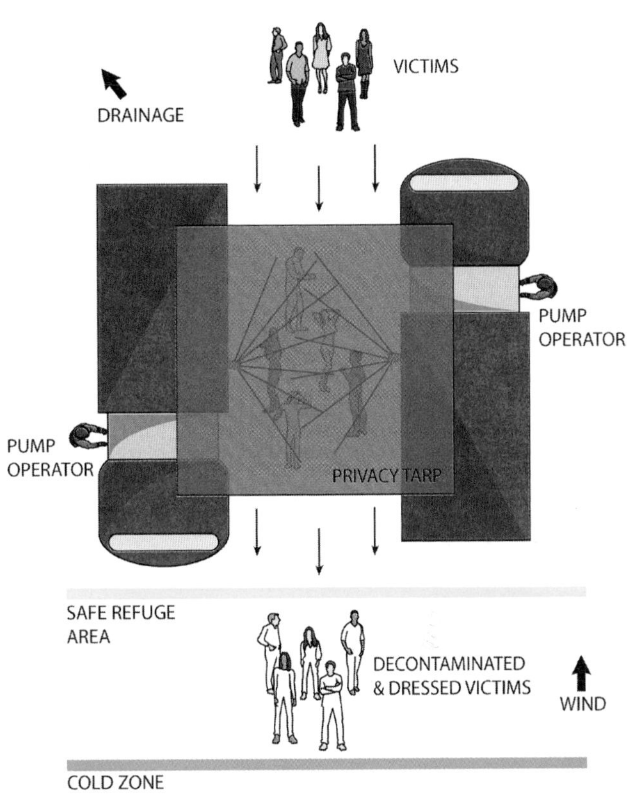

FIGURE 12-6 Diagram of a fire engine alley. Notice that the fire engines are positioned so that the pump operator is located away from the decontamination water and runoff. A tarp can be suspended between the two fire engines to provide privacy for the victims. *Art by David Heskett.*

Preconfigured Tents and Trailers

Preconfigured tents and trailers make the mass decontamination process very efficient and very effective, which increases safety for first responders as well as the victims who are being decontaminated. Figure 12-4 illustrates several different models of mass decontamination tents and trailers. Typical features of such equipment include prearranged shower lanes, nonambulatory decontamination corridors, preplumbed hot and cold water, preplumbed soap, and air heating and cooling capabilities. This type of equipment is typically easy and fast to set up with proper training. One major advantage of using this equipment is the increase in throughput and the increase in effectiveness of the decontamination procedure itself. Jurisdictions with a higher risk for terrorist incidents or hazmat incidents that could involve mass casualties should seriously consider using preconfigured tents and trailers as part of their response capability.

Preplumbed Decontamination Corridors at Fixed-Site Facilities

Hospital and clinic emergency rooms are increasingly being equipped with preplumbed decontamination corridors. Such fixed-site facilities are very helpful at locations where victims requiring decontamination are expected to congregate. Although the healthcare professionals at the facility are typically trained in the use of their preplumbed decontamination corridors, they may require the assistance of local first responders such as the fire department or hazmat response team to decontaminate a large number of victims effectively.

ambulatory ▪ Able to walk on one's own.

triage ▪ The process of prioritizing patient care based on available resources.

Improvised Equipment and Methods

Mass decontamination can also be carried out using improvised methods with the resources already on hand in the community. Swimming pools, both municipal and private, provide excellent sources of water. The chlorine or other disinfectants in the water are often efficient at breaking down chemicals, especially some chemical warfare agents and most biological agents. Shower facilities at pool complexes and gyms can be used as improvised decon corridors. Other excellent sources of decontamination are the emergency showers located at industrial facilities, research facilities, and laboratory teaching facilities. These facilities have reliable sources of water, and the runoff is channeled away from the victims and responders through the shower's drainage system. The owners and managers of these facilities will usually be very open to helping first responders during emergencies. However, plans should be in place for using these types of facilities in times of need. Additionally, coordination and training should be conducted with facility staff to ensure a smooth operation during an incident.

Decontamination Techniques

There are many different philosophies regarding decontamination. The type and amount of contaminants will dictate the type of decontamination required by the victim or entry team. The two primary types of technical decontamination are wet decontamination and dry decontamination. Some hazmat teams always use wet decon with soap and water, while others prefer dry decon with minimal use of water. The exact decontamination methods, equipment, and layout are less important than effectively removing the contaminants and verifying that decontamination is complete.

First, it must be determined whether the victim or entry team member has been exposed or contaminated. As described previously, there is a big difference between the two. Contamination must be removed, but exposed persons do not have hazardous materials on them or their equipment and likely do not need decontamination.

DRY DECONTAMINATION

Dry decontamination is often the method of choice for technical decontamination (Figure 12-7). Dry decon has a number of advantages compared with wet decon; however, dry decon is not appropriate for all hazardous materials. When the hazardous material has a moderate vapor pressure and is skin absorptive, it needs to be removed using wet decontamination. Many hazardous materials have a very high vapor pressure or a reasonably low vapor pressure and are not skin absorptive. For example, chlorine is a gas with a high vapor pressure and will normally completely evaporate before the entry team personnel reach the decontamination line, whereas sulfuric acid has an extremely low vapor pressure and will not generate significant vapors during the decontamination process. These two chemicals are excellent candidates for dry decontamination. Dry decontamination may also be chosen when monitoring equipment indicates the entry team has not been contaminated with the hazardous material. Box 12-3 explains the dry decontamination process.

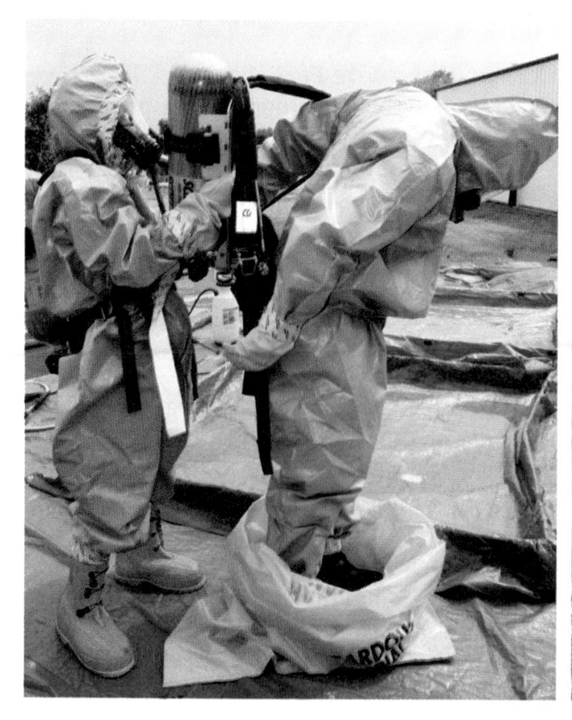

FIGURE 12-7 Wet decontamination (right) and dry decontamination (left). *Photos courtesy of Chris Weber, Dr. Hazmat, Inc.*

BOX 12-3 HAZMAT HOW TO: DRY DECONTAMINATION

1. Remove outermost gloves (third layer) and place in bag.
2. Step into bag.
3. Nonencapsulating PPE: Remove SCBA from back while staying on air.
4. Have partner unzip suit.
5. Carefully peel suit back from zipper. DO NOT allow outside of suit to contact clothing or skin. DO NOT allow the outer gloves to contact the inside of the suit or body.
6. Peel suit down to the boots.
7. Step out of the bag and move towards the cold zone.
8. Go off air and remove face piece (mask) with inner gloves (typically latex or nitrile).
9. Remove inner gloves.

Resource Central

View an illustrated step-by-step guide on Dry Decontamination

WET DECONTAMINATION

Wet decontamination supplies include at minimum water. Wet decontamination supplies usually also include soap, brushes, buckets, tarps, wash and rinse pools, sprayers, and trash cans. Figure 12-3 illustrates a common technical decontamination line schematic, and Figure 12-7 shows examples of wet and dry decontamination.

A standard decontamination line includes a tool drop area in the hot zone. The tool drop area allows multiple entry teams to reuse potentially contaminated tools to mitigate a more complex hazardous materials or WMD incident. This area may also include a boot rinse or overboot/booty removal area.

The first station located just inside the warm zone is the primary wash and rinse station. Decon personnel first wash the entry team personnel, using brushes if necessary, and rinse them using sprayers or hoses. This is the location where the majority of the contaminants are removed. Monitoring equipment should be used to determine the level of contamination prior to decontamination, to document potential exposures, and after decontamination, to determine the effectiveness using the chosen method. If a significant improvement is not achieved, the decon method must be modified or an entirely different method must be chosen.

The next station is the secondary wash and rinse station. This station may or may not be necessary based on the results of the monitoring equipment at the first station. After the secondary wash and rinse station the PPE is doffed appropriately for the level of protection and the type of CPC being used. Respiratory protection is the final piece of PPE that is removed. This is done at the station closest to the cold zone because it protects the most efficient route of entry—inhalation—from the contaminant. An optional field wash station may be included in the cold zone. It is a great idea to have the entry team members' footwear handy after the final PPE is removed, especially during inclement weather. Box 12-4 illustrates the wet decontamination process.

DECONTAMINATION OF ENTRY TEAM MEMBERS

Technical decontamination is primarily a hot zone support function. Decontamination activities should be closely coordinated with entry team operations to allow optimal decontamination line layout, to ensure that the appropriate decontamination methods are selected, and to ensure that the entry team members are familiar with the decontamination line layout.

Entry team members will often come into contact with hazardous materials in the hot zone while performing product control operations. Decontamination of entry team members will have unique challenges depending on the various levels of PPE. For example, with Level C PPE you will have to avoid getting the APR cartridges wet, since most tend to clog with excessive moisture. And with level B PPE, the SCBA can be challenging to decontaminate. Remember to always ensure that decontamination is complete using detection equipment.

View an illustrated step-by-step guide on Wet Decontamination

BOX 12-4 HAZMAT HOW TO: WET DECONTAMINATION

1. Have the entry team drop their tools in a bucket at the edge of the hot zone.
2. Visually inspect the entry members by having them rotate 360° and lift their feet.
3. If possible, evaluate the extent of contamination using monitoring instruments. (If the personnel are not contaminated, they don't need wet decon. They can go straight to dry decon to doff their PPE.)
4. Remove the outermost layer of gloves (third layer) and boot covers and place in first recovery drum or trash can.
5. Wash and scrub contaminated areas in the first pool and rinse. Start from the top and work your way down.
6. Step into the second pool and wash and rinse.
7. Remove the outer clothing as per the dry decon procedure (Box 12-3).

BOX 12-5 HAZMAT HISTORY: LITIGATION OVER PRIVACY DURING DECONTAMINATION

In October and November of 2001, in the wake of the anthrax letters that were mailed to Boca Raton, New York City, and Washington, most jurisdictions were faced with a rash of "white powder" calls. Most of these calls were false alarms, and many were malicious hoaxes. Most hazardous materials response teams were initially at a loss as to exactly what to do. Some hazmat teams always carried out full wet decontamination on anyone who may have been exposed to the substance.

Samples were routinely collected and given either to the FBI or local or state public health departments. These agencies were quickly overwhelmed with the volume of samples to analyze and store as evidence. Therefore, the FBI issued guidelines regarding a risk-versus-threat analysis before submitting samples. This helped the situation tremendously.

Eventually, hazardous materials response teams found their bearing and balanced caution with pragmatism. However, initially some of the people that were decontaminated felt they were not treated with dignity and respect. Some of the people were exposed in the nude to prying eyes, cameras, and the media because hazmat teams did not ensure modesty screening. Other people objected because they were decontaminated by first responders of the opposite sex. This led to lawsuits, which some emergency response agencies lost. Let's take a look at one of these cases.

EAST LANSING, MI (2001)

On October 12, 2001, 15 women were decontaminated after having been exposed to a threatening letter. The women all worked in Linton Hall on the Michigan State University (MSU) campus. One of the women felt a "burning sensation down her throat" after receiving a letter from a sender wanting to stop animal cruelty. MSU had already been the target of anti-animal-research terrorists. In the previous case a laboratory was vandalized with sulfuric acid, and laboratory research animals were freed. Therefore, authorities took this threat letter seriously. Because of the complaints of the woman that had opened the letter, university officials determined the best course of action would be complete wet decontamination and transport of the women to the hospital for evaluation. Unfortunately, the decontamination activities were carried out by male firefighters, with minimal privacy considerations, in full view of some of the women's coworkers inside the building. At least one woman was told she would be arrested if she did not submit to wet decontamination.

Some entry team members will require special consideration. Law enforcement officers may be contaminated and will also need to move through the decontamination line. They may have weapons that need to be decontaminated and handled in a safe manner according to their agency's standard operating procedures (SOP). This is often a bigger challenge than may at first appear. Many law enforcement SOPs prohibit law enforcement officers from turning over their weapons to anyone other than a sworn law enforcement officer and mandate that they maintain control of their weapons at all times (Figure 12-8). You should address this situation with the law enforcement agencies in your area during the planning stage to avoid any surprises at the scene of a hazmat or WMD incident. Other entry team members that require special consideration are canines and bomb squad members.

Entry team members will also carry many tools into the hot zone, including air-monitoring instrumentation, sampling tools, product control tools and equipment, and sample identification equipment. Some of this equipment is disposable, such as shovels, but most of this equipment is expensive and is certainly not disposable, such as air-monitoring instrumentation and sample identification equipment. Your decontamination plan should address how you will decontaminate not only PPE but also the equipment that is used in the hot zone. Always follow the manufacturer's guidelines when decontaminating equipment. This may include making special preparations before the equipment is brought into the hot zone—such as placing an air monitor in a plastic bag with the probe sticking out.

If the hazmat incident is a crime scene, such as the site of a WMD incident or an environmental crime, evidence will have to be decontaminated. Evidence poses a challenge because it may easily be destroyed—such as biological samples. Typically, most evidence is double or triple bagged and should be relatively easy to decontaminate; however, the chain of custody must be maintained. This may require the presence of a law enforcement officer in the warm zone who can accompany the evidence through the decontamination process.

FIGURE 12-8 Provisions must be made for the decontamination of firearms when law enforcement officers are involved in the emergency. Law enforcement officers may be required to enter the hot zone during operations, or law enforcement officers may have become victims themselves. *Courtesy of Chris Weber, Dr. Hazmat, Inc.*

DECONTAMINATION OF AMBULATORY AND NONAMBULATORY VICTIMS

Victims are among the most challenging people to decontaminate. Victims are not emergency responders and are not familiar with the decontamination line layout or the decontamination methods and procedures. You will have to explain what you are about to do and likely will have to calm them down if they feel they have been exposed or are contaminated and are showing signs and symptoms of exposure. This adds time to the decontamination process. The good news is that ambulatory patients (the "walking wounded") may be able to assist with their own decontamination. Box 12-6 illustrates the ambulatory patient decontamination process.

Nonambulatory patients, however, require specialized equipment and a significantly increased amount of time for adequate decontamination (Figure 12-9). Specialized equipment such as sawhorses, backboards, and rollers will be required for nonambulatory patient decontamination. Decontamination personnel must clean the face first, especially around the nose and mouth, to ensure that hazardous materials do not continue to be inhaled or ingested. Next, the top portion of the victim, which in the supine position is the chest, groin, and top of the legs, is decontaminated. The nonambulatory patient

nonambulatory ■ Not able to walk on one's own.

BOX 12-6 HAZMAT HOW TO: AMBULATORY PATIENT DECONTAMINATION

1. From a distance, instruct the patient to move away from the hazardous material.
2. From a distance, instruct the patient to remove contaminated clothing.
3. Wash the patient using lukewarm or cool water. Cooler water closes pores, but be careful not to induce hypothermia.
4. Rinse the patient with cool or lukewarm water.
5. Repeat until the patient is contaminant free.
6. Re-dress the patient in a commercially available re-dress kit, gown, a disposable Tyvek suit, a black plastic trash bag, or other suitable clothing.

View an illustrated step-by-step guide on Ambulatory Patient Decontamination

FIGURE 12-9 Nonambulatory patient decontamination in a tent. Rollers make moving patients easier, and preplumbed water and soap increase the efficiency of the decontamination operation. *Courtesy of Chris Weber, Dr. Hazmat, Inc.*

then must be rolled onto his or her side, and the back, buttocks, and back of the legs are decontaminated. This process is time- and resource-intensive. Each nonambulatory patient decontamination station will require a minimum of two decon personnel. Box 12-7 explains the nonambulatory patient decontamination process.

Several parts of the body tend to collect and hold contaminants more than others, such as the hair. Skin and hair should be flushed for 3 to 5 minutes with water or soap and water. The eyes are especially sensitive and absorb chemicals very effectively. The eyes should be flushed for at least 5 minutes with water or sterile saline solution. The Morgan Lens is a device designed to facilitate the decontamination of the eyes by continuous irrigation. It consists of two contact lenses with tubing to which a bottle of sterile saline solution can be attached. Prescription and vanity contact lenses should be removed whenever possible, because they tend to trap contaminants beneath them, causing more damage and absorption of the hazardous material. Once again, ensure that you use detection equipment to verify that decontamination has been accomplished.

Evaluating Decontamination Effectiveness

Before the decontaminated entry team members can leave the warm zone, the effectiveness of the decontamination process must be evaluated (Figure 12-10). The following are several ways to do this:

1. Visually inspect the PPE to determine if the product has been removed. This is the simplest technique but is rarely completely effective. Most chemicals, and certainly CBRNE agents, are toxic well below the level they can be visualized. This is not the recommended method for evaluating decontamination effectiveness. Relying on visual inspection will almost certainly lead to secondary contamination and possible injuries to the entry team member, to other first responders, and to the public.
2. Use detection equipment to analyze the rinse solution. This is a better method, and depending on the sensitivity of the monitoring equipment that is used, can be very effective. The drawback of this technique is that each entry team member or patient is not monitored individually, and spot contamination may go undetected. The advantage of this technique is that it is not resource-intensive. Large numbers of decon personnel and monitoring equipment are not necessary.

BOX 12-7 HAZMAT HOW TO: NONAMBULATORY PATIENT DECONTAMINATION

1. Evaluate ABCs. If possible, open the airway, assist ventilations, and secure the cervical spine (c-spine) (in the case of trauma). Depending on the hazard of the product and route of exposure, some lifesaving measures may need to be delayed briefly while rapid decontamination is performed on targeted areas (such as the face).
2. Remove clothing.
3. Wash patient.
4. Rinse patient.
5. Repeat until patient is contamination free (determined using appropriate monitoring instruments).
6. Re-dress the patient in a disposable Tyvek suit or other suitable clothing.

SOLVED EXERCISE 12-2

Draw the decontamination setup you would use for the opening scenario and justify your answer.

Solution: The decontamination line shown is laid out for emergency decon. The patients have severe signs and symptoms and must be decontaminated and treated as quickly as possible. A minimal decon line with water, or preferably soap and water, as a decontamination solution will accomplish this goal. A more extensive decon line may be set up later as the mitigation needs of the incident warrant.

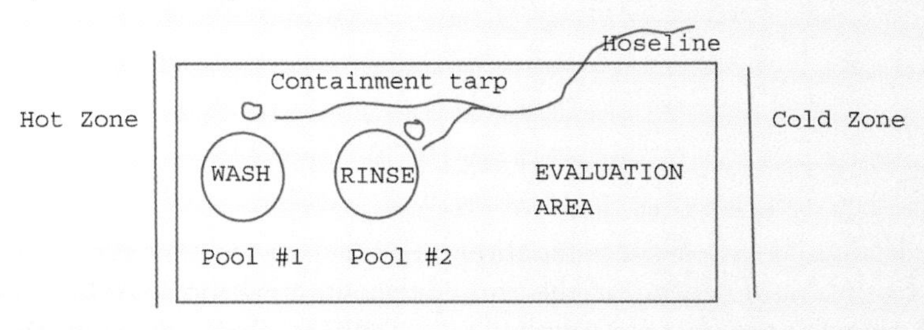

Courtesy of Chris Weber, Dr. Hazmat, Inc.

3. Use air-monitoring, radiation monitoring, or solid and liquid sample identification equipment to analyze each entry team member or victim directly. This is the safest method to use. If the resources on scene permit, each entry team member or victim should be monitored individually to confirm decontamination has been achieved.

Patient Treatment in the Warm Zone

EMS personnel should be prepared to treat victims in the warm zone immediately after decontamination or partial decontamination, as their medical condition warrants. Ideally, transfer of care will occur from the hot zone. The decontamination line should be arranged to allow EMS access while decontamination is taking place. EMS personnel must don the appropriate PPE and have the appropriate training to safely perform medical treatment in the warm zone (such as described in NFPA 473). Also, remember that patients who have

FIGURE 12-10 Decontamination efforts should always be verified using the appropriate detection equipment—in this case a photoionization detector (PID) is being used. *Courtesy of Chris Weber, Dr. Hazmat, Inc.*

SOLVED EXERCISE 12-3

How would you verify that decontamination has been successfully completed in the opening scenario?

Solution: The exact identity of the cleaning agent is presently unknown; however, many strong cleaning agents contain corrosive materials. Therefore, pH paper should initially be used to attempt to obtain a reading on the contents of the original container. If a reading is obtained, pH paper can then be used to verify that the patients have successfully been decontaminated. If this process proves unsuccessful, and the appropriate detection equipment is on-site, a variety of detection equipment may be used to investigate the identity of the cleaning solution.

TABLE 12-1 Relative Contaminant Absorption Efficiency of Body Parts

BODY PART	RELATIVE ABSORPTION EFFICIENCY
Forearm	1.0
Palm of the hands	1.3
Bottom of the feet	1.6
Forehead	4.2
Ear canal	5.4
Genitals	11.8

just emerged from a mass decontamination line may still be contaminated. The degree of product removal will depend on the type of decontamination solution, first responder oversight during the process, the number of patients that pass through, the speed with which they pass through, and whether decon verification could be completed before they entered the secondary triage site (unlikely).

Ideally, life-saving patient treatment will have begun in the hot zone. If the ABCs—airway, breathing, and circulation—were not attended to quickly there, the patient will not be viable by the time he or she arrives in the warm zone. The victim may not even be contaminated and may have injuries completely unrelated to any chemical exposure. If basic lifesaving measures were started in the hot zone, continue them uninterrupted even before decontamination is complete. Although ascertaining vital signs in full CPC may initially appear impossible, breathing may be assessed by watching for chest movement, and perfusion may be assessed by observing capillary refill. Potentially further contaminating the patient is the lesser worry if he or she is not breathing or does not have a pulse. After the patient has been decontaminated, more advanced treatment is possible. Box 12-8 summarizes some of the warm zone patient treatment guidelines.

Patients who have been contaminated with hazardous materials should never be flown. Cases have been documented in which pilots were overcome by noxious fumes off-gassing from a patient's clothing or even from the skin. For example, dimethyl sulfoxide (DMSO) is very skin absorptive, and even if surface decontamination is performed, the absorbed DMSO off-gases at varying rates. When patients are transported by road, arrangements should be

BOX 12-8 HAZMAT HOW TO: WARM ZONE PATIENT TREATMENT

1. Protect yourself, using proper PPE.
2. Maintain airway, breathing, and circulation (ABCs). Even if the patient is contaminated, ensure adequate perfusion.
   a. Quickly remove as much contamination from the face (nose and mouth area) as possible.
   b. If trauma is suspected, protect the c-spine.
   c. Assess for respirations. If the patient is not breathing, open the airway.
   d. Assess for respirations. If the patient is still not breathing, start artificial ventilation using an approved device (such as a bag valve mask). Never perform mouth-to-mouth ventilation on a potentially contaminated patient!
   e. Perform cardiopulmonary resuscitation (CPR) if the patient is pulseless.
3. Decontaminate the patient as soon as possible. Continue ventilations and CPR during the decontamination process. Field decontamination is necessary to prevent cross-contamination of medical personnel, the ambulance, and the hospital. Rapid field decontamination limits further injury to the patient from the hazardous material.
4. After wet decontamination is complete ensure that the patient is warm and has not become hypothermic.

SOLVED EXERCISE 12-4

What type of PPE should the EMS workers wear before decontamination is complete while tending to the patient in the opening scenario?

Solution: According to the patient's signs and symptoms, the chemical appears to pose primarily a contact hazard. Therefore, at a bare minimum, EMS personnel should wear chemical protective clothing including eye protection and gloves. Since the material is unknown, the highest level of respiratory protection is also warranted, especially indoors.

made to protect the ambulance crew. Dressing the patient in a Tyvek suit may prevent some off-gassing. In addition, the hospital must be notified that it is receiving a victim of a hazmat exposure, and pertinent chemical information must be relayed, preferably in person.

Crowd Management and Communications

Crowd control is an essential skill to master when responding to **mass casualty incidents** (Figure 12-11). Imagine a large sports stadium or convention center and all the occupants exiting the facility at one time after a hazmat incident or terrorist attack. You don't know which people have been exposed, which have been contaminated, or which people are suffering from chemical or warfare agent exposure. Your first priority must be to contain the crowd and prevent it from dispersing across the area to homes, clinics, and hospitals. Your second priority must be to determine which people have been contaminated and need field decontamination. Your third priority must be to determine which people need medical treatment and transport to a hospital for definitive care. Accomplishing these tasks relies heavily on crowd control.

mass casualty incident ■ An emergency incident that presents more patients than the responding agency can handle at the time.

Crowd control must be initiated immediately on arrival at a mass casualty incident. Law enforcement has extensive experience in crowd control and will be helpful during the planning process, during training, and during the actual emergency. Larger events that

FIGURE 12-11 During mass decontamination operations crowd control is very important. In this case, cones, barrier tape, and bilingual signs are being used to direct the crowd. *Courtesy of Chris Weber, Dr. Hazmat, Inc.*

have high attendance already have their own routine crowd control methods in place. Integrate your emergency plans with those of the facility. Typically, law enforcement and the public works department work together closely for routine crowd control procedures. These systems will be critical during an emergency. With proper planning and training, the existing crowd control infrastructure can quickly be converted to direct people to the mass decontamination and triage areas.

The mass decontamination corridor needs to be set up quickly in the path of the moving crowd. First responders must clearly and concisely communicate with the victims, or they will disperse and create a larger problem. Key information that must be quickly disseminated includes the following:

1. Location of the triage area
2. What will happen in the triage area
3. How to perform decontamination
4. What will happen to their personal effects
5. When they will get medical treatment
6. How they can find loved ones from whom they were separated on-site during the emergency
7. How they can contact loved ones off-site

This information can be disseminated using preprinted fliers, loudspeakers, bullhorns, or electronic signs. Crowd control resources, equipment, and procedures should be determined well in advance of large events. These same resources and procedures can be used for mass casualty incidents even when they are not part of a larger organized event.

SPECIAL CONSIDERATIONS FOR HOSPITALS

Hospitals will be the second "incident" in mass casualty incidents. It is almost inevitable that many people will self-transport to the hospital emergency department after a significant hazmat or WMD incident. The sarin attack in the Tokyo subway system is a stark example of that fact: five thousand people reported to local hospitals within several hours of the attack. This led to the cross-contamination of many hospital personnel and several emergency rooms. Some of these people were contaminated; most were not. How would your hospital handle that many potentially contaminated victims? It is imperative that contaminated people are not permitted to enter the hospital.

Hospitals should have a mass casualty incident (MCI) plan that includes hospital security, isolation of possibly contaminated patients, decontamination capabilities, and triage outside of the emergency room. These plans must be exercised regularly and should involve local emergency responders including EMS, the fire department, law enforcement, and hazardous materials response teams. Whenever possible, the decontamination runoff should be controlled. The only exception is when life safety is at stake and patient care would be compromised.

evidence preservation ■ The process of minimizing damage to evidence while performing other essential duties. At crime scenes, evidence preservation includes minimizing the number of entry personnel, disturbing as little as possible while performing work, good documentation through photography and videography, and maintaining a chain of custody.

Evidence Preservation and Processing

During criminal or terrorist incidents special care must be taken with regard to **evidence preservation** during the decontamination process. Traces of chemical or biological substances on victim clothing or equipment will be important for laboratory identification and as prosecutorial evidence (Figure 12-12). Decontamination of the deceased should be carried out only after consultation with the law enforcement agency having jurisdiction. Harsh decontamination methods such as chemical degradation, disinfection, sterilization, and neutralization will destroy the evidence and should be avoided when a criminal or terrorist incident is suspected.

Care should be taken when decontaminating samples. The samples should be double bagged to minimize contact with the decontamination solution. In addition, care must be taken while handling the secondary container to avoid puncturing or otherwise damaging it. The chain of custody of these samples must remain intact. This may require the presence of a law enforcement officer in or near the warm zone. For this process to work smoothly SOGs must be written and training should be conducted well before the incident.

Mass casualty incidents caused by criminal or terrorist events pose additional dangers and problems. One danger is that the perpetrators may be among the apparent victims. They may be contaminated, or uncontaminated and posing as victims. They may still be extremely dangerous at this point, and of course need to be apprehended. If it is feasible, record the mass decontamination operation on video to help law enforcement identify suspects and to help EMS with patient accountability.

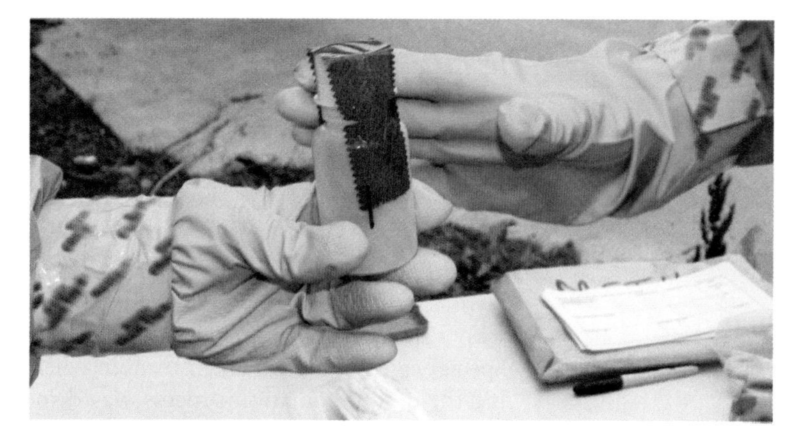

FIGURE 12-12 The chain of custody must be maintained when evidence is decontaminated. *Courtesy of Chris Weber, Dr. Hazmat, Inc.*

Decontamination Management

Sound management of decontamination operations is crucial. Technical decontamination requires a lot of equipment and may be complicated to set up. The decontamination methods and procedures may also be complicated and require training and a good entry team briefing to convey the intricacies of the decontamination plan to both the entry team members and the decon personnel. Therefore, the decontamination group supervisor should be an experienced member of your agency.

Because of the large number of victims and the large amount of resources needed for mass decontamination, effective management of mass decon is also critical. It is very important to place the mass decontamination facility in a practical and safe location. As mentioned earlier, it will be critical to funnel as many of the victims as possible into the mass decontamination lanes. If the location of the mass decontamination line is not chosen properly, the contaminated crowd will quickly scatter and endanger themselves and others. A sufficient amount of resources must be made available for mass decontamination efforts to minimize wait time. If wait times are extensive, particularly if victims know or strongly believe they have been contaminated, they will scatter and find help elsewhere. If sufficient resources cannot be obtained, as is typical for mass casualty events, it will be crucial to triage victims properly to prioritize the decontamination order and ensure others they are safe.

INCIDENT MANAGEMENT

The decontamination operation will be managed by a decontamination team leader under a NIMS-compliant incident management system. Decontamination operations can be equipment- and personnel-intensive. Thus, resource management is very important to successfully carry out both technical and mass decontamination. Resources must be ordered early and used effectively.

Technical decontamination efforts should be closely coordinated with entry team operations to ensure everyone's safety. The entry team members must understand the decontamination line layout and be able to confidently pass through the individual stations. The decontamination group supervisor and the decon personnel should attend the entry team briefing to ensure that both the decon personnel and the entry team members understand the procedure.

The mass decontamination operation will typically be managed by a mass decontamination branch director or mass decontamination group supervisor under a NIMS-compliant incident management system. As you have seen, mass decontamination operations are even more equipment- and personnel-intensive than technical decontamination. Thus, resource management is even more important to successfully manage mass decontamination events. It is very important to rapidly notify area hospitals that they should expect a large influx of people self-transporting to the emergency department and that you may need to transport a large number of patients to their location. Always ask what their capacity is to avoid overloading their emergency department.

REPORTS AND SUPPORTING DOCUMENTATION

The decontamination team leader should include a description of the decontamination line layout and the decontamination methods in the incident report. In addition, all personnel who were involved in decontamination efforts must be listed, as well as their job functions. Special care should be taken to note when and how the hazardous waste generated during the decontamination process was disposed of. The hazardous waste manifest should become part of the incident documentation. Remember, hazardous waste is tracked from cradle to grave per RCRA.

Personnel Exposure Records

Personnel exposure records must be kept for all first responders involved in decontamination efforts. Exposure records should include the name of the first responder, the employing agency, the employee identification number, the location, length of time, and the job function the first responder was performing at the incident. In addition, routine air-monitoring readings at the scene should be recorded as well as any documented chemical exposures first responders may have received. These exposure records must be kept by the employer for 30 years after termination of employment per HAZWOPER (29 CFR 1910.120).

Activity Log

An activity log consists of a list of all the actions performed during the decontamination operation. The activity log should include a description of the activity, the time it was started and completed, and the personnel who performed the activity. Examples of activities that are logged include setting up and performing decontamination, donning and doffing personal protective equipment (including on-air times), and chain-of-custody transfers of samples. This list is certainly not all-inclusive, and any significant activities should be logged.

Filing Documentation and Maintaining Records

Any documentation generated during the decontamination process should be filed with the rest of the incident documentation. Some records, such as personnel exposure reports, are required to be kept for a lengthy time by federal regulatory agencies. Other documentation, such as costs and equipment used, will be required by federal agencies and state agencies for reimbursement. Also, at criminal or terrorist incidents, most chemical samples and personal effects of victims will become evidence that will make or break a criminal prosecution. This is another reason documentation and tracking are crucial.

DISPOSAL OF CONTAMINATED WASTE

Wastewater from decontamination operations, used PPE, and any other disposable equipment must be treated as hazardous waste unless tests can be performed that exclude the decontamination waste from being regulated under RCRA. The EPA or state regulatory agencies may impose strict penalties for violations of RCRA. Usually, this is not a particularly large problem for hazmat response teams, since cleanup contractors are almost always called for hazardous materials spill remediation. But you must be aware that the decontamination wastewater and other contaminated PPE and equipment must be treated as hazardous waste and be disposed of by a licensed hazardous waste hauler. Documentation proving lawful disposal should be retained for a minimum of 3 years. If you are treating these materials as regular solid waste, make sure you have the appropriate documentation proving they are not RCRA-regulated waste (see Chapter 2 for more information on the RCRA regulation).

Write a decontamination plan for the incident described in the opening scenario.

Solution:

INCIDENT: 565 Oak Street DATE: 2/5/2013

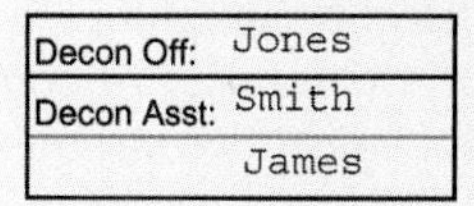

Decon Off:	Jones
Decon Asst:	Smith
	James

Hazardous Materials Response Team

DECON OFFICER Checklist

1. **Consult Operations Officer and Safety Officer for decon zone placement.**
2. **Consult Research Officer for product hazards, proper level of PPE, proper decontamination methods, and proper detection equipment.**

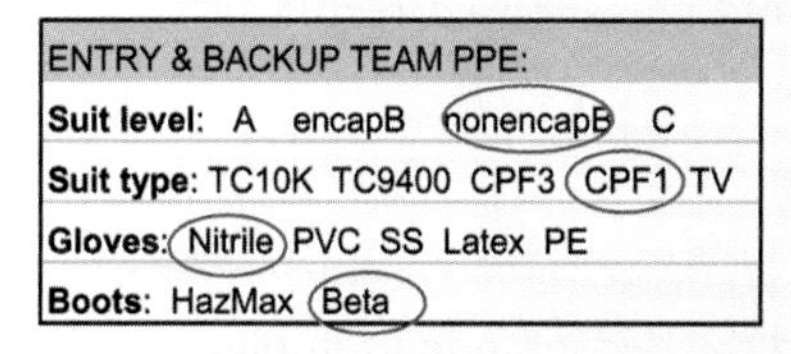

ENTRY & BACKUP TEAM PPE:
Suit level: A encapB (nonencapB) C
Suit type: TC10K TC9400 CPF3 (CPF1) TV
Gloves: (Nitrile) PVC SS Latex PE
Boots: HazMax (Beta)

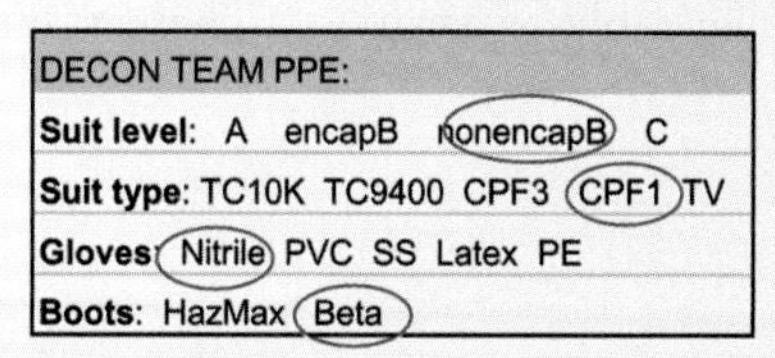

DECON TEAM PPE:
Suit level: A encapB (nonencapB) C
Suit type: TC10K TC9400 CPF3 (CPF1) TV
Gloves: (Nitrile) PVC SS Latex PE
Boots: HazMax (Beta)

4 GAS (PID)	GEIGER	
DRAEGER TUBE	Tube type:	Pumps:
HAZCAT KIT		
(pH PAPER)	SPILFYTER STRIPS	

3. **Determine appropriate decontamination method and decontamination line setup. Draw a diagram:**

Unknown chemical is involved therefore will use soap and water. Decontamination verification will consist of pH paper and PID until chemical identity is known.

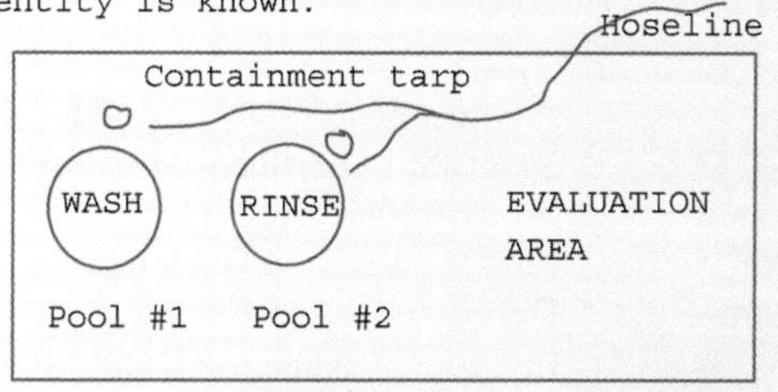

4. **Request (from Operations Officer) and assign personnel to setup decon line.**
5. **Request (from Operations Officer) and assign personnel to Decon Team.**

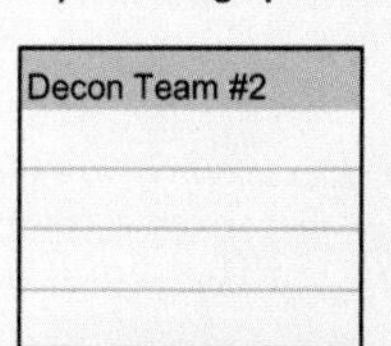

Decon Team #1
Winston
Alberts

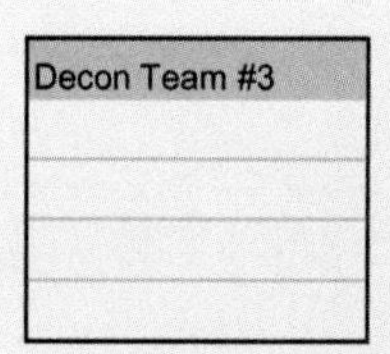

Decon Team #2

Decon Team #3

6. **Arrange Decon Team member dressout with Dressout Officer.**
7. **Verify that decon line setup is complete and effective.**
8. **Notify Operations Officer, Safety Officer, and Dressout Officer when decon line setup is complete.**
9. **Ensure Entry Team, patients, and Decon Team are properly and completely decontaminated. Verify decontamination is complete using the appropriate detection equipment.**
10. **Ensure equipment is appropriately decontaminated or disposed of properly.**

Courtesy of Chris Weber, Dr. Hazmat, Inc.

CHAPTER **REVIEW**

Summary

Decontamination is the process of removing chemical contaminants from personnel and equipment to prevent cross-contamination and the transfer of contaminants out of the hot zone. Technical decontamination is the process by which entry personnel are decontaminated, and mass decontamination is the process by which a large number of victims are rapidly decontaminated. There are significant differences between these two processes. In both cases, it is desirable to verify that decontamination has been successfully completed using monitoring equipment. In the case of mass decontamination, this may not always be practical. Hazardous materials response teams should have standard operating guidelines that address technical decontamination of entry team personnel and ambulatory and nonambulatory patients, as well as the rapid mass decontamination of a large number of victims. These SOGs should be exercised regularly.

Review Questions

1. What is the difference between mass decontamination and technical decontamination?
2. When must a decontamination plan be formulated?
3. What components must be covered in the decontamination plan?
4. When should the decontamination plan be revised?
5. What methods can be used for evaluating decontamination effectiveness?
6. How can preplanning events facilitate crowd control?
7. What is the importance of evidence preservation?

References

Currance, Phil and Alvin C. Bronstein (1999) Hazardous Materials for EMS: Practices and Procedures. Mosby.

Lake, William A.; Fedele, Paul D.; Marshall, Stephen M. (2000) SBCCOM Guidelines for Mass Casualty Decontamination during a Terrorist Chemical Agent Incident. Washington DC: U.S. Army.

Occupational Safety and Health Administration. (1990). 29 CFR 1910.120, *Hazardous Waste Operations and Emergency Response (HAZWOPER)*. Washington, DC: U.S. Department of Labor.

U.S. Army Soldier and Biological Chemical Command (2002) SBCCOM Guidelines for Cold Weather Mass Decontamination During a Terrorist Chemical Agent Incident. Washington DC: U.S. Army.

CHAPTER

13

Incident Command

Courtesy of Chris Weber, Dr. Hazmat, Inc.

KEY TERMS

command staff, *p. 426*
division, *p. 429*
emergency operations center (EOC), *p. 440*
emergency support functions (ESF), *p. 446*
finance/administration section, *p. 429*
general staff, *p. 428*
group, *p. 429*
incident commander (IC), *p. 426*
incident command post (ICP), *p. 439*
joint information center (JIC), *p. 440*
liaison officer, *p. 427*
logistics section, *p. 429*
operational period, *p. 430*
operations section, *p. 428*
planning P, *p. 436*
planning section, *p. 430*
public information officer (PIO), *p. 427*
safety officer, *p. 427*
staging area, *p. 440*
staging area manager, *p. 440*
strike team, *p. 429*
task force, *p. 429*
unified incident command (UIC), *p. 442*

OBJECTIVES

After reading this chapter, the student should be able to:

- Implement the incident command system (ICS) at a hazardous materials or WMD incident.
- Formulate response objectives at a hazardous materials or WMD incident.
- Describe the role of the emergency operations center (EOC) at hazardous materials incidents.
- Explain the utility of unified command at hazardous materials incidents.
- Describe the 10 components of a NIMS-compliant incident command system.
- Describe the role of the national response team (NRT) at hazardous materials releases.
- Describe the role of the federal on-scene coordinator (OSC) at hazardous materials releases.

Resource**Central** For additional review and practice tests, visit **www.bradybooks.com** and click on Resource Central to access book-specific resources for this text! To access Resource Central, follow directions on the Student Access Card provided with this text. If there is no card, go to **www.bradybooks.com** and follow the Resource Central link to Buy Access from there.

There is a large structure fire at one of the bigger agricultural supply companies in the area. Firefighters have been on the scene for approximately an hour when radio traffic of sick and injured firefighters is reported. The sickened firefighters are complaining of nausea, vomiting, headache, and a metallic taste in their mouth. You are responding as part of the county hazardous materials response team and are asking yourself the following questions:

- To whom do I report when I get there?
- How will the hazmat team fit into the incident command structure?
- How is scene safety being addressed?
- This is a resource-intensive incident. How are they getting adequate resources to the scene?
- Are public health agencies and the hospital involved?
- Are environmental agencies involved?
- Will the person in charge listen to the hazmat team's advice?

Let's see if we can get some answers to these questions and more.

All emergency incidents, especially hazardous materials incidents, require an organized approach when responding to them. The complexity of hazardous materials incidents, along with the technical expertise needed to safely mitigate them, makes this an even more critical requirement. The incident management system is a standardized method of managing emergency incidents safely and efficiently. It has been standardized as the National Incident Management System (NIMS) at the federal, state, tribal, and local level. All agencies that respond to emergency incidents should be NIMS-compliant to ensure interoperability and to be eligible to receive federal funding.

Implementation of an incident management system at all hazardous materials incidents, including criminal and terrorist incidents using weapons of mass destruction (WMD), is important to ensure responder safety and an efficient and effective mitigation of the emergency. In addition, HAZWOPER (29 CFR 1910.120) requires the use of an incident management system at all hazardous materials incidents. An incident commander and a safety officer must be appointed at every hazmat incident.

What does NIMS-compliant mean? In 2003, Homeland Security Presidential Directive 5 (HSPD 5) standardized an incident management system for all federal, state, and local responses to both natural and man-made disasters. Hazardous materials technicians must operate within the framework of these statutes and directives. The implementation of NIMS means that participating agencies will use common terminology, interoperable communications, a defined span of control, unity of command, and a unified command approach—especially when multiple agencies and/or multiple jurisdictions are involved. One component of NIMS is the use of the standardized incident command system (ICS). Figure 13-1 is a sample incident command chart for an incident or event using NIMS standardized ICS terminology. ICS is designed to be able to expand and contract in a modular fashion to accommodate the smallest to the largest incident. We cover basic ICS concepts in this chapter, but we do not cover detailed ICS concepts. As an emergency responder you should have received or should receive further NIMS-compliant ICS training appropriate to your job duties and rank.

Lines of authority and pecking orders are as old as humanity itself. These rank structures were presumably formalized as our ancestors organized into tribes and clans and became the lines of authority used during hunting and fighting. The first modern uses of command structures appeared in the military. These command systems were very rigid and resembled the strict military command and control system used today.

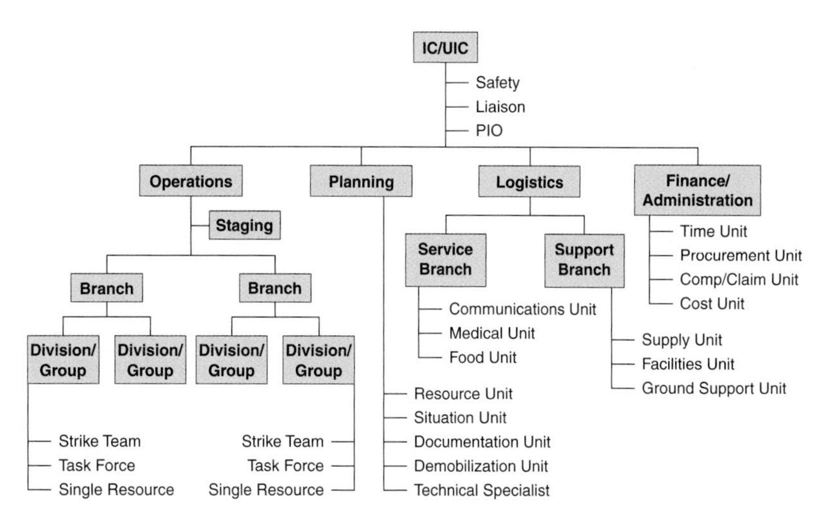

FIGURE 13-1 A National Incident Management System (NIMS)-compliant incident command system.

Management systems have also been used in the business community to maximize efficiency and profit margins. The civilian management systems are also formalized like their military counterparts, but they are far more flexible and adaptable to different business atmospheres and cultures.

The incident management system in use by first responders, including hazardous materials teams, is a mixture of these styles. Recently, most organizations have made their incident command systems NIMS-compliant to ensure that federal funding is available for their agency and jurisdiction.

Incident Command System Characteristics

An incident command system (ICS) performs several important functions at emergency incidents and larger events. Primarily, use of ICS greatly improves the safety of first responders and, secondarily, greatly improves the efficiency and effectiveness of the emergency response by optimizing the following:

- personnel accountability
- resource use and allocation
- short-term and long-term planning
- communications

A well-developed ICS structure is essential for effective resource management and personnel accountability.

The following are the core features of any incident command system:

1. Common terminology
2. Modular organization
3. Integrated communications
4. Unity of command
5. A single command structure
6. Unified command
7. Span of control
8. A single incident action plan (IAP)
9. Incident facilities
10. Resource management

First responder safety should be the highest priority of the incident commander. The ICS organizational chart allows for effective management of on-scene personnel by accounting for the human resources that have been assigned to the incident. The location and job assignment of all personnel should be recorded on the ICS chart, at a minimum by crew or unit assignment. Each agency should have a roster that is made available to the incident commander. This serves not only to enhance personnel accountability and scene safety but also facilitates cost estimates during the cost recovery and billing phase of the incident.

Resource management, both human and equipment, involves proper allocation of manpower and equipment at the incident. Effective resource management ensures that appropriate strategies and tactics can be carried out in a timely fashion. The resource needs of an incident must be planned for. Ideally, resource needs are accurately predicted, allowing resources to be ordered in a timely manner and staged near the incident before they are needed.

Effective management of resources depends heavily on good management practices. First and foremost, the management structure must encompass a reasonable span of control. *Span of control* refers to the number of people the average leader can effectively manage. The optimal span of control is five direct subordinates, with a range of three to seven being acceptable.

Establishing an Incident Command System

incident commander (IC) ▪ The person in charge of the emergency incident.

Incident command must be established immediately on arrival of the first emergency responders, whether they are firefighters, law enforcement, EMS personnel, or an industrial facility emergency response team. An **incident commander (IC)** should remain on scene until the last emergency response personnel have left the scene. Command is often transferred at incidents of long duration, or those that have multiple operational periods, so that the IC has a chance to rest. Command is also transferred as more qualified individuals arrive on the scene. Transfer of command simply means appointing a new IC in place of the previous one.

A single IC is in charge of any emergency incident. The IC has the responsibility for the safe and effective mitigation of the incident, which typically includes accomplishing the following tasks:

- analyzing the incident (scene size-up)
- ensuring the safety of all emergency responders and the public
- determining strategies and tactics
- developing an incident action plan (IAP)
- appointing positions within the ICS framework
- ordering and managing necessary resources (logistics)
- disseminating information to the public
- documenting the incident (for legal and financial purposes)

This is a complex set of tasks, which may be difficult for a single person to complete at even a small incident, let alone a large incident, without a significant amount of practice and experience. At larger incidents the incident commander therefore usually delegates many of these tasks to others using the ICS framework described next.

COMMAND STAFF

command staff ▪ In an ICS, the safety officer, liaison officer, and public information officer (PIO). These positions report directly to the incident commander.

Command staff comprises the incident commander (IC), the safety officer, the public information officer (PIO), and the liaison officer. The command staff can be viewed as the support staff—or office staff—of the incident commander and report directly to the IC. At smaller incidents these functions are typically handled by the IC directly. These positions are appointed by the IC when the complexity of the emergency incident starts to become unmanageable. Additional command staff may be appointed depending on the needs of the incident (such as legal counsel, an intelligence officer, or other technical experts).

Safety Officer

HAZWOPER requires the appointment of a **safety officer** during all hazmat and WMD incident responses. The safety officer's responsibilities include reviewing the incident action plan, maintaining an overview of incident operations, and ensuring that personnel on the scene are properly trained and capable of safely performing their assigned job function. At larger incidents or geographically dispersed incidents the safety officer may have one or more assistants to cover the area effectively. These personnel should be experienced hazardous materials responders and must be able to generate a site safety plan and evaluate strategies, tactics, and actions for their safety and effectiveness. The safety officer also typically delivers the safety briefing immediately before any personnel enter the hot zone.

safety officer ■ In an ICS, the person responsible for ensuring the emergency is mitigated in a safe manner. The safety officer has the authority to unilaterally stop unsafe actions as they occur.

The safety officer continuously monitors incident operations during the response phase and advises the IC of any safety issues. The safety officer must coordinate closely with the IC and the general staff to ensure that he or she understands the incident objectives and the strategies and tactics that are being used. The safety officer and his or her assistants have the authority to bypass the chain of command and stop unsafe actions as they occur. Incident command must be notified as soon as possible when unsafe actions have been stopped so that the incident action plan is not compromised. This is a significant amount of responsibility, and inappropriate decisions can adversely affect the incident. Thus, the safety officer should be one of the most experienced and disciplined personnel on the scene.

Assistant Safety Officer, Hazmat At larger emergency incidents or events, where hazardous materials are one piece of a much larger response, hazardous materials operations will be carried out by the hazmat branch or hazmat group (see the operations section). Owing to the dangers and complexity of hazardous materials operations, one or more additional safety officers will likely be appointed specifically for the hazmat branch, termed the assistant safety officer, hazmat.

Public Information Officer (PIO)

The **public information officer (PIO)** is responsible for informing the general public and the news media regarding the incident or event. For example, the PIO may keep the community apprised of road closures, smoke and vapor cloud location, and evacuation areas. The PIO gathers pertinent incident information such as geographic size, impacted areas, resources committed, and estimated duration. This information must be current and accurate and is released only after approval of the IC. At larger incidents the PIO may have one or more assistants to help gather and process the information.

public information officer (PIO) ■ In an ICS, the person responsible for informing the general public and news media of incident developments as instructed by the Incident commander (IC).

The PIO briefs the media and other agencies regarding the status and nature of the incident. He or she often writes the press release based on a brief outline issued by the IC. The PIO position is extremely important at hazmat incidents, since it facilitates the rapid dissemination of accurate information. Evacuation and in-place sheltering orders can be facilitated through use of the media to rapidly notify residents and businesses. Rumors are minimized and controlled when the media have access to accurate and timely incident information. Traffic backups and other inconveniences due to road closures are minimized when the community has timely access to information, since the public usually has the ability to adjust their traffic routes to avoid impacted areas.

Liaison Officer

The **liaison officer** is the point of contact for representatives from other agencies that are involved in responding to the incident. The liaison officer is like the incident receptionist. He or she will direct arriving agencies to the appropriate locations. These outside agencies

liaison officer ■ In an ICS, the person who serves as the primary contact with supporting agencies and organizations at an emergency incident.

SOLVED EXERCISE 13-1

A suspicious white powder is found in a threatening letter addressed to the governor of your state. Employees in the mail room that came in contact with the letter are feeling ill and have red and swollen hands. There are also a few cases of irritated eyes. Why would an intelligence officer be beneficial as part of the ICS at this incident?

Solution: An intelligence officer with the appropriate clearance level, especially a law enforcement officer, could gather information from federal agencies (such as the FBI and DHS), other jurisdictions (such as neighboring states), and other agencies (such as public health, hospitals, and EMS systems) regarding other similar incidents, other patients with similar symptoms, or current threats and terrorist activity.

may include federal, state, tribal, and local government agencies, nongovernmental organizations (NGOs) such as the Red Cross, and private companies such as cleanup contractors. The liaison officer must have the authority to speak for the incident commander after consultation and appropriate direction has been given.

Intelligence Officer (Intel)

An intelligence officer may be appointed at incidents that are information intensive or that involve classified or secure information such as during WMD incidents. It is the responsibility of the intelligence officer to gather the necessary information and distribute it to the appropriate personnel. When an incident requires the dissemination of classified information, it is the responsibility of the intelligence officer to ensure that the appropriate clearances are maintained and only personnel with operational requirements receive the sensitive information.

GENERAL STAFF

general staff ▪ In an ICS, the operations section chief, the logistics section chief, the finance/administration section chief, and the planning section chief.

Large or complex incidents require more significant support structures. As the incident grows in scope and complexity the needed resources necessary to ensure a positive outcome increases. Four sections can be implemented to meet these needs. **General staff** consists of the operations section chief, the logistics section chief, the finance/administration section chief, and the planning section chief. As with the command staff, general staff is appointed when the incident becomes too complex for the IC to handle all the incident management tasks by him- or herself. The IC is still responsible for all aspects of the incident outcome irrespective of what authority is delegated.

Operations Section

operations section ▪ In an ICS, the personnel tasked with accomplishing the incident objectives. The operations section is typically the largest section at an incident and can number in the hundreds of people.

As a hazardous materials technician you will typically operate within the operations section, in the hazardous materials branch or the hazmat group. The **operations section** is the workhorse of any incident. It is where the incident objectives are accomplished. The operations section is responsible for carrying out the incident objectives by formulating and executing the tactics that support the safe and successful completion of the strategies, and managing and assigning the appropriate resources to accomplish them. That is not to say that the other sections are not important. They will support your efforts to bring the situation under control, but they are in large part there to support the work that gets done by the personnel assigned to operations. The operations section is managed by the operations section chief.

Branches, divisions, and groups are implemented when the number of personnel and resources at an incident becomes too large for a single individual to manage directly. This is referred to as *exceeding the span of control.* As mentioned previously, under ideal circumstances a single individual is in charge of five subordinates. An acceptable span of control is three to seven subordinates. Reasonable span of control optimizes the use of resources by not overwhelming managers. As the number of personnel and resources grows at a scene, the number of managers must increase. This is accomplished by adding branches, divisions, and groups.

Branches Different responsibilities, functions, or agencies—such as hazmat, fire, law enforcement, and EMS—are often designated as their respective "branch" at larger incidents. There may be a hazmat branch, EMS branch, fire branch, and law enforcement branch at a complex incident involving not only hazardous materials response but also fire suppression, law enforcement, and patient treatment. Branches are managed by branch directors, and they may have deputy branch directors as necessary.

Divisions and Groups Divisions and groups are an additional layer of management directly below the branch level that helps organize resources and maintain a viable span of control. Divisions and groups are managed by supervisors who report to their respective branch director. **Divisions** are entities that cover a geographical area, for example there might be a tank farm division and separation unit division at a hazardous materials incident at a refinery. **Groups,** in contrast, cover functional areas, such as a mass decontamination group or a research group. There may be both divisions and groups within any particular branch at any given incident.

division ▪ In a NIMS–compliant incident command system, a resource designation based on location or geography. Examples of divisions are the tank farm A division and the tank farm B division.

Resources Resources may be organized into single resources, strike teams, and task forces. Single resources are used at every emergency incident, especially hazmat incidents. Units and crews are common single resources. *Units* are a piece of equipment and one to seven personnel that operate that piece of equipment. Units may be fire trucks, police cars, ambulances, or even entire hazmat teams at extremely large incidents. A *crew* consists of two to seven personnel who are managed by a crew leader. Crews and units are used to maintain proper span of control.

group ▪ In an ICS, a collection of resources assigned to perform a certain function, generally irrespective of the location within the emergency incident.

Strike teams are a grouping of the same kind and type of resource. **Task forces** are a grouping of different resources combined to gain functionality as a group. They are used when the incident is extremely large, and resources must be grouped together to maintain span of control. Strike teams and task forces must also adhere to the normal span-of-control limits. Strike teams are formed when a larger number of the same resource is needed at an incident; for example, ground crews at a wildland fire, or perimeter air-monitoring crews at a hazmat incident. Task forces are created when a variety of different skills are necessary simultaneously at multiple locations. For example, during a large snowstorm that renders roads impassable, task forces consisting of a snowplow, police cruiser, fire engine, and an ambulance may be formed. These task forces then respond to emergency calls, since multiple tasks must be performed for the mission to succeed (such as clearing of route, scene security, EMS, and fire suppression), yet no single resource can accomplish the task alone.

strike team ▪ In an ICS, a set of identical resources grouped together for a specific purpose.

task force ▪ In an ICS, a set of different resources grouped together for a specific purpose.

Logistics Section

A logistics section chief may be appointed to handle the supply issues associated with incident operations. As more and more responders arrive at the scene the number of support and service functions increases. The **logistics section** is in charge of procuring resources in the form of personnel, equipment, and services. They provide the resources needed to accomplish the response objectives in the IAP. The logistics section is tasked with making sure that there is enough water, food, and sanitary facilities to accommodate all the responders. The logistics section also ensures that needed supplies and services are acquired in a timely fashion. For example, if the supply of respirators or CPC is running low at a hazmat incident, the logistics section ensures that more are obtained. Without the logistics section, the operations section could not carry out the incident objectives.

logistics section ▪ In an ICS, the personnel responsible for ensuring that the needed service and support resources are ordered and available at the incident.

Finance/Administration Section

The **finance/administration section** keeps track of purchased services and equipment and personnel hours at complex incidents, which ultimately entails tracking incident costs. This section is responsible for ensuring that vendors get paid after the incident, and that cash flow is adequate when vendors must be paid immediately. In turn, the section

finance/administration section ▪ In an ICS, the personnel in charge of keeping track of expenditures such as salaries and orders, and paying for needed equipment and supplies.

is accountable for making sure that all applicable federal and state reimbursements are received. Hazardous materials incidents are very costly owing to the required equipment and services; these costs can sometimes reach millions of dollars. Therefore, individuals from the finance or accounting department of the agency or jurisdiction in charge typically staff the roles in the finance/administration section, since they possess the requisite knowledge, skills, and experience.

planning section ■ In an ICS, the personnel responsible for planning the next operational period of the incident action plan (IAP).

operational period ■ A defined increment of time—which may vary from hours to days, but is typically 12 hours—after which new personnel are rotated in and the incident is formally reassessed, and the incident action plan (IAP) is reassessed and modified.

Planning Section

The **planning section** is appointed when incidents become complex and/or last for days, weeks, or months. Because no individual can function for extended time periods, the incident is divided into operational periods. **Operational periods** are defined increments of time—which may vary from hours to days, but is typically 12 hours—after which new personnel are rotated in and the incident is formally reassessed, and the IAP is updated and modified. Therefore, the planning section must gather incident information from the operations section chief and determine which resources will be needed in the next operational period and how the incident objectives need to be updated and/or modified. The planning chief then notifies the logistics section chief of the needed resources for the next operational period.

BOX 13-1 HAZMAT HISTORY: PEPCON EXPLOSION IN HENDERSON, NV (1988)

On May 4, 1988, a massive explosion in Henderson, Nevada, occurred at the Pacific Engineering Production Company of Nevada (PEPCON) that killed two people and injured almost 400. The explosion was caused by maintenance activities and the improper storage of 4000 tons of ammonium perchlorate in HDPE plastic containers. PEPCON produced large quantities of ammonium perchlorate for the space shuttle program which was currently on hold due to the 1986 *Challenger* accident. Therefore, the market for ammonium perchlorate dried up almost instantly.

A fire was caused by welding on the steel support for a fiberglass tank that had been damaged in an earlier wind storm. The fiberglass started to burn, and the fire rapidly grew in intensity due to the oxidizing properties of the ammonium perchlorate residue on the structure. The fire then spread to the large quantity of plastic 55 gallon drums containing ammonium perchlorate. Employees tried unsuccessfully to douse the flames with hoses. The first explosions start to occur about 15 minutes after the fire started. Fortunately employees had already started to evacuate and about 75 employees escaped without harm.

The City of Henderson Fire Chief was the first to respond to the scene. In one of the initial explosions the glass was blown out of his vehicle and upon advice from evacuating plant employees he turned around to retreat to a safe distance. In all, several members of the Henderson fire department were injured by flying glass and several vehicles were damaged. Several other area fire departments responded, but all units stayed at a safe distance and allowed the fire to consume the dangerous ammonium perchlorate. The decision to operate in a non-intervention mode likely saved many first responder lives. The Nevada Highway Patrol and the Las Vegas Metro police department closed nearby roads and led the evacuation of a 5 mile radius around the PEPCON plant.

Different agencies and jurisdictions had an excellent working relationship prior to this incident which made cooperation much easier and contributed to the successful outcome. In addition a multijurisdictional command center was successfully used for interagency coordination. This included the hospitals. Due to advance notification, hospitals were able to set up triage points outside of their facilities to receive the large numbers of injured people, thereby preventing the emergency rooms from becoming overwhelmed.

The blast caused approximately $100 million worth of damage, affected a large portion of the Las Vegas metro area (which is only about 10 miles away), and caused several agencies to activate their disaster plans. Structural damage was recorded as far as 10 miles away. The explosions also ruptured a 16 inch natural gas main underneath the facility. The fire was finally brought under control after all of the ammonium perchlorate was consumed and the 16 inch gas main was shut off at a valve about a mile away. Hazardous materials teams were needed at the plant to deal with residual anhydrous ammonia and other corrosives left after the explosion and fire. Close coordination between multiple types of agencies from multiple jurisdictions was vital to a successful outcome.

BUILDING A COMMAND STRUCTURE

The incident command system is built from the bottom up in the sense that a small number of operational units are first dispatched—which form the initial incident command structure—and more management units are dispatched as needed—and the incident command structure grows. The most senior member of the first arriving unit typically assumes the role of incident commander. He or she performs a quick scene size-up, determines incident priorities, decides on the overall strategy to use and the tactics that will support the chosen strategy, and assigns available personnel to perform the highest priority tasks first. He or she also determines what additional resources are needed and ensures they are requested. At this point the incident commander typically manages all personnel directly, since there are fewer than seven people on scene. As more qualified personnel arrive on scene, command may be transferred.

The incident command system is modular in nature. Incident management systems are capable of handling extremely large and complex incidents when fully implemented, but all components of the system need not, and should not, be implemented during smaller incidents. The ICS chart or organizational structure reflects the number of resources employed at any given incident.

The Hazardous Materials Group or Branch

A hazmat branch director or hazmat group supervisor typically manages the hazmat-specific functions at more complex incidents. The hazmat branch director manages several groups corresponding to the common functions that must be carried out, for example, a research group, a decontamination group, an air-monitoring group, a product control group, an evidence collection group, and a medical group. The hazardous materials group generally implements the following positions:

- hazardous materials group supervisor (or hazmat branch director)
- assistant safety officer, hazmat
- site access control leader
- technical specialist–hazmat reference (research team leader)
- entry team leader
- decontamination team leader
- safe refuge area manager
- medical team leader

Each of these positions will have specific roles and responsibilities at the hazmat incident. Figure 13-2 illustrates an incident command chart for a hypothetical hazardous

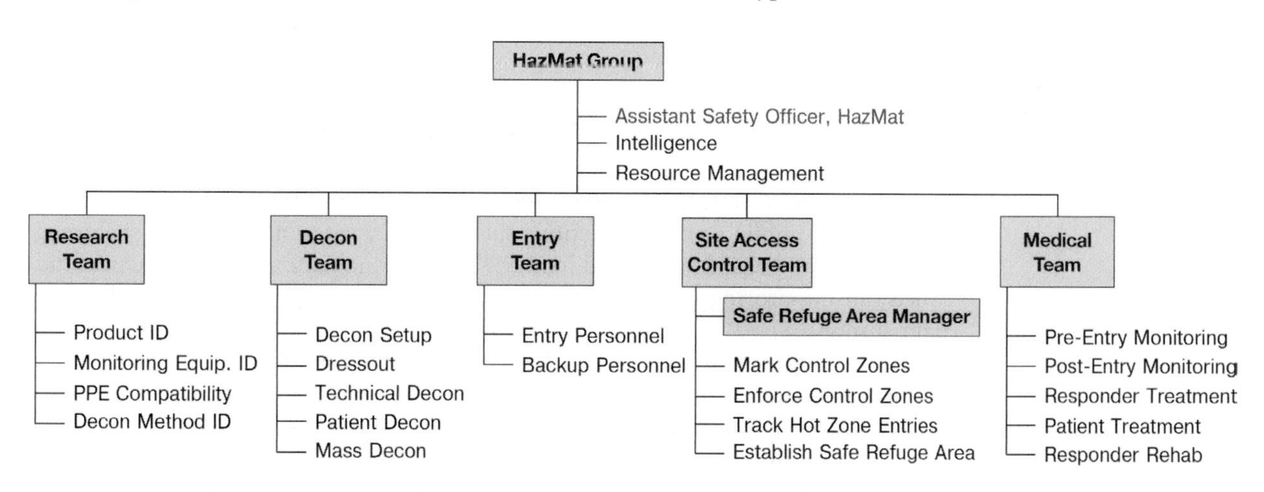

FIGURE 13-2 An incident command system flowchart for the hazardous materials group. *Courtesy of Chris Weber, Dr. HazMat, Inc.*

materials incident. The hazmat team you will eventually join and work with may or may not use exactly the same terminology; however, each of these core functions will be represented one way or another. For example, FIRESCOPE in California uses the technical specialist–hazmat reference, entry team, decontamination team, site access control leader, and safe refuge area manager positions. Other teams may have a medical team and/or a dress-out team. Let's explore the functional responsibilities of some of these positions.

ASSISTANT SAFETY OFFICER, HAZMAT

The Assistant Safety Officer, Hazmat (ASO-HM) reports to the incident safety officer, and not directly to the hazardous materials branch director or hazardous materials group supervisor. Typically the ASO-HM operates in the vicinity of the hazardous materials group and interacts very closely with the hazmat group supervisor and his or her designated team leaders. The ASO-HM has the discretion to immediately stop any unsafe acts if they are observed and then must inform the incident commander and the hazmat group supervisor of this action as quickly as possible. The following are some of the key responsibilities of the ASO-HM:

- Receive an incident briefing from the incident safety officer and the hazmat group supervisor
- Assist with the preparation of, review, and approve the hazardous materials site safety plan (SSP)
- Ensure the safety of all personnel assigned to the hazardous materials group
- Review the established site access control zones
- Assist in the identification of life safety hazards
- Review the decontamination plan
- Review PPE selection and use
- Review all hot zone activities—including victim rescue, reconnaissance, air monitoring, and product control actions
- Immediately stop any unsafe actions and notify the hazardous materials group supervisor
- Ensure that hazardous materials group personnel obtain appropriate medical care
- Ensure that medical records are maintained for hazardous materials group personnel

Thus, the ASO-HM must be a very experienced hazardous materials responder to competently perform all these functions.

HAZMAT GROUP SUPERVISOR

The hazmat group supervisor is responsible for managing the hazardous materials group. In this capacity he or she must obtain and assign appropriate resources, establish lines of authority within the group, direct operations, and ensure the safety of group members. The following are some of the key responsibilities of the hazmat group supervisor:

- Receive an incident briefing from the branch manager or operations section chief
- Reevaluate and/or establish control zones, isolation areas, and boundary markings
- Recommend public protection actions such as evacuations and in-place sheltering
- Determine site hazards using pre-incident plans, visual clues such as placards and labels, air monitoring, and reference materials such as shipping papers and safety data sheets
- Notify the appropriate agencies such as the National Spill Response Center, state and local environmental agencies, hospitals, and public health agencies
- Obtain current weather data and weather predictions
- Develop a written site safety plan
- Select and use appropriate PPE for hot zone entries
- Conduct pre-entry safety briefings

- Track the use and movement of all equipment and resources
- At larger incidents, participate in the development of the IAP

When the hazmat group supervisor cannot efficiently and competently perform all these (and often additional) critical functions, one or more of the following team leaders must be appointed (resources permitting).

SITE ACCESS CONTROL TEAM LEADER

The site access control team leader is responsible for enforcing the control zone boundaries. This means keeping all unauthorized personnel—both first responders and the public—from entering the hot zone or exclusion zone to prevent injury or death. He or she must also enforce the warm zone or contamination reduction zone boundary to prevent cross-contamination. The site access control team leader may need one or more team members on geographically dispersed hazmat incidents that are not easily controlled. He or she typically works closely with law enforcement as well. The following are some of the key responsibilities of the site access control team leader:

- Receive an incident briefing from the hazmat group supervisor
- Coordinate with the entry team leader and the decontamination team leader
- Determine the number of personnel needed to enforce control zone boundaries
- Consult with law enforcement agencies when necessary
- If not already completed, place control zone boundaries
- Station personnel at key entrance and egress points to prevent unauthorized access and cross-contamination
- Track all personnel entering and leaving the warm zone and hot zone
- Establish the safe refuge area inside a contamination reduction zone and appoint a safe refuge area manager
- Ensure that all personnel and victims who enter the cold zone have been decontaminated

Safe Refuge Area Manager

The safe refuge area manager is appointed by the site access control leader to establish and oversee the safe refuge area located in the contamination reduction zone, typically next to the decontamination line. The safe refuge area is a victim isolation and triage area designed to prevent cross-contamination of the decontamination line and cold zone by uncontrolled movement of victims. The safe refuge area manager does not necessarily have to work within the safe refuge area. The following are some of the key responsibilities of the safe refuge area manager:

- Receive an incident briefing from the site access control leader
- Coordinate with the entry team leader and the decontamination team leader
- Establish the safe refuge area within the contamination reduction zone
- Monitor the safe refuge area for contamination
- Evaluate victims in the safe refuge area for contamination
- Evaluate the medical condition of victims
- Track victim information (such as name, address, telephone number, previous location in the hot zone, contamination status, and medical status)
- Manage victim movement from the safe refuge area to either the decontamination line when contaminated, into the cold zone if uncontaminated, or to the appropriate personnel for scene-specific informational interviews if the victim is knowledgeable about key aspects of the incident

DECONTAMINATION TEAM LEADER

The decontamination team leader is appointed by the hazmat group supervisor to establish and oversee the decontamination line. Therefore, he or she is responsible for the successful decontamination of victims, entry personnel, and equipment. The decontamination team leader does not necessarily have to work in the warm zone. The following are some of the key responsibilities of the decontamination team leader:

- Receive an incident briefing from the hazmat group supervisor
- Coordinate with the site access control team leader, entry team leader, the technical specialist–hazmat reference, and medical team leader
- Determine the appropriate decontamination method with input from the research team leader
- Determine the appropriate PPE for decontamination team members
- Establish the decontamination line
- Brief the decontamination team members on the chosen decontamination methods and decontamination line layout
- Identify contaminated people and equipment entering the decontamination line using detection devices
- Supervise the operation of the decontamination line
- Maintain control of the access points to the decontamination line
- Ensure that victims receive medical care during decontamination
- Maintain control of contaminated waste and wastewater for proper disposal.

ENTRY TEAM LEADER

The entry team leader is appointed by the hazmat group supervisor to oversee the entry team while they are in the hot zone. In essence, they are responsible for all rescue and mitigation activities in the hot zone. The following are some of the key responsibilities of the entry team leader:

- Receive an incident briefing from the hazmat group supervisor
- Coordinate with the site access control leader, safe refuge area manager, decontamination team leader, and the technical specialist–hazmat reference
- Track all personnel entering and leaving the hot zone
- Recommend mitigation actions to the hazmat group supervisor
- Ensure that the appropriate level of PPE is chosen for the entry team
- Ensure that the entry team receives pre-entry medical checks
- Brief the entry team on their hot zone function
- Ensure that the entry team understands the site map, knows the route to take to their work location, and knows the location of safe refuge areas within the hot zone
- Supervise entry team operations—including victim rescue, air monitoring, reconnaissance, and mitigation activities
- Track entry and egress times of personnel entering the hot zone
- Ensure that the entry team receives post-entry medical checks and rehabilitation

TECHNICAL SPECIALIST–HAZMAT REFERENCE

The technical specialist–hazmat reference, or research team leader, is appointed by the hazardous materials group supervisor to gather all pertinent information regarding the chemicals, facilities, and containers involved in the hazardous materials release. In addition, the research team leader aids in the preparation of air-monitoring and detection instruments, and selection of PPE, decontamination methods, and mitigation techniques. The research team leader should be one of the most knowledgeable members of the team. The following are some key responsibilities of the research team leader:

- Receive an incident briefing from the hazmat group supervisor
- Coordinate with the entry team leader, medical team leader, and the decontamination team leader

- Determine the chemical and physical properties of all chemicals involved in the hazardous materials incident
- Select and prepare air-monitoring instruments
- Determine action levels for air-monitoring instrument readings
- Select appropriate PPE for the entry team, decontamination team, and backup team
- Select an appropriate decontamination method
- Perform plume modeling calculations
- Brief the entry team regarding technical aspects of their work functions

MEDICAL TEAM LEADER

The medical team leader is appointed by the hazardous materials group supervisor to ensure that pre- and post-entry medical monitoring is carried out for all personnel wearing PPE. In addition, the medical team leader determines the appropriate treatment of victims of hazardous materials exposure and ensures that receiving hospitals are notified of patient injuries, exposures, and decontamination methods and verification of decontamination. The medical team leader should have significant medical training. The following are some of the key responsibilities of the medical team leader:

- Receive an incident briefing from the hazmat group supervisor
- Coordinate with the technical specialist–hazmat reference and decontamination team leader
- Perform pre-entry medical monitoring for personnel wearing PPE
- Perform post-entry medical monitoring for personnel wearing PPE
- Ensure that personnel receive the appropriate rehabilitation and rehydration after hot zone entries and decontamination line work
- Determine appropriate medical treatment for victims of chemical exposure
- Notify receiving EMS agencies and hospitals of victim injuries, chemical exposure, and decontamination methods and verification of decontamination

DRESS-OUT TEAM LEADER

The dress-out team leader is appointed by the hazardous materials group supervisor to coordinate donning and doffing of PPE. Some hazardous materials response teams find it helpful to simultaneously dress out the entry team, backup team, and decontamination team. When PPE donning is coordinated, the hot zone entry often runs more smoothly, because all personnel are ready simultaneously, do not go on air unnecessarily, and do not languish in hot PPE waiting for other teams to get dressed, and therefore are ready for the entry briefing at the same time. This can significantly reduce hot zone entry preparation times. Not all hazmat teams coordinate the donning of PPE in this way, but many teams find this to be a very efficient and effective method. The following are some of the key responsibilities of the dress-out team leader:

- Receive an incident briefing from the hazmat group supervisor
- Coordinate with the entry team leader, decontamination team leader, and technical specialist–hazmat reference
- Assist with the determination of the appropriate PPE for the entry team, decontamination team, and backup team
- Coordinate the simultaneous donning of PPE for the entry team, backup team, and decontamination team

For large entry operations, the dress-out team leader will require multiple personnel to act as dress-out assistants. Each person donning PPE should have his or her own dress-out assistant for optimal efficiency.

The Planning Process

An effective plan is the cornerstone of an effective response. The better the plan, the better the response outcome will be. There are five distinct phases in the NIMS planning process:

1. Understanding the situation
2. Establishing incident objectives and strategy
3. Developing the IAP
4. Preparing and disseminating the IAP
5. Executing, evaluating, and revising the plan

It is critical to understand as much about a hazardous materials or WMD incident you are faced with as quickly as possible. Situational awareness involves assessing the scene—such as through air monitoring and reconnaissance, accessing information from pre-incident plans, and contacting other agencies and jurisdictions that may have helpful information (such as the building department or the public works department). You should also know the capabilities and extent of your own resources and those available through mutual-aid partnerships. Thus, effective networking, training, and pre-incident planning are essential activities for any hazardous materials response team.

planning P ■ A process developed by the U.S. Coast Guard by which an incident or event can be successfully managed in the long term.

THE PLANNING P

The **planning P** was developed by the U.S. Coast Guard in response to difficulties they faced at complex incidents. The planning P is a graphical representation of the key steps in the operational period planning cycle (Figure 13-3). The incident starts at the foot of the P, then progresses into the loop, where each revolution around the loop represents one operational period. During each operational period, incident objectives, strategies, and tactics are developed for the next operational period, a planning meeting is held, the new plan is approved by the IC, there is an operational briefing, and the new plan is executed during the next operational period. Then, during the next operational period, the planning section works on the IAP for the subsequent operational period using the latest information provided by the operations section.

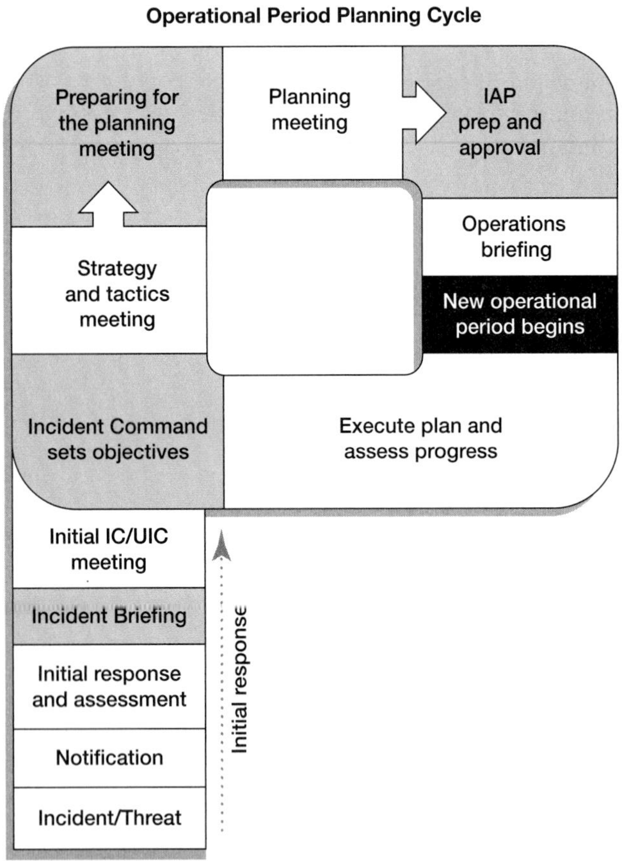

FIGURE 13-3 The planning P illustrates the planning process for long-term incidents. *Courtesy of FEMA.*

THE INCIDENT ACTION PLAN

A critical component of any incident management system (IMS) is the incident action plan (IAP). Every incident has an IAP, although only complex incidents or incidents of long duration typically have written IAPs. Written IAPs are used to record incident objectives, strategy and tactics, and the distribution of resources for incidents that are complex and/or span multiple operational periods. A written IAP facilitates the dissemination of information from one shift to another across operational periods and to large numbers of first responders from different agencies or geographical areas at an incident. Incident objectives should be measurable, flexible, and time-sensitive. This allows incident objectives to be evaluated for effectiveness and modified if they are not.

Part of the IAP is a safety analysis of the incident, often called the site safety plan (SSP). Every hazmat incident must have a written SSP per HAZWOPER. The following are key components of the SSP according to the EPA:

- site description and map, including control zone boundaries (hot, warm, and cold)
- response objectives and entry team objectives
- on-site incident command system organization
- on-site personnel (by control zones)
- hazard evaluation, including confined spaces, trenches, and other hazards
- personal protective equipment
- on-site work plan
- communication procedures
- decontamination procedures
- site safety and health plan

Safety briefings should be given to all personnel at hazmat and WMD incidents. The briefing provides a site safety overview and methods for contacting the incident commander and the safety officer in the event of an emergency, and includes incident communication details such as radio frequencies in use and procedures for relaying information up the chain of command. It is especially important to hold a detailed safety briefing, called the *entry team briefing*, for all responders entering the hot zone (Figure 13-4).

INCIDENT DOCUMENTATION

Accurate and complete documentation is an important component of hazmat and WMD incident response. Good documentation improves communication between personnel; improves the quality of the communicated information; provides a rationale for decision making, thereby improving scene safety; facilitates after-action reviews; and provides solid evidence during legal proceedings. Documentation may take many forms, including handwritten notes, photographs and video recordings, standardized response forms

FIGURE 13-4 The safety briefing should be conducted immediately before the entry team enters the hot zone. *Courtesy of Chris Weber, Dr. HazMat, Inc.*

(such as the IAP), air-monitoring and sample identification results, electronic data output from instrumentation, and recorded radio transmissions. After the incident has been terminated, the incident reports, medical surveillance, and critiques also become part of the incident documentation.

At large or complex incidents, document control is often difficult. To avoid this problem, a person—such as an incident recorder—is assigned to keep track of all incident documentation. Document control procedures may include the following:

- listing all documents in a document inventory
- collecting and collating documents at regular intervals during extended operations
- keeping copies of documents in a centralized, secure location
- assigning a control number to all documents

Good documentation has many advantages: When response objectives are written, it is easier for the safety officer to accurately review them. When strategies and tactics are written, it is easier for other agencies and subordinates to accurately and safely accomplish them. Finally, after the incident is over, it is much easier to infer lessons learned and improve agency operations when the documentation accurately reflects the actions that were taken under the given circumstances.

Incident Command Forms

ICS forms have been developed by many agencies to help facilitate the seamless cooperation of local, state, tribal, and federal agencies. Many ICS forms have been standardized in NIMS and follow a logical format (Figure 13-5). There are at least two dozen different forms that when completed form the basis of the IAP. The following forms constitute the core of the IAP:

ICS 201: Incident briefing
ICS 202: Incident objectives
ICS 203: Organization assignment list
ICS 204: Assignment list
ICS 205: Radio communications plan
ICS 205A: Communications list
ICS 206: Medical plan
ICS 207: Incident organization chart
ICS 208: Safety message/plan
ICS 209: Incident status summary
ICS 210: Resource status change
ICS 211: Incident check-in list
ICS 213: General message
ICS 214: Activity log
ICS 215: Operational planning worksheet
ICS 215A: IAP safety analysis

SOLVED EXERCISE 13-2

When should a written incident action plan be completed at a hazardous materials incident?

Solution: A written incident action plan should be completed as soon as possible, and a rudimentary plan listing site hazards, chemical hazards, PPE levels, and entry team objectives should be completed before anyone enters the hot zone. This plan may be a simple one- or two-page document containing bullet lists of the key components and a copy of a facility map with hazard zones indicated.

INCIDENT ACTION PLAN SAFETY ANALYSIS (ICS 215A)

1. Incident Name:		2. Incident Number:	
3. Date/Time Prepared: Date: Time:	4. Operational Period: Date From: Time From:		Date To: Time To:
5. Incident Area	6. Hazards/Risks	7. Mitigations	

8. **Prepared by** (Safety Officer): Name: ____________ Signature: ____________

Prepared by (Operations Section Chief): Name: ____________ Signature: ____________

ICS 215A Date/Time: ____________

FIGURE 13-5 The NIMS-compliant ICS 215A Incident Action Plan Safety Analysis form. *Courtesy of FEMA.*

Incident Command Facilities

At larger incidents the large number of personnel and resources will require more substantial facilities in which to operate effectively.

incident command post (ICP) ■ The physical location of the command functions and where the incident commander or the unified incident command is located.

INCIDENT COMMAND POST

All emergency incidents, both large and small, will have an **incident command post (ICP)**. This is where command and control functions take place and where the IC, most of the

FIGURE 13-6 A designated incident command post provides a quiet and secure location for incident command activities. In addition, a well-equipped command post has additional communications and computer facilities.
Courtesy of Chris Weber, Dr. HazMat, Inc.

command staff, and most of the general staff are located (if they are appointed) (Figure 13-6). The ICP can quickly become crowded at larger incidents if it is not chosen wisely. It must be located in a clean area that is upwind, uphill, and upstream of the hot zone. Still, it should be as close to the incident as safety considerations and practical conditions dictate. The ICP should also be secured from unauthorized access by the public and even by other first responders. This is especially important at criminal incidents, such as WMD and terrorist incidents.

At smaller incidents the ICP is typically the vehicle in which the incident commander arrived. At larger incidents the ICP may be upgraded to a mobile command post that has superior communications, lighting, electrical, and sanitary capabilities. At extremely large incidents the ICP may be moved to a fixed location such as a building—possibly remote from the site. In this case the ICP should have good visual and verbal communication with the incident location via radios and cameras.

emergency operations center (EOC) ▪ A permanent ICS installation that facilitates the deployment of resources by acting as a logistics clearinghouse at larger emergency incidents. Most cities and counties have an EOC that is maintained by the local emergency management agency (EMA).

EMERGENCY OPERATIONS CENTER

The **emergency operations center (EOC)** is a fixed facility run and maintained by the local emergency management agency. It has excellent communications resources that can accommodate representatives from stake-holding agencies involved in the incident. The EOC responds to resource requests from the incident commander. The EOC and emergency manager are a resource to the incident commander but do not assume command of the incident. When the logistics section has been established, it is usually in close communication, possibly even colocated, with the EOC.

staging area ▪ In an ICS, the location where incoming resources wait until they are needed. The staging area is located far enough away from the hot zone so as not to interfere with incident operations but close enough to permit rapid deployment.

staging area manager ▪ In an ICS, the individual in charge of the staging area.

STAGING AREA

The **staging area** is where resources report to when they arrive on scene. This keeps them from interfering with ongoing operations and causing traffic jams and other access and egress problems. The staging area is managed by a **staging area manager**, who reports directly to the IC or to the operations section chief (when appointed). To maintain the staged resources in a state of readiness, the staging area manager may need to request resources from the logistics section.

An efficient IC should ensure that resources arrive before they are actually needed. The staging area should be located in a safe location, out of the immediate work zone, yet close enough to allow rapid incident access. On arrival the responding units receive an incident briefing and, when needed, report to the appropriate manager on scene.

joint information center (JIC) ▪ In an ICS, the facility at which PIOs from different agencies and jurisdictions can brief the public with a unified voice.

JOINT INFORMATION CENTER (JIC)

At larger incidents, or at multijurisdictional incidents, a **joint information center (JIC)** may be formed by multiple agencies or jurisdictions to coordinate the release of mutually consistent information. For example, at a hazmat incident involving a large toxic plume, environmental agencies such as the EPA or their contractors may monitor the plume remotely while the local public health department conducts contamination surveys, EMS services treat victims, and the hazmat team assesses and controls the leak. Information from all four agencies must be coordinated to give the public consistent and accurate information, thereby ensuring that the released information advances the incident objectives (such as evacuation versus in-place sheltering advice).

BOX 13-2 HAZMAT HISTORY: MASSIVE CHLORINE RELEASE IN HENDERSON, NEVADA (1991)

On May 6, 1991, almost 70 tons of liquid chlorine was released in Henderson, Nevada, from the Pioneer Chlor Alkali Company. This facility produced chlorine gas through the electrolysis of table salt. The liquefied chlorine gas was stored in eight 150-ton storage tanks. The chlorine was shipped to customers in railcars and highway cargo tankers for use in water treatment processes. At any given time up to 300 tons of liquefied chlorine was stored on-site.

The release started when water mixed with the liquid chlorine, creating a very corrosive solution (hydrochloric acid) that created a pinhole leak in a transfer pipe during lading of rail cars from one of the massive storage tanks. Initially, company workers believed they could handle the apparently minor leak themselves. However, the pinhole leak grew relatively rapidly in size to a 1-inch-diameter liquid leak. Company employees believed they had stopped the pinhole leak, closed the appropriate valves, and were waiting for the situation to stabilize as residual chlorine leaked out of the transfer piping.

Therefore, the fire department was never notified by the company. Instead, the fire department was notified by concerned members of the public approximately 45 minutes after the release started and arrived at the plant through its own investigation. At that time, several plant employees needed medical attention from the chlorine exposure. Within a few minutes of arrival a Clark County battalion chief was also overcome by the chlorine cloud as it expanded to the front gate of the plant. At this point the plant was evacuated, and a command post was established half a mile away, uphill from the plant. Owing to the incapacitation of the Clark County battalion chief, a Henderson battalion chief assumed command. At this time several firefighters and plant employees were also transported to the hospital. Later, a Clark County assistant fire chief assumed command and reassigned the Henderson battalion chief as the operations section chief owing to his more extensive hazmat experience. This is an excellent example of the need for and use of transfer of command during a hazmat incident.

The Pioneer Chlor Alkali plant employees were actually members of the regional CHLOREP response team and were quite qualified to deal with liquid chlorine releases. Therefore, a nonintervention approach was taken initially with the hope that the vapor cloud would rapidly dissipate (since the release had theoretically been stopped).

Hazmat teams attempted to monitor and track the chlorine cloud using air-monitoring equipment. However, it was becoming obvious that the cloud was not dissipating, and the situation was becoming worse. In fact, the chlorine cloud expanded to such a degree that it overwhelmed the initial command post even though it was located uphill and a half mile from the release. Almost 2-1/2 hours after the initial release, and nearly 2 hours after fire department arrival, the command post and staging area had to be evacuated. In fact, the command post had to be relocated twice during this incident and was ultimately located several miles from company property, illustrating the importance of proper scene size-up in determining isolation zones. It is very important to consider not only current conditions but also worst-case scenarios when setting up the initial zones. In this case, because of the enormous size of the storage tanks, a much larger isolation area was necessary.

Since the situation was not improving, the fire department hazmat team made joint entries with the facility hazmat team to investigate and mitigate the leak. Plant employees did the plugging and patching on the first entry and installed a Teflon-coated flange during the second entry. Installation of the Teflon-coated flange ultimately stopped the liquid chlorine release. As the sun came up several hours later, increasing temperatures and wind started to disperse the chlorine vapors. However, it took several more hours for the released chlorine gas to completely dissipate.

During the incident, more than 200 people were taken to the hospital, and 30 were hospitalized. More than 700 people were evacuated to shelters, and 2000 to 7000 people were evacuated elsewhere. Evacuations were accomplished using joint fire department and law enforcement assets. Law enforcement agencies from multiple jurisdictions evacuated areas that were believed to be chlorine free, while firefighters using SCBA evacuated areas that potentially contained chlorine. Nevertheless, several law enforcement officers were exposed to chlorine during traffic control or evacuation duties and needed medical treatment. The Clark County School District made school buses available for the evacuation and allowed firefighters to drive them—an excellent example of resource coordination with outside agencies.

Unified Incident Command

unified incident command (UIC) ■ The joint command of an incident by incident commanders from different agencies and/or jurisdictions.

At larger or more complex incidents, multiple agencies or jurisdictions may form a **unified incident command (UIC)**. The UIC is composed of entities that have significant responsibilities for successful termination of the incident. Terrorist incidents are usually managed using the UIC concept owing to the overlapping legal responsibilities and complementary response functions of law enforcement agencies, fire departments, and EMS agencies at the federal, state, tribal, and local level. The technical knowledge required to successfully manage and mitigate complex incidents such as these is usually beyond the capabilities of a single agency or person and requires a unified command structure.

Hazardous materials incidents that are complex or affect multiple jurisdictions can be very difficult to manage. Often, there are conflicts between different jurisdictions and different agencies. Unified command is a way to minimize political differences between jurisdictions as well as to bridge technical capabilities between different agencies. For example, the discovery of an illicit laboratory involves hazardous materials in which the fire department may possess the emergency response expertise; however, the lead agency will be law enforcement, since it involves a crime. Incidents like this are well suited to a unified command involving both the fire department and the police department. In this way the technical capabilities of both agencies can be brought to bear at the command level and help ensure that no critical components are overlooked. Sometimes, unified command is chosen for jurisdictional reasons. For example, at a large fire at a chemical warehouse at the border of two municipalities, a unified command will likely be formed consisting of the fire chiefs of both municipalities.

Unified command is also often formed when incidents require resources from several distinct entities, for example, from different agencies such as the fire department, law enforcement, EMS, public works, and private industry; or different municipalities; or different levels of government, such as local, state, tribal, and federal levels. When a UIC is formed, it is much easier to determine all the essential incident objectives, select optimal strategies, and use all the available resources optimally.

Area Command

An area command is used when the incident affects multiple geographic areas simultaneously. When an emergency is spread over a large geographic area, it may be difficult to prioritize and assign resources in the most effective manner. For example, a natural disaster such as an earthquake might lead to multiple hazardous materials incidents across a substantial portion of the state. Each geographic area has its own needs, and the ICs require critical resources. Under an area command each separate hazardous materials incident, such as a refinery explosion, a train derailment, and multicar accident on the freeway, retains its own incident command system including an incident commander. However, all resource requests pass through an area command, which assigns resources based on the needs of each individual incident. This type of system allows for optimal resource allocation when resources in a particular geographic area are stretched thin.

SOLVED EXERCISE 13-3

Under what conditions would you conceive using an area command at a hazardous materials or WMD incident?

Solution: An area command may be necessary under a regional emergency, such as extensive flooding or a hurricane, in which there are many hazardous materials incidents and a limited number of resources. Resources can more efficiently be distributed using an area command and thereby accomplish more with those resources. The management of terrorist incidents that involve multiple attacks at multiple locations would also benefit from the implementation of an area command.

The Role of the Local and State Emergency Operations Centers

As discussed earlier, emergency operation centers are designed to more effectively obtain and manage resources during emergency responses. The EOC is made up of representatives from key agencies with available resources that may be needed at large emergency incidents. Because of the complexity of hazmat incidents, the local EOC is often activated. The EOC does not assume command of the incident; however, sometimes the EOC and the incident command post are colocated for convenience.

The state EOC (SEOC) is very similar in concept to the local EOC except that it is activated during emergencies that involve substantial regions of the state. When the SEOC is activated, representatives from state agencies attend and contribute their respective resources, which are available for request by local ICs through their respective local EOCs.

National Incident Management System

The National Incident Management System (NIMS) was developed by the federal government in the wake of the terrorist attacks of September 11, 2001, and the Space Shuttle *Columbia* disaster to respond effectively and efficiently to significant domestic incidents such as natural disasters and terrorist attacks. Homeland Security Presidential Directive 5 (HSPD-5), *Management of Domestic Incidents*, dated February 28, 2003, directed the Secretary of Homeland Security to develop a national incident management system. NIMS is a standardized management system designed to ensure efficient and effective control of resources at the scene of emergencies. It is designed to effectively manage a wide range of local, tribal, state, and federal resources at events of national significance. However, NIMS is also capable of being used at smaller incidents because of its modular nature.

Common terminology allows NIMS implementation by a wide variety of response agencies and both traditional and nontraditional first responders. Indeed, there is a push to train not only emergency responders but also state and federal regulatory agencies, nongovernmental organizations (NGOs) such as the Red Cross, as well as the private sector in the use and implementation of NIMS. The aim of broad NIMS incident command training is to facilitate the integration of resources and responders from a wide variety of backgrounds at future emergencies and events.

Communications and the broader concept of new technology are also addressed by NIMS. The terrorist attacks of 9-11 painfully illustrated the lack of effective and efficient communications between and within agencies. Other technology-intensive shortcomings were also identified. These issues are specifically addressed in HSPD-5.

Several IMS components also are dealt with by HSPD-5 and are implemented in NIMS. These components are designed to ensure that all responders use the same command structure and common terminology to manage the incident, and continue to incorporate new technology in the future. Specifically, the NIMS components are as follows:

1. Command and Management
2. Preparedness
3. Resource Management
4. Communications and Information Management
5. Supporting Technologies
6. Ongoing Management and Maintenance

Let's take a look at each of these components in greater detail. You will see that many of these components overlap and build on each other.

COMMAND AND MANAGEMENT

NIMS-compliant incident management systems comprise four key organizational components: the incident command system, multiagency coordination systems, public information systems, and publication management.

As discussed earlier, ICS is the organizational framework used to manage an incident. The operating characteristics, such as single command versus unified command, play a critical role in the outcome of most incidents. Typically, the operating characteristics change over time as the incident increases in complexity or scope, or becomes less severe as the incident is mitigated. Usually, an incident begins as a single command involving one or two agencies. As the severity of the incident is recognized or grows, other jurisdictions and other agencies are called, and the incident may transition into a unified command. As the incident is brought under control it will likely transition back to a single command involving fewer resources.

The incident command structure and organizational chart will naturally vary from incident to incident depending on the nature of the incident, the location of the incident, the available resources, the training level of the responding personnel, and the incident commander. Owing to the modular and flexible nature of ICS, many different organizational charts may be appropriate for any given incident. The implementation and design of the incident command organization is up to the incident commander and depends on such factors as the type of agency, training level and discipline of responders, and the capability of the incident commander. Even a suboptimal ICS organizational structure is very beneficial, because personnel and resource accountability are greatly enhanced through the use of an ICS.

Multiagency coordination systems provide a framework in which local, state, tribal, and federal agencies can work together in an efficient manner. Mutual-aid agreements and compacts should be developed during emergency response plan development and pre-incident planning. This concept is especially important for hazardous materials incidents. Jurisdictions and agencies generally have a limited supply of trained personnel and specialized equipment necessary to respond to hazmat incidents. Hazardous materials teams should therefore have agreements with neighboring jurisdictions to supplement equipment and manpower. They should also have agreements with other agencies, such as law enforcement SWAT teams, that complement their capabilities. How would your hazmat team deal with a barricaded gunman in a facility containing hazardous materials? Even more important, how would your local law enforcement deal with this situation? The time to find out whether a SWAT team will make an entry into a possibly chemically contaminated area is not during an incident!

Public information systems are vital to transmitting critical information quickly and accurately to the public during a crisis or emergency. Communication involves a multifaceted approach including technological tools such as reverse 911, cable alert systems, and paging systems. It also includes managing the media and using their capabilities to protect the public. Establishing a single PIO early on during an incident is vital to building trust in the community and effectively carrying out evacuation and in-place sheltering operations.

PREPAREDNESS

NIMS emphasizes community preparedness to ensure the availability of trained personnel and appropriate resources to manage emergency incidents effectively when they inevitably occur. Preparedness involves planning, training, exercising, qualifying and certifying personnel, acquiring and certifying equipment, and forming mutual-aid agreements, and each of these activities must be ongoing. Training is a never-ending task owing to personnel turnover, changing techniques and practices, and new equipment acquisitions. All these activities should be modified if deficiencies are found during exercises or actual events (either emergency incidents or large public events).

Planning involves analyzing your community and predicting the types of emergencies that could occur, assessing the vulnerability of the community, determining the resource level of the community, and using hazard- and risk-based analysis to establish what measures can be taken to eliminate the vulnerabilities or minimize their effect. Planning is a never-ending task given the rapid growth and development of most communities, evolving threats, and constant changes in technology.

Mutual-aid agreements are an important component of good resource management. Hazardous materials incidents are labor- and equipment-intensive. No single hazmat response team can prepare for all conceivable emergencies. Strong ties should be formed with neighboring hazmat response teams, as well as with other regional, state, tribal, and federal response teams. In addition, other organizations such as public health departments, EMS agencies, law enforcement agencies, the military, school systems, and public works departments may offer key resources during hazardous materials incidents.

Training is one of the most critical components of hazardous materials preparedness. Hazmat response teams must be able to function as a cohesive unit, safely and effectively use all the tools at their disposal, and work with other emergency responders from outside agencies. These goals can be accomplished only through regular training that incorporates classroom, skill station, and exercise-based training. Training should also be conducted at hazardous materials facilities in your jurisdiction. The time to get to know facility managers, key employees, and the facility layout is not during a hazmat emergency.

Publications management refers to the standardization, dissemination, and operational security of written materials. Examples include, but are not limited to, standard operating procedures or guidelines, command forms for incident management, and common terminology used during radio communications and between agencies. These documents must be used to train personnel, both within and outside the promulgating agency, and revised when they prove inadequate or when technology, training, or equipment changes are made.

All these activities must be performed in a competent fashion. Any one activity may be the weak link in the chain, and if any one of these activities is overlooked, tragedy can strike. Hurricane Katrina is a prime example. The New Orleans area ran a tabletop exercise of a large hurricane dubbed "Pam." The foresight of the planning team was tremendous; they essentially modeled Hurricane Katrina only a few months before it hit. However, at many levels and in many agencies, the lessons learned were not implemented, and the consequences were tragic—more than a thousand people lost their lives in the storm and its aftermath. Planning and preparedness activities must be implemented in the field.

RESOURCE MANAGEMENT

NIMS defines standardized mechanisms and processes to describe, inventory, dispatch, track, and recover resources before, during, and after an incident. Resource management is used in combination with planning, personnel management and training, mutual-aid agreements, and equipment acquisition to ensure that appropriate resources are available for use during an emergency and appropriately trained personnel are available to operate the resources.

COMMUNICATIONS AND INFORMATION MANAGEMENT

NIMS requires a standardized framework for communications, information management, and information sharing. Organizations must be able to communicate effectively across agencies and jurisdictions using interoperable communications equipment, procedures, and systems. Information management includes the collection, analysis, and dissemination of information at all levels of incident management. Sometimes, this information includes classified or sensitive materials. Operational security (Op Sec) must be maintained for this type of information.

SUPPORTING TECHNOLOGIES

NIMS provides a mechanism for ensuring that new technology is incorporated into the incident management toolbox as it becomes available. Often, this technology relates to communications systems and networks, but supporting technologies may also include other aspects of ICS such as security-related equipment and resource tracking and accountability systems. Technological developments play an important role in maintaining efficient and effective incident management systems. Several technological deficiencies were noted in the after-action analysis of 9-11 that may have cost many lives.

ONGOING MANAGEMENT AND MAINTENANCE

NIMS will continuously be refined as weaknesses and deficiencies are discovered. Local implementations of NIMS must likewise be updated, exercised, and revised as necessary owing to new technology, new operating procedures, or other changes. Incident management systems are only as effective as the people that use them. Exercising and training are essential for all personnel who will operate under the ICS.

National Response Framework

emergency support functions (ESF) ■ Common types of assistance provided by the federal government to local, state, tribal, and across different federal agencies.

The *National Response Framework (NRF)* establishes a comprehensive, national, all-hazards approach to disasters and emergencies. It is designed to seamlessly coordinate the response of all federal agencies to emergency incidents. The *NRF* assigns federal lead agencies depending on the type of emergency, or which **emergency support function (ESF)** is needed (Table 13-1). ESFs are essentially common categories or types of assistance

TABLE 13-1 Emergency Support Functions as Established in the National Response Framework

ESF #	ESF NAME	COORDINATING AGENCY
ESF #1	Transportation	DOT
ESF #2	Communications	DHS
ESF #3	Public Works and Engineering	DOD/Army Corps of Engineers
ESF #4	Firefighting	USDA/U.S. Forest Service
ESF #5	Emergency Management	DHS/FEMA
ESF #6	Mass Care, Emergency Assistance, Housing, and Human Services	DHS/FEMA
ESF #7	Logistics Management and Resource Support	GSA & DHS/FEMA
ESF #8	Public Health and Medical Services	DHHS
ESF #9	Search and Rescue	DHS/FEMA
ESF #10	**Oil and Hazardous Materials Response**	**EPA with DHS/USCG**
ESF #11	Agriculture and Natural Resources	USDA
ESF #12	Energy	DOE
ESF #13	Public Safety and Security	DOJ
ESF #14	Long Term Community Recovery	DHS/FEMA
ESF #15	External Affairs	DHS/FEMA

provided by the federal government to local, state, tribal, and other federal agencies. Each ESF has a coordinating agency, one or more primary agencies, and usually many supporting agencies within the federal government.

There are three levels of contingency plans under the national response system: the National Contingency Plan (NCP), Regional Contingency Plans (RCP), and Area Contingency Plans (ACP).

NATIONAL OIL AND HAZARDOUS SUBSTANCES POLLUTION CONTINGENCY PLAN

The National Oil and Hazardous Substances Pollution Contingency Plan, commonly called the National Contingency Plan (NCP), describes the federal response structure for significant hazardous materials releases in the United States. The NCP provides guidance for federal agencies, such as the EPA, involved in hazardous materials response under CERCLA. The industry Tier I and Tier II plans received by the local emergency planning committees (LEPC) and reviewed by the state emergency response commission (SERC) must follow the NCP.

Three fundamental kinds of activities are performed pursuant to the NCP:

1. Preparedness planning and coordination for response to a discharge of oil or release of a hazardous substance, pollutant, or contaminant
2. Notification and communications
3. Response operations at the scene of a discharge or release

National Response Team

National planning and coordination is the function of the National Response Team (NRT), which consists of representatives from many agencies involved in hazardous materials response, from boots on the ground to financing and logistics. Figure 13-7 shows its organizational framework. The NRT is chaired by a representative from the EPA, and the vice-chair is a representative of the U.S. Coast Guard.

Regional planning and coordination of preparedness and response actions is accomplished through regional response teams (RRTs). The two principal components of the RRT mechanism are a standing team, which consists of designated representatives from each participating federal agency, state government, and local government (as agreed on by the states); and incident-specific teams formed from the standing team when the RRT is activated for a response. Participation by the RRT member agencies on incident-specific teams will depend on the technical nature of the incident and its geographic location.

The National Response Center (NRC), located at USCG Headquarters, is the national communications center and is continuously staffed for handling activities related to response actions. The NRC acts as the single point of contact for all pollution incident reporting, and as the NRT communications center. Notice of discharges and releases must be made telephonically through a toll free number (1-800-424–8802). The NRC receives and immediately relays telephone notices of discharges or releases to the appropriate designated federal on-scene coordinator (OSC). The telephone report is distributed to any interested NRT member agency or federal entity that has established a written agreement or understanding with the NRC. The NRC evaluates incoming information and immediately advises FEMA of a potentially major disaster situation.

Federal On-Scene Coordinator

The federal on-scene coordinator (OSC) is the representative from the lead federal agency (such as EPA, DOE, or DHS) who is responsible for coordinating response of federal government assets with the local, state, or tribal incident commander. The OSC is also responsible for overseeing the development of the area contingency plan (ACP) in the

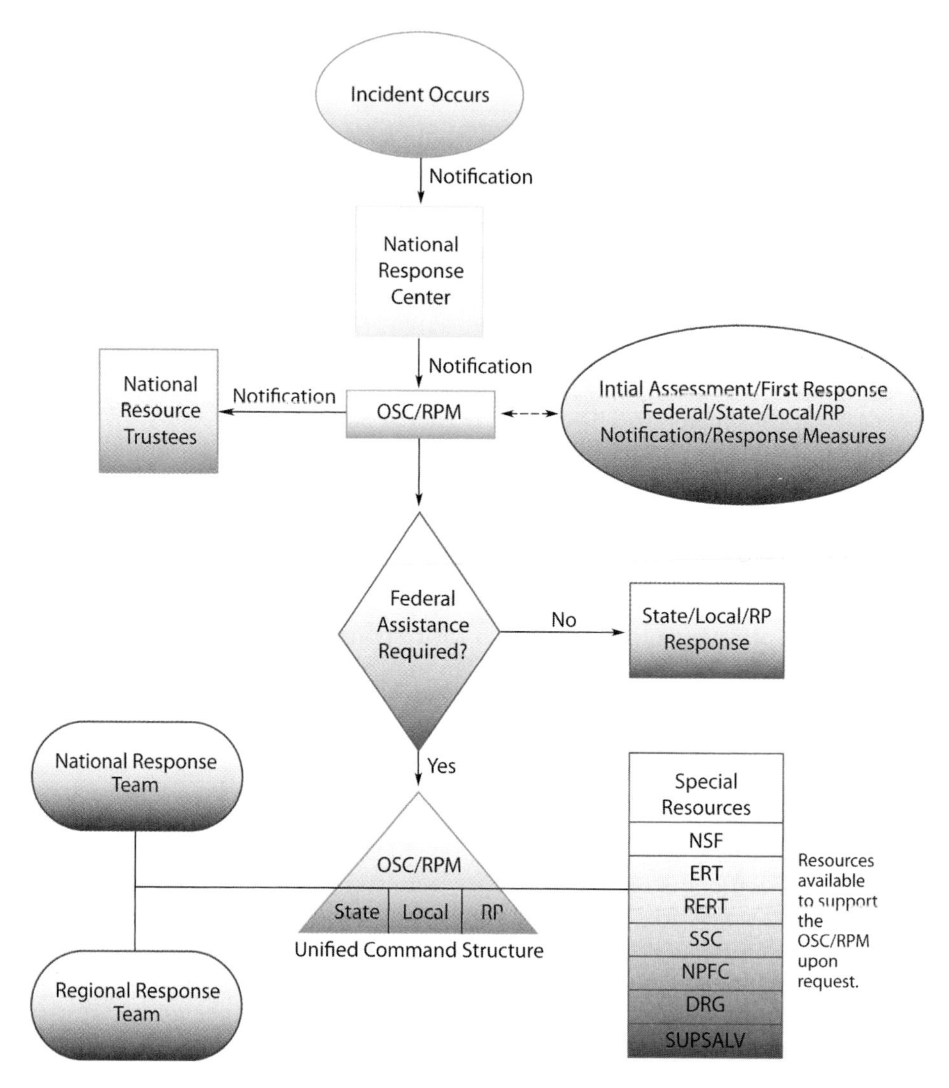

FIGURE 13-7 The response process when federal agencies are involved. *Art by David Heskett.*

area of the OSC's responsibility. During contingency planning and removal, the OSC coordinates, directs, and reviews the work of other agencies, area committees, responsible parties, and contractors to assure compliance with the NCP, decision documents, consent decrees, administrative orders, and lead agency–approved plans applicable to the response.

Summary

Responding to hazardous materials and weapons of mass destruction incidents can be dangerous, and first responder safety should always be a top priority. A NIMS-compliant incident command system must always be implemented at hazmat and WMD incidents to comply with federal regulations (HAZWOPER). ICS is an effective management tool for ensuring essential tasks are completed while maintaining an effective span of control and personnel accountability. Unified incident command is often used at hazardous materials and WMD incidents owing to their complexity and multi-jurisdictional nature. Training and exercising with partner agencies helps ensure that essential functions can be safely and quickly accomplished during an emergency.

Review Questions

1. What are the 10 core features of any incident command system (ICS)?
2. Why is resource management important at hazardous materials incidents?
3. What are the members of the command staff?
4. What are the responsibilities of the safety officer?
5. What are the members of the general staff?
6. What are the functional components of the Hazmat Group?
7. How do the incident command post (ICP) and the emergency operations center (EOC) relate to each other?
8. What are the components of NIMS?

References

National Fire Protection Association. (2008). NFPA 472, *Standard for Competence of Responders to Hazardous Materials/Weapons of Mass Destruction Incidents.* Quincy, MA: Author.

Noll, Hildebrand and Yvorra. (2005) *Hazardous Materials Managing the Incident*, 3rd ed., Annapolis, Maryland: Red Hat Publishing.

Occupational Safety and Health Administration. (1990). 29 CFR 1910.120, *Hazardous Waste Operations and Emergency Response (HAZWOPER).* Washington, DC: U.S. Department of Labor.

U.S. Fire Administration/Technical Report Series, Fire and Explosions at Rocket Fuel Plant in Henderson, Nevada; USFA-TR-021/May 1988.

U.S. Fire Administration/Technical Report Series, Massive Leak of Liquefied Chlorine Gas in Henderson, Nevada; USFA-TR-052/May 1991.

U.S. Department of Homeland Security. (2004, March). *National Incident Management System.* http://www.fema.gov/emergency/nims.

CHAPTER

14

Response Organization

Courtesy of Chris Weber, Dr. Hazmat, Inc.

KEY TERMS

Knox box, *p. 458*
postincident analysis (PIA), *p. 476*
safety briefing, *p. 475*
secondary device, *p. 459*
site safety plan (SSP), *p. 471*

OBJECTIVES

After reading this chapter, the student should be able to:

- Institute scene control measures at a hazardous materials or WMD incident.
- Formulate response objectives at a hazardous materials or WMD incident.
- Explain the differences between strategies and tactics.
- Describe the difference between life safety, incident stabilization, and property and environmental conservation objectives.
- Explain the importance of pre-entry medical monitoring.
- List the essential components of a safety briefing.
- Describe the required documentation for hazardous materials incidents.
- Explain proper incident termination procedures.

ResourceCentral For additional review and practice tests, visit **www.bradybooks.com** and click on Resource Central to access book-specific resources for this text! To access Resource Central, follow directions on the Student Access Card provided with this text. If there is no card, go to **www.bradybooks.com** and follow the Resource Central link to Buy Access from there.

Your hazmat team is on the scene of a rolled-over gasoline tanker. The tank is leaking a moderate amount of gasoline from the manway covers. Based on your previous training you know that the apartment complex 100 feet away will need to be evacuated, the air will need to be monitored for flammable atmospheres, the leaks will need to be stopped with a combination of catch basins and dome clamps, the gasoline will need to be off-loaded using stingers, and then the tanker will need to be righted. You are however wondering where to start and what preparations need to be made and asking yourself the following questions:

- How do we make sure all essential tasks are performed?
- How do we make sure all hazards are accounted for and the scene is safe for first responders and the public?
- Have I thought of all response objectives?
- In what order should the response objectives be carried out?
- Do we need a written site safety plan?
- What information should the site safety plan contain?
- What types of scene control procedures should be implemented?
- What information does the entry team need to know before they enter the hot zone?
- How do we legally turn the scene over to the cleanup contractor?

Let's see if we can get some answers to these questions and more.

All emergency incidents require an organized, consistent approach when responding. The complexity of hazardous materials incidents, and the technical expertise needed to mitigate them safely, makes this an even more critical requirement. The first step is to approach the scene safely—including obtaining accurate dispatch information and approaching the scene from upwind, uphill, and upstream.

Initial Information

Most agencies classify the severity of hazardous materials incidents into different levels during dispatch that require different response capabilities, including varied equipment, diverse personnel training levels, and various additional support agencies. Classifying hazardous materials incidents in such a way helps organize and expedite the response. Dispatchers should be trained to ask the correct questions when taking 911 calls, including the following:

- What is the nature of the emergency?
- What is the location of the emergency?
- What is the best way to access the emergency?
- Is anybody hurt?
- Is anyone contaminated?
- When did the incident start?
- What type of occupancy or vehicle is involved?
- What type of container is involved?
- Is the container damaged?
- What chemicals are involved? Spell out the chemical name to responding units.
- What quantity of chemical is involved?
- Has an explosion or fire started?
- Is the release ongoing or has it stopped?
- Is runoff entering the sanitary sewer system, storm sewer system, or any other body of water?

- Is the incident escalating, or has it stabilized?
- Is a knowledgeable individual available? Have him or her meet the first arriving units in a safe location (upwind, uphill, upstream, at a safe isolation distance).
- Determine current weather conditions and any changes that are forecast.

Based on this information, the scope of the hazardous materials incident will be classified as follows:

Level 1: Incidents that can typically be handled by one or two responding units
Level 2: Incidents that can typically be handled by a trained local hazmat team
Level 3: Incidents that require regional, state, or federal assets to mitigate

NIMS has a five-tiered classification system based on the scope and complexity of the incident. The following is a brief summary of the five categories:

Type 5: Incidents that can be handled by one or two responding units
Type 4: Incident that requires multiple resources to safely mitigate
Type 3: Incidents that will likely extend into multiple operational periods
Type 2: Incident that is beyond the capabilities of local resources
Type 1: Incident that requires national resources to safely manage

Whichever classification system your agency uses should inform responders of the type of incident they will face and the resources they will likely need. It is helpful to have memoranda of understanding and/or automatic mutual-aid agreements in place with other response agencies such as law enforcement, fire, EMS, and neighboring jurisdictions.

Once dispatched, first responders should approach hazardous materials incidents from upwind, uphill, and upstream whenever possible (Figure 14-1). This will minimize the chances of inadvertent exposure to gases, vapors, and/or hazardous runoff.

Systematic Incident Management

Using a system or process to manage hazmat and WMD incidents will make the response much safer and less stressful to all involved, especially the incident commander. Most ICs, fire company officers, paramedics, and law enforcement officers have their own system in place to handle routine calls. Employing a familiar system during responses ensures that key steps are not missed and that preventable accidents don't happen. The same is true for hazmat incidents, which, fortunately, for most first responders are rare. Mnemonics can therefore be very helpful.

A very basic awareness-level mnemonic is RAIN: recognize, avoid, isolate, and notify. At obvious hazmat emergencies, these steps should already have been performed by plant personnel trained to the awareness level, or by the first arriving law enforcement officers, or at the latest by operations level trained firefighters. However, often, the dispatch is not specifically to a hazmat incident but, rather, to a domestic disturbance that reveals an illicit lab in the kitchen, or an unresponsive patient in a vehicle who has committed suicide using hydrogen sulfide, or a garage fire that turns out to be a mini chemical storage warehouse. You always need to stay on your toes and recognize

FIGURE 14-1 Approach a hazardous materials incident from upwind, uphill, and upstream whenever possible. *Art by David Heskett.*

BOX 14-1 HAZMAT HISTORY: EXPLOSIVES STORAGE MAGAZINE AT KANSAS CITY, MISSOURI, CONSTRUCTION SITE (1988)

In November 1988 six firefighters were killed instantly when they responded to a reported fire at a highway construction site. Unbeknownst to the firefighters, the fire—which was later shown to be arson—was consuming an explosive storage magazine. The magazine contained Maynes mix, a blasting agent consisting of ammonium nitrate, diesel fuel, and powdered aluminum. Eventually, two magazines exploded, leaving one crater 80 feet in diameter and 8 feet deep and another 20 feet in diameter and 6 feet deep. The fire department was not aware of any explosive storage at the construction site, since it was not involved in the blasting permit process. The dispatcher was made aware that explosives were stored somewhere on the site by two security guards who called 911, and relayed this information to responding engine companies. The explosive storage magazines were not required to be placarded or to display signage at the time.

This incident illustrates the importance of pre-incident planning for hazardous materials, good communications with other agencies such as building departments and permitting departments, and comprehensive hazardous materials training that stresses hazmat recognition and the dangers of explosives.

hazardous materials incidents early, before you become too committed or, worse yet, contaminated, exposed, or injured. Even for a hazardous materials technician it is always prudent to verify that these basic steps have been completed.

There are several mnemonics that can be helpful while responding at the hazardous materials technician level. There is the DECIDE process developed by Ludwig Benner in 1970s; the Eight Step Process© developed by Greg Noll, Mike Hildebrand, and Jim Yvorra; and the AFIRMED process developed by the author. These acronyms are designed to jog your memory to perform critical steps at hazmat and WMD incidents.

Ludwig Benner was a renowned accident investigator with the NTSB for many years. He formulated the DECIDE process after noticing patterns in the accidents and the emergency responses he investigated. The DECIDE process consists of the following steps:

Detect: Detect the presence of hazardous materials.
Estimate: Estimate the likely harm that would occur without intervention.
Choose: Choose your response objectives.
Identify: Identify the action options with available resources and training.
Do: Do the best option.
Evaluate: Evaluate your progress.

The DECIDE process stresses recognition, analysis of the problem in conjunction with the available resources, implementation of the appropriate actions, and a continuous evaluation of the effectiveness of your actions. The strength of this system is that it guides your decision-making process toward your ability to improve the outcome of the incident. There is no point in acting when your actions will make the incident worse, endanger yourself, other first responders, or the public. The 1970s were especially tragic with respect to emergency response to flammable liquefied compressed gases (see Box 5-5 for details).

The Eight Step Process© is a management process that addresses the tactical and strategic goals at hazardous materials incidents. It is designed to address the important aspects of hazardous materials response. The Eight Step Process consists of the following:

Step 1: Site management and control
Step 2: Identification of the problem
Step 3: Hazard and risk assessment
Step 4: Selection of the proper level of PPE
Step 5: Coordination of information and resources

Step 6: Selection and implementation of response objectives
Step 7: Decontamination procedures
Step 8: Termination procedures

This process is designed for use at recognized hazardous materials incidents and is excellent for managing the response from beginning to end. The strength of this management process is the level of detail with which the actions specifically needed at a hazmat incident are addressed.

The AFIRMED process developed by the author consists of the following:

Assess: Assess the situation prior to committing resources.
Formulate: Formulate an IAP consistent with your level of training, the available resources, and the current situation.
Initiate: Initiate actions that will support the response objectives and IAP.
Reassess: Reassess the situation frequently to ensure a positive outcome.
Modify: Modify the incident action plan based on the reassessment.
Extended Ops: Continue with extended operations to mitigate the incident.
Demob: Demobilize all resources safely and with accountability.

This management process addresses the key incident command concepts needed at more complex incidents, including hazardous materials and terrorist incidents. The strength of this process lies in that specific incident tactical considerations are kept to a minimum while the overall incident management philosophy is stressed from start to finish in a NIMS-compliant way.

Choose one or more of these incident management processes to help guide your routine responses, and the "big one" won't seem quite as big after all. It is important to have a consistent and comprehensive response system in place that you routinely use when responding to all hazmat and WMD incidents.

Scene Control Procedures

The immediate area of the release should be isolated first. If some information is known about the hazardous material or the mode of transportation, the isolation distances can be obtained from the current edition of the *Emergency Response Guidebook*. If little information is known, the size of the hazard area must be estimated by considering the type of release, the size of the release, witness accounts, readouts from air-monitoring equipment, or usually a combination of all four. The general rule of thumb is to initially isolate the area 75 feet for solids, 150 feet for liquids, and 300 feet for gases at a minimum. Many chemicals are extremely toxic, and the initial isolation zone may have to be increased significantly based on other factors. Approach the area slowly from upwind, uphill, and upstream. Stop well away from the hazard and make a preliminary assessment using binoculars. Once determined, the hazard area is known as the exclusion zone or the hot zone. This zone may need to be increased, especially downwind, if the release cannot be stopped in a timely manner or is expected to get worse.

CONTROL ZONES

Ultimately, there will be three control zones: the cold zone, the warm zone, and the hot zone (Figure 14-2). Initially, the incident is divided into only two zones: the hot zone, which is the initial isolation zone, and the cold zone. The hot zone is any area that is currently contaminated or can reasonably be expected to be contaminated at any time before the incident is mitigated. Everyone should be evacuated from the hot zone and undergo decontamination; only personnel properly trained and equipped with PPE should be allowed to enter it. The use of personal protective equipment was covered in Chapter 7. People can be alerted to evacuate by announcements over public address systems, by activating alarms, and by searching room to room if necessary.

The uncontaminated area is called the cold zone, or support zone. A warm zone, or decontamination zone, is carved out of the cold zone on the upwind and uphill side of the release. Because the warm zone originates from the cold zone, it is initially not contaminated and maybe set up without the use of PPE. However, once entry team personnel or contaminated patients enter the warm zone, it is considered contaminated, and access points must be controlled. Emergency decontamination is also carried out in the warm zone.

FIGURE 14-2 The three control zones at hazardous materials and WMD incidents are the cold zone (green), the warm zone (yellow), and the hot zone (red). *Art by David Heskett.*

The control zone boundaries should be clearly marked (Figure 14-3). Using existing control features such as fences and roadways can simplify the marking of the hot zone boundary. When this is not feasible, barrier tape, cones, or temporary fencing can be used. It is very important that all emergency response personnel on the scene, including fire department, EMS, and law enforcement personnel, understand the location of the hot zone/cold zone boundary and what it means. It is equally important that all bystanders and the public understand how dangerous the hot zone is and where the hot zone perimeter is located. An outer perimeter should be enforced to keep the public and onlookers at a safe distance and prevent them from interfering with emergency operations. It is helpful to have law enforcement personnel on scene to manage crowd control. It is imperative to properly brief all personnel who are on scene—including law enforcement—of hazards.

EVACUATION VERSUS IN-PLACE SHELTERING

Evacuation is the removal of people from a defined area based on a hazard-risk assessment. In contrast, *in-place sheltering* refers to the protection of occupants at their current location. Evacuation is a time-consuming and resource-intensive process. Operations level personnel may need to be recruited to evacuate an area quickly. In the case of hazmat incidents, evacuating people from a building may make them more vulnerable to high-level exposure via a passing vapor cloud. In-place sheltering is often an attractive option when

FIGURE 14-3 Always clearly mark control zone boundaries. *Courtesy of Chris Weber, Dr. Hazmat, Inc.*

the hazardous materials release is outside the occupied building, or in large buildings, at a safe distance from the affected population.

Downwind Evacuation

The decision to evacuate or shelter in-place depends on the following factors:

Nature of the emergency: The vapor cloud may already have passed, making evacuation unnecessary, or occupants will have to pass through the vapor cloud, dramatically increasing their exposure.

Nature of the chemical: Evacuation is preferred in the case of flammable vapor clouds owing to the risk of explosion, although evacuating people through a flammable vapor cloud is even more dangerous. If occupants are sheltered in-place, advise them to stay away from windows and glass in case of explosion.

Resources: It may be impractical to completely evacuate the designated area. Targeted evacuation from the most dangerous area may be the most effective solution.

When the decision to evacuate has been made, ensure that occupants are evacuated into a safe atmosphere. The external atmosphere and evacuation route should be assessed using air-monitoring equipment. If it does become necessary to evacuate occupants through a vapor cloud, ensure that the rescuers and the occupants are properly protected before exiting the structure. Use radio, television, and reverse-911 systems to notify residents of what to do. If first responders are sent into the protective action zone (evacuation zone), ensure that they are properly trained and equipped with air-monitoring equipment and PPE (if necessary).

The downwind evacuation distance can be estimated using the green section of the *ERG*. Table 1 of the green section lists downwind isolation distances for large and small spills as well as for daytime and nighttime spills for all highlighted materials in the blue and yellow sections. Remember, consult the green section for highlighted chemicals only when the chemical is *not* on fire. If the product is on fire, most of the vapors are being consumed and will most likely not pose a significant downwind exposure hazard. The values found in Table 1 can be applied in the following way:

1. Identify the chemical by name or UN/NA identification number. If it is highlighted and not on fire, locate the appropriate entry in Table 1 using the identification number. Numbers are listed in numerical order.
2. Determine if the spill is large or small. A small spill is from one or more containers that have an aggregate capacity of approximately 55 gallons (200 L) or less. A small leak from a large container may be classified as a small spill under appropriate conditions. However, keep in mind that if the leak is getting worse, or can reasonably be expected to get worse, estimate accordingly. All other spills are classified as large.
3. Isolate the area in all directions using the appropriate isolation distance listed in Table 1 of the green section of the *ERG* (for a small or large spill)
4. Determine the appropriate downwind evacuation distance, depending on the time of day. Daytime evacuation distances are shorter than the nighttime evacuation distances. Daytime is classified as the period between sunrise and sunset. During the day, atmospheric mixing from sun-induced wind currents disperses the vapor cloud and dilutes the airborne chemical vapor concentration. At night, more stable atmospheres allow concentrated vapor clouds to drift farther downwind. If the incident started at night, or is likely to continue into the night, use the nighttime protective action distances.
5. The protective action zone extends downwind from the spill the distance indicated in the appropriate column (depending on spill size and time of day), and extends half the distance to either side. This creates a square whose left side is centered at the release point, and extends downwind (see Figure 14-4). Start the evacuation as

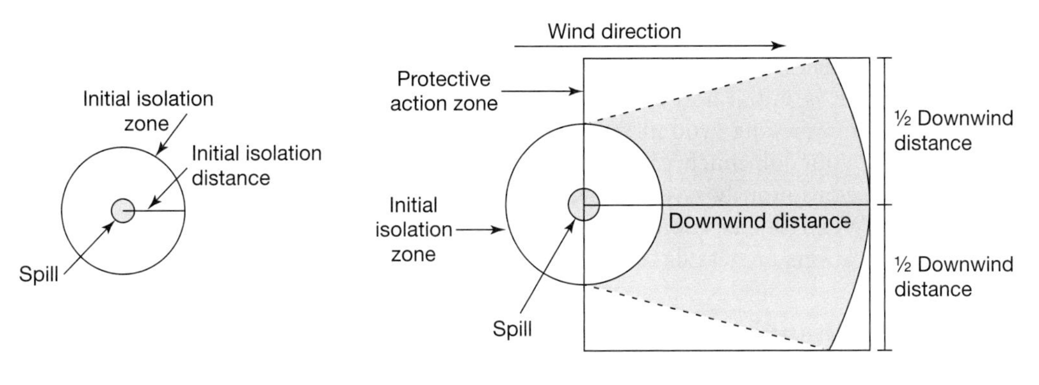

FIGURE 14-4 Determination of the downwind evacuation distance according to the DOT *Emergency Response Guidebook. Courtesy of the U.S. Department of Transportation Pipeline and Hazardous Materials Safety Administration.*

close to the spill site as safely possible, and continue downwind as far as resources permit. Call in additional resources if evacuation or in-place sheltering cannot be carried out in a timely fashion.

This method is also described on pages 290–291 of the 2012 ERG.

Sheltering In-Place

In-place sheltering, or sheltering in-place, should be considered when a gas or vapor cloud will move through the area relatively quickly and is not expected to linger. This is especially true when it is not feasible to evacuate owing to a lack of resources or time. Sheltering in-place may *not* be appropriate in the following situations:

- Vapors are flammable.
- A lingering vapor cloud will infiltrate the building.
- Buildings cannot be tightly sealed.
- Explosives are involved.

In-place sheltering is begun by notifying the occupants of the building to close all their windows and doors and turn off their air-handling (HVAC) system for a specified amount of time. The specified amount of time should be long enough to allow the vapor cloud to safely pass the sheltered-in-place occupancy. Contact should be maintained with the occupants if at all possible, keeping them updated of changing conditions and when the in-place sheltering order is lifted. Air monitoring should be conducted when resources permit.

In-place sheltering is often used when occupancies are difficult or dangerous to evacuate, such as hospitals and nursing homes. When a large number of occupancies need to be notified, reverse-911 systems or call centers can be very helpful. As a last resort, responders may enter the area to notify occupants in person or use vehicle-mounted public address systems. However, this may put those responders at risk, and they should wear the appropriate PPE.

Site Characterization and Scene Size-Up

As discussed in the last chapter, HAZWOPER requires that the incident command system be used at hazmat incidents. Therefore, the first step after arrival on scene is to establish incident command. After command is established, the scene must be sized-up and the site characterized to determine the extent of the hazards and the risk to responders and the public.

Site characterization, also known as *scene assessment,* is an important component of a safe and effective response to a hazmat incident. The size-up is an evaluation of critical incident factors that will help determine incident objectives, strategies, and tactics. Site characterization should be a continuous process through the duration of the incident, since many aspects of hazmat incidents change over time. If these subtle changes are not noticed and addressed in a timely fashion, the incident may abruptly change for the worse.

One of the positive consequences of an accurate size-up is the prevention or early detection of potentially dangerous site condition changes. A very important part of site characterization is initial and continuous air monitoring of the hot zone coupled with perimeter air monitoring around the exclusion zone. This will prevent injuries and save lives, making your job much easier. All personnel at hazardous materials or WMD releases should continuously size up the incident and immediately report key observations to superiors. When all hazardous materials personnel take responsibility for scene size-up, critical observations are less likely to be missed.

Site Information Acquisition and Management

After arriving at the hazardous materials or WMD incident and establishing command or reporting to the established incident command post, you should gather information as rapidly as possible. Such information may include the following:

- types of containers involved
- identification markings on containers
- quantity and/or capacity of containers
- type of hazardous materials involved
- type of release
- size of the release
- movement of the release
- exposures
- weather
- stage of the hazardous materials incident
- available resources

Knox box ■ A storage device for building keys, maps, and facility emergency plans to which emergency responders within a jurisdiction have access through a common key.

FIGURE 14-5 A Knox box can be used to store building keys and maps for rapid emergency responder access. Local emergency responders have a universal key to all Knox boxes in their jurisdiction. *Courtesy of Chris Weber, Dr. Hazmat, Inc.*

If operations level responders are already on scene, they should have gathered this information already and should be ready to brief you on your arrival. If not, attempt to determine this information from a distance, such as using binoculars, and make an initial site assessment.

Information may also be obtained from facility plans located at your station or the facility emergency response plan, which may be located in a **Knox box** at the facility, or can be obtained from the responsible party or the local emergency management agency. Knox boxes are small to relatively large wall-mounted safes that contain building keys, plans, and maps (Figure 14-5). They allow the fire department, emergency medical services, and law enforcement to gain access to the building using a jurisdiction-wide universal key. This limits the amount of forcible entry required during emergencies. Additional information can be obtained from shipping papers, witnesses, drivers, or workers at the facility. Once the product has been identified, technical experts such as CHEMTREC or reference materials—such as those we discussed in Chapter 6—can be consulted to obtain detailed information. The chemical and physical properties gleaned from the reference materials will allow you to understand how the product will behave, to estimate the harm it may cause, and to predict where it might go and what it might do.

It is important to form an accurate picture of the nature of the emergency to be able to set optimal incident objectives. Witnesses, building occupants, employees, and equipment operators should be interviewed as quickly as possible. The following are key questions to ask:

Are there any victims?
Is anyone contaminated?
Has a release occurred?
Where is the product going?
Is the situation getting worse?
Could a BLEVE or a catastrophic failure occur?
Do I have the appropriate and necessary resources on scene?

In the case of terrorist or criminal incidents, you should consider additional hazards, such as **secondary devices**, armed resistance and use of weapons, booby traps, as well as secondary contamination from handling victims and patients.

secondary device ▪ An explosive device designed to injure first responders and people evacuating the site of a previous attack.

You should also note the surrounding conditions, such as the following:

- topography
- land use
- accessibility
- weather conditions
- bodies of water
- public exposure potential
- discarded drums or containers
- dead vegetation or animals
- unusual odors
- overhead and underground wires and pipelines
- storm sewer drains
- possible ignition sources
- adjacent land usage such as rail lines, highways, and airports
- nature and extent of injuries
- building information

On-site weather conditions should be monitored using a weather station to obtain the most accurate and up-to-date information. An advantage of using a portable on-site weather station is that it can be interfaced with plume modeling software that allows you to generate real-time chemical plume predictions. If an on-site weather station is not available, contacting the closest airport is the next best option, since it will have accurate weather stations. You should choose the airport closest to the release site for the most accurate results.

As soon as feasible, you should verify the information you have received. Witnesses can be mistaken, reference materials have misprints, and information and numbers can be transcribed incorrectly. Information can be verified by using multiple reference sources and interviewing multiple witnesses to corroborate the information. It is advisable to use at least three independent sources and use the most conservative information when the data conflict. If left uncorrected, such errors and misconceptions can lead to catastrophic results.

PRELIMINARY SITE EVALUATION

A preliminary site evaluation can often be performed without even nearing the release area. For example, pre-incident plans can be used to determine the types of hazardous materials on site, witnesses and employees who have evacuated the site can be interviewed to determine the nature of the problem, and even news helicopter footage can be reviewed

BOX 14-2 HAZMAT HISTORY: DOMESTIC TERRORISM

CENTENNIAL PARK, ATLANTA (1996)

On July 27, 1996 Eric Robert Rudolph detonated a bomb in Centennial Olympic Park in Atlanta, Georgia during the 1996 Olympics. Two people were killed and 111 people were injured. The explosive device consisted of a 40 pound pipe bomb, a directional device consisting of a steel plate, and nails used as shrapnel in a knapsack. The explosive material consisted of dynamite.

The number of casualties was greatly reduced by an observant security guard named Richard Jewell. Before the satchel charge could detonate, he along with another security guard, evacuated people from the immediate area and called 911. It is important to regularly screen the building, surroundings, and attendees at events that could be terrorist targets. The personnel assigned to these tasks should have experience with that facility or area, and be able to tell what is normal and what is not. Personnel should ideally be assigned to that role from start to finish, including set up and during the event itself. This way they will have a baseline for comparison.

ABORTION CLINIC BOMBINGS, ATLANTA (1997–1998)

On January 16, 1997 Eric Robert Rudolph targeted an abortion clinic in Sandy Springs which is a suburb of Atlanta, Georgia. Then, on January 29, 1998 he targeted an abortion clinic in Birmingham, Alabama which killed a police officer and critically injured a nurse. Similarly to the Centennial Park bombing these bombs were constructed of dynamite.

The disturbing aspect of the Sandy Springs bombing was that more than an hour and a half into the response, a secondary device targeting first responders was detonated and injured seven first responders. The device was placed next to a dumpster which was intended to act as a directional device. Luckily a fire truck was parked between the secondary device and the first responders which helped to minimize injuries. This was the first time in decades that a secondary device was used to specifically target first responders. It is important to thoroughly search the area for secondary devices when responding to suspected terrorist incidents. It is also good practice to vary your routine responses to potential terrorist targets. False alarms at these locations may be used to gather intelligence on your agency's response protocols.

OTHERSIDE LOUNGE BOMBING, ATLANTA (1997)

On February 21, 1997 Eric Robert Rudolph targeted a gay bar in Atlanta, Georgia and injured five people. Similarly to the Centennial Park bombing the bomb was constructed of dynamite. In this case – as with the abortion clinic bombing, a secondary device was found in some shrubbery next to the parking lot, but fortunately the bomb squad was able to render it safe. Eric Robert Rudolph was motivated by the Christian Identity movement and strong anti-gay and antiabortion viewpoints.

to gain a better understanding of what is happening on site before personnel are committed to a reconnaissance mission. The preliminary evaluation is important for determining the appropriate level and type of PPE, the air-monitoring equipment to be used, the entry objectives, and the part of the site or facility to be investigated first. This evaluation minimizes the risk to personnel entering the hot zone.

Information can also be obtained from a distance using various methods and equipment such as a temperature gun or a TIC to determine the product level in containers (Figure 14-6). Assuming a worst-case scenario that a container was full, the amount of spilled product can be estimated by subtracting the volume of liquid remaining in the container from the maximum container capacity. Air-monitoring equipment should also be used to determine the extent of gas and vapor cloud migration (see Chapter 8).

RISK ASSESSMENT

You should be able to use all this information to accurately and realistically assess the hazards and risks of any given hazardous materials incident. To make a risk assessment you must evaluate the hazards—which are constant, and the risks—which vary and

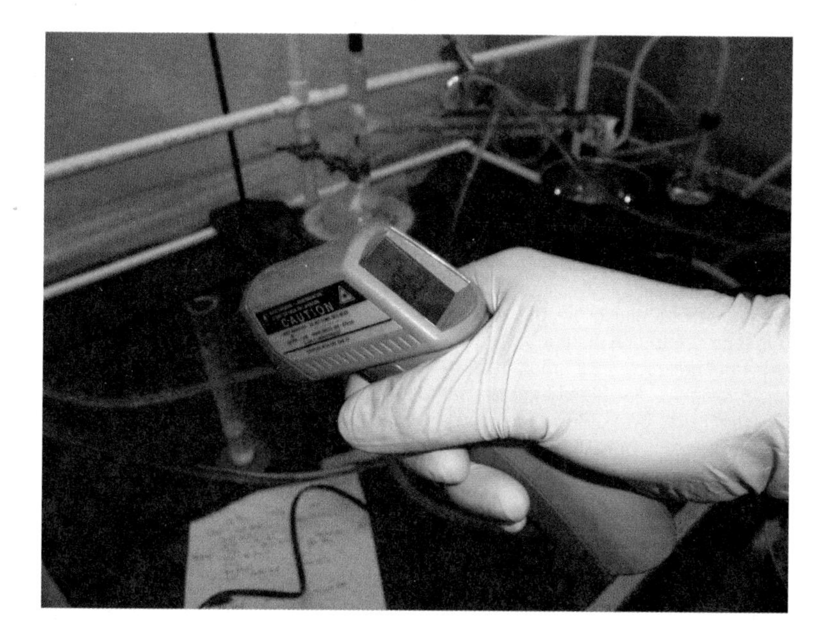

FIGURE 14-6 A temperature gun or TIC can be used to determine whether the container is warmer or colder than the ambient temperature. A colder temperature may indicate that a pressurized container is leaking, and a warmer temperature may signal that a chemical reaction is occurring.
Courtesy of Chris Weber, Dr. Hazmat, Inc.

BOX 14-3 TACTICS: ILLICIT LABORATORY RESPONSE

Illicit, or clandestine, laboratories can be some of the most dangerous hazmat incidents to which you may respond. By definition, these are not only hazardous materials incidents but also crime scenes. Consequently, you must be extremely careful not only because of the presence of hazardous atmospheres, chemicals, and equipment but also because of the potential presence of booby traps, secondary devices, and armed and dangerous suspects. Therefore, responses to clandestine laboratories are typically multiagency coordinated responses.

RECOGNITION AND IDENTIFICATION

There are many types of illicit laboratories that may produce many different final products. Recognition of the type of laboratory is essential for your own safety, the safety of the public, notification of the appropriate AHJ, and successful evidence collection and ultimate prosecution. Currently, the most

Courtesy of Chris Weber, Dr. Hazmat, Inc.

(continued)

common types of laboratories encountered are clandestine drug laboratories (clan labs), especially methamphetamine-producing labs.

Clandestine labs may be stationary or mobile. Stationary illicit laboratories may be found just about anywhere: hotel and motel rooms, homes, storage units, apartments, rural areas, and urban areas. Mobile illicit laboratories may be found in cars, trucks, vans, RVs, campers, and boats. Because of the illicit nature of the clandestine laboratory operation, most lab operators will try to find a relatively secluded area for their "business."

A wide range of chemicals and other hazardous materials may be present at illicit laboratories. Each of these chemicals plays a vital role in generating the final product. The most important chemicals are the *precursors*, which are the raw materials that will be converted into the final product. Without the appropriate precursor in a sufficient quantity, the clandestine laboratory will not be able to produce a final product. A common precursor is pseudoephedrine, used in the synthesis of methamphetamine. *Reagents* are the chemicals that facilitate the conversion of the precursor to the final product. Reagents may or may not become part of the finished product. An example of a reagent in the synthesis of methamphetamine is lithium metal. Some reagents are *catalysts*, which accelerate the rate of a chemical reaction but do not become part of the finished product. Catalysts are required in small amounts, since they are not used up during the chemical reaction. An example of a catalyst in the synthesis of methamphetamine is red phosphorus. *Solvents* are chemicals used to dissolve the reagents and precursors so that the chemical reaction can occur. Solvents are used in large amounts and are usually the cause of fires and explosions in clan labs, so it is essential to use a CGI during any illicit laboratory reconnaissance, during the chemical and hazard assessment phase, while gathering evidence, and during the remediation phase. An example of a solvent in the synthesis of methamphetamine is diethyl ether.

Illicit laboratories may contain much exotic glassware and equipment, or they may contain a number of normal items hooked up into funny looking contraptions. All this equipment is designed to facilitate one or more chemical reactions. Exotic glassware and equipment is commercially designed for use in industrial laboratories or research laboratories, and may be small- or large-scale depending on the particular use. The contraptions constructed of normal household items are improvised laboratory glassware and equipment, which are often used in mom-and-pop labs.

Methamphetamine Drug Labs

Methamphetamine is an extremely addictive stimulant that can easily be produced from either ephedrine or pseudoephedrine in one chemical step—a reduction reaction. This is one reason there are thousands of small mom-and-pop meth labs all across the country. There are several different synthetic routes for generating methamphetamine from the precursor. Recipes typically contain more than one step because of additional separation and purification processes.

Each meth lab typically starts with a tablet or liquid extraction step using a solvent and/or acids and bases (depending on the precursor source). You may find coffee grinders, mixers, or mortars and pestles to grind up the tablets. You may find hot plates, a stove, or heating mantles to evaporate the solvent. A filtering system is used to remove the undissolved solids. Filters may be as simple as a coffee filter and funnel, or more complex glassware such as a vacuum filtration system. The meth oil is produced according to different processes and then dissolved in a readily available solvent such as Coleman fuel or ether. Methamphetamine is subsequently *salted out* of solution using an HCl gas

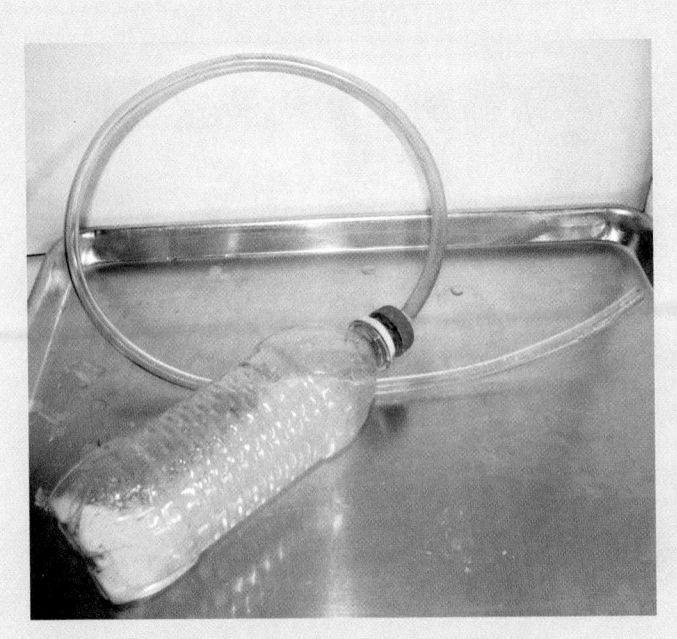

Courtesy of Chris Weber, Dr. Hazmat, Inc.

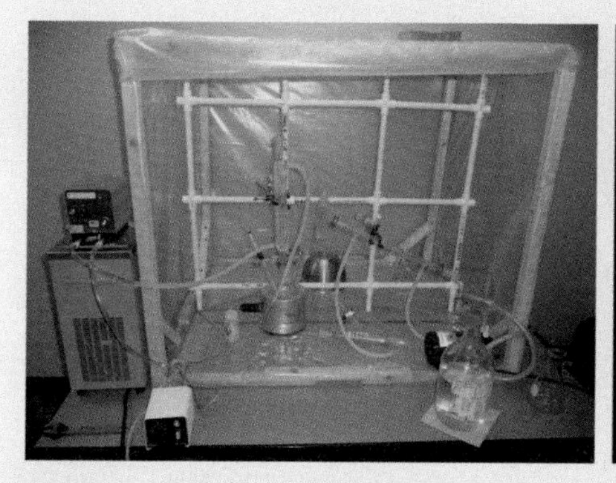

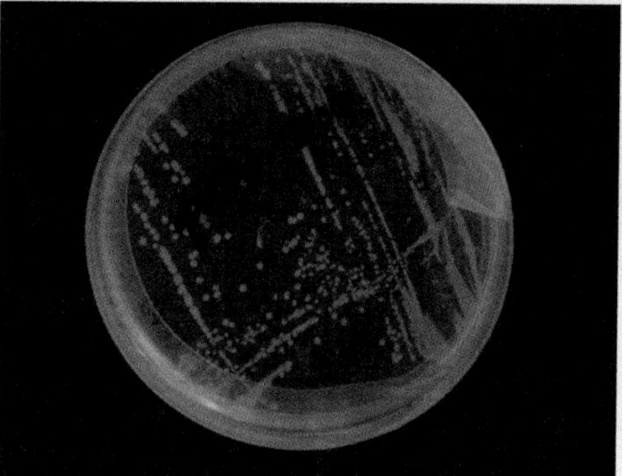

Courtesy of Chris Weber, Dr. Hazmat, Inc.

generator to convert the meth oil to the meth salt methamphetamine hydrochloride. Common constituents of HCl generators are sulfuric acid, muriatic acid, rock salt, or aluminum foil. If you find a gas can with a long hose attached, be especially careful, as highly corrosive HCl gas may be emitted unexpectedly. Even if the HCl generator does not look active, moving the container may mix the reagents inside and restart the HCl generation process. Since the HCl generator is often reused, the tip of the hose is an especially good place to find residual methamphetamine that can be used as evidence, even if the bulk of the methamphetamine that was produced has already been removed from that location. After salting out, the solid methamphetamine final product is then filtered and possibly washed with clean solvent.

Explosives Labs

Explosives are the most commonly used WMDs in the United States and worldwide. Traditionally, commercially available and easily produced explosives such as ANFO have been used. This is the explosive that was used in the Alfred P. Murrah Federal Building bombing in Oklahoma City in 1995. Recently, terrorists have been producing *organic peroxide*–based explosives such as TATP and HMTD. The former is an explosive commonly used by al-Qaeda operatives. For security reasons, we are omitting details about the chemicals and the processes. If you suspect you have encountered one of these laboratories, it is imperative that you exit the laboratory immediately, notify the FBI and your local bomb squad, and consult with technical experts as soon as possible.

Explosives laboratories also will contain typical chemical laboratory glassware and improvised glassware. There is a wide variety of chemical synthesis routes for the common explosives, including ANFO, TNT, RDX, PETN, urea nitrate, TATP, and HMTD. The unique hazards you will encounter are the extremely sensitive nature of the final product, strong oxidizers, corrosives, and flammable liquids. The following are common names associated with oxidizers you may find in an explosives laboratory:

- peroxide (such as hydrogen peroxide or methyl ethyl ketone peroxide)
- nitrate (such as nitric acid, ammonium nitrate, and potassium nitrate, or saltpeter)
- perchlorate (such as ammonium perchlorate)
- chlorate (such as sodium chlorate)
- hypochlorite (such as calcium hypochlorite, or bleach)
- permanganate (such as potassium permanganate)
- chromate (such as sodium chromate and chromic acid)
- dichromate (such as potassium dichromate)
- iodate (such as lead iodate)
- periodate (such as potassium periodate)

Powdered metals such as aluminum and magnesium may also be used as fuels in an improvised explosive.

Chemical Warfare Agent Labs

CWA labs, although exceedingly uncommon, can be deadly owing to the extreme toxicity of the final product being produced. Therefore you should have a basic understanding of the types of chemical warfare agents and their hazards. Again, for security reasons, we are omitting details about the chemicals and the processes. If you suspect you have encountered one of these laboratories, it is imperative that you exit the laboratory immediately, notify the FBI, and consult with technical experts as soon as possible. The state WMD–Civil Support Teams are very well trained hazardous materials responders and are technical experts in CWAs. They are a vital resource to have on site when confronted with this type of laboratory.

(continued)

Biological Warfare Agent Labs

Biological warfare agent laboratories, although uncommon, are easy to set up and operate. Anyone who has brewed beer at home has set up a biological laboratory, namely, growing the yeast that produces the alcohol in the beer. Furthermore, biological laboratories and the associated reagents and equipment are extremely common in industry and academia, owing to the commercial and scientific usefulness of advanced molecular biological techniques. Biological warfare agent laboratories may be set up using commercially and industrially available reagents and equipment, or relatively simple home-made or home-brew type of equipment. Generally, the most difficult part of setting up a biological warfare agent laboratory is obtaining a starter culture of the target organism.

Three primary types of biological agents are commonly produced: *bacteria, viruses,* and *toxins.* Each of these different production methods uses different reagents and equipment, which allows them to be distinguished from one another with a little practice and experience.

Biological laboratories use much unique glassware and equipment designed to keep the bacteria and viruses alive and thriving. Arguably, the most important of this equipment and glassware are the *incubators.* Incubators come in many different sizes and shapes depending on the type of organism they are designed to culture, and the amount of product they are designed to produce. Small-scale incubators can be as simple as an *Erlenmeyer flask* containing *culture media* in a temperature-controlled orbital shaker. Large-scale incubators are fancy contraptions with many different inlets and outlets that permit the addition of essential gases such as oxygen and carbon dioxide, the addition of culture media and other nutrients, sampling, and harvesting. The type of culture media helps distinguish the different classes and types of organisms being grown.

RESPONSE

An illicit laboratory may be discovered in one of three distinct phases:

- operational (active)
- nonoperational (inactive)
- boxed (inactive)

Operational labs are fully functional and actively producing materials, nonoperational labs are fully or partially set up but not active (pre-or postproduction), and boxed labs are in storage. As you can imagine, the most dangerous phase in which to find an illicit lab is the functional and actively producing

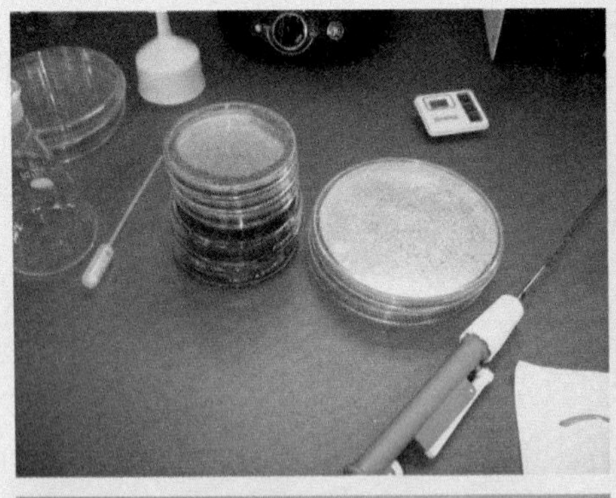

Courtesy of Chris Weber, Dr. Hazmat, Inc.

phase. During this phase all the necessary chemicals, reagents, solvents, precursors, and products are present, and most of the equipment is energized and poses significant hazards such as a potential ignition source, electrocution hazard, thermal hazard, or explosion hazard. Actively producing laboratories must be treated with the utmost care! Inactive labs are generally the least dangerous type of lab to encounter. However, caution is required around the inactive labs, even the boxed ones, since dangerous chemical residue may remain on the glassware and equipment.

During a response to any illicit laboratory the following must be accomplished:

1. Coordinate crime scene operations with law enforcement agencies.
2. Ensure that the law enforcement agency secures and preserves the crime scene.
3. Develop a site safety plan.
4. Consider performing the first entry and reconnaissance jointly with SWAT, EOD, and hazardous materials teams to avoid booby traps, secondary devices, and hazardous atmospheres.
5. Continuously monitor the air during the operation.
6. Have a decontamination plan in place that can effectively and safely deal with all the deployed assets, including SWAT, EOD, and K-9 units.
7. Mitigate the immediate hazards while preserving evidence.
8. Coordinate post–crime scene remediation operations.
9. Thoroughly document all completed operations.

Before the illicit laboratory is entered, a thorough hazard assessment should be made based on available information such as dispatch information, interviews of witnesses and neighbors, a visual overview of the exterior, and thermal imaging data from the exterior. Always be aware of the potential for booby traps inside the building itself as well as the exterior yard and grounds.

depend on the situation. Examples of hazards include the chemical and physical properties of a material such as anhydrous ammonia, the type of container in which a material is stored such as a gas cylinder, and weather conditions such as high heat and humidity. Weather considerations must be addressed, especially when long-term hazmat incidents are involved. A damaged pressure container that may be stable overnight when the temperature is 20°F will likely not be stable when the sun rises and temperatures reach 70°F (see Box 5-5, which describes the Waverly, Tennessee, train derailment).

However, the risk that any material poses varies greatly from one incident to another. For example, the risk from a properly packaged and stored 1-ton cylinder of chlorine on a 95° day is minimal. You can walk around that cylinder all day long without PPE with virtually zero risk. However, if the same 1-ton cylinder of chlorine falls off the flatbed truck carrying it and cracks, releasing its contents on a 95° day, it presents significant risk to responders and the public. Now, this hazard must be managed.

DETERMINING RESPONSE OBJECTIVES

After all necessary information has been gathered and analyzed—that is, a risk assessment has been made, the next step is to determine the incident response objectives. At most emergency incidents responders have limited resources, at least initially. It is therefore extremely important to correctly prioritize incident objectives. Lives will depend on it!

The following are some key questions to ask when determining the response objectives:

- How many victims can be saved with the available resources?
- Who is in most need of intervention?
- Is there a potential for secondary devices or a secondary attack?
- How big is the leak?
- Can we control the leak safely and effectively with the available resources?
- How are business and commerce being affected?
- What type of environmental damage is occurring?
- How will this affect the community in the near and the long term?
- Is nonintervention the safest and least damaging option?

FIGURE 14-7 Life safety—such as rescuing victims—should be the first priority in any hazardous materials incident. *Courtesy of Chris Weber, Dr. Hazmat, Inc.*

Asking these questions is part of a risk-versus-benefit analysis and is vital to making sound strategic and tactical decisions at a hazmat incident. It is very important to examine the risks associated with all actions, as well as the benefits to be gained from them. If the risks outweigh the benefits, a different response objective should be chosen.

Response objectives to any incident, including hazmat and WMD incidents, should be prioritized as follows:

1. Life safety
2. Incident stabilization
3. Property and environmental conservation
4. Social normalcy

Life Safety

Life safety refers to first-responder safety as well as the safety of the public. Life safety objectives in hazmat incidents include isolating the immediate area, controlling the scene, removing nonambulatory victims from the hot zone, prioritizing emergency medical care of the victims (triage), evacuating downwind, and sheltering in place (Figure 14-7). If you determine that there are objectives that you do not have the resources or training to meet, it's very important that you notify the appropriate agencies that do. For example, most hazmat teams will not be able to safely address life safety in a hostage standoff at a chemical facility. This will require close cooperation and mutual support from a SWAT team.

It is also important to remember there is a difference between affected populations and endangered populations. Endangered people are in close proximity to IDLH levels of the hazardous material or are in the predicted path of a chemical plume—they are at imminent risk of exposure. Affected people are in the vicinity of the hazardous materials release but are not yet at imminent risk of exposure to health-endangering levels of the hazardous material. Prioritizing life safety must take this difference into account and ensure that

BOX 14-4 TACTICS: VICTIM RESCUE

The feasibility of performing victim rescue and recovery operations should be analyzed at the pre-incident planning stage as well as at the start of any hazardous materials incident. In determining the feasibility of victim rescue and recovery, the benefit to the victim must be balanced against the risk to the responders. For example, if there will be little risk to the responders performing victim rescue, if the patient outcome will likely be positive, and if the appropriate PPE and rescue equipment are available, a victim rescue should almost certainly be attempted. However, if the risk to the responders is high, and the patient outcome is highly doubtful, a victim rescue should almost certainly not be attempted. These decisions are highly subjective and are based largely on agency preparation, the IC's experience and training, and evaluation of tactical considerations of the hazmat incident.

VICTIM CONSIDERATIONS

One of the primary incident response considerations should be the victim: What is the victim's condition? What is his or her location? The answers to these two questions, along with the available agency resources, will determine the feasibility of a victim rescue operation. There are four victim possibilities:

- The victim is *ambulatory* and you have line-of-sight contact with him.
- The victim is *nonambulatory* but you have line-of-sight contact with her.
- The victim is ambulatory but you do not have line-of-sight contact with her.
- The victim is nonambulatory and you do not have line-of-sight contact with him.

When other conditions are factored in, these facts will determine whether the operation will be a rescue or a recovery. The difference is, rescue operations are time-sensitive and must be performed as rapidly and safely as possible, whereas recoveries of bodies are typically not time-critical operations. For rescue operations some risk is incurred based on the perceived benefit to the victim. For body recoveries no appreciable risk should be incurred.

When patients are ambulatory, the benefits of rescue are high. Depending on how long the victim has been in the chemical environment, something is known about the danger of the hazardous material. The higher the exposure level and the longer the patient remains viable, the less danger that hazardous

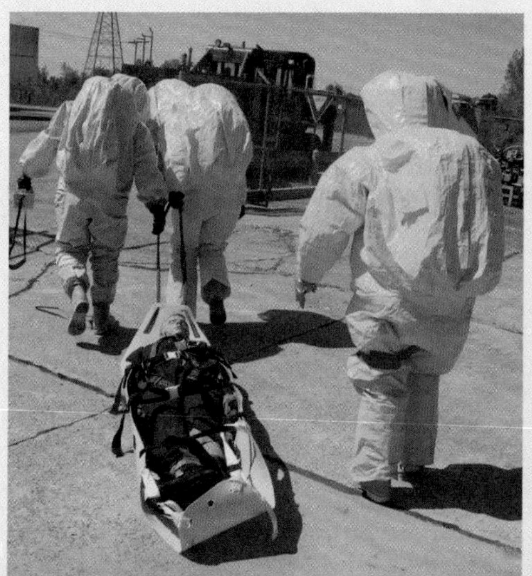

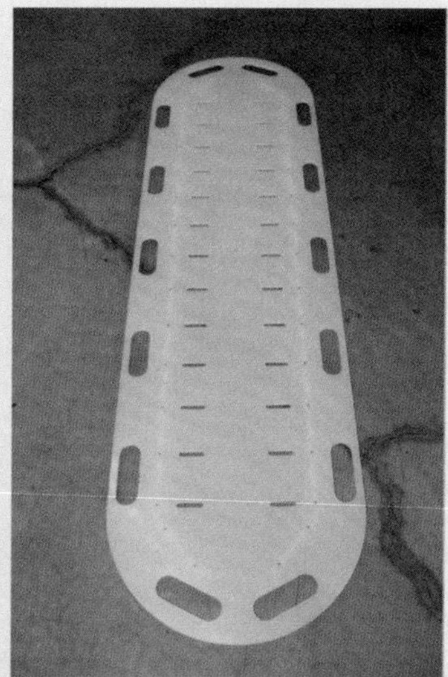

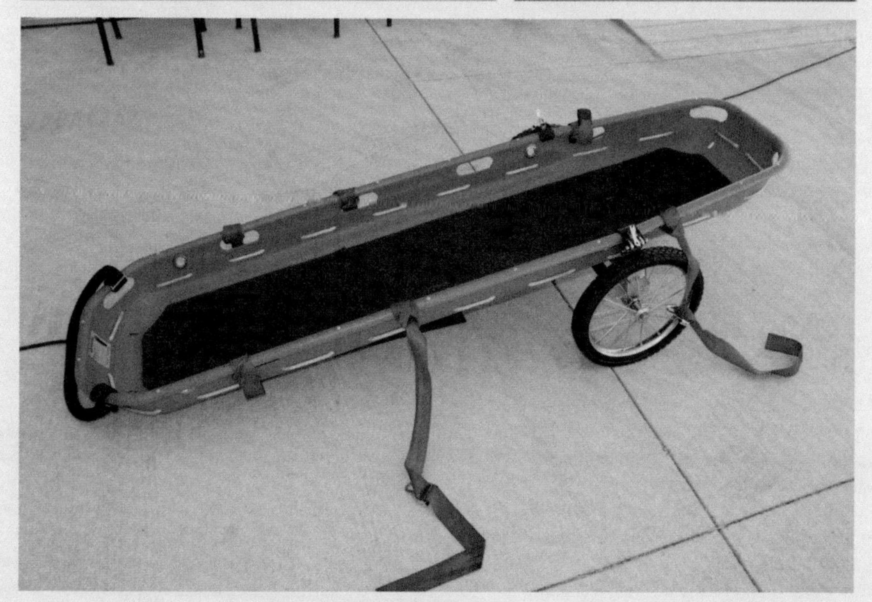

Courtesy of Chris Weber, Dr. Hazmat, Inc.

(continued)

material poses in the short term. Conversely, when patients are nonambulatory, the benefits of rescue may be lower. It is also not known whether the patient's condition is due to an unrelated emergency such as a preexisting medical condition or trauma unrelated to the chemical exposure. However, barring additional information, it must be assumed that the chemical may have caused the injuries and is hazardous. Nonambulatory patients also require more personnel and equipment to be rescued. For example, a Sked board or stretcher must be used for each victim who is to be removed from the hot zone. This requires at least two rescuers per patient. Conversely, ambulatory patients can usually self-extricate from the hot zone, which requires just one rescuer with a bullhorn.

The complexity and risk of a rescue is much lower when the victim can be seen, that is, is in the line of sight. This makes assessment of the patient easier, makes communication with the patient easier if he or she is conscious, and permits the command staff and backup team to have direct visual contact with the rescue team. When the victim can't be seen, patient assessment and communications become much more difficult, and rescue operations become more dangerous. These factors must be considered in the hazard-risk assessment. Depending on the situation, nonambulatory victims with whom there is no line-of-sight communication may be deemed recovery operations.

VICTIM RESCUE TECHNIQUES AND EQUIPMENT

The victim rescue techniques and equipment employed will depend primarily on the location and condition of the victim.

Drags and Carries

The clothing drag is an effective way to quickly remove a nonambulatory victim who is wearing normal clothing. The victim is essentially grabbed by the collar and dragged out. The advantage of the clothing drag is that it is simple and requires no extra equipment. The disadvantage of this drag is that the victim's body creates friction, which makes victim removal more difficult.

The blanket drag is an effective way of removing a victim wearing flimsy clothing, such as a robe or nightgown. A blanket or tarp is placed beneath the victim, and the blanket is used as a support to drag the victim out. The advantage of the blanket drag is that it is simple, and a blanket or other suitable material can usually be readily found. The disadvantage of this drag is that the victim's body on the blanket creates friction, which makes victim removal more difficult. However, if the "blanket" is chosen wisely, such as a plastic tarp, the friction can be reduced dramatically. In addition, most fire department engine companies and truck companies have tarps readily available. These can be quickly folded to a suitable size, or can be cut into suitable sizes when many rescues need to be made.

The sling drag is another way to remove a victim wearing flimsy clothing. A sling of webbing, rope, or long clothing is placed underneath the armpits of the victim and is used to drag the victim out. The wider the sling, the more comfortable it is for the victim and the less chance there is of causing injury or aggravating a preexisting injury. Of course, if the victim is in an IDLH atmosphere, minor secondary injuries are of minimal concern. The advantage of the sling drag is that it can be quickly and easily performed with minimal equipment. The disadvantage of this drag is that the victim's body creates friction, which makes victim removal more difficult.

Victim Removal Equipment

Patient backboards can be used to remove victims from the hot zone. The victim can be strapped to the backboard and carried out by two people, or dragged by one person. Using a rigid backboard has the advantage of reducing the friction to a small surface area where the backboard edge contacts the ground. The advantage of using backboards is that most ambulances, medic units, and fire departments carry two or more backboards at any given time. The disadvantage of backboards is that they are bulky, and strapping the patient onto the backboard takes time.

Litters and *Stokes baskets* have advantages and disadvantages similar to those of patient backboards. A Stokes basket is essentially a reinforced wire basket capable of supporting the weight of a person. Stokes baskets are often used for vertical lifting and lowering operations. The victim is strapped into these devices using webbing or rope and can be carried by two or more people or dragged by a single person. One advantage they offer is an attachment point for ropes and lifting, which may be necessary at complicated rescues at hazmat incidents. The disadvantage of these devices is the amount of time it takes to strap the person into the litter or basket.

Sked boards are often used in confined space rescues owing to their narrow profile and lifting capabilities (if so rated). These boards are also extremely useful at hazmat and WMD incidents. The Sked board is made of pliable plastic yet is rigid enough to easily support the weight of the victim. The plastic also provides much lower friction than the victim's clothing, a blanket, or a Stokes basket when dragged. The Sked board also provides a point to tie off a rope, allowing the victim to be dragged from a distance when there is a clear line of sight. In fact, the victim can even be dragged using a simple mechanical advantage system, and the boards can be attached to multiple attachment points on a rope to remotely evacuate multiple patients rapidly.

appropriate resources are used to rescue or remove the endangered populations first.

Incident Stabilization

Incident stabilization objectives cover a wide variety of actions (Figure 14-8). Any objective that keeps an incident from getting worse is considered an incident stabilization objective. Examples of such objectives are product control, diking and damming, building overflow and underflow dams in rivers and streams, setting up water curtains to control gases and vapors, applying chlorine kits, and extinguishing hydrocarbon fires.

FIGURE 14-8 Incident stabilization—such as stopping leaks—should be the next priority after life safety. *Courtesy of Chris Weber, Dr. Hazmat, Inc.*

Property and Environmental Conservation

After life safety has been addressed and the incident has been stabilized, it is time to consider how to minimize property damage and the environmental impact of the incident (Figure 14-9). Property conservation may include preventing runoff from affecting neighboring areas, decontaminating property and equipment, and transferring residual material inside leaking tanks to undamaged tanks. Environmental conservation may include preventing runoff from affecting lakes and streams, neutralizing residual corrosive materials, and preventing groundwater contamination by excavating contaminated soil.

Social Normalcy

Hazardous materials and WMD incidents are complex and resource-intensive emergencies that tend to disrupt the normal daily routine of communities. Large incidents often disrupt communities for days or weeks. Disruptions include closed freeways and interstates, in-place sheltering orders, and evacuations. ICs should strive to restore social normalcy as quickly as possible, or not to disrupt it in the first place. Unnecessary road closures and evacuations can have lethal effects, such as causing fatal automobile accidents, or they can adversely affect businesses and tourism. However, keep in mind that this objective is last on the list and should never be placed above life safety or incident stabilization priorities.

STRATEGIES AND TACTICS

After response objectives are determined, they must be accomplished. Objectives can best be executed by forming an overall strategy and then subdividing it into manageable tactics. The essential difference between a *strategy* and a *tactic* is that the tactics support the execution of the strategy. Strategies are broad goals that serve to accomplish the response objectives. Strategies must be measurable so that progress can be evaluated, which allows for strategic and tactical corrections when necessary. The following are examples of strategies at hazardous materials incidents:

- isolation of the impacted area
- victim rescue in the hot zone
- downwind public protective actions
- gathering information (reconnaissance)
- confinement
- containment
- mass decontamination of victims

FIGURE 14-9 Property and environmental conservation should be a priority after the incident has been stabilized. *Courtesy of Chris Weber, Dr. Hazmat, Inc.*

Each one of these strategies may have multiple tactics that must be executed to get the job done. For example, to protect the public, you may need to isolate the immediate area 300 feet in all directions, evacuate two nearby residences, block two roads, and demarcate the hot zone with barrier tape. Those four tactics accomplish the strategy of public protection in the immediate area of the release. Similarly, the other strategies listed can be accomplished using specific tactics that will vary depending upon a given hazmat incident. Notice that the tactics tend to be quite incident specific, while strategies tend to be relatively nonspecific and common to most incidents.

MANAGING RESOURCES

Hazardous materials and WMD incidents are resource-intensive operations. Small incidents typically require the presence of the fire department, EMS units, and the hazardous materials response team (HMRT). Large incidents that affect the wider community and require more resources typically additionally require activation of the local EOC. Many different agencies are usually needed at these larger hazmat incidents, including EMS personnel, the fire department, law enforcement, public works, utility workers, cleanup contractors, the health department, and state and federal environmental agencies. The EOC is an ideal tool for requesting these resources. Special resources such as the bomb squad, the meth team, or other technical experts may be needed at particularly challenging incidents. Network with representatives from these agencies now, before an incident occurs, and have their contact information handy. These notifications should be made as early in the incident as possible.

Releases that are above the RQ require a minimum of three notifications: one each at the local, state, and federal level. At the local level it is usually the local emergency management agency or LEPC that must be notified, and this should occur automatically through the 911 dispatch system. At the state level, it is typically the DEQ or the Department of Natural Resources that must be notified. At the federal level, the NRC must be notified at 1-800-424-8802. By law it is the responsibility of the responsible party to make these notifications, but it's wise to double-check that they have been made. Notification of the NRC is especially important, since it can provide many resources, equipment, personnel, and money.

EVALUATING THE PLANNED RESPONSE

After the response objectives have been formulated and are starting to be executed, it is important to evaluate the effectiveness of these measures. How well have the needs of the incident been anticipated? Have all life safety considerations, including downwind populations been taken care of? Is the incident being stabilized? Has the release been stopped or slowed? Or is the release increasing in scope or size? If the answers to these questions are not favorable, the IAP must be modified. If the situation becomes too dangerous, responders may have to be withdrawn to a safe distance.

Generally there are four modes of operation at any emergency incident, including hazmat incidents:

- nonintervention
- defensive
- offensive
- combination

The operational mode is carefully chosen using a risk-based response philosophy. Nonintervention is sometimes the best option at a hazmat incident, as when good offensive options are not available or when the necessary resources are not available to safely accomplish the tasks. For example, when a situation is extremely hazardous, the best strategy may be to evacuate the area and allow the situation to run its course—such as when a flammable liquefied compressed gas container is in imminent danger of a BLEVE.

Defensive mode involves using incident resources to protect exposures and minimize damage without active intervention. Defensive mode is appropriate when active intervention in the hazardous materials incident would pose a significant danger to first responders or when appropriate resources are not available to handle the release. Examples of defensive mode include deploying unmanned monitors to cool a flame-impinged pressurized tank containing a flammable liquefied compressed gas that is not showing imminent signs of failure.

Offensive mode involves using incident resources to actively mitigate the hazard. Offensive mode is appropriate for a manageable hazardous materials release that will not significantly endanger first responders, and the appropriate resources are available to handle it. Appropriate resources means properly trained personnel and the appropriate equipment. For example, extinguishing a gasoline tanker fire when the necessary quantities of foam are on scene and a good water supply has been established.

In most hazmat incidents a combination of modes is used—often at different times during the incident. For example, at a gasoline tanker fire, a nonintervention mode is initially used while waiting for a water supply and foam. Once a water supply has been established, but foam is not yet available, a defensive mode may be used to cool nearby exposures such as a bridge or building. Then, once a sufficient quantity of foam has arrived, an offensive mode is employed to finally extinguish the tanker fire. In any case, resources should be staged so that a quick withdrawal is possible if incident conditions change rapidly—such as owing to weather conditions, wind direction, or catastrophic container failure.

Site Safety Plan

The **site safety plan (SSP)**, often also called a *health and safety plan* (HASP), is developed at the hazmat incident and is designed to ensure the safety of personnel operating at hazardous materials releases. During the initial emergency operations, the SSP may initially be formed primarily from agency SOPs, facility pre-incident plans, and local and facility contingency plans. These documents detail how the hazmat response team should perform its duties, contain maps of the facility as well as documents the chemicals contained in the storage locations, list emergency contact numbers, and outline strategies and tactics for dealing with several different types of anticipated emergencies. Collectively, these documents are essentially a roadmap to the hazmat response.

site safety plan (SSP) ■ A comprehensive document that enumerates the hazards and risks associated with responding to a hazmat incident and describes the methods that will be used to minimize the risk to on-site personnel and outlines safe work practices for mitigation of the incident.

However, each hazardous materials release is different and has its own safety issues and response difficulties. As more information is gathered, such as the nature of the release and which chemicals are involved, the SSP becomes specific to the hazmat incident at hand. The initial information from emergency response plans and pre-incident plans should be verified using facility personnel and reconnaissance as time and resources permit.

Most hazmat response teams use standardized forms to ensure that no essential actions are overlooked and all pertinent information is gathered and documented. These forms should become part of the SSP of the hazmat incident. FIRESCOPE has developed a standardized site safety and control plan form titled ICS 208 HM which has also been adopted by OSHA (Figure 14-10). This form contains the following 12 sections:

1. Site information
2. Organization
3. Hazard/risk analysis
4. Hazard monitoring

SITE SAFETY AND CONTROL PLAN ICS 208	1. Incident Name:	2. Date Prepared:	3. Operational Period: Time:

Section I. Site Information

4. Incident Location:

Section II. Organization

5. Incident Commander:	6. HM Group Supervisor:	7. Tech. Specialist - HM Reference:
8. Safety Officer:	9. Entry Leader:	10. Site Access Control Leader:
11. Asst. Safety Officer - HM:	12. Decontamination Leader:	13. Safe Refuge Area Mgr:
14. Environmental Health:	15.	16.

17. Entry Team: (Buddy System) Name:	PPE Level	18. Decontamination Element: Name:	PPE Level
Entry 1		Decon 1	
Entry 2		Decon 2	
Entry 3		Decon 3	
Entry 4		Decon 4	

Section III. Hazard/Risk Analysis

19. Material:	Container type	Qty.	Phys. State	pH	IDLH	F.P.	I.T.	V.P.	V.D.	S.G.	LEL	UEL

Comment:

Section IV. Hazard Monitoring

20. LEL Instrument(s):	21. O_2 Instrument(s):
22. Toxicity/PPM Instrument(s):	23. Radiological Instrument(s):

Comment:

Section V. Decontamination Procedures

24. Standard Decontamination Procedures:	YES:	NO:

Comment:

Section VI. Site Communications

25. Command Frequency:	26. Tactical Frequency:	27. Entry Frequency:

Section VII. Medical Assistance

28. Medical Monitoring:	YES:	NO:	29. Medical Treatment and Transport In-place:	YES:	NO:

Comment:

ICS 208 Page 1 of 3 3/98

FIGURE 14-10 OSHA requires a site safety plan at all hazardous materials incidents. *Courtesy of FIRESCOPE*

Section VIII. Site Map					
30. Site Map:					
Weather ❑	Command Post ❑	Zones ❑	Assembly Areas ❑	Escape Routes ❑	Other ❑

Section IX. Entry Objectives
31. Entry Objectives:

Section X. SOP'S and Safe Work Practices		
32. Modifications to Documented SOP's or Work Practices:	YES:	NO:
Comment:		

Section XI. Emergency Procedures
33. Emergency Procedures:

Section XII. Safety Briefing	
34. Asst. Safety Officer - HM Signature:	Safety Briefing Completed (Time):
35. HM Group Supervisor Signature:	36. Incident Commander Signature:

ICS 208 Page 2 of 3 3/98

FIGURE 14-10 *(Continued)*

5. Decontamination procedures
6. Site communications
7. Medical assistance
8. Site map
9. Entry objectives
10. SOPs and safe work practices
11. Emergency procedures
12. Safety briefing

Although this two-page form is not strictly a NIMS form, it can be very helpful at hazmat incidents. Standardized forms typically contain a sketch of the area including control zones, research information regarding the release of hazardous material, an accounting of all personnel at the scene, incident objectives and the strategies and tactics that will be used to accomplish them, and the type of PPE that will be used in the hot zone and the warm zone. All personnel at the scene should be familiar with the SSP, have access to the information it contains, and receive a verbal safety briefing.

SITE MAP

A site map is a critical component of the SSP. A good site map lays the foundation for good site access control by clearly marking all control zone boundaries, shows the location of all life safety hazards, shows the placement of all containers containing hazardous materials, and serves as a guide map for the entry team in the hot zone. The site map should contain all the following critical information:

- hot zone, warm zone, and cold zone boundaries
- inner and outer perimeter
- nearby exposures (such as high population density, critical infrastructure, hospitals, schools, businesses, and residential areas)
- location of all life safety hazards
- location of all containers containing hazardous materials
- leaking containers and processes (highlighted)
- decontamination line layout and location
- safe refuge area layout and location
- emergency escape routes
- wind direction
- surface topography
- bodies of water

The site map should be drawn to scale whenever possible. At geographically large incidents a minimum of two maps should be used: one map provides an overview including control zone boundaries and key incident command facilities; one or more additional maps provide details of critical areas—such as the location of rail cars in a derailment or the location of storage tanks in a tank farm. Maps do not need to be drawn from scratch. A ready source of maps are the Internet—such as at Google Earth, and pre-incident plans and emergency response plans for incidents at fixed-site facilities.

The site map is updated as necessary and as site conditions change, and at a minimum each operational period. The most common updates to site maps are due to changes in weather, in life safety hazards, and in conditions of containers of hazardous materials.

Backup Team

According to HAZWOPER, a backup team must be available for immediate rescue whenever personnel operate in the hot zone. The backup team should have their PPE almost completely donned, with the exception of their respirator face piece. The backup team should not be on air while standing by. They must conserve their air in case they are needed for emergency assistance. The PPE level of the backup team should at a minimum be the same as that of the entry team, if not one level higher (when applicable). If the entry team goes down because of a suit breach or chemical incompatibility issue, it does not help for the backup team to be wearing the same PPE. However, many other hot zone emergencies can occur—such as a medical crisis or a slip and fall. The appropriate balance must be struck using a risk-based analysis of the incident hazards.

The backup team, much like a fire department RIT, should be well trained and have victim extrication tools at their immediate disposal. Remember, the backup team is the entry team's last line of defense.

Decontamination Line

A decontamination plan must be in place before anyone enters the hot zone, according to HAZWOPER. At a minimum, emergency decontamination in the form of a charged hose line or emergency shower should be available. Ideally, a fully staffed technical decontamination line will be in place and the decontamination line will be completely set up with all water sources pressurized and ready to go.

Whenever possible, decon team members should don their PPE at the same time as the entry team and the backup team. The decon team should be at the same level of PPE or one level below that of the entry team per the EPA. However, the decon team members should go on air only as the first entry team begins to exit the hot zone, to avoid running out of air during the decontamination process.

Safety Briefing

After a thorough hazard assessment has been conducted, the hazards have been evaluated and appropriate countermeasures have been instituted, and the SSP has been written, the entry team, backup team, and decontamination team must be briefed before anyone enters the hot zone (Figure 14-11). The **safety briefing** advises the team members of the dangers present in the hot zone and the actions they are expected to perform. When complicated incident objectives are to be carried out, such as air monitoring, valve repositioning, or controlling a leak, the safety briefing may be longer than when a hot zone rescue must be performed. In either case, the briefing should cover only the needed information at a level appropriate to the task at hand. A 20-minute briefing will not be remembered.

safety briefing ■ The oral description of the hot zone safety hazards, instituted safety precautions, and work expectations given to the entry team prior to entry.

The safety briefing should at a minimum address the following key points:

- incident objectives
- health hazards
- site layout
- PPE use and limitations
- equipment use and limitations
- radio frequencies
- contact frequency and stay times
- decontamination line location and procedures
- emergency signals
- a reminder to all members to stay alert and observe their surroundings

The safety briefing should be carried out in the presence of all personnel entering the hot zone, the decontamination line personnel, and the hazmat group team leaders. It can also be helpful to have facility representatives at the safety briefing to verify facility information such as access routes and mitigation tactics.

The safety briefing is conducted by the IC, the hazmat group supervisor, the research team leader, or the safety officer. Exactly who delivers the safety briefing will depend on the size of the incident, the complexity of the incident, and technical expertise of the relevant personnel.

FIGURE 14-11 The safety briefing should cover important information such as the safety hazards entry personnel can expect to encounter as well as the job functions they are expected to carry out. *Courtesy of Chris Weber, Dr. Hazmat, Inc.*

Entry Team Debriefing

A debriefing is conducted immediately on the return of the entry team from the hot zone. A good debriefing is critical to furthering the IAP and planning the next hot zone entry. The debriefing should include the entry team members, the hazmat group supervisor, the safety officer, and the hazmat group team leaders. The debriefing should consist of the pertinent hot zone conditions, the nature of the problem, the results that were accomplished during the hot zone entry, any problems that were encountered, and a list of the tasks that could not be accomplished and the reasons why they could not be accomplished. The debriefing can be much more effective when the entry team documents the hot zone and their actions using video and/or still photography. Visual documentation will also help with the postincident analysis (PIA), discussed later.

Termination of the Incident

The last task that must be accomplished at a hazmat incident is proper termination of the incident. Hazmat incidents tend to generate hazardous waste, including contaminated equipment, PPE, and runoff that must be disposed of properly. The EPA requires that runoff from decontamination activities be properly disposed of as hazardous waste and may require proof of proper waste disposal at a later time. This is one reason it vital that the incident be properly documented. Proper documentation is also important if personnel may have been exposed to hazardous materials, to ensure proper medical monitoring in the future, as well as resolution of possible workers' compensation claims.

TRANSFERRING INCIDENT RESPONSIBILITY

At the end of emergency operations the hazardous materials incident is transferred to another agency, typically a cleanup contractor for hazmat incidents and a law enforcement agency for crime scenes. Depending on the nature of the emergency this could be one of any number of different public or private entities:

1. accident investigation teams
2. a cleanup contractor
3. the EPA or state DEQ
4. the property owner
5. a local or state law enforcement agency
6. the FBI

It is very important that any remaining hazards be clearly indicated in writing to the agency assuming responsibility for the scene. This will help prevent injuries and illnesses owing to hazardous materials exposure, the use of improper cleanup and mitigation techniques, and unsafe disposal of hazardous waste.

TERMINATION PROCEDURES

postincident analysis (PIA) ■ A comprehensive examination of an emergency incident after it has been completed designed to improve future responses through a detailed deconstruction of the incident.

A debriefing, **postincident analysis (PIA)**, and critique should be carried out after every hazmat or WMD incident. Hazardous materials incidents are uncommon occurrences for most agencies. Consequently, there are usually many lessons to be learned from the response that can make future responses safer and more effective. If many different departments or agencies were involved in the hazardous materials incident, it is also prudent to have an after-action review with members of these different entities.

Debriefing

The debriefing is an on-scene review that occurs after the emergency incident has ended and the scene has been released but before units leave the scene. The authority having jurisdiction (AHJ) as well as any mutual-aid organizations should take part in the debriefing. The following are key topics that should be discussed:

- the chemicals that were present
- signs and symptoms of chemical exposure
- method of hazardous waste disposal
- any unsafe site conditions
- assessment of need for critical incident stress debriefing (CISD)
- any equipment needs, such as repairs, reorders, and refills
- assignment of personnel to gather information for reports, the PIA, and the critique

All the key players, including the hazmat group team leaders, entry team members, and the decontamination team members, should have a few minutes to report on and discuss their role in the hazmat incident. Specifically, the parts of the response that went well and those parts that could use improvement should be addressed. These topics will then be addressed in further detail in the PIA and the critique.

Postincident Analysis (PIA)

The postincident analysis occurs after the hazardous materials incident has been completed and is designed to improve future responses through a detailed deconstruction of the incident. Completing a rigorous PIA helps ensure the following:

- The incident is properly documented and reported.
- There is a clear description of the emergency response for evaluation.
- A foundation for a formal investigation exists (if necessary).
- The responsible party can be identified.

The PIA should address several key topics:

1. The command-and-control structure
2. The tactics that were used
3. The resources that were employed
4. Support services that were needed
5. The plans and procedures
6. The training that response members received

Critique

The critique is the final component of the incident evaluation, and it should generate recommendations for improvement. The critique is usually carried out several days to weeks after the incident. This delay allows for completion of the PIA and reporting documents. It also allows members to reflect on their actions and how they affected the outcome of the incident. Unlike the debriefing, which took place on scene and included everybody, the critique is by invitation only and generally includes only the key players. However, three distinct types of critiques can take place:

1. Participant-level critique
2. Operation-level critique
3. Group-level critique

CHAPTER **REVIEW**

Summary

Responding to hazardous materials incidents and weapons of mass destruction incidents can be dangerous, and safety should always be the top priority. Hazmat and WMD incident response starts with accurate dispatch information. Incident sites should always be approached from upwind, uphill, and upstream. HAZWOPER requires the use of an incident command system at hazmat incidents. Life safety, incident stabilization, and property and environmental conservation are the mainstays of response objectives, which are determined by a good site assessment involving reconnaissance and air monitoring. Strategies and tactics must be carefully chosen to support reasonable and feasible response objectives. HAZWOPER requires a written site safety plan at all hazardous materials incidents. Hazardous materials response involves a comprehensive risk-based philosophy to keep first responders and the public safe.

Review Questions

1. What are the three control zones called at hazardous materials incidents?
2. In what control zone are support functions carried out?
3. In what control zone are decontamination functions carried out?
4. What are the key components of a site safety plan at a hazardous materials incident?
5. What are three factors to consider when prioritizing response objectives?
6. Why is resource management important at hazardous materials incidents?
7. In which control zone would you expect to find chemical contamination?
8. What are the key components of a pre-entry safety briefing?
9. How is a hazardous materials incident properly terminated?

References

Benner, Ludwig, Jr. (1975) DECIDE In Hazardous Materials Emergencies. Fire Journal p13–18.

Byers, M, M. Russell, and D. J. Lockey. (2008). Clinical care in the "Hot Zone." *Emergency Medicine Journal* 25:108–112.

Lesak, David. (1999) *Hazardous Materials Strategies and Tactics*. Upper Saddle River, NJ: Pearson/Brady.

Michigan State Police Emergency Management and Homeland Security Training Center. (2008). *Hot Zone Rescue for the Operational Level Responder* (Student Manual). Lansing, MI: Author.

National Fire Protection Association. (2008). NFPA 472, *Standard for Competence of Responders to Hazardous Materials/Weapons of Mass Destruction Incidents*. Quincy, MA: Author.

Noll, Hildebrand and Yvorra. (2005) *Hazardous Materials Managing the Incident*, 3rd ed., Annapolis, Maryland: Red Hat Publishing.

Occupational Safety and Health Administration. (1990). 29 CFR 1910.120, *Hazardous Waste Operations and Emergency Response (HAZWOPER)*. Washington, DC: U.S. Department of Labor.

Weber, Chris. (2011) *Hazardous Materials Operations*. Upper Saddle River, NJ: Pearson/Brady.

Weber, Chris. (2007). *Pocket Reference for Hazardous Materials Response*. Upper Saddle River, NJ: Pearson/Brady.

Weber, Chris, and John Meyers. (2009). *Chemical Warfare Agent–Biological Warfare Agent Illicit Laboratory Response Course*. Longmont, CO: Dr. Hazmat, Inc.

York, Kenneth J. and Gerald L. Grey. (1989) *Hazardous Materials/Waste Handling For the Emergency Responder*. New York: Fire Engineering.

ACRONYMS

AAR	Association of American Railroads
AAS	atomic absorption spectroscopy
ABCs	airway, breathing, circulation
ACGIH	American Conference of Governmental Industrial Hygienists
ACP	area contingency plan
AEGL	acute exposure guideline level
AFFF	aqueous film-forming foam
AHJ	authority having jurisdiction
AIHA	American Industrial Hygiene Association
ALARA	as low as reasonably achievable
ALOHA	Areal Locations of Hazardous Atmospheres
amu	atomic mass unit
ANFO	ammonium nitrate–fuel oil
ANSI	American National Standards Institute
API	American Petroleum Institute
APR	air-purifying respirator
ARF	alcohol resistant foam
ASME	American Society of Mechanical Engineers
ASO	assistant safety officer
ASO-HM	Assistant Safety Officer, HazMat
ASTM	ASTM International
ATF	Bureau of Alcohol, Tobacco, Firearms and Explosives
atm	atmosphere
ATR	attenuated total reflectance
ATSDR	Agency for Toxic Substances and Disease Registry
BAL	British anti-lewisite
BLEVE	boiling-liquid, expanding vapor explosion
BME	betamercaptoethanol
BQ	becquerel
BTEX	benzene, toluene, ethyl benzene, xylene
BVM	bag valve mask
BWA	biological warfare agent
C	ceiling
CAA	Clean Air Act
CAMEO	Computer-Aided Management of Emergency Operations
CAS	Chemical Abstract Services
CBRN	chemical, biological, radiological, and nuclear
CBRNE	chemical, biological, radiological, nuclear, and explosive
CDC	Centers for Disease Control and Prevention
CERCLA	Comprehensive Environmental Response, Compensation, and Liability Act
CFC	chlorofluorocarbon
CFR	Code of Federal Regulations
CGA	Compressed Gas Association
CGI	combustible gas indicator
CHEMTREC	Chemical Transportation Emergency Center
CHLOREP	CHLORine Emergency Plan
CHRIS	Chemical Hazard Response Information System
Ci	curie
CIH	certified industrial hygienist
CISD	critical incident stress debriefing
CNS	central nervous system
CO	carbon monoxide
COLIWASA	composite liquid waste sampler
CPC	chemical protective clothing
cpm	counts per minute
CPR	cardiopulmonary resuscitation
CRC	contamination reduction corridor
CSB	U.S. Chemical Safety and Hazard Investigation Board
CST	civil support team
CWA	chemical warfare agent; Clean Water Act
DEA	Drug Enforcement Administration
DECIDE	detect, estimate, choose, identify, do, evaluate
DEP	Department of Environmental Protection
DEQ	Department of Environmental Quality
DHHS	Department of Health and Human Services
DHS	Department of Homeland Security
DMSO	dimethyl sulfoxide
DNA	deoxyribonucleic acid
DOD	Department of Defense
DOE	Department of Energy
DOJ	Department of Justice
DOT	Department of Transportation
DRD	direct reading dosimeter
EKG	electrocardiogram
EMA	emergency management agency
EMS	emergency medical services
EMT	emergency medical technician
EOC	emergency operations center
EOD	explosive ordnance disposal
EPA	Environmental Protection Agency
EPCRA	Emergency Planning and Community Right-to-Know Act
EPD	electronic pocket dosimeter; electronic personal dosimeter
EPDM	ethylene propylene diene monomer
ERCV	emergency response containment vessel
ERG	*Emergency Response Guidebook*
ERP	emergency response plan
ERPG	Emergency Response Planning Guideline

ESF	emergency support function
ESLI	end of service life indicator
EU	European Union
eV	electron-volt
FAA	Federal Aviation Administration
FBI	Federal Bureau of Investigation
FEMA	Federal Emergency Management Agency
FEPCA	Federal Environmental Pesticide Control Act
FFPE	firefighter protective ensemble
FID	flame ionization detector
FIFRA	Federal Insecticide, Fungicide, and Rodenticide Act
FIRESCOPE	FIrefighting RESources of California Organized for Potential Emergencies
FM	Factory Mutual Research Corporation
FPD	flame photometric detection
FTIR	Fourier transform infrared spectroscopy
GC/MS	gas chromatography/mass spectrometry
GHS	Globally Harmonized System of Classification and Labeling of Chemical Substances
GI	gastrointestinal
GSA	General Services Administration
Gy	gray
HASP	health and safety plan
HAZWOPER	Hazardous Waste Operations and Emergency Response (29 CFR 1910.120)
HCl	hydrogen chloride
HF	hydrogen fluoride
HD	distilled mustard agent (vesicant)
HDPE	high-density polyethylene
HEPA	high efficiency particulate air
HM	hazardous materials
HME	homemade explosive
HMIS	Hazardous Materials Identification System
HMR	Hazardous Materials Regulations
HMRT	hazardous materials response team
HMTD	hexamethylene triperoxide diamine
HSLA	high-strength low-alloy
HSPD 5	Homeland Security Presidential Directive 5
HVAC	heating, ventilating, and air-conditioning
IAP	incident action plan
IBC	intermediate bulk container
IC	incident commander
IC_{50}	incapacitating concentration, 50% (by inhalation)
ICP	incident command post
ICS	incident command system
ID	identification
ID_{50}	incapacitating dose, 50% (by injection, skin contact, or ingestion)
IDLH	immediately dangerous to life or health
IED	improvised explosive device
IM	intermodal; intramuscular
IMO	International Maritime Organization
IMS	incident management system; ion mobility spectroscopy
IP	ionization potential
ISP	Internet service provider
IV	intravenous
JIC	joint information center
K-9	canine
kg	kilogram
KI	potassium iodide
KOOROC	keep out of reach of children
LC_{50}	lethal concentration, 50% (by inhalation)
LD_{50}	lethal dose, 50% (by injection, skin contact, or ingestion)
LEL	lower explosive limit
LEPC	local emergency planning committee
LOC	level of concern
LOX	liquid oxygen
LPG	liquefied petroleum gas
LRN	Laboratory Response Network
LSA	low specific activity
LSD	lysergic acid diethylamide
MAWP	maximum allowable working pressure
MC	motor carrier
MCI	mass casualty incident
MEKP	methyl ethyl ketone peroxide
MIC	methyl isocyanate
mJ	millijoule
mmHg	millimeters of mercury
MOPP	mission-oriented protective posture
MSDS	material safety data sheet
MSST	maximum safe storage temperature
MTBE	methyl tertiary-butyl ether
MUC	maximum use concentration
MW	molecular weight
NCP	National Contingency Plan
NEC	National Electric Code
NFPA	National Fire Protection Association
NGO	nongovernmental organization
NIMS	National Incident Management System
NIOSH	National Institute for Occupational Safety and Health
NIST	National Institute of Standards and Technology
NLM	National Library of Medicine
NOAA	National Oceanic and Atmospheric Administration
NOS	not otherwise specified
NPIC	National Pesticide Information Center
NPL	National Priorities List
NQT	nonquenched and tempered
NRC	National Response Center
NRF	National Response Framework
NRT	National Response Team
NTSB	National Transportation Safety Board
OJP	Office of Justice Programs
OPA	Oil Pollution Act

OSC	on-scene coordinator (federal)
OSHA	Occupational Safety and Health Administration
PAC	protective action criteria
PAPR	powered air-purifying respirator
PASS	personal alert safety system
PCB	polychlorinated biphenyl
PCR	polymerase chain reaction
PDF	portable document file
PEL	permissible exposure limit (OSHA)
PET	positron emission tomography
PHMSA	Pipeline and Hazardous Materials Safety Administration
PIA	post-incident analysis
PID	photoionization detector
PIH	poison inhalation hazard
PIO	public information officer
PNS	peripheral nervous system
ppb	parts per billion
PPE	personal protective equipment
PPLS	Pesticide Product Labeling System
ppm	parts per million
PPT	parts per trillion
PRD	pressure relief device
PRV	pressure relief valve
PSI	pounds per square inch
PSIA	pounds per square inch absolute
PSIG	pounds per square inch gauge
PVC	polyvinyl chloride
QT	quenched and tempered
rad	radiation absorbed dose
RAIN	recognize, avoid, isolate, notify
RCP	regional contingency plan
RCRA	Resource Conservation and Recovery Act
RDD	radiological dispersion device
REL	recommended exposure limit (NIOSH)
rem	roentgen equivalent man
RGasD	relative gas density
RIT	rapid intervention team
RQ	reportable quantity
RRT	regional response team
RSC	Royal Society of Chemistry
SABA	supplied-air breathing apparatus
SADT	self-accelerating decomposition temperature
SAR	supplied-air respirator
SARA	Superfund Amendments and Reauthorization Act
SAW	surface acoustic wave
SBCCOM	U.S. Army Soldier and Biological Chemical Command
SCBA	self-contained breathing apparatus
SCO	surface-contaminated object
SDS	safety data sheet
SEB	staphylococcal enterotoxin B
sec	second
SEOC	State Emergency Operations Center
SERC	State Emergency Response Commission
SLUDGEM	salivation, lacrimation, urination, defecation, gastrointestinal upset, emesis, and miosis
SME	subject-matter expert
SOG	standard operating guideline
SOP	standard operating procedure
SP	special permit
SSP	site safety plan
START	simple triage and rapid treatment
STCC	Standard Transportation Commodity Code
STEL	short-term exposure limit
STP	standard temperature and pressure
Sv	sievert
SWAT	special weapons and tactics
TATP	triacetone triperoxide
TB	tuberculosis
TCDD	2,3,7,8-tetrachlorodibenzodioxin
TDI	toluene diisocyanate
TDS	time, distance, shielding
TEEL	temporary emergency exposure limit
THF	tetrahydrofuran
TI	transport index
TIC	toxic industrial chemical; thermal imaging camera
TIH	toxic inhalation hazard
TIM	toxic industrial material
TLD	thermoluminescent detector
TNT	trinitrotoluene
TSCA	Toxic Substances Control Act
TSD	treatment, storage, and disposal
TWA	time-weighted average
UEL	upper explosive limit
UIC	unified incident command
UL	Underwriters Laboratories, Inc.
UN/NA	United Nations/North America
USAMRIID	U.S. Army Medical Research Institute for Infectious Diseases
USCG	United States Coast Guard
USDA	United States Department of Agriculture
UV	ultraviolet
VOC	volatile organic compound
VRD	victim retrieval device
WISER	Wireless Information System for Emergency Responders
WMD	weapon of mass destruction

INDEX

A

B

C

D

E

F

G

N

O

P

Q

R

S

T

U

V

W

X

Y